CRC Series in
ADVANCES in FISHERIES SCIENCE

PATHOBIOLOGY of
MARINE and ESTUARINE ORGANISMS
Edited by John A. Couch and John W. Fournie

ADVANCES in FISHERIES SCIENCE

PATHOBIOLOGY of MARINE and ESTUARINE ORGANISMS

Edited by
John A. Couch
John W. Fournie
U.S. Environmental Protection Agency
Environmental Research Laboratory
Office of Research and Development
Gulf Breeze, Florida

CRC Press
Taylor & Francis Group
Boca Raton London New York

CRC Press is an imprint of the
Taylor & Francis Group, an **informa** business

CRC Press
Taylor & Francis Group
6000 Broken Sound Parkway NW, Suite 300
Boca Raton, FL 33487-2742

Visit the Taylor & Francis Web site at
http://www.taylorandfrancis.com

and the CRC Press Web site at
http://www.crcpress.com

SERIES PREFACE

ADVANCES IN FISHERIES SCIENCE

Advances in Fisheries Science is a book series published with the objective of providing in-depth treatment of the diverse subject matter that, taken together, forms the scope of fisheries science. Areas of emphasis within the series will include, but not be limited to, aquaculture, fishery management, fishing methods, descriptions of vertebrate and invertebrate fisheries, taxonomy and evolution of commercially important aquatic organisms, policy making with respect to fisheries, and relationships between fisheries and both natural and perturbed environments. Many additional topics of sufficient scope for a book in the series are encompassed within those broad headings. Some of the additional topics are genetics, molecular biology, nutrition, pathology, reproduction, behavior, and the general ecology of aquatic animals and, in some cases, plants.

Books in the series are designed to meet the needs of fisheries professionals and, in most instances, will be appropriate for use in upper division undergraduate and graduate courses in fisheries science.

ACKNOWLEDGMENTS

Many people contributed to the development of this book. We gratefully acknowledge authors of individual chapters. Informal comments on technical aspects of each chapter were provided by experts in relevant fields, whom we thank for their time and thoughtful critiques. Betty Jackson, editor for the U.S. EPA Laboratory, Gulf Breeze, Florida, was responsible for copy editing and organization of manuscripts. We thank Valerie Coseo, of Computer Science Corporation, for production of camera-ready text and Tom Poe, of Computer Science Corporation, for graphics. Susan Martin, of Technical Resources Incorporated, Gulf Breeze, assisted in reading page proofs. Cover illustration was designed by Lee A. Courtney of the EPA Laboratory, Gulf Breeze.

PREFACE

With the support of the United States Environmental Protection Agency's Environmental Monitoring and Assessment Program (EMAP), the Gulf Breeze Environmental Research Laboratory has created a Center for Marine and Estuarine Disease Research. As it develops the Center will focus on research questions dealing with the impact of xenobiotics and other anthropogenic stress factors on disease processes in marine and estuarine organisms. The Center will bring together research activities in this area of science within the Environmental Protection Agency, other federal agencies, universities, and independent research institutes. In addition to conventional research activities, the Center might organize response teams to investigate and identify the causes of major incidents involving mortality of marine or estuarine organisms with the objective of establishing the cause of the incident, either directly or through research which would ensue under the auspices of the Center.

The Center's first activity was to convene the *Gulf Breeze Symposium on Marine and Estuarine Disease Research* in October, 1990. The purpose of the symposium was to assess the state of the science in marine and estuarine disease research. Experts in the various aspects of diseases of aquatic mammals, fishes, and invertebrates comprehensively and critically reviewed the specific areas of their expertise. These presentations have been converted into the chapters of this book. The book is, however, an up-to-date compendium of what we know about diseases of marine and estuarine organisms at this time, not simply a compilation of the papers presented at the symposium. We believe it will serve students of marine pathobiology well as a resource for years to come.

The editors and authors hope that this book will serve as a stimulus for research in this important area of environmental science. In addition to the obvious economic importance of fish and shellfish, the organisms living in the marine and estuarine environment serve as useful indicators of the health of the nation's coastal waters. Their disease state is of critical importance.

Robert E. Menzer, Director
U.S. EPA
Environmental Research Laboratory
Gulf Breeze, Florida

THE EDITORS

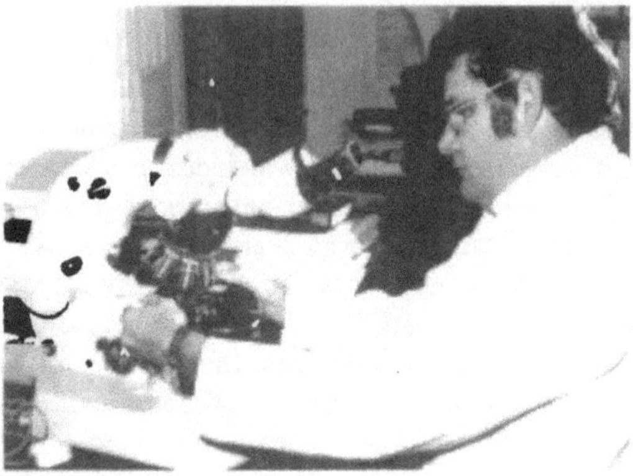

JOHN A. COUCH, Senior Research Scientist, U.S. Environmental Protection Agency (EPA), Environmental Research Laboratory, Gulf Breeze, FL, received a M.S. in Parasitology in 1964 and a Ph.D. in Cell Biology and Parasitology in 1971 from Florida State University. Dr. Couch has completed graduate and special course work in comparative pathology at Johns Hopkins School of Medicine, U.S. Armed Forces Institute of Pathology, and Harvard Medical School. His research interests are in toxicological pathology, experimental oncology with aquatic animals, and pathology of microbial disease agents in aquatic species. He has published extensively on infectious and noninfectious diseases of aquatic animals, and planned and coordinated major research programs on pathology and on lower animals as cancer research models.

Dr. Couch is a leading authority on diseases of molluscs and crustacea, and a pioneer in the pathology and experimental study of cancer in poikilothermic animals. He helped establish the emerging discipline of toxicological pathology of fishes. He originated and managed a four-year, multimillion dollar research program for the EPA and National Cancer Institute on the use of aquatic species in cancer research. He organized and managed the Pathobiology Branch at the EPA Research Laboratory in Gulf Breeze, FL, and originated the concept and original plan for implementation of the center for Marine and Estuarine Disease Research to resolve cause-and-effect problems in coastal and marine organisms' mass mortalities and diseases.

JOHN W. FOURNIE is a Fish Pathologist in the Pathobiology Branch at the EPA Center for Marine and Estuarine Disease Research in Gulf Breeze, Florida. He obtained B.A. and M.S. degrees in Biology from St. Louis University in 1974 and 1979. Dr. Fournie obtained a Ph.D. in Biology from the University of Mississippi and the Gulf Coast Research Laboratory in 1985 where his research emphasized fish parasitology and tumor pathology. He was a post doctoral fellow at the laboratory from 1985-1986, conducting research in small fish carcinogenesis.

Dr. Fournie has attained international recognition as a fish parasitologist for his work on coccidian parasites of fishes. He discovered that a true intermediate host is necessary in the life cycle of an extraintestinal fish coccidian, *Calyptospora funduli*, and reported on its infectivity, life stages, ultrastructure, and pathogenicity to a number of fish species. His work led to the redescription of this group, the creation of a new genus (*Calyptospora*) and a description of a new species. Dr. Fournie's more recent work has expanded into tumor pathology of fish and the use of small fish as carcinogenesis models. He has published numerous papers on various aspects of neoplastic diseases of fish, including detailed descriptions of ocular and exocrine pancreatic neoplasms. Research with carcinogen-induced pancreatic neoplasms in guppies led to an invitation to participate in the International Pancreatic Cancer Study Group meeting in Verona, Italy in 1988. Dr. Fournie has authored chapters on neoplasms of the exocrine pancreas and cardiovascular system, and co-authored a chapter on neoplasms of bone, cartilage, and the soft tissues.

AUTHORS

Robert S. Anderson, Chesapeake Biological Laboratory, University of Maryland, Box 38, Solomons, MD 20688

Thomas C. Cheng, Department of Anatomy and Cell Biology, Medical University of South Carolina, 171 Ashley Avenue, Charleston, South Carolina 29425

John A. Couch, U.S. Environmental Protection Agency (EPA), Environmental Research Laboratory, Sabine Island, Gulf Breeze, FL 32561

John W. Fournie, U.S. EPA, Environmental Research Laboratory, Sabine Island, Gulf Breeze, FL 32561

John L. Fryer, Department of Microbiology, Nash Hall 200, Oregon State University, Corvallis, OR 97331

George R. Gardner, U.S. EPA, Environmental Research Laboratory, 27 Tarzwell Drive, Narragansett, RI 02882

Romona J. Haebler, U.S. EPA, Environmental Research Laboratory, 27 Tarzwell Drive, Narragansett, RI 02882

John C. Harshbarger, Registry of Tumors in Lower Animals, National Museum of Natural History, Smithsonian Institution, Washington, DC 20560

David E. Hinton, Department of Medicine, School of Veterinary Medicine, University of California, Davis, CA 95616

Michael L. Kent, Pacific Biological Station, Nanaimo, BC Canada V9R 5K6

Donald V. Lightner, Department of Veterinary Science, University of Arizona, Tucson, AZ 85721

Robert B. Moeller, Jr., U.S. Armed Forces Institute of Pathology, Washington, DC 20306

Edward J. Noga, College of Veterinary Medicine, 4700 Hillsborough Street, Raleigh, NC 27606

Robin M. Overstreet, Gulf Coast Research Laboratory, P.O. Box 7000, Ocean Springs, MS 39564

Frank O. Perkins, Virginia Institute of Marine Science, P.O. Box 1346, Gloucester Pt, VA 23062

Esther C. Peters, Tetra Tech, Inc., 10306 Eaton Place, Suite 340, Fairfax, VA 22030

John A. Plumb, Department of Fisheries and Allied Aquaculture, Auburn, University, Auburn, AL 36849

John S. Rohovec, Department of Microbiology, Nash Hall 200, Oregon State University, Corvallis, OR 97331

Carl J. Sinderman, National Marine Fisheries Service, Northeast Fisheries Center, Oxford, MD 21654

Albert K. Sparks, U.S. Department of Commerce, NOAA, Seattle, WA 98115

Phyllis M. Spero, Registry of Tumors in Lower Animals, National Museum of Natural History, Smithsonian Institution, Washington, DC 20560

James E. Stewart, Habitat Ecology Division Biological Sciences Branch, Dept. of Fisheries & Oceans, P.O. Box 1006, Bedford Institute of Oceanography, Dartmouth, Nova Scotia, Canada B2Y 4A2

Norman M. Wolcott, Registry of Tumors in Lower Animals, National Museum of Natural History, Smithsonian Institution, Washington, DC 20560

CONTENTS

Authors
Preface
Acknowledgments

1 Importance of Marine Fish Diseases -- An Overview 1
Michael L. Kent and John W. Fournie

2 Viral Diseases of Marine Fish 25
John A. Plumb

3 Bacterial Diseases of Fish 53
John L. Fryer and John S. Rohovec

4 Fungal Diseases of Marine and Estuarine Fish 85
Edward J. Noga

5 Parasitic Diseases of Fishes and Their Relationship with Toxicants and Other Environmental Factors 111
Robin M. Overstreet

6 Neoplasms in Wild Fish from the Marine Ecosystem Emphasizing Environmental Interactions 157
John C. Harshbarger, Phyllis M. Spero, and Norman M. Wolcott

7 Toxicologic Histopathology of Fishes: A Systemic Approach and Overview 177
David E. Hinton

8 Pathobiology of Selected Marine Mammal Diseases 217
Romona Haebler and Robert B. Moeller, Jr.

9 Invertebrate Diseases -- An Overview 245
Albert K. Sparks

10 Infectious Diseases of Molluscs 255
Frank O. Perkins

11 Noninfectious Diseases of Marine Molluscs 289
Thomas C. Cheng

12 Infectious Diseases of Marine Crustaceans 319
James E. Stewart

13 Noninfectious Diseases of Crustacea With an Emphasis on Cultured Penaeid Shrimp 343
Donald V. Lightner

14 Chemically Induced Histopathology in Aquatic Invertebrates 359
George R. Gardner

15 Diseases of Other Invertebrate Phyla: Porifera, Cnidaria, Ctenophora, Annelida, Echinodermata 393
Esther C. Peters

16 Interactions of Pollutants and Disease in Marine Fish and Shellfish 451
Carl J. Sindermann

17 Modulation of Nonspecific Immunity by Environmental Stressors 483
Robert S. Anderson

18 Observations on the State of Marine Disease Studies 511
John A. Couch

Index 531

1

Importance of Marine Fish Diseases--An Overview

Michael L. Kent
John W. Fournie

INTRODUCTION

Diseases and pathogens may impact commercially important fish species, which include food, sport, and ornamental fishes. In addition to causing morbidity and mortality, diseases may cause poor flesh quality and undesirable aesthetic changes. Furthermore, some fish pathogens may infect humans or domestic animals. Fish diseases may also be useful as bioindicators of anthropogenic contamination (AC) or to elucidate the pathologic effects of specific toxicants. Fish disease research has generally evolved in two directions. One area of research has concentrated on diseases in economically important species, especially fishes reared in government or private aquaculture facilities. The other area involves diseases of fishes as indicators of anthropogenic contamination, either in natural populations from potentially polluted waters or in experimental populations exposed to toxicants. However, as with most areas of research, boundaries are arbitrary and overlap exists.

The numbers of researchers and programs in fish pathology have increased significantly in recent years. Fish farming is growing rapidly on a worldwide basis. As a result, research on diseases of these commercial stocks is increasing in importance. Research on diseases of wild fishes from potentially polluted waters has also increased due to greater concern about xenobiotics in the aquatic environment and an elevated awareness that fish may be useful sentinels of xenobiotic effects. In addition, the use of certain fish diseases as models in biomedical research is a new and growing field because fish have some advantages over mammalian models for toxicity and carcinogenicity testing (Dawe and Couch 1984).

Contribution No. 751, Environmental Research Laboratory, Gulf Breeze, FL

0-8493-8662-4/93/$0.00 + $.50

Other chapters in this book assess fish diseases according to etiologic agents (e.g., viruses, bacteria, and parasites) or specific pathological changes (e.g., neoplasms). This chapter reviews the major diseases of marine and estuarine fishes in three categories: 1) those affecting wild fishes, 2) those affecting captive fishes (e.g., fishes reared in fish farms or aquaria), and 3) those used as models for biomedical research. The three categories reflect the two major research areas of fish pathology. Research on infectious diseases is performed primarily on captive fishes, whereas research on fish as sentinels of xenobiotic effects primarily utilizes wild fish but may use small aquarium species for biomedical models.

Diseases of Wild Fishes

It is often difficult to assess the importance of diseases in wild fishes because moribund fish may be quickly eaten by predators or otherwise lost from the population. This is particularly true for chronic infectious diseases. The best documentation of infectious diseases in the wild are those which have resulted in massive, acute mortalities, yet there are undoubtedly many infectious diseases that occur in wild fish that have gone unrecorded. Capture methods may also be biased. Bait and lures tend to catch healthy fish, whereas trawls and nets may be biased toward catching moribund fish. Determining the effects of disease is complicated by the lack of knowledge on normal histological features and hematological parameters for most wild species. As a result, it may be difficult to determine the overall effect of a disease or pathogen on an individual or population. For example, it is often difficult to assess the significance of heavy parasite infections or the presence of *Mycobacterium* tubercles in an individual fish that appears otherwise normal, let alone the impact of such infections on the population.

Diseases Caused by Physical and Chemical Changes

Disease and mortality in wild fishes from coastal marine environments may be caused by changes in physical or chemical conditions of the water, including AC. Dramatic changes in water temperature have caused fish mortalities in various coastal areas throughout the world (Möller and Anders 1986). A few examples of massive fish kills associated with excessively cold water include kills in Mississippi (Overstreet 1974), Georgia (Dahlberg and Smith 1970), and the Gulf of St. Lawrence (Templeman 1965). Mortalities associated with a rapid decrease in temperature may be dramatic, and an estimated 30,000 tons of fish died due to cold shock in Texas during the spring of 1935 (Gunter 1952). In most cases, cold shock does not induce any rapidly recognizable pathological changes, but some affected fishes have exhibited skin lesions (Möller and Anders 1986). Cold water temperatures may also cause increased susceptibility to infectious disease by inducing immunosuppression (Avtalion *et al.* 1976; Ellis 1982), or by altering natural host-parasite relationships. For example, Gulf killifish, *Fundulus grandis,* that are infected with the coccidian, *Calyptospora funduli,* can be severely impacted by exposure to low temperatures. Solangi *et al.* (1982) showed that temperatures of 7-10°C inhibited development and caused abnormalities in developmental stages of *C. funduli* in the Gulf killifish. Fournie (1985) showed that exposure of experimentally infected killifish to 3-4°C resulted in mortalities in infected fish, whereas uninfected fish survived. In addition, this cold temperature killed the parasite in the few surviving infected fish. The parasite is highly prevalent in wild killifish, and this observation could explain the low abundance of this fish reported by Solangi and Overstreet (1980) in years that followed cold winters.

There are examples of direct mortality in fishes due to increases in water temperature, particularly in planktonic stages of deep sea species (Möller and Anders 1986). These mortalities usually occur in fish trapped in abnormal water conditions. Fish trapped in the thermal effluent of power plants may be exposed to excessively high water temperatures. Young (1974) reported four separate incidents of high mortality in Atlantic menhaden, *Brevoortia tyrannus,* exposed to warm water from power plant outlets.

Low dissolved oxygen levels have caused fish kills in fjords and bays in Europe (Möller and Anders 1986), in estuaries of the southeastern United States (May 1973), and in Namibia (Brongersma-Sanders 1957). Reduced dissolved oxygen levels are often caused by decomposing organic material following plankton blooms or by upwelling of nutrient-enriched deep waters.

In addition to their importance as food and recreational sources, wild fishes may be useful as indicators of environmental contamination. There are numerous reports associating diseases and acute fish kills with AC of coastal waters, and there are several reviews on this subject (Overstreet and Howse 1977; Hodgins *et al.* 1977; Sindermann 1979; Sindermann 1980; Möller 1985; Mix 1986; Harshbarger and Clark 1990; Dawe 1990). In addition to acute mortalities, AC can cause non-neoplastic and neoplastic lesions, immunosuppression, increased susceptibility to infectious agents, and may affect reproduction. Contaminants may also affect fish by changing the composition of available food sources and may cause behavioral changes, such as avoidance of polluted waters. Möller and Anders (1986) suggested that changes in fish behavior caused by exposure to toxic wastes (e.g., emigration of sensitive species) may be a more important factor than disease induction in changing the composition of resident fish populations.

A few examples of acute mortality in fishes caused by AC in coastal waters include phosphorus pollution in Placentia Bay, eastern Canada (Iangaard 1970), copper sulfate pollution in the Netherlands (Roskam 1965), and parathion in waters along the eastern coast of Denmark (Boetius 1968).

Chronic pollution can cause histopathological changes, including neoplasms. Epizootics of neoplastic diseases in estuarine fishes in several locations have been associated with xenobiotics (Smith at al. 1979; Couch and Harshbarger 1985; Murchelano and Wolke 1985; Mix 1986; Baumann *et al.*, 1987; Harshbarger and Clark 1990; Myers *et al.* 1990; Vogelbein *et al.* 1990). In addition, AC has been cited as the cause of fin lesions, skin ulcers (Balouet and Laurencin 1983), skeletal anomalies (Couch *et al.* 1977), gill lesions (Couch 1975), hepatic megalocytosis (Myers *et al.* 1987; Peters *et al.* 1987), and preneoplastic liver lesions (Malins *et al.* 1987; Myers *et al.* 1987; Baumann *et al.* 1990; Hayes *et al.* 1990; Vogelbein *et al.* 1990).

However, the occurrence of these lesions, as well as many frank neoplasms, are not necessarily pathognomonic indicators of exposure to AC. Some of the diseases and lesions that have been attributed to AC are poorly substantiated (Mix 1986). Fin and skin lesions have been associated with AC, but they can also be induced by physical trauma, often with concurrent *Cytophaga-Flexibacter* bacterial infections (Anderson and Conroy 1969; Hikida *et al.* 1979; Kent *et al.* 1988). Epithelial separation of the gills has often been associated with exposure to contaminants, but this change can also be induced within minutes of death if the gills are not preserved immediately in appropriate fixatives (Speare and Ferguson 1989). Although hepatic megalocytosis is commonly observed following exposure to anthropogenic hepatotoxicants, this change is also observed in captive salmon (Kent 1990) and in certain species of wild fish from apparently pristine waters. Lastly, epizootics of hepatic and epidermal neoplasms have been shown to be strongly influenced by AC; however, others such as hemic, neural, and connective tissue neoplasms seem unrelated to environmental pollution exposure. In fact, there is evidence that some hemic and neural neoplasms of fishes are caused by infectious agents (Sonstegard 1976; Schmale and Hensly 1988; Kent and Dawe 1990).

Less obvious changes caused by pollution include impairment of defense mechanisms. Weeks and Warinner (1984) and Weeks *et al.* (1986) reported reduction in chemotactic efficiency and phagocytic activity of macrophages in fishes from the heavily polluted Elizabeth River, Virginia. These studies are substantiated by immunosuppression observed in laboratory exposure of fishes to specific chemicals (Anderson *et al.* 1984). Immunosuppressive chemicals include cadmium (Schreck and Lorz 1978; O'Neil 1981), copper (Donaldson and Dye 1975; Anderson *et al.* 1989), tributyltin (Wishkovsky *et al.* 1989), dioxin (Spitsbergen *et al.* 1986), and aflatoxin (Arkoosh and Kaattari 1987).

Immunosuppressive changes induced by toxic chemicals may lead to increased susceptibility to disease caused by infectious agents, particularly those caused by opportunistic pathogens (Rodsaether *et al.* 1977; Hetrick *et al.* 1979; Knittel 1981; MacFarlane *et al.* 1986; Khan 1990).

There have been several reports of detrimental effects on fish reproduction (e.g., ovarian development and spawning) in fish exposed to AC. Stott *et al.* (1983) found altered ovarian development in plaice, *Pleuronectes platessa*, exposed to crude oil from an oil spill; Cross and Hose (1986) reported increased atresia in oocytes in white croakers, *Genyonemus lineatus*, from contaminated sites near Los Angeles, California; Sloof and DeZwart (1983) reported decreased gonadosomatic indices in bream, *Abramis brama*, from polluted sites in the Rhine river; and Johnson *et al.* (1988) reported inhibition of ovarian development in English sole, *Parophrys vetulus*, from polluted sites in Puget Sound, Washington.

Detrimental Algal Blooms

Algal blooms caused by several dinoflagellate species have caused fish kills in coastal waters throughout the world (Möller and Anders 1986; White 1987). Worldwide, the natural production of toxins by algae is probably more important for fish health than AC (Möller and Anders 1986). However, detrimental algal blooms have recently increased and AC may play a role. Nutrient enrichment of coastal waters by runoff of inorganic fertilizer and sewage has caused eutrophication of coastal waters (Prakash 1975; Jingzhong *et al.* 1985; Kimura 1988; Anderson 1989; Smayda 1990). Probably the most dramatic examples of this phenomenon are the blooms of *Chattonella antiqua* in the Seto Inland Sea, Japan, which killed large numbers of pen-reared yellowtail, *Seriola quinqueradiata* (Smayda 1990; Watanabe *et al.* 1990). Although the algae responsible for fish kills and the associated oceanographic conditions are often well documented, the actual mechanisms by which fish are killed are not always understood. There are four known mechanisms by which algal blooms may kill fish (Black *et al.* 1991): 1) physical damage to the gills by the spines of diatoms such as *Chaetoceros* spp.; 2) asphyxiation caused by oxygen depletion; 3) gas-bubble trauma due to extreme oxygen saturation from algal photosynthesis; and 4) direct chemical toxicity from algal toxins.

The pathological effects of algal ichthyotoxins have been best described for *Chattonella antiqua* and *C. marina* that were associated with losses of yellowtail in Japan. Fish exposed to the algae have been reported to show impaired osmoregulation (Shimida *et al.* 1983; Toyoshima *et al.* 1985), reduction in oxygen pressure in the blood (Kobayashi 1978), acidosis (Yamaguchi *et al.* 1981) due possibly to a reduction in gill carbonic anhydrase activity (Sakai *et al.* 1986), and electrophoretic changes in gill mucus (Kobayashi *et al.* 1989). Histologically, affected gills showed degeneration of mucous goblet cells and edema (Shimada *et al.* 1983). All of these may contribute to the ultimate cause of death, which is considered to be asphyxiation.

The following are other examples of phytoplankton that have caused significant fish kills in wild populations in marine waters: *Pytchodiscus brevis* (= *Gymnodinium breve*) occurs throughout the Gulf of Mexico and produces at least two neurotoxins that have been incriminated in fish kills of various species. In 1986, an estimated 20 million fishes were killed by these algae off the coast of Texas (White 1987). *Protogonyaulax tamarensis* has caused mortality in herring, *Clupea harengus*, and other fishes due to accumulation of neurotoxins (saxitoxin) in the planktonic food chain (White 1980, 1981; Gosselin *et al.* 1989). A little known algal species, *Chrysochromulina polylepis*, caused massive mortality in finfish and shellfish in Sweden and Norway in 1988 (Lindahl and Dahl 1990; Johnsen and Lømsland 1990). Apparently, mortalities were caused by a toxin that destroyed the gill epithelium (Saunders 1988). This alga also produces a hemolysin (Edvardsen *et al.* 1990). *Gyrodinium aureolum* is a naked dinoflagellate that has caused mortalities in wild and farmed fishes in western Europe and Japan (Potts and Edwards 1987; Dahl and Tangen 1990). The alga causes necrosis of gill epithelium, but the presumed toxin produced by the organism has not been identified (White 1987).

Infectious Diseases in Wild Fishes

Biological causes of morbidity and mortality are caused by infectious agents, ranging from viruses to metazoan parasites, algal blooms, or predators. When mortalities are associated with parasitic infections, they are usually chronic and they seldom induce mass mortalities. It is difficult, therefore, to assess the impact of parasitic diseases on wild fishes because deaths of affected fish probably go unrecorded, thus the impact of parasitic infections on wild populations is probably underestimated. Sindermann (1990) reviewed a number of cases of such mortalities caused by protozoan and metazoan parasites.

Fish parasites may also cause poor flesh quality or unpleasant aesthetic appearance, resulting in rejection of affected fish by wholesalers, retailers, and consumers. Myxosporeans of the genera *Kudoa*, *Hexacapsula*, *Unicapsula*, and *Henneguya* that infect muscle tissue often do not cause clinical disease. However, several are of economic importance because they reduce the market quality of the flesh (Arai and Matsumoto 1953; Lom 1970; Kabata and Whitaker 1981, 1986; Patashnik *et al.* 1982; Tsuyuki *et al.* 1982; Boyce *et al.* 1985). Probably the best documented example of soft, unmarketable flesh is in Pacific hake, *Merluccius productus*, caused by infections of *K. paniformis* and *K. thyrsitis*.

After the fish dies, proteolytic enzymes produced by the parasites diffuse into the flesh and soften its texture (Patashnik *et al.* 1982; Tsuyuki *et al.* 1982).

Some fish parasites are also important because they are capable of infecting man. Examples include anisakine nematodes, heterophyid trematodes, and the cestode, *Diphyllobothrium latum*. Smith and Wootten (1975), Williams and Jones (1976), Margolis (1977), Ruitenberg *et al.* (1979), Deardorff and Overstreet (1990), and Sindermann (1990) have prepared reviews on the general subject of zoonoses due to fish parasites.

Although better documented in captive fishes, diseases caused by fungi, bacteria, and viruses have been reported in wild marine fishes (Sindermann 1980, 1990; Möller and Anders 1986). The fungus *Ichthyophonus hoferi* has been reported to cause massive disease in herring (Sindermann 1980, 1990) and ulcerative lesions associated with fungi occur in high prevalences in Atlantic menhaden and other fishes from coastal waters of North Carolina (Noga and Dykstra 1986; Levine *et al.* 1990).

Among bacterial diseases, vibriosis, caused by *Vibrio anguillarum* and other *Vibrio* species, has probably had the greatest impact on wild marine fishes. Infections have been reported in numerous species throughout the world (Egidius 1987). *Vibrio anguillarum*, like many other Gram-negative bacteria, is ubiquitous in the environment and often acts as a secondary pathogen in fishes compromised by other infections or adverse environmental conditions.

Pasteurella piscicida is a Gram-negative bacterium that has caused high mortality in marine and estuarine fishes, but within a more restricted host and geographic range. This bacterium has caused mortalities in wild striped bass, *Morone saxatilis*, white perch, *M. americanus*, striped mullet, *Mugil cephalus*, and Atlantic menhaden, *Brevoortia tyrannus*, along the Gulf of Mexico and Atlantic coasts of the United States (Snieszko *et al.* 1964; Lewis *et al.* 1970). One epizootic of pasteurellosis caused approximately 50% mortality in white perch in Chesapeake Bay in the summer of 1963 (Snieszko *et al.* 1964).

Gram-positive bacteria suspected of causing disease in wild marine fishes include *Streptococcus* sp., *Renibacterium salmoninarum*, and *Mycobacterium marinum*. *Streptococcus* sp. has been reported in several fish species in the Gulf of Mexico (Plumb *et al.* 1974; Baya *et al.* 1990). *Renibacterium salmoninarum*, the causative agent of bacterial kidney disease (BKD) in salmonid fishes, has been detected in wild Pacific salmon (Evelyn *et al.* 1973; Banner *et al.* 1986). This disease can be devastating in captive salmonids both in fresh and sea water (Evelyn 1987), and it is suspected that many wild salmon die at sea from BKD after they are released from hatcheries. Granulomatous lesions caused by *Mycobacterium marinum* are often observed in wild fishes, but we know of no reports of epizootics in wild fish. However, infections are prevalent in some species and have been associated with reduced growth rate in concurrence with the severity of the lesions (Bucke 1980; Hastings *et al.* 1982). This bacterium is of concern because it can infect humans who handle infected fish (e.g., aquarist or fish processors) or swim in infected waters (Good 1985).

Numerous viruses have been reported from wild marine fishes (Wolf 1988), and a few have been associated with epizootics. A birnavirus similar to infectious pancreatic necrosis virus of salmonids caused a spinning disease in Atlantic menhaden (Stephens *et al.* 1980), and this virus was implicated as the cause of disease in other fishes from the Pamlico Sound on the east coast of the United States (McAllister *et al.* 1984). An iridovirus and a rhabdovirus similar to viral hemorrhagic necrosis virus of salmonids were associated with the "ulcus syndrome" in Atlantic cod, *Gadus morhua* (Wolf 1988). However, the etiology of the disease is unclear and appears to be multifactorial, with pollution and pathogenic vibrios being important co-factors (Jensen and Larsen 1982; Larsen and Jensen 1982; Jensen 1983). Jensen and Larsen (1982) suggested that the iridovirus was the primary etiologic agent; they were able to experimentally transmit the disease to cod by injection of the iridovirus grown in culture. Iridoviruses that infect erythrocytes, collectively referred to as viral erythrocytic necrosis viruses, infect a wide variety of marine fishes and have been reported to cause anemia in Atlantic cod, *Gadus morhua*, and herring, *Clupea harengus* (Reno *et al.* 1986; MacMillan *et al.* 1989).

Diseases of Captive Fishes

Compared to wild fishes, there have been many more reports and a wider variety of pathogens described from captive fishes. Some important reasons follow: Captive fishes are reared in crowded conditions that enhance transmission of water-borne pathogens. Crowding can also cause immunosuppression and increased susceptibility to disease (Fagerlund *et al.* 1981). Furthermore, moribund fish in captive environments probably live longer than wild individuals due to lack of predation; it is much easier to observe clinical signs in captive fishes.

Fish farming is a rapidly growing industry worldwide and diseases are often the most important limiting factor for the success of these operations. With substantial growth in aquaculture, there has been a corresponding increase in the reported disease outbreaks. Concern about diseases affecting captive fishes is therefore growing. The increasing importance of disease in aquaculture may be due to several factors: more fish and new fish species are being reared, new methods of aquaculture are being employed, new geographic areas are being exploited for aquaculture, and more researchers are involved in the field. In the past, most workers in aquaculture diseases have been fishery biologists, microbiologists, or zoologists in governmental laboratories or universities. More recently, with the rapid growth of private aquaculture, veterinarians are entering the field.

It is beyond the scope of this chapter to comprehensively describe all the important diseases of finfish mariculture. Diseases of various cultured marine fishes have been reviewed: mullets, gilthead sea bream, *Sparus aurata*, and other Red Sea fishes (Paperna 1975; Paperna *et al.* 1977; Paperna and Overstreet 1981; Paperna 1983), yellowtail, *Seriola quinqueradiata*, and other cultured fishes in Japan (Egusa 1983; Sano and Fukuda 1987); and Pacific salmonids in netpens (Kent and Elston 1987; Hicks 1989). Instead, we present a discussion of how aquaculture practices contribute to the occurrence of disease in captive fishes, and we describe a few specific examples. The method by which fish are reared and the culture environment greatly influence the types and severity of diseases. Aquaculture can be divided into two broad categories: extensive and intensive culture. Extensive culture involves rearing fish in ponds or netpens, and intensive culture involves rearing in aquaria, tanks and raceways.

There are advantages and disadvantages to each culture method, many of which pertain directly to disease problems. Extensive systems have the advantage of lower capital and operating costs, and the fish are less crowded than in intensive systems. A major disadvantage of extensive systems is that the control of water flow, and therefore of water quality, is often minimal or nonexistent. This relates directly to many of the disease problems encountered in extensive aquaculture. It is also difficult to administer external chemotherapeutics in ponds and especially in netpens. The potential, therefore, for severe, uncontrollable, and lethal infections by external parasites exists in these systems. For example, parasitic copepods, *Lepeophtheirus salmonis* (Fig. 1-1) and *Caligus* spp., are very serious problems in netpen salmon farms (Pike 1989; Jones *et al.* 1990); the only available treatment involves use of large quantities of organophosphates. These chemicals are difficult to apply to the systems and are dangerous to both fish culturists and the environment (Ross 1989). Monogenean infections have also caused serious diseases in pen-reared yellowtail (Eto *et al.* 1976) and in cultured Red Sea fishes (Paperna *et al.* 1984).

Facultative parasites can also proliferate unchecked in extensive aquaculture systems. For example, *Paramoeba pemaquidensis* is normally a free-living marine amoeba that has recently been associated with gill disease in pen-reared coho salmon, *Oncorhynchus kisutch*, in Washington State (Kent *et al.* 1988b). A similar, if not identical, amoeba has caused high mortality in pen-reared salmonids in Tasmania (Roubal *et al.* 1989). Fish in extensive systems will often feed on naturally-occurring food organisms in addition to the commercial diet provided. These organisms may not normally be ingested by the fish and, when eaten, may result in unusual heteroxenous parasite infections. Kent *et al.* (1991) observed mortality in chinook salmon smolts with eye lesions due to infections

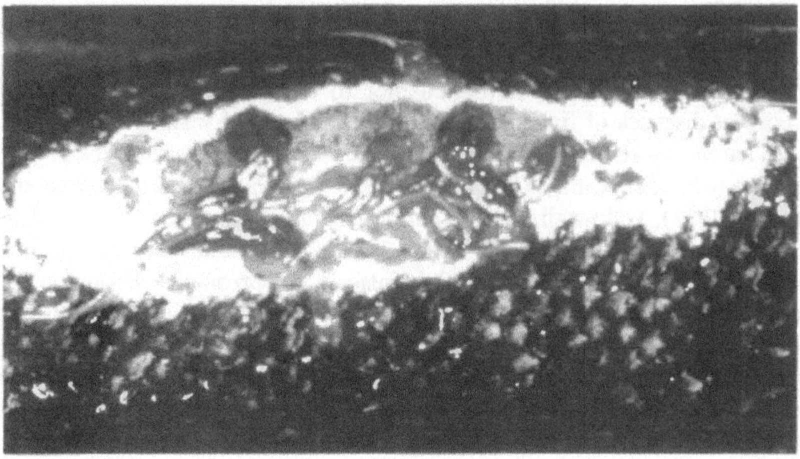

Fig. 1-1. Sea lice, *Lepeophtheirus salmonis*, on the head of a sockeye salmon, *Oncorhynchus nerka*. (By courtesy of Dr. S. Johnson)

of metacestodes of *Gilquinia squali*. The spiny dogfish, *Squalus acanthias*, is the definitive host for the parasites, and the North Sea whiting, *Merlangius merlangus*, is the normal second intermediate host. The first intermediate host, presumably a crustacean, is unknown. Infections have not been reported in wild salmon; thus it was concluded that the pen-reared salmon were feeding on an invertebrate that is not normally part of the fish's diet.

Mixing of year classes at netpen sites, or not drying and disinfecting ponds between growing cycles, allows for horizontal transmission of pathogens between year classes. This transmission is most likely responsible for the widespread prevalence and severity of bacterial kidney disease in salmon. In addition to water-borne pathogens, noxious algae may enter or proliferate in ponds and netpens. For salmon and yellowtail cultured in sea water, serious losses have been caused by toxin-producing algae such as *Heterosigma akashiwo* (Fig. 1-2) (Okaichi 1983; Black *et al.* 1991; Gaines and Taylor 1986; Rensel *et al.* 1989; Chang *et al.* 1990), *Gonyaulax excavata* (Mortensen 1984), *Gyrodinium aureolum* (Tangen 1977; Jones *et al.* 1982), *Chrysochromulina polylepis* (Saunders 1988), and *Chattonella* sp. (Yanagida 1980; Gaines and Taylor 1986). Blooms of *Prymnesium parvum* have caused massive mortalities in brackish water ponds in Israel (Sarig 1971). Mortalities in pen-reared salmon have been caused by *Chaetoceros* spp. and *Corethron sp.* (Fig. 1-3) diatoms due to physical damage to the gills (Bell *et al.* 1974; Farrington 1988; Bruno *et al.* 1989; Horner *et al.* 1990; Speare *et al.* 1989).

In addition to providing an economic way of raising fish, these extensive aquaculture systems, particularly netpens in coastal waters, may offer the fish

Fig. 1-2. Mortality of pen-reared chinook salmon, *Oncorhynchus tshawytscha*, due to bloom of *Heterosigma akashiwo*. (By courtesy of Dr. J.N.C. Whyte)

pathologist an opportunity to elucidate diseases that occur in wild marine fishes. Several new diseases caused by marine pathogens have been described in pen-reared salmonids. It is possible that these diseases are affecting wild salmon but are going undetected for reasons stated earlier.

Species utilized in extensive aquaculture systems may also be useful as sentinels for AC. In contrast to wild fish, fish in these confined environments cannot avoid noxious chemicals in the water, and the duration and exact location of exposure can usually be determined. Furthermore, the diseases, normal histology, and physiology of the species employed in these systems are often better known than those of wild fishes, and moribund fish are less likely to escape attention because they are not removed from the population by predators. Most fish farms are intentionally established in areas with little risk of exposure to AC. Although specific examples are difficult to identify, degradation of coastal waters due to AC was considered to have a significant impact on aquaculture in Southeast Asia (Iwashita 1982; Okaichi 1983; Eng *et al.* 1989). A primary concern with AC in coastal aquaculture systems is the enhancement of

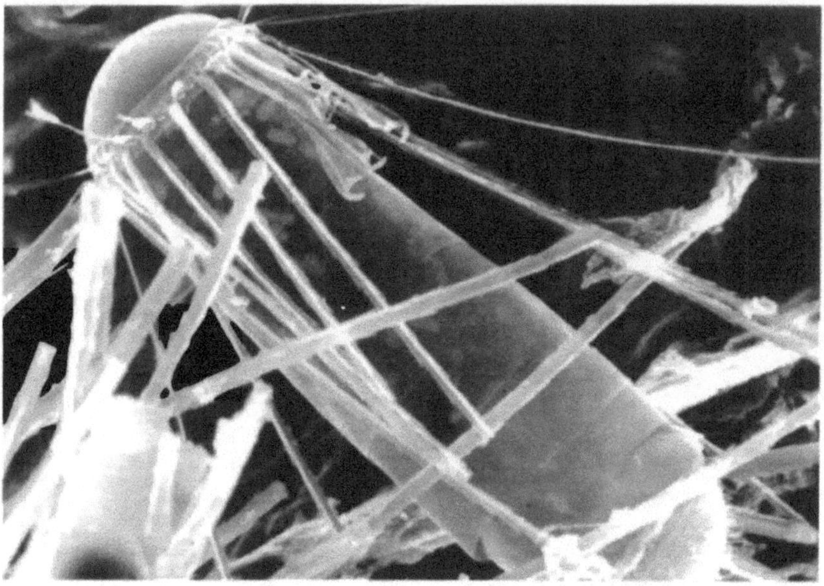

Fig. 1-3. Scanning electron micrograph of the diatom *Corethron sp.* from the gills of a coho salmon, *Oncorhynchus kisutch*. (From *Speare et al.* 1989).

eutrophication, which can cause blooms of toxic algae (Anderson 1989; Smayda 1990).

In intensive aquaculture systems, there is usually good control of the water supply. These operations often use pathogen-free well water or disinfect the incoming water with ultraviolet or ozone sterilization. In addition, administration of external chemotherapeutics is relatively easy to carry out. This all leads to potentially better control and treatment of diseases, and certain pathogens, particularly external parasites, are more easily eradicated or controlled in such systems. However, due to the high densities in these systems, such infections left unchecked can cause rapid, massive mortality. For example, two ectoparasitic protozoans, *Amyloodinium ocellatum* and *Cryptocaryon irritans*, are probably the two most important pathogens in marine aquaria (Wilkie and Gordin 1969; Spotte 1973; Lawler 1977; Sindermann 1990). Both of these parasites infect the skin and gills, causing respiratory and osmoregulatory difficulties when fish are allowed to become heavily infected. Several species of monogeneans have also caused epizootics in marine aquaria and other intensive marine systems when treatments were not applied promptly (Spotte 1973; Lawler and Cave 1978). Bacterial diseases, such as vibriosis, are of concern in intensive marine systems, but can usually be controlled with antibiotics if detected early enough.

Viral diseases are also important in intensive systems because they are untreatable with chemotherapeutics and spread rapidly in confined systems. Viral diseases are recognized as serious pathogens in freshwater intensive systems (Wolf 1988), but with growth in marine aquaculture, the list of viral diseases in cultured marine fishes is rapidly expanding (Lupiani *et al.* 1989; Miyazaki *et al.* 1989; Bloch *et al.* 1991; Iida *et al.* 1991).

Nutritional problems are of concern in intensive systems because the fish are often reared on artificial diets. Although adequate artificial diets have been developed for salmonid fishes, such diets are not readily available for many other marine fishes, particularly the larval forms. Proper diets for fish larvae are, therefore, usually the main obstacle for successful finfish aquaculture. A number of pathological conditions in cultured marine fishes have been attributed to diet, including a systemic granulomatous disease in gilthead sea bream (Paperna 1987), liver anomalies in sea bass, *Dicentrarchus labrax* (Mosconi-Bac 1987), intestinal degeneration in milkfish, *Chanos chanos* (Segner *et al.* 1987), and ceroid deposition, retinal degeneration, and accumulation of calcium oxalate crystals in the kidney of anemonefish, *Amphiprion ocellaris* (Blazer and Wolke 1983). Another potential problem associated with diet in marine fish culture is the introduction of pathogens into confined systems via natural diets.

Mere presence of certain pathogens, even in the absence of any signs of disease, can seriously affect aquaculture operations due to governmental restrictions on transport of infected fishes or to policies that require destruction of infected stocks. In fact, many more apparently healthy fish from populations infected with *Myxobolus cerebralis* or infectious hematopoietic virus have been deliberately destroyed by hatchery personnel than have been killed by the pathogens.

DISEASES OF FISHES IN BIOMEDICAL RESEARCH

From the human perspective, not all diseases of fishes are considered deleterious because some have provided important insights into the basic mechanisms involved in disease development. For example, three types of chemically induced neoplasms in fishes have been sufficiently characterized to be considered as useful models. These include retinal neoplasms of medaka, *Oryzias latipes* (Hawkins *et al.* 1986); exocrine pancreatic neoplasms in the guppy, *Poecilia reticulata* (Fournie *et al.* 1987); and hepatic neoplasms in rainbow trout, *Oncorhynchus mykiss* (Hendricks *et al.* 1984); medaka (Aoki and Matsudaira 1977, 1984; Hinton *et al.* 1988); sheepshead minnow, *Cyprinodon variegatus* (Couch and Courtney 1987); guppy (Simon and Lapis 1984; Hawkins *et al.* 1989), and *Poeciliopsis lucida* (Schultz and Schultz 1985).

Retinal neoplasms in medaka provide one of the few animal models for intraocular neoplasms caused by systemic carcinogen exposure and illustrate the relationship between dysplasia and neoplasia in retinal tumor induction (Hawkins *et al.* 1986). The methylazoxymethanol-acetate- (MAM-Ac) induced medaka medulloepitheliomas demonstrate the sensitivity of a fish sensory organ to a carcinogenic stimulus and the potential for various cellular elements in the eye to become neoplastic. Further work with this model may aid in understanding the histogenesis of intraocular neoplasms.

Exocrine pancreatic neoplasms are important causes of tumor-related death in humans. Although these lesions have been experimentally induced in rodents, cellular origin and oncogenesis of pancreatic tumors are not well known. It is possible that MAM-Ac-induced guppy tumors might provide a model to study the origin of pancreatic tumors. Results to date have suggested that the acinar cells are more significant in the histogenesis of pancreatic carcinomas than previously believed.

Multiple types of hepatic neoplasms have been induced in a number of fish species, but a complete picture of hepatocarcinogenesis in fishes is still lacking. Some piscine hepatic neoplasms and putative precursor lesions share many characteristics with their mammalian counterparts; however, additional work is still necessary for better descriptions and classifications. Further work will hopefully lead to a well-defined standardized classification scheme for liver tumors and associated precursor lesions induced in experimental fishes and a complete understanding of histogenetic pathways involved in hepatic neoplasia. Models of liver neoplasia in fishes are also environmentally significant because of several documented cases of epizootics of liver cancer in wild fish populations from polluted areas (Harshbarger and Clark 1990).

Melanoma in platyfish, *Xiphophorus maculatus*/swordtail, *X. helleri*, hybrids has provided comparative oncologists with a model to investigate the role of oncogenes in the development of melanoma in fish and possibly a means for elucidating the role of oncogenes in carcinogenesis in general. These hybrids and F_1 backcrosses with swordtails have developed melanomas, and it has been demonstrated that a 'tumor regulatory gene complex' (*Tu-complex*) codes for neoplastic transformation in these fish (Anders *et al.* 1984; Zechel *et al.* 1988; Clauss *et al.* 1989). Southern blot analysis has shown that cellular homologs of at least 15 oncogene probes are present in both platyfish and swordtails (Anders *et al.* 1987), and a v-erb B-related oncogene is probably responsible for the transformation from the normal to the neoplastic state of the pigment cells (Zechel *et al.* 1988; Anders *et al.* 1990). The Tu-complex is probably also responsible for the large variety of neurogenic, epithelial and mutagenic tumors that can be induced by carcinogens in these hybrids (Clauss *et al.* 1989).

Infectious agents cause some neoplasms in fishes, especially those of hemic origin (Mulcahy and O'Leary 1970; Sonstegard 1976; Papas *et al.* 1976; Harada *et al.* 1990; Kent *et al.* 1990; Kent and Dawe 1990). Certain neoplasms of fish, therefore, may also be useful as models of neoplasms induced by infectious agents (e.g., oncogenic viruses). Agents that evoke chronic inflammation have been associated with plasma cell tumors in mice and humans (Potter 1972), but their relationship to each other is unclear. A plasmacytoid leukemia of netpen-reared chinook salmon (Kent *et al.* 1990) may be a good model for elucidating this relationship. Although not the primary cause of the plasmacytoid leukemia, many affected fish have concurrent infections of *Renibacterium salmoninarum*. This bacterium causes diffuse, granulomatous lesions, and may be an important co-factor in the disease, the primary cause of which may be a retrovirus. The hemic neoplasm may also be a good model because affected and unaffected salmon are readily available, the neoplasm occurs in experimentally exposed fish 8-12 weeks after injection of tissue homogenates, and large amounts of tumor tissue and purified neoplastic cells can easily be obtained.

A neurofibroma in the bicolor damsel fish is a potential model for human neurofibromatosis (Schmale *et al.* 1983; 1986). This neoplasm is apparently caused by an infectious agent and can be readily transmitted in the laboratory with tissue homogenates (Schmale and Hensley 1988) or cell-free filtrates (Schmale, University of Miami, Florida, personal communication). Presently, there are few good higher vertebrate models for this disease.

Additional research is needed on these and other potential models; new fish models should be developed to study behavior of corresponding or similar mammalian neoplasms or other diseases. Establishment of fish neoplasm models further strengthens the usefulness of fish species in carcinogen testing, as well as the use of small fish as environmental monitors and indicators.

CONCLUSION

Use of fish as sentinels or models for effects of AC has become a well-established field. In the future, emphasis likely will increase on the more cryptic or indirect effects of AC on fish and use of gross and histopathological endpoints. There is an increased interest in effects of AC on immune systems of fishes and relationships between AC and infectious diseases. As research on effects of pollution on fish attempts to determine physiological, biochemical, and immunological consequences of pollution-related disease, aquaculture research will continue to focus on infectious diseases. Further, as aquaculture in coastal areas continues to grow, particularly with netpen systems, research on aquaculture diseases will have more interactions and overlap with the study of potential effects of AC on fish. Thus, two specialties in fish pathology are beginning to merge into a more cohesive, interdependent discipline.

REFERENCES

Anders, F., T. Gronau, M. Schartl, A. Barnekow, G. Yaenel-Dess, and A. Anders. 1987. Cellular oncogenes as ubiquitous genomic constituents in the animal kingdom and as a fundamental in melanoma formation. In *Cutaneous Melanoma: Status of Knowledge and Future Perspective*, ed. U. Veronesi, N. Cascinelli, and M. Santinami, pp. 351-371. New York: Academic Press.

Anders, A., F. Anders, C. Zechel, U. Schleenbecker, and A. Smith. 1990. Attempts to analyze the initiation of initiating processes of carcinogenesis in the *Xiphophorus* melanoma models. *Arch. Geschwulstforsch.* 60: 249-263.

Anders, F., M. Schartl, A. Barnekow, and A. Anders. 1984. *Xiphophorus* as an *in vivo* model for studies on normal and defective control of oncogenes. *Adv. Cancer Res.* 42:191-275.

Anderson, D.M. 1989. Toxic algal blooms and red tides: a global perspective. In *Red Tides: Biology, Environmental Science, and Toxicology*, ed. T. Okaichi, D. M. Anderson, and T. Nemoto, pp. 11-16. New York: Elsevier Science Publishing Co., Inc.

Anderson, D.P., W.B. van Muiswinkel, and B.S. Roberson. 1984. Effects of chemically induced immune modulation on infectious diseases of fish. *Progr. Chem. Biolog. Res. Series* 161:187-211.

Anderson, D.P., O.W. Dixon, J.E. Bodammer, and E.F. Lizzio. 1989. Suppression of antibody-producing cells in rainbow trout spleen sections exposed to copper *in vitro*. *J. Aquat. Anim. Health* 1:57-61.

Anderson, J.I.W. and D.A. Conroy. 1969. The pathogenic myxobacteria with special reference to fish diseases. *J. Appl. Bacteriol.* 32:30-39.

Aoki, K., and H. Matsudaira. 1984. Factors influencing methylazoxy-methanol acetate initiation of liver tumors in *Oryzias latipes*: Carcinogen dosage and time of exposure. *Natl. Cancer Inst. Monogr.* 65:345-351.

Arai, Y. and K. Matsumoto. 1953. On a new sporozoa, *Hexacapsula neothunni* g.n. and sp.n. from the muscle of yellowfin tuna, *Neothunnus macropterus*. *Bull. JPN Soc. Sci. Fish.* 18:293-298.

Arkoosh, M.R. and S.L. Kaattari. 1987. Effect of early aflatoxin B1 exposure on *in vivo* and *in vitro* antibody responses in rainbow trout, *Salmo gairdneri*. *J. Fish. Biol.* 31:19-22.

Avtalion, R.R., E. Weiss, and T. Moalem. 1976. Regulatory effects of temperature upon immunity in ectothermic vertebrates. In *Comparative Immunology*, ed. J.J. Marchalonis, pp. 227-238. Oxford: Blackwell Scientific Publications.

Balouet, G. and F. Baudin Laurencin. 1983. Penetrating ulcers of mullet and possible relationships with petroleum. *Rapp. P.-v. Reun. Cons. Int. Explor. Mer* 182:87-90.

Banner, C.R., J.J. Long, J.L. Fryer, and J.S. Rohovec. 1986. Occurrence of sa.lmonid fish infected with *Renibacterium salmoninarum* in the Pacific Ocean. *J. Fish Dis.* 9:273-275.

Baumann, P.C., W.D. Smith, and W.K. Parland. 1987. Tumor frequencies and contaminant concentrations in brown bullheads from an industrialized river and a recreational lake. *Trans. Am. Fish. Soc.* 116:79-86.

Baumann, P.C., J.C. Harshbarger, and K.J. Hartman. 1990. Relationship between liver tumors and age in brown bullhead populations from two Lake Erie tributaries. *Sci. Total Environ.* 94:71-87.

Baya, A.M., B. Lupiani, F.M. Hetrick, B.S. Roberson, R. Lukacovic, E. May, and C. Poukish. 1990. Association of *Streptococcus* sp. with fish mortalities in the Chesapeake Bay and its tributaries. *J. Fish Dis.* 13:251-253.

Bell, G.R., W. Griffioen, and O. Kennedy. 1974. Mortalities of pen-reared salmon associated with blooms of marine algae. pp. 58-60. In *Proceedings of the Northwest Fish Culture Conference*, 25th Anniversary, 4-6 December, Seattle, Washington.

Black, E.A., J.N.C. Whyte, J.W. Bagshaw and N.G. Ginther. The effects of *Heterosigma akashiwo* on juvenile *Oncorhynchus tshawytscha* and its implications for fish culture. *J. Appl. Ichthyol.* (in press).

Blazer, V.S. and R.E. Wolke. 1983. Ceroid deposition, retinal degeneration and renal calcium oxalate crystals in cultured clownfish, *Amphiprion ocellaris*. *J. Fish Dis.* 6:365-376.

Bloch, B.K. Gravningen, and J.L. Larsen 1991. Encephalomyelitis among turbot associated with a picornavirus-like agent. *Dis. Aquat. Organ.* 10:65-70.

Boetius, J. 1968. Toxicity of waste from a parathion industry at the Danish North Sea coast. *Helgol. Wiss. Meeresunters.* 17:182-187.

Boyce, N.P., Z. Kabata, and L. Margolis. 1985. Investigations of the distribution, detection, and biology of *Henneguya salminicola* (Protozoa, Myxozoa), a parasite of the flesh of Pacific salmon. *Can. Tech. Rep. Fish. Aquat. Sci.* No. 1405.

Brongersma-Sanders, M. 1957. Mass mortality in the sea. *Geol. Soc. Am. Mem.* 67:941-1010.

Bruno, D.W., G. Dear, and D.D. Seaton. 1989. Mortality associated with phytoplankton blooms among farmed Atlantic salmon, *Salmo salar* L., in Scotland. *Aquaculture* 78:217-22.

Bucke, D. 1980. A note on acid-alcohol-fast bacteria in mackerel, *Scomber scombrus* L. *J. Fish Dis.* 3: 173-175.

Chang, F.H., C. Anderson, and N.C. Boustead. 1990. First record of a *Heterosigma* (Raphidophyceae) bloom with associated mortality in cage-reared salmon in Big Glory Bay, New Zealand. *New Zealand J. Mar. Freshwater Res.* 24:461-469.

Clauss, G., J. Lohmeyer, C.V. Hamby, S. Ferrone, and F. Anders. 1989. Melanoma-associated antigens in *Xiphophorus* fish. In *Human Melanoma From Basic Research to Clinical Application*, ed. S. Ferrone, pp. 74-86. New York: Springer-Verlag.

Couch, J.A., and L.A. Courtney. 1987. N-nitrosodiethylamine-induced hepatocarcinogenesis in estuarine sheepshead minnow, *Cyprinodon variegatus*: neoplasms and related lesions compared with mammalian lesions. *J. Natl. Cancer Inst.* 79:297-321.

Couch, J.A., and J.C. Harshbarger. 1985. Effects of carcinogenic agents on aquatic animals: An environmental and experimental overview. *Environ. Carcinog. Rev.* 3:63-105.

Couch, J.A. 1975. Histopathological effects of pesticides and related chemicals on the livers of fishes. In *The Pathology of Fishes*, ed. W.E. Ribelin and G. Migaki, pp. 559-584. Madison: The University of Wisconsin Press.

Couch, J.A., J.T. Winstead, and L.R. Goodman. 1977. Kepone-induced scoliosis and its histological consequences in fish. *Science* 197:585-587.

Cross, J.N., and J.E. Hose. 1986. Reproductive impairment in white croaker from contaminated areas off Los Angeles, p. 48-49. In *Southern California Coastal Waters Research Project Annual Report 1986*.

Dahl, E., and K. Tangen. 1990. *Gyrodinium aureolum* bloom along the Norwegian coast in 1988. In *Toxic Marine Phytoplankton*, ed. E. Granéli, B. Sundström, L. Edler, and D.M. Anderson, pp. 123-127. New York: Elsevier.

Dahlberg, M.D. and F.G. Smith. 1970. Mortality of estuarine animals due to cold on the Georgia coast. *Ecology* 33:165-210.

Dawe, C.J. and J. A. Couch. 1984. Debate: Mouse versus minnow: the future of fish in carcinogenicity testing. *Nat. Cancer Inst. Monogr.* 65: 223-238.

Dawe, C.J. 1990. Implications of aquatic animal health for human health. *Environ. Health Perspect.* 86:245-255.

Deardorff, T.L., and R.M. Overstreet. 1990. Seafood transmitted zoonoses in the United States: the fishes, the dishes, and the worms. In *Microbiology of Marine Food Products*, ed. D. Ward and C.R. Hackey, pp. 211-265. New York: Van Nostrand Reinhold.

Donaldson, E.M., and H.M. Dye. 1975. Corticosteroid concentrations in sockeye salmon (*Oncorhynchus nerka*) exposed to low concentrations of copper. *J. Fish. Res. Board Can.* 32:533-539.

Edvardsen, B., F. Moy, and E. Paashe. 1990. Hemloytic activity in extracts of *Chrysochromulina polylepis* grown at different levels of selenite and phosphate. In *Toxic Marine Phytoplankton*, ed. E. Granéli, B. Sundström, L. Edler, and D.M. Anderson, pp. 284-289. New York: Elsevier.

E. Granéli, B. Sundström, L. Edler, and D.M. Anderson, pp. 284-289. New York: Elsevier.

Egidius, E. 1987. Vibriosis: pathogenicity and pathology. A review. *Aquaculture* 67:15-28.

Egusa, S. 1983. Disease problems in Japanese yellowtail, *Seriola quinqueradiata*, culture: a review. *Rapp. P.-v. Reun. Cons. Int. Explor. Mer* 182:10-18.

Ellis, A.E. 1982. Difference between the immune mechanisms of fish and higher vertebrates. In *Microbial Diseases of Fish*, ed. R.J. Roberts, pp. 1-29. New York: Academic Press.

Eng, C.T., J.N. Paw, and F.Y. Guarin. 1989. The environmental impact of aquaculture and the effects of pollution on coastal aquaculture development in Southeast Asia. *Mar. Pollut. Bull.* 20:335-343.

Eto, A., S. Sakamoto, M. Fukii and Y. Yone. 1976. Studies on an anemia of yellowtail parasitized by a trematode, *Axine (Heteraxine) heterocerca*. *Rep. Fish. Res. Lab., Kyushu Univ.* 3:45-51.

Evelyn, T.P.T. 1987. Bacterial kidney disease in British Columbia, Canada: comments on its epizootiology and on methods for its control on fish farms. In: *AQUA NOR 87 Trondheim International Conference, Norske Fiskeoppdretternes Forening-Fiskeoppdretternes Salgslag A/L*, Trondheim, Norway, pp. 51-57.

Evelyn, T.P.T., G.E. Hoskins, and G.R. Bell. 1973. First record of bacterial kidney disease in an apparently wild salmonid in British Columbia. *J. Fish. Res. Board Can.* 30:1578-1580.

Fagerlund, U.H.M., J.R. McBride, and E.T. Stone. 1981. Stress-related effects of hatchery rearing density on coho salmon. *Trans. Am. Fish. Soc.* 110:644-649.

Farrington, C.W. 1988. Mortality and pathology of juvenile chinook salmon (*Oncorhynchus tshawytscha*) and chum salmon (*Oncorhynchus keta*) exposed to cultures of the marine diatom *Chaetoceros convolutus*. Master's thesis, University of Alaska-Southeast, Juneau. 80 pp.

Fournie, J. W. 1985. Biology of *Calyptospora funduli* (Apicomplexa) from Atheriniform Fishes. University of Mississippi. Doctoral dissertation. 100 pp.

Fournie, J.W., W.E. Hawkins, R.M. Overstreet, and W.W. Walker. 1987. Exocrine pancreatic neoplasms induced by methylazoxymethanol acetate in the guppy (*Poecilia reticulata*). *J. Nat. Cancer Inst.* 78:715-725.

Gaines, G., and F.J.R. Taylor. 1986. A Mariculturist's Guide to Potentially Harmful Marine Phytoplankton of the Pacific Coast of North America. Information Report No. 10, Marine Resources Section, Fisheries Branch, Ministry of Environment, Vancouver, British Columbia.

Good, R.C. 1985. Opportunistic pathogens in the genus *Mycobacterium*. *Annu. Rev. Microbiol.* 39:347-369.

Gosselin, S., L. Fortier, and J.A. Gagné. 1989. Vulnerability of marine fish larvae to the toxic dinoflagellate *Protogonyaulax tamarensis*. *Mar. Ecol. Progr. Ser.* 57:1-10.

Gunter, G. 1952. The importance of catastrophic mortalities for marine fisheries along the Texas coast. *J. Wildl. Manage.* 16:63-69.

Harada, T., J. Hatanaka, S.S. Kubota, and M. Enomoto. 1990. Lymphoblastic lymphoma in medaka, *Oryzias latipes* (Temminck et Schlegel). *J. Fish Dis.* 13:169-173.

Harshbarger, J.C., and J.B. Clark. 1990. Epizootiology of neoplasms in bony fish of North America. *Sci. Total Environ.* 94:1-32.

Hastings, T.S., K. MacKenzie and A.E. Ellis. 1982. Presumptive mycobacteriosis in mackerel (*Scomber scombrus* L.). *Bull. Eur. Assoc. Fish Pathol.* 2: 19-21.

Hawkins, W.E., J.W. Fournie, R.M. Overstreet, and W.W. Walker. 1986. Intraocular neoplasms induced by methylazoxymethanol acetate in Japanese medaka (*Oryzias latipes*). *J. Nat. Cancer Inst.* 76:453-465.

Hawkins, W.E., W.W. Walker, J.S. Lytle, T.F. Lytle, R.M. Overstreet. 1989. Carcinogenic effects of 7,12-dimethylbenz[a]anthracene on the guppy, *Poecilia reticulata*. *Aquat. Toxicol.* 15:63-82.

Hayes, M.A., I.R. Smith, T.H. Rushmore, T.L. Crane, C. Thorn, T.E. Kocal, and H.W. Ferguson. 1990. Pathogenesis of skin and liver neoplasms in white suckers from industrially polluted areas in Lake Ontario. *Sci. Total Environ.* 94:105-23.

Hetrick, F.M., M.D. Knittel, and J.L. Fryer. 1979. Increased susceptibility of rainbow trout to infectious hematopoietic necrosis virus after exposure to copper. *Appl. Environ. Microbiol.* 37:198-201.

Hicks, B. 1989. *British Columbia Salmonid Disease Handbook.* Victoria, British Columbia: Province of British Columbia, Ministry of Agriculture and Fisheries.

Hikida, M., H. Wakabayashi, S. Egusa, and K. Masumura. 1979. *Flexibacter* sp., a gliding bacterium pathogenic to some marine fishes in Japan. *Bull. JPN Soc. Sci. Fish.* 45:421-28.

Hinton, D.E., J.A. Couch, S.J. Teh, and L.A. Courtney. 1988. Cytological changes during progression of neoplasia in selected fish species. *Aquat. Toxicol.* 11:77-112.

Hodgins, H.O., B.B. McCain, and J.W. Hawkes. 1977. Marine fish and invertebrate diseases, host disease resistance, and pathological effects of petroleum. In *Effects of Petroleum on Arctic and Subarctic Marine Environments and Organisms*, ed. D.C. Malins, pp.95-173. New York: Academic Press, Inc.

Horner, R.A., J.R. Postel, and J.E. Rensel. 1990. Noxious phytoplankton blooms in western Washington waters. A Review. In *Toxic Marine Phytoplankton*, ed. E. Granéli, B. Sundström, L. Edler, and D.M. Anderson, pp. 171-176. New York: Elsevier.

Iangaard, P.M. 1970. The role played by the Fisheries Research Board of Canada in the "red" herring phosphorus pollution crisis in Placentia Bay, Newfoundland. *Fish. Res. Board Can. Circ.* 1:1-20.

Iida, Y., T. Nakai, and K. Masumura. 1991. Histopathology of a herpesvirus infection in larvae of Japanese flounder *Paralichthys olivaceus. Dis. Aquat. Organ.* 10:59-63.

Iwashita, M. 1982. Present situation and some problems of marine fish propagation in Japan. *Proceedings of the North Pacific Aquaculture Symposium, Alaska Sea Grant Report Alaska Sea Grant Program Alaska University.* pp. 203-208.

Jensen, N.J. 1983. The ulcus syndrome in cod (*Gadus morhua*): a review. *Rapp. P.-v. Reun. Cons. Int. Explor. Mer* 182: 58-64.

Jensen, N.J., and J.L. Larsen. 1982. The ulcus syndrome in cod (*Gadus morhua*) IV. Transmission experiments with two viruses isolated from cod and *Vibrio anguillarum. Nord. Veterinaermed.* 24: 136-142.

Jingzhong, Z., D. Liping and Q. Baoping. 1985. Preliminary studies on eutrophication and red tide problems in Bohai Bay. *Hydrobiologia* 127:27-30.

Johnsen, T.M., and E.R. Lømsland. 1990. The culmination of the *Chrysochromulina polylepis* (Manton & Parke) bloom along the western coast of Norway. In *Toxic Marine Phytoplankton*, ed. E. Granéli, B. Sundström, L. Edler, and D.M. Anderson, pp. 177-182. New York: Elsevier.

Johnson, L.L., E. Casillas, T.K. Collier, B.B. McCain, and U. Varanasi. 1988. Contaminant effects on ovarian development in English sole (*Parophrys vetulus*) from Puget Sound, Washington. *Can. J. Fish. Aquat. Sci.* 45:2133-2146.

Jones, K.J., P. Ayres, A.M. Bullock, R.J. Roberts, and P. Tett. 1982. A red tide of *Gyrodinium aureolum* in sea lochs of the Firth of Clyde and associated mortality of pond-reared salmon. *J. Mar. Biol. Assoc. U.K.* 62:771-782.

Jones, M.W., C. Sommerville, and J. Bron. 1990. The histopathology associated with the juvenile stages of *Lepeophtheirus salmonis* on the Atlantic salmon, *Salmo salar* L. *J. Fish Dis.* 13:303-310.

Kabata, Z. and D.J. Whitaker. 1981. Two species of *Kudoa* (Myxosporea: Multivalvulida) parasitic in the flesh of *Merluccius productus* (Ayres, 1855) (Pisces: Teleostei) in the Canadian Pacific. *Can. J. Zool.* 59:2085-2091.

Kabata, Z., and D.J. Whitaker. 1986. Distribution of two species of *Kudoa* (Myxozoa: Multivalvulida) in the offshore population of the Pacific hake, *Merluccius productus* (Ayres, 1855). *Can. J. Zool.* 64:2103-2110.

Kent, M.L. 1990. Netpen liver disease (NLD) of salmonid fishes reared in sea water: species susceptibility, recovery, and probable cause. *Dis. Aquat. Organ.* 8:21-28.

Kent, M.L., and S.C. Dawe. 1990. Experimental transmission of a plasmacytoid leukemia of chinook salmon, *Oncorhynchus tshawytscha*. *Can. Res*. 50:5679s-5681s.

Kent, M.L. and R.A. Elston. 1987. Diseases of seawater reared salmon in Washington state. *Proceedings, Fourth Alaska Aquaculture Conference, Alaska Sea Grant Report*. 88 4:161-163.

Kent, M.L., J.M. Groff, G.S. Traxler, J.G. Zinkl, and J.W. Bagshaw. 1990. Plasmacytoid leukemia in seawater reared chinook salmon *Oncorhynchus tshawytscha*. *Dis. Aquat. Organ*. 8: 199-209.

Kent, M.L., L. Margolis, and J.W. Fournie. 1991. A new eye disease in pen-reared chinook salmon caused by metacestodes of *Gilquinia squali* (Trypanorhyncha). *J. Aquat. Anim. Health* 3:134-140.

Kent, M.L., C.F. Dungan, R.A. Elston, and R.A. Holt. 1988a. *Cytophaga* sp. (Cytophagales) infection in seawater pen-reared Atlantic salmon *Salmo salar*. *Dis. Aquat. Organ*. 4:173-179.

Kent, M.L., T.K. Sawyer, and R.P. Hedrick. 1988b. *Paramoeba pemaquidensis* (Sarcomastigophora: Paramoebidae) infestation of the gills of coho salmon *Oncorhynchus kisutch* reared in seawater. *Dis. Aquat. Organ*. 4: 163-169.

Khan, R.A. 1990. Parasitism in marine fish after chronic exposure to petroleum hydrocarbons in the laboratory and to the Exxon Valdez Oil Spill. *Bull. Environ. Contam. Toxicol*. 44:759-763.

Kimura, I. 1988. Aquatic pollution problems in Japan. *Aquat. Toxicol*. 11:287-301.

Knittel, M.D. 1981. Susceptibility of steelhead trout *Salmo gairdneri* Richardson to redmouth infection *Yersinia ruckeri* following exposure to copper. *J. Fish Dis*. 4:33-40.

Kobayashi, H. 1978. Sakana no Kokyu to Junkan. In *Respiration and Circulation in Fish*, JPN. Society of Scientific Fisheries, Susisangaku Ser. 244:111-124. Tokyo: Koseisha-Koseikaku.

Kobayashi, H., Y. Takahashi, and T. Itami. 1989. Changes in the electrophoretic pattern of mucous solution extracted from gills of the yellowtail *Seriola quinqueradiata* affected by red tide. *Nippon Suisan Gakkaishi* 55:577.

Larsen, J.L., and N.J. Jensen. 1982. The ulcus-syndrome in cod (*Gadus morhua*). V. Prevalence in selected Danish marine recipients and a control site in the period 1976-1979. *Nord. Veterinaermed*. 34: 303-312.

Lawler, A.R. and R.N. Cave. 1978. Deaths of aquarium-held fishes caused by monogenetic trematodes. I. *Aspinatrium pogonae* (MacCallum, 1913) on *Pogonius chromis* (Linnaeus). *Drum and Croaker* 18:31-33.

Lawler, A.R. 1977. Dinoflagellate (*Amyloodinium*) infestation of pompano. In *Disease Diagnosis and Control in North American Marine Aquaculture*, ed. C.J. Sindermann, pp. 257-261. Amsterdam: Elsevier.

Levine, J.F., J.H. Hawkins, M.J. Dykstra, E.J. Noga, D.W. Maye, and R.S. Cone. 1990. Epidemiology of ulcerative mycosis in Atlantic menhaden in Tar-Pamlico River Estuary, North Carolina. *J. Aquat. Anim. Health* 2:162-171.

Lewis, D.H., L.C. Grumbles, S. McConnell, and A.I. Flowers. 1970. *Pasteurella*-like bacteria from an epizootic in menhaden and mullet in Galveston Bay. *J. Wildl. Dis*. 6:160-162.

Lindahl, O., and E. Dahl. 1990. On the development of the *Chrysochromulina polylepis* bloom in the Skagerrak in May - June 1988. In *Toxic Marine Phytoplankton*, ed. E. Granéli, B. Sundström, L. Edler, and D.M. Anderson, pp. 189-194. New York: Elsevier.

Lom, J. 1970. Protozoa causing diseases in marine fishes. *Am. Fish. Soc. Spec. Pub*. 5:101-123.

Lupiani, B., C.P. Dopazo, A. Ledo, B. Fouz, J.L. Barja, F.M. Hetrick, and A.E. Taranzo. 1989. New syndrome of mixed bacterial and viral etiology in cultured turbot *Scophthalmus maximus*. *J. Aquat. Anim. Health* 1:197-204.

MacFarlane, R.C., G.L. Bullock, and J.J.A. McLaughlin. 1986. Effects of five metals on susceptibility of striped bass to *Flexibacter columnaris*. *Trans. Am. Fish. Soc*. 115:227-231.

MacMillan, J.R., D. Mulcahy, and M.L. Landolt. 1989. Cytopathology and coagulopathy associated with viral erythrocytic necrosis in chum salmon. *J. Aquat. Anim. Health* 1:255-262.

Malins, D.C., B.B. McCain, D.W. Brown, M.S. Myers, M.M. Krahn, and S.L. Chan. 1987. Toxic chemicals, including aromatic and chlorinated hydrocarbons and their derivatives, and liver lesions in white croaker (*Genyonemus lineatus*) from the vicinity of Los Angeles. *Environ. Sci. Technol*. 21:765-770.

Margolis, L. 1977. Public health aspects of 'codworm' infection: a review. *J. Fish. Res. Board Can.* 34:887-898.

May, E.B. 1973. Extensive oxygen depletion in Mobile Bay, Alabama. *Limnol. Oceanogr.* 18:353-366.

Mix, M.C. 1986. Cancerous diseases in aquatic animals and their association with environmental pollutants: a critical literature review. *Mar. Environ. Res.* 20(1,2):1-141.

Miyazaki, T., K. Fujiwara, J. Kobara, N. Matsumoto, M. Abe, and T. Nagano. 1989. Histopathology associated with two viral diseases of larval and juvenile fishes: epidermal necrosis of the Japanese flounder *Paralichthys olivaceus* and epithelial necrosis of black sea bream *Acanthopagrus schlegeli. J. Aquat. Anim. Health* 1:85-93.

McAllister, P.E., M.W. Newman, J.H. Sauber, and W.J. Owens. 1984. Isolation of infectious pancreatic necrosis virus (serotype Ab) from diverse species of estuarine fish. *Helgol. Wiss. Meeresunters.* 37:317-328.

Möller, H. and K. Anders. 1986. *Diseases and Parasites of Marine Fishes.* Kiel, Germany: Verlag Möller. 365 pp.

Möller, H. 1985. A critical review on the role of pollution as a cause of fish diseases. In *Fish and Shellfish Pathology*, ed. A.E. Ellis, pp. 169-182. London: Academic Press.

Mortensen, A.M. 1984. Massive fish mortalities in the Faroe Islands caused by a *Gonyaulax excavata* red tide. In *Toxic Dinoflagellates*, ed. D.M. Anderson, A.W. White, and D.G. Baden, pp. 165-70. New York: Elsevier.

Mosconi-Bac, N. 1987. Hepatic disturbances induced by an artificial feed in the sea bass (*Dicentrarchus labrax*) during the first year of life. *Aquaculture* 67:93-99.

Mulcahy, M.F. and A. O'Leary. 1970. Cell-free transmission of lymphosarcoma in northern pike *Esox lucius* L. (Pisces, Esocidae). *Experientia* 26:891.

Murchelano, R.A., and R.E. Wolke. 1985. Epizootic carcinoma in the winter flounder, *Pseudopleuronectes americanus. Science* 228:587-589.

Myers, M.S., L.D. Rhodes, and B.B. McCain. 1987. Pathologic anatomy and patterns of occurrence of hepatic neoplasms, putative preneoplastic lesions, and other idiopathic hepatic conditions in English sole (*Parophrys vetulus*) from Puget Sound, Washington. *J. Nat. Can. Inst.* 78:333-363.

Myers, M.S., J.T. Lendahl, M.M. Krahn, L.L. Johnson, and B.B. McCain, 1990. Overview studies on liver carcinogenesis in English sole from Puget Sound; Evidence for a xenobiotic chemical etiology I: Pathology and epizootiology. *Sci. Total Environ.* 94: 33-50.

Noga, E.J. and M.J. Dykstra. 1986. Oomycete fungi associated with ulcerative mycosis in menhaden, *Brevoortia tyrannus* (Latrobe). *J. Fish Dis.* 9:47-53.

O'Neill, J.G. 1981. Effects of intraperitoneal lead and cadmium on the humoral immune response of *Salmo trutta. Bull. Environ. Contam. Toxicol.* 27:42.

Okaichi, T. 1983. Red tides and fisheries damages. *Eisei Kagaku* 29:1-4.

Overstreet, R.M. and H.D. Howse. 1977. Some parasites and diseases of estuarine fishes in polluted habitats of Mississippi. *Ann. N.Y. Acad. Sci.* 298:427-442.

Overstreet, R.M. 1974. An estuarine low-temperature fish-kill in Mississippi, with remarks on restricted necropsies. *Gulf Res. Rep.* 4:328-350.

Papas, T.S., J.E. Dahlberg and R.A. Sonstegard. 1976. Type C virus in lymphosarcoma in northern pike (*Esox lucius*). *Nature* 261:506-508.

Paperna I., A. Diamant, and R.M. Overstreet. 1984. Monogenean infestations and mortality in wild and cultured Red Sea fishes. *Helgol. Meeresunter.* 37:445-462.

Paperna, I. and R.M. Overstreet. 1981. Diseases of mullets (Mugilidae). In *Aquaculture of Grey Mullets*, ed. O.H. Oren, pp. 411-493. Cambridge: Cambridge University Press.

Paperna, I. 1975. Parasites and diseases of the grey mullet (Mugilidae) with special reference to the seas of the near east. *Aquaculture* 5:65-80.

Paperna, I. 1983. Review of diseases of cultured warm-water marine fish. *Rapp. P.-v. Reun. Cons. Int. Explor. Mer* 183:44-48.

Paperna, I. 1987. Systemic granuloma of sparid fish in culture. *Aquaculture* 67:53-58.

Paperna, I., A. Colorni, H. Gordin, and G.Wm. Kissil. 1977. Diseases of *Sparus aurata* in marine culture at Eilat. *Aquaculture* 10:195-213.

Patashnik, M., H.S. Groninger Jr., H. Barnett, G. Kudo, and B. Koury. 1982. Pacific whiting, *Merluccius productus*: I. Abnormal muscle texture caused by myxosporidian-induced proteolysis. *Mar. Fish. Rev.* 44:1-12.

Peters, N., A. Kohler, and H. Kranz. 1987. Liver pathology in fishes from the lower Elbe as a consequence of pollution. *Dis. Aquat. Organ.* 2:87-97.

Pike, A.W. 1989. Sea lice -- major pathogens of farmed Atlantic salmon. *Parasitol. Today* 5(9):291-297.

Plumb, J.A., J.H. Schachte, J.L. Gaines, W. Peltier, and B. Carroll. 1974. *Streptococcus* sp. from marine fishes along the Alabama and northwest Florida coast of the Gulf of Mexico. *Trans. Am. Fish. Soc.* 2:358-361.

Potter, M. 1972. Immunoglobulin-producing tumors and myeloma proteins of mice. *Physiol. Rev.* 52:631-719.

Potts, G.W., and J.M. Edwards. 1987. The impact of a *Gyrodinium aureolum* bloom on inshore young fish populations. *J. Mar. Biol. Assoc. U.K.* 67:293-297.

Prakash, A. 1975. Dinoflagellate blooms - an overview. In *First International Congress on Toxic Dinoflagellate Blooms*, pp. 1-6. Wakefield, MA: Massachusetts Science and Technology Foundation.

Rensel, J.E., R.A. Horner, and J.R. Postel. 1989. Effects of phytoplankton blooms on salmon aquaculture in Puget Sound, Washington: initial research. *Northwest Environ. J.* 5(1):53-69.

Reno, P.W., D.V. Serreze, S.K. Hellyer, and B.L. Nicholson. 1985. Hematological and physiological effects of viral erythrocytic necrosis (VEN) in Atlantic cod and herring. *Fish Pathol.* 20:353-360.

Rodsaether, M.C., J. Olafsen, J. Raa, K. Myhre, and J.B. Steen. 1977. Copper as an initiating factor of vibriosis (*Vibrio anguillarum*) in eel (*Anguilla*). *J. Fish Biol.* 10:17-21.

Roskam, R.T. 1965. A case of copper pollution along the Dutch shore. *Inter. Counc. Explor. Sea.* C. M. E, 40.

Ross, A. 1989. Nuvan use in salmon farming: the antithesis of the precautionary principle. *Mar. Pollut. Bull.* 20:372-374.

Roubal, F.R., R.J.G. Lester, and C.K. Foster. 1989. Studies on cultured and gill-attached *Paramoeba* sp. (Gymnamoebae: Paramoebidae) and the cytopathology of paramoebic gill disease in Atlantic salmon, *Salmo salar* L., from Tasmania. *J. Fish Dis.* 12:481-492.

Ruitenberg, E.J., F. Van Knapen, and J.W. Weiss. 1979. Food-borne parasitic infections -- a review. *Vet. Parasitol.* 5:1-10.

Sakai, T., K. Yamamoto, M. Endo, A. Kuroki, K. Kumanda, K. Takeda, and T. Aramaki. 1986. Changes in the gill carbonic anhydrase activity of fish exposed to *Chattonella marina* red tide, with special reference to mortality. *Bull. JPN Soc. Sci. Fish.* 52:1351-1354.

Sano, T., and H. Fukuda. 1987. Principal microbial diseases of mariculture in Japan. *Aquaculture* 67:59-69.

Sarig, S. 1971. *Book 3: The Prevention and Treatment of Diseases of Warmwater Fishes Under Subtropical Conditions, with Special Emphasis on Intensive Fish Farming.* In *Diseases of Fishes*, ed. S.F. Snieszko and H.R. Axlerod, 127 pp. Neptune, New Jersey: T.F.H. Publications.

Saunders, R.L. 1988. Algal catastrophe in Norway. *World Aquacult.* 10(2):11-12.

Schmale, M.C., and G.T. Hensley. 1988. Transmissibility of a neurofibromatosis-like disease in bicolor damselfish. *Can. Res.* 48:3828-3833.

Schmale, M.C., G. Hensley, and L.R. Udey. 1983. Multiple schwannomas in the bicolor damselfish, *Pomacentrus partitus* (Pisces, Pomacentridae). *Am. J. Pathol.* 112:238-241.

Schmale, M.C., G.T. Hensley, and L.R. Udey. 1986. Neurofibromatosis in the bicolor damselfish (*Pomacentrus partitus*) as a model of von Recklinghausen neurofibromatosis. *Ann. N.Y. Acad. Sci.* 486:386-402.

Schreck, C.B., and H.W. Lorz. 1978. Stress response of coho salmon (*Oncorhynchus kisutch*) elicited by cadmium and copper and potential use of cortisol as an indicator of stress. *J. Fish. Res. Board Can.* 35:1124-1129.

Schultz, M.E., and R.J. Schultz. 1985. Transplantable chemically-induced liver tumors in the viviparous fish *Poeciliopsis*. *Exp. Mol. Pathol.* 42: 320-330.

Segner, H., P. Burkhardt, E.M. Avila, J.V. Juario, and V. Storch. 1987. Nutrition-related histopathology of the intestine of milkfish *Chanos chanos* fry. *Dis. Aquat. Organ.* 2:99-107.

Shimada, M., T.M. Murakami, T. Imahayashi, H.S. Ozaki, T. Toyoshima, and T. Okaichi. 1983. Effects of sea bloom, *Chattonella antiqua*, on gill primary lamellae of the young yellowtail, *Seriola quinqueradiata*. *Acta Histochem. Cytochem.* 16:232-244

Simon, K., and K. Lapis. 1984. Carcinogenesis studies on guppies. *Natl. Cancer Inst. Monogr. 65:71-81.*

Sindermann, C.J. 1979. Pollution-associated diseases and abnormalities of fish and shellfish: a review. *Fish. Bull.* 76:717-749.

Sindermann, C.J. 1980. The use of pathological effects of pollutants in marine environmental monitoring programs. *Rapp. P.-v. Reun. Cons. Int. Explor. Mer* 179:129-134.

Sindermann, C.J. 1990. *Principal Diseases of Marine Fish and Shellfish.* New York: Academic Press, Inc. 521 pp.

Slooff, W., and D. DeZwart. 1983. The growth, fecundity and mortality of bream (*Abramis brama*) from polluted and less polluted surface waters in the Netherlands. *Sci. Total Environ.* 27:149-162.

Smayda, T. J. 1990. Novel and nuisance phytoplankton blooms in the sea: evidence for global epidemic. In *Toxic Marine Phytoplankton*, ed. E. Granéli, B. Sundström, L. Edler, and D.M. Anderson, pp. 29-40. New York: Elsevier.

Smith, C.E., T.H. Peck, R.J. Klauda, and J.B. McLaren. 1979. Hepatomas in Atlantic tomcod *Microgadus tomcod* (Walbaum) collected in the Hudson River estuary in New York. *J. Fish Dis.* 2:313-319.

Smith, J.W., and R. Wootten. 1975. Experimental studies on the migration of *Anisakis* sp. larvae (Nematoda: Ascaridida) into the flesh of herring, *Clupea harengus* L. *Int. J. Parasitol.* 5:133-136.

Snieszko, S.F., G.L. Bullock, E. Hollis, and J.G. Boone. 1964. *Pasteurella* sp. from an epizootic of white perch (*Roccus americanus*) in Chesapeake Bay tidewater areas. *J. Bacteriol.* 88:1814-1815.

Solangi, M.A. and R.M. Overstreet. 1980. Biology and pathogenesis of the coccidium *Eimeria funduli* infecting killifishes. *J. Parasitol.* 66:513-526.

Solangi, M.A., R.M. Overstreet and J.W. Fournie. 1982. Effect of low temperature on development of the coccidium *Eimeria funduli* in the Gulf killifish. *Parasitology* 84:31-39.

Sonstegard, R.A. 1976. Studies of the etiology and epizootiology of lymphosarcoma in *Esox* (*Esox lucius* L. and *Esox masquinongy*). *Prog. Exper. Tumor Res.* 20:141-155.

Speare, D.J., J. Brackett, and H.W. Ferguson. 1989. Sequential pathology of the gills of coho salmon with a combined diatom and microsporidian gill infection. *Can. Vet. J.* 30:571-575.

Speare, D.J., and H.W. Ferguson. 1989. Fixation artifacts in rainbow trout (*Salmo gairdneri*) gills: a morphometric evaluation. *Can. J. Fish. Aquat. Sci.* 46:780-785.

Spitsbergen, J.M., K.A. Schat, J.M. Kleeman, and R.E. Peterson. 1986. Interactions of 2,3,7,8-tetrachlorodibenzo-p-dioxin (TCDD) with immune responses of rainbow trout. *Vet. Immunol. Immunopathol.* 12:263-280.

Spotte, S.H. 1973. *Marine Aquarium Keeping: The Science, Animals and Art.* New York: Wiley (Interscience). 171 pp.

Stephens, E.B., M.W. Newman, A.L. Zachary, and F.M. Hetrick. 1980. A viral aetiology for the annual spring epizootics of Atlantic menhaden, *Brevoortia tyrannus*, (Latrobe) in Chesapeake Bay. *J. Fish Dis.* 3:387-398.

Stott, G.G., W. Haensly, J. Neff, and J. Sharp. 1983. Histopathologic survey of ovaries of plaice, *Pleuronectes platessa* L., from Aber Wrac'h and Aber Benoit, Brittany, France oil spills. *J. Fish Dis.* 6:429-437.

Tangen, K. 1977. Blooms of *Gyrodinium aureolum* (Dinophyceae) in north European waters, accompanied by mortality in marine organisms. *Sarsia* 63:123-133.

Templeman, W. 1965. Lymphocystis disease in American plaice of the eastern Grand Bank. *J. Fish. Res. Board Can.* 22:1345-1356.

Toyoshima, T., H.S. Ozaki, M. Shimada, T. Okaichi, and T.H. Murakami. 1985. Ultrastructural alterations on chloride cells of the yellowtail, *Seriola quinqueradiata*, following exposure to the red tide species *Chattonella antiqua*. *Mar. Biol.* 88:101-108.

Tsuyuki, H., S.N. Williscroft, Z. Kabata, and D.J. Whitaker. 1982. The relationship between acid and neutral protease activities and the incidence of soft cooked texture in the muscle tissue of Pacific hake *Merluccius productus* infected with *Kudoa paniformis* and/or *K. thyrsitis*, and held for varying times under different pre-freeze chilled storage conditions. *Fish. Aquat. Sci. Can. Tech. Rep. No. 1130.*

Vogelbein, W.K., J.W. Fournie, P.A. Van Veld, and R.J. Huggett. 1990. Hepatic neoplasms in the mummichog *Fundulus heteroclitus* from a creosote-contaminated site. *Can. Res.* 50:5978-5986.

Watanabe, M., K. Kohata, and M. Kunugi. 1990. Nitrogen and phosphate accumulation by *Chattonella antiqua* during diel vertical migration in a stratified microcosm. In *Toxic Marine Phytoplankton*, ed. E. Granéli, B. Sundström, L. Edler, and D.M. Anderson, pp. 171-176. New York: Elsevier.

Weeks, B.A., and J.E. Warinner. 1984. Effects of toxic chemicals on macrophage phagocytosis in two estuarine fishes. *Mar. Environ. Res.* 14:327-335.

Weeks, B.A., J.E. Warinner, P.L. Mason, and D.S. McGinnis. 1986. Influence of toxic chemicals on the chemotactic response of fish macrophages. *J. Fish Biol.* 28:653-658.

White, A.W. 1980. Recurrence of kills of Atlantic herring (*Clupea harengus harengus*) caused by dinoflagellate toxins transferred through herbivorous zooplankton. *Can. J. Fish. Aquat. Sci.* 27:2262-2265.

White, A.W. 1981. Sensitivity of marine fishes to toxins from the red-tide dinoflagellate *Gonyaulax excavata* and implications for fish kills. *Mar. Biol.* 65:255-260.

White, A.W. 1987. Blooms of toxic algae worldwide: their effects on fish farming and shellfish resources. In: *AQUA NOR 87 Trondheim International Conference, Norske Fiskeoppdretternes Forening-Fiskeoppdretternes Salgslag A/L*, Trondheim, Norway, p. 9-14.

White, A.W., O. Fukuhara, and M. Anraku. 1989. Mortality of fish larvae from eating toxic dinoflagellates or zooplankton containing dinoflagellate toxins. In: *Red Tides: Biology Environmental Science and Toxicology*, ed. T. Okaichi, D.M. Anderson, and T. Nemoto pp. 395-398. New York: Elsevier.

Wilkie, D.W. and H. Gordin. 1969. Outbreak of cryptocaryoniasis in marine aquaria at Scripps Institution of Oceanography. *Calif. Fish Game* 55:227-236.

Williams, H. H. and A. Jones. 1976. *Marine Helminths and Human Health*. Commonwealth Institute of Helminthology Publication No. 3. Berkhamsted Hets, England: Clunbury Cottrell Press. 47 pp.

Wishkovsky, A.,E.S. Mathews, and B.A. Weeks. 1989. Effect of tributyltin on the chemiluminescent response of phagocytes from three species of estuarine fish. *Arch. Environ. Contam. Toxicol.* 18:826-31.

Wolf, K. 1988. *Fish Viruses and Fish Viral Diseases*. London: Comstock Publishing Associates. 476 pp.

Yamaguchi, K., K. Ogawa, N. Takeda, K. Hashimoto, and T. Okaichi. 1981. Oxygen equilibria of hemoglobins of cultured sea fishes, with special reference to red tide-associated mass mortality of yellowtail. *Bull. JPN Soc. Sci. Fish.* 47:403-409.

Yanagida, T. 1980. *The Red Tide*. Tokyo: Kodansha Scientific.

Young, J.S. 1974. Menhaden and power plants - a growing concern. *Mar. Fish. Rev.* 36(10): 19-23.

Zechel, C., U. Schleenbecker, A. Anders, and F. Anders. 1988. v-erb B-related sequences in *Xiphophorus* that map to melanoma determining Mendelian loci and overexpression in a melanoma cell line. *Oncogene* 3:605-617.

2

Viral Diseases of Marine Fish

John A. Plumb

INTRODUCTION

Known viral diseases of fish date back 30 to 35 years, as the first definitive virus of fish was demonstrated in the late 1950's (Wolf *et al.* 1959). However, if one considers clinical signs and descriptive pathology throughout the fish pathology literature, it is clear that viruses are not new to fish. Wolf (1988) provides an excellent history of viral diseases of fish, and it is clear that viruses are among the most devastating infectious agents that afflict fish. Further, it is difficult to separate viruses that infect only marine fish, not freshwater fish, from those that infect both. Thus, this paper includes viruses of teleosts that infect marine and freshwater fish with the emphasis on their effects on fish in the marine environment (Table 2-1).

The vast majority of literature related to viral diseases of fish emphasizes those viruses that affect primarily cultured species and populations (Wolf 1988). Thus, the effect of viruses on wild fish populations is not clearly understood, but it is conceivable that viruses in natural marine fish populations could reduce available recruitment into harvestable-size fish by killing large numbers of juveniles. Tumors resulting from viruses may make harvestable-size fish aesthetically unacceptable and increase the vulnerability of smaller fish to gill nets, as the case with lymphocystis (Wolf 1988). Viral diseases of cultured marine fishes are more easily assessed because of daily contact and observation at the more highly susceptible small size or young age. Poor understanding of viruses and viral diseases of marine fishes is further impaired because of the inaccessability of the most susceptible size and age (fry or small fingerlings). As mariculture becomes more intensive throughout the world, and previously uncultured species are adapted to domesticated spawning, rearing, and culture, additional specific viral diseases of fish will be discovered. Whether their discovery will result from increased awareness, aquaculture intensification, or both, is difficult to ascertain; however, history indicates that intensification of fish culture has resulted in the

0-8493-8662-4/93/$0.00+$.50

Table 2-1. Viral Diseases of Marine Fish and Their Environments.

Disease	Virus classification	How known		Environment	
		Isolation	Electron microscopy	Marine	Marine & freshwater
Infectious Pancreatic Necrosis (IPN)	Birnovirus	X	X	–	X
Viral Hemorrhagic Septicemia (VHS)	Rhabodovirus	X	X	–	X
Pacific Salmon Anemia Virus (EIBS)	———	–	–	–	X
Viral Erythrocytic Necrosis (VEN)	Iridovirus	–	X	X	–
Chum Salmon Reovirus	Reovirus	X	X	X	–
Chinook Salmon Paramyxovirus	Paramyxovirus	X	X	X	X
Lymphocystis	Iridovirus	X	X	–	X
Atlantic Menhaden Spinning Disease	?	X	X	X	X
Atlantic Cod Adenovirus	Adenovirus	–	X	X	–
Atlantic Cod Ulcus Syndrome	Iridovirus	–	X	X	–
Atlantic Salmon Papillomatosis	?	–	X	–	–
Atlantic Salmon Swim Bladder Sarcoma Virus	Reovirus	–	X	X	–
Pacific Cod Herpesvirus	Herpesvirus	–	X	X	X
Pleuronectid Papilloma	?	–	–	X	–
Winter Flounder Papilloma	?	–	–	X	–

appearance of previously unknown diseases. Viral diseases of marine and estuarine fishes can be grouped in various ways; in this paper, they are divided into 1) salmonid virus diseases, 2) virus diseases of other species, and 3) viruses known only by electron microscopy.

EFFECTS OF POLLUTION ON FISH

Degradation of the marine environment through industrial, municipal, and petroleum contamination, eutrophication, increased siltation, and human encroachment on coastal and estuarine habitats is a significant concern. Fish population shifts in age classes and other population parameters can be measured by annual or periodic sampling, but the effects of environmental insults on the basic health of fish populations are difficult to assess.

Researchers have associated external lesions, tumors, and other detrimental responses to environmental changes in coastal waters; and some have been specific in the chemical or environmental factor responsible for such changes (Christmas and Howse 1970; Couch and Nimmo 1974; Gardner 1975, Giles *et al.* 1978; Mearns and Sherwood 1977; Murchelano and Ziskowski 1982; Overstreet 1988).

Möller (1985) addressed the role of pollution as a precipitator of fish diseases and extensively discussed these relationships in the marine environment. There is much positive evidence to correlate environmental pollutants and diseases, but several reports refute this relationship. Studies have implicated increased skin, fin, and gill lesions, including ulcers, tumors, and deformities, on chemical contamination of waters (Bengtsson 1975; Gardner 1975; Weis and Weis 1976; Minchew and Yarbrough 1977; Giles *et al.* 1978; McCain *et al.* 1978; Couch *et al.* 1979; Solangi and Overstreet 1982).

Field studies in coastal waters have implicated pollution with fish diseases (Stich *et al.* 1976; Hard *et al.* 1979; McCain *et al.* 1979; Mearns and Sherwood 1977; Murchelano and Ziskowski 1982; Maurin and McArdle 1983). Some significant contradictory reports are by Wellings *et al.* (1976) and Phillips *et al.* (1976).

Several studies correlated viral infections of marine animals to pollutants, some involving shrimp (Couch and Nimmo 1974; Couch and Courtney 1977) and fish (Christmas and Howse 1970; Overstreet and Howse 1977; Hill 1984, Reiersen and Fugelli 1984; Overstreet 1988).

The role of pollution in the marine environment and resulting stress on fish is not clearly known. Sindermann (1988) stressed that fish, among other aquatic animals, could serve as biological indicators for environmental degradation and pollutants in estuarine and coastal habitats. Sindermann (1979) noted that environmental stress in fish may come from a variety of pollutants that affect various life stages of animals differently. Such stressors may result in infectionsby facultative pathogens or the conversion of latent to active infections. Overstreet (1988) concentrated on effects of pollution on diseases of marine fish in the southeastern U.S. He noted that light or moderate parasitic infections that may not be harmful under non-polluted conditions are damaging in heavily polluted waters. He also pointed out that the stalked protozoan *Epistylis* sp. occurred more heavily on fish in polluted waters than in non-polluted waters. Concerning virus diseases of fish, Overstreet (1988) found no conclusive evidence that lymphocystis was more prevalent in polluted versus non-polluted waters, but Reiersen and Fugelli (1984) reported a higher incidence of lymphocystis in polluted waters. There is little doubt that environmental conditions play an important role in incidence of diseases in fishes, whether fish are freshwater or saltwater inhabitants, but the extent of their role in viral diseases is not clear.

VIRAL DISEASES OF SALMONIDS

Perhaps the best known virus diseases of fish are those that afflict salmonids because several of their agents, notably infectious pancreatic necrosis virus (IPNV), infectious hematopoietic necrosis virus (IHNV), and viral hemorrhagic septicemia virus (VHSV), can cause high mortality. Other viral diseases of salmonids are less virulent, less frequent, or produce tumors or papillomas. These viruses infect juvenile life stages in fresh water where they are the most devastating, but also occur in marine environments where their effect is not easily measured.

Infectious Pancreatic Necrosis Virus

Infectious pancreatic necrosis virus (IPNV) is generally considered a viral pathogen of freshwater fish (Wolf 1988). However, because IPNV infects juvenile salmonids, it has an impact on recruitment of fish into free swimming stocks of salmonids in the marine environment. Snieszko *et al.* (1957) first named and described IPN, but Wolf *et al.* (1959) and Wolf *et al.* (1960) established its viral etiology, thereby reporting the first known viral disease of fish. However, M'Gonigle (1941) previously reported similar clinical signs in rainbow trout, *Oncorhynchus mykiss*, and named the disease "catarrhal enteritis." The virus is a non-enveloped icosahedron with a single capsid and diameter of 55 to 75 nm (Chang *et al.* 1978). The type strain is designated VR-299 (ATCC). IPNV is a bisegmented, double-stranded RNA genome that belongs to the birnavirus group. Hill and Way (1988), in standardizing serological classification of aquatic birnaviruses, divided viruses into Serogroup I, which contains 9 serotypes including VR-299, and Serogroup II with 6 serotypes.

A wide variety of fish are susceptible to IPNV: Wolf (1988) lists 16 different species of salmonids from which IPNV has been isolated, including Atlantic salmon (*Salmo salar*), chinook salmon (*O. tshawytscha*), chum salmon (*O. keta*), coho salmon (*O. kisutch*), sockeye salmon (*O. nerka)*, and steelhead rainbow trout. IPNV, or an IPN-like virus, has been isolated from an additional 20 families, including some in the marine environment: lampreys, herrings, eels, true bass, flounders, halibuts, and soles. For the most part, IPN-like viruses isolated from non-salmonids have not been proven pathogenic to salmonids. In addition to the wide species susceptibility to IPNV, or IPNV-like, the virus has a world-wide distribution, having been reported from 16 countries in North America, South America, Europe, the British Isles, the Mediterranean region, and Asia.

IPNV causes a systemic, lethal disease of young salmonids. IPNV-infected fish become darkly pigmented, are bilaterally exophthalmic, have distended abdomens, and pale to pink gills (Fig. 2-1) (Wolf *et al.* 1959). They swim in an erratic corkscrew fashion. Internally, the body cavity is filled with a clear yellow fluid; the gut is pale and void of food, but contains a gelatinous mucous plug in the anterior portion of the intestine. This plug will survive formalin fixation of the fish. A general hyperemic appearance of the viscera is characteristic, especially in the pyloric caecae region where pancreatic tissue is embedded in adipose tissue. The liver and kidney are pale to hyperemic. Mortality patterns range from an acute 80% mortality in 30- to 50-day-old fish (less than 4 cm) to more chronic mortality rates of less than 10% in 100- to 120-day-old fish. Fish over 15 cm are seldom adversely affected. These mortality rates are also influenced by water temperature, strain and species of fish, and strain of virus (Dorson and Torchy 1981; McAllister and Owens 1988).

Fish that survive an IPNV epizootic become active carriers of the virus, and these fish are capable of transmitting the virus to their offspring either on the spermatozoa or in the egg (Bullock *et al.* 1976; Ahne and Neyele 1985; Dorson and Torchy 1985). Therefore, any fish that survive an IPNV epizootic and are stocked into waters, fresh or salt, pose a threat to future generations. Yamamoto and Kilistoff (1979), however, showed that over a period of years, if subsequent fish stockings are virus free, the incidence of carrier fish decreases over time. In the case of infected anadromous salmonid adults, vertical transmission has not been reported to be a problem.

It has been shown that in some cases IPNV in non-salmonids can be horizontally transmitted to salmonids. McAllister and McAllister (1988) transmitted IPNV from striped bass, *Morone saxatilis*, to brook trout, *Salvelinus*

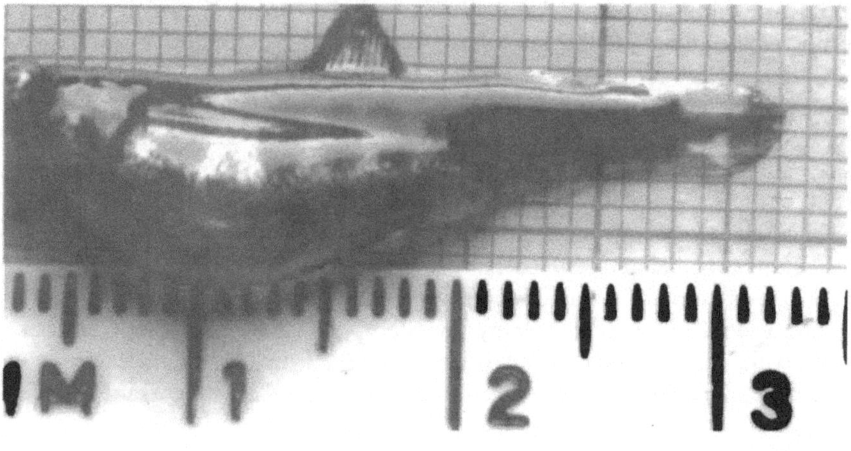

Fig. 2-1. Rainbow trout infected with infectious pancreatic necrosis virus. Note dark pigmentation, abdominal distention, and exophthalmia.

fontinalis. IPNV was also experimentally transmitted to crayfish, *Astacus astacus*, and then transmitted to rainbow trout (Holder and Ahne 1988).

There are no reports of IPNV causing mortality among wild spawned-fish populations, but it must be recognized that in such cases the severely affected small fish could unknowingly be killed and escape detection.

Viral Hemorrhagic Septicemia

Viral hemorrhagic septicemia virus (VHSV) was first reported in Europe by Altara (1963) and Jensen (1963). VHS for many years has been the most severe viral disease of European salmonid culture and was confined to Europe until the virus was isolated in 1989 and 1990 from adult salmon returning on spawning runs in Washington state, U.S. (Hooper 1989; Brunson *et al.* 1989). VHSV is a member of the *Rhabdoviridae* (Zwellenberg *et al.* 1965) that has an RNA genome. The bullet-shaped virion measures approximately 70 by 180 nm.

VHSV infects juvenile, subadult, and adult fish, but the smaller and younger fish are more susceptible and suffer a high mortality. Survivors of VHSV epizootics become carriers of the virus, but shedding seems to be intermittent (de Kinkelin 1983) and vertical transmission has not yet been proven. Reservoirs of the virus are carriers in the water supply, clinically diseased fish, and possibly the intestines of fish-eating birds (Bregenballe 1981, from Wolf 1988).

Although rainbow trout is the primary and most susceptible species of fish to VHSV, Wolf (1988) lists 10 additional species and one hybrid salmonid as susceptible to the virus. Three of these are of marine origin, including Atlantic salmon, sea bass, *Dicentrarchus labrax*, and turbot, *Scophthalmus maximus*. Wolf (1988) further listed nine species of fish that are refractive to VHSV. Ironically, the list of resistant species includes chinook salmon (Ord *et al.* 1976) and coho salmon (de Kinkelin *et al.* 1974), two species of Pacific salmon from which VHSV was isolated in Washington state (U.S.) in 1989. The vast majority of VHSV epizootics have been reported from freshwater rainbow trout populations, but Bovo and Giorgetti (1983, from Wolf 1988) isolated VHSV from rainbow trout that had been stocked into brackish water 5 months prior to the onset of disease. Horlyck *et al.* (1984) described a similar epizootic in rainbow trout that had been in brackish water for 18 months. Castric and de Kinkelin (1980) reported an 80% mortality in rainbow trout 1 month after they were moved from fresh water to full-strength sea water. Thus the ground work has been laid for the possibility of VHSV to occur in mariculture salmonids in brackish or sea-water environments.

The recent detection of VHSV in coho and chinook salmon adults in the state of Washington showed no signs of disease (Hooper 1989; Brunson *et al.* 1989), thus possibly indicating a potentially wider geographical range than previously thought. Winton *et al.* (1989) showed that the Washington (WVHSV) isolates were similar serologically and biochemically to the European (EVHSV) isolates. However, J.R. Winton (U.S. Fish and Wildlife Service, Seattle, WA, personal communication) showed that the Washington strain is significantly less virulent than the European. The WVHSV does not kill any of eight different historically EVHSV susceptible species of salmonids by either water-borne or injection exposure. This fact would appear to confirm the reports of Ord *et al.* (1976) and de Kinkelin *et al.* (1974) that chinook and coho salmon were resistant to VHSV. J.R. Winton (personal communication) also showed that virulence of WVHSV was reduced 5-fold in sea water compared to a 200-fold loss of virulence in fresh water. In spite of attempts to eliminate VHSV from the continental United States in 1989 by destruction of infected populations and disinfection of facilities, the virus was again isolated from coho and chinook salmon early in 1990 (J.R. Winton, personal communication).

Typically infected individuals do not feed, and their swimming behavior ranges from lethargic to hyperactive. Clinically, VHSV-infected salmonids are distinctly darker than normal and may have unilateral or bilateral exophthalmia. Gills are usually pale, but have petechial hemorrhaging (Fig. 2-2). Hemorrhage may also occur in eyes and fins. Internally, there is a wide variation of lesions. In severe cases, internal organs and visceral cavity show profuse bleeding, with the liver being a chalky gray and the kidney a deeper red than normal. The spleen is usually enlarged and dark red. Petechiae occur on the swim bladder, peritoneum, adipose tissue, and in the musculature (de Kinkelin 1983).

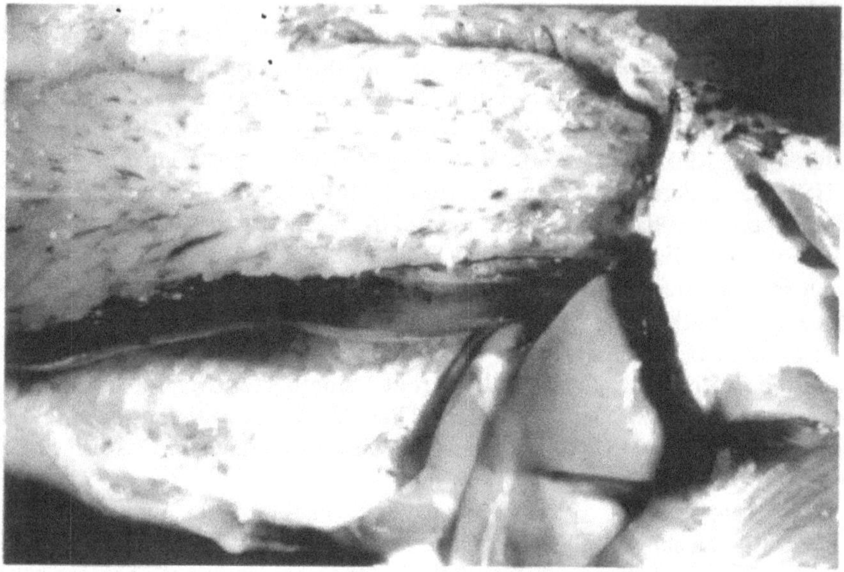

Fig. 2-2. Rainbow trout infected with viral hemorrhagic septicemia. Note pale gills and liver, and hemorrhage in the musculature and visceral organs. (Photograph by P. Ghittino).

The impact of VHSV on marine or estuarine fish populations is unclear, but the potential for this virus to adversely affect mariculture operations has not been demonstrated. As culture of salmonids, and possibly other fish species, develops in estuarine and sea waters, VHSV could become more important.

Infectious Hematopoietic Necrosis

Infectious hematopoietic necrosis virus (IHNV) produces infectious hematopoietic necrosis (IHN) primarily in Pacific salmon (Amend *et al.* 1969). IHN has also been called chinook salmon virus disease, Columbia River sockeye disease, Sacramento River chinook disease, and others (Wolf 1988). By definition, IHN is an acute, systemic disease caused by a rhabdovirus, usually in young cultured trout and salmon in the U.S. Pacific Northwest.

According to Wolf's (1988) historical discussion, IHN probably dates back to the late 1940's and 1950's when severe epizootics were reported in young salmon in some hatcheries in the Pacific salmon-producing region of North America (Rucker *et al.* 1953; Watson *et al.* 1954). It is generally accepted that these early reports of probable viral diseases were caused by the same agent that was later named infectious hematopoietic necrosis virus by Amend *et al.* (1969).

IHNV is a typical rhabdovirus that measures about 70 by 170 nm and has RNA genomic material. The virus is serologically distinct from other rhabdoviruses of fish, VHSV and spring viremia of carp virus, for example. Although IHNV is a serologically homogeneous virus, it can be separated into one variable and four distinct groups based on the relative size and molecular weights of their proteins (Leong *et al.* 1981). Winton *et al.* (1988) further classified IHNV isolates into four serological groups, using monocolonal antibody, but concluded that these groups were related to geographical origin rather than host species, virulence, or date of isolation.

The Salmonidae are the only known species susceptible to IHNV. Outbreaks have occurred in culture populations of chinook, pink (*O. gorbuscha*), and sockeye salmon and brown (*Salmo trutta*) and rainbow trout (Wolf 1988). Originally IHNV was confined to coastal rivers in the U.S. Pacific Northwest from California to Alaska (Amend and Wood 1972; Grischkowski and Amend 1976). As result of probable transport of IHNV-infected eggs, the virus was transmitted to Minnesota (Plumb 1972), Montana (Holway and Smith 1973), Virginia and West Virginia (Wolf *et al.* 1973), and New York (Carlisle *et al.* 1979). IHNV has also been reported in Japan, Europe, and possibly Australia (Wolf 1988).

IHNV is most severe in fry and fingerling fish where accumulated losses can approach 100%. The disease may occur in smolts, but mortality is not as great as in the younger fish. Most IHNV epizootics occur at 12°C or less, but some have been reported in water temperatures of 15°C (Wolf 1988). Mortality is arrested by increasing the water temperature to 18°C early in development of the disease, but survivors remain carriers (Amend 1970).

Infected fry and fingerlings have a general lethargic swimming behavior that is interspersed with erratic or spiral swimming (Amend *et al.* 1969). These fish maintain themselves with difficulty in any water current. Externally, infected fish are exophthalmic and appear darker than normal. Abdominal swelling and fecal casts are evident. Pale gills, hemorrhage at the base of fins, petechiae, and areas of excessive bleeding beneath the skin behind the head often characterize IHNV infections. Spinal deformities (scoliosis and lordosis) may occur among a small percentage of surviving fish (Amend *et al.* 1969). Internally, the body cavity often contains a clear, pale yellow fluid, and the peritoneum, adipose tissue, and other tissues show petechiation and general hemorrhage. Livers and kidneys are usually pale, but the hind gut may be hemorrhagic and void of food.

IHNV is a major limiting factor in the culture of some species of salmonids in western and Pacific Northwest areas of North America (Meyers *et al.* 1988). The majority of the problems with IHNV occur in anadromous Pacific salmon in their freshwater life stage. Although the virus apparently has no adverse effect on salmon once they migrate to sea, IHNV is capable of severe effects on young cultured salmonids and could reduce the fish available for smoltification.

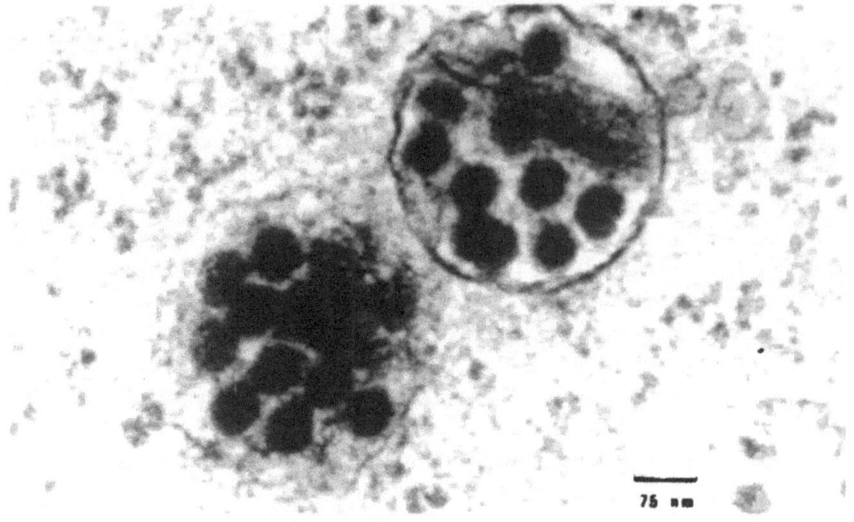

Fig. 2-3. Electron micrograph of Pacific salmon anemia virus (erythrocytic inclusion body syndrome [EIBS]). Virus measures 150 nm in diameter. (Photograph by J.R. Winton: reproduced by permission of Springer-Verlag.)

Erythrocytic Inclusion Body Syndrome

Erythrocytic inclusion body syndrome (EIBS), also known as Pacific salmon anemia virus, is a severe anemic condition primarily of cultured juvenile coho and chinook salmon (Holt and Rohovec 1984; Leek 1985; Leek 1987). EIBS is caused by a transmittable 80 nm icosahedral virus (Fig. 2-3) that has not been isolated in tissue culture but is considered a togavirus (Anakawa *et al.* 1989). The virus has been detected by electron microscopy in cytoplasmic inclusions in erythrocytes of anemic salmon. The EIBS virus is morphologically distinct from the iridovirus of viral erythrocytic necrosis.

Piacentini *et al.* (1989) demonstrated experimental transmission of EIBS in coho salmon and reported an incubation period for erythrocytic inclusion bodies at 12°C and hematocrit to drop to their lowest point (25%) in 28 days. Following the drop in hematocrit, fish returned to normal by day 45 post injection. Horizontal transmission was also achieved by cohabitation of infected fish with naive coho salmon. Experimental transmission also occurred in rainbow trout and cutthroat trout, *0. clarki*. Leek (1987) described a natural infection in juvenile spring chinook salmon reared in freshwater, but mortality due to the virus disease was only 2 to 3% above normal.

No behavioral, external, or internal clinical signs other than severe anemia are described. The gills are pale and hematocrits are low. Free swimming adult salmon in open seas are also affected, but EIBS impact on these fish is not known (J.R. Winton, personal communication). EIBS has been reported only from the Pacific Northwest, U.S. A significant problem associated with EIBS is that infected fish are more prone to bacterial kidney disease (*Renibacterium salmoninarum*) in salt water, enteric red mouth disease (*Yersinia ruckeri*), Cytophaga-like external bacteria in freshwater, and external fungi in both environments. EIBS-afflicted individuals are likely to recover if these secondary diseases do not kill them (J.R. Winton, personal communication).

Means of EIBS transmission are not known. Vertical transmission is questionable, and impact on freshwater and marine salmonids is currently minimal. However, EIBS is another viral disease that may become a greater problem as culture intensification of the susceptible species increases. How EIBS affects survivors in their marine environment is also unknown.

Herpesvirus Salmonis Disease

Herpesvirus salmonis disease is a viral infection of rainbow trout and steelhead trout originally described by Wolf and Taylor (1975). Their initial isolation was from rainbow trout adults in Washington, while the second was from steelhead in California (Hedrick *et al.* 1986).

The etiological agent is *Herpesvirus salmonis* (Wolf *et al.* 1978) that measures 90 to 95 nm. Hedrick *et al.* (1986) showed that the first and second isolates were similar, but distinguishable from each other. Hedrick *et al.* (1987) compared *Herpesvirus salmonis* isolated from North America to herpesvirus of salmonids from Japan and, due to serological differences, named the two North American isolates Type 1 and the Asian isolates Type 2 (Hedrick and Sano 1989).

Disease caused by *H. salmonis* occurs at 10°C or less. Adult fish from which the original isolate was taken (Wolf and Taylor 1975) showed only distinctly darkened pigmentation. Experimentally infected juvenile rainbow trout stopped eating after 2 weeks and became lethargic (Wolf and Smith 1981). The fish lay on the bottom of the tanks; some fish became dark, while most developed exophthalmia with hemorrhage in the eyes. Gills were pale and abdomens distended.

All of the population associated with the original *H. salmonis* virus was destroyed and the hatchery disinfected. The virus was presumed to be eliminated until Hedrick *et al.* (1986) isolated it again 10 years later. The importance of *H. salmonis* virus is minimal because it is a disease encountered infrequently; it possesses the potential to become a significant disease agent.

Oncorhynchus Masou Virus

Oncorhynchus masou virus (OMV) produces a disease in young members of several species of the genus *Oncorhynchus* (Kimura *et al.* 1981a). A lethal disease occurs in young fish while a neoplastic disease develops in older fish.

The causative agent is *Oncorhynchus masou* virus, which is a herpesvirus (Sano *et al.* 1983). The icosahedral nucleocapsid measures 100 to 115 nm in diameter, but with the envelope measures 220 to 240 nm (Tanaka *et al.* 1987). This description is specifically for the OMV agent that occurs only in Japan. The naturally susceptible species of fish are young himemasu (kokanee) (*0. masou*) and yamame (*0. yamame*); however, experimentally chum and coho salmon, and rainbow trout are susceptible (Yoshimizu *et al.* 1987).

OMV is an intriguing disease, because the virus manifests itself as a lethal infection at one point and a neoplastic infection at another. The fish with lethal infections from 50- to 150-days-old lose their appetite, while some fish show exophthalmia and petechiae on the body surface, especially under the jaw. The liver of infected chum salmon fry is mottled with white areas; in advanced cases, the liver is totally white. The spleen may be swollen, but ironically the kidney appears overtly normal (Kimura *et al.* 1981a; 1983).

Neoplasms develop on surviving chum, coho, and masou salmon and rainbow trout about 4 months after exposure and persist at least 1 year. Tumors occur mainly around the mouth (Fig. 2-4) and may also be found on fins, gill cover, body surface, corneas of the eye, and rarely in the kidney (Kimura *et al.* 1981a; Sano *et al.* 1983; Yoshimizu *et al.* 1987).

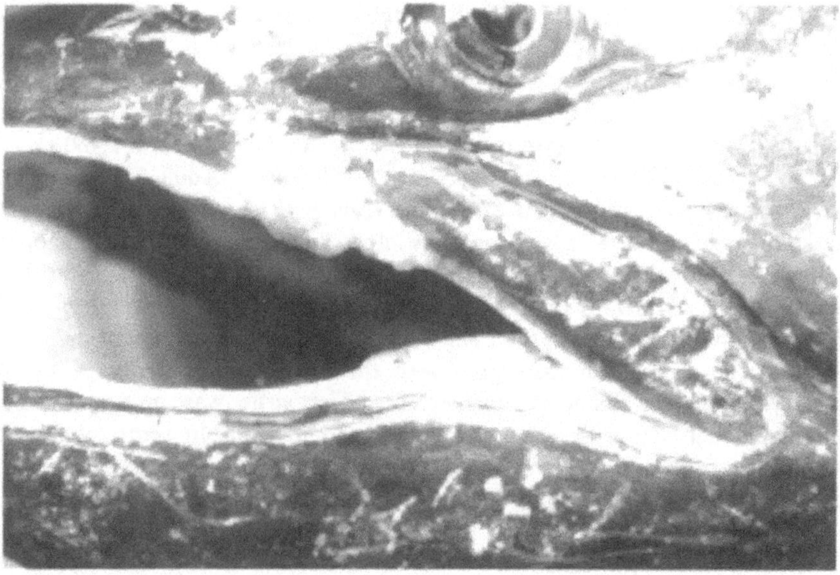

Fig. 2-4. *Oncorhynchus masou* virus (OMV) tumor on the mouth of wild masou salmon. (Photograph by M. Yoshimizu).

OMV was first discovered in normal land-locked masou salmon in 1978 (Kimura *et al.* 1981b), but can infect anadromous salmon as well. Most of the research has been done in Japan where the tumor-producing herpesviruses have been isolated from land-locked or sea pen-reared salmonids (Sano 1988).

Transmission of OMV is most certainly vertical through reproductive products and horizontally via the water. Very high percentages of mortality of young coho and masou salmon have been reported, but as fish get older they become more refractive (Kimura *et al.* 1981c; Tanaka *et al.* 1984).

The oncogenic nature of OMV provides a model for some interesting studies on virus induced tumors. Among masou survivors of lethal OMV epizootics, Kimura *et al.* (1981a,b) reported that 60% of the survivors had tumors and about 45% were around the mouth.

OMV is a major disease problem among land-locked salmon, particularly in northern Japan. A significant number of 1- to 5-month-old salmon may be killed, and survivors become carriers of the virus in the tumors and at least periodically shed virus. Since the virus was isolated from ovarian fluids of adult fish, vertical transmission is certain.

Chum Salmon Reovirus

Chum salmon reovirus (CSR) is a member of the family Reoviridae (Winton *et al.* 1980, 1981). Typically, CSR is not pathogenic when tested in Pacific salmon, although the virus was isolated from adult chum salmon returning to Hokkaido Island, Japan, on normal spawning runs. Isolation was obtained during routine virological examinations from overtly normal fish. There is no mention of any clinical signs that indicate infection, even in experimentally infected juvenile salmon.

VIRAL DISEASES OF NON-SALMONID FISHES

Viral diseases of non-salmonid marine fish are generally not as severe as those of salmonids in terms of causing large numbers of deaths; however, there are some exceptions. Many known viral agents of marine fish produce tumors, papillomas, or other types of growths on the skin and fins of afflicted fish.

Erythrocytic Necrosis Virus

Erythrocytic necrosis virus (ENV), causative agent of the disease viral erthrocytic necrosis (VEN), is a viremic condition in reptiles, amphibians, and fishes, characterized by cytoplasmic inclusion bodies in erythrocytes. In fish, VEN has also been known as piscine erythrocytic necrosis (PEN) (Laird and Bullock 1969); however, VEN holds precedent. Laird and Bullock (1969) first reported viral-like particles in cytoplasmic inclusions in fish erythrocytes, thus VEN was definitely considered to have a viral etiology (Fig. 2-5). Walker and Sherburne (1977) published electron micrographs showing distinct virions in the inclusions, thus conclusively demonstrating a viral etiology. For the most part, VEN of fish is known only from marine or anadromous species although Sherburne and Bean (1979) reported inclusions in erythrocytes of landlocked American smelt, *Osmerus mordax*. ENV has been reported from the hagfishes and lampreys and from members of at least 14 families of bony fishes, including anguillids, salmonids, clupeids, and others (Walker and Sherburne 1977; Kahn and Newmann 1982; and Wolf 1988). Geographical range of ENV includes the Northern Pacific Ocean, the Atlantic Ocean of North America, Greenland, the United Kingdom, and the Mediterranean Sea. VEN appears to be a more serious disease of young fishes, but presumably adult fish serve as reservoirs. Critical experimental age discrimination studies of VEN are lacking, mainly, because the causative agent cannot be isolated routinely in cell culture.

The primary overt clinical sign of VEN is anemia, characterized by pale gills and viscera. Wright-Giemsa stained blood smears of infected fish show small, round eosinophilic inclusions in the cytoplasm of erythrocytes that may bud from the nucleus. The percentage of infected erythrocytes in an individual fish may

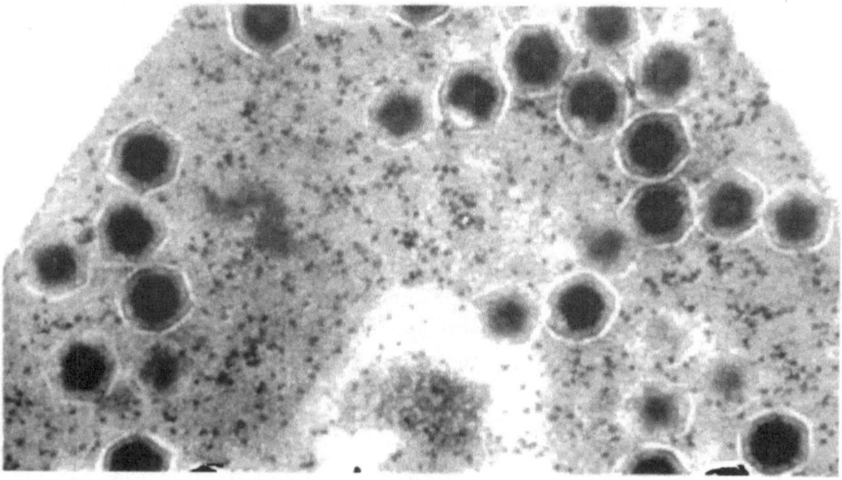

Fig. 2-5. Viral erythrocytic necrosis virus in cytoplasm of erythrocyte of Atlantic cod. Virus measures 300 nm. (Photograph by B. Nicholson).

range from 1% to 100% (Wolf 1988) and the inclusions measure 1 to 4 nm in diameter. Hematocrit values of ENV-infected fish may be as low as 2 to 10% (Evelyn and Traxler 1978).

VEN virus has not been identified and classified; however, electron micrographs indicate that the spheroidal causative agent measures from 150 to 400 nm in diameter. The size varies from fish species to species. It has been accepted that ENV is an iridovirus, but due to the various sizes of viral particles from different species of fish, there may be at least three different VEN-associated iridoviruses. For example, Reno *et al.* (1978) reported 145 nm diameter ENV in Atlantic herring. The ENV in Atlantic cod is about 360 nm (Appy *et al.* 1976). Evelyn and Traxler (1978) reported a viral particle of 190 nm in ENV-infected Pacific salmon.

The significance of VEN is unclear, but with the comparatively wide range of host susceptibility and geographical distribution, the potential for severity is present. Culture conditions may increase the incidence of VEN in infected populations. The incidence increased from 7 to 96% over a 2-week period in captive Atlantic herring, *Clupea harengus*, (Sherburne 1973) and (Philippon *et al.* 1977) and showed an increase of infectivity in Atlantic herring from 11% at capture to 90% 13 days later. VEN infection also has a tendency to increase the susceptibility of certain species of fish to vibriosis and oxygen depressions (MacMillan *et al.* 1980). Thus, VEN is a viral disease that could have major consequences on mariculture operations that use captured fish for seed or use wild brood fish for spawning. Also, MacMillan and Mulcahy (1979) successfully transmitted ENV from Pacific herring (*C. harengus pellasi*) to chum salmon, while Eaton (1990) showed that pink and sockeye salmon were also susceptible. In view of these reports ENV presents a significant threat to salmonid culture.

Lymphocystis

Lymphocystis virus produces a hypertrophic disease primarily of cells in the skin and fins of fish. It is perhaps the best known of all fish viral diseases due to its infectivity to a wide variety of fish species and extensive geographical range. Lymphocystis is also one of the oldest known viral diseases of fish.

Lymphocystis of fish was first recognized in 1874, and it was proposed that the etiology was a virus in 1914 (Weissenberg 1914, from Wolf 1988). However, it was not until 1964 that River's postulates were fulfilled for this virus and its virological nature fully appreciated (Weissenberg 1965). The virus is now considered an iridovirus. Since its early description, lymphocystis has had a long history in fish pathology. Until lymphocystis virus was isolated in cell cultures in the early 1960's (Wolf 1962), all investigations involved histopathology or electron microscopy, a trend which continues today.

Lymphocystis virus has the greatest host species range of all known fish viruses. The list of fish susceptibility includes at least 9 taxonomic orders, and over 34 families in which at least one species is susceptible (Wolf 1988); thus over 100 species are affected. While some of these species are freshwater inhabitants, many are of marine or brackish water environments and include anadromous and catadromous species. More common susceptible marine groups are flounders, soles, smelts, croakers, true basses, herrings, snappers, and drums.

Recognizing the extensive and unusual species susceptibility of lymphocystis, it is not surprising that the virus has essentially a world-wide geographical distribution. The disease has been reported from North, Central and South America, Europe, Africa, Asia, and nearly all major saline bodies of water. It is truly a cosmopolitan fish disease.

Lymphocystis has no distinct fish age preference, occurring in juvenile as well as adult fish. Incidence and transmission of lymphocystis are influenced by a number of factors from fish density to environmental conditions. In culture conditions, or confinement, a very high percentage of a population will be infected, probably reflecting the ease of horizontal transmission. Incidence may reach 100% in some intensively cultured fish (Paperna *et al.* 1982); however, incidences in wild populations of marine fishes are lower and range from 4 to 57%, depending on season of the year, with the higher incidence in the summer (Shelton and Wilson 1973; Reierson and Fugelli 1984).

Transmission and incubation depend upon host species, virus, and temperature. Also, lymphocystis virus from one species of fish may not be infective to a different species (Overstreet and Howse 1977). In fact, due to DNA cleavage patterns by restrictions enzymes, there may be different types of lymphocystis virus, depending on the species of fish infected. Flugel (1985) subdivided the virus into two groups: FLDV-1 in flounder and plaice and FLDV-2 in dabs. Lymphocystis virus disease is more prevalent in cool weather than warm weather, but at 10 to 15°C, incubation may take up to 6 weeks for lesions to develop, compared to 5 to 12 days at 20 to 25°C (Wolf 1962; Cook 1972; Roberts 1975). Transmission is facilitated by increased fish density, trauma during spawning (Ryder 1961), and from netting or tagging practices (Olson 1958; Clifford and Applegate 1970). Even some external parasites (copepods or isopods) that disrupt the protective mucous layer may increase rate of transmission (Nigrelli 1950; Lawler *et al.* 1974). It has been suggested that a higher incidence of lymphocystis is influenced by pollution (Christmas and Howse 1970; Hill 1984). Reierson and Fugelli (1984) described higher incidence of lymphocystis in flounder in polluted waters than in less polluted waters of Norway, but cautioned against definitive conclusions because of the complexity of the disease/environmental relationship.

Lymphocystis manifests itself by producing hypertrophied cells in the connective tissue of the integument of the body surface and the fins (Fig. 2-6).

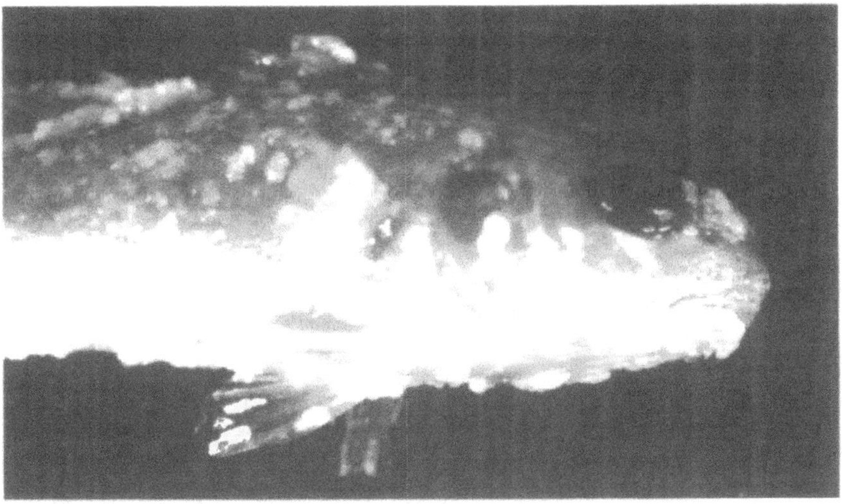

Fig. 2-6. *Lymphocystis* on Atlantic croaker.

Overstreet (1988) reported that naturally infected croakers had lymphocystis lesions on 60 to 80% of their body surface, but this degree of infection is unusual. Large grayish to white cells measuring up to 2 mm in diameter are easily seen without magnification. These cells may occur singly or be grouped together in "grape-like" clusters, giving a tumor appearance. However, lymphocystis is not a true tumor. Only rarely do lymphocystis-induced cells occur in visceral organs. There are no distinct behavioral patterns associated with lymphocystis infections.

The hypertrophied cells may be 20 to 5,000 times larger than normal cells and are easily visualized and identified by histopathology. These cells have an enlarged nucleus and Feulgen positive basophilic, ribbon shaped, cytoplasmic inclusion bodies (Walker 1965). A hyaline capsule, or matrix, surrounds the cells (Walker 1962). The icosahedral virus is classified as an iridovirus that measures 145 to 300 nm in diameter (Kelly and Robertson 1973). Transmission presumably takes place horizontally when lesions slough and cells lyse, releasing virions into the environment.

Significance of lymphocystis is not from the lethality of the virus, but from an aesthetic point of view. Infected fish are not retained for food by fishermen, although lesions are transient. In spawning and seed fish operations, entire populations of fingerlings or juveniles may be rejected for further use if they are heavily infected with lymphocystis. Although lymphocystis-infected fish do not die as result of infection, they are of reduced economic value.

Atlantic Menhaden Spinning Disease

Atlantic menhaden spinning disease (AMSD) occurs almost annually among populations of Atlantic menhaden, *Brevoortia tyrannus*, along the Atlantic coastal waters of the United States during spring and summer (Newman 1980). According to Wolf (1988), IPN-like virus has been isolated from fish in some of these epidemics, but not all; therefore IPN-like virus is not considered the universal pathogen of AMSD. Sindermann (1970) suggested a viral aetiology of AMSD, that a virus that closely resembled IPNV biophysically and serologically was not isolated until the late 1970's (Stephens *et al.* 1980). When juvenile menhaden were immersed in a virus solution, the menhaden spinning disease was reproduced. The disease normally infects young-of-the-year Atlantic menhaden, but older individuals have occasionally been involved. Most epidemics have occurred in the Chesapeake Bay area.

Clinical signs of AMSD were reported by Newman (1980). Gross signs include abnormal swimming (corkscrew fashion), hyperemia at the base of fins, and hemorrhage in the eye and vent. Internally, the spleen and brain are congested, and the lumen of the intestine often contains a bloody fluid.

Evidence points to a viral aetiology with IPN-like virus a possibility, but a virus other than IPNV is the suspected principal cause. Atlantic menhaden spinning disease must be considered a minor problem based on present knowledge.

Atlantic Cod Ulcus Syndrome

Atlantic cod ulcus syndrome is a condition of 1.5- to 2.5-year-old Atlantic cod, *Gadus morhua*, taken from organically polluted coastal waters of Denmark (Jensen and Larsen 1976, from Wolf 1988). The disease has been recognized for over 50 years and was considered of bacterial origin, primarily *Vibrio anguillarum*, until Jensen *et al.* (1979) isolated a virus from cod with ulcus syndrome. Jensen and Larsen (1982) demonstrated viral transmission of the disease, using an iridovirus from the naturally infected fish. Jensen and Larsen (1979) described the disease in five stages, ranging from an initial papulovesicular stage to an intermediate ulcerative stage and finally healing. These signs include initial papules 2-8 mm in diameter and 1-3 mm high forming on the skin followed by crater-like erosions that progress into large, red ulcers followed by healing of these ulcers. Atlantic cod ulcus disease is most prevalent in autumn. Larsen and Jensen (1982) also have suggested that the incidence of the disease increased where waters were polluted by organic effluents from fruit and vegetable processing plants.

Hirame Rhabodovirus

Hirame rhabdovirus (HRV) was isolated from cultured Japanese flounder (hirame), *Paralichthys olivaceus*, and ayu, *Plecoglossus altivelis*, in Japan in the mid-1980's (Kimura *et al.* 1986). HRV was responsible for a severe hemorrhagic disease in fish that showed pale gills, abdominal distention, ascites, and hemorrhage in the fins and skeletal muscle. Internally gonadal congestion and hemorrhaged organs were evident. Rainbow trout were susceptible to experimental inoculation, but chum, coho, and masu salmon were not. Effects of HRV on aquaculture are not fully known, but have been reported only from Japan.

VIRUSES KNOWN ONLY BY ELECTRON MICROSCOPY

Several diseases of marine fishes are known to be associated with viruses only through electron microscopy and generally have limited known geographical range. Some are suspected to be capable of killing fish, while others cause unsightly neoplasia or tumors.

Atlantic Cod Adenovirus

Atlantic cod adenovirus was seen in nuclei of hyperplastic epidermal plaque-like lesions on the body surface of wild Atlantic cod (Jensen and Bloch 1980, from Wolf 1988). The skin lesions are slightly raised, transparent plaques that occur on various areas of the body surface. The virus has no known lethal capability.

Atlantic Salmon Papillomatosis

This disease is a chronic proliferation of epidermal cells producing plaque-like lesions on various regions of parr and occasionally young adult Atlantic salmon. The disease has been reported in Scandinavia (Chronwall 1976) and Scotland (Roberts and Bullock 1979). Wolf (1988), in a historical review of Atlantic salmon papillomatosis, cites references that suggest this disease has been known at least from the 1950's. Outbreaks occur among salmon primarily in fresh water during summer and peak in September and October (Johansson 1977, from Wolf 1988). A near 100% incidence is seen in some freshwater populations, but incidence is reduced in salt water to 1% if it is present at all. Infected fish develop multiple skin lesions anywhere on the body. These lesions are smooth or rough and several millimeters thick and up to 40 mm in diameter. Attempts to isolate virus from the lesions have generally failed, but Carlisle (1977) noted virus-like particles in electron micrographs of tissues from infected salmon. Significance of Atlantic salmon papilloma is not known but it occurs in cultured and wild stocks.

Atlantic Salmon Swim Bladder Sarcoma Virus

This neoplasm is associated with a C-type retrovirus (Duncan 1978). Atlantic salmon swim bladder sarcoma is a massive tumor in the swim bladder of subadult Atlantic salmon that was associated with mortality (McKnight 1978). The disease occurred in 1-year-old smolts and 2-year-old subadults in cages in a Scottish loch. Massive tumors, 1.5-3 cm in diameter, developed on the swim bladder and filled much of the visceral cavity, killing about 3.4% of the fish. Although only one incidence has been reported, this disease could become a significant problem.

Pacific Cod Herpesvirus

The Pacific cod herpesvirus has been seen in giant cells in some lesions of Pacific cod, *Gadus macrocephalus*, in the Bering Sea. This disease was first reported by McArn *et al.* (1978), who found that about 4% of Pacific cod examined had lesions, some of which were hypertrophied cells containing herpesvirus-like particles. The lesions were circular, pale, 1-15 mm in diameter, and contained giant cells embedded in a hyaline extracellular coat. McCain *et al.* (1979) described the lesions as ring-like and large (10-50 mm).

Pleuronectid Papilloma

According to Wolf (1988), pleuronectid papilloma is a collective term used for a group of similar growths that occur in epidemic proportions on the skin of several species of Pacific flatfishes. The etiology of these tumors, which have been recognized for many years, has not been precisely determined, but viruses have been seen in cells in these lesions (Nigrelli *et al.* 1965; Cooper and Keller 1969). Papillomas occur anywhere on the body, but most often on the pigmented side. The early, 1- to 2-mm, smooth, raised, pink growths enlarge to several centimeters and become rough. These epidermal papillomas may not be pigmented although they are on the pigmented skin surface (Wellings *et al.* 1976). Generally, younger pleuronectids are more severely affected and incidence declines with age, inferring that the tumors may result in death (Wellings *et al.* 1976; Peters and Watermann 1979). In a global literature survey, Stich *et al.* (1977) found 18 species of pleuronectids with epidermal papillomas, all of which came from waters in the northern hemisphere.

Winter Flounder Papilloma

Winter flounder papilloma is a benign neoplasm in the skin of winter flounder *Pseudopleuronectes americanus*, with affected cells containing cytoplasmic virus-like particles measuring 30 nm. The only incidence of this disease was reported from Newfoundland (Emerson *et al.* 1985). Lesions are distinct areas of raised tissue on the pigmented skin. The papillomas resemble lymphocystis, but the cells lack hypertrophy. This disease has very little significance.

Turbot Herpesvirus

Turbot herpesvirus disease is a condition of turbot characterized by giant cells on the skin and gill epithelium of young-of-the-year fish (Buchanan and Madeley 1978). In the first report of this disease, a large number of young turbot in a fish farm were killed. The disease occurs in wild fish as well as cultured fish. To date the only cases have been reported from Scotland.

Overt external lesions or internal clinical signs are lacking in infected fish (Buchanan *et al.* 1978); however, affected fish become anorexic and lethargic. The giant cells contain large numbers of herpesvirus-like viral particles that are circular or hexagonal and measure approximately 100 nm in diameter (Buchanan and Madeley 1978). Cytoplasmic enveloped virions measure 200 nm. Richards and Buchanan (1978) postulated that health problems arise when fish are stressed during transport, handling, and temperature fluctuations, or when fish are exposed to other stressful environmental conditions.

RESEARCH NEEDS IN MARINE FISH VIRAL DISEASES

Research needs for virus diseases of marine finfish should follow several basic avenues. From an environmental point of view, the effects of water quality, chemical, heavy metal, and organic pollutants on the susceptibility of fish to viral agents need further elucidation. Sufficient data are available to indicate a potential relationship between environmental quality and viral infections, but definitive correlations are lacking. Environmental research must be carried out in culture situations as well as natural fish populations. Estuarine areas of marine waters are the most vulnerable to pollution; therefore, research should concentrate on such habitats and include studies on the effect of viruses on survival, growth, and quality of selected, economically important fish species.

A second area of viral research needs to examine effects of known viruses on wild and cultured fish populations. Projects should include epidemiological studies, such as reservoir and transmission studies; pathogenic and pathogenesis studies; and refinement and development of virus detection methods both *in vivo* in fish and *in vitro* in the environment. Fate of cultured salmonids released in the marine environment after surviving epizootics of IHNV, IPNV, EIBS, VHSV, or ENV is unknown. Investigations into their survival would be of great importance in disposition of surviving populations. Such studies could be correlated with projects that concentrate on the role of the environment (pollutants) on the incidence of viral diseases. Since a few viral diseases of fish are neoplastic or hypertrophic, the opportunity exists to measure the relationships of viruses and chemicals in neoplastic diseases.

An effort should be made to further develop measures to prevent and control viral diseases of marine fishes. These control studies should be in cooperation with detection and identification studies. Also investigations of currently developing vaccines and new vaccines should be pursued. Success of any research effort on virus diseases of marine fish can be achieved only through a long-term commitment by research and funding agencies.

REFERENCES

Ahne, W., and R.D. Negele. 1985. Studies on the transmission of infectious pancreatic necrosis virus via eyed eggs and sexual products of salmonid fish. In: *Fish and Shellfish Pathology*. ed. A.E. Ellis, pp. 261-269. London: Academic Press.

Altara, J. 1963. Resolutions on Item A, viral hemorrhagic septicemia. Permanent Commission for the Study of Diseases of Fish, Office International des Epizooties. *Bull. Off. Int. Epizoot.* 59:298-299.

Amend, D.F. 1970. Control of infectious hematopoietic necrosis virus disease by elevating the water temperature. *J. Fish. Res. Board Can.* 27:265-270.

Amend, D.F., and J.W. Wood. 1972. Survey of infectious hematopoietic necrosis (IHN) virus in Washington salmon. *Prog. Fish-Cult.* 34:143-147.

Amend, D.F., W.T. Yasutake, and R.W. Mead. 1969. A hematopoietic virus disease of rainbow trout and sockeye salmon. *Trans. Am. Fish. Soc.* 98:796-804.

Anakawa, C.K., D.A. Hursh, C.N. Lannan, J.S. Rohovec, and J.R. Winton. 1989. Preliminary characterization of a virus causing infectious anemia among stocks of salmonid fish in the United States. In *Viruses of Lower Vertebrates*, ed. W. Ahne and E. Kurstak, pp. 442-450. Berlin, Heidelberg: Springer-Verlag.

Appy, R.G., M.D. B. Burt, and T.J. Morris. 1976. Viral nature of piscine erythrocytic necrosis (PEN) in the blood of Atlantic cod (*Gadus morhua*). *J. Fish. Res. Board Can.* 33:1380-1385.

Bengtsson, B.E. 1975. Vertebral damage in fish induced in pollutants. In *Sublethal Effects of Toxic Chemicals on Aquatic Animals*, ed. J.H. Koeman and J.J. Sikes. Amsterdam: Elsevier.

Bovo, G., and G. Giorgetti. 1983. Episodio di septicemia hemorragica virale trote iridea (*Salmo gairdneri*) allevate in acqua salmastra. *Atti Soc. Ital. Sci. Vet.* 37:688-689.

Bregenballe, F. 1981. Kan Fiskehenjren overfore Egtved virus? *Med. Forsogsdambruget* No. 64.

Brunson, R., R. True, and J. Yancey. 1989. VHS virus isolated at Makah National Fish Hatchery. *Am. Fish. Soc. Fish Health Section Newsl.* 17 (2):3.

Buchanan, J.S., and C.R. Madeley. 1978. Studies on *Herpesvirus scophthalmi* infection of turbot (*Scophthalmus maximus* L.): ultrastructural observations. *J. Fish Dis.* 1:283-295.

Buchanan, J.S, R.H. Richards, C. Sommerville, and C. R. Madeley. 1978. A herpes-type virus from (*Scophthalmus maximus* L). *Vet. Rec.* 2:527-528.

Bullock, G.L., R.R. Rucker, D. Amend, K. Wolf, and M.H. Stuckey. 1976. Infectious pancreatic necrosis: transmission with iodine treated and nontreated eggs of brook trout (*Salvelinus fontinalis*). *J. Fish. Res. Board Can.* 33:1197-1198.

Carlisle, J.C. 1977. An epidermal papilloma of the Atlantic salmon II: ultrastructure and etiology. *J. Wild. Dis.* 13:235-239.

Carlisle, J.C., K.A. Schat and R. Elston. 1979. Infectious pancreatic necrosis in rainbow trout *Salmo gairdneri* Richardson in a semi-closed system. *J. Fish Dis.* 2:511-517.

Castric, J., and P. de Kinkelin. 1980. Occurrence of viral hemorrhagic septicemia in rainbow trout *Salmo gairdneri* Richardson reared in seawater. *J. Fish Dis.* 3:21-27.

Chang, N., R.D. MacDonald, and T. Yamamoto. 1978. Purification of infectious pancreatic necrosis (IPN) virus and comparison of polypeptide composition of different isolates. *Can. J. Microbiol.* 24:19-27.

Christmas, J.Y., and H.D. Howse. 1970. The occurrence of lymphocystis in *Micropogon undulatus* and *Cynoscion arenarius* from Mississippi estuaries. *Gulf Res. Rep.* 3:131-154.

Chronwall, B. 1976. Epidermal papillomas in *Salmo salar* L.: a histological description. *Zoonoses* 4:109-114.

Clifford, T.J., and R.L. Applegate. 1970. Lymphocystis disease in tagged and untagged walleyes in a South Dakota lake. *Prog. Fish-Cult.* 32:177.

Cook, D.W. 1972. Experimental infection studies with lymphocystis virus from Atlantic croaker. In *Proceedings of the 3rd Annual Workshop of the World Mariculture Society*, ed. J. W. Avault, E. Boudreaux, and E. Jaspers, pp. 329-335.

Cooper, R.C., and C.A. Keller. 1969. Epizootiology of papillomas in English sole, *Parophrys vetulus*. In *Neoplasms and Related Disorders of Invertebrates and Lower Vertebrate Animals*. ed. C.J. Dawe and J.C. Harshbarger, *Natl. Cancer Inst. Monogr.* 31.

Couch, J.A., and D.R. Nimmo. 1974. Detection of interactions between natural pathogens and pollutants in aquatic animals. In *Proceedings Gulf Coast Regional Symposium on Disease of Aquatic Animals*, ed. R.L. Amborski, M.M. Hood, and R.R. Miller, LSU-SG-7405, pp. 261-268. Baton Rouge, LA: Center for Wetlands Resource, Louisiana State University.

Couch, J.A., and L. Courtney. 1977. Interaction of chemical pollutants and virus in a crustacean: A novel bioassay system. *Ann. NY Acad. Sci.* 298:497-504.

Couch, J.A., J. Winstead, T. Hansen, and L.R. Goodman. 1979. Vertebral dysplasia in young fish exposed to the herbicide trifluralin. *J. Fish Dis.* 2: 35-42.

de Kinkelin, P. 1983. Viral haemorrhagic septicemia. In *Antigens of Fish Pathogens: Development and Production for Vaccines and Serodiagnostics*. ed. D.P. Anderson, M. Dorson, and P. Dubourget, pp. 51-62. Lyon, France: Collection Foundation Marcel Merieux.

de Kinkelin, P., M. Le Berre, A. Meurillon, and M. Calemsl. 1974. Septicémie hémorragique virale: démonstration de l'état réfractaire du saumon coho (*Oncorhynchus kisutch*) et de la truite fario (*Salmo trutta*). *Bull. Fr. Piscic.* 253:166-176.

Dorson, M., and C. Torchy. 1981. The influence of fish age and water temperature on mortalities of rainbow trout (*Salmo gairdneri* Richardson) caused by a European strain of infectious pancreatic necrosis virus. *J. Fish Dis.* 4:213-221.

Dorson, M., and C. Torchy. 1985. Experimental transmission of infectious pancreatic necrosis virus via the sexual products. In *Fish and Shellfish Pathology*, ed. A. E. Ellis, pp. 251-260. London: Academic Press.

Duncan, I.B. 1978. Evidence for an oncovirus in swim bladder fibrosarcoma of Atlantic salmon *Salmo salar* L. *J. Fish Dis.* 1:127-131.

Eaton, W.D. 1990. Artificial transmission of erythrocytic necrosis virus (ENV) from Pacific herring in Alaska to chum, sockeye, pink salmon. *J. Appl. Ichthyol.* 6:136-141.

Emerson, C.J., J.F. Payne, and A.K. Bal. 1985. Evidence for the presence of a viral non-lymphocystis type disease in winter flounder, *Pseudopleuronectes americanus* (Walbaum), from the north-west Atlantic. *J. Fish Dis.* 8:91-102.

Evelyn, T.P.T., and G.S. Traxler. 1978. Viral erythrocytic necrosis: natural occurrence in Pacific salmon and experimental transmission. *J. Fish. Res. Board Can.* 35:903-907.

Flugel, R.M. 1985. Lymphocystis disease virus. *Curr. Top. Microbiol. Immunol.* 116:134-150.

Gardner, G.R. 1975. Chemically induced lesions in estuarine or marine telosts. In *The Pathology of Fishes*, ed. W.E. Ribelin and G. Magaki, pp. 657-693. Madison: University Wisconsin Press.

Giles, R.C., L.R. Brown, and C.D. Minchew. 1978. Bacteriological aspects of erosion in mullet exposed to crude oil. *J. Fish Biol.* 13:113-117.

Grischkowski, R.S. and D.F. Amend. 1976. Infectious hematopoietic necrosis virus: prevalence in certain Alaskan sockeye salmon, *Oncorhynchus nerka*. *J. Fish. Res. Board Can.* 33:186-188.

Hard, G.C., R. Williams, and J. Lee. 1979. Histopathology of superficial fish "tumours" found during a cancer survey of demersal fish in Port Philip Bay, Victoria, Australia. *J. Fish Dis.* 2:455-467.

Hedrick, R.P., and T. Sano. 1989. Herpesvirus of fishes. In *Viruses of Lower Vertebrates*, ed. W. Ahne and E. Kurstak, pp. 161-170. Berlin: Springer-Verlag.

Hedrick, R.P., T. McDowell, W.D. Eaton, L. Chan, and W. Wingfield. 1986. *Herpesvirus salmonis* (HPV): first occurrence in anadromous salmonids. *Bull. Eur. Assoc. Fish Pathol.* 6:66-68.

Hedrick, T.P., T. McDowell, W.D. Eaton, T. Kimura, and T. Sano. 1987. Serological relationship of five herpesviruses isolated from salmonid fishes. *J. Appl. Ichthyol.* 3:87-92.

Hill, B.J. 1984. Lymphocystis disease of fish. In *Fiches d'Identification des Maladies et Parasites des Poissons, Crustacés et Mollusques*. ed. C.0. Sindermann, *Cons. Int. Explor. Mer* 2:5.

Hill, B.J., and K. Way. 1988. Proposed standardization of the serological classification of aquatic birnavirus. *Fish Health Section/American Fisheries Society International Fish Health Conference*, Vancouver, British Columbia, Canada, July 19-21, 1988.

Holder, M., and W. Ahne. 1988. Freshwater crayfish *Astacus astacus* - a vector for infectious pancreatic necrosis virus (IPNV). *Dis. Aquat. Organ.* 4:205-209.

Holt, R., and J. Rohovec. 1984. Anemia of coho salmon in Oregon. *Am. Fish. Soc. Fish Health Sec. Newsl.* 12:4.

Holway, J.E., and C.E. Smith. 1973. Infectious hematopoietic necrosis of rainbow trout in Montana: a case report. *J. Wildl. Dis.* 9:287-290.

Hooper, K. 1989. The isolation of VHSV from chinook salmon at Glenwood Springs, Orcas Island, Washington. *Am. Fish. Soc. Fish Health Sec. Newsl.* 17(2):1.

Horlyck, V., S. Mellergard, I. Dalsgaard, and P.E. Vestergard Jorgensen. 1984. Occurrence of VHS in Danish maricultured rainbow trout. *Bull. Eur. Assoc. Fish Pathol.* 4:11-13.

Jensen, M.H. 1963. Preparation of fish tissue cultures for virus research. *Bull. Off. Int. Epizoot.* 59:131-134.

Jensen, N.J., and J.L. Larsen. 1976. Research project concerning the possible relationship between pollution factors, especially carbohydrate pollution and marine vibrios. *Dan. Vertinaertidsskr.* 59:521-524.

Jensen, N.J., and J.L. Larsen. 1979. The ulcus-syndrome in cod (*Gadus morhua*). I. A pathological and histological study. *Nord. Veterinaermed.* 31:221-228.

Jensen, N.J., and B. Bloch. 1980. Adenovirus-like particles associated with epidermal hyperplasia in cod (*Gadus morhua*). *Nord. Veterinaermed.* 32:173-175.

Jensen, N.J., and J.L. Larsen. 1982. The ulcus-syndrome in cod (*Gadus morhua*). IV. Transmission experiments with two viruses isolated from cod and *Vibrio anguillarum*. *Nor. Veterinaermed* 34:136-142.

Jensen, N.J., B. Bloch, and J.L. Larsen. 1979. The ulcus-syndrome in cod (*Gadus morhua*). III. A preliminary virological report. *Nord. Veterinaermed.* 31:436-442.

Johannsson, N. 1977. Studies on diseases in hatchery-reared Atlantic almon (*Salmo salar*) L. and sea trout (*Salmo trutta* L.) in Sweden. *Abstracts of Uppsala Dissertations From the Faculty of Science*, 401, Acta University Uppsaliensis, Sweden.
Kelly, D.C., and J.S. 1973. Icosahedral cytoplasmic deoxyriboviruses. *J. Gen. Virol.* 20 (Suppl.):17-14.
Kahn, R.A., and M.W. Newman. 1982. Blood parasites from fish of the Gulf of Maine to Cape Hatteras, Northwest Atlantic Ocean, with notes on the distribution of fish hematozoa. *Can. J. Zool.* 60:396-402.
Kimura, T., M. Yoshimizu and M. Tanaka. 1981a. Studies on a new virus (OMV) from *Oncorhynchus masou*. II. Oncogenic nature. *Fish Pathol.* 5:149-153.
Kimura, T., M. Yoshimizu, and M. Tanaka. 1981b. Fish viruses: tumor induction in *Oncorhynchus keta* by the OMV herpesvirus. In *Phyletic Approaches to Cancer*,. ed. C.J. Dawe, J.C. Harshbarger, T. Sugimara, S. Takamatsu, pp. 59-68. Tokyo: Japanese Scientific Society Press.
Kimura T., M. Yoshimizu, M. Tanaka, and H. Sannohe. 1981c. Studies on a new virus (OMV) from *Oncorhynchus masou*. I. Characteristics and Pathogenicity. *Fish Pathol.* 15:149-153.
Kimura, T., M. Yoshimizu, and M. Tanaka. 1983. Susceptibility of different fry stages of representative salmonid species to *Oncorhynchus masou* virus (OMV). *Fish Pathol.* 17:251-258.
Kimura, T., M. Yoshimizu, and S. Gorie. 1986. A new rhabdovirus isolated in Japan from cultured hirame (Japanese flounder) *Paralichthys olivaceus* and ayu *Plecoglossus altivelis*. *Dis. Aquat. Organ.* 1:209-217.
Laird, M., and W.L. Bullock. 1969. Marine fish hematozoa from New Brunswick, New England. *J. Fish. Res. Board Can.* 26:1075-1102.
Larsen, J.L., and N.J. Jensen. 1982. The ulcus-syndrome in cod (*Gadus morhua*). V. Prevalence in selected Danish marine recipients and a control site in the period 1976-1979. *Nord. Veterinaermed* 34:303-312.
Lawler, A.R., H.D. Howse, and D.W. Cook. 1974. Silver perch, *Bairdiella chrysura*: new host for lymphocystis. *Copeia* 1974:266-269.
Leek, S.L. 1985. Artificial transmission of erythrocytic virus. *Am. Fish. Soc. Fish Health Sec. Newsl.* 13:4.
Leek, S.L. 1987. Viral erythrocytic inclusion body syndrome (EIBS) occurring in juvenile spring chinook salmon (*Oncorhynchus tshawytscha*) reared in freshwater. *Can. J. Fish. Aquat. Sci.* 44:685-688.
Leong, J.C., Y. L. Hsu, H.M. Engelking, and D. Mulcahy. 1981. Strains of infectious hematopoietic necrosis (IHN) virus may be identified by structural protein differences. *Dev. Biol. Stand.* 49:43-55.
MacMillian, J.R., and D. Mulcahy. 1979. Viral erythrocytic necrosis (VEN) disease: incidence in Puget Sound, artificial transmission and species susceptibility. *J. Fish. Res. Board Can.* 36:1097-1101.
MacMillan, J.R., D. Mulcahy, and M. Landolt. 1980. Viral erythrocytic necrosis: some physiological consequences of infection in chum salmon (*Oncorhynchus keta*). *Can. J. Fish. Aquat. Sci.* 37:799-804.
Maurin, C., and J. McArdle. 1983. Report of the working group "Pathology and Diseases in Marine Organisms." *Int. Counc. Explor. Sea* 23.
McAllister, P.E., and W.J. Owens. 1988. Infectious pancreatic necrosis virus: protocol for a standard challenge in brook trout. *Trans. Am. Fish. Soc.* 115:466-470.
McAllister, K.W., and P.E. McAllister. 1988. Transmission of infectious pancreatic necrosis from carrier striped bass to brook trout. *Dis. Aquat. Organ.* 4:101-104.
McArn, G.E., B. McCain, and S.R. Wellings. 1978. Skin lesions and associated virus in Pacific cod (*Gadus macrocephalus*) in the Bering Sea. *Fed. Proc.* 37:937.

McCain, B.B., M.S. Myers, W.D. Gronlund, S.R. Wellings, and C.E. Alpers. 1978. The frequency, distribution, and pathology of three diseases of demersal fishes in the Bering Sea. *J. Fish Biol.* 12:267-276.

McCain, B.B., W.D. Gronlund, M.S. Myers, and S.R. Wellings. 1979. Tumors and microbial diseases of marine fishes in Alaskan waters. *J. Fish Dis.* 2:111-130.

McKnight, I.J. 1978. Sarcoma of the swim bladder of Atlantic salmon (*Salmo salar* L.). *Aquaculture* 13:55-60.

Mearns, A.J., and M.J. Sherwood. 1977. Changes in the prevalence of fin erosion of Los Angeles and Orange Counties. *South. Calif. Coastal Water Res. Proj. Annu. Rep.* 143-145.

Meyers, T.R., J. Thomas, J. Follett, and R. Saff. 1988. IHNV: Trends in prevalence and the "farming around" approach in Alaskan sockeye salmon culture. *Fish Health Section/American Fisheries Society International Fish Health Conference.*, Vancouver, B.C., Canada, July 19-21, 1988.

M'Gonigle, R.H. 1941. Acute catarrhal enteritis of salmonid fingerlings. *Tran. Am. Fish. Soc.* 70:297-303.

Minchew, C.D., and J. Yarbrough. 1977. The occurrence of fin rot in mullet (*Mugil cephalus*) associated with oil contamination of an estuarine pond-ecosystem. *J. Fish Biol.* 10:319-323.

Möller, H. 1985. A critical review on the role of pollution as a cause of fish diseases. In *Fish and Shellfish Pathology*, ed. A. E. Ellis, pp. 169. London: Academic Press.

Murchelano, R.A., and J. Ziskowski. 1982. Fin rot disease in the New York Bight (1973-1977). In *Ecological Stress and New York Bight: Science and Management*, ed. G. F. Mayer, Columbia Estuarine Research Foundation.

Newman, M.W. 1980. IPN virus disease of clupeid fishes. *Int. Counc. Explor. Sea, Special Meeting on Diseases of Commercially Important Marine Fish and Shellfish*, No. 24.

Nigrelli, R.F. 1950. Lymphocystis disease and ergasilid parasites in fishes. *J. Parasitol.* 36:36.

Nigrelli, R.F., K.S. Ketchen, and G.D. Ruggieri. 1965. Studies on virus diseases of fishes: Epizootiology of epithelial tumors in the skin of flatfishes of the Pacific coast, with special reference to the sand sole (*Psettichthys melanosticus*) from northern Hecate Strait, British Columbia, Canada. *Zoologica* 50:115-122.

Olson, D.E. 1958. Statistics of a walleye sport fishery in a Minnesota lake. *Trans. Am. Fish. Soc.* 87:52-72.

Ord, W.M., M. Le Berre, and P. de Kinkelin. 1976. Viral hemorrhagic septicemia: comparative susceptibility of rainbow trout (*Salmo gairdneri*) and hybrids (*S. gairdneri* x *Oncorhynchus kisutch*) to experimental infection. *J. Fish. Res. Board Can.* 33:1205-1208.

Overstreet, R.M. 1988. Aquatic pollution problems, southeastern U.S. coasts: histopathological indicators. *Aquat. Toxicol.* 11:213-239.

Overstreet, R.M., and H.D. Howse. 1977. Some parasites and diseases of estuarine fishes in polluted habitats of Mississippi. *Ann. NY Acad. Sci.* 298:427-462.

Paperna, I., I. Sabnai, and A. Colorni. 1982. An outbreak of lymphocystis in *Sparus aurata* L. in the Gulf of Aqaba, Red Sea. *J. Fish Dis.* 5:433-437.

Peters, N., and B. Watermann. 1979. Three types of skin papillomas of flatfishes and their causes. *Mar. Ecol. Prog. Ser.* 1:269-276.

Philippon, M.B., L. Nicholson, and S.W. Sherburne. 1977. Piscine erythrocytic necrosis (PEN) in the Atlantic herring (*Clupea harengus harengus*): evidence for a viral infection. *Am. Fish. Soc. Fish Health News* 6:6-10.

Phillips, M.L., N.E. Warner, and H.W. Puffer. 1976. Oral papillomas in *Genyonemus lineatus* (white croakers). *Prog. Exper. Tumor Res.* 20:108-112.

Piacentini, S.C., J.S. Rohovec, and J.L. Fryer. 1989. Epizootiology of erythrocytic inclusion body syndrome. *J. Aquat. Anim. Health* 1:173-179.

Plumb, J.A. 1972. A virus-caused epizootic of rainbow trout (*Salmo gairdneri*) in Minnesota. *Trans. Am. Fish. Soc.* 101:121-123.

Reiersen, L.O., and K. Fugelli. 1984. Annual variation in lymphocystis infection frequency in flounder, *Platichthys flesus* (L.). *J. Fish Biol.* 24:187-191.

Reno, R.W., M. Phillippon-Fried, B.L. Nicholson, and S.W. Sherburne. 1978. Ultrastructure studies of piscine erythrocytic necrosis (PEN) in Atlantic herring (*Clupea harengus harengus*). *J. Fish. Res. Board Can.* 35:148-154.

Richards, R.H., and J.S. Buchanan. 1978. Studies on Herpesvirus scophthalmi infection of turbot *Scophthalmus maximus* (L.): histopathological observations. *J. Fish Dis.* 2:75-77.

Roberts, R.J. 1975. Experimental pathogenesis of lymphocystis in the plaice (*Pleuronectus platessa*). In *Wildlife Disease*, ed. L.A. Page, pp. 431-441. New York: Plenum Press.

Roberts, R.J., and A.M. Bullock. 1979. Papillomatosis in marine cultured rainbow trout *Salmo gairdneri* Richardson. *J. Fish Dis.* 2:75-77.

Rucker, R.R., W.J. Whipple, J.R. Parvin, and C.A. Evans. 1953. A contagious disease of salmon possibly of virus origin. *U.S. Fish Wildl. Ser. Fish. Bull.* 54:35-46.

Ryder, R.A. 1961. Lymphocystis as a mortality factor in a walleye population. *Prog. Fish. Bull.* 23:183-186.

Sano, T. 1988. Characterization, pathogenicity and oncogenicity of herpesvirus in fish. *Fish Health Section/American Fisheries Society International Fish Health Conference.* Vancouver, British Columbia, Canada. July 19-21, 1988.

Sano, T.H. Fukuda, N. Okamoto, and F. Kaneko. 1983. Yamame tumor virus: lethality and oncogenicity. *Bull. JPN Soc. Sci. Fish.* 49:1159-1163.

Shelton, R.G.J., and K.W. Wilson. 1973. On the occurrence of lymphocystis, with notes on other pathological conditions, in the flatfish stocks of the north-east Irish Sea. *Aquaculture* 2:395-410.

Sherburne, S.W. 1973. Erythrocyte degeneration in the Atlantic herring. *Clupea harengus harengus* L. *U.S. Nat. Mar. Fish. Ser. Bull.* 71:125-134.

Sherburne, S.W., and L.L. Bean. 1979. Incidence and distribution of piscine erythrocytic necrosis and the microspordian, *Glugea hertwigi*, in rainbow smelt, *O. mordax*, from Massachusetts to the Canadian Maritimes. *U.S. Nat. Mar. Fish. Ser. Fish. Bull.* 77:503-509.

Sindermann, C.J. 1970. Principal diseases of marine fish and shellfish. New York: Academic Press.

Sindermann, C.J. 1979. Pollution-associated diseases and abnormalities of fish and shellfish: a review. *U.S. Dept. Comm. Fish. Bull.* 76:717-749.

Sindermann, C.J. 1988. Biological indicators and biological effects of estuarine/coastal pollution. *Water Resour. Bull.* 24(5):931-939.

Snieszko, S.F., E.M. Wood, and W.T. Yasutake. 1957. Infectious pancreatic necrosis in trout. *Arch. Pathol.* 63:229-233.

Solangi, M.A., and R.M. Overstreet. 1982. Histopathological changes in two estuarine fishes, *Menidia beryllina* (Cope) and *Trinectes maculatus* (Bloch and Schneider), exposed to crude oil and its water soluble fractions. *J. Fish Dis.* 5:13-35.

Stephens, E.B., M.W. Newman, A.L. Zachary, and F.M. Hetrick. 1980. A viral aetiology for the annual spring epizootics of Atlantic menhaden *Brevoortia tyrannus* (Latrobe) in Chesapeake Bay. *J. Fish Dis.* 3:387-398.

Stich, H.F., A.B. Acton, and C.R. Forrester. 1976. Fish tumors and sublethal effects of pollutants. *J. Fish. Res. Board Can.* 33:1193-2001.

Stich, H.F., A.B. Acton, B.P. Dunn, K. Oishi, F. Yamazaki, T. Harada, G. Peters, and N. Peters. 1977. Geographic variations in tumor prevalence among marine fish populations. *Int. J. Cancer* 20:780-791.

Tanaka, M., M. Yoshimizu, and T. Kimura. 1984. *Oncorhynchus masou* virus: Pathological changes in masou salmon (*Oncorhynchus masou*), chum salmon (*O. keta*) and coho salmon (*O. kisutch*) fry infected with OMV by immersion method. *Bull. JPN Soc. Sci. Fish.* 50:431-437.

Tanaka, M.,M. Yoshimizu, and T. Kimura. 1987. Ultrastructures of OMV infected RTG-2 cells and hepatocytes of chum salmon *Oncorhynchus keta*. *Nippon Suisan Gakkaishi*. 53:47-55.

Walker, R. 1962. The structure of lymphocystis virus of fish. *Virology* 18: 503-505.

Walker, R. 1965. Viral DNA and cytoplasmic RNA in lymphocystis virus of fish. *Ann. NY Acad. Sci.* 126:386-395.

Walker, R., and S.W. Sherburne. 1977. Piscine erythrocytic necrosis virus in Atlantic cod, *Gadus morhua*, and other fish: ultrastructure and distribution. *J. Fish. Res. Board Can.* 34:1188-1195.

Watson, S.W., R.W. Guenther, and R.R. Rucker. 1954. A virus disease of sockeye salmon: interim report. *U.S. Fish Wildl. Ser. Spec. Sci. Rep. on Fish.* 138.

Weis, P., and J.S. Weis. 1976. Abnormal locomotion associated with skeletal malformations in the sheepshead minnow, *Cyprinodon variegatus*, exposed to malathion. *Environ. Res.* 12:196-200.

Weissenberg, R. 1914. Uber infekfiose zellhypertrophie bei fischen (Lymphocystiserkrankung). *Sitzungsber. Kg Preuss Akad. Wiss. Physiks Math.* 30:792-804.

Weissenberg, R. 1965. Fifty years of research on the lymphocystis virus disease of fishes (1914-1964). *Ann. NY Acad. Sci.* 126:362-374.

Wellings, S.R., B.B. McCain, and B.S. Miller. 1976. Epidermal papillomas in Pleuronectidae of Puget Sound, Washington. *Prog. Exper. Tumor Res.* 20:55-74.

Winton, J.R., C.N. Lannan, J.I. Fryer, and T. Kimura. 1980. Isolation and characterization of a new reovirus from chum salmon. In *Proceedings of the North Pacific Aquaculture Symposium*, Anchorage, Alaska, August 18-21, 1980. pp. 359-367.

Winton, J.R., C.N. Lannan J.I. Fryer, and T. Kimura. 1981. Isolation of new reovirus from chum salmon in Japan. *Fish Pathol.* 15:155-162.

Winton, J.R., C.K. Arakawa, C.N. Lannan, and J.L. Fryer. 1988. Neutralizing monoclonal antibodies recognize antigenic variants among isolates of infectious hematopoietic necrosis virus. *Dis. Aquat. Organ.* 4:199-204.

Winton, J.R., W.N. Batts, T. Nishizawa, and C.M. Stehr. 1989. Characterization of the first North American isolates of viral hemorrhagic septicemia virus. *Am. Fish. Soc. Fish Health Sect. Newsl.* 17(2):2-3.

Wolf, K. 1962. Experimental propagation of lymphocystis disease of fishes. *Virology* 18:249-256.

Wolf, K. 1988. Fish viruses and fish viral diseases. 476 pp. Ithaca: Cornell University Press.

Wolf, K., and C.E. Smith. 1981. *Herpesvirus salmonis*: pathological changes in parentally infected rainbow trout, *Salmo gairdneri* Richardson, fry. *J. Fish Dis.* 4:445-457.

Wolf, K., and W.G. Taylor. 1975. Salmonid viruses: a syncytium-forming agent from rainbow trout. *Fish Health News* 4:3.

Wolf, K., S.F. Snieszko, and C.E. Dunbar. 1959. Infectious pancreatic necrosis, a virus-caused disease of fish. *Excerpta Med.* 13:228.

Wolf, K., S.F. Snieszko, C.E. Dunbar, and E. Pyle. 1960. Virus nature of infectious pancreatic necrosis in trout. *Pro. Soc. Exp. Biol. Med.* 104: 105-108.

Wolf, K., M.C. Quimby, L.L. Pettijohn, and M.L. Landolt. 1973. Fish viruses: isolation and identification of infectious hematopoietic necrosis in eastern North America. *J. Fish. Res. Board Can.* 30:1625-1627.

Wolf, K., R.W. Darlington, W.G. Taylor, M.C. Quimby, and T. Nagabayashi. 1978. *Herpesvirus salmonis*: characterization of a new pathogen of rainbow trout. *J. Virol.* 27:659-666.

Yamamoto, T. and J. Kilistoff. 1979. Infectious pancreatic necrosis virus: quantification of carriers in lake populations during a 6-year period. *J. Fish. Res. Board Can.* 36:562-567.

Yoshimizu, M., M. Tanaka, and T. Kimura. 1987. *Oncorhynchus masou* virus (OMV): incidence of tumor development among experimentally infected representative salmonid species. *Fish Pathol.* 22:7-10.

Zwellenberg, L.O., M.H. Jensen, and H.H.L. Zwellenberg. 1965. Electron microscopy of the virus of viral hemaorrhagic septicemia of rainbow trout (Egtved virus). *Arch. Virusforsch* 17:1-19.

3

Bacterial Diseases of Fish

John L. Fryer
John S. Rohovec

INTRODUCTION

Considering the large number of species of fish and their divergent environments, it is surprising to find that only about 37 bacteria have been reported to be pathogenic for such animals. Most of the bacteria presently recognized were not isolated from wild populations, but were from fish used in artificial propagation. Fish maintained in this manner are frequently crowded and under conditions that impose additional stress. This enhanced the ease with which pathogenic bacteria are transmitted from one individual to another as imposed stress tends to alter (increase) the relative susceptibility of individuals within the population. Conditions of propagation also provide an opportunity to observe animals for specific alterations resulting from disease and for potential mortality. Such conditions are seldom available with wild populations of fish. Much of what is known about the bacteria that infect fish and the extent of the diseases they cause has resulted from man's efforts to artificially propagate selected species and, almost without fail, the species selected for propagation are food fish. Examples of nonfood fish might include aquarium and certain species of bait fish. The bacteria that exist as pathogens among cultured fish were almost certainly derived from carriers used to initiate culture of the species. Transmission by water-borne infection from wild populations to cultured fish also occurs. We believe these bacteria did not evolve as pathogens among hatchery stocks but were transmitted from wild to hatchery fish. Bacteria recovered from and associated with diseased fish are shown in Table 1. The important and best researched of these pathogens listed are the subject of this report.

0-8493-8662-4/93/$0.00 + $.50

Table 3-1. Bacterial Pathogens of Fish

Bacterium	Disease
Gram-Negative Pathogens	
Vibrio anguillarum[1,2]	Vibriosis
Vibrio ordalii[1,2]	Vibriosis
Vibrio salmonicida[1,2]	Vibriosis
Vibrio alginolyticus	Vibriosis
Vibrio damsela	Vibriosis
Vibrio cholera (non-01)	Vibriosis
Vibrio vulnificus (Biogroup 2)	Vibriosis
Aeromonas salmonicida[1,2]	Furunculosis
subsp. *salmonicida*	
subsp. *achromogenes*	
subsp. *masoucida*	
Aeromonas hydrophila[1,3]	Motile Aeromonad Septicemia
Pasteurella piscicida[1]	Pasteurellosis
Providencia rettgeri[3]	Bacterial Hemorrhagic Septicemia
Edwardsiella tarda[3]	Edwardsiellosis
Edwardsiella ictaluri[1]	Enteric Septicemia
Serratia plymuthica	*Serratia* Septicemia
Yersinia ruckeri[1]	Enteric Redmouth Disease
Acinetobacter sp.	Acinetobacterosis
Pseudomonas anguilliseptica	*Pseudomonas* Septicemia
Pseudomonas chlororaphis	*Pseudomonas* Septicemia
Pseudomonas fluorescens[3]	*Pseudomonas* Septicemia
Flexibacter psychrophilus[1]	Bacterial Coldwater Disease
Flexibacter columnaris[1]	Columnaris
Flexibacter maritimus	Flexibacterosis
Flavobacterium branchiophila	Bacterial Gill Disease
Sporocytophaga sp.	Saltwater Columnaris
Rickettsiales[1,2]	*Rickettsiosis*
Renibacterium salmoninarum[1,2]	Bacterial Kidney Disease
Eubacterium tarantellus	Eubacterial Meningitis
Carnobacterium piscicola	Pseudokidney Disease
Vagococcus salmoninarum	Pseudokidney Disease
Gram-Positive Pathogens	
Lactococcus piscium	Pseudokidney Disease
Staphylococcus sp.	Staphylococcal Septicemia
Streptococcus sp.	Streptococcal Septicemia
Streptoverticillium	Streptomycosis
Clostridium botulinum	Botulism

Table 3-1 (continued)

Bacterium	Disease
Acid Fast Pathogens	
Mycobacterium marinum[3]	Mycobacteriosis
Mycobacterium fortuitum[3]	Mycobacteriosis
Mycobacterium chelonei[3]	Mycobacteriosis
Nocardia asteriodes[3]	Nocardiosis
Nocardia seriolae[1,2]	Nocardiosis

[1]Species of bacteria considered major or important pathogens of finfish.
[2]Species of bacteria considered obligate pathogens of finfish.
[3]Species of bacteria which are also associated with human disease.

Edwardsiella

The genus *Edwardsiella* is comprised of three species, two of which are pathogens of fish. The type species, *Edwardsiella tarda*, is an opportunistic pathogen of many species of animals and is the cause of disease in at least 15 different species of fish (Bullock and Herman 1985). Although *Edwardsiella ictaluri* has been isolated from several species of diseased fish (Bullock and Herman 1985), it is most commonly found in channel catfish, *Ictalurus punctatus*, and is currently the most important bacterial pathogen affecting the catfish aquaculture industry of the southeastern USA (Meyer and Bullock 1973; Hawke *et al.* 1981).

Edwardsiella tarda and *E. ictaluri* are both Gram-negative rods and oxidase negative. Both organisms are peritrichously flagellated, but *E. ictaluri* is only weakly motile at 20-30°C and not motile at 37°C. Within each species there is biochemical homogeneity of the isolates from fish, even among those from different species and different geographical locations (Austin and Austin 1988). At least four serotypes (A, B, C, D) of *E. tarda* have been described as pathogens of fish (Park *et al.* 1983). The isolates of *E. ictaluri* are also serologically homogeneous (Plumb and Vinitnantharat 1989). *Edwardsiella tarda* differs from *E. ictaluri* in several respects: it is motile at 37°C, produces indole from tryptone broth, is methyl red positive, utilizes tartrate, and grows on Christensen's citrate.

Isolation of these bacterial pathogens from lesions or internal organs can be accomplished with bacteriological media, such as brain heart infusion agar

(BHIA) or tryptic soy agar (TSA). On both media, incubation for 2-3 days at 22-26°C results in small (0.2-0.5 mm diameter), transparent, circular, raised, and slightly convex colonies that grow best on blood agar. *Edwardsiella ictaluri* is more fastidious, grows more slowly, and produces smaller colonies. Inoculation of kidney tissue into thioglycollate broth and incubation at 22°C for 4 days prior to subculture on BHIA have been reported to improve recovery of *E. tarda*.

Fish infected with *E. tarda* sometimes become lethargic and orient vertically at the surface of the water or swim erratically in a spiral pattern. External signs in channel catfish may begin as small cutaneous lesions (3-5 mm in diameter) and progress to deep abscesses of the body musculature in the postero-lateral region. Abscesses may develop into gas-filled hollow areas that if punctured, emit a foul odor.

Internally, the most common gross lesions are light-colored nodules on the spleen, kidney, and liver. Histologically, the lesions are focal necrotic areas, often filled with the bacterium.

Channel catfish infected with *E. ictaluri* have behavioral signs similar to those with *E. tarda*. Fish are lethargic, anorexic, and may swim with spiral movements with occasional erratic bursts. The disease caused by *E. ictaluri* is known as enteric septicemia of catfish (ESC) and is characterized by hemorrhages around the mouth, on the lateral and ventral portions of the body and on the fins. There may be pale gills and exophthalmia. Open lesions on the skull are the basis for a second common name, "hole-in-the-head disease."

Internally, petechiae occur throughout the visceral organs and body musculature. Ascites may be present and there is commonly swelling of the kidney, liver, and spleen. Histologically, there is inflammation of the intestinal mucosa and submucosa and diffuse necrosis in the visceral organs.

Control of disease caused by species of *Edwardsiella* is by therapeutic administration of antimicrobials. Both Terramycin® and a potentiated sulfonimide, Romet®, have been used successfully. There is, however, bacterial resistance to both compounds, and efforts are being made to develop an effective immunogen against *E. ictaluri*.

Yersinia

The genus *Yersinia* contains seven species, of which one, *Yersinia ruckeri*, is an important pathogen of salmonid fish. This bacterium causes the disease enteric redmouth (ERM) and affects salmonids in North and South America, Europe, and Australia. The disease was first recognized as the cause of a major economic problem of the rainbow trout industry of southern Idaho, and the bacterium was described first by Ross *et al.* (1966). The disease primarily affects salmonids, but there have been a few reports of *Y. ruckeri* in non-salmonid

(Daley *et al.* 1986), two serotypes (I and II) are the primary fish pathogens (O'Leary *et al.* 1982).

Isolates of *Y. ruckeri* have characteristics of the enterics and are Gram-negative, peritrichously flagellated rods (Fig. 3-1) which are oxidase negative and catalase positive, ferment glucose, and reduce nitrates to nitrites. There are reports that serotype II can be differentiated from serotype I by its ability to ferment sorbitol. *Yersinia ruckeri*, unlike other species of the genus, has lysine decarboxylase and gelatinase enzymes.

The pathogen can be readily isolated from kidneys of diseased fish on TSA or BHIA which is incubated at 20-25°C. Resulting bacterial colonies are small, smooth, circular, slightly raised, and appear after 48 h incubation.

The disease may occur in peracute, acute, or chronic form. In the acute form, disease signs are similar to those found in other Gram-negative septicemias of fish. The name of the disease, ERM, is descriptive of one of the most common signs, which is reddening of the mouth with inflammation and erosion of the jaws, palate, and opercula. Other external signs include: darkening of the body coloration, hemorrhaging at the base of the fins, and bilateral exophthalmia. Internally, petechial hemorrhaging occurs in the visceral fat, musculature, and

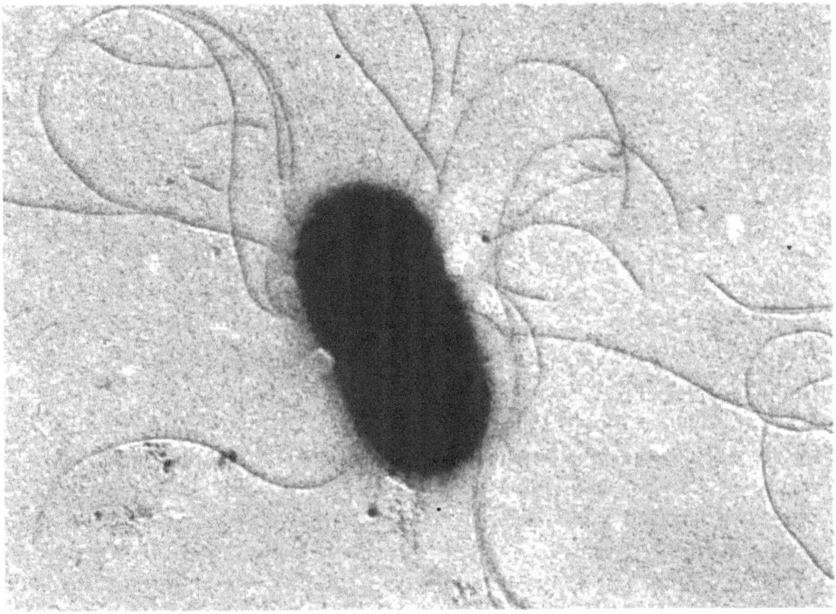

Fig. 3-1. Electron micrograph of *Yersinia ruckeri*.

species. Although there is serological heterogeneity among isolates of *Y. ruckeri* intestine, which may contain yellow fluid. Splenomegaly and enlargement of the kidney may also occur.

At one time, ERM caused significant mortality among cultured rainbow trout, and control of the disease depended on antibiotic therapy. However, effective, easily administered vaccines have been developed to control this disease (Tebbit *et al.* 1981; Amend *et al.* 1983; Ellis 1988).

Vibrio

Vibriosis, caused by serotypes of *Vibrio anguillarum*, *V. ordalii*, and *V. salmonicida*, is the most common, and potentially the most devastating, bacterial disease of cultured marine fish. Control of this disease is an absolute necessity for the culture of salmonids and other fish in saltwater environments where it can be responsible for mortality of 90% or greater (Cisar and Fryer 1969). Although vibriosis is primarily a disease that occurs in salt water, it has also caused serious losses among fish in fresh water, especially in Japan (Egusa 1969). The etiology of the disease consists of a diverse group of 10 known serotypes of *V. anguillarum* (Ezura *et al.* 1980; Kitao *et al.* 1983; Larsen and Mellengaard 1984), *V. ordalii* (Schiewe *et al.* 1981), and *V. salmonicida* (Egidius *et al.* 1986).

Like other members of the genus *Vibrio*, these organisms are Gram-negative, slightly curved rods that are motile by polar flagella; *V. anguillarum*, and *V. ordalii* have a single flagellum (Fig. 3-2), but *V. salmonicida* has a tuft of nine. Vibrios are noncapsulated and do not produce spores. These facultative anaerobes are oxidase and catalase positive and ferment carbohydrates anaerogenically. They typically are sensitive to the vibriostatic agent, 0/129.

In culture, typical strains of *V. anguillarum* produce entire colonies which are round, raised, convex, and cream-colored. Colonies produced by *V. ordalii* are smaller and whitish in color, and those of *V. salmonicida* are small and greyish. *Vibrio anguillarum* and *V. ordalii* are easily isolated from diseased fish, but *V. salmonicida* may take several days before visible colonies are detected and blood agar may be required for growth.

A wide variety of both warm- and cold-water fish are susceptible to infection by *V. anguillarum* (Anderson and Conroy 1970). The bacterium is highly invasive, producing septicemia and usually acute disease. External pathologic changes are similar to those caused by other Gram-negative fish pathogens and include erythema at the base of the fins and within the mouth, hemorrhaging of the vent, and petechiae in the musculature. As the disease progresses, necrotic lesions develop in the body musculature. Internally, there can be congestion of the liver and spleen and swelling of the kidney. In advanced cases, internal organs may become necrotic and liquified.

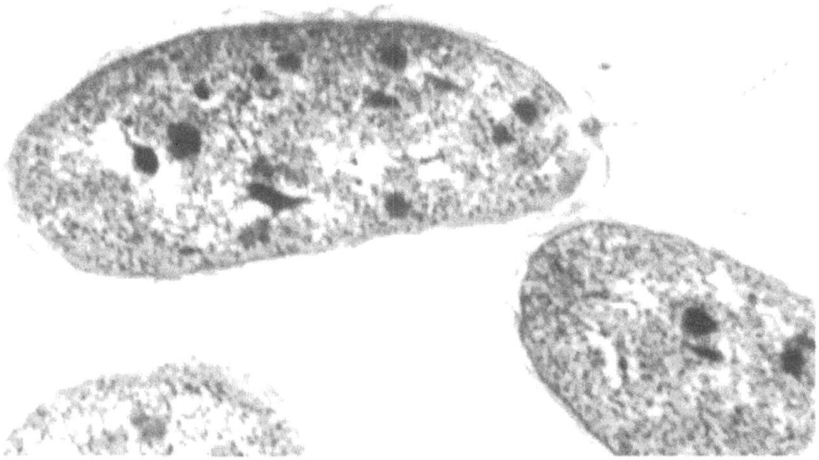

Fig. 3-2. Ultrathin section of *Vibrio anguillarum* depicting the cell, the irregular outer layer and single polar flagellum.

Ransom *et al.* (1984) compared the histopathology resulting from *V. anguillarum* and *V. ordalii*. They showed that *V. anguillarum* was most easily observed and caused pathologic changes in blood, blood-forming tissues (kidney and spleen), and in loose connective tissue and gills. Bacteria were found to be rather evenly distributed throughout these tissues. *Vibrio ordalii* formed distinct colonies within fish tissue, and skeletal and cardiac muscle were the areas most affected. Both organisms caused necrosis in the lower gastrointestinal tract.

Vibrio salmoninarum has occurred primarily in Atlantic salmon, *Salmo salar*, cultured in Norway and Scotland. The disease is characterized by hemorrhaging in the integument surrounding the internal organs and by marked anemia. There is a generalized septicemia with large numbers of bacterial cells in the blood.

Control of vibriosis has been accomplished, using chemotherapeutants, but this method has not been entirely satisfactory because drug resistant forms can occur (Aoki *et al.* 1974). However, effective vaccines have been developed against all three species and are commonly used as polyvalent preparations (Hayashi *et al.* 1964; Fryer *et al.* 1978; Amend and Johnson 1981; Holm and Jorgensen 1987).

Aeromonas

Aeromonas salmonicida, the causative agent of furunculosis, is perhaps the best-studied of the bacterial fish pathogens. Although it is primarily a cause of disease and extensive mortality among fish in fresh water, it has also been responsible for losses in salt water. Fish that succumb to *A. salmonicida* in salt water probably are initially infected in fresh water and carry the organism into salt water where the bacterium continues to proliferate to lethal levels. However, there are documented observations of marine fish becoming infected with *A. salmonicida*. In most cases the fish resided in close proximity to salmonids cultured in sea-water net pens (Sindermann 1977). They may also become infected after consuming diseased salmon (Evelyn 1977a). It has been shown that *A. salmonicida* is capable of surviving in salt water and infecting susceptible salmonids (Scott 1968).

Aeromonas salmonicida is a Gram-negative coccoid rod. The bacterium is nonsporulating and nonencapsulated. Like other members of the genus, it is

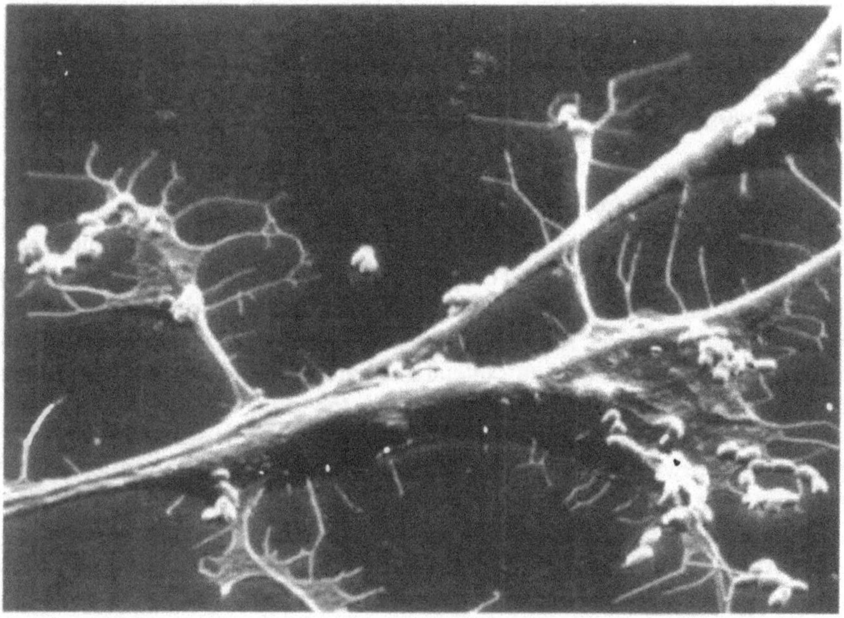

Fig. 3-3. Virulent *Aeromonas salmonicida* adhering to CHSE-214 cells *in vitro*.

oxidase positive, but differs because it is nonmotile and most strains do not produce gas when carbohydrates are fermented. The organism can be grown on a wide variety of bacteriological media; on TSA, the colonies are small, round, raised, convex, and entire. Most strains produce a water- soluble brown pigment when cultivated in the presence of tyrosine or phenylalanine. Growth *in vitro* is best between 18 and 25°C, occurs slowly at 4°C, and is absent at 37°C.

Aeromonas salmonicida is primarily a pathogen of salmonids, but is capable of infecting other freshwater and marine fish. The disease is typical of those associated with Gram-negative septicemias. Gross pathology may include focal necrosis of musculature, hemorrhaging at the base of pelvic and pectoral fins, splenomegaly, and congestion of the kidney which can progress to a liquefactive necrosis of most internal organs. Virulent forms of the organism autoagglutinate. This characteristic is a result of an additional outer or "A" layer on virulent cells (Udey and Fryer 1978; Munn *et al.* 1982), allowing them to adhere to host cells (Fig. 3-3). A leucocidin which acts to increase pathogenicity has also been described (Fuller *et al.* 1977). In addition, at least two proteases and two hemolysins have been characterized, and glycerophospholipid:cholesterol acyltransferase complexed with lipopolysaccharide was reported to be a virulence factor of *A. salmonicida* (Lee and Ellis 1990).

Furunculosis has been controlled with chemotherapeutants, such as oxytetracycline, sulfamerazine, and oxolinic acid. Vaccines have received much attention and some efficacious preparations have been described; however, none are in common use.

Motile aeromonads, especially *Aeromonas hydrophila*, can cause septicemias in a wide variety of freshwater fish and this bacterium has been isolated from marine fish. There is some disagreement concerning the role of *A. hydrophila* as a fish pathogen. Some workers contend that it is a secondary pathogen of a compromised host and possesses limited invasive qualities. The bacterium is ubiquitous in freshwater; when it does cause disease, it produces a hemorrhagic septicemia. This is characterized by petechiae in the body musculature, hemorrhaging in the gills and vent, and accumulation of ascites in the peritoneal cavity.

Isolation of motile aeromonads is easily accomplished by streaking kidney material of diseased animals onto BHIA or TSA. A selective and differential medium, Rimler-Shotts (Shotts and Rimler 1973), has also been described for cultivation of *A. hydrophila*. On nonselective medium, the organism produces cream, round, raised entire colonies in 24 h at incubation temperatures of 20-25°C. Unlike *A. salmonicida*, *A. hydrophila* can grow at 37°C.

Aeromonas hydrophila is a Gram-negative bacterium, which is motile by a single polar flagellum. It is a rod-shaped organism (0.8-1.0 x 1.0-3.5 μm) which is fermentative and oxidase and catalase positive.

Pasteurella

During an epizootic among white perch, *Morone americanus*, in Chesapeake Bay, a Gram-negative bacterium was isolated from infected fish (Snieszko *et al.* 1964). Morphological and physiological characteristics indicated the isolate was a member of the genus *Pasteurella*. Janssen and Surgalla (1968) determined that the organism had unique properties and proposed the species name *Pasteurella piscicida*. Although this bacterium has not caused subsequent mortality among fish in the U.S., it has caused serious problems among cultured yellowtail, *Seriola quinqueradiata*, in Japan (Kubota *et al.* 1970; Kusuda and Yamaoka 1972). Because of the prominent white granulomas which are present in organs of infected fish, the disease caused by *P. piscicida* is called pseudotuberculosis.

Strains isolated from white perch in the U.S. and from yellowtail in Japan have been compared; all appear to be identical in morphological, physiological and biochemical characteristics (Kusuda and Yamaoka 1972; Koike *et al.* 1975). The bacterium is a Gram-negative rod which shows bipolar staining and some pleomorphism. The nonmotile cells range in size from 0.6-1.2 by 0.7-2.6 μm, are not encapsulated, and do not produce spores. On solid media, *P. piscicida* forms entire, convex, translucent colonies which are viscid, especially when grown on BHIA. These facultative anaerobes grow at temperatures between 20 and 30°C, but do not grow at 15 or 35°C.

Characteristics of the organism from fish were compared to other species of *Pasteurella* (Koike *et al.* 1975). There were similarities, but negative reactions of fish isolates in the methyl red, Voges-Proskauer, and arginine decarboxylase tests were unique. These tests, plus the inability to grow at 37°C, were the basis for establishing this new species of *Pasteurella*.

Pasteurella piscicida has been associated with disease in several species of marine fish. It has been isolated from white perch, yellowtail, striped bass (*Morone saxatilis*), menhaden (*Brevoortia tyrannus*), and mullet (*Mugil cephalus*).

The disease can be either acute or chronic and seems to be more severe at higher water temperatures (>23°C) and when water quality is poor. Gross pathological signs resemble other Gram-negative septicemias of fish and include darkening of body coloration and hemorrhaging around the opercula and fins. Internally there may be miliary lesions in the kidney and spleen; these lesions, when observed histopathologically, are tubercle-like and composed of masses of bacterial cells, epithelial cells, and fibroblasts.

Diagnosis of pseudotuberculosis is usually accomplished by isolating and characterizing the organism by morphological and biochemical properties. A fluorescent antibody test is also available (Kitao and Kimura 1974).

Chemotherapeutants have been used to control *P. piscicida* infections; however, their effectiveness has become limited in Japan because of the appearance of drug-resistant strains. Effective vaccines have been tested (Fukuda and Kusuda 1981; Kitao *et al.* 1981); however, they are not yet commonly used.

Rickettsiales

Rickettsiales-like (also chlamydiales-like) organisms have been observed in many species of aquatic animals (Hoffman *et al.* 1969; Harshbarger *et al.* 1977; Johnson 1984). These reports are descriptive in nature and center on the ultrastructure of the agents and the pathologic changes produced in their hosts. The *in vitro* culture of these intracellular microorganisms, observed in fish and shellfish, has been limited (Wolf 1981; Bradley *et al.* 1988). Ozel and Schwanz-Pfitzner (1975) observed a rickettsia-like agent in RTG-2 cells that had been inoculated with tissue from rainbow trout infected with VHS virus. Work with this agent was limited to examination and description by light and electron microscopy. The intracellular microorganism observed from rainbow trout by the authors is no longer available.

During 1989, coho salmon, *Oncorhynchus kisutch*, cultured in sea-water net pens in Chile experienced an epizootic of unknown etiology. The disease was first observed in the vicinity of Puerto Montt (the location of many salmon farms). The peak mortality occurred in May and continued into November, causing losses up to 90% at certain farms (Bravo and Campos 1989; Cvitanich *et al.* 1990). The disease was most severe in coho salmon; however, chinook, *Oncorhynchus tshawytscha*, Atlantic salmon, *Salmo salar*, and rainbow trout, *Oncorhynchus mykiss*, were also infected.

Disease signs in moribund fish included lowered hematocrit, swollen kidney, enlarged spleen, and occasionally, mottled liver (Fig. 3-4). No infectious agents were isolated by Bravo and Campos (1989) who first described the disease, but they did observe an unidentified parasite in the blood and internal organs of infected fish by both light and electron microscopy. The disease is not known to occur naturally in fresh water.

To determine the cause of this epizootic, Lannan *et al.* 1984 inoculated kidney tissue from diseased coho onto an established chinook salmon cell line, CHSE-214 (ATCC CRL 1681). A rickettsiales-like agent was isolated in these cell cultures (Fig. 3-5). Specifically coho salmon (900 to 1200 g in weight) were collected from a saltwater net pen at a salmon farm near Puerto Montt, Chile, where an epizootic was in progress. Tissues were taken for bacteriological and virological analysis. At the same time, kidney tissue was aseptically removed and inoculated directly into 25-cm tissue culture flasks containing a monolayer of CHSE-214 cells in antibiotic-free Eagle's Minimum Essential Medium with Earle's salts (MEM-10) (Automod, Sigma Chemical Co., St. Louis, MO) and supplemented with 10% fetal bovine serum (FBS) (Hyclone Laboratories, Inc.,

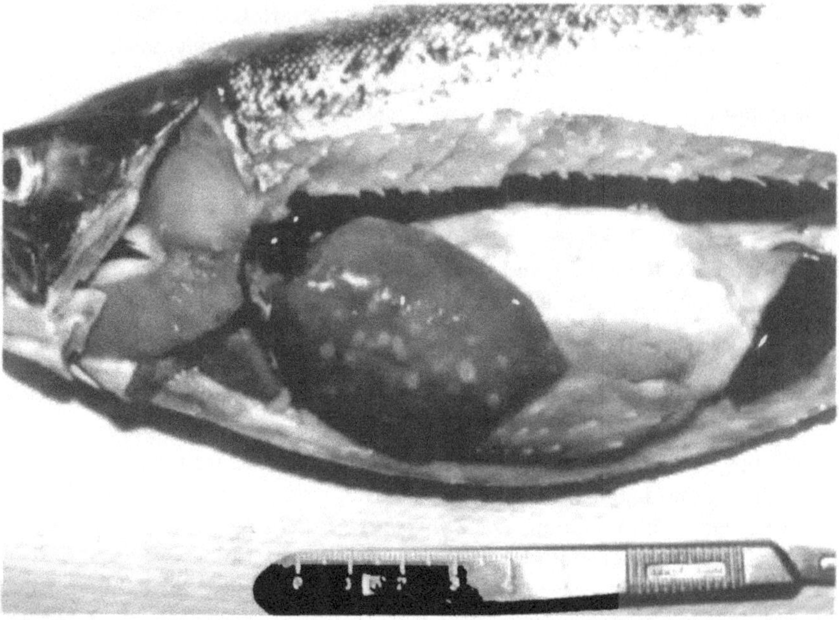

Fig. 3-4. Coho salmon naturally infected with rickettsia show gross signs of disease. Note lesions in liver, pale gills, enlarged spleen. Photo provided by Sandra Bravo, SalmoLab S.A. Puerto Montt, Chile.

Logan, UT). Cultures were kept at 4°C for transport to Oregon State University Hatfield Marine Science Center Fish Disease Laboratory, Newport, Oregon. At the laboratory, cultures were incubated at 15°C. When cytopathic effect (CPE) was observed in these cultures, aliquots of spent culture medium were transferred to fresh CHSE-214 cultures and incubation continued.

The microorganism associated with the Chilean coho salmon disease was isolated, and preliminary characterization made in cell culture. The organism was an obligately intracellular parasite and replicated in cultured salmonid fish cells, but failed to grow in standard bacteriological media. The optimum temperature for growth was 15-18°C. Growth of the microorganism was inhibited by streptomycin, gentamicin, and tetracycline (Fryer *et al.* 1990).

Giemsa-stained smears of fluid from infected cultures contained large numbers of darkly-staining microorganisms. These microorganisms were pleomorphic, occurring as coccoid or ring forms, and frequently in pairs. They varied in diameter from approximately 0.5 to 1.5 μm. Microscopic examination of fixed and stained cell cultures showed that microorganisms replicated within cytoplasmic inclusions in infected cells (Fig. 3-5).

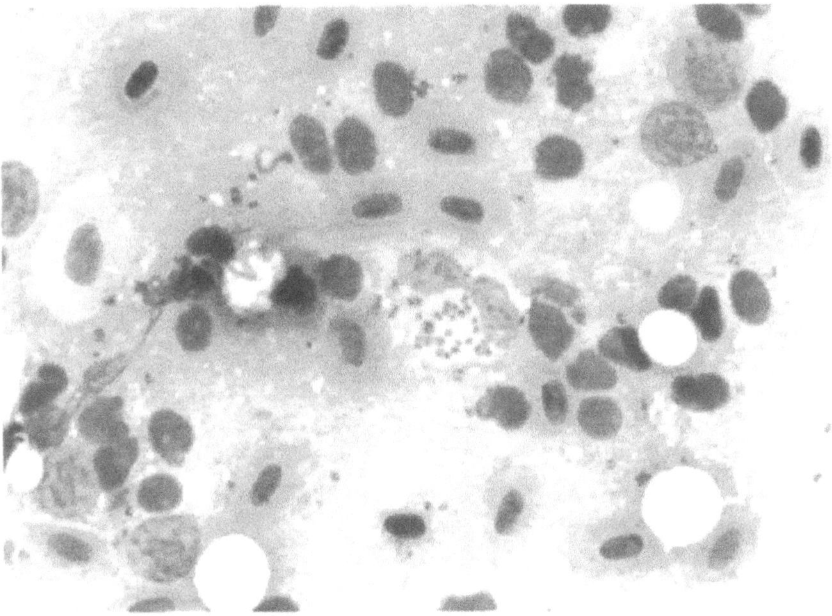

Fig. 3-5. Rickettsial organism within a cytoplasmic inclusion, CHSE-214 cells four days post inoculation (May Greenwald-Giemsa stain).

Transmission electron microscopy revealed individual or paired organisms enclosed in membrane-bound vacuoles (Fig. 3-6). Organisms were bound by three membrane layers, an undulate outer membrane, a cell wall, and a closely opposed inner membrane. Electron-dense areas containing ribosome-like structures were concentrated near the plasma membrane, and fibrillar DNA-like material was localized in the central region. Many organisms contained one or more electron-lucent spherical structures. Organisms apparently undergoing binary fission were frequently observed (Fig. 3-6).

Infected cells examined by scanning electron microscopy after 24 h incubation showed irregular coccoid organisms of approximately 1 μm in diameter attached to the exterior surfaces of host cells. These organisms had highly folded outer membranes and varied in size and morphology (Fig. 3-7).

Exact taxonomic placement of this marine rickettsia is yet to be determined. It is an obligate intracellular parasite, replicating *in vitro* within cytoplasmic inclusions in host cells. It does not react with the monoclonal antibody against the group-specific LPS chlamydial antigen and, although polymorphic, does not develop small infectious elementary bodies or large replicating initial bodies characteristic of chlamydia. It possesses the rippled cell wall and electron-lucent

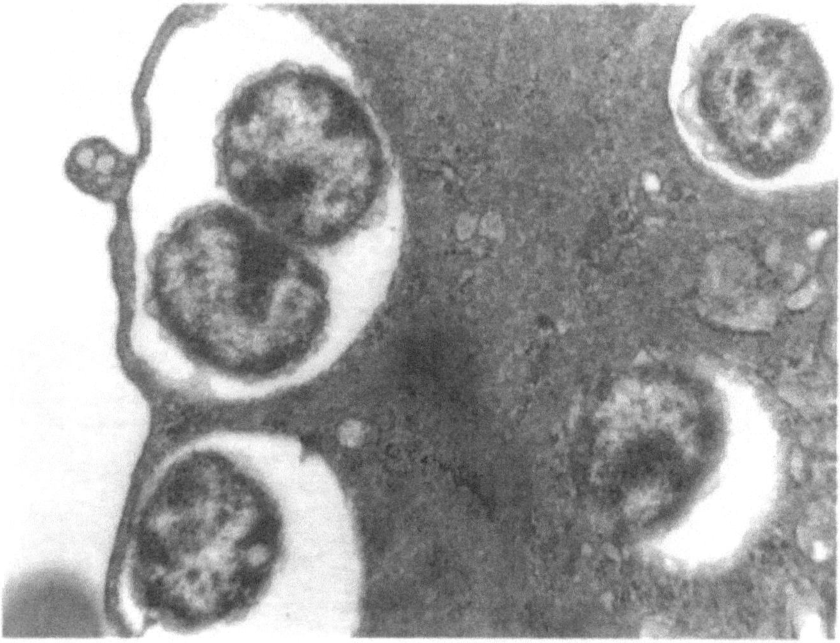

Fig. 3-6. Ultrathin section of an infected CHSE-214 cell. Rickettsia lie within membrane bound vacuoles. Note the ripple cell wall and the electron lucent spherical structures.

spherical structures described for certain rickettsial species (Anderson *et al.* 1965). Therefore, we have suggested the organism be associated with the order Rickettsiales (Fryer *et al.* 1990).

Clostridium

Cells of *Clostridium botulinum* and the other clostridia are similar in shape to the bacilli. Anaerobic growth, absence of catalase activity, and the formation of endospores are characteristics that permit identification of *Clostridium*. Isolates of *C. botulinum* have been obtained from the intestinal contents and tissues of fish by strict anaerobic culture at 25-30°C on enriched media, e.g., egg yolk and blood agar and trypticase, peptone, glucose liquid medium (Eklund *et al.* 1967; Eklund *et al.* 1984). Types B, E, and F of *C. botulinum* grew and produced toxin at temperatures as low as 3.3°C. Sediments from ponds in which a type E toxin-induced fish mortality occurred have been used to culture this bacterium. Botulinum toxin caused severe losses of fish in earthen ponds that allowed anaerobic conditions to prevail in the sediments. This same condition also has occurred in salt water and resulted in loss of fish (Henley and Lewis 1976).

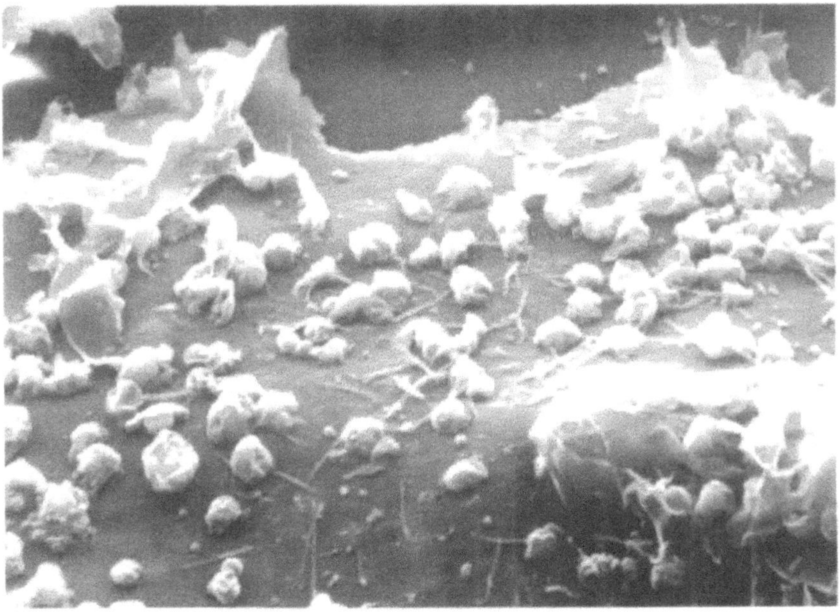

Fig. 3-7. Scanning electron micrograph of rickettsial organisms released from CHSE-214 cells, eight days post inoculation. Note the variation in size among these organisms.

Colony characteristics are somewhat variable among the types of *C. botulinum*; in general, they are circular to slightly raised, translucent to semi-opaque, and have a matte to semi-glossy surface. Types B and E have been described as producing a pearly (iridescent) layer covering and surrounding each colony.

The presence of *C. botulinum* type E toxin can be demonstrated by intraperitoneal injection of mice or fish with supernatant fluids from tissue homogenates or culture filtrates. Injected animals develop signs of botulism and death. Further confirmation can be obtained by mixing these preparations with type E antitoxin and again injecting test animals. Failure to develop botulism further corroborates presence of type E toxin in the original preparation. Similar procedures can be of value to detect other toxins produced by *C. botulinum.*

The Lactic Acid Bacteria

The lactic acid bacteria are considered as a group here rather than by genus and species. Members of this group are Gram-positive, non-motile, non-spore forming rods and cocci that occur singly or in chains. Lactobacilli have been identified as part of the normal flora of both marine and freshwater fish. In marine fish, *Lactobacillus* species have been isolated from the mackerel, *Scombridae*, hake, *Merluccius gayi*, and Atlantic cod, *Gadus morhua*, and *Lactobacillus plantarum* from the saithe, *Gadus virens*. Several freshwater species of the Salmonidae have been reported to harbor *Lactobacillus* sp. in internal organs such as the heart, liver, and kidney.

In this genus, only the species, *Lactobacillus piscicola*, is known to cause disease in fish (Hiu *et al.* 1984). *Lactobacillus piscicola* has been transferred to a new genus and is now called *Carnobacterium piscicola* (Collins *et al.* 1987). In adult salmonids, *C. piscicola* is often observed shortly after spawning and is thought to be stress-related (Cone 1982; Hiu *et al.* 1984). The bacterium has been isolated from juvenile salmonids; however, the mortality was negligible.

Strains of *C. piscicola* can be isolated on BHIA and TSA from internal organs, principally the kidney, or lesions on diseased fish. Incubation at 25°C for 2 days produces nonpigmented (~2 mm in diameter), white, round, entire colonies. Morphologically, individual rods measure about 0.5 by 1-1.5 μm. Short chains of 2 to 3 cells are often seen. Young cultures are Gram-positive but those more than 24-h-old frequently become Gram-variable. The majority of lactic acid bacteria isolated from diseased fish have been identified as *C. piscicola*.

Other lactic acid bacteria isolated from diseased fish include *Vagococcus salmoninarum* (Wallbanks *et al.* 1990) and *Lactococcus piscium* (Williams *et al.* 1990). These bacteria were recently reported but little is known about their occurrence in fish. *Carnobacterium piscicola* is regarded as the most important. It is possible that all three bacteria form part of the normal flora of the fish.

Renibacterium salmoninarum

Renibacterium salmoninarum, the etiological agent of bacterial kidney disease (BKD), usually infects and causes mortality of anadromous salmonid fish in fresh water (Fryer and Sanders 1981), but can also cause high mortality of salmonids reared in salt water (Earp *et al.* 1953; Bell 1961; Banner *et al.* 1983). Wood and Wallis (1955) first advanced the hypothesis that salmonid fish infected with *R. salmoninarum* in fresh water continue to die during the saltwater phase of their life cycle. Observations by Ellis *et al.* (1978) and unpublished data from our laboratory support this hypothesis. Bacterial kidney disease is responsible for economic loss among salmonids reared in mariculture and can also cause significant mortality among hatchery-reared fish released into the ocean.

All isolates of *R. salmoninarum* seem to be relatively homogeneous in their characteristics. The organisms are Gram-positive rods which often occur in pairs (Fig. 3-8). These nonmotile cells are 0.3-1.5 μm by 0.1-1.0 μm in dimension, and they do not form spores. Young and Chapman (1978) have presented data on the ultrastructure of selected isolates of *R. salmoninarum* which indicate that,

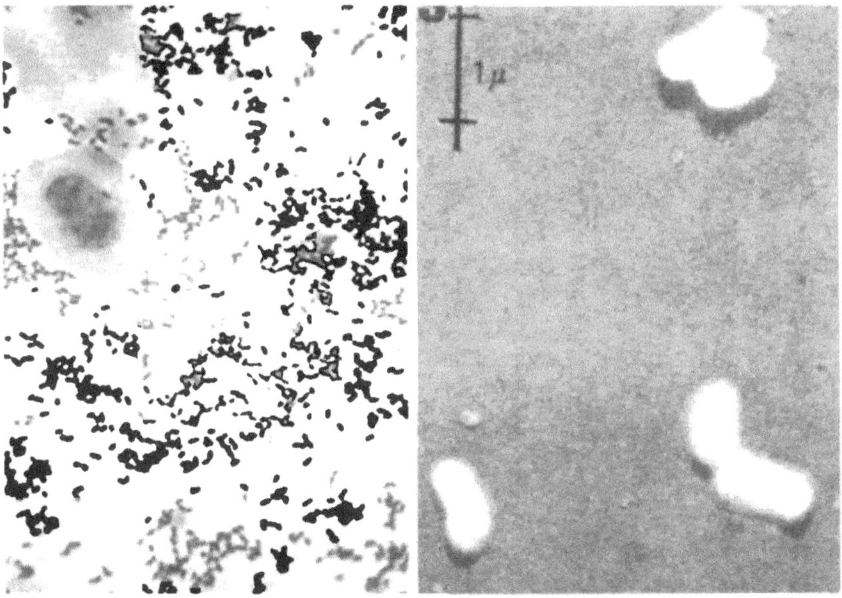

Fig. 3-8. Blood smear and electron micrograph showing *Renibacterium salmoninarum.*

in vitro, the organism is consistently rod-shaped; however, some irregularity in morphology of the bacterium was observed in host tissue. When grown on KDM-2 medium (Evelyn 1977b), white to yellowish colonies are circular, convex, and of varying size. The variation in colony size may result from auto-aggregation of the cells, the colony-forming units consisting of varying numbers of bacteria.

A unique physiological characteristic of *R. salmoninarum* is a strict requirement for cysteine. The organism is proteolytic but does not utilize carbohydrates. The aerobic bacterium grows optimally at 15°C (range 5-22°C) with a generation time of approximately 24 h. It produces catalase but not cytochrome oxidase. The guanine plus cytosine content of bacterial DNA is 53% mol in the range of the *Corynebacterium*; however, cell wall biochemistry was sufficiently different from members of the Corynebacteria to place the organism in a separate genus (Sanders and Fryer 1980).

The difficulty of culturing *R. salmoninarum* stimulated development of diagnostic procedures that do not require growth of the organism. Several serologically based techniques have been developed. Those most frequently used are fluorescent antibody techniques (Bullock *et al.* 1980; Banner *et al.* 1982) and enzyme-linked immunosorbent assays (Pascho and Mulcahy 1987; Turaga *et al.* 1987a,b).

The only natural hosts of *R. salmoninarum* are members of the family Salmonidae (Fryer and Sanders 1981). Chinook salmon, *Oncorhynchus tshawytscha*, and pink salmon, *O. gorbuscha*, are probably the most susceptible and rainbow trout seem to be most resistant. The disease is a slowly progressing, chronic infection that results in external signs which can include exophthalmia, lesions in the eyes, swollen abdomen as a result of ascites accumulation and blood-filled blisters on body surfaces. Internally, the classic lesions are white or greyish pustules in the kidney. The histopathology of BKD has been described (Bruno 1986) and is classified as a diffuse, chronic granulomatous inflammatory reaction that frequently involves the reticulo-endothelial system. Focal necrosis and proliferating macrophages and fibroblasts are common features.

Renibacterium salmoninarum is an obligate pathogen of salmonids. Although it can survive for short periods in the environment, replication outside of the host has not been demonstrated. Reservoirs of infection are other infected salmonids that can transmit the bacterium to uninfected hosts. The pathogen is also transmitted vertically from infected, spawning females via the ovum to resulting fry (Evelyn *et al.* 1984; Lee and Evelyn 1989). Once fish are infected, the disease can persist for a long time. Anadromous fish which survive initial infections may carry the pathogen for their entire life cycle and return from the ocean heavily infected (Banner *et al.* 1986).

The pathogenesis of BKD is not well described, but there are several characteristics of *R. salmoninarum* that might contribute to its virulence. Daly and Stevenson (1987; 1990) have shown that the cell is hydrophobic and gives the bacterium the capability of adhering to host tissue. An extracellular protein, the F antigen (Getchell *et al.* 1984), is produced in large quantities during infection and the amount produced can be correlated with the severity of BKD (Turaga *et al.* 1987b). This protein probably functions as a virulence factor of *R. salmoninarum* (Turaga *et al.* 1987b; Bruno 1986). Recently it has been demonstrated that *R. salmoninarum* is encapsulated (Dubreuil *et al.* 1990), and although its role in the pathogenesis of BKD is not determined, the presence of a capsule on other bacterial pathogens contributes to their virulence. Perhaps the aspect that contributes most to pathogenicity of *R. salmoninarum* is its apparent ability to survive phagocytosis and perhaps to even replicate within macrophages (Young and Chapman 1978). This capability allows the pathogen to avoid the most important defense mechanism of the host and sequester the bacterium so that it is not susceptible to the action of many antimicrobial agents used for therapy.

Although BKD is difficult to treat and drug therapy often is unsuccessful, erythromycin has long been considered the antibiotic of choice (Wolf and Dunbar 1959); its effectiveness against *R. salmoninarum* has been supported by more recent investigations (Austin 1985). The drug has been administered to adult salmon to prevent vertical transmission of *R. salmoninarum* (Evelyn *et al.* 1986) and has been fed as a dietary component both prophylactically and as therapy after BKD has been diagnosed. Although erythromycin has been used extensively in salmonids for disease control, it is not currently an approved fisheries chemical (Schnick *et al.* 1986).

Mycobacterium spp.

Acid-fast organisms from fish were first described by Bataillon *et al.* (1897), who studied mycobacterial lesions of carp, *Cyprinus carpio*. Since this early description, numerous workers have reported mycobacterial infections of fish, and the diversity of hosts and bacterial strains isolated has led to confusion in the classification of these organisms (Reichenbach-Klinke 1972). The etiological agents of mycobacteriosis of fish are considered to be *Mycobacterium marinum* in marine fishes and *Mycobacterium fortuitum* in freshwater and brackish-water fishes. *Mycobacterium chelonei* has been reported to infect aquarium fishes and salmonids (Kubica *et al.* 1972; Arakawa and Fryer 1984).

An acid-fast organism was frequently observed among Pacific salmon in the 1950's but was never cultivated artificially and, therefore, was not characterized (Ross 1970). The epizootics caused by this organism were brought under control by discontinuing the practice of feeding raw, unpasteurized salmon carcasses to juvenile fish. Today relatively few epizootics of mycobacteriosis occur in cultured salmonids.

There have been many descriptions of *Mycobacterium* spp. isolated from fish. In general, these organisms are long, Gram-positive, acid-fast, and rod-shaped. They are nonmotile, do not produce spores, are relatively fastidious, and grow slowly at temperatures between 18 to 25°C. They usually will not grow at 37°C.

These organisms have a wide geographic and host range (Parisot 1958), but disease caused by *Mycobacterium* rarely becomes epizootic unless environmental conditions are unfavorable for the fish. Mycobacteriosis is a slow-developing chronic disease and may not be evident externally. Lethargy and anorexia may occur, and in some cases skin ulcerations are present. Internally there are granulomatous, caseous, or necrotic lesions of the body organs, especially in the kidney, liver, and spleen. Diagnosis of mycobacteriosis depends on these disease signs and identification of the bacterial pathogen.

There is no effective treatment for the disease and control depends on good sanitation. Mycobacteriosis has been almost eliminated from populations of Pacific salmon by including pasteurized fish products in the diets of hatchery-reared fish. Control can also be accomplished by restricting movement of infected animals into areas where the disease is not enzootic.

Nocardia

There have been relatively few reports of nocardial infections among fish. In most instances, the etiological agent has been tentatively identified as *Nocardia asteroides*. However, a bacterium associated with mortality of cultured yellowtail in Japan has been referred to as *Nocardia kampachi*. This organism was first described by Kariya *et al.* (1968) and Kubota *et al.* (1968). The name of this bacterium has now been formally established as *Nocardia seriolae* (type strain JCM 3360) (Kudo *et al.* 1988).

Nocardia seriolae is a nonmotile, Gram-positive rod with a beaded appearance. The weakly acid-fast cells are of varying lengths and can occur in filaments. They are nonsporulating and produce aerial hyphae. On Ogawa's medium, the bacterium produces flat wrinkled growth after 10 days incubation at 25°C.

Biochemically, *N. seriolae* is similar to *N. asteroides*. Catalase, hydrogen sulfide production, starch hydrolysis, and nitrate reduction are positive. Oxidase, urease, indole production, and gelatin liquefaction are negative. The organisms were differentiated on whether they were able to grow at 37°C and survive at 50°C for more than 4 h. Both characteristics are properties of *N. asteroides*, not of *N. seriolae*. All strains of the bacterium isolated from yellowtail have very similar characteristics to the original isolate (Kusuda and Taki 1973; Kusuda, 1975). Kitao *et al.* (1989) has reported a *Nocardia* species isolated from diseased giant gourami *Osphronemus goramy* in fresh water.

Nocardiosis primarily affects cultured yellowtail (Kusuda and Nakagawa 1978) and occurs only sporadically and in isolated regions of Japan. It is a chronic disease that can result in high mortality. The clinical signs may include emaciation, inactivity, and skin discoloration. In later stages of infection, tubercles may appear on the body surface. Large whitish tubercles are sometimes found on the gills of older fish suffering from this infection, which led some investigators to call the disease *gill tuberculosis* (Kusuda *et al.* 1974).

Internally, creamy white solid nodules up to 4 mm in diameter can commonly be seen in spleen, swim bladder, and kidney, and less frequently in the liver and pericardium. Histopathological examination of these lesions shows that they are tubercular in nature and characterized by large masses of bacteria surrounded by concentrically arranged fibrous tissue.

Flexibacter

Flexibacter psychrophilus (formerly *Cytophaga psychrophila*), *F. columnaris*, and *F. maritimus* are gliding bacteria in the order Cytophagales. The taxonomy of this group of organisms is not well-defined, but the nomenclature suggested by

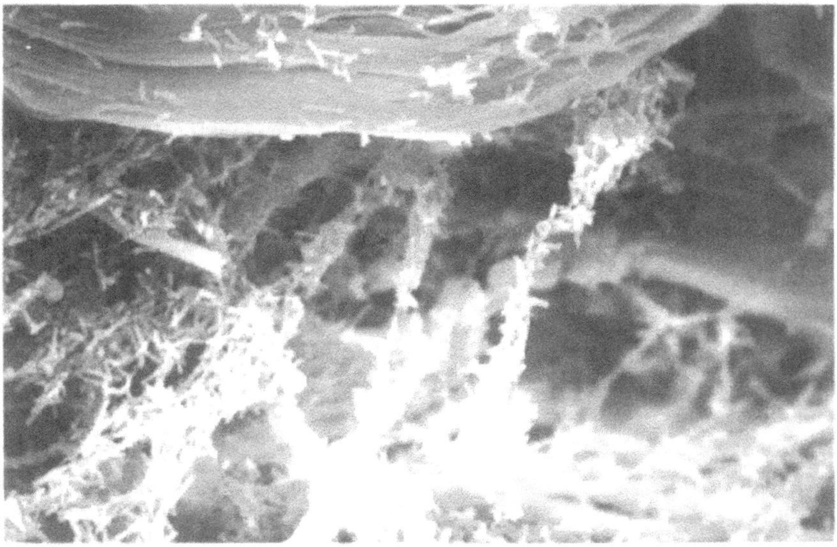

Fig. 3-9. Scanning electron micrograph showing a lesion containing *Flexibacter psychrophilus*, the causative agent of bacterial cold water disease.

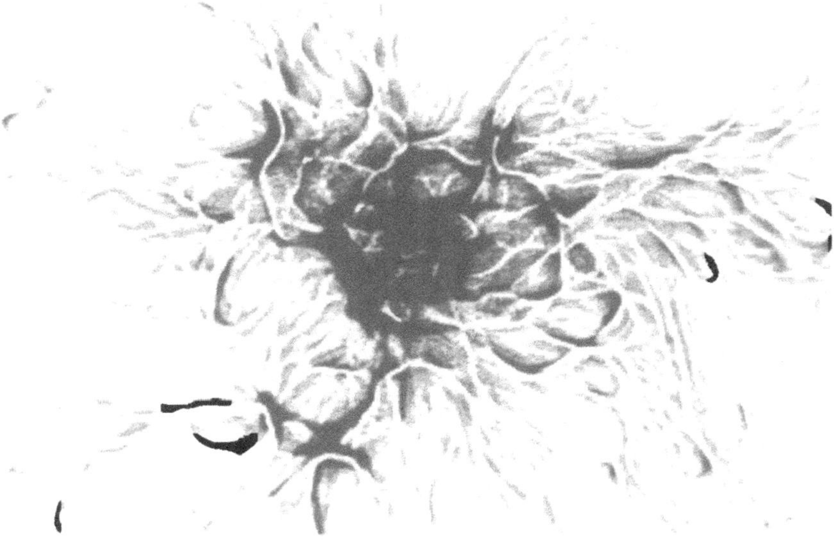

Fig. 3-10. Colony of *Flexibacter columnaris*. Note the rough, irregular appearance of surface and edges of the colony.

Bernardet and Grimont (1989) is used here. These bacteria are morphologically similar and composed of long, thin, Gram-negative rods.

Flexibacter psychrophilus is the etiological agent of bacterial cold-water disease of salmonids. It can cause external lesions (Fig. 3-9); and internally the spleen and kidney are the best sources for isolation of the microorganism. Cytophaga agar is widely used for cultivation of this bacterium. This medium is nonselective and will support the growth of a wide variety of bacteria; hence, knowledge of cell and colony morphology is essential. Colonies of *F. psychrophilus* form after 3-5 days incubation at 15-18°C and are bright yellow with a raised convex center and a thin spreading irregular periphery revealing a "fried egg" appearance. Occasional strains may produce colonies with an entire edge. *Flexibacter psychrophilus* grows at temperatures from 5-20°C but rarely at 25°C (Holt *et al.* 1989; 1991).

Flexibacter columnaris is a pathogen of many species of both cold- and warm-water fishes. The disease, columnaris, caused by this bacterium is characterized by yellowish external lesions on the gills and body surfaces. The bacterium seen on bits of tissue collected from these lesions is arranged in column-like or "haystack" masses of cells. Internally, the bacterium can be isolated from the kidneys, but often only after the disease has reached an advanced stage. On cytophaga agar, colonies of *F. columnaris* that develop after 3-5 days incubation at 25°C are yellow with spreading, convoluted centers, rhizoid edges and adhere to the agar surface (Fig. 3-10). Atypical colonies that may occasionally be isolated are yellow with rhizoid edges, but differ by having mucoid centers and not adhering to the agar surface. *Flexibacter columnaris* grows at temperatures from 10 to 35°C with some strains also growing at 37°C (Holt *et al.* 1975; Becker and Fujihara 1978).

Flexibacter maritimus has been isolated from several marine fishes in Japan (Wakabayashi *et al.* 1989). Diseased fish have eroded mouths and frayed fins. Infected tissue can appear pale yellow. Some affected fish may have shallow skin lesions. The bacterium will not grow on cytophaga agar unless at least 30% sea water is used in the medium. Both KCl and NaCl are required for growth. The yellow colonies produced are flat and thin and have uneven edges. Optimum growth temperature is approximately 30°C with a range between 15 and 34°C.

MISCELLANEOUS BACTERIAL PATHOGENS OF FISH

Flavobacterium

Bacteria in this genus are slender, Gram-negative, non-motile rods that are strict aerobes and catalase, oxidase and phosphatase positive. Except for their inability to glide or swarm on agar surfaces, they are taxonomically similar to members of the family Cytophagaceae (the genera *Cytophaga* and *Flexibacter*). Flavobacteria and *Cytophaga*-like bacteria are commonly isolated from the external gill surfaces of cultured salmonids and may form part of their normal flora. These bacteria are abundant when fish show signs of the condition referred to as *bacterial gill disease*; however, their role in the disease process is not clear and Koch's postulates have never been completed. Wakabayashi *et al.* (1980) examined 15 strains of flavobacteria isolated from fish with clinical signs of bacterial gill disease in Japan and the U.S. These authors concluded that they appeared to represent a new species of *Flavobacterium*; however, no species designation was proposed. These bacteria were isolated by streaking infected gill tissue onto either cytophaga agar or very dilute (20 fold) TSA. After 5 days incubation at 18°C, colonies appeared as round (0.5-1 mm in diameter), transparent, smooth, and light yellow in color. Growth occurred over a temperature range of 10-25°C.

Pseudomonas

This large and complex genus consists of approximately 27 recognized species, two of which have been isolated from diseased fish. A third proposed species, *Pseudomonas anguilliseptica*, has been isolated from diseased eels in Japan (Wakabayashi and Egusa 1972; Muroga *et al.* 1977). *Pseudomonas fluorescens* has been shown to cause disease, although mortalities are limited and occur only after stress or injury to the fish. The bacterium is not believed to be highly invasive.

Staphylococcus

There are few available reports describing staphylococcal infections of fish in detail. *Staphylococcus epidermidis* was reported associated with disease outbreaks in 1976 and 1977 in yellowtail (*Seriola quinqueradiata*) and red sea bream (*Chrysophrys major*) in Japan (Kusuda and Sugiyama 1981). After incubation for 48 h at 25°C these isolates developed colonies on BHIA which were circular, convex, entire, and white to whitish yellow in color. Individual cells ranged from 0.6 to 1.8 μm in diameter and replicated to form irregular clusters. Staphylococcal infection was observed in fish but may require specific and severe stress imposed by the environment.

Streptococcus

All the streptococci that have been isolated from diseased fish share similar morphological characteristics. The organisms are Gram-positive, ovoid cells with a diameter of 0.6 to 0.9 μm. These nonmotile bacteria are nonsporulating and do not produce capsules. Colonies on agar plates are small (approximately 0.5 mm in diameter), white, and translucent. They are entire, smooth, and slightly raised. When grown on rabbit-blood agar, most freshwater isolates are ß hemolytic (Jo 1982). For the most part, isolates that cause serious problems in yellowtail culture are nonhemolytic (Kusuda and Kawai 1982; Kitao 1982); however, there are reports of ß hemolytic streptococci occurring in salt water (Minami *et al.* 1976; Nakatsugawa 1983). Todd-Hewitt or brain heart infusion supplemented with 0.5% glucose support the growth of isolates from yellowtail at temperatures ranging from 10 to 45°C with an optimum between 20 to 37°C (Kusuda and Kawai 1982).

Biochemical characteristics of the *Streptococcus* sp. from diseased fish vary among isolates (Kusuda and Komatsu 1978). Kusuda *et al.* (1976) and Kitao (1982) have indicated that the organisms which they isolated from yellowtail are very closely related biochemically to *S. faecalis* and *S. faecium*. Streptococci have been isolated from fish only in Japan and the U.S. However, they have been recovered from both marine and freshwater fish and appear to have a broad host range. Hoshina *et al.* (1958) isolated streptococci from naturally infected rainbow trout undergoing an epizootic, causing 0.3% mortality per day. The dying animals were 10 to 20 cm in length. The researchers were able to artificially infect carp, *Cyprinus carpio*, goldfish, *Carassius auratus*, eels, *Anguilla japonicus*, frogs, *Rana nigromaculata*, and mice with this isolate from trout. In an epizootic in estuarine areas in the southern U.S., tens of thousands of fish were estimated to have died (Plumb *et al.* 1974). The primary species affected was menhaden, *Brevoortia patronus*, but a variety of other marine fish, including sea catfish, *Arius felis*, striped mullet, *Mugil cephalus*, and silver trout, *Cynoscion nothus*, also died during the epizootic. The bacterial isolates from this disease outbreak were infective for channel catfish, *Ictalurus punctatus*, when challenged artificially. Robinson and Meyer (1966) isolated streptococci from naturally infected golden shiner, *Notemigonus crysoleucas*, and were able to artificially infect bluegills, *Lepomis macrochirus*, green sunfish, *L. cyanellus*, and American toads, *Bufo americanus*. Streptococcal infections have been a problem in cultured yellowtail because they can cause high mortality in two-year-old fish which are at a marketable age. Sea bream, *Parius major*, flounder, *Paralichthys olivaceus*, and other cultured marine fish are also susceptible.

Signs of the disease may include lethargic or erratic swimming. External manifestations are hemorrhagic lesions of the skin, especially around the opercula, mouth, anus, and at the base of the fins. Lesions are also found on the caudal fin. Exophthalmia with hemorrhaging around the eye is a common sign. Internally, the peritoneal cavity may be filled with a bloody fluid, and there may be congestion of the internal organs (Miyazaki 1982), including the brain (Shiomitsu 1982).

SUMMARY

Although somewhat limited in number, there is an interesting assemblage of pathogenic bacteria associated with fish. Most of these organisms have been isolated from populations of cultured fish experiencing mortality resulting from disease. Some are unique, exhibiting a degree of specificity (e.g., *Renibacterium salmoninarum*); others are opportunistic (e.g., certain aeromonads). Some may even be members of the normal flora and only assume the role of pathogens when the host has been subjected to sufficient insult (e.g., the lactic acid bacteria). The environment exerts an important influence over the various pathogenic bacteria known to occur in fish. It is possible to generalize by stating that as temperature increases, the replication rate of the bacteria increases, and, therefore, the disease problem frequently worsens. There are exceptions. For example, the bacteria replicate within a rather narrow range of temperature. No doubt many other environmental factors play an important role in the physiological and pathogenesis of bacteria of fish. However, probably none are more important than temperature.

ACKNOWLEDGEMENTS

The authors thank the Oregon State University Sea Grant College Program, supported by NOAA Office of Sea Grant, U.S. Department of Commerce under Grant No. NA89AA-D-SG108 for assistance. They thank Carlene Pelroy for typing this manuscript, Oregon Agricultural Experiment Station Technical Paper No. 9424.

REFERENCES

Amend, D.F., and K.A. Johnson. 1981. Current status and future needs of *Vibrio anguillarum* bacterins. *Dev. Biol. Stand.* 49:403-417.

Amend, D.F., K.A. Johnson, T.R. Croy, and P.H. McCarthy. 1983. Some factors affecting *Yersinia ruckeri* bacterins. *J. Fish Dis.* 6:337-344.

Anderson, D.R., H.E. Hopps, M.F. Barile, and B.C. Bernheim. 1965. Comparison of the ultrastructure of several rickettsiae, ornithosis virus, and mycoplasma in tissue culture. *J. Bacteriol.* 90:1387-1404.

Anderson, J.I. W., and D.A. Conroy. 1970. Vibrio disease in marine fishes. In *A symposium of diseases of fishes and shellfishes*, ed. S.F. Snieszko. Am. Fish. Soc., Washington D.C. Special Publ. pp. 266-272.

Aoki, T., S. Egusa, and T. Arai. 1974. Detection of R factors in naturally occurring *Vibrio anguillarum* strains. *Antimicrob. Agents Chemother.* 6:534-538.

Arakawa, C.K., and J.L. Fryer. 1984. Isolation and characterization of a new subspecies of *Mycobacterium chelonei* infectious for salmonid fish. *Helog. Meeresunters.* 37:329-342.

Austin, B. 1985. Evaluation of antimicrobial compounds for the control of bacterial kidney disease in rainbow trout, *Salmo gairdneri* Richardson. *J. Fish Dis.* 8:209-220.

Austin, B., and D.A. Austin. 1988. *Bacterial fish pathogens: disease in farmed and wildfish.* Chichester: Halsted Press, John Wiley and Sons.

Banner, C.R., J.J. Long, J.L. Fryer, and J.S. Rohovec. 1986. Occurrence of salmonid fish infected with *Renibacterium salmoninarum* in the Pacific Ocean. *J. Fish Dis.* 9:273-275.

Banner, C.R., J.S. Rohovec, and J.L. Fryer. 1982. A rapid method for labeling rabbit immunoglobulin with fluorescein for use in detection of fish pathogens. *Bull. Eur. Assoc. Fish Pathol.* 2:35-37.

Banner, C.R., J.S. Rohovec, and J.L. Fryer. 1983. *Renibacterium salmoninarum* as a cause of mortality among chinook salmon in salt water. *J. Wildl. Maricult. Soc.* 14:236-239.

Bataillon, E., L. Dubard, and L. Terre. 1897. Un nouveau type de tuberculose. *C.R. Seánces Soc. Biol.* 49:446.

Becker, C.D. and M.P. Fujihara. 1978. The bacterial pathogen *Flexibacter columnaris* and its epizootiology among Columbia River fish. *Am. Fish. Soc. Monograph* No. 2.

Bell, G.R. 1961. Two epidemics of apparent kidney disease in cultured pink salmon (*Oncorhynchus gorbuscha*). *J. Fish. Res. Board Can.* 18:559-562.

Bernardet, J.F., and P. A.D. Grimont. 1989. Deoxyribonucleic acid relatedness and phenotypic characterization of *Flexibacter columnaris* sp. nov., nom. rev., *Flexibacter psychrophilus* sp. nov., nom. rev., and *Flexibacter maritimus* Wakabayashi, Hikida, and Masumura. 1986. *Int. J. Syst. Bacteriol.* 39:346-354.

Bradley, T.M., C.E. Newcomer, and K.O. Maxwell. 1988. Epitheliocystis associated with massive mortalities of cultured lake trout *Salvelinus namaycush*. *Dis. Aquat. Org.* 4:9-17.

Bravo, S., and M. Campos. 1989. Coho salmon syndrome in Chile. *Fish Health Sec. Am. Fish. Soc. Newsl.* 17:3.

Bruno, D.W. 1986. Histopathology of bacterial kidney disease in laboratory infected rainbow trout, *Salmo gairdneri* Richardson, and Atlantic salmon, *Salmo salar* L., with reference to naturally infected fish. *J. Fish Dis.* 9:523-537.

Bullock, G.L., B.R. Griffin, and H.M. Stucky. 1980. Detection of *Corynebacterium salmoninus* by direct fluorescent antibody test. *Can. J. Fish. Aquat. Sci.* 37:719-721.

Bullock, G.L., and R.L. Herman. 1985. *Edwardsiella* infections of fishes. *U.S. Fish Wildl. Ser. Fish Dis. Leaflet* 71.

Cisar, J.O., and J.L. Fryer. 1969. An epizootic of vibriosis in chinook salmon. *Bull. Wildl. Dis. Assoc.* 5:73-76.

Collins, M.D., J.A.E. Farrow, B.A. Phillips, S. Ferusu, and D. Jones. 1987. Classification of *Lactobacillus divergens*, *Lactobacillus piscicola*, and some catalase-negative asporogeneous, rod-shaped bacteria from poultry, in a new genus, *Carnobacterium*. *Int. J. Syst. Bacteriol.* 37:310-316.

Cone, D.K. 1982. A *Lactobacillus* sp. from diseased female rainbow trout, *Salmo gairdneri* Richardson, in Newfoundland, Canada. *J. Fish Dis.* 5:479-485.

Cvitanich, J., O. Garate, and C.E. Smith. 1990. Etiological agent in a Chilean coho disease isolated and confirmed by Koch's postulates. *Fish Health Sect. Am. Fish. Soc. Newsl.* 18:1-2.

Daly, J.G., B. Lindvik, and R.M.W. Stevenson. 1986. Serological heterogeneity of recent isolates of *Yersinia ruckeri* from Ontario and British Columbia. *Dis. Aquat. Org.* 1:151-153.

Daly, J.G., and R.M.W. Stevenson. 1987. Hydrophobic and hemagglutinating properties of *Renibacterium salmoninarum*. *J. Gen. Microbiol.* 133:3575-3580.

Daly, J.G., and R.M.W. Stevenson. 1990. Characterization of the *Renibacterium salmoninarum* hemagglutinin. *J. Gen. Microbiol.* 136:949-953.

Dubreuil, D., R. Hallier, and M. Jacques. 1990. Immunoelectron microscopic demonstration that *Renibacterium salmoninarum* is encapsulated. *FEMS Microbiol. Lett.* 66:313-316.

Earp, B.J., C.H. Ellis, and E.J. Ordal. 1953. Kidney disease in young salmon. *Spec. Rep. Ser. Wash. Dept. Fish.* 1:1-72.

Egidius, E., R. Wiik, K. Andersen, K.A. Hoff, and B. Hjeltnes. 1986. *Vibrio salmonicida* sp. nov., a new fish pathogen. *Int. J. Syst. Bacteriol.* 36:518-520.

Egusa, S. 1969. *Vibrio anguillarum*, a bacterium pathogenic to salt water and freshwater fishes. *Fish Pathol.* 4:31-44.

Eklund, M.W., F.T. Poysky, M.E. Petersen, L.W. Peck, and W.D. Brunson. 1984. Type E botulism in salmonids and conditions contributing to outbreaks. *Aquaculture* 41:293-309.

Eklund, M.W., F.T. Poysky, and D.I. Wieler. 1967. Characteristics of *Clostridium botulinum* type E isolated from Pacific Coast of the United States. *Appl. Microbiol.* 15:1316-1323.

Ellis, A.E. (ed.) 1988. *Fish Vaccination.* San Diego, CA: Academic Press.

Ellis, R.W., A.J. Novotny, and L.W. Harrell. 1978. Case report of kidney disease in a wild chinook salmon (*Oncorhynchus tshawytscha*) in the sea. *J. Wildl. Dis.* 14:121-123.

Evelyn, T.P.T. 1977a. An aberrant strain of the bacterial fish pathogen *Aeromonas salmonicida* isolated from a marine host, the sablefish (*Anoplopoma fimbris*) and from two species of cultured Pacific salmon. *J. Fish. Res. Board Can.* 28:1629-1634.

Evelyn, T.P.T. 1977b. An improved growth medium for the kidney disease bacterium and some notes on using the medium. *Bull. Off. Int. Epizoot.* 87:511-513.

Evelyn, T.P.T., J.E. Ketcheson, and L. Prosperi-Porta. 1984. Further evidence for the presence of *Renibacterium salmoninarum* in salmonid eggs and the failure of povodine-iodine to reduce the intra-ovum infection in water hardened eggs. *J. Fish Dis.* 7:173-182.

Evelyn, T.P.T., J.E. Ketcheson, and L. Prosperi-Porta. 1986. Use of erythromycin as a means of preventing vertical transmission of *Renibacterium salmoninarum*. *Dis. Aquat. Org.* 2:7-11.

Ezura, Y., K. Tajima, M. Yoshimizu, and T. Kimura. 1980. Studies on the taxonomy and serology of causative organisms of fish vibriosis. *Fish Pathol.* 14:167-179.

Fryer, J.L., J.S. Rohovec, and R.L. Garrison. 1978. Immunization of salmonids for control of vibriosis. *Mar. Fish. Rev.* 40(3):20-23.

Fryer, J.L., and J.E. Sanders. 1981. Bacterial kidney disease of salmonid fish. *Ann. Rev. Microbiol.* 35:273-298.

Fryer, J.L., C.N. Lannan, L.H. Garces, J.J. Larenas, and P.A. Smith. 1990. Isolation of a Rickettsiales-like organism from diseased coho salmon (*Oncorhynchus kisutch*) in Chile. *Fish Pathol.* 25:107-114.

Fukuda, Y., and R. Kusuda. 1981. Efficacy of vaccination for pseudotuberculosis in cultured yellowtail by various routes of administration. *Bull. JPN Soc. Sci. Fish.* 47:141-150.

Fuller, D.W., K.S. Pilcher, and J.L. Fryer. 1977. A leucocytolytic factor isolated from cultures of *Aeromonas salmonicida*. *J. Fish. Res. Board Can.* 34:1118-1125.

Getchell, R.G., J.S. Rohovec, and J.L. Fryer. 1984. Comparison of *Renibacterium salmoninarum* isolates by antigenic analysis. *Fish Pathol.* 20:149-159.

Harshbarger, J.C., S.C. Chang, and S.V. Otto. 1977. Chlamydiae (with phages), mycoplasmas, and rickettsiae in Chesapeake Bay bivalves. *Science* 196:666-668.

Hawke, J.P., A.C. McWhotter, A.G. Steigerwalt, and D.J. Brenner. 1981. *Edwardsiella ictaluri* sp. nov., the causative agent of enteric septicemia of catfish. *Int. J. Syst. Bacteriol.* 31:396-400.

Hayashi, K., S. Kobayashi, T. Kamata, and H. Ozaki. 1964. Studies on vibrio disease of rainbow trout (*Salmo gairdneri irideus*). II. Prophylactic vaccination against vibrio-disease. *J. Fac. Fish. Pref. Univ. Mie-Tsu* 6:181-191.

Henley, M.W., and D.L. Lewis. 1976. Anaerobic bacteria associated with epizootic in grey mullet (*Mugil cephalus*) and redfish (*Sciaenops ocellata*) along the Texas Gulf Coast. *J. Wildl. Dis.* 12:448-453.

Hiu, S.F., R.A. Holt, N. Sriranganathan, R.J. Seidler, and J.L. Fryer. 1984. *Lactobacillus piscicola*, a new species of salmonid fish. *Int. J. Syst. Bacteriol.* 34:393-400.

Hoffman, G.L., C.E. Dunbar, K. Wolf, and L.O. Zwillenberg. 1969. Epitheliocystis, a new infectious disease of bluegill (*Lepomis machrochirus*). *Antonie Leeuwenhoek J. Microbiol. Serol.* 35:146-158.

Holm, K.O., and T. Jorgensen. 1987. A successful vaccination of Atlantic salmon, *Salmo salar* L., against 'Hitra disease' or coldwater vibriosis. *J. Fish Dis.* 10:85-90.

Holt, R.A., J.E. Sanders, J.L. Zinn, J.L. Fryer, and K.S. Pilcher. 1975. Relation of water temperature to *Flexibacter columnaris* infection in steelhead trout (*Salmo gairdneri*), coho (*Oncorhynchus kisutch*) and chinook (*O. tshawytscha*) salmon. *J. Fish. Res. Board Can.* 32:1553-1559.

Holt, R.A., A. Amandi, J.S. Rohovec, and J.L. Fryer. 1989. Relation of water temperature to bacterial cold-water disease in coho salmon, chinook salmon, and rainbow trout. *J. Aquat. Anim. Health* 1:94-101.

Holt, R.A., J.S. Rohovec, and J.L. Fryer. 1991. Bacterial cold-water disease of salmonid fish. *Proceedings of Science in Aquaculture Conference*. Stirling, Scotland (in press).

Hoshina, T., T. Sano, and Y. Mommoto. 1958. A streptococcus pathogenic to fish. *J. Tokyo Univ. Fish.* 44:57-68.

Janssen, W.A., and M.J. Surgalla. 1968. Morphology, physiology and serology of a *Pasteurella* species pathogenic for white perch. *J. Bacteriol.* 96:1606-1610.

Jo, Y. 1982. Streptococcal infections of cultured freshwater fishes. *Fish Pathol.* 17:33-37.

Johnson, P.T. 1984. A rickettsia of the blue king crab, *Paralithodes platypus*. *J. Invert. Pathol.* 44:112-113.

Kariya, T., S. Kubota, Y. Nakamura, and K. Kira. 1968. Nocardial infection in cultured yellowtails (*Seriola quinqueradiata* and *S. purpurascens*). I. Bacteriological study. *Fish Pathol.* 3:16-23.

Kitao, T., and M. Kimura. 1974. Rapid diagnosis of pseudotuberculosis in yellowtail by means of the fluorescent antibody test. *Bull. JPN Soc. Sci. Fish.* 40:889-893.

Kitao, T., T. Aoki, and M. Kanda. 1981. Immune response of marine and freshwater fish against *Pasteurella piscicida*. *Dev. Biol. Stand.* 49:355-368.

Kitao, T. 1982. The methods for detection of *Streptococcus* sp.,causative bacteria of streptococcal disease of cultured yellowtail (*Seriola quinqueradiata*) especially, their cultural, biochemical, and serological properties. *Fish Pathol.* 17:17-26.

Kitao, T., T. Aoki, M. Fukudome, K. Kawano, Y. Wada, and Y. Mizuno. 1983. Serotyping of *Vibrio anguillarum* isolated from diseased freshwater fish in Japan. *J. Fish Dis.* 6:175-181.

Kitao, T., L. Ruangpan, and M. Fukudome. 1989. Isolation and classification of a *Nocardia* species from diseased giant gourami *Osphronemus goramy*. *J. Aquat. Anim. Health* 1:154-162.

Koike, Y., A. Kuwahara, and H. Fjuiwara. 1975. Characterization of *Pasteurella piscicida* isolated from white perch and cultivated yellowtail. *JPN J. Microbiol.* 19:241-247.

Kubica, G.P., I. Baess, R.E. Gordon, P.A. Jenkins, J.B.G. Kwapinski, C. McDurmont, S.R. Pattyn, H. Saito, V. Silcox, J.L. Stanford, K. Takeya, and M. Tsukamura. 1972. A cooperative numerical analysis of rapidly growing mycobacteria. *J. Gen. Microbiol.* 73:55-70.

Kubota, S., T. Kariya, Y. Nakamura, and K. Kira. 1968. Nocardial infection in cultured yellowtails (*Seriola quinqueradiata* and *S. purpurascens*). II. Histological study. *Fish Pathol.* 3:24-33.

Kubota, S., M. Kimura, and S. Egusa. 1970. Studies on bacterial tuberculosis of yellowtail. I. Symptomology and histopathology. *Fish Pathol.* 4:111-118.

Kudo, T., K. Hatai, and A. Seino. 1988. *Nocardia seriolae* sp. nov. causing nocardiosis of cultured fish. *Int. J. Syst. Bacteriol.* 38:173-178.

Kusuda, R. and M. Yamaoka. 1972. Etiological studies on bacterial pseudotuberculosis in cultured yellowtail with *Pasteurella piscicida* as the causative agent. I. On the morphological and biochemical properties. *Bull. JPN Soc. Sci. Fish.* 38:1325-1332.

Kusuda, R., and H. Taki. 1973. Studies on a nocardial infection of cultured yellowtail. I. Morphological and biological characteristics of *Nocardia* isolated from diseased fishes. *Bull. JPN Soc. Sci. Fish.* 39:937-943.

Kusuda, R., H. Taki, and T. Takeuchi. 1974. Studies on a nocardial infection of cultured yellowtail. II. Characteristics of *Nocardia kampachi* isolated from a gill-tuberculosis of yellowtail. *Bull. JPN Soc. Sci. Fish.* 40:369-373.

Kusuda, R. 1975. Nocardial infection in cultured yellowtails. In *Proceedings of the Third U.S.-Japan Meeting on Aquaculture.* Special Publication of Fishery Agency, Japan. pp. 63-66.

Kusuda, R., K. Kawai, T. Toyoshima, and I. Komatsu. 1976. A new pathogenic bacterium belonging to the genus *Streptococcus* isolated from an epizootic of cultured yellowtail. *Bull. JPN Soc. Sci. Fish.* 42:1345-1352.

Kusuda, R., and A. Nakagawa. 1978. Nocardial infection of cultured yellowtail. *Fish Pathol.* 13:25-31.

Kusuda, R., and I. Komatsu. 1978. A comparative study of fish pathogenic *Streptococcus* isolated from saltwater and freshwater fishes. *Bull. JPN Soc. Sci. Fish.* 44:1073-1078.

Kusuda, R., and A. Sugiyama. 1981. Studies on the characters of *Staphylococcus epidermidis* isolated from diseased fishes. I. On the morphological, biological and biochemical properties. *Fish Pathol.* 16:15-24.

Kusuda, R., and K. Kawai. 1982. Characteristics of *Streptococcus* sp. pathogenic to yellowtail. *Fish Pathol.* 17:11-16.

Lannan, C.N., J.R. Winton, and J.L. Fryer. 1984. Fish cell lines: establishment and characterization of nine cell lines from salmonids. *In Vitro* 20:671-676.

Larsen, J.L., and S. Mellengaard. 1984. Agglutination typing of *Vibrio anguillarum* isolates from diseased fish and the environment. *Appl. Environ. Microbiol.* 47:1261-1265.

Lee, E.G.H., and T.P.T. Evelyn. 1989. Effect of *Renibacterium salmoninarum* levels in the ovarian fluid of spawning chinook salmon on the prevalence of the pathogen in their eggs and progeny. *Dis. Aquat. Org.* 7:179-184.

Lee, K.K., and A.E. Ellis. 1990. Glycerophospholipid:cholesterol acyltransferase complexed with lipopolysaccharide (LPS) is a major lethal exotoxin and cytolysin of *Aeromonas salmonicida*: LPS stabilized and enhances toxicity of the enzyme. *J. Bacteriol.* 172:5382-5393.

Meyer, F.P., and G.L. Bullock. 1973. *Edwardsiella tarda*, a new pathogen of channel catfish (*Ictalurus punctatus*). *Appl. Microbiol.* 25:155-156.

Minami, T., M. Nakamura, Y. Ikeda, and H. Ozaki. 1976. A beta-hemolytic *Streptococcus* isolated from cultured yellowtail. *Fish Pathol.* 14:33-38.

Miyazaki, T. 1982. Pathological study on streptococcus histopathology of infected fishes. *Fish Pathol.* 17:39-47.

Munn, C.B., E.E. Ishiguro, W.W. Kay, and T.J. Trust. 1982. Role of surface components in serum resistance of virulent *Aeromonas salmonicida*. *Infect. Immunol.* 36:1069-1075.

Muroga, K., T. Nakai, and T. Sawada. 1977. Studies on red spot disease of pond cultured cells. IV. Physiological characteristics of the causative bacterium *Pseudomonas anguilliseptica*. *Fish Pathol.* 12:33-38.

Nakatsugawa, T. 1983. A streptococcal disease of cultured flounder. *Fish Pathol.* 17:281-285.

O'Leary, P.J., J.S. Rohovec, J.E. Sanders, and J.L. Fryer. 1982. Serotypes of *Yersinia ruckeri* and their immunogenic properties. In *Sea Grant College Program Publication* ORESU-T-82-001, Oregon State Univ., Corvallis, Oregon, pp. 1-15.

Özel, M., and I. Schwanz-Pfitzner. 1975. Comparative studies by the electron microscope of Rhabdoviruses of plant and of animal origin. III. Egtved virus (VHS) of the rainbow trout (*Salmo gairdneri*) and Rickettsia-like organisms. *Zbl. Bakt. Hyg., I. Abt. Orig.* A230:1-14.

Park, S.J., H. Wakabayashi, and Y. Watanabe. 1983. Serotype and virulence of *Edwardsiella tarda* isolated from eel and their environment. *Fish Pathol.* 18:85-89.

Parisot, T.J. 1958. Tuberculosis in fish. *Bacteriol. Rev.* 22:240-245.

Pascho, R.J., and D. Mulcahy. 1987. Enzyme-linked immunosorbent assay for a soluble antigen of *Renibacterium salmoninarum*, the causative agent of salmonid bacterial kidney disease. *Can. J. Fish. Aquat. Sci.* 44:183-191.

Plumb, J.A., J.H. Schachte, J.L. Gaines, W. Pelter, and B. Carroll. 1974. *Streptococcus* sp. from marine fishes along Alabama and northwest Florida coast of the Gulf of Mexico. *Trans. Am. Fish. Soc.* 103:358-361.

Plumb, J.A., and S. Vinitnantharat. 1989. Biochemical, biophysical, and serological homogeneity of *Edwardsiella ictaluri*. *J. Aquat. Anim. Health* 1:51-56.

Ransom, D.P., C.N. Lannan, J.S. Rohovec, and J.L. Fryer. 1984. Comparison of histopathology caused by *Vibrio anguillarum* and *Vibrio ordalii* in three species of Pacific salmon. *J. Fish Dis.* 7:107-115.

Reichenbach-Klinke, H.H. 1972. Some aspects of mycobacterial infections in fish. In *Diseases of Fish*, ed. L.E. Mawdesley-Thomas, *Symp. Zool. Soc. Lond.* 30. pp. 17-24.

Robinson, J.A. and F.P. Meyer. 1966. Streptococcal fish pathogen. *J. Bacteriol.* 92:512.

Ross, A.J. 1970. Mycobacteriosis among Pacific salmonid fishes. In *A symposium on diseases of Fishes and Shellfishes*, ed. S.F. Snieszko, (Spec. Publ. Fish. Soc. 5). pp. 279-283.

Ross, A.J., R.R. Rucker, and W.H. Ewing. 1966. Description of a bacterium associated with redmouth disease of rainbow trout (*Salmo gairdneri*). *Can. J. Microbiol.* 12:763-770.

Sanders, J.E., and J.L. Fryer. 1980. *Renibacterium salmoninarum* gen. nov., sp. nov., the causative agent of bacterial kidney disease in salmonid fishes. *Int. J. Syst. Bacteriol.* 30:496-502.

Schiewe, M.H ., T.J. Trust, and J.H. Crosa. 1981. *Vibrio ordalii* sp. nov.: A causative agent of vibriosis of fish. *Curr. Microbiol.* 6:343-348.

Schnick, R.A., F.P. Meyer, and D.L. Gray. 1986. A guide to approved chemicals in fish production and fishery resource management. University of Arkansas Cooperative Extension and U.S. Fish and Wildlife Service. MP 241, 24 pp.

Scott, M. 1968. The pathogenicity of *Aeromonas salmonicida* (Griffin) in sea and brackish waters. *J. Gen. Microbiol.* 50:321-327.

Shiomitsu, K. 1982. Isolation of *Streptococcus* sp. from the brain of cultured yellowtail. *Fish Pathol.* 17:27-31.

Shotts, E.B., and R. Rimler. 1973. Medium for the isolation of *Aeromonas hydrophila*. *Appl. Microbiol.* 26:550-553.

Sindermann, C.J. 1977. *Disease Diagnosis and Control in North American Marine Aquaculture*. New York:Elsevier.

Snieszko, S.F., G.L. Bullock, E. Hollis, and J.G. Boone. 1964. *Pasteurella* species from an epizootic of white perch (*Roccus americanus*) in Chesapeake Bay tidewater areas. *J. Bacteriol.* 88:1814-1815.

Tebbit, G.L., J.D. Erickson, and R.B. Van de Water. 1981. Development and use of *Yersinia ruckeri* bacterins to control enteric redmouth disease. *Develop. Biol. Stand.* 49:395-401.

Turaga, P., G.D. Weins, and S. Kaattari. 1987a. Bacterial kidney disease: the potential role of soluble antigen(s). *J. Fish Biol.* 31(supplement A):191-194.

Turaga, P.S.D., G.D. Weins, and S.L. Kaattari. 1987b. Analysis of *Renibacterium salmoninarum* antigen production *in situ*. *Fish Pathol.* 22:209-214.

Udey, L.R., and J.L. Fryer. 1978. Immunization of fish with bacterins of *Aeromonas salmonicida*. *Mar. Fish. Rev.* 40(3):12-17.

Wakabayashi, H., and S. Egusa. 1972. Characteristics of a *Pseudomonas* sp. from an epizootic of pond-cultured cells (*Anguilla japonica*). *Bull. JPN Soc. Sci. Fish.* 35:577-587.

Wakabayashi, H., S. Egusa, and J.L. Fryer. 1980. Characteristics of filamentous bacteria isolated from a gill disease of salmonids. *Can. J. Fish. Aquat. Sci.* 17:1499-1504.

Wakabayashi, H., M. Hikida, and K. Masumura. 1989. *Flexibacter maritimus* sp. nov., a pathogen of marine fishes. *Int. J. Syst. Bacteriol.* 36:396-398.

Wallbanks, S., A.J. Martiniz-Murcia, J.L. Fryer, B.A. Phillips, and M.D. Collins. 1990. 16SrRNA sequence determination for members of the genus *Carnobacterium* and related lactic acid bacteria and description of *Vagococcus salmoninarum* sp. nov. *Int. J. Syst. Bacteriol.* 40:224-230.

Williams, A.M., J.L. Fryer, and M.D. Collins. 1990. *Lactococcus piscium* sp. nov. a new *Lactococcus* species from salmonid fish. *FEMS Microbiol. Lett.* 68:109-114.

Wolf, K. 1981. Chlamydia and rickettsia of fish. *Fish Health News* 10:1-5.

Wolf, K., and L.E. Dunbar. 1959. Test of 34 therapeutic agents for control of kidney disease in trout. *Trans. Am. Fish. Soc.* 88:117-124.

Wood, J.W., and J. Wallis. 1955. Kidney disease in adult chinook salmon and its transmission by feeding to young chinook salmon. *Res. Briefs Fish Comm. Ore.* 6:32-40.

Young, C.L., and G.B. Chapman. 1978. Ultrastructural aspects of the causative agent and renal histopathology of bacterial kidney disease in brook trout (*Salvelinus fontinalis*). *J. Fish. Res. Board Can.* 35:1234-1248.

4

Fungal Diseases of Marine and Estuarine Fishes

Edward J. Noga

INTRODUCTION

Fungi are generally considered to be opportunistic, relatively weak pathogens that are problematic only when hosts are exposed to some stressful conditions or otherwise have reduced defenses (Noga 1990). Thus, they are potentially attractive candidates for monitoring programs that are intended to reflect the physiological health state of an aquatic population. While fish-pathogenic fungi are relatively uncommon in marine or estuarine ecosystems, several diseases have caused considerable morbidity and mortality: *Ichthyophonus*, ulcerative mycosis, and red spot disease/epizootic ulcerative syndrome.

ICHTHYOPHONUS

General Characteristics

Ichthyophonus hoferi is a fungus-like agent of uncertain taxonomy that causes a systemic infection in many marine fish. Mainly a pathogen of cold-water fishes, it has been most commonly observed in populations in the northeast and northwest Atlantic Ocean and the North and Baltic Seas, but has also been recorded from the Mediterranean, Australia, and Japan.

Pathology depends upon the host, dosage of pathogen, and possibly strain/species of fungus. Gross clinical signs include behavioral changes (loss of equilibrium or "staggers"), abnormal pigmentation, abdominal distension, muscle atrophy, spinal curvature, roughening of skin, and skin ulcers. Internally, white nodules or elongated foci may be present on various tissues, especially highly vascular organs (Fig. 4-1) (e.g., kidney, spleen, heart, liver). Hyperpigmentation is common in lesions in some fish. Areas of muscle liquefaction may be seen in acute cases (Sindermann 1990).

0-8493-8662-4/93/$0.00 + $.50

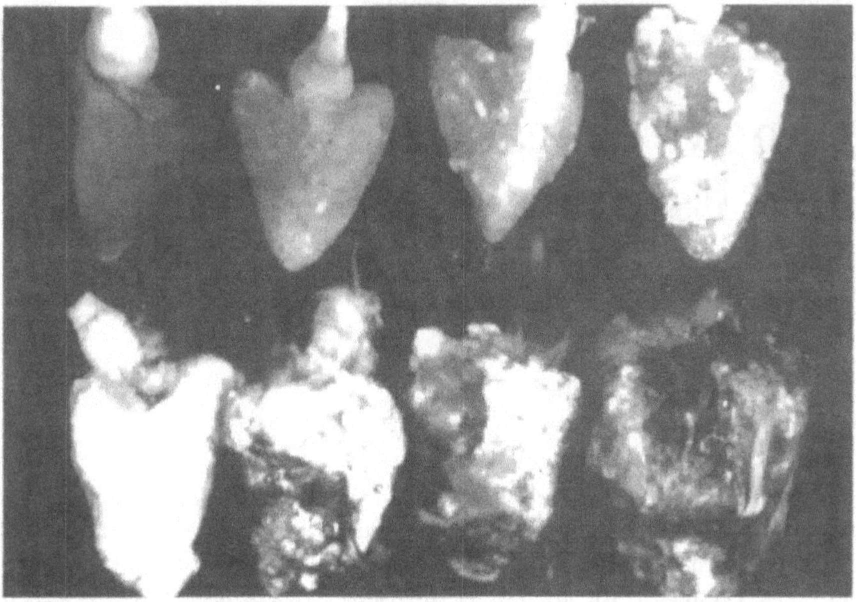

Fig. 4-1. Hearts of normal (upper left) and *Ichthyophonus*-infected Atlantic herring (*Clupea harengus*), with increasing extent of invasion (from Sindermann 1990).

In the early stages of infection, there is usually a pronounced mononuclear infiltrate in reponse to the presence of spores. In the chronic form, this progresses to a more advanced chronic inflammatory response, including epithelioid cells and a fibrous capsule. Some spores within this inflammatory lesion may degenerate, especially in resistant hosts. In the acute, more virulent form of the disease, there is often little inflammation. Host susceptibility appears to be a primary determinant of pathogenesis. More resistant hosts usually develop the chronic form of the disease, with chronic inflammatory foci developed in reponse to the various life stages of the fungi. Mortality often does not occur for at least several months. More susceptible hosts develop an acute form, which is associated with massive tissue invasion, necrosis, and death, usually within a few weeks. Both forms of the disease have been experimentally reproduced in herring, with a high dose of pathogen needed to induce acute lesions (Sindermann 1963). McVicar (1979) showed that plaice develop serum precipitins to *Ichthyophonus*; however, this was not associated with increased resistance to disease.

Ichthyophonus is clinically very similar to other chronic inflammatory diseases, such as mycobacteriosis. Thus, a definitive diagnosis requires the identification

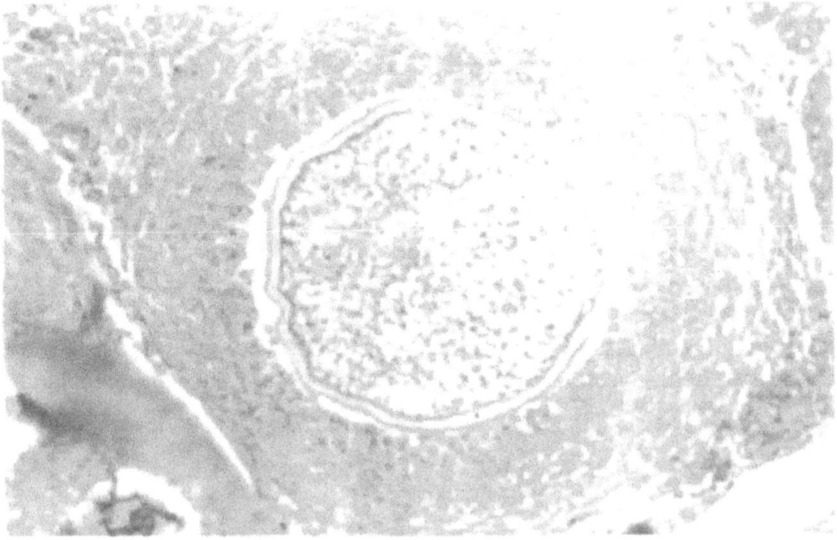

Fig. 4-2. Histopathology of a resting spore of *Ichthyophonus* surrounded by chronic inflammation.

of characteristic life stages in clinical material (Wolke 1975). This includes the presence of the typical thick-walled "resting spore" in wet mounts or histological sections (Fig. 4-2). Characteristic features of the *Ichthyophonus* resting spore include variable size (10 to 200 μm in diameter), a thick, PAS-positive, GMS-positive, double wall and a vacuolated, weakly basophilic, PAS-positive, weakly argyrophylic, multinucleated cytoplasm. The germinating spore is flask-shaped, with the neck consisting of the hypha that breaks though the outer wall, with the resulting cytoplasm contained only by the inner wall. Elongate hyphae that have developed from germinating spores are rarely seen in histological sections.

In wet mounts, it is often advisable to confirm structures as *Ichthyophonus* spores by observing germination, which usually occurs within minutes of the host's death (McVicar 1982). Spores also readily germinate in various culture media, such as Sabourad's dextrose agar with 1% serum, Hagem's medium, or minimum essential medium (McVicar 1982). The taxonomy of *Ichthyophonus* is poorly studied, and whether all diseases are due to the same pathogen is unknown. This is largely due to a lack of culture studies on isolates described from cases. McVicar (1982) and others (Egusa 1983; Sindermann and Scattergood 1954) have described several stages in *Ichthyophonus*'s life cycle, which include vegetative and proliferative stages. Asexual reproduction has usually been the only observed method of propagation, although Sproston (1944) reported sexual stages in an isolate from mackerel (*Scomber scombrus*).

Ichthyophonus is an obligate pathogen. Experimental transmission is easily accomplished by feeding spores or infected fish. The latter has been responsible for outbreaks of *Ichthyophonus* in freshwater fishes (Wood *et al.* 1955; Egusa 1983). Water-borne infections have been reported, but ingestion is believed to be the primary route of infection, with the release of amoeboid stages from the resting spore in the intestine. The amoeboid stage then penetrates the mucosa, entering the bloodstream to subsequently lodge in various organs. Spores are formed, which then produce further infective stages. The pathogen is eventually released via damaged body tissues (e.g., skin ulcers) or after death of the fish. Laboratory-held spores have remained infectious in seawater for up to 6 months.

Ecological/Economic Impact

Ichthyophonus is the only fungal pathogen that has caused significant epidemics in purely marine feral fish populations (Noga 1990). Most notable include outbreaks in Atlantic herring, *Clupea harengus,* and yellowtail flounder, *Limanda ferruginea*, of the northwestern Atlantic Ocean (Sindermann 1963; Ruggieri *et al.* 1970), haddock, *Melanogrammus aeglefini*, and plaice, *Pleuronectes platessa,* of the northeastern Atlantic Ocean (McVicar 1979) and cod, *Gadus morhua,* of the Baltic Sea (Moller 1974).

Ichthyophonus may be an important limiting factor to some feral fish populations. Epidemics in Atlantic herring have been associated with rapidly rising prevalence followed by a crash in the fish population and subsequent rapid decline in fungal prevalence. This has been associated with large reductions in herring abundance (Tibbo and Graham 1963). In other fish, such as plaice, disease losses take on a more chronic nature (McVicar 1986). Nonetheless, *Ichthyophonus*-associated chronic mortality is estimated to be as high as 50% in some populations, such as plaice in the North Sea (McVicar 1986). In other cases (e.g., Baltic cod), while disease prevalence is high, population effects are not discernible (Moller 1974). Some feel that resistant hosts may be important reservoirs of infection between epidemics (Sindermann 1990). The consistently high levels of *Ichthyophonus* infections in some marine fish populations suggest that endemic levels in some populations may be very high.

Ichthyophonus can render fillets less marketable, causing what is known as greasers in haddock, which causes a foul odor and poor flesh texture (McVicar 1982). Acute infections in herring cause degeneration and necrosis of body muscle, making them unsuitable for smoking or pickling (Sindermann 1958). Infected fillets can also contaminate normal fillets by contact (McVicar 1982).

Future Research Needs

There is a need for basic research to determine whether more than one species of *Ichthyophonus* exists as well as a need to clarify the life cycle. This research may also help to explain differences in host susceptibility and how potential reservoir hosts contribute to maintenance of the disease.

Tank studies with experimentally infected Atlantic herring have mimicked the course of epidemics seen in the wild (Sindermann 1990). While high population densities and unknown decreases in host resistance have been postulated as possible causes of outbreaks, few experimental studies confirm or deny the proposed theories. There is no evidence for the importance of pollution in the development of ichthyophonosis epidemics. Field studies by McVicar (personal communication) suggested that the highest disease prevalence in plaice was at unpolluted sites. This does not mean that pollution may not increase risk of disease in some cases, but if so, it is not readily discernible. Research on environmental risk factors may help to define conditions responsible for epidemic and endemic disease.

ULCERATIVE MYCOSIS

General Characteristics

Pathogenesis

One of the most common diseases presently affecting fishes in estuarine ecosystems of the northwest Atlantic Ocean is ulcerative mycosis (UM), a deep, ulcerative fungal infection, primarily affecting Atlantic menhaden, *Brevoortia tyrannus*. Ulcerative mycosis is distinguished by very deep penetrating lesions which are so aggressive that they commonly perforate the body wall, exposing the internal organs. When the dead tissue is sloughed off, a crater-shaped lesion is left, which is usually the most common stage seen during epidemics. Besides the prominent fungal component, lesions are also infected with many different types of bacteria and protozoa. Some of these other agents appear to be important in lesion development (see below).

Ulcerative mycosis appears to have a high mortality rate, since often few fish with evidence of previous infection are seen after an outbreak subsides. Many fish with advanced lesions have systemic bacterial infections (Noga and Dykstra 1986; Noga, unpublished data). Thus, the large numbers of microorganisms present in advanced cases combined with the osmotic stress caused by the large ulcers probably contribute to the death of the fish.

Fig. 4-3. Early ulcerative mycosis lesion in Atlantic menhaden (*Brevoortia tyrannus*)(from Noga *et al.* 1988).

Examination of spontaneous lesions suggests that UM begins as a skin infection that progresses to involve underlying muscle and viscera (Noga *et al.* 1988). Skin lesions begin as flat, up to 5 mm, red or yellow-red foci, having primarily a mononuclear infiltrate. Fungal infection appears to begin near the surface of the skin (Fig. 4-3), with invasion of underlying muscle. In advanced lesions (Fig. 4-4), large open ulcers contain masses of fungal hyphae, bacteria, and necrotic muscle. This necrotic mass eventually sloughs, leaving a crater-shaped cavity (Fig. 4-5). Healing lesions are usually small, nonulcerated areas of tissue loss that are most common after an outbreak subsides. Barely visible with hematoxylin and eosin stain, fungi are easily seen with silver stains (Fig. 4-6).

For presently unknown reasons, over half of all lesions occur in the anal area. Fungal chemotaxis towards intestinal contents, anatomical differences in local immunity, and trauma have been hypothesized as possible causes (Noga *et al.* 1988). Oomycete zoospores are attracted to menhaden mucus, scales, skin, and fecal material (D. Celio, personal communication, cited in Shafer *et al.* 1990); menhaden schools have been seen to disturb the mud substrate, which may stir up spores present in sediments (J. Merriner, personal communication).

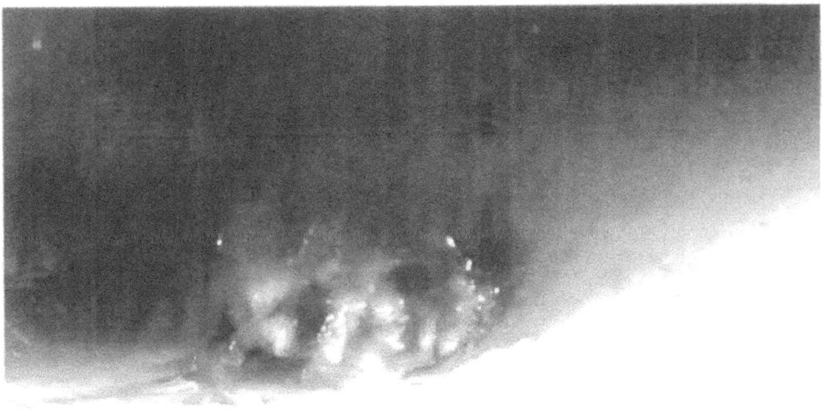

Fig. 4-4. Advanced stage ulcerative mycosis lesion in Atlantic menhaden, *Brevoortia tyrannus*. A large, white,edematous mass of friable necrotic muscle and hyphae fills the center of the lesion (from Noga *et al*. 1988).

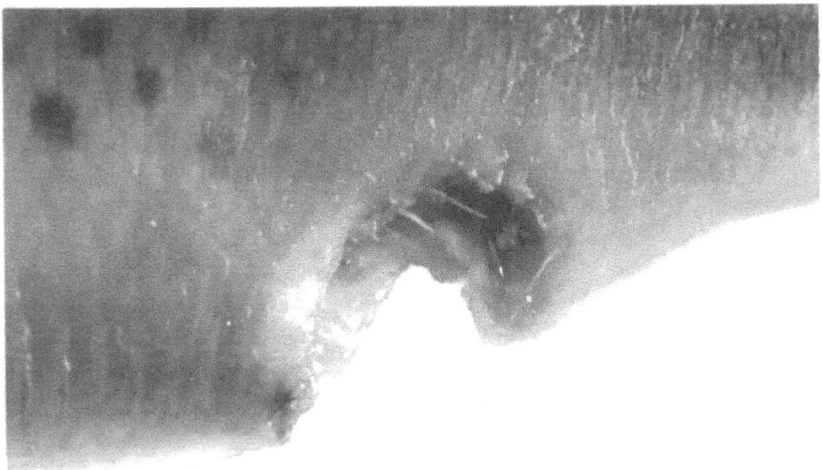

Fig. 4-5. End-stage ulcerative mycosis lesion in Atlantic menhaden, *Brevoortia tyrannus*. Note the deeply indented appearance (from Noga *et al*. 1988).

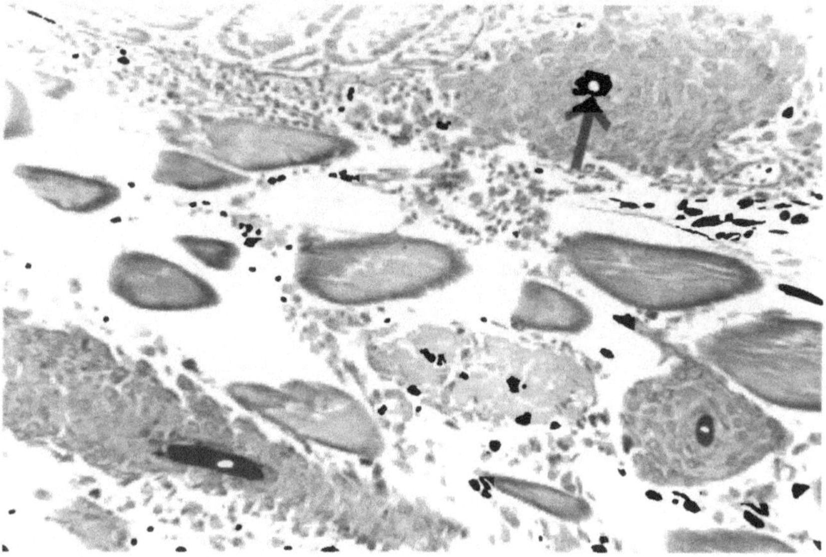

Fig. 4-6. Histopathology of advanced ulcerative mycosis lesion in Atlantic menhaden, *Brevoortia tyrannus*, showing hyphae within solid granulomas (arrow) Gomori methenamine silver, X 300 (from Noga *et al.* 1988).

The fungi in UM lesions are saprolegniaceous water molds (Oomycetes), including primarily *Aphanomyces*, but *Saprolegnia* has occasionally been isolated (Noga and Dykstra 1986; Dykstra *et al.* 1986; Dykstra *et al.* 1989). There may be more than one species of *Aphanomyces* involved; at least one isolate is taxonomically similar to *Aphanomyces laevis*. Previously, these saprolegniaceous water molds were considered to be exclusively freshwater pathogens and were not reported to cause disease in estuarine fishes.

Saprolegniaceous water molds are common freshwater inhabitants that usually form fuzzy, cottony growths on the skin of freshwater fishes. In contrast with UM, such lesions usually do not penetrate deeply into the body. The fish's inflammatory response to UM is also unusually severe; this may reflect the fact that the fungus grows aggressively into the tissue.

The growth and sporulation of an *Aphanomyces* isolate from menhaden with ulcerative mycosis were enhanced in the presence of low concentrations of salt (Dykstra *et al.* 1986). This is very unusual, for saprolegniaceous oomycetes are usually inhibited by salt and are not isolated from salinities greater than 3 ppt (TeStrake 1959). The ability of this *Aphanomyces* sp. isolate to sporulate in up to 20 ppt salinity is unique for a saprolegniaceous oomycete (Dykstra *et al.* 1986).

Similarly, Shafer *et al.* (1990) found that a *Saprolegnia* sp. isolated from a UM lesion in menhaden was relatively unstressed by exposure to mesohaline (10-15 ppt) sea water and germinated in large numbers in such conditions in the presence of high nutrients.

Salinity tolerance correlates with the highest UM prevalence occurring in waters of low to moderate salinity. Salinity tolerance may also explain how these fungi can penetrate deep into fish tissue, which has a high salt content. While reduction in host immunity has been speculated to be a contributor to fish developing UM, Hargis *et al.* (in press) observed increased immunoreactivity to mitogen stimulation in UM-affected menhaden. Values returned to normal in healed fish. This does not necessarily mean that UM-affected fish have a heightened immune response, but instead may reflect a response to the massive antigenic stimulation associated with microbial agents in UM lesions. More specific indicators of susceptibility to specific UM pathogens may be needed to discern immune mechanisms (see Experimental Model of UM).

Epidemiology

Ulcerative mycosis was first recognized in April 1984 from estuaries in North Carolina. First seen in the Pamlico River, it was soon reported from the Neuse River, New River, and Albemarle Sound, North Carolina. Within a few months, UM epidemics were recognized in the Chesapeake Bay (Maryland) and St. Johns River (Florida) estuaries (Dykstra *et al.* 1989)(Fig. 4-7). However, whether UM is actually a new disease is questionable. Interestingly, Kroger and Guthrie (1972) observed Atlantic and Gulf menhaden (*Brevoortia patronus*) with lesions strikingly similar to the healing stage of UM. Other chronic inflammatory mycoses were reported in other parts of the world around the same time (Table 4-1).

Since 1984, UM has continued to cause repeated outbreaks that in some instances has resulted in up to 100% infection rates in randomly sampled Atlantic menhaden schools in the Pamlico River (Levine *et al.* 1991). The overwhelming majority of fish that acquire the infection are young-of-year (0+) Atlantic menhaden. In North Carolina, outbreaks are most common in the Pamlico River; the severity of epidemics in other estuarine systems in the state appears to parallel the severity of the condition in the Pamlico River. During especially virulent episodes, it appears as if millions of fish may die of the infection (J. Hawkins, personal communication). Lesions that are pathologically similar to UM have been observed on many other estuarine species (Table 4-2). The prevalence of skin lesions on these other species seems to be greatest during the peaks of the menhaden epizootics (Levine *et al.* 1990), suggesting a spillover effect, as has been seen during *Ichthyophonus* epidemics (Sindermann 1990).

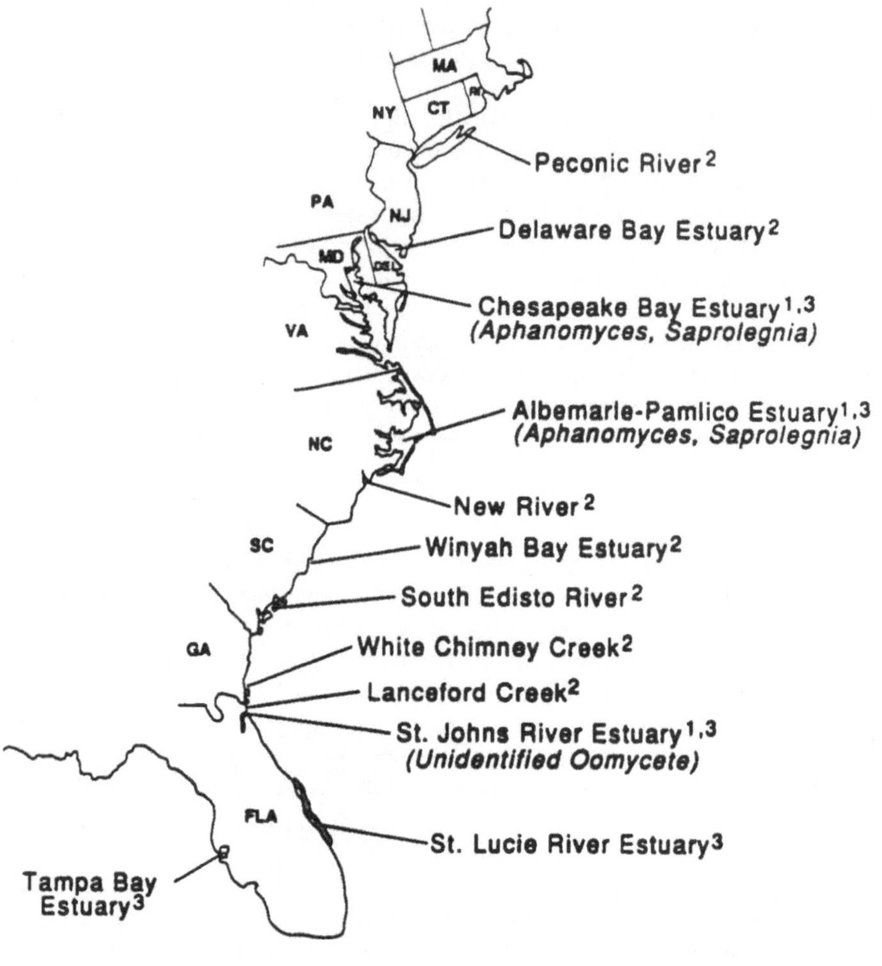

Fig. 4-7. Documented occurrences of (1) ulcerative mycosis in Atlantic menhaden, (2) ulcerative mycosis-like lesions in Atlantic menhaden, and (3) ulcerative mycosis-like lesions in estuarine fishes besides Atlantic menhaden in northwestern Atlantic estuaries. Identification of ulcerative mycosis is based upon the presence of deep skin ulcers having chronic inflammation in response to oomycete fungi. Isolated fungi are indicated in parentheses (primarily from data of Dykstra *et al.* 1986; Dykstra *et al.* 1989; TeStrake and Lim 1987). Ulcerative mycosis-like lesions are clinically similar lesions, but oomycete culture was not attempted (primarily from data of Ahrenholz *et al.* 1987; Grier and Quintero 1987; TeStrake and Lim 1987; Couch, unpublished data; Noga *et al.* 1991). Fish with ulcerative mycosis-like lesions have also been referred to as having "ulcerative disease syndrome" (Hargis 1985).

Table 4-1. Occurrences of chronic inflammatory mycoses associated with oomycete water molds.

Disease	Geographic Range	Oomycete Present	Primary Host range	Reference
Red-spot (Bundaberg disease)	Australia New Guinea	*Aphanomyces*	Barramundi (*Lates calcalifer*)* Mullet (*Mugil cephalus*) Yellowfin bream (*Acanthopagrus australis*) Luderick (*Girella tricuspidata*)	McKenzie and Hall (1976) Callinan (1988)
Australian epizootic ulcerative syndrome	Australia	*Saprolegnia?*	Grunters (*Teraponidae*)* Mullet (*Mugilidae*) Rainbow fish (*Melanotaenidae*)	Pearce (undated)
Asian epizootic ulcerative syndrome (Ulcerative disease)	Malaysia Indonesia Thailand Burma Lao Par Sri Lanka India	*Achlya*	Catfish (*Clarias* sp.)* Snakehead (*Channa* sp.)	Roberts *et al.* (1986) Chinabut and Limsuan (1983)
Ulcerative mycosis	U.S.	*Aphanomyces*	Atlantic menhaden (*Brevoortia tyrannus*)	Noga and Dykstra (1986) Dykstra *et al.* (1989)
(Ulcerative disease syndrome)		*Saprolegnia*	Gizzard shad (*Dorosoma cepedianum*)	
Mycotic granulomatosis	Japan	*Aphanomyces*	Goldfish (*Carassius auratus*) Ayu (*Plecoglossus altivelis*) Bluegill (*Lepomis macrochirus*)	Miyazaki and Egusa (1972) Hatai *et al.* (1986)

* Many other fish species have been reported with similar lesions. See Callinan (1988), Pearce (undated), and Roberts *et al.* (1986) for details.

Table 4-2. Estuarine fishes affected by ulcerative mycosis (UM) or similar lesions in North America[1]

BOTHIDAE	Southern flounder (*Paralichthys lethostigma*) Ocellated flounder (*Ancylopsetta dilecta*)[2]
CLUPEIDAE	Atlantic menhaden (*Brevoortia tyrannus*)[3] Gizzard shad (*Dorosoma cepedianum*)[3]
CYPRINODONTIDAE	Topminnow (*Fundulus* sp.)[2] Topminnow (*Cyprinodon* sp.)[2]
PERCICHTHYIDAE	Striped bass (*Morone saxatilis*) White perch (*Morone americana*)
POMATOMIDAE	Bluefish (*Pomatomus saltatrix*)
SCIAENIDAE	Atlantic croaker (*Micropogonias undulatus*) Weakfish (*Cynoscion regalis*) Red drum (*Sciaenops ocellata*)[4] Spot (*Leiostomus xanthurus*) Silver perch (*Bairdiella chrysura*) Black drum (*Pogonias cromis*)[2,3]
SOLEIDAE	Hogchoker (*Trinectes maculatus*)
SPARIDAE	Pinfish (*Lagodon rhomboides*)

[1] all data from Noga *et al.* (1991) and Noga (unpublished) unless otherwise noted
[2] data from TeStrake and Lim (1986)
[3] Oomycetes cultured from lesions; all other diagnoses based upon morphological identification of oomycete-like fungi in deep skin ulcers
[4] data from Grier and Quintero (1986)

There is evidence that fish infected with UM are a dead-end for the fungus. Most of the fungi present in advanced and end-stage lesions are dead (Dykstra *et al.* 1989) and sporulating mycelia have never been seen in lesions. Thus, infected fish may not be a source of contagion, at least for the fungal component of the disease.

In the Pamlico River, where UM has been best studied, outbreaks exhibit a bimodal annual cycle, with peaks in disease prevalence usually between April to June and September to December (Levine *et al.* 1991). The highest prevalences occur in oligo- to mesohaline (about 2-8 ppt) waters. Both observations of lay observers, as well as empirical evidence from sampling surveys (Levine *et al.* 1991), strongly support that the fish probably acquired the infection in this part of the estuary. This is also supported by the spatial shifting of the focus for high concentrations of infected fish from downriver in the spring-summer epidemic, to farther upriver during the fall-winter peak, corresponding to rising salinities at the end of the year (Levine *et al.* 1991). Finally, clinically normal menhaden held in tanks in low salinity sites develop the disease (Noga *et al.*, unpublished data).

Ecological/Economic Impact

Concerns about ulcerative mycosis have included the impact that this disease may have upon the fishery's productivity. Massive epidemics have been reported in the Pamlico River and other estuaries. Many epidemics have been associated with fish kills; the relatively few fish with healing lesions after some epidemics suggest that the disease is frequently fatal. This may be due to direct effects of infection or greater susceptibility of sick individuals to predation.

Furthermore, the reproductive competence of fish that have recovered from UM is unknown. Lesions often extend to internal organs and the most common site of infection is at the anal area, which often involves the site of the genital openings. The reproductive capabilities of these fish are unknown, but if a significant number of reproductively incompetent fish are present in a population, it may have a strong negative influence on reproductive success, especially since such fish would be competing with their cohorts for food and other resources without contributing to the gene pool. For menhaden, reproductive impacts would have to be affecting fish at age 2-3 years, when they become sexually mature.

Another concern has been the socioeconomic impact that this disease has had upon both the commercial fishery and the affected riverine communities. Visibly diseased fish are "unsalable in-the-round." If the fillet is affected, they may be totally rejected, thus excluding them from the commercial fishery, regardless of their potential human health risk. When ulcerative lesions affected flounder during one quite severe epidemic, flounder from North Carolina (both healthy and diseased) were temporarily rejected by out-of-state wholesalers, which caused considerable economic hardship to North Carolina fishermen. These episodes can also affect consumer confidence that extends well beyond the market for North Carolina seafood products. In addition, people do not like to see sick fish. Thus, appearance may reduce the attractiveness of fishing in affected areas, which could have considerable impact on a valuable recreational fishery. Negative impacts on land values and other economic factors have not been assessed.

While water molds in UM lesions are not a zoonotic threat, many bacteria in UM lesions and organs of affected fish are potential human pathogens (Noga and Dykstra 1986). Thus ulcerated fish may pose a human health threat if consumed raw.

Future Research Needs

Accurate Assessment of UM's Impacts

Atlantic menhaden constitute by weight the largest commercial fishery in the coastal United States (U.S. National Marine Fisheries Service 1989). Besides being an important commercial source of fish meal, they provide an important food source for organisms higher up in the food chain, such as flounder, striped bass, and sea trout, all of which support important sport and commercial fisheries.

While massive mortalities of menhaden have occurred during some UM outbreaks, the impact of the disease on fishery stocks is uncertain. This is largely due to the the great natural variation in Atlantic menhaden populations. Vaughan *et al.* (1986) have estimated that as much as a 70% reduction in age 0+ year class abundance would be needed to statistically detect a decrease in population abundance due to a catastrophic event. Such an event has not yet been detected in a 0+ year class of menhaden. Obviously, more accurate methods of determining the effect of catastrophic events is needed. Some possible solutions are discussed by Vaughan *et al.* (1986).

Of potentially greater concern is the impact of long-term mortalities due to UM. Since 1984, UM has caused repeated disease outbreaks in northwest Atlantic estuaries. The effect of smaller-scale chronic losses is often less dramatic and even more difficult to measure over the short term, but may have much more profound effects. For example, mathematical models of fishery stocks have suggested that as little as a 0.50% decline in fishery recruitment over a 30-year period could result in a 40% reduction in fishery biomass (Merriner and Vaughan 1987). For reasons stated previously, such dramatic decreases in biomass may not be noticed until serious and possibly irreversible damage has occurred.

Experimental Model of UM

The study of UM, as with most naturally occurring diseases, is severely hampered by the frequent inavailability of experimental animals. Challenge of menhaden with *Aphanomyces* has failed to induce UM (Noga *et al.* 1989). However, fish challenged with lesion material develop typical UM lesions (Noga and Burke, unpublished data), suggesting that some other pathogen(s) may be needed in conjunction with the fungus to cause disease.

Agents that interact to cause UM should be identified to develop a standardized, quantifiable, model for testing the effects of water quality on disease resistance; a rational plan for determining how host immunity affects disease development; and a surveillance program for identifying where the pathogens responsible for UM reside in the environment (e.g., in water, only on fish, etc.).

Data Linking Water Quality to UM

Since its dramatic appearance in 1984, ulcerative mycosis has remained prevalent in the Pamlico and Rappahanock Rivers. While many factors have the potential of making fish more susceptible to UM, potential risk factors must be prioritized because limited resources prevent the examination of all potential factors that may affect fish health.

There is very little data on the toxicant burdens in menhaden from the Pamlico and and Rappahannock Rivers, but the latter watersheds are not highly impacted by xenobiotics compared to many other estuaries in the U.S. (Rader *et al.* 1987). However, over the years, there has been a large increase in farming activity and population growth in their watersheds, with a subsequent increase in nutrient and agricultural chemical loading (Rader *et al.* 1987). These estuarine systems have a relatively low flushing rate compared to other smaller estuarine systems along the northwest Atlantic coast, suggesting that incubation of substances that come from the watershed may somehow be responsible for water quality changes that lead to UM.

While comparable data are not available for all years, it appears that the largest disease outbreaks in the Pamlico River were in 1984, 1989, and possibly 1988 (Levine *et al.* 1991). Interestingly, 1984 and 1989 were unusually wet years, resulting in below average salinities. While depressed salinities may be associated with increased risk of disease outbreaks, this is obviously by no means the only factor that is required.

A direct causal relationship between an anthropogenic substance and its effect on aquatic animal health may not exist; instead, the pollutant may act indirectly on fish health by causing other changes in environmental quality. More difficult to assess, yet potentially very important, are ecological changes that may affect trophic dynamics, and thus food quality or quantity or the presence of various chemical constituents in the water (oxygen levels, pH, various toxins, etc.).

Conditions that could merit special consideration as possible risk factors for ulcerative mycosis include low dissolved oxygen and phytoplankton-produced toxins. Hypoxia is responsible for acute mortalities of finfish in areas having UM (Officer *et al.* 1984; Rader *et al.* 1987). We have identified a highly ichthyotoxic dinoflagellate associated with fish kills in the Pamlico River (Smith *et al.* 1988,

Hobbs *et al.* unpublished data). The presence of such acute events suggests that sublethal effects leading to increased disease susceptibility may also occur. If this is true, changes in the quality and/or quantity of phytoplankton communities will require further investigation. Physical/chemical factors which might be important in influencing these end results include the volume and temporal/spatial distribution of water entering impacted estuaries.

Based upon the historical anthropogenic inputs that appear to be most significant in the Pamlico River watershed, prospective and retrospective studies have examined the relationship between selected water quality factors and UM prevalence. Some studies have failed to demonstrate any correlation between UM and either temperature, dissolved oxygen, salinity, or pH (Levine *et al.* 1987). However, later studies which involved more temporally intense sampling, revealed evidence associating outbreaks with low salinity, decreasing temperatures and/or low oxygen levels (Levine *et al.* 1989; Levine *et al.* 1991).

More research is needed to substantiate links between suspected water quality factors and ulcerative mycosis. Attempts should also be made to determine how UM may be linked to other deleterious changes in the environment, such as increased turbidity and loss of submerged aquatic vegetation (Rader *et al.* 1987), which may have adverse effects on fishery productivity. Thus, there needs to be a commitment to determine the impact of UM on estuarine ecosystems, so that rational judgments can be made before irreversible changes ensue. Finally, an international effort should be directed at determining if similar mycotic diseases reported worldwide (Table 4-1) are part of one larger problem.

RED SPOT DISEASE/EPIZOOTIC ULCERATIVE SYNDROME

General Characteristics

Since the early 1970's, several epidemic fungal diseases sharing similar clinico-pathological features have occurred in the Indo-Pacific region (Table 4-2). In 1972, an ulcerative skin disease known as Bundaberg disease or red spot disease (RSD) was reported in estuarine fishes of southern Queensland, Australia (McKenzie and Hall 1976). Subesquently, RSD epidemics have occurred in many economically important fishes in New South Wales and Western Australia, as well as New Guinea (Callinan 1988). In 1980, an epidemic skin disease referred to as ulcer disease or epizootic ulcerative syndrome (EUS) was reported in Malaysia (Boonyaratpalin 1985). It soon spread throughout southeast Asia and has now been reported as far westward as India (Roberts 1989b). Primarily affecting freshwater fishes, epidemics have also been reported from estuarine species in Thailand (Tonguthai 1985) and Sri Lanka (Costa and Wijeyaratne

1989). In 1986, an epidemic skin disease, also called epizootic ulcerative syndrome, was seen in many freshwater and oligohaline (less than 1 ppt) riverine areas of the Northern Territory, Australia (Pearce, undated). Since then, epidemics have usually recurred annually.

The similarity of these conditions to ulcerative mycosis is striking, as observed in many fishes, often including many estuarine species (e.g., RSD). Oomycetes play an important role in lesion development. *Aphanomyces*, a relatively uncommon fish pathogen, has been isolated from both RSD and UM lesions (Noga and Dykstra 1986; Callinan 1990). Oomycete-like fungi are a prominent component of Asian and Australian EUS lesions and oomycetes have been isolated from some of these lesions (Chinabut and Limsuwan 1983; Pearce, undated). These diseases cause deep skin ulcers that have a severe chronic inflammatory infiltrate in response to the invading fungi. In Australian fish with EUS, fungal infections may also involve the eye (Pearce, undated).

Lesions appear to begin on the skin surface and develop into deep ulcers via extension. One feature that distinguishes RSD and EUS from UM is the apparent lack of fungal involvement in early stages. Earliest stages of Asian EUS are epithelial edema, loss of surface epithelium, and dermal edema with

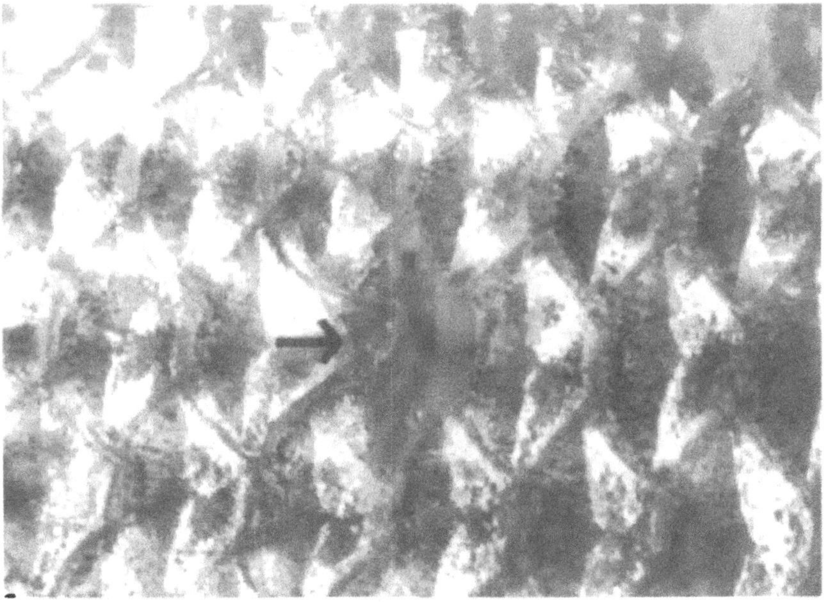

Fig. 4-8. Red spot disease in mullet, *Mugil cephalus*: Erythematous dermatitis affecting the posterior margin of a scale (from Callinan *et al.* 1989).

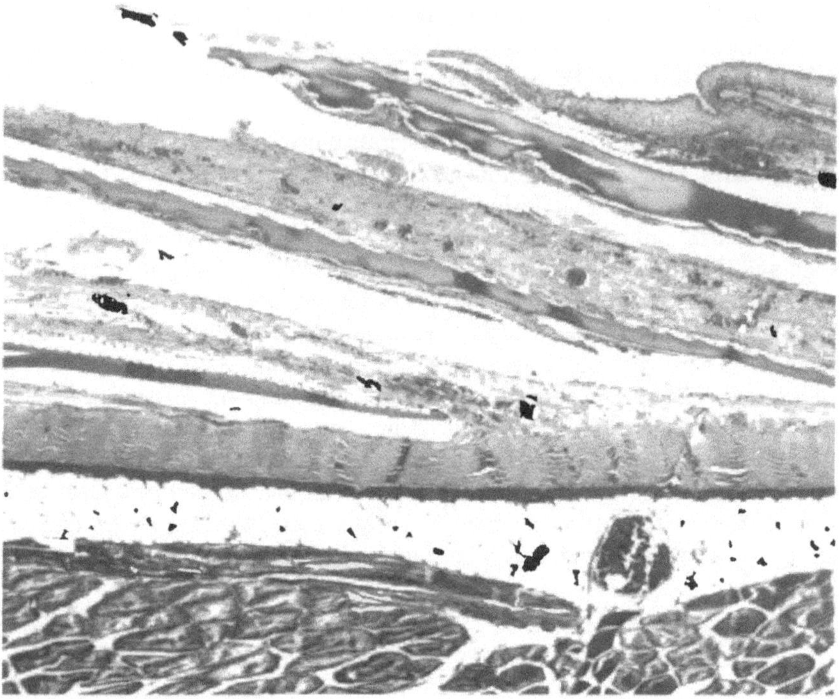

Fig. 4-9. Red spot disease in mullet, *Mugil cephalus*: Histological section of erythematous dermatitis showing ulceration, edema and cellular infiltration of the stratum spongiosum. Hematoxylin and eosin, X 38 (from Callinan *et al.* 1989).

hyperemia and hemorrhage. There is also extravasation of leucocytes forming perivascular cuffs (Roberts *et al.* 1986). Earliest stages of Australian EUS are similar but also include damage to dermal vasculature and perivascular accumulations of eosinophilic granular cells (Pearce, undated). Red spot disease lesions may evolve in one of two ways (Callinan *et al.* 1989). Erythematous dermatitis lesions have mild ulcerations or an intact albeit damaged epithelium, mild to severe, chronic-active dermatitis, and no fungi (Fig. 4-8, 4-9). In contrast, necrotizing dermatitis lesions are more extensively ulcerated (Fig. 4-10) with a pronounced mononuclear infiltrate in response to many fungal hyphae; they progress to a deep ulcer (Fig. 4-11), exposing skeletal muscle and showing severe, chronic myositis. Callinan *et al.* (1989) felt that erythematous dermatitis lesions spontaneously healed and thus did not progress to necrotizing dermatitis. However, they did not observe the early lesions that lead to necrotizing dermatitis.

Fig. 4-10. Red spot disease in mullet, *Mugil cephalus*: Necrotizing dermatitis lesions showing dermal swelling and early sloughing of dermis (from Callinan *et al.* 1989).

While myriad infectious agents have been isolated from RSD and EUS lesions, the primary cause(s) of these diseases remains uncertain. Rogers and Burke (1981) first suggested *Vibrio* as the initiator of RSD lesions in Queensland fish. Later, they argued that adult digenean trematodes may also be responsible for lesion initiation (Rogers and Burke 1988). However, a lack of histopathological data and limited microbiological studies make the linkage of these agents to RSD unclear. In a more extensive study, Callinan and Keep (1989) found some vibrios but also many other types of bacteria in all stages of RSD; no single bacterium was consistently present in all lesions. Trematodes were not associated with lesions.

Viral involvement has been suggested, but not substantiated. Tissue samples of several fish species with EUS from the Northern Territory, Australia, produced cytopathic effects in cell culture associated with rhabdovirus-like particles (Humphrey and Langdon 1986). Some of the early histopathological changes (e.g, eosinophilic intranuclear inclusions) present in Australian EUS lesions are also suggestive of a viral infection (Pearce, undated). While a rhabdovirus was reportedly isolated from a luderick, *Girella tricuspidata*, with RSD from New South Wales (Roberts, unpublished data cited in Callinan *et al.* 1989), doubts have been raised regarding the validity of this report (Callinan,

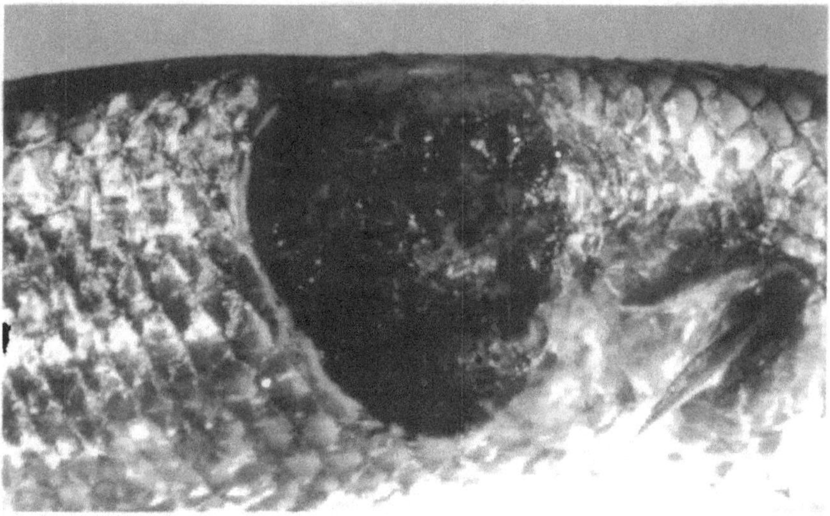

Fig. 4-11. Red spot disease in mullet (*Mugil cephalus*): Dermal ulcer exposing muscle (from Callinan *et al.* 1989).

personal communication). Virus isolation attempts from mullet with early RSD lesions have been unsuccessful (Callinan *et al.* unpublished data cited in Callinan *et al.* 1989). Birnaviruses (Hedrick *et al.* 1986; Saitanu *et al.* 1986; Wattananavijarn *et al.* 1988) and rhabdoviruses (Frierichs *et al.* 1986) have been isolated from Asian fish with EUS, but their importance to disease development is unproven.

The absence of any pathogen consistently present in early lesion stages has lead to the hypothesis that initial epidermal damage may be due to a rapid (adverse) change in water quality which, if severe enough, may allow fungal colonization leading to the large ulcers typical of advanced lesions (Callinan *et al.* 1989). Outbreaks of RSD have occurred when water temperatures drop in fall and also when the rainy season begins, depressing salinities. Precipitous and prolonged drops in dissolved oxygen and pH have also preceded RSD outbreaks (Callinan 1988). Dropping temperatures have also been associated with Asian and Australian EUS epidemics (Roberts *et al.* 1986; Pearce, undated). Interestingly, hypoxia and temperature change are also possible risk factors in UM development.

A primary noninfectious insult would help explain the presence of diverse, often opportunistic, infectious agents that might capitalize on a compromised host. Alternatively, their widespread occurrence of these diseases makes it possible that various outbreaks may be initiated by different infectious agents at the opportune time. In any case, the end result (i.e., deep mycosis) suggests that at least localized immunosuppression is occurring.

Ecological/Economic Impact

The impact of RSD or EUS on fishery stocks has not been determined, although impacts similar to those described for UM might be expected. Prevalences as high as 30% of commercial catches have been reported during RSD outbreaks in mullet (Callinan and Keep 1989). Callinan (personal communication) has estimated that during significant RSD outbreaks that occurred during 1984, 1985, 1987, 1988, and 1989, as much as 10% of the catch in affected rivers may have been discarded, resulting in a total loss of about $1,000,000/year to the Australian east coast commercial fishery. There are also concerns about the disease's esthetic/economic impact on the valuable recreational fishery. Of considerable concern is the pandemic nature of these chronic inflammatory mycoses. Red spot disease, Australian EUS, Asian EUS and ulcerative mycosis are clinico-pathologically similar to mycotic granulomatosis, a disease recognized in Japan in the early 1970's (Miyazaki and Egusa 1972; Hatai *et al.* 1984). Asian EUS is considered by some to be the most serious disease affecting finfish aquaculture in southeast Asia (Frierichs *et al.* 1989) and it appears to be spreading rapidly (Roberts 1989b). Interestingly, some feel that it may have originated in Australia as RSD. The spread of Asian EUS epizootics has even been likened to that of crayfish plague (coincidentally also caused by an *Aphanomyces* sp.), which was introduced into European crayfish populations by the importation of latently infected North American crayfish. Spread of Asian EUS could have occurred from the unrestricted transport of fish throughout the region (Callinan, personal communication). While spread of an infectious agent remains to be proven, the expeditious establishment of regulations that control the international transfer of aquatic stocks at risk is certainly justified.

Future Research Needs

Most basic questions about ulcerative mycosis are also applicable to these other fungal dermatopathies. First is the need for an assessment of the impact of these diseases on native fish populations. The same constraints are present in answering this question as with UM. There is also a need to develop experimental models of RSD and EUS. A defined experimental model implies the fulfilment of Koch's postulates and de facto knowledge of the infectious and noninfectious pathogenesis. Claims of reproducing Asian EUS with rhabdovirus challenge (Saitanu *et al.* 1986) have not yet been substantiated. Miyazaki and Egusa (1972) reproduced mycotic granulomatosis by simply implanting

Aphanomyces hypae under the skin. The mixed infections that are characteristic of RSD and EUS suggest that future studies should examine the possible synergism of multiple pathogens. There is also a need for determining the water quality conditions that facilitate these outbreaks. Close similarities among these chronic inflammatory mycoses suggest that common risk factors may be important. While the relationship of these fungal diseases to each other is uncertain, prevalence of such problems on a worldwide scale is of major concern.

MISCELLANEOUS PATHOGENS

Fish entering freshwater from marine environments are at risk of developing classical oomycete skin infections, consisting of relatively superficial foci of cottony hyphae (Neisch and Hughes 1980). In salmonids, some oomycete epidemics have been associated with ulcerative dermal necrosis (UDN). Early UDN lesions consist of focal epithelial necrosis; these ulcerate and are often secondarily invaded by Oomycetes, especially members of the *Saprolegnia diclina-parasitica* complex. Identified from Irish, British, French and Swedish estuaries (Stuart and Fuller 1968; Roberts 1989a), the cause of UDN is unknown, although viruses and toxins are suspects.

Exophiala and related pathogens have been responsible for epidemic disease in salmonids cultured in sea cages (Richards *et al.* 1978) and in aquaria (Blazer and Wolke 1979), but no epidemics have been reported from wild populations.

Most other fungal agents reported from marine fishes have been seen as single cases that do not have any known impact upon fishery populations. These include species of *Aureobasidium* in a stingray (*Trygon pstonacea*) (Alderman 1982), *Sarcinomyces* in black sea bream (*Spondiliosoma cantharus*)(Todaro *et al.* 1983), *Cladosporium* in cod (*Gadus morhua*) (Reichenbache-Klinke 1956), and *Fusarium* in triggerfish (*Melichthys vidua*) (Ostland *et al.* 1987) and bonnethead sharks (*Sphyrna tiburo*)(Muhvich *et al.* 1989).

SUMMARY AND CONCLUSIONS

Although relatively uncommon compared to other microbial diseases, fungal infections have been increasingly recognized as problems in marine and estuarine fish populations. *Ichthyophonus* infections in marine fish and relatively superficial oomycete skin lesions in marine species migrating into fresh water have been observed for many years. Recently, deep, chronic inflammatory, oomycete mycoses have caused widespread epidemics in estuarine fishes. The true role of fungi in initiating these chronic inflammatory mycoses is unclear; a number of other infectious agents have been implicated in the pathogenesis of these diseases. Of concern is the pandemic nature of the chronic inflammatory oomycete infections, their dramatic appearance and rapid dissemination over a wide geographic area. There is evidence that they may be associated with poor water quality. Future research should not only determine risk factors for individual diseases but also how similar ecological conditions may be predisposing to these problems.

ACKNOWLEDGEMENTS

I thank J.H. Merriner and colleagues (U.S. National Marine Fisheries Service, Beaufort, North Carolina), M. Pearce (Dept. of Primary Industry and Fisheries, Darwin, Northern Territory, Australia), and especially R.B. Callinan (North Coast Agricultural Institute, Wollongbar, New South Wales, Australia) for comments and information. I also acknowledge many colleagues who have participated in research discussed in this review, especially J.H. Hawkins, J.F. Levine, M.J. Dykstra, and J.F. Wright. Some research described in this report was supported by grants from the University of North Carolina Water Resources Research Institute, the North Carolina Division of Marine Fisheries, U.S. Environmental Protection Agency, and University of North Carolina Sea Grant College.

REFERENCES

Ahrenholz, D.W., J.F. Guthrie, and R.M. Clayton. 1987. Observations of ulcerative mycosis on Atlantic menhaden (*Brevoortia tyrannus*). *NOAA Adm. Tech. Memo*. NMFS-SEFC-196.

Alderman, D.J. 1982. Fungal diseases of aquatic animals. In *Microbial Diseases of Fish*, ed. R.J. Roberts, pp. 189-242. New York: Academic Press.

Blazer, V.S., and R.E. Wolke. 1979. An *Exophiala*-like fungus as the cause of a systemic mycosis of marine fish. *J. Fish Dis*. 2:145-152.

Boonyaratpalin, S. 1985. Fish disease outbreak in Burma. FAO Spec. Rep. TCP/BUR/4402.

Callinan, R.B. 1988. Diseases of native Australian fishes. In *Fish Diseases*, Post Graduate Committee in Veterinary Science, pp. 459-472. Sydney: University of Sydney.

Callinan, R.B., and J.A. Keep. 1989. Bacteriology and parasitology of red spot disease in sea mullet (*Mugil cephalus*). *J. Fish Dis*. 12:349-356.

Callinan, R.B., G.C. Fraser, and J.L. Virgona. 1989. Pathology of red spot disease in sea mullet *Mugil cephalus* from eastern Australia. *J. Fish Dis*. 2:467-479.

Callinan, R.B., G.C. Fraser, and J.L. Virgona. 1990. Pathogenesis of red spot disease, a cutaneous ulcerative syndrome of estuarine fish in Australia. Sym. on Dis. in Asian Aquacult. Asian Fish Soc./Bali Indonesia.

Chinabut, S., and C. Limsuwan. 1983. Histopathological changes in some freshwater fishes found during the disease outbreak: 1982-1983. *The Fisheries Gazette* 36: 281-289.

Costa, H.H., and M.J.S. Wijeryaratne. 1989. Epidemiology of the epizootic ulcerative syndrome occurring for the first time among fish in Sri Lanka. *J. Appl. Ichthyol*. 1:48-52.

Dykstra, M.J., J.F. Levine, E.J. Noga, J.H. Hawkins, P. Gerdes, W.J. Hargis, Jr., H.J. Grier, and D. TeStrake. 1989. Ulcerative mycosis: A serious menhaden disease of the southeast coastal fisheries of the United States. *J. Fish Dis.* 12:125-127.

Dykstra, M.J., E.J. Noga, J.F. Levine, J.H. Hawkins, and D.F. Moye. 1986. Characterization of the *Aphanomyces* species involved with ulcerative mycosis (UM) in menhaden. *Mycologia* 78:664-672.

Egusa, S. 1983. Disease problems in Japanese yellowtail, *Seriola quinqueradiata*, culture: A review. *Rapp. P.V. Reún. Cons. Int. Explor. Mer* 182:10-18.

Faisal, M., and Hargis, W.J., Jr. (In Press). Evaluation of the in vitro lymphoproliferative responses to mitogens of Atlantic menhaden *Brevoortia tyrannus* with ulcer disease syndrome. *Fish and Shellfish Immunology*. Vol. 1.

Frierichs, G.N., S.D. Millar, and R.J. Roberts. 1986. Ulcerative rhabdovirus in fish in southeast Asia. *Nature* 322: 216.

Frierichs, G.N., B.J. Hill, and K. Way. 1989. Ulcerative disease rhabdovirus: Cell line susceptibility and serological comparison with other fish rhabdoviruses. *J. Fish Dis.* 12: 51-56.

Grier, H., and I. Quintero. 1987. *A Microscopic Study of Ulcerated Fish in Florida*. Fl. Bur. Mar. Res. Rep. WM-164.

Hargis, W.H. 1985. Quantitative effects of marine diseases on fish and shellfish populations. *Trans. N. Am. Wildl. Nat. Resour. Conf.* 50:608-640.

Hatai K., S. Takahashi, and S. Egusa. 1984. Studies on the pathogenic fungus of mycotic granulomatosis-IV. Changes of blood constituents in ayu, *Plecoglossus altivelis*, experimentally inoculated and naturally infected with *Aphanomyces piscicida*. *Fish Pathol.* 19:17-23.

Hedrick, R.P., W.D. Eaton, J.L. Fryer, W.D. Groberg, and S. Boonyaratapalin. 1986. Characteristics of a birnavirus isolated from a cultured sand goby (*Oxeleotris marmoratus*). *Dis. Aquat. Org.* 1:219-225.

Humphrey, J.D., and J.S. Langdon. 1986. *Ulcerative Disease in Northern Territory Fish*. Benalla, Victoria: Australian Fish Health Reference Laboratory, Regional Veterinary Laboratory. Internal Report.

Kroger, R.L., and J.F. Guthrie. 1972. Effect of predators on juvenile menhaden in clear and turbid estuaries. *Mar. Fish. Rev.* 34:78-80.

Lauckner, G. 1984. Agents: Fungi. In *Diseases of Marine Animals*. Vol. IV Part I., pp. 89-113. Hamburg: Biologische Anstalt Helgoland.

Levine, J.F., R.S. Cone, N. Sandukas, K. Meier, L.C. Lee, and J. Hand. 1987. An Integrated Database Management Program for Documentation of Ulcerative Disease Syndrome. The Ulcerative Disease Regional Database. Florida Dept. of Environmental Regulation Rept. SP131.

Levine, J.F., M.J. Dykstra, J.T. Camp, R.S. Cone, A. Clark, N. Sandukis, N.M. Markwardt, and T. Wenzel. 1989. Temporal distribution of ulcerative lesions on menhaden in the Tar-Pamlico River. Report of Contract #88-0670. North Carolina Division of Marine Fisheries.

Levine, J.F., J.H. Hawkins, M.J. Dykstra, E.J. Noga, D.W. Moye, and R.S. Cone. 1990. Species distribution of ulcerative lesions on finfish in the Tar-Pamlico River Estuary, North Carolina, May 1985-April 1987. *Dis. Aquat. Org.* 8:1-5.

Levine J.F., J.H. Hawkins, M.J. Dykstra, E.J. Noga, D.W. Moye, and R.S. Cone. 1991. Epidemiology of Atlantic menhaden affected with ulcerative mycosis in the Tar-Pamlico River Estuary. *J. Aquat. Anim. Health* 2:162-171.

McKenzie, R.A., and W.T.K. Hall. 1976. Dermal ulceration of mullet (*Mugil cephalus*). *Aust. Vet. J.* 52:230-231.

McVicar, A.H. 1979. *Ichthyophonus* in haddock and plaice in Scottish waters. *Inter. Coun. Explor. Sea Doc.* C.M.1979/G:48.

McVicar, A.H. 1982. *Ichthyophonus* infections of fish. In *Microbial Diseases of Fish*, ed. R.J. Roberts, pp. 243-269. New York: Academic Press.

McVicar, A.H. 1986. Fungal infection in marine fish and its importance in mariculture. *Spec. Publ.-Eur. Aquaculture Society* 9:189-196.

Merriner, J.V., and D. Vaughan. 1987. Ecosystem and fishery implications of ulcerative mycosis. In *The Proceedings of the Workshop on Fishery Diseases for the Albemarle-Pamlico Estuarine Study*, pp. 39-46, University of North Carolina Water Resources Research Institute.

Miyazaki, T., and S. Egusa. 1972. Studies on mycotic granulomatosis in freshwater fishes. I. The goldfish. *Fish Pathol.* 7:15-25 (In Japanese).

Moller, H. 1974. *Ichthyosporidium hoferi* (Plehn et Muslow)(fungi) as parasite in the Baltic cod (*Gadus morhua*). *Kiel. Meeresforsch.* 30:37-41.

Muhvich, A.G., R. Reimschuessel, M.M. Lipsky, and R.O. Bennett. 1989. *Fusarium solani* isolated from newborn bonnethead sharks *Sphyrna tiburo* (L.). *J. Fish Dis.* 12:57-62.

Neish, G.A., and G.C. Hughes. 1980. *Fungal Diseases of Fishes*. Neptune, NJ: TFH Publications.

Noga, E.J. 1990. A synopsis of mycotic diseases of marine fishes and invertebrates. In *Pathology in Marine Science*, pp. 143-160, ed. F.O. Perkins and T.C. Cheng. New York: Academic Press.

Noga, E.J., and M.J. Dykstra. 1986. Oomycete fungi associated with ulcerative mycosis in Atlantic menhaden. *J. Fish Dis.* 9:47-53.

Noga, E.J., J.F. Levine, M.J. Dykstra, and J.H. Hawkins. 1988. Pathology of ulcerative mycosis in Atlantic menhaden *Brevoortia tyrannus*. *Dis. Aquat. Org.* 4:189-197.

Noga, E.J., J.F. Levine, and M.J. Dykstra. 1989. *Fish Diseases of the Albemarle-Pamlico Estuary*. University of North Carolina Water Resources Research Institute Report # 238.

Noga, E.J., J. Wright, J.F. Levine, M.J. Dykstra, and J.H. Hawkins. 1991. Dermatological diseases affecting fishes of the Tar-Pamlico Estuary, North Carolina. *Dis. Aquat. Org.* 10:87-92.

Officer, C.B., R.B. Biggs, J.L. Taft, L.E. Cronin, M.A. Tyler, and W.R. Boynton. 1984. Chesapeake Bay anoxia: Origin, development and significance. *Science* 223:22-27.

Ostland, V.E., H.W. Ferguson, R.D. Armstrong, A. Asselin, and R. Hall. 1987. Case report: Granulomatous peritonitis in fish associated with *Fusarium solani*. *Vet. Rec.* 121:595-596.

Epizootic Ulcerative Syndrome Technical Report Dec. 1987-Sept. 1989. Northern Territory (Australia) Department of Primary Industry and Fisheries Fishery Report No. 22.

Rader, D.N., L.F. Loftin, B.A. McGee, J.P. Dorney, and J.T. Clements. 1987. *Surface Water Quality Concerns in the Tar-Pamlico River Estuary*. Water Quality Technical Report No. 87-04, North Carolina Division of Environmental Management.

Reichenbache-Klinke, H.H. 1956. Uber einige bisher ubekannte Hyphomyceten bei Verscheidenen Suwasser und Meeresfischen. *Mycopath. Mycol. Appl.* 7:333-368.

Richards, R.H., A. Holliman, and S. Helagson. 1978. *Exophiala salmonis* infection in Atlantic salmon, *Salmo salar* L. *J. Fish Dis.* 1:357-368.

Roberts, R.J. 1989a. Miscellaneous non-infectious diseases. In *Fish Pathology*, ed. R.J. Roberts, pp. 363-373. London: Bailliere Tindall.

Roberts, R.J. 1989b. Rugged species near the top in farming. *Fish Farming Int.* (August) p. 6.

Roberts, R.J., D.J. MacIntosh, K. Tonguthai, S. Boonyaratpalin, N. Tayaputch, M.J. Phillips, and S.D. Millar. 1986. *Field and Laboratory Investigation into Ulcerative Fish Diseases in the Asia-Pacific Region*. FAO Tech. Rept. Proj. TCP/RAS/4508, Bangkok.

Rogers, L.J., and J.B. Burke. 1981. Seasonal variation in the prevalence of red spot disease in estuarine fish with particular reference to the sea mullet, *Mugil cephalus* L. *J. Fish Dis.* 4:297-307.

Rogers, L.J., and J.B. Burke. 1988. Aetiology of 'red spot' disease (vibriosis) with special reference to the ectoparasitic digenean *Prototransversotrema steeri* (Angel) and the sea mullet, *Mugil cephalus* L. *J. Fish Biol.* 32:655-663.

Ruggieri, G.D., R.F. Nigrelli, P.M. Powles, and D.G. Garnett. 1970. Epizootics in yellowtail flounder, *Limanda ferruginea* Storer, in the western North Atlantic caused by *Ichthyophonus*, an ubiquitous parasitic fungus. *Zoologica (N.Y.)* 55:57-62.

5

Parasitic Diseases of Fishes and Their Relationship with Toxicants and Other Environmental Factors

Robin M. Overstreet

INTRODUCTION

A wide variety of environmental factors influence symbiotic infections and diseases in fishes. Some factors, both natural and anthropogenic, enhance infections, whereas others reduce, interrupt, or exclude them. In all cases, an equilibrium becomes established among symbiont, host, and environment. Many freshwater habitats provide conditions where hosts cannot avoid either the symbiont (see comments on that term and the term "parasite") or the environmental restriction. Estuarine and marine habitats are less likely to offer such confinement, but additional factors in such habitats can complicate both infections and infectious diseases, especially when they constitute stresses that stimulate shift of an equilibrium from infection to disease condition.

My emphasis for this chapter is twofold. First, it addresses effects of natural environmental factors, allowing a better understanding of the host-parasite relationship, effect of the parasite on a fish stock, and general ecology of a system. Second, it addresses effects of anthropogenic environmental factors on the host-parasite relationship as both a condition in nature and a model system capable of providing bioindication of both specific and general contamination or influence.

In the first portion, I provide examples of parasitic diseases and disease-agents that can serve as either models or indicators of environmental stresses. Parasitic diseases are not well-appreciated in the natural environment, especially the marine and estuarine habitats. Diseases probably affect larval and postlarval stages much more dramatically than they affect juveniles or adults. They

0-8493-8662-4/93/$0.00 + $.50

certainly affect old adults, but relatively few of these cases occur in a stock, and the effect on the stock is spread over a long period of time. Unless a parasite is introduced into a new geographical area, an abnormal host, or a vital site in a host, the infection seldom develops into a disease with subsequent death or debilitation. That nonpathogenic relationship, however, depends on minimal or no stress and on strong to moderate host resistance. Because environmental factors can create stress and reduce host resistance, diseases occur.

The numerous factors that influence infections interrelate in such a way that the observed dynamic effects produce a complex web of interactions, in many cases nearly impossible to delineate or to observe duplicated in a natural setting. Most short-lived parasites in estuarine and marine habitats in Mississippi exhibit seasonality, and the seasonal dynamics of those infections differ, sometimes drastically, every year. Prevalence of diseases presents an even more unpredictable occurrence, probably because diseases often are heavily dependent on environmental factors.

Most fish have infections of one or more symbionts simultaneously, and one of those can influence others. Symbiotic agents may range from one that associates with an animal without harming it, and depending on it minimally, to a parasite that strongly depends on and harms its host. Some "parasites" depend on hosts for survival and harm them only when hosts undergo certain stresses. In some cases, a symbiont with a complex life cycle and a series of stages may harm one of a series of corresponding multiple hosts, usually an intermediate host. Harm in those cases usually increases the prospect for completion of the cycle and perpetuation of the population. In other cases, the degree of "harm" or even rendering of harm to the host by the agent has not been investigated and is therefore unknown. Consequently, one may avoid confusion by considering all agents as symbionts and describing the known relationship between the symbiont and the host in detail. Nevertheless, I often use the term "parasite" interchangeably with "symbiont," as have most others in the literature.

Related to the matter of harm to the host caused by an agent is the difference between a "disease" and an "infection." When a parasite infects a fish, that condition constitutes an infection. For purposes of this treatment (and definitions differ considerably among workers and fields of study), if the infection is restricted to that resulting from an external agent, it will be termed an "infestation" and if from an internal agent or a combination of internal and external ones, it will be termed an "infection." When either infection or infestation develops to a point that the host is subjectively considered significantly impaired, the condition can be considered a "disease." The reason for the disease may reflect numbers of agents, general health of host, or either direct or indirect interaction of agent with an internal or external factor. This chapter will emphasize mostly diseases.

TYPE OF PARASITE

The prevalence and intensity of an infection as well as other aspects may depend on whether the parasite is external or internal, whether it has a direct life cycle or an indirect one, and whether it can have an adverse effect on its fish host. A fish can be an intermediate, paratenic, reservoir, accidental, abnormal, definitive host and a passive vector. From a public health standpoint, the fish can be 1) infected by a parasite commonly infecting a human definitive host, 2) infected with a species that could accidentally infect either a normal or an immunosuppressed human, or 3) serve as a passive vector for an egg or some other infective stage of a human parasite acquired from human or domestic animal contamination. The direct cycle of a parasite can involve a free living stage or not. Environmental factors usually affect stages differently. In an indirect cycle, two to four or more hosts can be involved, and a toxicant or other factor can readily eliminate or reduce one of those hosts, resulting in a break in the cycle and a reduction or exclusion of the corresponding parasitic infection. On the other hand, an influencing factor can also increase the stock size of an intermediate host or increase that host's susceptibility to an infection.

The next sections provide examples of factors and interrelationships among factors that have an effect on the host and on the symbiont, sometimes in a deleterious manner and sometimes in a manner that is useful from a human point of view.

HOST EFFECTS

A special relationship exists between a symbiont and its host. This relationship depends on various host factors such as the host immunological system, stage of development, age, sex, nutrition, genetic history, and migratory behavior, including home range. These and other factors, such as secondary bacterial, fungal, or viral infections, all interact with external environmental factors.

Fishes exhibit an immunological response to some protozoans, helminths, and other metazoans, in addition to microbial agents. Most research on piscine immunology has been conducted with bacteria as a tool to evaluate and dissect the immune system, a cellular and humoral system requiring considerable additional attention (e.g., Stolen *et al.* 1986; Manning and Secombes 1987). Whereas the system differs and can be modified differently among various species of fishes, several factors significantly influence host immunity. These factors include stress, diet, toxicants, temperature, and length of day (Ellis 1988). Typically, multiple factors act simultaneously.

Larval and postlarval fishes are probably the most vulnerable stages to disease in the wild as well as those most difficult to assess. Parasitic copepods are more conspicuous than most symbionts because they are relatively large, are external, and therefore are obvious to ichthyologists. A single copepod chalimus stage can kill a Spanish sardine or bay anchovy when the fish is <12mm long (Overstreet 1983a). Survivors can be assessed. For example, unidentified copepods on goby postlarvae infested 4.4% of 1530 fish, and ranged from 5 to 39% of the goby's body length (Felley *et al.* 1987). Most certainly, infections with microbial agents, protozoans, and metazoans have a significant effect on abundance of local stocks and of populations. As indicated above, these relationships are difficult to assess in nature. Laboratory studies are necessary but can either significantly overestimate or underestimate importance of each effect; detrimental effects are heavily dependent on both host and environmental stressors.

The sex or developmental stage of a host can have a bearing on infections. Often, these differences in infections respond to behavioral differences in host migration, schooling, or feeding. Differences also respond to hormonal differences or other sexual differences in physiology.

Nutrition and dietary sources have a strong effect on parasites and diseases. A researcher needs to treat these factors when assessing infections, especially when establishing either parasites or diseases as bioindicators. Obviously, what constitutes the diet of a fish will dictate some of the parasites the fish can acquire. Because these prey organisms serve as intermediate hosts, the appropriate prey and other food sources must be eaten by the fish for it to obtain those parasites. Moreover, the fish must be susceptible to the agent -- a condition that can also be influenced by nutrition. Food can also contain toxicants that can interfere with infections. Lack of food or specific ingredients also can reduce both the cell-mediated and humoral immune responses (Landolt 1989).

Parasites also affect feeding. *Gasterosteus aculeatus* infected with the metacestode (plerocercoid) of *Schistocephalus solidus* died before noninfected ones, when experimentally placed on a restricted diet. Moreover, such fish required a higher feeding rate to maintain their total body weight (Pascoe and Mattey 1977). When infected by that cestode, fish on a restricted diet had a suppressed fright response compared with that of well-fed individuals (Giles 1987). "Starved" fish consequently foraged earlier than well-fed counterparts during that period when predators were still likely to be present, making the starved fish more vulnerable to predation, but more capable of perpetuating the life cycle of the worm.

Parasitic infections are intricately involved with host migration and home-range. As a fish enters a nursery area or develops into its next stage, it necessarily changes its diet and consequently its complement of parasites.

Migration can include foraging, other daily movements, spawning migrations, or seasonal migrations. During such trips, a fish may encounter a variety of prey that serve as intermediate hosts, feed on different infective stages not incorporated into hosts, and locate itself in a habitat containing infective agents that invade it (e.g., cercariae of digeneans). To acquire an infection, the prey must contain the agent, the agent must have reached its infective stage, the agent must be infective to the specific fish, the time of day or period of year must coincide for both parasite and host, and other intricate criteria must be met. On the other hand, a fish could lose parasites after entering an area with different environmental conditions. Direct contact with such factors or host alterations could initiate the loss of the parasite or its reproductive capabilities.

NATURAL ENVIRONMENTAL EFFECTS

Regardless of the symbiont or habitat being considered, infections will exhibit a dynamic pattern responding to several environmental factors. For example, in estuarine and marine environments, salinity, temperature, oxygen concentration, and pH of water, as well as light and pressure, contribute heavily to that dynamic pattern as do various chemical factors (e.g., Bauer 1959). These natural environmental factors, as well as various anthropogenic ones to be discussed later, serve to control the equilibrium between pathogen (or other symbiont) and host as diagrammed in other chapters. Moreover, this intricate balance among symbiont, host, and environment applies to a stock, population, and ecosystem as well as an individual. For example, a disease of a dominant fish host species can reduce the numbers of that species to a density level at which the species is permanently replaced by one or more other fish species.

Temperature and salinity are the most studied natural environmental conditions in regard to symbiont biology. They affect the growth, development, reproduction, susceptibility, and behaviors of symbionts, as well as the effect of both hosts and other natural and anthropogenic environmental conditions. Each symbiotic relationship is different and has to be examined separately. Temperature significantly affects immunological response in fishes and, in regard to interactions with contaminants, alters the host metabolism that may serve to detoxify compounds or produce toxic metabolites, and alters oxygen concentration in the environment and solubility of toxicants. As with temperature, salinity affects each host and each stage of symbiont differently. Salt buffers water and therefore may neutralize or otherwise react with some toxic compounds. Consequently, data from studies conducted in freshwater cannot necessarily be applied to saltwater conditions.

I present some examples of symbionts, first to illustrate how they relate to disease, and how they interact with natural environmental factors. That framework should permit an appreciation for later examples which introduce an interaction of the symbiont with anthropogenic toxicants.

Symbionts with Direct Life Cycles

Amyloodinium spp. are external flagellate protozoans with a direct life history (they have no intermediate host). The dinoflagellate *Amyloodinium ocellatum* has been studied in considerable detail because it infests and can readily kill a large number of fishes, both in aquaculture and in the aquarium industry. Mortality is at least partially related to number of organisms per gill filament. Host specificity is broad. In fact, of 79 local species tested experimentally in Mississippi, only 8 did not succumb (Lawler 1980).

The life cycle of the dinoflagellate *A. ocellatum* involves a trophont stage (Fig. 5-1) that typically infests the gills of its marine and estuarine fish hosts but can

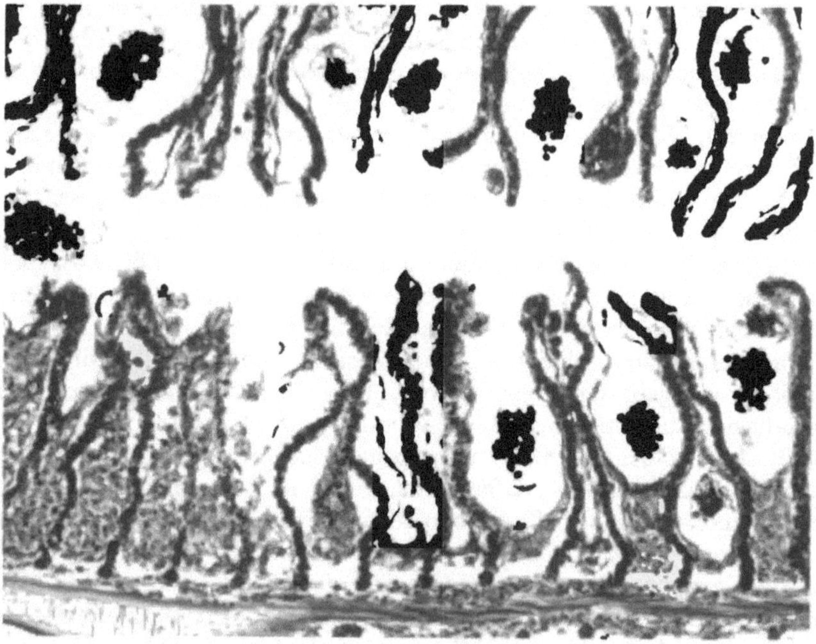

Fig. 5-1. Histological section showing small to large sized trophonts of heavy infestation of the dinoflagellate *Amyloodinium ocellatum* on the gills of the spot, *Leiostomus xanthurus*. Note hyperplasia, fused secondary lamellae, and epithelium separated from basement membrane.

involve other external epithelia. Heavy infestations (Fig. 5-1) can produce extensive hyperplasia of the gills with subsequent fusing of adjacent lamellae, degeneration of the inner epithelial layer, and loss of mucous cells (Paperna 1980). Infestations on larval fishes typically occur on the skin rather than on the gills. A freshwater counterpart, *Oodinium limneticum*, typically infests the epithelium of the skin, thus being responsible for the disease commonly known as "velvet disease."

Once one of the many trophonts of *A. ocellatum* reaches a size unable to remain attached to the host, it undergoes its reproductive cycle. The trophont retracts its rhizoids from host epithelium, detaches from the host, develops a thin cyst wall, and undergoes a series of synchronous divisions forming tomites by binary fission. Those tomites sporulate and form free-living infective "gymnodinium-like" dinospores, with the ultimate number dependent on initial size of the trophont as well as temperature, salinity, and other factors. Paperna (1984) observed small trophonts less than 24-h-old, that dislodged, sporulated without dividing, and formed two mobile dinospores. Paperna determined for a Red Sea isolate that temperature and salinity affected the viability of trophonts, onset of divisions, sequence of divisions, number of divisions before sporulation, synchrony of development, and survival of stages. In Mississippi estuaries, the intensity of infestation of the dinoflagellate on the Atlantic croaker, *Micropogonias undulatus*, was influenced by environmental conditions. It exhibited a positive relationship (based on partial correlation coefficients) with precipitation values (but not inversely with salinity values) over a 20-month period in a relatively pristine habitat. It also showed a similar relationship with average monthly temperature during the same period in a nearby polluted area about 20 km away (Overstreet 1982).

The trophont of *A. ocellatum* can detach and initiate its reproductive process when stressed by any of a variety of environmental factors such as rapid changes in temperature, salinity, or oxygen. Such conditions can initiate production of a large number of available infective dinospores. Consequently, in a confined area like a pond, mortalities ensue.

Whereas infestations of *A. ocellatum* and related species are known from fishes in their natural habitat (e.g., Overstreet 1968, 1982), mortalities from natural sites caused by such agents have not been reported (e.g., Bower and Turner 1988). I report here a mass mortality caused at least in part by the dinoflagellate infestation.

A mass mortality of the spot, *Leiostomus xanthurus*, occurred at Orange Beach Marina and the adjoining Shotgun Canal in Baldwin County, Alabama, on 31 October and 1 November 1984. Members of the State of Alabama Department of Conservation and Natural Resources (DCNR, John Hawke and R. Vernon Minton) and Department of Environmental Management (Brad Gane and Ray

Marchman) investigated the mortality after most of the fish had died, but while some fish were still dying. Samples of dead fish to count and of live fish to examine were collected along transect lines. Approximately 42,000 dead fish were collected from the Marina area and an additional 4,800 dead fish from Shotgun Canal. About 3,500 kg of these fish were removed from the area in containers. Of these, spot measuring 15-20 cm long accounted for nearly all specimens. Others killed included *Micropogonias undulatus* (6 specimens), *Arius felis* (4 specimens), and *Cynoscion arenarius* (3 specimens). Some living specimens of *Mugil curema*, *Brevoortia patronus*, *Cyprinodon variegatus*, and *Synodus foetens* were sampled but did not harbor *A. ocellatum*. Those samples of spot, *M. undulatus*, and *Lagodon rhomboides* had heavy infestations on the gills and light to moderate ones on the skin. I identified samples of the agent as *A. ocellatum*.

As indicated above, disease and subsequent mass mortalities from parasites usually result from an interaction with the pathogen and some stress factor. In this case, the factor could have been a low concentration of oxygen in the water. Water taken from four involved sites at 1700 on 31 October (18 h after the mortality began) had bottom and surface concentrations of oxygen ranging from 1.6 and 3.6 mg/l, respectively, to 4.3 and 7.0 mg/l, with salinity measuring 25-30 ppt and temperature 26-28°C. By 1300 the next day, values from five similar sites ranged from 2.4 and 5.4 mg/l to 6.0 and 8.2 mg/l; 22-26 ppt; and 32-34°C. Whereas the depleted oxygen could have resulted partially from the deteriorating fish, cloudy, overcast weather experienced at the location on 27-28 October could have initiated the depletion in the basin and canal historically known as being poorly flushed. The reported conditions could readily increase substantially the rate of infestation of the dinoflagellate. Moreover, during this same approximate period (15 October-early November), the DCNR experienced heavy losses of the striped bass, *Morone saxatilis*, being cultured in two ponds at nearby Claude Peteet Mariculture Center (CPMC). An additional complicating factor in that case was an infection by the bacterium, *Pasteurella piscicida* (see Hawke et al. 1987). Nevertheless, the intensity of infestations of *A. ocellatum* seen in the Marina area has been observed to cause chronic mortality at CPMC ponds and constitute the most serious pathogen at that site (Hawke and Minton, personal communication; personal observations), and in experimentally infested fish.

Different strains of *A. ocellatum* throughout the world exhibit different tolerances to temperature and salinity. The strain in Mississippi coastal water, as in most areas, cannot tolerate fresh water. Consequently, a freshwater dip may serve as a means to dislodge trophonts and help control the disease in aquaculture or aquaria systems.

The freshwater ciliate *Ichthyophthirius multifiliis* and its marine counterpart *Cryptocaryon irritans* have life histories similar to *A. ocellatum*. The trophonts, rather than being attached to the gills, live under the skin and gill epithelium of their fish hosts. Under stress or after an appropriate feeding period, trophonts detach from the host, encyst on a substratum, undergo binary fission, and produce many motile tomites. Both species occur throughout much of the world, causing disease and mortality in ponds and aquaria, and providing good models to assess environmental conditions. Other ciliates, trichodinids (Fig. 5-2), can also serve as indicators. In addition to naturally infesting gills and skin of both freshwater and saltwater fishes, they also infect renal tubules of wild and laboratory-reared fishes, increasing the value of the ciliates as indicators.

Free-living protozoans, including specific ciliates and amebas, can also cause gill-disease, including mortality, under fitting environmental conditions; some may be good indicators. The ameba *Paramoeba pemaquidensis* or a similar species has been associated with disease in cultured salmonids from different parts of the world in sea water by Kent *et al.* (1988) and Roubal *et al.* (1988). Infestations appear to be dependent on temperature, presence of bacteria, and host stress. The freshwater counterpart *Thecamoeba hoffmani* also has been reported from cultured salmonids (Sawyer *et al.* 1974).

Dynamics of numerous metazoan symbionts with direct cycles also have been investigated in regard to effects of natural environmental and host factors. On most occasions, relatively few specimens of a monogenean infest the natural host, and they cause only a slight host response. In confined and stressed unnatural habitats, such as fish ponds, a higher intensity of these agents cause severe disease and kill their hosts. Examples of mortality in wild populations stressed

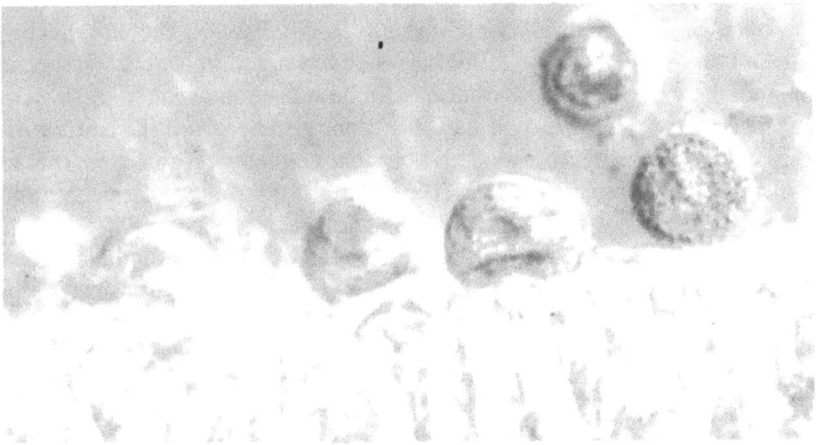

Fig. 5-2. Trichodinid infestation on gill of juvenile red drum, *Sciaenops ocellatus*, from Mississippi estuary. (From Overstreet 1983b).

by natural environmental conditions also exist. The monogenean *Benedenia monticelli* on the mullet *Liza carinata* in the coastal water of A-tor Bay and in El Bilaim Lagoon, Gulf of Suez, feeds heavily on host epithelium near the fish's fins, eyes, and mouth. Occasionally during some winter months with low temperature, many infested fish succumb, perhaps those with an inhibited immunological response and a secondary bacterial infection (Paperna *et al.* 1984).

Symbionts with Indirect Life Cycles

Trypanoplasma bullocki is an internal flagellate protozoan with an indirect life cycle influenced by an interaction of salinity, temperature, and host immunity. This hemoflagellate has hosts including the leech *Calliobdella vivida* and a variety of fishes. In at least the summer flounder, *Paralichthys dentatus*, and the hogchoker, *Trinectes maculatus*, infection can result in anemia, ascites, splenomegaly, and death (Overstreet 1982; Burreson and Zwerner 1984). Relatively high salinity in an estuary from a decrease in freshwater runoff or strong winds can result in prospective juvenile fish hosts persisting in inshore water during autumn. These conditions superimpose those optimal for the leech vector (15 to 22 ppt, <18°C). Burreson and Zwerner (1984) experimentally infected the juvenile flounder in November using both the leech vector and a syringe, killing all infected fish within 11 weeks. Because of the progressive disease signs consistent with parasite-induced mortalities observed in wild infected fish from north of but not south of Cape Hatteras in March, those authors assumed natural mortalities occurred in the northern sites. Dead flounder were also trawled from the lower York River at the same time that experimentally infected fish were dying. In Mississippi, the hogchoker was abundant and noninfected in 1978, but when examined in May 1979, specimens exhibited infections and signs of disease in both liver and spleen. A subsequent sharp decrease in abundance of local fish stock also occurred (Overstreet 1982).

An immuno-blot assay revealed a serum antibody response to the blood flagellate in the summer flounder (Burreson and Frizzell 1986). The antibody titer varied inversely with temperature, and the intensity of infection varied inversely with titer. Fish died from infections in water <5°C, and antibody titer increased sharply when temperature approached 10°C. Infections cleared from peripheral blood of both immunized and control fish in May when the antibody titer peaked. That study explained mortalities indicated above. Recovered fish were apparently immune to homologous challenge for at least one year.

Other trypanosomes produce mortalities and show similar relationships with disease and both the environment (e.g., *Cryptobia salmositica* by Bower and Margolis 1985) and host immune system (e.g., *T. borreli* by Steinhagen *et al.*

1989). Some fish-trypanosome relationships are more difficult to study in the laboratory because infected fish require larger holding facilities, and more time is necessary to establish appropriate infections. However, such studies can permit an understanding of offshore, coastal, and freshwater diseases. A notable, later example is discussed below in conjunction with hydrocarbon contamination.

Coccidian infections can serve as systems to investigate disease and to indicate environmental effects. *Calyptospora funduli* infects a variety of atheriniform fishes, mostly species in the genus *Fundulus*. This coccidian should be considered a prime model to assess disease in nature, to evaluate environmental effects on the disease, and, by utilizing its synchronous development and its formation of macrophage aggregates (MAs) and granulomas in an abnormal host, to assess relationships of toxicants on the piscine cellular response.

Calyptospora funduli is an internal tissue dwelling apicomplexan protozoan with an indirect cycle. Unlike the direct cycles of *Eimeria* spp. in homothermous hosts, the life cycle of this piscine species includes an intermediate host as probably do many but not all coccidian species in poikilothermic vertebrates. This parasite utilizes palaemonid shrimps as a true intermediate host in which development of the sporozoite stage occurs before being able to infect a relatively large number of cyprinodontid fishes (Fournie and Overstreet 1983). In a normal killifish host, an infection with this parasite can replace >85% of the fish's hepatic parenchyma and pancreatic tissue (Fig. 5-3).

The relatively small amount of functional liver and pancreas in fishes heavily infected with *C. funduli* does not support the complete needs of the host during stress and is probably a major contributor to production of disease. When held in aquaria under nonstressful conditions, heavily infected wild and experimentally infected fish usually exhibit no significant signs of poor health. At least one important stress is low temperature. Solangi and Overstreet (1980) reported a conspicuous drop in the local natural population of usually abundant *F. grandis* as probably associated with the combination of infections and low temperature; Overstreet (1982) indicated selective weakening and death of infected compared with noninfected individuals after a few days at 4°C. Further, Solangi *et al.* (1982) demonstrated that low temperature also affected the parasite and the host leucocytic response to it. Temperatures of 7 and 10°C, when maintained for 20 days, inhibited and reduced intensity of infection and delayed normal granulocytic leucocyte response. In fact, 7°C even eliminated early stages of infection after 5 days of exposure. The infected host also develops a premunition and perhaps even a stronger humoral response to inhibit concurrent infections and this immunity probably is affected also by low temperature. Consequently, the synchronous development is facilitated and interaction of temperature and infection is both complicated and dynamic.

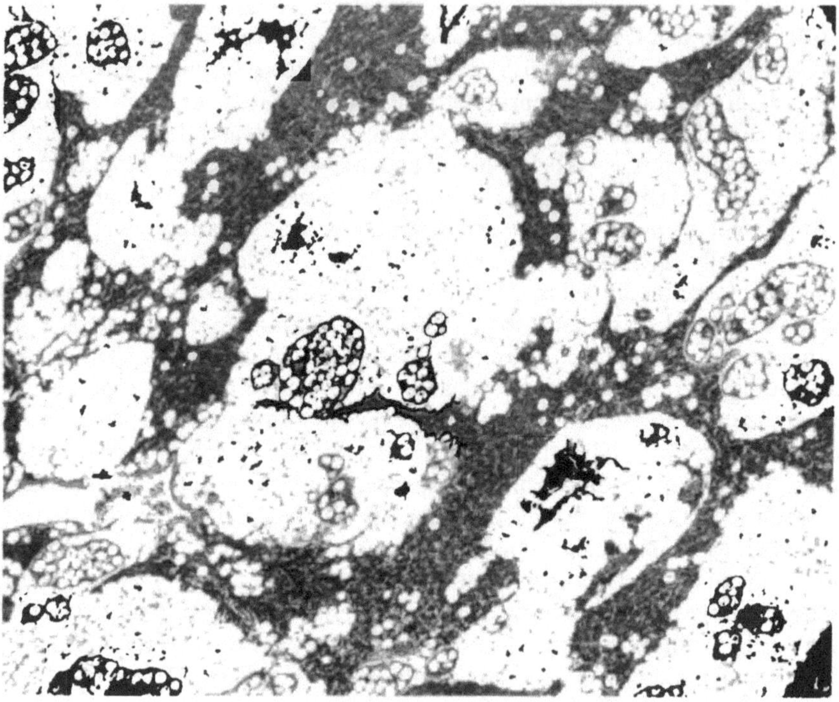

Fig. 5-3. Gross histological view of liver of Gulf killifish, *Fundulus grandis*, heavily infected by the coccidian *Calyptospora funduli*. Oocysts are single, in large groups, and encapsulated in relatively small groups; the darkly staining cells among large groups are normal hepatocytes and some of those inside the groups are remaining pancreatic acinar cells.

In addition to being indirectly responsible for disease and host death during appropriate periods of host stress, infections of *C. funduli* can provide a model system to assess some host and environmental stresses. For example, infections in the abnormal host *Rivulus marmoratus* induce at predictable times both MAs and granulomas (Fig. 5-4) (Vogelbein *et al.* 1987). Infections in *R. marmoratus*, as well as in normal hosts, typically develop synchronously. That is, individual organisms in various developmental stages typically progress from one stage to another within a short time period of each other, even though the exact timing of a stage depends on temperature and other factors. Typically, a mild granulocytic response, consisting of eosinophilic granulocytes and heterophils, develops about 6 days after the host is infected; this cellular response reacts to tissue destruction and increases during the next week, with an abundance of large and small mononuclear cells as secondary generation merozoites that cause necrosis

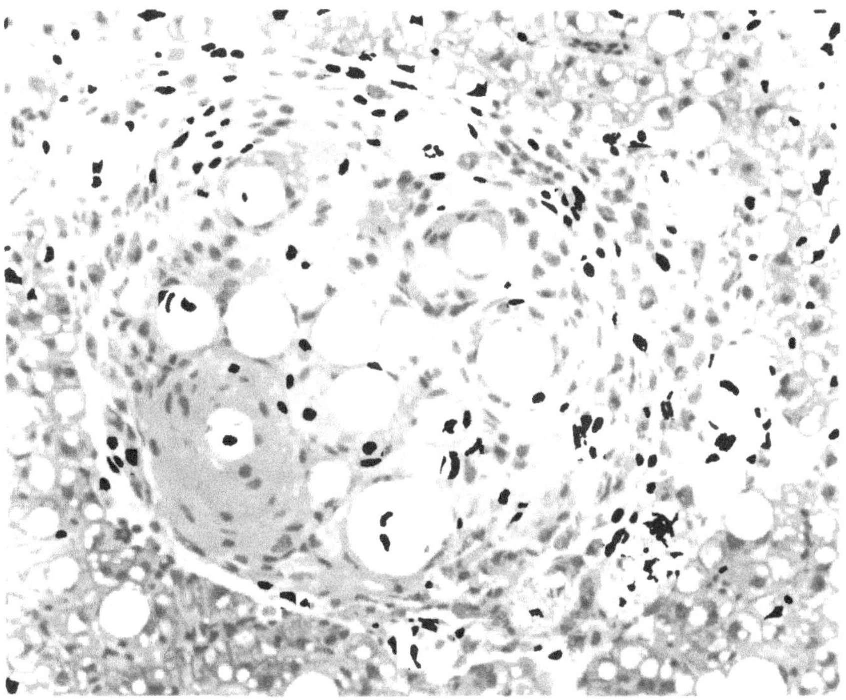

Fig. 5-4. Mature epithelioid granuloma in liver of *Rivulus marmoratus* sequestering oocysts of the coccidian *Calyptospora funduli* at 100 days after initiation of infection. (From Vogelbein *et al.* 1987.)

and degeneration of some host hepatic cells. At about 3 weeks after infection, the leucocytic infiltrate organizes as focal MAs responding to both cellular destruction and degenerating parasites. By about 6 weeks, a granulolomatous response becomes associated with oocyst sequestration. Because timing for the development of these various inflammatory responses can be predicted, components of the responses can be investigated and monitored and the responses can be altered by introduction of various chemicals, especially those that affect host cellular responses. Vogelbein *et al.* (1987; personal communication) are presently evaluating the effect of cadmium exposure on the responses of fishes. Depending on the rationale of an experiment, a researcher could use a normal killifish host with leucocytic and MA responses and with fibrotic encapsulation of oocysts (Solangi and Overstreet 1980; Hawkins *et al.* 1981) or *R. marmoratus* with a slightly different granulomatous response and granuloma formation. A researcher also could induce specific pathological alterations with experimental infections in a normal host model system for use as a biomonitor for specific types of toxicants or natural environmental conditions.

Other coccidians should also make good models. Whereas the cycles of some species are direct, those of others, in addition to *C. funduli* also require an intermediate host. Other life cycles of fish coccidians are more complicated. For example, Steinhagen and Körting (1990) could infect carp with *Goussia carpelli* directly by fecal contamination, but fish that had recovered from a previous infection required sporozoites incorporated into intestinal cells of tubificid oligochaetes. Sporozoites in the two tested species of tubificids could survive for at least 57 days; feeding crustaceans and a chironomid insect larva fed the coccidian did not produce infections in the carp.

Life cycles of most protozoa of fishes, especially those with multiple hosts or complex stages, are not yet elucidated. As indicated for trypanosomes and coccidians, describing the components and determining the interrelationships of various protozoans with the numerous dynamic influencing environmental factors can help answer many questions regarding the role of parasites in an ecosystem. For example, why does the microsporidan, *Ichthyosporidium giganteum*, infect the spot, *Leiostomus xanthurus*? What has been reported as the same species has been reported from two other fishes along the Atlantic coast of Europe and in the Black Sea (Canning *et al.* 1986). If all those infectious agents are conspecific, why does only the spot acquire infections (Fig. 5-5) in the U.S. or are infections present but less conspicuous in other hosts?

This conspicuous protozoan infection in spot consists of hyperplastic host connective tissue involving both small multilocular cysts made up of fibrous capsules (Fig. 5-6) and large, thick- to thin-walled xenomas (Fig. 5-7). A xenoma is the elaborate complex in which the host cell is altered to become physiologically and morphologically integrated with a parasite. Why is an xenoma formed? According to Canning *et al.* (1986), this entity apparently provides suitable conditions for parasitic proliferation, while masking it with host components to avoid host attack. The complex also benefits the host by confining the parasite rather than allowing its spread throughout the host. Its development is poorly understood. Typically, a chronic inflammatory response of histiocytes, lymphocytes, and solitary multinucleate cells surrounds the small cysts.

In spot, large-lobed xenomas massed with spores occasionally occur within the ovary, ceasing reproduction. Are the obvious infected individuals (Fig. 5-5) more susceptible to infection than other spot or are other individuals or other species infected but expressing the infection in a different manner? Do larval and postlarval spot become infected, and, if so, do these infections significantly reduce spot population? Is there an adverse effect on spot population based on loss of fecundity by infected adult females? Being able experimentally to infect spot and evaluate pathogenicity could begin to answer these and more questions.

Fig. 5-5. The spot, *Leiostomus xanthurus*, with a growth produced by the microsporan *Ichthyosporidium giganteum* (top photo). (From Overstreet and Howse 1977.)

On the basis of known microsporidan life cycles, infections of some species can be transmitted directly without another host but others cannot.

Future discovery of how to transmit the myxosporidan *Myxobolus lintoni*, to the sheepshead minnow, *Cyprinodon variegatus*, should provide an intriguing model to test environmental effects on infection (Figs. 5-8, 5-9, and 5-10). The infection superficially appears as a malignant neoplasm; the agent aggressively invades various tissues (Fig. 5-10) and also spreads the infection to distant sites (Fig. 5-9) in the host (see Overstreet and Howse 1977). If the infection requires a particular stress as suggested by Overstreet and Howse (1977) and by Overstreet (1988), who have reported infected fish as being restricted to specific polluted or stressed habitats, the ability to conduct experimental infections would allow one to assess different types of environmental stresses in relationship to both infection and disease as well as to assess the effect of the parasite on different stages of the fish. Perhaps infection with that myxosporidian is common, but the corresponding disease is stress-dependent.

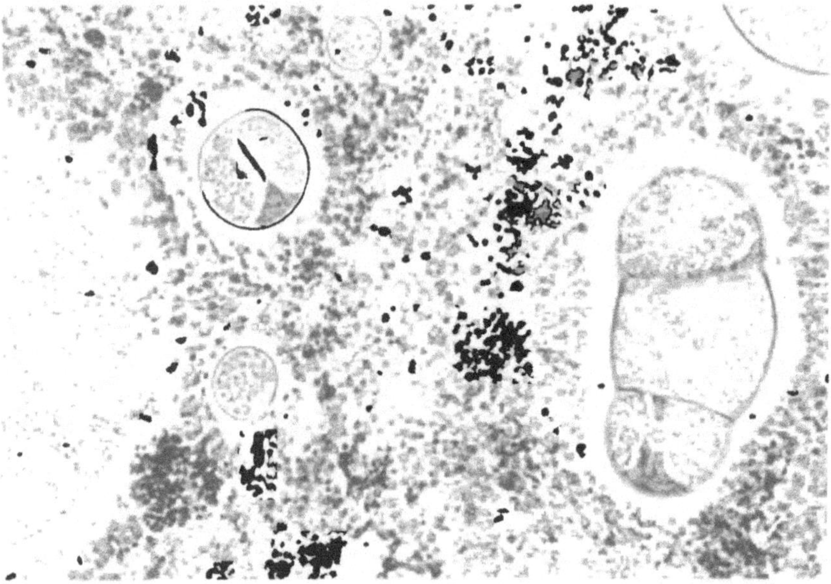

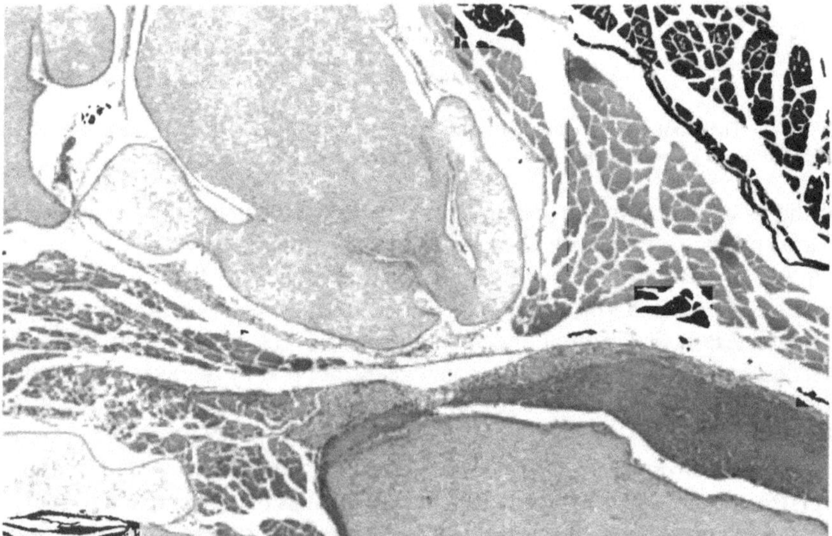

Fig. 5-6. Encapsulated cystic stages of the microsporan *Ichthyosporidium giganteum* surrounded by vegetative stages and extensive chronic inflammatory reaction; in growth from *Leiostomus xanthurus*. (top photo). **Fig. 5-7.** Lobulated mature xenomas of *Ichthyosporidium giganteum*;, some measure at least 4 mm across and comprise a dominant feature in some large growths from *Leiostomus xanthurus*.

Some myxosporidans are considered to thrive only in clean water. Narasimhamurti and Kalavati (1984) examined *Channa punctatus* for *Henneguya waltairensis* in a relatively clean reservoir tank used for domestic consumption and in a nearby one that received city drainage and wastes from humans and other animals. They found infections in fish from the clean site, mostly during winter (18-25°C) rather than summer (35-42°C), but not in the contaminated one (high alkalinity, high turbidity, and low oxygen concentration). Iva Dyková (personal communication) has also noted that myxosporidan infections occur much more readily in cultured fish from freshwater ponds in Europe with good water quality when compared with those from ponds with poor water quality. The discrepancy of infections that are more abundant in both stressed and clean environments may be explained by different possibilities. Some or all myxosporidans have intermediate hosts; El-Matbouli and Hoffmann (1989) have confirmed that at least two species of *Myxobolus* require tubificid oligochaete intermediate hosts, and Kent *et al.* (1990) have infected a third species, using a lumbriculid oligochaete intermediate host. Consequently, various "pollutants" can either eliminate or provide a nutritional source for the "intermediate" hosts. Specific toxicants can also increase susceptibility or reduce resistance of hosts to myxosporidan infections. Both microsporan and myxosporan infections, whether species with or without intermediate hosts, can help indicate areas and periods of stress, as indicated in relatively long-term studies conducted in the North Sea (van Banning 1987).

There are numerous other examples of protozoan as well as metazoan parasites with intermediate hosts, but most parasites seldom cause mortalities in natural habitats. Those parasites, however, can serve as sensitive monitors of dynamic environmental conditions. Some can also serve as indicators of characteristic symbiotic relationships involving a variety of complex stages, each reacting to environmental conditions differently and each stage and its host serving as a component of a larger dynamic ecosystem. The literature contains an overwhelming number of cases that exemplify how different normal environmental parameters influence differently the two to five or more hosts that can constitute a life cycle. Examples illustrating regulation over those parameters by anthropogenic influences will be presented later.

The more complicated the life cycle and the greater the variation in stages, the more a cycle can be influenced by the environment. Some nematodes provide good examples. Ascaridoid nematodes are internal helminths, some of which have a free-living juvenile stage in addition to the typical parasitic juvenile stage in an intermediate host (and sometimes also one or more paratenic hosts), plus adult and preadult stages in a vertebrate definitive host.

Hosts of the ascaridoids can provide a partial buffer to external environmental factors, making the internal worms less sensitive to the environment than the

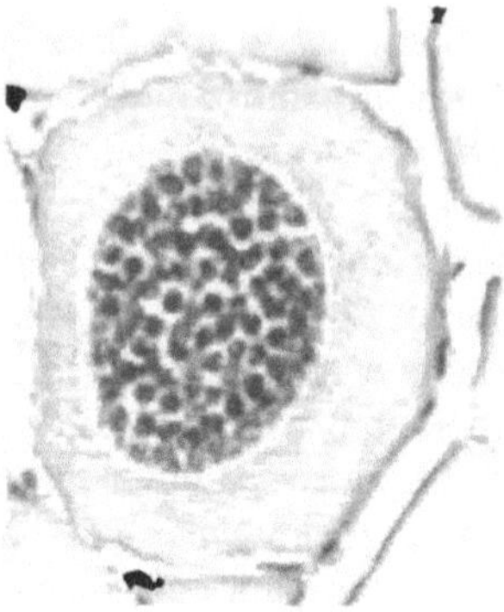

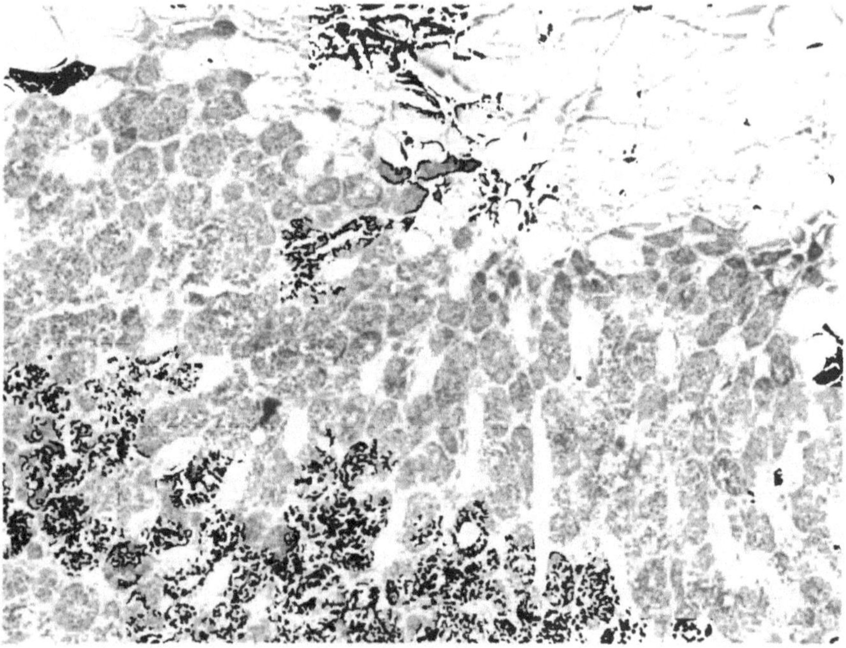

Fig. 5-8. The sheepshead minnow, *Cyprinodon variegatus*, with growths produced by the myxosporidan, *Myxobolus lintoni* (top, left). (From Overstreet and Howse 1977.) Fig. 5-9. Skeletal muscle bundle of *Cyprinodon variegatus* infected with the myxosporidan *Myxobolus lintoni* spread from a distant site (top, right). Fig. 5-10. Invasion of skeletal muscle of *Cyprinodon variegatus* by the myxosporidan, *Myxobolus lintoni*.

host. The "free-living stage" of most species, however, is probably more sensitive to water conditions than the parasitic stages. Sensitivity differs among species. The free-living juveniles (second stage) of two species tested in three different salinities and two temperatures exhibited different ranges of tolerance (Deardorff and Overstreet 1981b). The first species, *Hysterothylacium reliquens*, matures in a variety of euryhaline fishes, whereas the second, *Iheringascaris inquies*, is restricted as an adult to the cobia (lemonfish), *Rachycentron canadum*. As free-living juveniles at 28°C, about 75% of the specimens of *H. reliquens* survived 12 days when in either 15 or 30 ppt compared with most in 0 ppt dying at day 2. On the other hand, the juveniles of *I. inquies* did not survive as long. They survived about 6-7 days at 15 and 30 ppt but 3-4 days at 0 ppt, even though the fish host harboring the adult seldom entered low salinity areas. When reared at 4°C, however, most juveniles of *I. inquies* in sea water lived 3-4 weeks; and a few survived for over 5 weeks. Some individuals in fresh water survived more than 2 weeks. Some cobia live in cool, relatively deep water along the continental shelf of the northern Gulf of Mexico during winter, so juvenile nematodes are likely to survive long enough to perpetuate their stock in that and other presumably cool habitats throughout the extensive circumtropical and subtemperate geographic range of the host. On the other hand, most juveniles of *H. reliquens* at 4°C died after 2 days, with some survival, especially at 15 ppt until day 13. Intermediate and definitive hosts of *H. reliquens* typically occur in water of moderate to relatively high salinity (Deardorff and Overstreet 1981a,b).

In the case of the two example ascaridoids, perhaps the tolerance of the free-living stage controls the abundance of the adult populations. I observed thousands of specimens of *I. inquies* in each of about 10 adult cobia examined during a 1987 fishing tournament and in most of the ten or so fish examined irregularly during 15 years prior to that. Since 1987, few specimens (sometimes <10 and often none per fish) infected the approximately 200 cobia examined. Only a quarter of the fish had any specimens (Gabrielle Meyer and James Franks, personal communication; personal observations). Cobia migrate seasonally near shore in Mississippi in the Gulf of Mexico. The salinity in Mississippi Sound during spring 1987 was relatively high and has been somewhat lower thereafter.

In regard to infections of *H. reliquens*, I have seen up to 137 large adult specimens in the intestine and pyloric caeca of the sheepshead, *Archosargus probatocephalus*, the local primary host for that species. There have been so many specimens that the specific name *reliquens* refers to the proclivity of the worm to actively migrate from mouth, opercular openings, or anus of dead or physiologically stressed individuals (Norris and Overstreet 1975). High numbers were observed in 1979 and a few earlier years, but the species was not looked for specifically during the next 7 years. No adult specimens have been seen in

Mississippi Sound during the last 4 years even though several hundred sheepshead have been examined extensively for that infection during that period. However, juveniles still occurred frequently in local penaeid shrimps and fishes. Also, its recent absence in the sheepshead is especially strange because this ascaridoid has one of the widest distributions of any near-shore piscine ascaridoid known; it occurs both north and south of the Republic of Panama along the coasts of the Pacific Ocean, Gulf of Mexico, and Atlantic Ocean (Deardorff and Overstreet 1981a).

Temperature, salinity, and other natural environmental factors can also affect adults and juvenile ascaridoid nematodes within eggs. Few studies have been conducted to demonstrate these effects on growth and survival. Möller (1978) conducted a few experiments with *H. aduncum* from the eelpout, *Zoarces viviparus*, in the Western Baltic Sea and determined larval development in the egg was normal at 5-20°C, with a period of survival reduced in fresh water and prolonged to 150 days when kept at 5°C at salinity higher than 12 ppt. When the third-stage juvenile was removed from the body cavity of its fish intermediate host and placed in water of various salinities for 50 h, it gradually took up water in 0 and 4 ppt solutions and gradually lost water at 16 and 32 ppt. It remained constant at 12 ppt (osmolarity of 300 m osmol). When placed in other water of different salinities, half the specimens (L_{T50}) lived for about 3 days in 2 to 28 ppt, but died soon after being placed in 0 and 32 ppt. Adults were observed from dead eelpout. After the fish died, the worm vacated its host from its mouth and anus, as indicated for *H. reliquens*. Emigration was most rapid at 5 and 10°C. Low temperature of 0°C reduced the activity but increased to 14 days the length of survival of half the specimens. At 25°C, less than 25% of the worms survived more than 20 h.

Few internal helminths cause mortality in their normal hosts when those hosts are juveniles or adults inhabiting their natural environment. As indicated earlier, when fish are in stages of larva to fry, they are especially vulnerable to disease. About 10% of the Atlantic herring larvae experimentally reared on wild plankton died from infections with a juvenile ascaridoid related to those discussed above, a metacestode stage of a tetraphyllidean cestode, and two ectoparasitic copepods (Rosenthal 1967).

Of helminths that cause mortality, several include species that as adults inhabit sites other than the alimentary tract. One group of parasites that has a potential to produce disease and mortality consists of blood flukes. This potential is evident from mortalities of fish confined in culture conditions. Several cases involving different freshwater salmonids with *Cardicola klamathensis* and *C. davisi*, and *Cyprinus carpio* with *Sanguinicola inermis*, have resulted in mass mortality. Mortalities of the marine carangid, *Seriola purpurascens*, cultured in floating net cages in Japan appeared to Ogawa *et al.* (1989) to be caused by one or two species of *Paradeontacylix*. The digeneans produced hyperplasia of gills, encapsulation of eggs in the gills and ventricle, and papillae formation associated with endothelial proliferation in the afferent branchial arteries. Unlike freshwater associations reported by several authors, there was no necrosis in any organs and no lesions in the kidney. Infections were reported by Ogawa *et al.* (1989) as seasonal, with many deaths occurring in March, apparently because of reduced respiratory function of gills and heart. Infections in survivors cleared up, and such fish seemed to be partially immune to additional infections.

Different species of blood flukes have different cycles, infect varied sites in blood vessels and other spaces, and induce different host-parasite relationships. Density of infection in vulnerable tissues and immunological response by the host probably dictate severity of disease. Overstreet and Thulin (1989) investigated *Pearsonellum corventum* in serranid fishes and suggested that this marine parasite could cause mortalities of fishes in culture conditions. In addition to the fibrotic encapsulation of eggs also reported from other host-blood fluke relationships, that relationship involving *P. corventum* in *Plectropomus leopardus* and other serranids included MAs in the ventricle (Fig. 5-11), which sequesters eggs deposited by the worm. Such aggregates of macrophages are normal in the liver, kidneys, and spleen of most fishes, including serranids. I do not know if uninfected-naive *P. leopardus* have MAs in the heart, but MAs were not

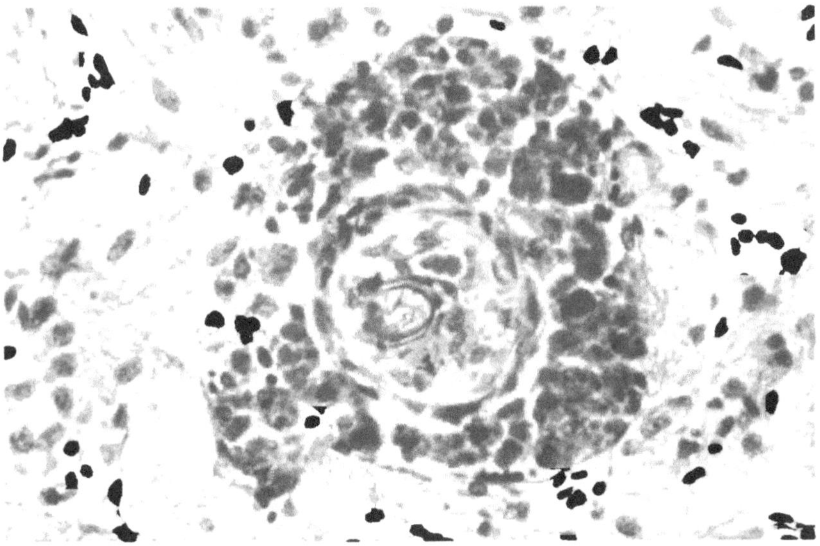

Fig. 5-11. Early granuloma containing egg fragments of *Pearsonellum corventum*, surrounded by a macrophage aggregate in ventricle of *Plectropomus leopardus*. (From Overstreet and Thulin 1989.)

observed in the serranid, *Epinephelus guttatus,* infected by a different blood fluke (Fig. 5-12) or in 11 species in 9 families other than Serranidae in the Gulf of Mexico (Overstreet and Thulin 1989). On the other hand, MAs (Fig. 5-13) were present in the ventricle of a different uninfected serranid (black grouper, *Mycteroperca bonaci*) off Florida in the Gulf of Mexico.

The MAs in the heart as well as related inflammatory processes consisting of fibrotic encapsulation with associated melanistic involvement may have evolved to allow many serranids (and some lutjanids) to sequester excess digenean eggs as well as other parasites and large foreign bodies in tissues and the body cavity. Overstreet and Thulin (1989) provided some cases. Another question is, did blood flukes stimulate perfection of this inflammatory system to the point that the serranids and miscellaneous others were evolutionarily successful in a harsh environment or was the system already established, allowing the fluke to be successful in serranids?

Different parasites and different conditions induce different types of harmful effects on their hosts. For example, a digenean not related to blood flukes also affects blood flow. Metacercariae of *Apatemon gracilis* within the pericardial cavity of its salmonid intermediate host reduce the cardiac output of that fish host. Apparently the effect on the heart relates to lesions resulting from the host inflammatory response towards the parasite and not directly to the presence of the encysted parasite or to adhesions that bind chambers of the heart to surrounding pericardial walls, restricting their movement (Tort *et al.* 1987).

The dynamics of helminth life cycles, just like protozoan cycles, are typically moderated by host factors such as the immunological response. Such a response apparently maintains low numbers of infections of the long-living *Poecilancistrium caryophyllum* in the musculature of the spotted seatrout, *Cynoscion nebulosus*, and other sciaenid fishes. The total number of metacestodes present appears to depend on the number acquired when the host first feeds on infective intermediate hosts, which in turn is dictated directly by salinity (Overstreet 1977) and should be useful for indicating long-term trends in salinity. Typically, about half the seatrout in estuaries of Mississippi and Louisiana each averaged about two worms, and nearly all of those from an area of higher salinity (Apalachee Bay, Florida) had about four, still not enough to harm them. In the relatively polluted Tampa Bay area, only about 10% of the stock was infected, and then with an average of about one worm. Life history and pathogenesis studies are necessary to evaluate the deleterious effect of this cestode on juvenile seatrout and, if the effect is serious, the effect on year-class of fish should prove significant.

Cestodes and other helminths can be useful indicators of environmental conditions in oceanic environments as well as estuarine ones indicated above. MacKenzie (1987) followed metacestodes of *Grillotia angeli* in the Atlantic

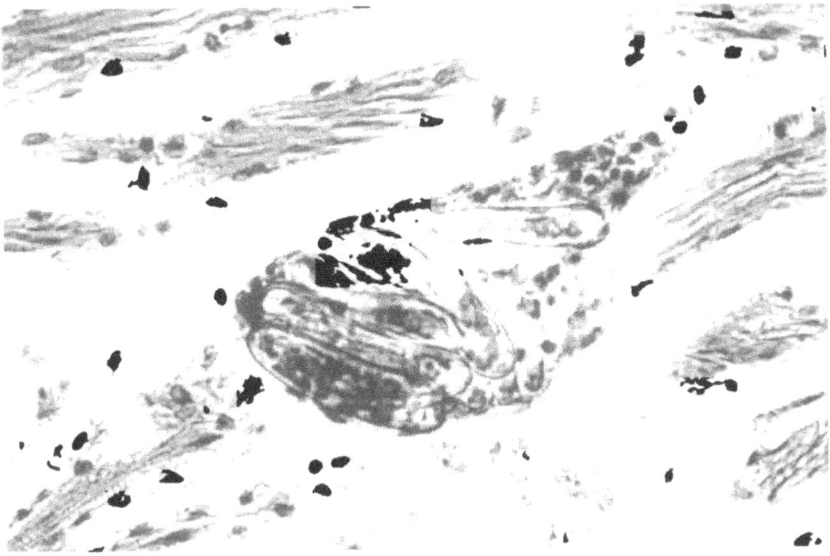

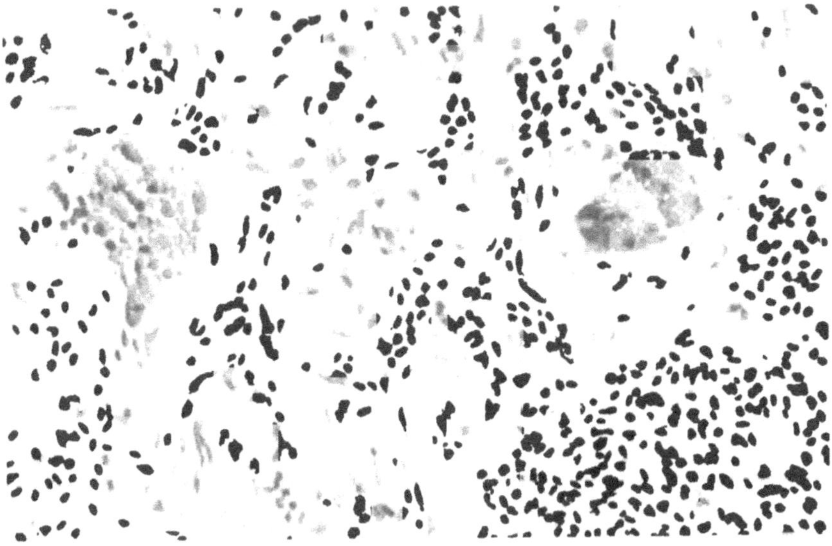

Fig. 5-12. Blood fluke eggs (top) in ventricle of the red hind (serranid), *Epinephelus guttatus*, surrounded by inflammatory cells but not macrophage aggregates. No such aggregates were observed in the heart. Fig. 5-13. Macrophage aggregates in ventricle of the black grouper, *Mycteroperca bonaci*, with no associated signs of an infection with a blood fluke.

mackerel, *Scomber scombrus*, and of *Lacistorhynchus* sp. in the Atlantic herring, *Clupea harengus*, irregularly over a 13-year period. Data collected for both host-parasite relationships corresponded with each other and to the "mid-70's salinity anomaly."

Numerous crustacean symbionts, with their diversity in life history strategies, can influence a fish population. Under normal environmental conditions, losses to the population amount to few fish and sometimes can be evaluated statistically. Weakened or otherwise compromised individuals become prey for other organisms and seldom succumb directly to the infestation. Those that seek shelter from aquatic predators provide interesting objects for study. Were sluggish spotted gars, *Lepisosteus oculatus*, in the boatslip at the Gulf Coast Research Laboratory and elsewhere that had an abundance of the isopod *Anilocra acuta*, the branchiuran, *Argulus nobilis*, and associated lesions (Overstreet and Howse 1977) debilitated by the crustaceans? Or, were these fish that could be picked out of the water by hand or by dip net weakened by a contaminant, microbial agent, natural environmental factor, or some combination of factors, allowing the crustaceans to seek out the host and to reproduce readily?

Isopods probably play a more important role in natural mortality than assumed by a casual observer witnessing a host-parasite relationship between a cymothoid isopod and its fish host. Maxwell (1982) provided good evidence that the cymothoid *Ceratothoa imbricatus* can remain with its pelagic jack mackerel, *Trachurus declivis*, host for 9 years. Adlard (1990) studied the cymothoid *Anilocra pomacentri* on the dorsum of the reef fish *Chromis nitida* near the midline just behind either eye. In the Great Barrier Reef, infested fish had a 6% loss of body condition, a difference in estimated mortality between infested and noninfested recruits as 26%, and, most important, 88% fewer eggs produced in infested than in noninfested counterparts. In laboratory aquaria, 78% of the small infested fish and 28% of the large infested ones died. In five reefs studied, *C. nitida* was the only host (except for larval isopod infestations on some other fishes), but other fishes could be infested in the laboratory for a short duration. Individuals of some species were all killed (also personal observations), and the isopod died in other tests.

Uncertainty exists as to whether *Lironeca ovalis* has an effect on highly fecund fishes like the red drum and seatrouts in the Northern Gulf of Mexico (Overstreet 1983a,b) and white perch in the Delaware River (Sadzikowski and Wallace 1974) similar to that of *A. pomacentri* on its reef hosts. In U.S. cases, infested fish averaged smaller than noninfested counterparts, and the isopod did not occur on older fish. The isopod also appears able by means of damaging host gill filaments of the silver perch, *Bairdiella chrysoura*, to allow the hematogenous spread of a strain of lymphocystis virus from that host's gills to its visceral organs and other internal tissues (Howse *et al.* 1977). As indicated above, the most influential effect by copepods on fish populations occurs when hosts are larvae or postlarvae and consequently are difficult to observe and evaluate. An equilibrium involving some loss of those early-stage fish is established, dependent on environmental conditions. Nevertheless, some copepods (Fig. 5-14) also adversely affect juvenile and adult hosts in their unaltered environment. Kabata and Forrester (1974) and Kabata (1984) reported 10.7% of young sampled specimens of the flatfish *Atheresthes stomias* off British Columbia to exhibit infestations of the copepod *Phrixocephalus cincinnatus* in both eyes, ultimately rendering the host blind. Fish with monocular infections did not survive well; those with adult copepods in both eyes disappeared from the stock, presumably eaten or starved to death. The parasite anchors into the choroid, feeding on blood. As it develops, it penetrates the iris and cornea, and its trunk containing the reproductive organs emerges from the eye, causing blindness. Infections diminish with host age and depth of water.

ANTHROPOGENIC ENVIRONMENTAL EFFECTS

A pollutant serves as just one more type of environmental factor that can increase or decrease intensity of a parasite or increase or decrease susceptibility of the piscine host to the infection. Action of the pollutant, however, may be unique because of its suddenness or abruptness in terms of lack of exposure to naive hosts. It can directly influence infections of either internal or external parasites as well as indirectly influence infections by affecting the host response or an intermediate stage.

Pollutants can be grouped into a wide variety of categories (such as temperature, sedimentation, and organic material); some are introduced into ecosystems naturally as well as from human intervention. Contaminants can be hydrophilic (watersoluble) or hydrophobic (relatively insoluble), and the difference can have a great bearing on the effect on both host and symbiont. Relatively few investigations involving parasites and their relationship with these toxicants or polluting conditions provide decisive cause-and-effect results. Consequently, results are usually considered as pollution-associated rather than pollution-caused. Several research and review papers on pollution have covered some effects of pollution associated parasitic interactions (e.g., Overstreet and Howse 1977; Sindermann 1979; Couch 1985; Möller 1985; 1986; Overstreet 1988).

Most authors dismiss parasites as a rather insignificant factor in natural host mortality. However, as implied above, I think that some parasites have or, under appropriate circumstances, may have enormous influence on populations.

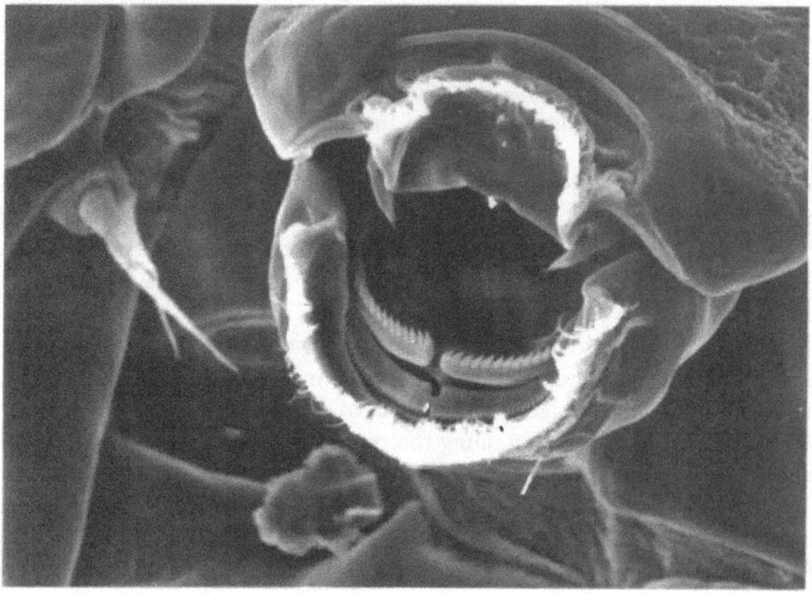

Fig. 5-14. Orifice of mouth cone of caligid copepod, presumed to be *Caligus mutabilis*, from skin, gills, and branchial chamber of Florida pompano, *Trachinotus carolinus*, in Mississippi. Infestations of this and related copepods usually have the most serious effects on young individuals and those in culture conditions. Note the two pairs of tooth-bearing structures in the lower portion. Fine teeth on the lower pair (strigil) saw the fish tissue, whereas the upper pair (mandible) accumulate that tissue and convey it into the copepod's buccal cavity.

Parasites appear to play an important compounding role in complex interactions among host factors, natural environmental factors, and anthropogenic factors that result in disease and mortality. Death, when observed, is often attributed to secondary infections. One seldom observes natural mortalities, whether from parasites, microbial agents, toxicants, or interaction of those causes, especially in the marine environment because "victims" are readily eaten by predators and diseases in larvae and fry are especially difficult to detect and assess. Because some research has provided data or hypotheses delineating a portion of this complex web (including parasites), I have arbitrarily grouped some pollutants and present some available examples.

Heavy Metals

Heavy metals have various effects on the immune system and other systems of fish. Because of the complex nature of their effects, heavy metals, including alterations in blood chemistry, tissue morphology, growth, and reproduction, probably influence parasites in a variety of ways. Consequently, parasitic infections can respond in a variety of ways.

The ciliate parasite, *Ichthyophthirius multifiliis*, has proved to be a good organism to assess effects of toxic metals. When the channel catfish, *Ictalurus punctatus*, was exposed to the ciliate, copper sulfate (which at specific levels serves as a chemotherapeutic agent), or a combination of both, it exhibited a response complementary in nature when both stress factors were present (Ewing *et al.* 1982). Gills of fish responded to both parasite and copper-exposure, but cystic spaces, epithelial hyperplasia, and especially lamellar fusion occurred more often when the combination was present. When the carp, *Cyprinus carpio*, was exposed to different levels of cadmium for up to 10 days and then exposed to the ciliate, the intensity of infestation significantly increased at >50 μg/l but not at levels <25 μg/l (Mohan and Sommerville 1989).

The marine counterpart to the agent of the disease "ich" is *Cryptocaryon irritans*. It occurs in wild populations but is usually present on a few different species in small numbers. In confined ponds and aquaria in the U.S. and throughout much of the world, however, it causes mass mortalities (e.g., Overstreet 1990). Its survival and reproduction are regulated strongly by temperature and salinity (Cheung *et al.* 1979), and that ciliate, as well as *I. multifiliis*, can also be controlled in culture systems with copper and other chemotherapeutants.

Levels of both helminth and protozoan infections can be reduced by metal exposure. A series of six heavy metals were exposed singly and in combination for 28 days to the fish, *Channa punctatus*, naturally infected with the nematode *Spinicauda spinicauda*. The prevalence of infection was less in all groups tested than in unexposed controls, and the effectiveness of the metals were ranked (Jana and Ghosh 1987; Ghosh and Jana 1988).

In at least one case, an internal parasite contained less metal than the tissues of its exposed host. The metacestode *Schistocephalus solidus* in the perivisceral cavity of the three-spined stickleback, *Gasterosteus aculeatus*, contained less cadmium than the whole body minus the worms (Pascoe and Mattey 1977). Those authors suggested that the finding may explain why the worms are normally alive in the body cavity of fish dying from heavy metal poisoning, including cadmium.

Some metals, such as free iron, can affect bacteria, which in turn can influence concurrent parasitic infections. Nakai *et al.* (1987) found that ferric and ferrous compounds injected into the Japanese eel, *Anguilla japonica*, but not the ayu, *Plecoglossus altivelis*, greatly enhanced the virulence of the bacterium *Vibrio anguillarum*.

Parasitic infections also have an influence on susceptibility of fishes to heavy metals. Median period of survival for the three-spined stickleback infected with metacestodes of *S. solidus* was notably shorter than that for noninfected controls experimentally exposed to four of five different levels of dissolved cadmium chloride (Pascoe and Cram 1977). The period was even shorter when the additional stress of dietary restriction was added (Pascoe and Woodworth 1980). On the other hand, fish subjected to the stresses of cadmium and food restriction did not die before those subjected to food restriction alone. Perhaps cadmium plays a protective disinfecting role by its toxic effect on the parasite. As noted with cadmium exposures, the cumulative percent mortality of sockeye salmon smolts infected with the intestinal cestode *Eubothrium salvelini* demonstrated that infected individuals were significantly more susceptible to 1 mg/l dissolved zinc than uninfected controls. Unexposed, infected individuals did not die (Boyce and Yamada 1977). Experimental infections of the striped bass, *Morone saxatilis*, with juveniles of the marine mammal nematode *Anisakis* sp. did not affect the uptake of a combination of zinc and benzene, but the infection may have reduced the host's hematocrit and antibody titer (Sakanari *et al.* 1984).

The above findings clearly demonstrate either that laboratory-reared, parasite-free fishes should be used for toxicity studies or that all potentially infected wild or cultured fish used in tests should be evaluated for infections. Care should be taken to exclude parasites from laboratory-reared fishes.

Infections can also affect the tolerance of invertebrate hosts to heavy metals, but as with fish, the effect can be complicated. Two species of blood flukes that infect the snail *Lymnaea stagnalis* increase host tolerance to acutely lethal doses of zinc. One species had more effect than the other, but only in the later stage of development and at high concentrations of zinc (Guth *et al.* 1977). Of course, any alterations of the larval stage of these or other parasites necessarily affect the degree of infections in fish or other vertebrate definitive hosts.

Kraft Mills

Effluents from kraft mills that contain oxygen-binding inorganic and organic substances deplete oxygen from the water, causing stress which can lead to mortality. Effluents that produce bleached pulp are known to have additional adverse effect on biological systems. The effects on parasitic infections have been investigated for several years in the Scandinavian region. In one study (Lehtinen *et al.* 1984), specimens of the flounder, *Platichtys flesus*, were collected from unpolluted brackish water in Sweden and placed into five tanks, each receiving clean water in a flow-through arrangement. Four tanks additionally received effluents from three pulp mills that used different processes; three received concentrations of 2.5% v/v from each mill, one received a lower concentration of 1% of the conventional chlorine bleach effluent, and one served as control.

The ciliate *Trichodina* sp. was a sensitive monitor of stress induced by the kraft mill effluents mentioned above. Intensity of infestation seemed related to the effect on gills. Heaviest infestation occurred in additional wild fish sampled near the discharge from a mill that used chlorine bleaching with a pre-bleaching step with oxygen (infestation was 200 times that for the control). The experimental group receiving effluent from the same mill exhibited the highest degree of infestation of four groups that received effluents. The least degree of infestation, next to that on the control, was on fish receiving unbleached pine pulp.

Ciliate infestations indicated above appear to be related to stress effects in the gills (Lehtinen *et al.* 1984). Fish receiving effluent with pulp pre-bleached with oxygen exhibited gill epithelium which was detached from the basement membrane, similar to that occurring in zinc- and copper-exposed fish. Gills of other fish exhibited less extensive pathological alterations. Under nonstressful conditions, few ciliates infest a fish, and these symbionts could be considered nonparasitic because they feed on bacteria, algae, and organic material in water. However, when epithelium becomes detached and otherwise effected along with stimulated mucus production, the ciliates then feed on the host tissues and secretions. These parasitic forms reproduce to take advantage of the increased food supply, damaging the gills by attaching as well as by feeding on them.

Gills of the perch, *Perca fluviatilis*, experimentally exposed to different concentrations of bleached kraft pulp mill waste in Finland demonstrated an "oodinium" infestation, but no agents occurred on gills of the controls (Lehtinen and Oikari 1980). Fishes examined in a brackish water area receiving effluent from a kraft mill producing bleached pulp in the Gulf of Bothnia had the same endoparasites. As the distance from the mill increased, however, frequencies of the ectoparasitic monogeneans *Paradiplozoon homoion* and *Dactylogyrus* sp. on gills of the roach, *Rutilus rutilus*, increased (Thulin *et al.* 1988). Those authors also reported ulcers and other pathological alterations of the fins, gills, vertebrae, and head in fish from the area.

Parasites from two species of fishes from four lakes in central Finland were investigated for differences among infections. Of those lakes, one received pulp waste, two were eutrophic, and one was oligotrophic (Valtonen *et al.* 1987). Most notable was the low number of strigeoid eye flukes in the lake receiving the kraft effluent because of the low number or absence of the snail intermediate hosts that were abundant elsewhere.

Petroleum Hydrocarbons

When dealing with petroleum hydrocarbons, one typically deals with a variety of components. Some components of petroleum are more toxic than others, some are aromatic and short lived in nature, and some have different and sometimes contrary effects on biological systems. Histopathological alterations in fishes exposed to crude oil and water-soluble fractions have been documented (e.g., Solangi and Overstreet 1982; Khan and Kiceniuk 1984). Also, a series of studies by Khan and Kiceniuk have clearly shown that mixtures of petroleum hydrocarbons affect several parasites and host-parasite relationships.

Fishes with trichodinid ciliate infestations typically demonstrate a pathological gill response to the association of protozoan and hydrocarbons as they do to protozoan and kraft effluent. Khan (1990) studied such infestations on two species of adult fish from the Atlantic Ocean. He exposed the longhorn sculpin, *Myoxocephalus octodecemspinosus*, to 12 weeks of 50 to 100 μg/l of water-soluble fractions and the Atlantic cod, *Gadus morhua*, to oil-contaminated sediment (2,200 μg/g). He let the groups depurate for 14 and 20 weeks, respectively, before examining their gills for infestations. The dramatic difference in both prevalence and intensity of ciliate infestation compared similarly with what he observed on wild fish from the Pacific Ocean. Taking advantage of the *Exxon Valdez* spill in Alaska, Khan (1990) examined field samples of the tidepool sculpin, *Oligocottus maculosus*, that he collected about 5 months after the spill from an oiled beach on the Pye Islands and a noncontaminated site in the Seward area.

Monogeneans show a similar relationship as trichodines to exposure to oil. Epithelial and mucus cell hyperplasia, capillary dilation, and lamellar fusion relate directly with hydrocarbon concentration. However, Khan and Kiceniuk (1988) reported no statistical difference in intensity of the monogeneans *Gyrodactylus* spp. on gills of control Atlantic cod compared with that for infestations on cod exposed to oil fractions. They then exposed a group of cod to Venezuelan oil fractions for 12 weeks and waited 16 weeks before examining the gills. The prevalence of infestation in the control group dropped from 70% to < 20% and the intensity was low. All exposed fish had an infestation, and the intensity was high (Fig. 5-15). Another group depurated for 20 weeks after a 13-week exposure showed similar results (Fig. 5-15).

Petroleum toxicants cause mortalities in exposed fish, but added effects of parasites need to be recognized. Venezuelan oil can produce severe tail rot, hypersecretion of mucus, and death of the winter flounder, *Pseudopleuronectes americanus* (see Khan 1987). Khan (1987) compounded these effects by experimentally infecting half a group of winter flounder with *Trypanosoma murmanensis* and exposing half of those and half the remainder with 3200 μg/g (decreasing to 2600 μg/g) total hydrocarbon concentration in oil-contaminated sediment. He demonstrated that separately both oil and infection caused mortalities after 6 weeks, but combined produced considerably more deaths. This mortality affected the juvenile group more than adult counterparts. The same trypanosome infection in Atlantic cod added to the effects of 50-100 μg/l

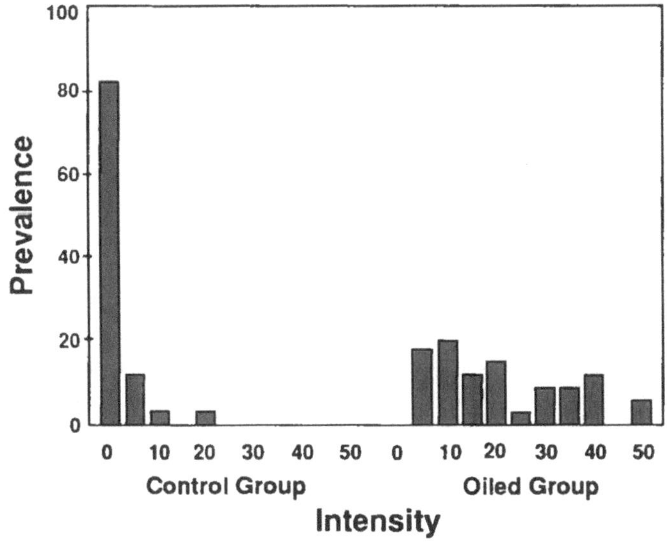

Fig. 5-15. Prevalence and intensity of the monogeneans *Gyrodactylus* spp. on the gills of the Atlantic cod, *Gadus morhua*, following a 16-week depuration period from a 12-week exposure to water-soluble fractions of Venezuelan crude oil in a flow-through system. Data to the left represent the control group not exposed to the oil fractions. (From Khan and Kiceniuk 1988.)

water-soluble fractions of the oil. Both prevalence and intensity of infections were greater in oil-exposed fish, and those fish had lower body condition and excessive mucus secretion from gills. Adults had retarded gonadal development.

The above trypanosome infection also illustrates another important aspect of parasitic infections. One parasite can potentiate the effect of another. A concurrent infection of the trypanosome and the blood-feeding adult copepod *Lernaeocera branchialis* in the Atlantic cod could cause mortality in juvenile but not adult fish (Khan and Lacey 1986). Fish surviving a dual infection had a lower body-condition than fish with either parasite alone or no parasites. The effect on cod by two or more adult copepods was more pronounced than that by one. The authors suggested that the low prevalence of the copepod infestation in cod from some offshore locations results from dual infections.

Hydrocarbons also can influence helminths in the digestive tract. Khan and Kiceniuk (1983) exposed flounder and cod to contaminated sediments or water-soluble fractions of crude oil, and examined the intestines for the digenean *Steringophorus furciger* in the flounder and the acanthocephalan *Echinorhynchus gadi* in cod. They found lower values of both prevalence and intensity of infections for both when fish were exposed to oil, whether in sediments or as water soluble fractions. Whether such an effect resulted from direct toxicity by drinking contaminated water or from a modification of host physiology was not established.

As with cases of exposure to some heavy metals and probably to many stresses, the presence of symbionts can influence the effect of aromatic hydrocarbons. In an extensive study by Moles (1980), the number of glochidia of the mussel *Anodonta oregonensis* present on the gills of coho salmon fry greatly influenced fish mortality when exposed to the toxicant. This larval molluscan stage must attach to fish gills within hours of being released by the female mussel and remains attached until it develops into a juvenile. Moles show that fry infested with 20 to 35 glochidia were significantly more sensitive to crude oil, toluene, or naphthalene than noninfested fish. This sensitivity increased linearly with an increase in number of symbionts. As indicated above, interpretation and application of results of toxicity tests need to assess the type and intensity of infection both in bioassay fishes and in wild stocks.

Thermal

Some problems involving thermal pollution are semantic and concern the definition of "thermal pollution." Consequently, they have to do with one's point of view. For example, water heated by some nuclear and conventional power plants is used for aquaculture; its elevated temperature promotes growth of finfishes and shellfishes and is considered a benefit. Also, the source of heat can be problematical. Normal temperature fluctuations can mask effects of an artificially heated discharge. The most conspicuous thermal damage from power plants usually is restricted to a short distance from the effluent, especially if the site is carefully chosen (e.g., Roessler 1977).

Temperature has considerable influence on toxic effects, diseases, and parasites. Cairns *et al.* (1975) pointed out that even with a vast amount of literature on effects of temperature on aquatic organisms, the knowledge was hard to apply. That is still true. Moreover, heated waste water is often located near discharges of toxic chemicals, intensifying or otherwise influencing effects of the chemicals.

Heated effluents alter the life cycles of some parasites and hosts, initiating or complicating infections or potential infections. For example, heated water from a nuclear power plant along the coast of the Bothnian Sea in Sweden attracted *Salmo trutta*, which in turn acquired a variety of parasites and abnormalities (Thulin 1987). At the same site, a temporal shift occurred in the life cycle of the monogenean *Paradiplozoon homoion* without altering its prevalence or intensity of infestation in *Rutilus rutilus* (see Höglund and Thulin 1989). Apparently increased recruitment of juveniles was neutralized by a decreased survival of adults in the heated area. Near another Swedish nuclear plant along the Baltic

Sea, 90% of a large sample of *Gadus morhua* contained the freshwater digenean *Diplostomum* sp. in the eyes (Thulin 1984). About 25% of these fish exhibited grey to cloudy lenses, and the eyes of these contained from 50 to 200 specimens of the digenean, mostly in the lens. Thulin (1984) was not sure if this pathogenic infection resulted from hot water discharge, a host reaction involving a freshwater worm in a marine fish, or a combination of those factors. Höglund and Thulin (1988) documented the cercariae of four strigeoid species of *Diplostomum* to shed a month earlier than expected when in the heated water. They found a relationship between experimentally induced fry mortality of *Alburnus alburnus* and number of cercariae; this relationship intensified with temperature. Bamber *et al.* (1983) studied the strigeoid *Neodiplostomum* sp. that caused visible black spots on the skin of fish near a Scottish power plant. Whereas the infected specimens were not more likely than uninfected ones to become impinged in the intake screen of the plant, the authors still reported that a crab with a rhizocephalan parasite and a fish with the copepod *Lernaeocera* sp. on it appeared more likely to be impinged than their uninfected counterparts.

Thermal stress from a nuclear power plant in South Carolina has been associated with depressed body condition and "red sore" disease of centrarchid fishes (Esch *et al.* 1976). That relationship has been attributed by Hazen *et al.* (1978) and Esch and Hazen (1980) to temperature stress and the bacterium *Aeromonas hydrophila* without intimate association with the peritrich ciliate, *Epistylis* sp. Whereas the bacterial disease can occur without the ciliate present both in the Carolinas and on fish in Mississippi estuaries, the ciliate appears to allow entrance of the bacterium in Mississippi (Overstreet and Howse 1977; Overstreet 1990). The ciliate can erode epithelium and underlying scales and then invade muscle tissue, allowing secondary infection. In Mississippi bayous (low salinity), the disease commonly occurs when there is an abundance of organic material. That material serves as the ciliate's direct food source and a source for bacteria, which also constitute food for the ciliate. In addition to centrarchids, a variety of estuarine species that encroach into low salinity areas rich with organic material are highly susceptible to the ciliate and to the disease. Consequently, red sore disease should not be considered as limited to areas of thermal stress, although it does indicate a stressed environment.

Several general surveys for parasites have been conducted in heated European lakes or reservoirs. An extensive one involved about 50 species of parasites found in 5 species of fish from 6 freshwater lakes, 5 of which were heated to different degrees by effluents from electric power plants in Poland (Pojmanska *et al.* 1980). Those authors found some parasites in the warmer lakes to be more abundant, but most were less abundant -- some because of absence of intermediate hosts. Not all parasites, fish hosts, or intermediate hosts responded similarly to the same increased temperature. Notable was a shift in host specificity of three monogeneans to a related fish in the two warmer lakes. That publication and associated papers provide a good understanding of the dynamics of different parasites, the life histories of some already having been determined.

Domestic Sewage and Waste

Parasites that use human definitive hosts can thrive near untreated sewage effluents. Artamoshin and Khodakova (1976) reported a relatively high prevalence of infection of *Diphyllobothrium latum* in perch fingerlings from the river Usolka in the Soviet Union downstream from villages without industrial waste. In contrast, few fish harbored the metacestodes in nearby rivers Glotikhi and Chernaya where chemical effluents mixed with the domestic wastes.

Human parasites other than *D. latum*, *Clonorchis sinensis*, and some heterophyids that utilize fish as intermediate hosts might also be vectored by fishes that feed on sewage, even though such species are not stipulated in WHO (1989) guidelines for use of wastewater in agriculture and aquaculture. The extensive guidelines consider inactivation of helminth eggs as important. Eggs of *Ascaris lumbricoides*, *Trichuris trichura*, and others persist for long periods and should be able to survive in fish. Moreover, with the increased numbers of immunodeficient people, fishes, especially those that feed on benthic detritus, could vector *Cryptosporidium parvum* and other parasites not typically considered as human parasites (Thomas Deardorff and Overstreet, unpublished data).

In the alpine Naini Tal Lake in Kumaun, India, fish kills commonly occur from December through February because of a combination of polluting sources, including domestic and industrial sewage, detergents, silt, and toxic plants (Das and Shrivastava 1985). Following mass mortalities attributed to eutrophication with its accompanying oxygen depletion and high ammonia concentration, a few fish species became heavily infected with *Chilodonella cyprini*, *Chloromyxum esocinum*, *Argulus japonicus*, and leeches. These fishes experienced die-offs during February and March. Along the shore zone in March, dying specimens of *Puntius* spp. still occurred. They exhibited grayish, mucus-laden gills clogged with infestations of *Chilodonella cyprini* and *Trichodina domerguli*.

Similar scenarios as described above can occur in rivers and estuaries. Oxygen depletion, stress-induced mucus, and lamellar lesions all support parasitic infestations on fishes, compounding an already stressful state. If domestic wastes are not excessive, they should promote a rich growth of phytoplankton available directly or indirectly to intermediate hosts of metazoan parasites. For example, fishkills in the Main River, Germany, have been attributed to metacercariae of *Bucephalus polymorphus* produced by a bivalve mollusc fed indirectly by "polluting wastes" (Schaefer and Hoffmann 1985). Each future survey of biota of "polluted areas" or analysis of mortalities should carefully evaluate parasitic infections relative to life cycles and host-parasite relationships.

The status of sewage sludge relative to finfishes has not been properly assessed. McVicar *et al.* (1988) examined fishes in a North Sea dumping area but found fewer diseased fish than in a control area. Those authors did not report on parasites. Hendrix (1985) examined monogeneans on fishes from the sewage-sludge, acid waste, cellar dirt, and dredge spoil dump sites in the New York Bight. He reported 16.4 specimens of *Bothitrema bothi* per fish on the pathologically altered gills of the windowpane (flounder), *Scophthalmus aquosus*, collected from the dump sites but only 8.2 per fish from the relatively clean adjacent areas where the gills did not exhibit extensive alterations. Murchelano (1982) reviewed pollution-associated diseases, stressing New York Bight, but he did not include protozoans or metazoans, with the possible exception of an organism suspected as an ameba associated with the epizootic epidermal papillomas of flatfishes. That disease has not been demonstrated to be related to a virus (e.g., Diamant *et al.* 1988), and the resulting atypical amebic-like cells are not of host origin. Murchelano and Azarovitz (1979), summarizing a groundfish survey of general fish diseases conducted in the Western North Atlantic, mentioned a few lesions that did not appear unusual or life-threatening, but were caused by parasites. They stated "the paucity of diseases noted may signify that the area surveyed is unimpacted by pollution." Oishi *et al.* (1976) associated epidermal papillomas of flatfishes near Hokkaido, Japan, with a philometrid nematode, but pointed out that at least three flatfish species with tumors inhabited areas free of either industrial or municipal discharges. A philometrid infection in two flatfishes in Puget Sound, Washington, did not appear to be related to METRO outfalls, but the prevalence did increase from north to south in the Sound (Miller *et al.* 1976). A fibroma in a flounder from Mississippi possibly related with a philometrid was not associated with pollution (Overstreet and Edwards 1976). Hard (1988) pointed out that at least a few different neoplasms used to suggest an association with pollution actually were pseudotumors induced by protozoans, a virus, and a digenean, and cautioned those assessing neoplasia to be aware of infectious agents.

Sedimentation

Turbidity can be beneficial to some species and harmful to others. Some fishes prefer turbid conditions (Cyrus and Blaber 1987). Fish that die in laboratory assays involving high levels of suspended sediments usually succumb from lack of oxygen due to sediment particles clogging the gills. Under natural conditions of prolonged suspension of sediments or presence of irritating substances, the gills of a fish unaccustomed to such exposure will produce excessive amounts of mucus and develop hyperplasia or other lesions. A variety of solids, depending on their concentration, size-distribution, organic content, and reactivity and on tolerance of exposed fish, can have lethal effects on fishes. O'Connor *et al.* (1977) studied sublethal physiology of seven estuarine species experimentally exposed to Fuller's earth suspension and described the effects on the gills of exposed white perch (*Morone americana*) in detail. Mucous cells increased, especially along the anterior margin of the filaments. Epithelium of the secondary lamellae separated from the pillar cells which occasionally become disrupted, and the epithelial cells underwent hypertrophy and hyperplasia. When Goldes *et al.* (1988) exposed juvenile rainbow trout to inert suspended clay kaolin for 64 days, lamellar proliferation was conspicuous at 16 and 32 days only in gills of fish infested with the symbiotic flagellate *Icthyobodo necator.* Apparently the kaolin induced mucus production, which in turn promoted an increase in flagellate infestation. The parasite then promoted hyperplasia and, ultimately by day 64, increased mucus production, an immune response, reduction in flagellates, turnover of branchial epithelium, and a return to normal gill architecture.

As indicated above for exposure to heavy metals and other toxicants, the pathological conditions in fish exposed to sediments provide an abundance of food for a variety of ectoparasites, such as trichodinids, flagellates, and monopisthocotylid monogeneans, supporting their growth and reproduction.

General Agricultural, Industrial, and Miscellaneous Toxicants

Only a few studies have treated specific chemicals as causing a specific effect on parasites and their hosts. Studies on parasites or diseases are not necessarily beneficial in delineating causative sources of general pollution. Nevertheless, they acknowledge that problems exist, provide baseline data for future comparisons, and encourage experimentation to solve specific problems.

Three fishes in Biscayne Bay, Florida, had nearly 100% prevalence of infestation by a monogenean on its gill, whether from the relatively pristine southeastern region or from the southwestern region polluted with relatively high concentrations of ammonia, trace metals, and pesticides (Skinner 1982). Gills of the species in the contaminated southwestern region, however, showed the pathological lesions described from areas laden with other toxicants: epithelial hyperplasia, excessive mucus secretion, fused lamellae, clubbed and fused filaments, and aneurysms. All three species of fish there harbored considerably heavier infestations than in the clean area.

Low rates of infections of some parasites utilizing intermediate hosts occur in regions where those hosts are affected by toxicants. Overstreet (1988) cited one case where both prevalence and intensity of an acanthocephalan utilizing amphipod intermediate hosts were less in the definitive Atlantic croaker host from the polluted than the unpolluted estuarine site. One site was downriver from and heavily influenced by effluents from a kraft pulp mill, chemical plant, and fish reduction plants.

The amphipod *Gammarus pulex* is susceptible to acid water and is used for assays. Some individuals in a study by McCahon *et al.* (1989) had an infection with the acanthocephalan *Pomphorhynchus laevis*, and those amphipods exhibited greater mortality than uninfected ones in various tests. Several studies have been conducted to assess the effect of pesticides, fungicides, and herbicides as a means to control parasites in culture facilities. These studies, however, also allowed an appreciation of the effect of those compounds as pollutants. Depending on the concentrations, most compounds affect differently a variety of symbionts (especially crustaceans), intermediate hosts, and final hosts. For example, Nair and Nair (1982) tested the effect of five organophosphates on the juvenile parasitic aegid isopod *Alitropus typus*. Organophosphates have advantages over chlorinated hydrocarbons and carbamates when used in culture systems. From the five compounds, authors found that the insecticide folithion (0,0-dimethyl 0-[4-nitro-m-tolyl] phosphorothioate) was most toxic (48-h LD_{50} = 0.001 ppm); however, another insecticide, nuvan, was probably more toxic to symbionts protected in the gill chamber. Dimecron (0-2-chloro-2-[diethylcarbamoyl]-1-methyl vinyl-0-0-dimethyl phosphate) was the least toxic (48-h LD_{50} = 5.2 ppm). About 1-4 ppm of folithion was necessary to similarly affect four fishes.

The sensitivity of host-symbiont associations to insecticides differs according to host, symbiont, and salinity. Paperna and Overstreet (1981) reported that whereas two tested compounds, dylox and malathion, killed parasitic copepods on mullets in brackish water, sensitivity of the fish to the compounds increased to that of the parasite when in fresh water. They also discussed other effects of salinity as they relate to controlling infestations in aquaculture systems.

FUTURE RESEARCH

Bioindicators

Parasitic surveys should become a routine part of any overall "base-line" or "follow-up" survey of polluted localities or localities suspected of being polluted. They should be conducted in conjunction with other disciplines involving pathobiology, biology, ecology, and associated fields. Because of the complex nature of the multihost life-histories, some parasites provide extremely sensitive indicators of both specific groups of toxicants and specific habitats and geographic areas. These should be utilized, and accumulation of all the parasitic data should provide an important aspect of any such surveys.

Infections by symbionts, categorized either as ectoparasites and endoparasites or as those with direct and indirect cycles, can be better, more sensitive indicators than some lesions -- a few of which take long periods to develop and cannot be related to specific conditions. Additional research can clarify relationships between a symbiont and its relationships with specific environmental conditions, thus increasing sensitivity of the indicator and creating a model.

Laboratory studies to back up field observations are needed, as are laboratory studies conducted for the sake of evaluating specific components of a complex system. Results from such studies, however, do not always express specific conditions of an environment receiving input from numerous natural and man influenced sources. Understanding differences between laboratory and field work constitutes important avenues to investigate.

In addition, parasitic infections also influence the sensitivity of hosts to toxicants and other environmental factors. Consequently, assessments of laboratory tests, using wild fish as well as laboratory-reared fish, should consider effects of parasites, if present. Standards for allowable concentrations of various chemicals can be established in error without such consideration.

Seasonal patterns in infections differ in response to numerous different environmental and host parameters, regardless of any compounding toxicity factors. Consequently, results may differ, depending on when collected. For example, diplostome infections in Sweden occurred earlier in thermally influenced habitats, but evened out over time (Höglund and Thulin 1989). Moreover, immunological factors take time to become established and to be modified. I should stress the importance of recognizing appropriate endpoints in assessing parasitic indicators for both field and laboratory studies. Examples have already been given regarding the need by Khan (1990) and Khan and Kiceniuk (1988) to allow 14- to 20-week periods after exposure to the toxicant to demonstrate differences in infestations of gill parasites.

When considering the relationships between parasitic infections and various environmental stresses, factors of time and endpoints have to be appreciated in ways other than indicated above. Whereas some factors might cause a short-term increase in infection, those same pollutants or other environmental factors may cause a decrease over a long period. Also, the factor of time may differ between results obtained from young fish or from young parasites and those from older counterparts. Model systems also might become obscured and the resulting data misconstrued, if examined after a long period of time. Consequently, the examination would not represent well the effects of toxicity.

An important approach to relating laboratory and field studies is to place laboratory-reared, uninfected fish in cages (e.g., Borthwick and Stanley 1985; Goodman and Cripe 1987) located both at point sources and at nearby control sites and then to compare infections from those groups with those in wild stocks from the same sites. Also, experimentally infected fish with model infections, such as *Calyptospora funduli*, can be placed into cages and the alterations in the cellular responses by the host to the parasite monitored by collection of serial histological samples from the reference site, control site, and laboratory-control.

Models should be established to be used in cages without fear of introducing exotic species that can provide usable parasitic data of a general or specific nature, and that can be used in the laboratory as well as in the field. If models do not consist of native fishes, some assurance should be made that neither fish nor their offspring would be introduced permanently into the tested environment.

The most important aspect of identifying and refining sensitivity of parasitic bioindicators is to delineate "cause-and-effect" relationships and be able to interpret them properly. Some relationships should show what contaminants and other environmental conditions are present and to what extent; perhaps these and others could be used to predict short-term as well as long-term effects on the ecosystem. I previously emphasized that such data were lacking in the southeast U.S. (Overstreet 1988) and scarce from other areas. Such data are still scarce and more necessary now than before.

Diseases

Because diseases, both infectious (parasitic and nonparasitic) and noninfectious, exert an important but often unappreciated role in the ecology of aquatic habitats, the need to understand those diseases and the corresponding disease processes is paramount. Much of what was indicated above as future needs to assess bioindicators also applies to assess disease. Information on parasites can provide explanations on changes in environmental conditions and in the ecosystem resulting from acute and chronic infections and help explain physiological variables, bioenergetics and reproduction of the hosts, and consequently, the structure of the population and community. On the other hand, not all indicators represent disease states, and several aspects of disease conditions require attention. Good models are difficult to maintain and assess, and each should be appreciated and aggressively investigated.

As indicated earlier, parasitic diseases often result from a disruption of the established equilibrium among host, symbiont, and environment. Stress upsets and shifts this equilibrium so that diseases can have a significant effect on the dynamics of many fishes. Also, a combination of parasites can have synergistic effects on the host. Some parasitic infections do not progress into a disease state without additional or concurrent secondary bacterial, fungal, or viral infections. Microbial infections probably constitute a greater threat to health of most fishes than parasites, but the synergistic effects as well as other relationships between the two require more attention. Because diseases usually cannot be visualized in nature, careful laboratory and field studies have to be designed to assess them. Because of the dynamic environmental conditions in an estuary, the estuary can either be considered the best or worst habitat to study the problems. If carefully monitored for perturbations related to environmental changes, the disease process can be dissected and various tests applied to the portions. Statistical analyses without good biological information may be difficult to obtain. For example, such analyses to determine mortality of fish from metacercariae of four diplostomatid digeneans from the eye, even with extensive biological data, were not conclusive (Kennedy 1984). Experiments should use cage studies of wild infections and laboratory-induced infections. Stocks in the field should be tagged and monitored. Laboratory studies should be designed to verify hypotheses established from the field observations.

At least some neoplasms relate to parasitic infections, and such associations should be clarified. Studies could be conducted to show the relationship between experimentally induced neoplasms and parasitic infections established both before and after exposure to the carcinogen.

Disease conditions should be critically examined from as many points of view as researchers in a group are willing to approach them. Team approaches should provide more information than individual studies. If possible, the relationships between and among disease and resistance, pathology, toxicology, genetics, nutrition, and fisheries biology should be simultaneously evaluated.

ACKNOWLEDGMENTS

I thank Teri Turner, Ronnie Palmer, Marie Wright, Nate Jordan, Jennifer Weber, Jim Franks, and Helen Gill of the Gulf Coast Research Laboratory for technical assistance. John Crites, retired from Ohio State University, graciously read the manuscript. I also thank John Hawke and Vernon Minton from the State of Alabama Department of Conservation and Natural Resources for specimens of fish infested with *Amyloodinium ocellatum* and permission to report the mass mortality. Lew Bullock, Department of Natural Resources, Division of Marine Resources, Florida Marine Research Institute, provided some fish material collected off the west coast of Florida. A portion of the study was conducted in cooperation with the U.S. Department of the Interior, Fish and Wildlife Service, Contract No. MS881027-002, Project No. F-91.

REFERENCES

Adlard, R.D. 1990. The effects of the parasitic isopod *Anilocra pomacentri* Bruce (Cymothoidae) on the population dynamics of the reef fish *Chromis nitida* Whitley (Pomacentridae). *University of Queensland, Australia. Dissertation*. 118 p.

Artamoshin, A.S., and V.I. Khodakova. 1976. The effect of sewage on the rate of infestation of fingerling fish by *Diphyllobothrium latum* plerocercoids in the Kama reservoir. *Gidrobiol. ZH.* 12(6):89-91. (English translation from Russian by Translation Bureau (WRi), Multilingual Services Division, Department of the Secretary of State of Canada, *Fisheries and Marine Service Translation Series* 4138:1-7.)

Bamber, R.N., R. Glover, P.A. Henderson, and A.W.H. Turnpenny. 1983. Diplostomiasis in the sand smelt, *Atherina presbyter* (Cuvier), population at Fawley Power Station. *J. Fish Biol.* 23(2):201-210.

Bauer, O.N. 1959. The influence of environmental factors on reproduction of fish parasites. *Vopr. Ekol. (Izdatelstvo Kievskogo Universiteta)* 3:132-141. (English translation from Russian by Z. Kabata, and L. Margolis, *Fisheries Research Board of Canada Translation Series* No. 1099:1-7).

Borthwick, P.W., and R.S. Stanley. 1985. Effects of ground ULV applications of fenthion on estuarine biota III. Response of caged pink shrimp and grass shrimp. *J. Fla. Anti-Mosq. Assoc.* 56(2):69-72.

Bower, C.E., and D.T. Turner. 1988. Drug and chemical effects on the behavior and survival *in vitro* of the dinoflagellate fish parasite *Amyloodinium ocellatum*. Paper read at International Association for Aquatic Animal Medicine, 22-26 May 1988, at Orlando, Florida.

Bower, S.M., and L. Margolis. 1985. Effects of temperature and salinity on the course of infection with the haemoflagellate *Cryptobia salmositica* in juvenile Pacific salmon, *Oncorhynchus* spp. *J. Fish Dis.* 8(1):25-33.

Boyce, N.P., and S.B. Yamada. 1977. Effects of a parasite, *Eubothrium salvelini* (Cestoda: Pseudophyllidea), on the resistance of juvenile sockeye salmon, *Oncorhynchus nerka*, to zinc. *J. Fish. Res. Board Can.* 34(5):706-709.

Burreson, E.M., and D.E. Zwerner. 1984. Juvenile summer flounder, *Paralichthys dentatus*, mortalities in the western Atlantic Ocean caused by the hemoflagellate *Trypanoplasma bullocki*: evidence from field and experimental studies. *Helgol. Meeresunters.* 37(1-4):343-352.

Burreson, E.M., and L.J. Frizzell. 1986. The seasonal antibody response in juvenile summer flounder (*Paralichthys dentatus*) to the hemoflagellate *Trypanoplasma bullocki*. *Vet. Immunol. Immunopathl.* 12:395-402.

Cairns, J. Jr., A.G. Heath, and B.C. Parker. 1975. The effects of temperature upon the toxicity of chemicals to aquatic organisms. *Hydrobiolgia* 47(1):135-171.

Canning, E.U., J. Lom, and I. Dyková. 1986. *The Microsporidia of Vertebrates*. Orlando, Florida: Academic Press, Inc.

Cheung, P.J., R.F. Nigrelli, and G.D. Ruggieri. 1979. Studies on cryptocaryoniasis in marine fish: effect of temperature and salinity on the reproductive cycle of *Cryptocaryon irritans* Brown, 1951. *J. Fish Dis.* 2(2):93-97.

Couch, J.A. 1985. Prospective study of infectious and noninfectious diseases in oysters and fishes in three Gulf of Mexico estuaries. *Dis. Aquat. Organ.* 1(1):59-82.

Cyrus, D.P., and S.J.M. Blaber. 1987. The influence of turbidity on juvenile marine fishes in estuaries. Part 2. Laboratory studies, comparisons with field data and conclusions. *J. Exp. Mar. Biol. Ecol.* 109(1):71-91.

Das, M.C., and A.K. Shrivastava. 1984. Fish mortality in Naini Tal Lake (India) due to pollution and parasitism. *J. Hydrobiol.* 20(4):60-64.

Deardorff, T.L., and R.M. Overstreet. 1981a. Review of *Hysterothylacium* and *Iheringascaris* (both previously = *Thynnascaris*) (Nematoda: Anisakidae) from the northern Gulf of Mexico. *Proc. Biol. Soc. Wash.* 93(4):1035-1079.

Deardorff, T.L., and R.M. Overstreet. 1981b. Larval *Hysterothylacium* (= *Thynnascaris*) (Nematoda: Anisakidae) from fishes and invertebrates in the Gulf of Mexico. *Proc. Helminthol. Soc. Wash.* 48(2):113-126.

Diamant, A., D.A. Smail, L. McFarlane, and A.M. Thomson. 1988. An infectious pancreatic necrosis virus isolated from common dab *Limanda limanda* previously affected with X-cell disease, a disease apparently unrelated to the presence of the virus. *Dis. Aquat. Organ.* 4(3):223-227.

Ellis, A.E. 1988. Optimizing factors for fish vaccination. In *Fish Vaccination*, ed. A.E. Ellis, pp. 32-46. San Diego, California: Academic Press, Inc.

El-Matbouli, M., and R. Hoffmann. 1989. Experimental transmission of two *Myxobolus* spp. developing bisporogeny via tubificid worms. *Parasitol. Res.* 75(6):461-464.

Esch, G.W., and T.C. Hazen. 1980. Stress and body condition in a population of largemouth bass: implications for red-sore disease. *Trans. Am. Fish. Soc.* 109(5):532-536.

Esch, G.W., T.C. Hazen, R.V. Dimock, Jr., and J. Whitfield Gibbons. 1976. Thermal effluent and the epizootiology of the ciliate *Epistylis* and the bacterium *Aeromonas* in association with centrarchid fish. *Trans. Am. Micros. Soc.* 95(4):687-693.

Ewing, M.S., S.A. Ewing, and M.A. Zimmer. 1982. Sublethal copper stress and susceptibility of channel catfish to experimental infections with *Ichthyophthirius multifiliis*. *Bull. Environ. Contam. Toxicol.* 28(6):674-681.

Felley, S.M., M. Vecchione, and S.G.F. Hare. 1987. Incidence of ectoparasitic copepods on ichthyoplankton. *Copeia* 1987(3):778-782.

Fournie, J.W., and R.M. Overstreet. 1983. True intermediate hosts for *Eimeria funduli* (Apicomplexa) from estuarine fishes. *J. Protozool.* 30(4):672-675.

Ghosh, K., and S. Jana. 1988. Effects of combinations of heavy metals on population growth of fish nematode *Spinicauda spinicauda* in aquatic environment. *Environ. Ecol.* 6(4):791-794.

Giles, N. 1987. Predation risk and reduced foraging activity in fish: experiments with parasitized and non-parasitized three-spined sticklebacks, *Gasterosteus aculeatus* L. *J. Fish Biol.* 31(1):37-44.

Goldes, S.A., H.W. Ferguson, R.D. Moccia, and P.-Y. Daoust. 1988. Histological effects of the inert suspended clay kaolin on the gills of juvenile rainbow trout, *Salmo gairdneri* Richardson. *J. Fish Dis.* 11(1):23-33.

Goodman, L.R., and G.M. Cripe. 1987. A cage for use with small aquatic animals in field studies. *J. Am. Mosq. Control Assoc.* 3(1):109-110.

Guth, D.J., H.D. Blankespoor, and J. Cairns, Jr. 1977. Potentiation of zinc stress caused by parasitic infection of snails. *Hydrobiolgia* 55(3):225-229.

Hard, G.C. 1988. Fish tumors and ecological surveillance: a cautionary example from Port Phillip Bay. *Water Res. Bull.* 24(5):975-980.

Hawke, J.P., S.M. Plakas, R. Vernon Minton, R.M. McPhearson, T.G. Snider, and A.M. Guarino. 1987. Fish pasteurellosis of cultured striped bass (*Morone saxatilis*) in coastal Alabama. *Aquaculture* 65(3/4):193-204.

Hawkins, W.E., M.A. Solangi, and R.M. Overstreet. 1981. Ultrastructural effects of the coccidium, *Eimeria funduli* Duszynski, Solangi and Overstreet, 1979 on the liver of killifishes. *J. Fish Dis.* 4(4):281-295.

Hazen, T.C., M.L. Raker, G.W. Esch, and C.B. Fliermans. 1978. Ultrastructure of red-sore lesions on largemouth bass (*Micropterus salmoides*): association of the ciliate *Epistylis* sp. and the bacterium *Aeromonas hydrophila*. *J. Protozool.* 25(3):351-355.

Hendrix, S.S. 1985. Monogenetic trematodes from fishes of the New York Bight. In *Annual Midwest Conference of Parasitology Presentations*, Ohio State University.

Höglund, J., and J. Thulin. 1988. Parasitangreep i ögon hos fisk, som leveri kylvatten fran kärnkraftsreaktorer. *Naturvardsverket Rapp.* 3539 11:47.

Höglund, J., and J. Thulin. 1989. Thermal effects on the seasonal dynamics of *Paradiplozoon homoion* parasitizing roach, *Rutilus rutilus* (L.). *J. Helminthol.* 63(2):93-101.

Howse, H.D., A.R. Lawler, W.E. Hawkins, and C.A. Foster. 1977. Ultrastructure of lymphocystis in the heart of the silver perch, *Bairdiella chrysoura* (Lacépède), including observations on normal heart structure. *Gulf Res. Rep.* 6(1):39-57.

Jana, S., and K. Ghosh. 1987. Effect of heavy metals on population growth of a fish nematode *Spinicauda spinicauda* in aquatic environment. *Environ. Ecol.* 5(4):811-813.

Kabata, Z. 1984. Diseases caused by metazoans: crustaceans. In *Diseases of Marine Animals*, ed. O. Kinne, Vol. 4, Part 1, pp. 321-399. Hamburg, Germany: Westholsteinische Verlagsdruckerei Boyens & Co.

Kabata, Z., and C.R. Forrester. 1974. *Atheresthes stomias* (Jordan and Gilbert 1880) (Pisces: Pleuronectiformes) and its eye parasite *Phrixocephalus cincinnatus* Wilson 1908 (Copepoda: Lernaeoceridae) in Canadian Pacific waters. *J. Fish. Res. Board Can.* 31(10):1589-1595.

Kennedy, C.R. 1984. The use of frequency distributions in an attempt to detect host mortality induced by infections of diplostomatid metacercariae. *Parasitology* 89(2):209-220.

Kent, M.L., T.K. Sawyer, R.P. Hedrick. 1988. *Paramoeba pemaquidensis* (Sarcomastigophora: Paramoebidae) infestation of the gills of coho salmon *Oncorhynchus kisutch* reared in sea water. *Dis. Aquat. Organ.* 5(3):163-169.

Kent, M.L., D.J. Whitaker, and L. Margolis. 1990. Experimental transmission of the myxosporean *Myxobolus arcticus* to sockeye salmon using an aquatic oligochaete, *Eclipidrilus* sp. (Lumbriculidae). *Fish Health Section-American Fisheries Society Newsletter* 18(4):4-5.

Khan, R.A. 1987. Effects of chronic exposure to petroleum hydrocarbons on two species of marine fish infected with a hemoprotozoan, *Trypanosoma murmanensis*. *Can. J. Zool.* 65(11):2703-2709.

Khan, R.A. 1990. Parasitism in marine fish after chronic exposure to petroleum hydrocarbons in the laboratory and to the Exxon Valdez oil spill. *Bull. Environ. Contam. Toxicol.* 44(5):759-763.

Khan, R.A., and J. Kiceniuk. 1983. Effects of crude oils on the gastrointestinal parasites of two species of marine fish. *J. Wildlife Dis.* 19(3):253-258.

Khan, R.A., and J. Kiceniuk. 1984. Histopathological effects of crude oil on Atlantic cod following chronic exposure. *Can. J. Zool.* 62(10):2038-2043.

Khan, R.A., and J.W. Kiceniuk. 1988. Effect of petroleum aromatic hydrocarbons on monogeneids parasitizing Atlantic cod, *Gadus morhua* L. *Bull. Environ. Contam. Toxicol.* 41(1):94-100.

Khan, R.A., and D. Lacey. 1986. Effect of concurrent infections of *Lernaeocera branchialis* (Copepoda) and *Trypanosoma murmanensis* (Protozoa) on Atlantic cod, *Gadus morhua*. *J. Wildlife Dis.* 22(2):201-208.

Landolt, M.L. 1989. The relationship between diet and the immune response of fish. *Aquaculture* 79(1-4):193-206.

Lawler, A.R. 1980. Studies on *Amyloodinium ocellatum* (Dinoflagellata) in Mississippi Sound: natural and experimental hosts. *Gulf Res. Rep.* 6(4):403-413.

Lehtinen, K.-J., and A. Oikari. 1980. Sublethal effects of kraft pulp mill waste water on the perch, *Perca fluviatilis*, studied by rotary-flow and histological techniques. *Ann. Zoolog. Fenn.* 17:255-259.

Lehtinen, K.-J., M. Notini, and L. Landner. 1984. Tissue damage and parasite frequency in flounders, *Platichtys flesus* (L.) chronically exposed to bleached kraft pulp mill effluents. *Ann. Zool. Fenn.* 21:23-28.

MacKenzie, K. 1987. Long-term changes in the prevalence of two helminth parasites (Cestoda: Trypanorhyncha) infecting marine fish. *J. Fish Biol.* 31(1):83-87.

McCahon, C.P., A.F. Brown, M.J. Poulton, and D. Pascoe. 1989. Effects of acid, aluminum and lime additions on fish and invertebrates in a chronically acidic Welsh stream. *Water Air Soil Pollut.* 45:345-359.

McVicar, A.H., D.W. Bruno, and C.O. Fraser. 1988. Fish diseases in the North Sea in relation to sewage sludge dumping. *Mar. Pollut. Bull.* 19(4):169-173.

Manning, M.J., and C.J. Secombes. 1987. Immunology and Disease Control Mechanisms of Fish. *Fisheries Society of the British Isles Symposium* 31(Supplement A):1-261.

Maxwell, J.G.H. 1982. Infestation of the jack mackerel, *Trachurus declivis* (Jenyns), with the cymothoid isopod, *Ceratothoa imbricatus* (Fabricus), in south eastern Australian waters. *J. Fish Biol.* 20(3):341-349.

Miller, B.S., B.B. McCain, R.C. Wingert, S.F. Borton, and K.V. Pierce. 1976. Puget Sound Interim Studies. Ecological and disease studies of demersal fishes near *metro* operated sewage treatment plants on Puget Sound and the Duwamish River. *Fisheries Research Institute Publication* UW-7608. 135 p.

Mohan, C.V., and C. Sommerville. 1989. Interaction between parasitic infection and metal toxicity in fish; an experimental study. *Proceedings of the XIV Symposium of the Scandinavian Society for Parasitology*, Information 20:62.

Moles, A. 1980. Sensitivity of parasitized coho salmon fry to crude oil, toluene and naphthalene. *Trans. Am. Fish. Soc.* 109(3):293-297.

Möller, H. 1978. The effects of salinity and temperature on the development and survival of fish parasites. *J. Fish Biol.* 12(4):311-323.

Möller, H. 1985. A critical review on the role of pollution as a cause of fish diseases. In *Fish and Shellfish Pathology*, ed. A.E. Ellis, pp. 169-182. London, England: Academic Press.

Möller, H. 1986. Pollution and parasitism in the aquatic environment. *Int. J. Parasitol.* 17(3):353-361.

Murchelano, R.A. 1982. Some pollution-associated diseases and abnormalities of marine fishes and shellfishes: a perspective for the New York Bight. In *Ecological Stress and the New York Bight*: Science and Management, ed. G.F. Mayer, pp. 327-346. Columbia, South Carolina: Estuarine Research Federation.

Murchelano, R., and T. Azarovitz. 1979. Fish disease surveys in the western North Atlantic. *Int. Counc. Explor. Sea* C.M.1979/E:24, 7 p.

Nair, A.G., and N.B. Nair. 1982. Effect of certain organophosphate biocides on the juveniles of the isopod, *Alitropus typus* M. Edwards (Crustacea: Flabellifera: Aegidae). *J. Anim. Morphol. Physiol.* [1983] 29(1-2):265-271.

Nakai, T., T. Kanno, E.R. Cruz, and K. Muroga. 1987. The effects of iron compounds on the virulence of *Vibrio anguillarum* in Japanese eels and ayu. *Fish Pathol.* 22(4):185-189.

Narasimhamurti, C.C., and C. Kalavati. 1984. Seasonal variation of the myxosporidian, *Henneguya waltairensis* parasitic in the gills of the fresh water fish, *Channa punctatus* Bl. *Arch. Protistenkd.* 128:351-356.

Norris, D.E., and R.M. Overstreet. 1975. *Thynnascaris reliquens* sp.n. and *T. habena* (Linton, 1900) (Nematoda: Ascaridoidea) from fishes in the northern Gulf of Mexico and eastern U.S. seaboard. *J. Parasitol.* 61(2):330-336.

O'Connor, J.M., D.A. Newmann, and J.A. Sherk, Jr. 1977. Sublethal effects of suspended sediments on estuarine fish. *U.S. Department of Commerce, National Technical Information Service, Technical Paper No.* 77-3:1-90.

Ogawa, K., K. Hattori, K. Hatai, and S. Kubota. 1989. Histopathology of cultured marine fish, *Seriola purpurascens* (Carangidae) infected with *Paradeontacylix* spp. (Trematoda: Sanguinicolidae) in its vascular system. *Fish Pathol.* 24(2):75-81.

Oishi, K., F. Yamazaki, and T. Harada. 1976. Epidermal papillomas of flatfish in the coastal waters of Hokkaido, Japan. *J. Fish. Res. Board Can.* 33(9):2011-2017.

Overstreet, R.M. 1968. Parasites of the inshore lizardfish, *Synodus foetens*, from South Florida, including a description of a new genus of Cestoda. *Bull. Mar. Sci.* 18(2):444-470.

Overstreet, R.M. 1977. *Poecilancistrium caryophyllum* and other trypanorhynch cestode plerocercoids from the musculature of *Cynoscion nebulosus* and other sciaenid fishes in the Gulf of Mexico. *J. Parasitol.* 63(5):780-789.

Overstreet, R.M. 1982. Abiotic factors affecting marine parasitism. In *Parasites -- Their World and Ours*, ed. D.F. Mettrick and S.S. Desser, *the Fifth International Congress of Parasitology, Proceedings and Abstracts*, Vol. 2, pp. 36-39. Toronto, Canada: World Federation of Parasitologists.

Overstreet, R.M. 1983a. Aspects of the biology of the spotted seatrout, *Cynoscion nebulosus*, in Mississippi. *Gulf Res. Rep.*, Supplement 1:1-43.

Overstreet, R.M. 1983b. Aspects of the biology of the red drum, *Sciaenops ocellatus*, in Mississippi. *Gulf Res. Rep.*, Supplement 1:45-68.

Overstreet, R.M. 1988. Aquatic pollution problems, Southeastern U.S. coasts: histopathological indicators. *Aquat. Toxicol.* 11(3,4):213-239.

Overstreet, R.M. 1990. Antipodean aquaculture agents. *Int. J. Parasitol.* 20(4):551-564.

Overstreet, R.M., and R.H. Edwards. 1976. Mesenchymal tumors of some estuarine fishes of the northern Gulf of Mexico. II. Subcutaneous fibromas in the southern flounder, *Paralichthys lethostigma*, and the sea catfish, *Arius felis*. *Bull. Mar. Sci.* 26(1):41-48.

Overstreet, R.M., and H.D. Howse. 1977. Some parasites and diseases of estuarine fishes in polluted habitats of Mississippi. In *Annals of the New York Academy of Sciences*, Vol. 298, ed. H.F. Kraybill, C.J. Dawe, J.C. Harshbarger, and R.G. Tardiff, pp. 427-462. New York: The New York Academy of Sciences.

Overstreet, R.M. and J. Thulin. 1989. Response by *Plectropomus leopardus* and other serranid fishes to *Pearsonellum corventum* (Digenea: Sanguinicolidae), including melanomacrophage centres in the heart. *Aust. J. Zool.* 37:129-142.

Paperna, I. 1980. *Amyloodinium ocellatum* (Brown, 1931) (Dinoflagellida) infestations in cultured marine fish at Eilat, Red Sea: epizootiology and pathology. *J. Fish Dis.* 3(5):363-372.

Paperna, I. 1984. Reproduction cycle and tolerance to temperature and salinity of *Amyloodinium ocellatum* (Brown, 1931) (Dinoflagellida). *Ann. Parasitol. Hum. Com.* 59(1):7-30

Paperna, I., and R.M. Overstreet. 1981. Parasites and diseases of mullets (Mugilidae). In *Aquaculture of Grey Mullets*, ed. O.H. Oren, pp. 411-493. Great Britain: Cambridge University Press.

Paperna, I., A. Diamant, and R.M. Overstreet. 1984. Monogenean infestations and mortality in wild and cultured Red Sea fishes. *Helgol. Meeresunter.* 37(1-4):445-462.

Pascoe, D., and P. Cram. 1977. The effect of parasitism on the toxicity of cadmium to the three-spined stickleback, *Gasterosteus aculeatus* L. *J. Fish Biol.* 10(5):467-472.

Pascoe, D., and D. Mattey. 1977. Studies on the toxicity of cadmium to the three-spined stickleback *Gasterosteus aculeatus* L. *J. Fish Biol.* 11(2):207-215.

Pascoe, D., and D. Mattey. 1977. Dietary stress in parasitized and non-parasitized *Gasterosteus aculeatus* L. *Z. Parasitenkd.* 51:179-186.

Pascoe, D., and J. Woodworth. 1980. The effects of joint stress on sticklebacks. *Z. Parasitenkd.* 62(2):159-163.

Pojmanska, T., B. Grabda-Kazubska, S.L. Kazubski, J. Machalska, and K. Niewiadomska. 1980. Parasite fauna of five fish species from the Konin lakes complex, artificially heated with thermal effluents, and from Goplo lake. *Acta Parasitol. Pol.* 27(38):319-357.

Roessler, M.A. 1977. Thermal additions in a tropical marine lagoon. In *Proceedings of the World Conference Towards a Plan of Actions for Mankind. Biological Balance and Thermal Modifications*, ed. M. Marois, Vol. 3, pp. 79-87. New York: Pergamon Press.

Rosenthal, H. 1967. Parasites in larvae of the herring (*Clupea harengus* L.) fed with wild plankton. *Int. J. Life in Oceans and Coastal Waters* 1(1):10-15.

Roubal, F.R., T.C. Jones, and R.J.G. Lester. 1988. Studies on the ultrastructure of cultured and gill-attached *Paramoeba* sp. and the cytopathology of amoebic gill disease in Atlantic salmon, *Salmo salar*. In *Society for Parasitology, Annu. Sci. Gen. Meeting, Program and Abstracts*, p. 30.

Sadzikowski, M.R., and D.C. Wallace. 1974. The incidence of *Lironeca ovalis* (Say) (Crustacea, Isopoda) and its effects on the growth of white perch, *Morone americana* (Gmelin), in the Delaware River near Artificial Island. *Chesapeake Sci.* 15(3):163-165.

Sakanari, J.A., M. Moser, C.A. Reilly, and T.P. Yoshino. 1984. Effects of sublethal concentrations of zinc and benzene on striped bass, *Morone saxatilis* (Walbaum), infected with larval *Anisakis* nematodes. *J. Fish Biol.* 24(5):553-563.

Sawyer, T.K, J.G. Hnath, and J.F. Conrad. 1974. *Thecamoeba hoffmani* sp.n. (Amoebida: Thecamoebidae) from gills of fingerling salmonid fish. *J. Parasitol.* 60(4):677-682.

Schaefer, W., and R. Hoffmann. 1985. Fish kills in Main River. *Fischer and Teichwirt* 36(7):199-202.

Sindermann, C.J. 1979. Pollution-associated diseases and abnormalities of fish and shellfish: a review. *Fish. Bull.* 76(4):717-749.

Skinner, R.H. 1981. The interrelation of water quality, gill parasites, and gill pathology of some fishes from South Biscayne Bay, Florida. *Fish. Bull.* 80(2):269-279.

Solangi, M.A., and R.M. Overstreet. 1980. Biology and pathogenesis of the coccidium *Eimeria funduli* infecting killifishes. *J. Parasitol.* 66(3):513-526.

Solangi, M.A., R.M. Overstreet, and J.W. Fournie. 1982. Effect of low temperature on development of the coccidium *Eimeria funduli* in the Gulf killifish. *Parasitology* 84(1):31-39.

Steinhagen, D., and W. Körting. 1990. The role of tubificid oligochaetes in the transmission of *Goussia carpelli*. *J. Parasitol.* 76(1):104-107.

Steinhagen, D., P. Kruse, and W. Körting. 1989. Effects of immunosuppressive agents on common carp infected with the haemoflagellate *Trypanoplasma borreli*. *Dis. Aquat. Organ.* 7(1):67-69.

Stolen, J.S., D.P. Anderson, and W.B. Van Muiswinkel. 1986. *Fish Immunology*. New York: Elsevier Science Publishing Company.

Thulin, J. 1984. The impact of some environmental changes on the parasite fauna of cod in Swedish waters. In *Abstracts of the Fourth European Multicolloquium of Parasitology*, pp. 239-240. Izmir, Turkey: Bilgehan Publishing House.

Thulin, J. 1987. Some diseases and parasites of trout (*Salmo trutta*) attracted to a hot water effluent. In *Parasites and Diseases in Natural Waters and Aquaculture in Nordic Countries*, ed. A. Stenmark, and G. Malmberg, p. 72. Stockholm, Sweden: Zoo-Tax.

Thulin, J., J. Höglund, and E. Lindesjöö. 1988. Diseases and parasites of fish in a bleached kraft mill effluent. *Water Sci. Technol.* 20(2):179-180.

Tort, L., J.J. Watson, and I.G. Priede. 1987. Changes in *in vitro* heart performance in rainbow trout, *Salmo gairdneri* Richardson, infected with *Apatemon gracilis* (Digenea). *J. Fish Biol.* 30(3):341-347.

Valtonen, E.T., M. Koskivaara, and H. Brummer-Korvenkontio. 1987. Parasites of fishes in central Finland in relation to environmental stress. *Biological Research Report University Jyvaeskylae* 10:129-130.

van Banning, P. 1987. Long-term recording of some fish diseases using general fishery research surveys in the south-east part of the North Sea. *Dis. Aquat. Organ.* 3(1):1-11.

Vogelbein, W.K., J.W. Fournie, and R.M. Overstreet. 1987. Sequential development and morphology of experimentally induced hepatic melano-macrophage centres in *Rivulus marmoratus*. *J. Fish Biol.* 31 (Supplement A):145-153.

WHO Scientific Group. 1989. Health guidelines for the use of wastewater in agriculture and aquaculture. In *World Health Organization Technical Report Series* 778, pp. 1-75, Geneva, Switzerland: World Health Organization.

6

Neoplasms in Wild Fish from the Marine Ecosystem Emphasizing Environmental Interactions

John C. Harshbarger
Phyllis M. Spero
Norman M. Wolcott

INTRODUCTION

The vast majority of neoplasms from cold-blooded vertebrates in the marine ecosystem submitted to the Registry of Tumors in Lower Animals (RTLA) or described in the published literature were from specimens collected in the estuarine and neritic zones. Nearly all types of tumors recognized in fish are represented, but many are isolated by space and time. However, a moderate number of marine, estuarine, and neritic cases have been found in clusters or epizootics with common denominators that provide clues to their etiology and environmental interactions. These tumor clusters have primarily originated from liver, epidermis, hematopoietic tissue, connective tissue, pigment cells, and peripheral nerve sheath cells.

Among these various clusters, liver neoplasms have a well established chemical etiology that is operative in the environment. Some of the epidermal neoplasms appear to have been caused or influenced by chemicals, while others appear to have been caused or influenced by viruses. Hematopoietic neoplasms appear to have been caused by viruses.

Clusters of connective tissue neoplasms have circumstantial evidence for a viral cause. Neoplasms arising from both peripheral nerve sheath cells and from pigment cells appear to have a genetic basis that can be influenced or activated by several different factors.

0-8493-8662-4/93/$0.00 + $.50

EPIZOOTIC LIVER NEOPLASMS

Epizootic liver neoplasms (Table 6-1) have been found in 12 fish species taken from 13 heavily polluted brackish or briny waterways of the Pacific North American coast from Vancouver to Southern California, the Atlantic North American coast from Boston to Norfolk, and the European North Sea. Epizootic liver neoplasms are unreported in the southeastern U.S. and Gulf of Mexico, but appropriate fish have not been adequately surveyed at point sources of pollution.

Along the Pacific coast, epizootic liver neoplasms occur in English sole, *Pleuronectes vetulus*, from Vancouver Harbor (Goyette *et al.* 1986); in English sole, *P. vetulus*, rock sole, *Pleuronectes bilineata*, starry flounder, *Platichthys stellatus*, and Pacific staghorn sculpin, *Leptocottus armatus*, from Commencement Bay, Everett Harbor, Elliott Bay, Duwamish waterway, and other pollution point sources in Puget Sound (McCain *et al.* 1977; Myers and Rhodes 1988); in white croaker, *Genyonemus lineatus*, from the Long Beach area of California, including the Hyperion sewage outfall (Malins *et al.* 1987).

Along the North Atlantic coast, epizootic liver neoplasms in winter flounder, *Pleuronectes americanus*, occur in Boston Harbor at the Deer Island sewage outfall (Murchelano and Wolke 1991); Quincy and New Bedford Harbors in Massachusetts; Gaspee Point in Narragansett Bay, Rhode Island; Black Rock Harbor at Bridgeport, Connecticut; and the Central Long Island Sound Dump Site for Black Rock Harbor dredge material (Gardner *et al.* 1989). Epizootic liver neoplasms occur in Atlantic tomcod, *Microgadus tomcod*, from the lower Hudson River (Smith *et al.* 1979; Cormier and Racine 1990); brown bullhead, *Ameiurus nebulosus*, and white perch, *Morone americana*, in the Delaware River from Philadelphia to Trenton (Anonymous 1988) (Lucké and Schlumberger 1941); white perch, *M. americana*, in the Chesapeake Bay (May *et al.* 1987; Bunton and Baksi 1988); oyster toadfish, *Opsanus tau*, from near the mouth of the York River, Virginia (Thiyagarajah and Bender 1988); and mummichog, *Fundulus heteroclitus*, from the Elizabeth River near Norfolk, Virginia (Vogelbein *et al.* 1990). The mummichog epizootic is especially relevant because 33% of mummichogs at a creosote plant had liver cancer compared to none of the mummichog along the opposite riverbank. In the southern North Sea, liver neoplasms ranged from 35%-38% of dab, *Limanda limanda*, captured on three annual cruises (Bucke *et al.* 1984; Kranz and Dethlefsen 1990).

In addition to this epidemiological association of fish liver cancer with environmental pollutants, other lines of evidence are consistent with a chemical etiology: 1) Experimental studies (Couch and Harshbarger 1985) show that approximately two dozen fish species treated with one or more of three dozen established carcinogens produced liver cancer. Some produced a variety

Table 6-1. Epizootic Liver Neoplasms in Marine Fish

Species	Site	Key References
Atlantic tomcod, *Microgadus tomcod*	Hudson River, NY	Smith *et al.* 1979; Cormier and Racine 1990
Brown bullhead, *Ameiurus nebulosus*	Delaware River, PA	Lucké and Schlumberger 1941; Anonymous 1988
White perch, *Morone americana*	Delaware River, PA	Anonymous 1988
White perch, *M. americana*	Chesapeake Bay	May *et al.* 1987; Bunton and Baksi 1988
Oyster toadfish, *Opsanus tau*	York River, VA	Thiyagarajah and Bender 1988
Mummichog, *Fundulus heteroclitus*	Elizabeth River, VA	Vogelbein *et al.* 1990
Winter flounder, *Pleuronectes americanus*	Boston Harbor, MA	Murchelano and Wolke 1991
Winter flounder, *P. americanus*	Quincy Bay, MA	Gardner *et al.* 1989
Winter flounder, *P. americanus*	Black Rock Harbor, Bridgeport, CT	Gardner *et al.* 1989
Winter flounder, *P. americanus*	New Bedford, MA	Gardner *et al.* 1989
Winter flounder, *P. americanus*	Central Long Island Sound Dump Site, NY	Gardner *et al.* 1989
Winter flounder, *P. americanus*	Gaspee Point, Narragansett Bay, RI	Gardner *et al.* 1989
English sole, *P. vetulus*	Vancouver Harbor, B.C., Canada	Goyette *et al.* 1986
English sole, *P. vetulus*	Puget Sound, WA	Myers and Rhodes 1988
Starry flounder, *Platichthys stellatus*	Puget Sound, WA	Myers and Rhodes 1988
Rock sole, *Pleuronectes bilineatus*	Puget Sound, WA	Myers and Rhodes 1988
Pacific staghorn sculpin, *Leptocottus armatus*	Puget Sound, WA	Myers and Rhodes 1988
White croaker, *Genyonemus lineatus*	Pacific Ocean, Los Angles, CA area	Malins *et al.* 1987
Dab, *Limanda limanda*	Southern North Sea	Bucke *et al.* 1984; Kranz and Dethlefsen 1990

of other types of cancer, but liver cancer resulted consistently. 2) Fish livers have the necessary enzyme system to metabolically activate indirect acting carcinogens (Stegeman and Lech 1991). 3) Extracts of sediment from the polluted Buffalo River, a freshwater habitat containing brown bullhead, *A. nebulosus*, with epizootic skin and liver neoplasms, were mutagenic by the Ames test and produced skin neoplasms in brown bullheads, *A. nebulosus*, following skin paintings and liver neoplasms following dietary exposure (Black 1983). Skin painting also produced skin cancer in mice. 4) When extracts of sediment from Hamilton Harbor in Lake Ontario, where white suckers, *Catostomus commersoni*, have epizootic liver cancer (Cairns and Fitzsimons 1988), were injected into rainbow trout, *Oncorhynchus mykiss*, sac fry they induced liver cancer (Metcalfe *et al.* 1988). 5) Gardner *et al.* (1991) demonstrated trophic transfer of toxins and possibly carcinogens by exposing blue mussels, *Mytilus edulis*, to suspended sediment from the Central Long Island Dump Site for Bridgeport, Connecticut, and feeding the mussels to winter flounder, *P. americanus*. Winter flounder, *P. americanus*, livers exhibited toxicosis and early stages of carcinogenesis similar to lesions in the winter flounder, *P. americanus*, collected at the dump site and at Black Rock Harbor in Bridgeport. 6) 32P-Postlabeling analyses show DNA adducts in fish liver cells from polluted areas (Dunn *et al.* 1987; Varanasi *et al.* 1989; Maccubbin *et al.* 1990). Previously, hydrolyzed aflatoxin derivatives had been shown to covalently bind to rainbow trout, *O. mykiss*, DNA (Croy *et al.* 1980) which is now linked to toxicity, regeneration and presumably carcinogenicity (Nunez *et al.* 1990). 7) The DNA in liver cancer cells of winter flounder, *P. americanus*, (McMahon *et al.* 1990), Atlantic tomcod, *M. tomcod*, (Wirgin *et al.* 1989) and rainbow trout, *O. mykiss*, (Mangold *et al.* 1991; Chang *et al.* 1991) contain activated *ras* oncogenes. Transfection of the winter flounder, *P. americanus*, tumor activated *ras* to NIH3T3 mouse fibroblast cells transformed those cells and caused them to produce tumors when injected into athymic mice. 8) There is no electron microscopic evidence or immunocytochemistry evidence of viruses associated with any fish liver neoplasms and no evidence of a familial basis. 9) Historically, the discovery of liver neoplasms in wild fish correlates with the increase of synthetic organic chemicals in the environment. From 1900 until World War II, annual production of synthetic organics held steady at about 15,000,000 pounds, then started increasing exponentially with a 7-8 year doubling time. Current annual production is about one billion pounds; there are 60,000 chemicals in common use and more than 100,000 in the environment. Epizootic liver tumors were first reported in 1 species of wild fish in 1964 compared to at least 16 species from at least 25 fresh and saltwater sites in North America today.

EPIZOOTIC EPIDERMAL NEOPLASMS

Epidermal neoplasms (Table 6-2) seem to be of several basic types including the plaque type where evidence favors a viral etiology and the smooth lobular type where evidence favors a chemical etiology except in Japanese salmonids. Grossly, the plaque type, also called fish pox, presents as multiple, low, white, or opaque lesions scattered over the skin and in the webbing of the fins. Histologically, plaques are well-differentiated epidermal thickenings with circumscribed epidermal pegs usually limited to the stratum spongiosum of the dermis. They have been reported since the mid 1500's in cultured common carp, *Cyprinus carpio*, and occur on a variety of other freshwater and marine fish species including the slippery dick, *Halichoeres bivittatus*, from Bimini, Bahama Islands (Lucké 1938). A herpesvirus has been repeatedly isolated from such lesions on a fancy strain of the common carp, *C. carpio*, by Sano *et al.* (1985). The virus was grown in the fathead minnow (FHM), *Pimephales promelas*, cell line, the epithelioma papillosum cyprini (EPC) cell line, and two other cell lines; virus-containing tumor plaques developed after five months in 3 of 10 carp injected with the virus.

Table 6-2. Epizootic Epidermal Neoplasms in Marine Fish

Species	Site	Key References
European smelt, *Osmerus eperlanus*	Baltic Sea	Breslauer 1916
Dab, *Limanda limanda*	North Sea (Danish)	Bloch *et al.* 1986
Northern pike, *Esox lucius*	Baltic Sea (Swedish)	Winqvist *et al.* 1968
Atlantic cod, *Gadus morhua*	Little Belt and Loge Bay, Denmark	Jensen and Bloch 1980
European eel, *Anguilla anguilla*	Baltic Sea	Christiansen and Jensen 1950
Slippery dick, *Halichoeres bivittatus*	Dry tortugas	Lucké 1938
Brown bullhead, *Ameiurus nebulosus*	Delaware River, PA	Lucké and Schlumberger 1941
White croaker, *Genyonemeus lineatus*	Los Angles, CA area	Russell and Kotin 1957
Cunner, *Tautogolabrus adspersus*	Sakonnet River	Harshbarger *et al.* 1976
Masu salmon, *Oncorhynchus masou*	Northern Japan	Yoshimizu *et al.* 1988a

Herpesvirus has been visualized in but not yet isolated from plaques of at least six other species, i.e., northern pike, *E. lucius* (Yamamoto *et al.* 1983), Pacific cod, *Gadus macrocephalus* (McArn *et al.* 1978), sheatfish, *S. glanis* (Bekesi *et al.* 1984), turbot, *S. maximus* (Buchanan *et al.* 1978), European smelt, *O. eperlanus*, (Anders and Möller 1985), and walleye, *S. vitreum* (Yamamoto *et al.* 1985). However, retrovirus was also visualized in walleye, *S. vitreum* (Walker 1969), northern pike, *E. lucius* (Winqvist *et al.* 1968), and a picornavirus in the European smelt, *O. eperlanus* (Ahne *et al.* 1990). In addition, adenovirus has been visualized in epidermal plaques on dab, *L. limanda* (Bloch *et al.* 1986), and Atlantic cod, *Gadus morhua* (Jensen and Bloch 1980). Thus it seems viruses are routinely present in the plaques; yet only Sano, working with carp pox, has actually transmitted a lesion.

One of the first epizootic epidermal papillomas of the globular type was recorded in the European eel, *Anguilla anguilla* in 1910. It was described by Christiansen and Jensen (1950) and a virus etiology was suggested by Schäperclaus (1953). Since then, there have been nine viral isolates (Ahne *et al.* 1987). Five, including the original EV (Berlin) isolate, appear to be Infectious Pancreatic Necrosis Virus (IPHN) and four appear to be eel, *Rhabdovirus anguilla* EVX. There is no evidence that any isolates cause the tumor. Since the stomatopapilloma occurs in high prevalence in polluted rivers and estuaries of Europe, a chemical etiology remains an attractive possibility.

Epizootic, globular epitheliomas of the lip and mouth of brown bullhead, *A. nebulosus*, were first reported in the lower Delaware River by Lucké and Schlumberger (1941). Their continued presence was confirmed more than four decades later by Edward Washuta, New Jersey Division of Fish, Game, and Wildlife, who collected *A. nebulosus* with skin neoplasms at a number of points in the Delaware River from Trenton to Philadelphia (Anonymous 1988). Specimens sent to the RTLA for diagnosis had epidermal papillomas on the jaws; some fish also had hepatocellular neoplasms (Harshbarger and Clark 1990). Brown bullheads have a high prevalence of papillomas and squamous carcinomas from a number of polluted freshwater habitats including lakes in the citrus areas of Florida (Harshbarger 1972), the Buffalo and Niagara Rivers in New York (Black 1983), the Black and Cuyahoga Rivers in Ohio (Baumann *et al.* 1990), the Fox River in Illinois (Brown *et al.* 1973) and Silver Stream reservoir, Newburgh, New York (Bowser *et al.* 1991).

Harshbarger and Wolf (Wolf 1988) were unable to demonstrate virus in tumors from the Florida fish. A 1965 meeting report of a virus isolate by R. Sullivan, R.R. Malsberger, and E.J. Benz has been proven to be unfounded (Wolf 1988). Subsequently, Edwards *et al.* (1977) published a photograph of 50 nm particles with a double envelope in a papilloma from a brown bullhead from Canaan Lake (Long Island, New York), but the tissue had been frozen prior to glutaraldehyde fixation so it cannot be determined if the particles were viruses or artifacts. On the basis of higher epidermal tumor prevalence in heavily polluted waterways compared to waterways believed to be relatively unpolluted, the failure to convincingly demonstrate virus in spite of repeated attempts and especially Black's experiments (Black *et al.* 1985) showing that extracts of polluted sediments induced both skin and liver neoplasms, indicates a chemical influence in the development of brown bullhead epidermal neoplasms.

Two other epizootic bilobular type epidermal neoplasms are associated with polluted environments. These are neoplasms arising from dental epithelium in cunner, *Tautogolabrus adsperus* collected from the Sakonnet River near Portsmouth, Rhode Island (Harshbarger *et al.* 1976). Other than proximity to an industrialized area, however, there is no additional information to support a chemical cause. Also from polluted environments are white croaker, *Geyonemus lineatus*, with papilloma of the lip from near the Santa Monica, California sewage outfall (Russell and Kotin 1957) from the Long Beach, California, harbor area (Young 1964), from near the Santa Ana, California sewage outfall (Harshbarger 1972) and from the Los Angeles, California harbor area (Phillips et al 1976). Russell and Kotin (1957) found no papillomas on white croakers, *G. lineatus*, from unpolluted water 50 miles from Santa Monica. In contrast to the previous examples favoring a chemical etiology for tumors that do not have a plaque type of gross morphology are invasive and metastatic epithelial neoplasms that developed on the mouth, cornea, and operculum of several species of Japanese salmon following experimental infection with herpesvirus isolates (Yoshimizu *et al.* 1987). Actually, there have been three herpesvirus isolates from Japanese salmon: one from unhealthy landlocked *Oncorhynchus nerka* (Sano 1976), one from unhealthy landlocked *Oncorhynchus masou* (Kimura *et al.* 1981a; Kimura *et al.* 1981b) and one from naturally occurring tumors on *Oncorhynchus masou* (Sano *et al.* 1983). However, all three isolates are very closely related if not nearly identical (Hedrick *et al.* 1987). The one most intensely used in experimental studies, *Oncorhynchus masou* virus (OMV), is widespread in rivers, hatcheries, lakes and fish farms of northern Japan (Yoshimizu *et al.* 1988a). By injection, OMV produced tumors in four species of *Oncorhynchus*: coho salmon, *O. kisutch*, chum salmon, *O. keta*, masu salmon, *O. masou*, and rainbow trout, *O. mykiss* (Yoshimizu *et al.* 1988b).

EPIZOOTIC HEMATOPOIETIC NEOPLASMS

While only one epizootic hematopoietic neoplasm (Table 6-3) is known in a wild fish species: northern pike, *E. lucius*, in the Swedish Baltic Sea (Ljungberg 1976) and the Finnish Baltic Sea (Thompson 1982; Bogovski 1988), a close parallel

Table 6-3. Epizootic Hematopoietic Neoplasms in Marine Fish

Species	Site	Key References
Northern pike, *Esox lucius*	Baltic Sea (Swedish) Baltic Sea (Finnish) Baltic Sea (eastern part)	Ljungberg 1976 Thompson 1982 Bogovski 1988
Chinook salmon, *Oncorhynchus tshawytscha*	Sechelt Inlet, B.C., Canada	Kent *et al.* 1990

occurs in chinook salmon, *Oncorhynchus tshawytscha*, raised in netpens in Sechelt Inlet, British Columbia (Kent *et al.* 1990).

The northern pike, *E. lucius*, lymphoma first reported as a single case from Wisconsin (Ohlmacher 1898) was soon reported to be prevalent in Ontario (Higgins and Hadwen 1909; cited by Sonstegard 1976). It appeared to spread among pike maintained in the same aquarium as if caused by an infectious agent (Nigrelli 1947), and was epizootic in rivers in Ireland (Mulcahy 1963).

Disease progression was similar in all populations to that described in Canadian pike. Tumors originated in skin, invaded underlying muscle, and ultimately metastasized to kidney, spleen, and liver in that order (Sonstegard 1976).

Histologically, tumors consist of large unidentified blast cells resembling immunoblasts (Mulcahy *et al.* 1970), but a combination of immunological markers with the fine structure suggest it is in the monocytic lineage (Thompson and Kostiala 1990). It would be consistent with its cutaneous origin to be a tissue histiocyte.

Based on isoenzyme studies (Healy and Mulcahy 1980), there is a low genetic variability among geographically diverse northern pike, *E. lucius*, populations. Likewise, karyotypes of transplanted lymphoma from the Baltic Sea (Ljungberg 1976) and primary lymphoma of North America (Whang-Peng *et al.* 1976) are similar, but both populations consistently had a number of chromosomal aberrations. Van Beneden *et al.* (1990) transfected NIH3T3 mouse fibroblasts with a DNA sequence from pike lymphoma which neoplastically transformed the mouse cells when assayed in athymic mice. However, no specific oncogene has been identified.

Several lines of evidence strongly suggest, but do not confirm, a viral etiology. Cell-free transmission has been reported repeatedly (Mulcahy and O'Leary 1970; Sonstegard 1976; Brown *et al.* 1980) although not all attempts have succeeded (Ljungberg 1980). Additionally, type C virus particles have been visualized in a cytoplasmic fraction of tumor tissue in a sucrose gradient (Papas *et al.* 1976), and this fraction also contained reverse transcriptase activity (Papas *et al.* 1977).

However, virus particles have never been actually visualized in pike tumor tissue, but electron microscopy has revealed concentric membranous RER bodies in the cytoplasm (Banfield *et al.* 1976) that resemble ribosome lamella complexes in some human leukemias and other types of tumor cells.

The chinook salmon, *O. tshawytscha*, tumor discovered and lost in a freshwater hatchery (Harshbarger 1982, 1984) and rediscovered in seawater netpens in Sechelt Inlet, British Columbia (Kent *et al.* 1990) causes mortality in up to 80% of the stock. Affected fish have exophthalmia, cutaneous petechiae, ascites and splenorenomegaly due to a widely disseminated, infiltrative population of immature "plasmacytoid" cells.

The disease was transmitted to chinook, *O. tshawytscha*; sockeye, *Oncorhynchus nerka;*, and Atlantic salmon, *Salmo salar*, but not rainbow trout, *O. mykiss*, or coho salmon, *Oncorhynchus kisutch*, by intraperitoneal injections of pooled homogenated kidneys from affected fish (Kent and Dawe 1990; Newbound and Kent 1991). This implies an infectious etiology because any viable tumor cells left in the homogenate would have been rejected in the interspecific and intergenera inoculations.

EPIZOOTIC CUTANEOUS FIBROCYTIC NEOPLASMS

The only epizootic fibrocytic neoplasms (Table 6-4) reported in the marine environment were in the striped mullet, *Mugil cephalus* from three locations. In the first report two striped mullet (*M. cephalus*), collected from the Galveston, Texas, beachfront and one striped mullet collected from Galveston Bay had multiple, exophytic, grayish-white, pedunculated, lobular masses up to 6.0 cm in greatest dimension in the head/operculum region (Lightner 1974). Histologically they originated in the dermis and consisted of well-differentiated, non-invasive fibrous tissue containing bony spicules. Diagnosis was ossifying fibroma.

Nine striped mullet, *M. cephalus*, with 1 to 10 histologically, nearly similar, dermal fibrocytic neoplasms were collected in Mississippi Sound during 1972 (Edwards and Overstreet 1976) and several more cases have been collected subsequently (Overstreet 1988). These tumors occurred on the trunk and apparently did not contain spicules of bone. They extended well into the stratum compactum but did not reach the underlying hypodermis or skeletal muscle.

Four additional cases reported in striped mullet, *M. cephalus*, from Lake Tunis, Tunisia, (Lopez and Raibaut 1981) were located both on the head and the trunk. They all originated from dermal fibrous tissue and one of the head tumors contained islands of bone. None was invasive.

Evidence for a chemical etiology is weak. While the three collecting sites are all near population centers, mullet migrate in schools and usually do not linger

Table 6-4. Epizootic Cutaneous Fibrocytic Neoplasms in Marine Fish

Species	Site	Key References
Striped mullet, *Mugil cephalus*	Galveston, TX	Lightner 1974
	Mississippi Sound, MS	Edwards and Overstreet 1976
	Tunis Lake, Tunisia	Lopez and Raibaut 1981

in one place. Additionally, no one has reported liver cancer or any of the other neoplasms that are strongly linked to chemical carcinogens in mullet; no one has experimentally induced fibromas with established carcinogens in any fish; and no one has found epizootic fibromas in any fish from the polluted sites where fish liver neoplasms are epizootic. Finally, fishermen have reported that approximately 5% of certain schools of mullet have fibroma-like lesions (Overstreet 1988), which suggests that the disease is endemic in certain populations. Either a small number of fish have a genetically susceptible gene combination, possibly augmented by pollutants or certain populations are infected with an oncogenic virus. This later possibility is made attractive by the reports that dermal sarcoma in the freshwater walleye, *Stizostedion vitreum*, contains retrovirus particles (Walker 1969; Martineau *et al.* 1991) and is transmissible by cell-free filtrate (Bowser *et al.* 1990).

EPIZOOTIC PIGMENT CELL NEOPLASMS

Pigment cell neoplasms (Table 6-5) occur in a wide range of fish species under the influence of genetics, chemicals, or age factors (Masahito *et al.* 1989). Because at least some neoplastic fish pigment cells are capable of expressing one or another of the several pigment types at different times, they are often called "chromatoblastomas" (Matsumoto *et al.* 1981).

The platyfish/swordtail hybrid provides an example of a genetic influence. Laboratory backcrosses that selectively eliminate genes that regulate a pigment gene allow melanophores to proliferate continuously (Anders *et al.* 1984).

Epizootic chromatophoromas in the nibe croaker, *Nibea mitsukurii* (Kimura *et al.* 1984; Kimura *et al.* 1989; Kinae *et al.* 1990), on the other hand are strongly influenced by chemicals. Prevalence ranged from 0-47% at 24 sites along the eastern coast of Japan. Highest prevalence was at a kraft pulp mill and sediment at the mill was mutagenic by the Ames test. Mutagens identified in sediment included 5 chloroacetones, tetrachlorocyclopentene-1,3-dione, and 2 alpha-

Table 6-5. Epizootic Pigment Cell Neoplasms In Marine Fish

Species	Site	Key References
Nibe croaker, *Nibea mitsukurii*	Japan (East Coast)	Kimura *et al.* 1984
Hawaiian butterflyfish, *Chaetodon multicinctus* and *Chaetodon miliaris*	Hawaii	Okihiro 1988
Deepwater redfish, *Sebastes mentella*	North Atlantic Ocean	Bogovski and Bakai 1989

dicarbonyl compounds. After removal of contaminated sediment, chromatophoroma prevalence declined to 20%. Experimental exposure of nibe croaker, *N. mitsukurii*, to 7,12-dimethyl- benz(a)anthracene, 1-methyl-3-nitro-1-nitrosoguanidine and nifurpirinol induces chromatophoromas.

Epizootic chromatophoromas in 2 of 22 species of the Hawaiian butterflyfish, *Chaetodon multicinctus* and *Chaetodon miliaris*, (Okihiro 1988) are histologically similar to the nibe croaker, *N. mitsukurii*, tumors and likewise can exhibit the production of mixed pigment patterns within a given lesion. Tumors range in prevalence up to 50% with higher prevalence in fish populations in close proximity to runoff from agricultural lands.

Epizootic chromatoblastomas in deepwater redfish, *Sebastes mentella*, from the North Atlantic (Bogovski and Bakai 1989), on the other hand, express a different histologic pattern and prevalence from the nibe croaker and the Hawaiian butterflyfish and have no apparent correlation with pollution.

Table 6-6. Epizootic Peripheral Nerve Sheath Cell Neoplasms in Marine Fish

Species	Site	Key References
Gray snapper, *Lutjanus griseus*	Dry Tortugas and Key West, FL	Lucké 1942
Dog snapper, *Lutjanus jocu*	Dry Tortugas and Key West, FL	Lucké 1942
Schoolmaster, *Lutjanus apodus*	Dry Tortugas and Key West, FL	Lucké 1942
Slippery dick, *Halichoeres bivittatus*	Bimini, Bahama Islands	Nigrelli 1949
Bicolor damselfish, *Pomacentrus partitus*	Florida Keys	Schmale *et al.* 1983

EPIZOOTIC PERIPHERAL NERVE SHEATH CELL TUMORS

Schwannomas and neurofibromas (Table 6-6) are among the most common neoplasms seen in wild fish. They exhibit a spectrum of patterns influenced by the nerve of origin and whether they are endoneurial, perineurial or epineurial. Because Schwann cells are closely related to pigment cells embryologically, both normal and neurofibroma Schwann cells have been shown in humans to express melanoma associated antigens (Ross *et al.* 1986) and in fish some pigmented Schwann cell tumors are hard to distinguish histologically from some pigment cell tumors.

Lucké (1942) reported neurilemmoma (schwannoma) in gray snapper, *Lutjanus griseus*, the dog snapper, *Lutjanus jocu*, and the schoolmaster, *Lutjanus apodus*, from the Florida Keys (near the Dry Tortugas and near Key West, Florida. The cause was not determined and transmission studies using the gray snapper were inconclusive.

Nigrelli (1949) reported a 1.57% prevalence of single and multiple invasive cutaneous neural neoplasms in 1,788 slippery dicks, *Halichoeres bivittatus*, collected near Bimini, Bahama Islands. The neoplasms ranged from a few millimeters to several centimeters, occurring in fins as well as being scattered over the skin. Many were pigmented, especially those developing in pigmented areas of the fish. While Nigrelli called them ganglioneuromas, his description was consistent with nerve sheath cell neoplasms described in other fish species.

Schmale *et al.* (1983) reported that epizootic schwannoma in bicolor damselfish, *Pomacentrus partitus*, also occurred in the Florida Keys among the various coral reefs. In fact schwannomas were present in bicolor damselfish, *P. partitus*, from 18 of 19 reefs surveyed with a stable prevalence per given reef ranging from 0.4% to 23.8% (Schmale 1991). While multiple neurofibromas have been experimentally induced in coho salmon, *O. kisutch*, with a carcinogen (Kocan and Landolt 1989), this distribution of tumor prevalence among the reefs, plus the clusters of tumors among subpopulations of bicolor damselfish at the same reef, was not consistent with a chemical influence. Rather, that data supported a genetic susceptibility, possibly mediated in part by immunity (Schmale and McKinney 1987). The additional discovery that the tumors can be experimentally transmitted 84% of the time with injections of homogenized tumor tissue (Schmale and Hensley 1988) strongly suggests a transmissible agent.

CONCLUSION

The presence of epizootic liver cancer in 12 fish species in the marine ecosystem of three continents and oceania is correlated with the proximity of the affected populations to point sources and other concentrations of chemical pollutants in their habitat compared to a virtually zero background rate in non-polluted habitats. Additionally, extracts of sediments from several of these areas with concentrated pollutants have induced tumors in fish when applied experimentally by feeding, painting, or egg injection. This is consistent with experimental carcinogenesis data showing that established carcinogens almost invariably induce liver cancer in fish, indicating that fish liver enzymes are efficient metabolizers of indirect acting carcinogens with the production of electrophilic intermediates that adduct DNA and activate oncogenes. This leads to the conclusion that fish liver cancer is an excellent indicator of the presence of carcinogens in the environment. Epizootic epidermal neoplasms of the plaque, or fish pox, type are strongly associated with viruses, and carp pox has a proven viral etiology.

The bulbous or globular type of epidermal neoplasm, including squamous papillomas, squamous carcinomas, and various odontogenic tumors, is prevalent in polluted environments but less clearly associated with point sources than liver neoplasms. However, evidence for an alternative viral etiology is weak for all globular tumors except those in several species of salmonids in Japan caused by the *Oncorhynchus masou* herpesvirus (OMV). Squamous carcinomas have been induced in brown bullhead, *A. nebulosus*, by skin painting of polluted sediment extracts. The above observations suggest that "fish pox" is viral in origin, while squamous neoplasms are chemical, but that the latter are less precise as indicators of the presence of environmental carcinogens than liver cancer. Epizootic hemic neoplasms are caused by retrovirus with little environmental influence.

Only one fish species with epizootic fibrocytic neoplasm is known from the marine ecosystem. Low numbers have been found in three polluted coastal areas as well as at low levels in some populations collected off coast. Therefore, evidence of a chemical etiology is weak vis-à-vis the alternative possibility of an endemic virus, which is supported by studies of epizootic fibrous neoplasms in walleye, *S. vitreum*, from freshwater habitats. Nevertheless, a chemical influence or a low level of a susceptible gene combination cannot be excluded.

Of three epizootic pigment cell neoplasms in the marine ecosystem, the nibe croaker, *N. mitsukurii*, is clearly associated with polluted environments and experimentally various carcinogens will produce the neoplasms. Hawaiian butterflyfish, *C. multicinctus* and *C. miliaris*, chromatophoromas are more prevalent in polluted habitats but experimental studies have not been done. The deepwater redfish, *S. mentella*, pigment cell tumors have no apparent correlation with pollution. In the freshwater platyfish/swordtail, hybridization can reduce the influence of pigment cell regulator genes on one or more oncogenes, resulting in a continuous proliferation of melanophores -- a process which can be promoted by chemical exposure. This suggests that any exogenous factors that can interact with the genome (i.e., chemicals, viruses, and radiation) can act through multiple points of genetic vulnerability to initiate or promote the formation of pigment cell tumors and that the more susceptible fish species, such as the nibe croaker, *N. mitsukurii*, could be useful sentinels for the presence of environmental carcinogens.

In the marine ecosystem, epizootic neurofibromas and schwannomas occur in bicolor damselfish, *P. partitus*, from coral reefs along the Florida Keys. Prevalence seems to be correlated with specific populations and subpopulations rather than chemical pollution. Experimental carcinogenesis studies have not been done in the bicolor damselfish, *P. partitus*, although neurofibromas have been chemically induced in coho salmon, *O. kisutch*. However, most bicolor damselfish, *P. partitus*, receiving injections of tumor homogenate develop schwannomas, and damselfish, *P. partitus*, with schwannomas have a reduced immune competence. As with pigment cell tumors, this suggests an endemic genetic vulnerability that could be influenced by several exogenous factors. There is strong evidence that one of the possible factors is a virus.

REFERENCES

Ahne, W., I. Schwanz-Pfitzner, and I. Thomsen. 1987. Serological identification of 9 viral isolates from European eels (*Anguilla anguilla*) with stomatopapilloma by means of neutralization tests. *J. Appl. Ichthyol.* 3: 30-32.

Ahne, W., K. Anders, M. Halder, and M. Yoshimizu. 1990. Isolation of picornavirus-like particles from the European smelt, *Osmerus eperlanus* (L.). *J. Fish Dis.* 13: 167-168.

Anders, F., M. Schartl, and A. Barnekow. 1984. *Xiphophorus* as an *in vivo* model for studies on oncogenes. In *Use of Small Fish in Carcinogenicity Testing*: Proceedings of a Symposium at Bethesda, MD, 1981, ed. K.L. Hoover. *Nat. Cancer Inst. Monogr.* 65: 97-109.

Anders, K., and H. Moller. 1985. Spawning papillomatosis of smelt, *Osmerus eperlanus* L., from the Elbe estuary. *J. Fish Dis.* 8: 233-235.

Anonymous. 1988. *Fish Health and Contamination Study*. Delaware Estuary Use Attainability Project, Delaware River Basin Commission, West Trenton, New Jersey (DEL USA Project Element 10), 120 pp.

Banfield, W.G., C.J. Dawe, C.E. Lee, and R. Sonstegard. 1976. Cylindroid lamella-particle complexes in lymphoma cells of northern pike (*Esox lucius*). *J. Nat. Cancer Inst.* 57(2): 415-420.

Baumann, P.C., J.C. Harshbarger, and K.J. Hartman. 1990. Relationship between liver tumors and age in brown bullhead populations from two Lake Erie tributaries. *Sci. Total Environ.* 94: 71-88.

Bekesi, L., E. Kovacs-Gayer, F. Ratz, and O. Turkovics. 1984. Skin infection of the sheatfish (*Silurus glanis* L.) caused by a herpes virus. *Symp. Biol. Hung.* 23: 25-30.

Black, J., H. Fox, P. Black, and F. Bock. 1985. Carcinogenic effects of river sediment extracts in fish and mice. In *Water Chlorination: Volume 5: Chemistry, Environmental Impact and Health Effects*, Proceedings 5th Conference, Williamsburg, June 1984, ed. R.L. Jolley, R.J. Bull, W.P. Davis, S. Katz, M.H. Roberts, Jr., and V.A. Jacobs, pp. 415-428. Chelsea, MI: Lewis Publishers.

Black, J.J. 1983. Field and laboratory studies of environmental carcinogenesis in Niagara River fish. *J. Great Lakes Res.* 9(2): 326-334.

Bloch, B., S. Mellergaard, and E. Nielsen. 1986. Adenovirus-like particles associated with epithelial hyperplasias in dab, *Limanda limanda* (L.). *J. Fish Dis.* 9(3): 281-285.

Bogovski, S.P. 1988. Malignant lymphomas of different histological types in pikes from the eastern part of the Baltic Sea. *Eksp. Onkol.* 10(4): 31-33.

Bogovski, S.P., and Y.I. Bakai. 1989. Chromatoblastomas and related pigmented lesions in deepwater redfish, *Sebastes mentella* (Travin), from North Atlantic areas, especially the Irminger Sea. *J. Fish Dis.* 12: 1-13.

Bowser, P.R., D. Martineau, and G.A. Wooster. 1990. Effects of water temperature on experimental transmission of dermal sarcoma in fingerling walleyes. *J. Aquat. Anim. Health* 2: 157-161.

Bowser, P.R., M.J. Wolfe, J. Reimer, and B.S. Shane. 1991. Epizootic papillomas in brown bullheads, *Ictalurus nebulosus* from Silver Stream Reservoir, New York. *Dis. Aquat. Org.* 11: 117-127.

Breslauer, T. 1916. Zur Kenntnis der Epidermoidalgeschwuelste von Kaltbluetern. Histologische Veranderungen des Integuments und der Mundschleimhaut beim Stint (*Osmerus eperlanus* L.). *Arch. Mikrosk. Anat.* 87(1): 200-264.

Brown, E.R., J.J. Hazdra, L. Keith, I. Greenspan, J.B.G. Kwapinski, and P. Beamer. 1973. Frequency of fish tumors found in a polluted watershed as compared to nonpolluted Canadian waters. *Cancer Res.* 33(2): 189-198.

Brown, E.R., L. Keith, J.J. Hazdra, P. Beamer, O. Callaghan, and V. Nair. 1980. Water pollution and its relationship to lymphomas in poikilotherms. In *Advances in Comparative Leukemia Research 1979*, ed. Yohn, D.S., B.A. Lapin, and J.R. Blakeslee, pp. 209-210. New York: Elsevier/North-Holland.

Buchanan, J.S., and C.R. Madeley. 1978. Studies on Herpesvirus scophthalmi infection of turbot *Scophthalmus maximus* (L.) ultrastructural observations. *J. Fish Dis.* 1: 283-295.

Bucke, D., B. Watermann, and S. Feist. 1984. Histological variations of hepato-splenic organs from the North Sea dab, *Limanda limanda*. *J. Fish Dis.* 7: 255-268.

Bunton, T.E., and S.M. Baksi. 1988. Cholangioma in white perch *Morone americana* from the Chesapeake Bay. *J. Wildl. Dis.* 24: 137-141.

Cairns, V.W., and J.D. Fitzsimons. 1988. The occurrence of epidermal papillomas and liver neoplasia in white suckers (*Catostomus commersoni*) from Lake Ontario. *Can. Tech. Rep. Fish Aquat. Sci.* No. 1607: 151-152.

Chang, Y.-J., C. Mathews, K. Mangold, K. Hendricks, J. Marien, G. Bailey. 1991. Analysis of *ras* gene mutations in rainbow trout liver tumors initiated by aflatoxin B1. *Mol. Carcinog.* 4: 112-119.

Christiansen, M., and A.J.C. Jensen. 1950. On a recent and frequently occurring tumor disease in eel. I. Occurrence of the disease in the various years and its distribution. II. Investigations of the tumors. *Report of the Danish Biological Station to the Ministry of Agriculture and Fisheries* 50: 31-44.

Cormier, S.M., and R.N. Racine. 1990. Histopathology of Atlantic tomcod: a possible monitor of xenobiotics in northeast tidal rivers and estuaries. In *Biomarkers of Environmental Contamination*, ed. J.F. McCarthy and L.R. Shugart, pp. 59-71. Boca Raton, FL: Lewis Publishers.

Couch, J.A., and J.C. Harshbarger. 1985. Effects of carcinogenic agents on aquatic animals: an environmental and experimental overview. *Environ. Carcinog. Rev.* 3(1): 63-105.

Croy, R.G., J.E. Nixon, R.O. Sinnhuber, and G.N. Wogan. 1980. Investigation of covalent aflatoxin B1-DNA adducts formed in vivo in rainbow trout (*Salmo gairdneri*) embryos and liver. *Carcinogenesis (Lond.)* 1(11): 903-909.

Dunn, B.P., J.J. Black, and A.E. Maccubbin. 1987. 32P-postlabeling detection of aromatic carcinogen: DNA adducts in liver from fish from polluted areas. *Cancer Res.* 47: 6543-6548.

Edwards, M.R., W.A. Samsonoff, and E.J. Kuzia. 1977. Papilloma-like viruses from catfish. *Fish Health News* 6: 94-95.

Edwards, R.H., and R.M. Overstreet. 1976. Mesenchymal tumors of some estuarine fishes of the northern Gulf of Mexico. 1. Subcutaneous tumors, probably fibrosarcomas in the striped mullet, *Mugil cephalus*. *Bull. Mar. Sci.* 26(1): 33-40.

Gardner, G.R., R.J. Pruell, and L.C. Folmar. 1989. A comparison of both neoplastic and non-neoplastic disorders in winter flounder (*Pseudopleuronectes americanus*) from eight areas in New England. *Mar. Environ. Res.* 28: 393-397.

Gardner, G.R., P.P. Yevich, J.C. Harshbarger, and A.R. Malcolm. 1991. Carcinogenicity of Black Rock Harbor sediment to the eastern oyster and trophic transfer of Black Rock Harbor carcinogens from the blue mussel to the winter flounder. *Environ. Health Perspect.* 90: 53-66.

Goyette, D., D. Brand, and M. Thomas. 1986. Prevalence of idiopathic liver lesions in English sole and epidermal abnormalities in flatfish from Vancouver Harbour, British Columbia. Environment Canada, Conservation and Protection, Environmental Protection, Pacific and Yukon Region, Regional Program Report 87-09.

Harshbarger, J.C. 1972. Work of the Registry of Tumours in Lower Animals with emphasis on fish neoplasms. *Symp. Zool. Soc. Lond.* 30: 285-303.

Harshbarger, J.C. 1982. Epizootiology of leukemia and lymphoma in poikilotherms. In *Advances in Comparative Leukemia Research 1981*, ed. D.S. Yohn and J.R. Blakeslee, pp. 39-46. Elsevier Biomedical: New York.

Harshbarger, J.C. 1984. Pseudoneoplasms in ectothermic animals. In *Use of Small Fish in Carcinogenicity Testing*: Proceedings of a Symposium at Bethesda, MD, 1981, ed. K.L. Hoover. *Nat. Cancer Inst. Monogr.* 65: 251-273.

Harshbarger, J.C., and J.B. Clark. 1990. Epizootiology of neoplasms in bony fish from North America. *Sci. Total Environ.* 94: 1-32.

Harshbarger, J.C., S.E. Shumway, and G.W. Bane. 1976. Variable differentiating oral neoplasms, ranging from epidermal papilloma to odontogenic ameloblastoma, in cunners [(*Tautogolabrus adspersus*) Osteichthyes; Perciformes: Labridae]. In *Tumors in Aquatic Animals*, ed. C.J. Dawe, D.G. Scarpelli, and S.R. Wellings, pp. 113-128. Basel: Karger.

Healy, J.A., and M.F. Mulcahy. 1980. A biochemical genetic analysis of populations of the northern pike *Esox lucius* L. from Europe and N. America. In *Advances in Comparative Leukemia Research 1979*, ed. D.S. Yohn, B.A. Lapin, and J.R. Blakeslee, pp. 211-212. Elsevier/North-Holland: New York.

Hedrick, R.P., T. McDowell, W.D. Eaton, T. Kimura, and T. Sano. 1987. Serological relationships of five herpesviruses isolated from salmonid fishes. *J. Appl. Ichthyol.* 3(2): 87-92.

Jensen, N.J., and B. Bloch. 1980. Adenovirus-like particles associated with epidermal hyperplasia in cod (*Gadus morhua*). *Nord. Veterinaermed.* 32: 173-175.

Kent, M.L., and S.C. Dawe. 1990. Experimental transmission of a plasmacytoid leukemia of chinook salmon, *Oncorhynchus tshawytscha*. *Cancer Res. (suppl.)* 50: 5679s-5681s.

Kent, M.L., J.M. Groff, G.S. Traxler, J.G. Zinkl, and J.W. Bagshaw. 1990. Plasmacytoid leukemia in seawater reared chinook salmon *Oncorhynchus tshawytscha*. *Dis. Aquat. Org.* 8: 199-209.

Kimura, I., N. Kinae, H. Kumai, M. Yamashita, G. Nakamura, M. Ando, H. Ishida, and I. Tomita. 1989. Environment: Peculiar pigment cell neoplasm in fish. *J. Invest. Dermatol.* 92: 248S-254S.

Kimura, I., N. Taniguchi, H. Kumai, I. Tomita, N. Kinae, K. Yoshizaki, M. Ito, and T. Ishikawa. 1984. Correlation of epizootiological observations with experimental data: Chemical induction of chromatophoromas in the croaker, *Nibea mitsukurii*. In *Use of Small Fish in Carcinogenicity Testing*: Proceedings of a Symposium at Bethesda, MD, 1981, ed. K.L.Hoover. *Nat. Cancer Inst. Monogr.* 65: 139-154.

Kimura, T., M. Yoshimizu, and M. Tanaka. 1981a. Studies on a new virus (OMV) from *Oncorhynchus masou*--II. Oncogenic nature. *Fish Pathol.* 15(3/4): 149-153.

Kimura, T., M. Yoshimizu, M. Tanaka, and H. Sannohe. 1981b. Studies on a new virus (OMV) from *Oncorhynchus masou*–I. Characteristics and pathogenicity. *Fish Pathol.* 15(3/4): 143-147.

Kinae, N., M. Yamashita, I. Tomita, I. Kimura, H. Ishida, H. Kumai, and G. Nakamura. 1990. A possible correlation between environmental chemicals and pigment cell neoplasia in fish. *Sci. Total Environ.* 94: 143-154.

Kocan, R.M., and M.L. Landolt. 1989. Survival and growth to reproductive maturity of coho salmon following embryonic exposure to a model toxicant. *Mar. Environ. Res.* 27: 177-193.

Kranz, H., and V. Dethlefsen. 1990. Liver anomalies in dab *Limanda limanda* from the southern North Sea with special consideration given to neoplastic lesions. *Dis. Aquat. Org.* 9: 171-185.

Lightner, D.V. 1974. Case reports of ossifying fibromata in the striped mullet. *J. Wildl. Dis.* 10: 317-320.

Ljungberg, O. 1980. Skin tumours in northern pike (*Esox lucius* L.): transmission and immunization studies. In *Advances in Comparative Leukemia Research 1979*, ed. D.S. Yohn, B.A. Lapin, and J.R. Blakeslee, pp. 213-214. Elsevier/North-Holland: New York.

Ljungberg, O. 1976. Epizootiological and experimental studies of skin tumours in northern pike (*Esox lucius* L.) in the Baltic Sea. In *Progress in Experimental Tumor Research*, vol. 20, ed. F. Homburger, *Tumors in Aquatic Animals*, ed. C.J. Dawe, D.G. Scarpelli, and S.R. Wellings, pp. 156-165. Basel: Karger.

Lopez, A. and A. Raibaut. 1981. Multiple cutaneous fibromas in a mullet, *Mugil cephalus cephalus* L. *J. Fish Dis.* 4(2): 169-174.

Lucké, B. 1938. Studies on tumors in cold-blooded vertebrates. *Annu. Rep. Tortugas Lab.*, Carnegie Inst., Washington, DC, 1937-1938: 92-94.

Lucké, B. 1942. Tumors of the nerve sheath in fish of the snapper family (Lutianidae). *Arch. Pathol.* 34(1): 133-150.

Lucké, B., and H.G. Schlumberger. 1941. Transplantable epitheliomas of the lip and mouth of catfish. I. Pathology. Transplantation to anterior chamber of eye and into cornea. *J. Exp. Med.* 74(5): 397-408 and Figure Plates 18-22.

Maccubbin, A.E., J.J. Black, and B.P. Dunn. 1990. 32P-postlabeling detection of DNA adducts in fish from chemically contaminated waterways. *Sci. Total Environ.* 94: 89-104.

Malins, D.C., B.B. McCain, D.W. Brown, M.S. Myers, M.M. Krahn, and S-L. Chan. 1987. Toxic chemicals, including aromatic and chlorinated hydrocarbons and their derivatives, and liver lesions in white croaker (*Genyonemus lineatus*) from the vicinity of Los Angeles. *Environ. Sci. Technol.* 21(8): 765-770.

Mangold, K., Y.-J. Chang, C. Mathews, K. Marien, J. Hendricks, and G. Bailey. 1991. Expression of *ras* genes in rainbow trout liver. *Mol. Carcinog.* 4: 97-102.

Martineau, D., R. Renshaw, J.R. Williams, J.W. Casey, and P.R. Bowser. 1991. A large unintegrated retrovirus DNA species present in a dermal tumor of walleye *Stizostedion vitreum*. *Dis. Aquat. Org.* 10: 153-158.

Masahito, P., T. Ishikawa, and H. Sugano. 1989. Pigment cells and pigment cell tumors in fish. *J. Invest. Dermatol.* 92: 266S-270S.

Matsumoto, J., T. Ishikawa, P. Masahito, A. Oikawa, and S. Takayama. 1981. Multiplicity in phenotypic expression of fish erythrophoroma and irido-melanophoroma cell *in vitro*. In *Phyletic Approaches to Cancer*, ed. C.J. Dawe, J.C. Harshbarger, S. Kondo, and S. Takayama, pp. 253-266. Tokyo: Japan Scientific Societies Press.

May, E.B., R. Lukacovic, H. King, and M. Lipsky. 1987. Hyperplastic and neoplastic alterations in the livers of white perch (*Morone americana*) from the Chesapeake Bay. *JNCI. J. Nat. Cancer Inst.* 79(1): 137-143.

McArn, G.E., B. McCain, and S.R. Wellings. 1978. Skin lesions and associated virus in Pacific cod (*Gadus macrocephalus*) in the Bering Sea. *Fed. Proc.* 37(3): 937.

McCain, B.B., K.V. Pierce, S.R. Wellings, and B.S. Miller. 1977. Hepatomas in marine fish from an urban estuary. *Bull. Environ. Contam. Toxicol.* 18(1): 1-2.

McMahon, G., L.J. Huber, M.J. Moore, J.J. Stegeman, and G.N. Wogan. 1990. Mutations in c-Ki-*ras* oncogenes in diseased livers of winter flounder from Boston Harbor. *Proc. Nat. Acad. Sci. U.S.A.* 87: 841-845.

Metcalfe, C.D., V.W. Cairns, and J.D. Fitzsimons. 1988. Experimental induction of liver tumours in rainbow trout (*Salmo gairdneri*) by contaminated sediment from Hamilton Harbour, Ontario. *Can. J. Fish Aquat. Sci.* 45: 2161-2167.

Mulcahy, M.F. 1963. Lymphosarcoma in the pike, *Esox lucius* L., (Pisces; Esocidae) in Ireland. *Proc. R. Ir. Acad. Sect. B Biol. Geol. Chem. Sci.* 63(7): 129.

Mulcahy, M.F., and A. O'Leary. 1970. Cell-free transmission of lymphosarcoma in the northern pike *Esox lucius* L. (Pisces; Esocidae). *Experientia (Basel).* 26: 891.

Mulcahy, M.F., G. Winqvist, and C.J. Dawe. 1970. The neoplastic cell type in lymphoreticular neoplasms of the northern pike, *Esox lucius* L. *Cancer Res.* 30(11): 2712-2717.

Murchelano, R.A., and R.E. Wolke. 1991. Neoplasms and nonneoplastic liver lesions in winter flounder, *Pseudopleuronectes americanus*, from Boston Harbor, Massachusetts. *Environ. Health Perspect.* 90: 17-26. (Symposium on Chemically Contaminated Aquatic Food Resources and Human Cancer Risk, September 29 and 30, 1988, National Institute of Environmental Health Sciences, Research Triangle Park, NC).

Myers, M.S., and L.D. Rhodes. 1988. Morphologic similarities and parallels in geographic distribution of suspected toxicopathic liver lesions in rock sole (*Lepidopsetta bilineata*), starry flounder (*Platichthys stellatus*), Pacific staghorn sculpin (*Lepatocottus armatus*), and Dover sole (*Microstomus pacificus*) as compared to English sole (*Parophrys vetulus*) from urban and non-urban embayments in Puget Sound, Washington. *Aquat. Toxicol. (Amst)* 11: 410-411.

Newbound, G.C., and M.L. Kent. 1991. Experimental interspecies transmission of plasmacytoid leukemia in salmonid fishes. *Dis. Aquat. Org.* 10: 159-166.

Nigrelli, R.F. 1947. Spontaneous neoplasms in fish. III. Lymphosarcoma in *Astyanax* and *Esox*. *Zoologica (N Y).* 32(2): 101-108.

Nigrelli, R.F. 1949. Studies on spontaneous neoplasma in fishes. IV. Ganglioneuroma in the marine fish, *Halichoeres bivittatus* (Bloch), from Bimini, B.W.I. *Cancer Res.* 9(10): 615 (abstr).

Nunez, O., J.D. Hendricks, and A.T. Fong. 1990. Inter-relationships among aflatoxin B1 (AFB1) metabolism, DNA-binding, cytotoxicity, and hepatocarcinogenesis in rainbow trout *Oncorhynchus mykiss*. *Dis. Aquat. Org.* 9: 15-23.

Ohlmacher, A.P. 1898. Several examples illustrating comparative pathology of tumors. I. Round-celled sarcomata in a fish. II. An osteo-sarcoma in a frog. *Bull. Ohio Hosp. Epileptics* 1: 223-239.

Okihiro, M.S. 1988. Chromatophoromas in two species of Hawaiian butterflyfish, *Chaetodon multicinctus* and *C. miliaris*. *Vet. Pathol.* 25: 422-431.

Overstreet, R.M. 1988. Aquatic pollution problems, Southeastern US coasts: histopathological indicators. *Aquat. Toxicol. (Amst).* 11: 213-239.

Papas, T.S., J.E. Dahlberg, and R.A. Sonstegard. 1976. Type C virus in lymphosarcoma in northern pike *Esox lucius*. *Nature (London)* 261(5560): 506-508.

Papas, T.S., T.W. Pry, M.P. Schafer, and R.A. Sonstegard. 1977. Presence of DNA polymerase in lymphosarcoma in northern pike (*Esox lucius*). *Cancer Res.* 37(9): 3214-3217.

Phillips, M.L., N.E. Warner, and H.W. Puffer. 1976. Oral papillomas in *Genyonemus lineatus* (white croakers). Etiological considerations. In *Progress in Experimental Tumor Research*, vol. 20, ed. F. Homburger, *Tumors in Aquatic Animals*, ed. C.J. Dawe, D.G. Scarpelli, and S.R. Wellings, pp. 108-112. Basel: Karger.

Ross, A.H., D. Pleasure, B.A. Sonnenfeld, B. Kreider, D.M. Jackson, E.S. Taff, R.P. Lisak, and H. Koprowski. 1986. Expression of melanoma-associated antigens by normal and neurofibroma Schwann cells. *Cancer Res.* 46: 5885-5892.

Russell, F.E., and P. Kotin. 1957. Squamous papilloma in the white croaker. *J. Nat. Cancer Inst.* 18(6): 847-861.

Sano, T. 1976. Viral diseases of cultured fishes in Japan. *Fish Pathol.* 10: 221-226.

Sano, T., H. Fukuda, and M. Furukawa. 1985. *Herpesvirus cyprini*: biological and oncogenic properties. *Fish Pathol.* 20(2/3): 381-388.

Sano, T., H. Fukuda, N. Okamoto, and F. Kaneko. 1983. Yamame tumor virus: lethality and oncogenicity. *Bull. JPN. Soc. Sci. Fish.* 49(8): 1159-1163.

Schaperclaus, W. 1953. Die Blumenkohlkrankheit der aale und anderer Fische der Ostee. *Z. Fisch.* 1(NS): 105-124.Schmale, M.C. 1991. Prevalence and distribution patterns of tumors in bicolor damselfish (*Pomacentrus partitus*) on South Florida reefs. *Mar. Biol. (Berl).* 109(2): 203-212.

Schmale, M.C., G. Hensley, and L.R. Udey. 1983. Neurofibromatosis, von Recklinghausen's disease, multiple schwannomas, malignant schwannomas. Multiple schwannomas in the bicolor damselfish, *Pomacentrus partitus* (Pisces, Pomacentridae). *Am. J. Pathol.* 112(2): 238-241.

Schmale, M.C., and G.T. Hensley. 1988. Transmissibility of a neurofibromatosis-like disease in bicolor damselfish. *Cancer Res.* 48: 3828-3833.

Schmale, M.C., and E.C. McKinney. 1987. Immune responses in the bicolor damselfish, *Pomacentrus partitus* and their potential role in the development of neurogenic tumors. *J. Fish Biol.* 31(Suppl A): 161-166.

Smith, C.E., T.H. Peck, R.J. Klauda, and J.B. McLaren. 1979. Hepatomas in Atlantic tomcod *Microgadus tomcod* (Walbaum) collected in the Hudson River Estuary in New York. *J. Fish Dis.* 2: 313-319.

Sonstegard, R.A. 1976. Studies of the etiology and epizootiology of lymphosarcoma in Esox (*Esox lucius* L. and *Esox masquinongy*). In *Progress in Experimental Tumor Research*, vol. 20, ed. F. Homburger, *Tumors in Aquatic Animals*, ed. C.J. Dawe, D.G. Scarpelli, and S.R. Wellings, pp. 141-155. Basel: Karger.

Stegeman, J.J., and J.J. Lech. 1991. Cytochrome P-450 monooxygenase systems in aquatic species: carcinogen metabolism and biomarkers for carcinogen and pollutant exposure. *Environ. Health Perspect.* 90: 101-109.

Thiyagarajah, A., and M.E. Bender. 1988. Lesions in the pancreas and liver of an oyster toadfish, *Opsanus tau* (L.), collected from the lower York River, Virginia, USA. *J. Fish Dis.* 11: 359-364.

Thompson, J.S. 1982. An epizootic of lymphoma in northern pike, *Esox lucius* L., from the Aland Islands of Finland. *J. Fish Dis.* 5: 1-11.

Thompson, J.S., and A.A.I. Kostiala. 1990. Immunological and ultrastructural characterization of true histiocytic lymphoma in the northern pike, *Esox lucius* L. *Cancer Res. (suppl.)* 50: 5668s-5670s.

Van Beneden, R.J., K.W. Henderson, D.G. Blair, T.S. Papas, and H.S. Gardner. 1990. Oncogenes in hematopoietic and hepatic fish neoplasms. *Cancer Res. (suppl.)* 50: 5671s-5674s.

Varanasi, U., W.L. Reichert, and J.E. Stein. 1989. P-Postlabeling analysis of DNA adducts in liver of wild English sole and winter flounder. *Cancer Res.* 49: 1171-1177.

Vogelbein, W.K., J.W. Fournie, P.A. Van Veld, and R.J. Huggett. 1990. Hepatic neoplasms in the mummichog *Fundulus heteroclitus* from a creosote-contaminated site. *Cancer Res.* 50: 5978-5986.

Walker, R. 1969. Virus associated with epidermal hyperplasia in fish. In *Neoplasms and Related Disorders of Invertebrate and Lower Vertebrate Animals. Nat. Cancer Inst. Monogr.* 31: 195-208.

Whang-Peng, J., R.A. Sonstegard, and C.J. Dawe. 1976. Chromosomal characteristics of malignant lymphoma in northern pike (*Esox lucius*) from the United States. *Cancer Res.* 36(10): 3554-3560.

Winqvist, G., O. Ljungberg, and B. Hellstroem. 1968. Skin tumours of northern pike (*Esox lucius* L.) II. Viral particles in epidermal proliferations. *Bull. Off. Int. Epizoot.* 69(7/8): 1023-1031.

Wirgin, I., D. Currie, and S.J. Garte. 1989. Activation of the K-*ras* oncogene in liver tumors of Hudson River tomcod. *Carcinogenesis (London)* 10: 2311-2316.

Wolf, K. 1988. *Fish Viruses and Fish Viral Diseases*, 476 pp. London and Ithaca: Comstock Publishing Associates, Cornell University Press.

Yamamoto, T., R.K. Kelly, and O. Nielsen. 1984. Epidermal hyperplasia of northern pike (*Esox lucius*) associated with herpesvirus and C-type particles. *Arch. Virol.* 79: 255-272.

Yamamoto, T., R.K. Kelly, and O. Nielsen. 1985. Epidermal hyperplasia of walleye, *Stizostedion vitreum vitreum* (Mitchill), associated with retrovirus-like type-C particles: prevalence, histologic and electron microscopic observations. *J. Fish Dis.* 19: 425-436.

Yoshimizu, M., T. Nomura, T. Awakura, and T. Kimura. 1988a. Incidence of fish pathogenic viruses among anadromous salmonid in northern part of Japan (1976-1986). *Sci. Rep. Hokkaido Salmon Hatchery* (42): 1-20.

Yoshimizu, M., M. Tanaka, and T. Kimura. 1988b. Histopathological study of tumors induced by *Oncorhynchus masou* virus (OMV) infection. *Fish Pathol.* 23: 133-138.

Yoshimizu, M., M. Tanaka, and T. Kimura. 1987. *Oncorhynchus masou* virus (OMV): Incidence of tumor development among experimentally infected representative salmonid species. *Fish Pathol.* 22: 7-10.

Young, P.H. 1964. Some effects of sewer effluent on marine life. *Calif. Fish Game.* 50(1): 33-41.

7

Toxicologic Histopathology of Fishes: A Systemic Approach and Overview

David E. Hinton

INTRODUCTION

Wild fish flesh is an important component of human nutrition worldwide (Report Committee on Aquaculture 1978). Many species spend their infancy in estuaries and coastal zones associated with densities of urban populations and industrial complexes (Möller 1985; McCain *et al.* 1988; O'Connor and Huggett 1988; Kimura 1988). Sediments of many estuaries are contaminated (Malins *et al.* 1984; 1985a,b) and composite xenobiotics may threaten human nutrition and health directly or indirectly. Directly, loss of the food source may arise through acute toxicity to early life stages of fishes or through chronic toxicity by impairment of reproductive potential or by cancer and premature death of older, productive fish. Indirectly, human contaminant loading may occur through consumption of tainted fish flesh. The application of pathobiologic approaches to monitor this threat and to prioritize abatement procedures is needed.

Environmental studies using fish as sentinel organisms are not new with investigations reported of varying magnitude (Wellings 1969; Sparks 1972; Hodgins *et al.* 1977; Brown *et al.* 1979; Sindermann 1979; Christensen 1980; Murchelano 1982). In each of these, analysis of histological alterations played a major role. Unfortunately, assessing either exposure to or effects of environmental contaminants in feral fishes is a process fraught with uncertainties. McCarthy and Shugart (1990a) reviewed difficulties in estimating the extent of exposure to toxic chemicals in the environment and attributing adverse health or ecological effects to such exposure. Due to the above, interest is growing in an approach that evaluates exposure and effects of environmental chemicals by the use of biological markers. These have been defined (McCarthy and Shugart 1990a,b) as "measurement of body fluids, cells, or tissues that indicate in

0-8493-8662-4/93/$0.00 + $.50

biochemical or cellular terms the presence of contaminants or the magnitude of the host response." Histopathology has a major role in this approach. Recently, Hinton *et al.* (1991a) evaluated various histopathologic biomarkers and categorized them as present or future. Present markers were produced in laboratory exposures of fishes to chemical toxicants and verified in field studies. In addition, Darrel Laurén and I reviewed this subject, using examples from liver (Hinton and Laurén 1990a) gill, kidney, spleen, skin, and skeletal systems (Hinton and Laurén 1990b). These and the excellent review by Meyers and Hendricks (1985) led to the organization and approach used herein.

Other reviews (Ribelin and Migaki 1975; McCarthy and Shugart 1990a; Ellis 1985; Murty 1986; Sindermann 1990; Adams 1990; Ferguson 1989) enabled a systemic overview and an appraisal of areas that require additional effort.

With a thorough prior knowledge of normal anatomy, the investigator uses histological analysis to detect alterations in tissues and organs (Hinton and Couch 1984) caused by exposure to toxicants. When concentration of a toxicant is sufficient to result only in cellular injury, not death, sublethal (adaptive) changes may be observed in affected cells. For a review of these alterations, see Trump *et al.* (1980). On the other hand, death of cells without death of the organism (as in some forms of acute toxicity) is followed by a series of cellular reactions and host responses. When tissues are properly fixed immediately after the animal is euthanized, toxicant-induced antemortem necrosis can be differentiated from postmortem changes (Trump *et al.* 1980) in the overall organ. Hinton (1990) described common histological techniques for fish tissues and detailed processing methodology and commonly used stains. A partial listing of major fish histology references includes: Bucke (1972), Ashley (1975), Grizzle and Rogers (1976), Ellis *et al.* (1978), Groman (1982), Kubota *et al.* (1982), Yasutake and Wales (1983), Hinton *et al.* (1984), Meyers and Hendricks (1985), and Ferguson (1989). The review of a necropsy-based approach to assess fish health and condition (Goede and Barton 1990) presents procedures and a computerized data storage and retrieval system for systemic gross evaluation. When followed by histopathologic assessment and data management, condition indices for populations from specific sites can be used to monitor environmental change.

Use of the histopathological approach in toxicologic pathology of fishes has a number of strengths. A number of important tissues may be simultaneously studied while maintaining *in situ* cellular, tissue, and organ system relationships. Maintenance of spatial relationships is required to appreciate biological effects associated with toxicity in localized portions of an organ and the subsequent derangement(s) in fluids, tissues, or cells at other locations. While homogenates of entire organs may be used to chemically detect and estimate levels of marker enzymes and reflect change in specific organelles, detection of the cell(s) responsible for such change is not possible due to the inclusion and mixing of various cell types within the homogenate (Bolender 1978). Contrast this with the histologic section retaining *in situ* relationships and, in the case of liver, revealing alteration within biliary cells without obvious change in hepatocytes and other cell types. These results steer the physiologist and biochemist to appropriate tests (e.g., biliary retention times or identification of metabolites or serum analyses of substances normally cleared in the bile). When such interdisciplinary integration is achieved in toxicologic investigations, relevant biomarker effects and mechanistic considerations are likely to follow (Hinton and Laurén 1990a,b).

Perhaps no other approach enables examination of many potential sites of injury so rapidly. This feature was particularly well illustrated by Wester and Canton (1986) in medaka, *Oryzias latipes*, after long-term exposure to β-hexachlorocyclohexane. This estrogen mimetic caused alterations in gonads, liver, kidney, pituitary, thyroid, spleen, and heart. Affected portions of organs were known targets of estrogen or involved in the metabolism of estrogen-stimulated products.

Histopathologic analysis is "user friendly" for the field investigator. In most applications, fixed tissues may be stored until it is convenient to process them further. The rapid detection of *in vivo* toxicity helps to prioritize sites for more detailed analysis. With fish not easily captured or reared, histopathology is the only viable alternative to other types of analysis. Histopathologic analysis yields data on a number of organ systems and permits localization of lesions within specific cell types. The other emerging importance of histopathologic methods relates to assessment of changes in animals too small to dissect for biochemical studies. In addition to eggs and larvae, adults of many small fish species that have value as sentinel organisms may be studied in this way. With conventional histopathology, it is routinely possible not only to evaluate target organ alteration but determine sex and reproductive status of affected animals.

Potential sources of error in feral fish histopathology exist. Specific criteria for classification of toxicologic alterations of fishes are emerging. Since studies range among a variety of approaches at varying levels of resolution and among lesions in an assortment of tissues, variation in inter-investigator diagnosis is possible. One of the purposes of this volume is to address infectious as well as toxicologic pathology. With focused workshops and the inclusion of additional fish histopathologic findings in the toxicologic pathology literature, a sufficient baseline will arise. Already, microcomputers and appropriate software acquire quantitative morphologic data (i.e., morphometry and stereology) within a reasonable timeframe obviating the subjective nature of conventional morphologic studies (Rohr *et al.* 1976; Weibel 1979, 1980; Reide and Reith 1980; Bolender 1981; Loud and Anversa 1984). When more of these approaches are incorporated, integration of morphologic with biochemical and physiologic approaches will strengthen the findings and our interpretation.

Seasonal and hormonal changes, i.e., inter- and intraspecific anatomical variations in the range of normality, should be considered to prevent errors in histopathologic analysis. For accurate diagnoses, considerations of normal seasonal variations, gender, and hormonal differences are important. However, these variations are reflected in only one or two levels of structural organization and do not negate acquisition of meaningful data at other levels. For example, glycogen and lipid levels in hepatocyte cytoplasm are likely to change seasonally or in relation to the reproductive cycle (van Bohemen *et al.* 1981). Despite this, the basic architectural pattern of the liver is not altered and detection of many important toxicologic lesions is not compromised.

Infectious disease and parasitism must be considered before implicating a chemical etiology. Viruses, bacteria, fungi, protozoa, and metazoan parasites may cause degenerative and necrotizing lesions in various organs. These potentially confounding issues have been reviewed (Hinton *et al.* 1991a). Fortunately, the causes of these lesions can often be determined histologically by visualization of the offending organism or the resultant inflammatory response. Some viral infections, however, can result in severe parenchymal and epithelial necrosis with minimal or no inflammation. Light microscopical examination for inclusions (nuclear and cytoplasmic) and electron microscopical (EM) examination for viral particles will help eliminate the latter as potential etiologic agents for necrotizing lesions.

Viruses induce certain neoplastic lesions causing epithelial tumors in masu salmon, *Oncorhynchus masou*, (Kimura *et al.* 1981a,b; Yoshimizu *et al.* 1987) and perhaps lymphoma in northern pike, *Esox lucius* (Mulcahy and O'Leary 1970; Papas *et al.* 1976, 1977). Examination of tumors may demonstrate a possible viral etiology. However, oncogenic retroviruses are often non-productive once infected cells have undergone neoplastic transformation; therefore lack of detection of viral particles does not necessarily negate viral causation. On the other hand, papillomas of eels, *Anguilla anguilla*, contained viruses that have not been shown to be oncogenic (McAllister *et al.* 1977; Ahne and Thomsen 1985). Assays for reverse transcriptase and transmission studies, in addition to EM, may rule out viruses as etiologic agents of specific neoplasms (Hinton *et al.* 1991a).

Two pseudoneoplastic conditions, lymphocystis and X-cell pseudotumors, can be confused with chemically induced epizootics in wild fish. The hypertrophied fibroblasts of lymphocystis are distinctive and the causative agent can be diagnosed with EM screening. X-cell pseudotumors can be differentiated from true neoplasms (Dawe 1981) and the accompanying inflammatory reaction. For a review of conditions resembling but not representing true neoplasms, see Harshbarger (1984).

Ectoparasites (ciliated protozoa and metazoa) induce lesions in fins, skin, and gills that mimic those induced by either water-borne or sediment-deposited chemicals. Cutaneous hyperplasia, erosion, and ulceration are caused by a variety of ectoparasites. Similarly, ectoparasites in the gill have been associated with mucous and epithelial cell hyperplasia, lamellar capillary aneurysms, and lamellar clubbing and fusion. Examinations of the skin and gill with wet mount preparations, at the time of collection, will help eliminate ectoparasites as the cause of noninflammatory cutaneous and gill lesions. Certain lesions, such as respiratory epithelial hyperplasia in gill filaments and lamellae, can be the result of *past* ectoparasitic infestations, e.g., by copepods, so that the absence of infectious or parasitic agents even as demonstrated in a wet mount does not fully eliminate ectoparasites as etiologic agents of the condition (Hinton et al. 1991a).

Residual lesions from infectious disease including parasitism are additional sources of uncertainty. Macrophage aggregates, commonly found in the kidney, spleen, liver, and heart of fish exposed to chemical toxicants, are also abundant in fish with healing bacterial, fungal, or protozoan infections. Microsporidian and myxosporidian infections (Hoffman *et al.* 1962) can cause musculoskeletal lesions that may mimic deformities induced by a number of different heavy metals and organophosphate pesticides.

A final source of uncertainty is disease (infectious or neoplastic) secondary to the debilitation and immunosuppression caused by a toxicant. Fish culturists are very aware that a variety of environmental stressors affect the immune system and result in disease outbreaks and mortality (Anderson 1990). In reviewing immunosuppression and disease, Anderson (1990) cites various challenge tests (metals and corticosteroid drugs) that demonstrated suppression of disease resistance in exposed fish. Anderson (1990) lists experiments necessary to make this linkage of exposure to anthropogenic toxicants and immunosuppression.

SYSTEMIC TOXICOLOGIC PATHOLOGY OF FISHES

Skin

Perhaps the largest teleost organ and an important interface between internal tissues and external environment, skin reflects changes after various conditions of exposure that may be used as an indicator of host injury and response. Certain skin lesions are classified as "pollution-associated diseases" (Sindermann 1990) but the co-association of opportunistic microorganisms makes it difficult to determine the specific etiologic agent and whether xenobiotics were the principal or simply a contributory etiologic factor.

Ulcerative lesions represent the breakdown of the skin as a protective organ. Such lesions may be somatic or may occur on the fins as "fin rot." The etiology for such lesions is probably multifactorial, involving immune system deficiencies, fungal and bacterial pathogens, and toxic contaminants (Wellings *et al.* 1976). Ulcers have been positively associated with contaminated marine environments in red hake, *Urophycis chuss*, from the New York Bight (Murchelano and Ziskowski 1979) cod, *Gadus morhua*, and dab, *Limanda limanda*, from the North Sea (Dethlefsen 1980) and in cod from Danish coastal waters (Jenson and Larsen 1978). Elevated frequencies of such lesions have also been noted in brown bullhead, *Ictalurus nebulosus*, from the Cuyahoga and Black Rivers, Ohio (Baumann, personal communication). Sindermann (1990) reviewed various laboratory exposures to verify etiology of fin erosion. Oil, Aroclor 1254, and lead produced these lesions in laboratory exposures of various fish.

Perhaps the most characteristic response of skin is the production and secretion of mucus. Physical trauma (Mittal and Munshi 1974; Pickering *et al.* 1982), exposure to toxic metals (Varanasi *et al.* 1975; Lock and van Overbeeke 1981; Miller and MacKay 1982), or to acid pH (Anthony *et al.* 1971; Daye and Garside 1976) lead to enhanced mucus production. The shape (spherical) and staining characteristics (magenta) of mucous granules (alcian blue-PAS reaction) lend themselves to rapid quantification. True cutaneous neoplasms, suspected of having a toxic etiology, include epidermal papillomas, squamous cell carcinomas, and chromatophoromas. Although initial reports of lip papillomas in white suckers, *Catostomus commersoni*, from the Great Lakes described the presence of virus (Sonstegard 1977), later studies failed to confirm these findings (Smith *et al.* 1989). The tumors are now believed to be associated with contaminated habitats. Papillomas occurred in 39% of white suckers from heavily polluted Hamilton Harbor in Lake Ontario and were found in only low prevalence at three reference sites (6% in eastern Lake Ontario, less than 1% in Lake Superior and Lake Huron) (V. Cairns, personal communication; Baumann and Whittle 1988). Hayes *et al.* (1990) studied epidermal neoplasms in white suckers from industrially polluted areas in Lake Ontario and compared them to lesions in fish from less polluted sites in the Great Lakes. High prevalence of epidermal papillomas have also been reported in white suckers from western Lake Erie and the Niagara River (Black 1983a).

Elevated prevalence of lip neoplasms (described as epidermoid carcinomas) were reported in brown bullhead, *Ictalurus nebulosus*, from polluted areas of the Schuylkill and Delaware Rivers, Pennsylvania (Lucke and Schlumberger 1941; Schlumberger and Lucké 1948). In addition, both lip and skin neoplasms have also been found in brown bullhead from the Black River, Ohio (Baumann *et al.* 1987). Approximately 60% of these lip neoplasms were diagnosed as epidermal papillomas and about 40% as papillary or squamous carcinomas. Although these neoplasms occurred in less than 1% of 2-year-old fish from the Black River (N=263), 4-year-old fish (N=50) had a 32% prevalence of lip neoplasms and an 18% prevalence of skin neoplasms. Brown bullhead 3 years or older (N=78) from Buckeye Lake, a reference location, had no skin tumors and only a 1.5% frequency of lip tumors (Baumann *et al.* 1987). Sediment in the Black River is highly contaminated with polynuclear aromatic hydrocarbons (PAH) (Fabacher *et al.* 1988). When extracted from this sediment, these PAH compounds induced skin papillomas when painted on the skin of mice (Black *et al.* 1985). Black (1983b) also induced skin papillomas by painting PAH-rich extract of Buffalo River, New York, sediment on the heads of brown bullhead catfish.

Black bullhead, *Ictalurus melas*, from a sewage treatment pond in Alabama were first reported by Grizzle *et al.* (1984) as having a high prevalence (73%) of oral papillomas. Caged black bullheads placed in the pond also developed neoplasms, and their aryl hydrocarbon hydroxylase activity was elevated compared to that of bullhead from a nearby pollution-free pond (Tan *et al.* 1981). While a precise etiology was not postulated, the prevalence of neoplasms declined significantly after chlorination had been reduced by approximately one-third (Grizzle *et al.* 1984).

Dermal pigment cell neoplasms described as chromatophoromas occurred in freshwater drum, *Aplodinotus grunniens*, from western Lake Erie and the Niagara River (Black 1983a). Drum from 5 polluted sites had a significantly ($P<0.05$) higher tumor frequency (8.8%, N=305) than did drum collected from two reference areas (2.2%, N=891). Variation in tumor frequency occurred among the 5 polluted sampling sites and weakened evidence for a chemical etiology.

Two marine species of drum nibe, *Nibea mitsukurii*, and koichi, *Nibea albiflora*, found off the Pacific coast of Japan, were also found to have elevated prevalence of chromatophoromas (Kimura *et al.* 1984). Tumor prevalence as high as 75% was associated with marine environments adjacent to pulp and paper factories, and extracts of sediment and wastewater were found to be mutagenic with both the Rec and Ames assays. Kimura *et al.* (1984) were able to chemically induce chromatophoromas in laboratory-reared nibe, using subcutaneous injections of PAH, 7,12-dimethyl-benz[a]anthracene (DMBA), oral administration of N-methyl-N'-nitro-N-nitrosoguanidine (MNNG), and bath immersion with MNNG or nifurpirinol. Kinae *et al.* (1990) induced melanophore hyperplasia in 70 to 100% of marine catfish, *Plotosus anguillaris*, and chromatophoroma in one *N. mitsukurii*, using effluent from a kraft pulp mill which has been epidemiologically linked with identical lesions in wild fish of the same species. This work is especially important because it is one of the few examples where the hypothesis that chromatophoromas in feral fish are caused by exposure to chemical carcinogens was tested, at least partially, under controlled laboratory conditions.

Chromatophoromas have also been found in two species of Hawaiian butterflyfish, *Chaetodon multicinctus*, and *C. miliaris*, at prevalances of 50% and 5%, respectively (Okihiro 1988). At least circumstantial evidence supported the possibility of a chemical etiology. Pacific rockfish were sampled from Cordell Bank, 37 kilometers of central California, from 1985 to 1990. Hyperplastic and neoplastic cutaneous lesions, involving dermal chromatophores, were observed in five species: yellowtail, *Sebastes flavidus*, bocaccio, *S. paucispinis*, olive, *S. serranoides*, widow, *S. entomelas*, and chilipepper rockfish, *S. goodei* (Okihiro *et al.* 1991). Electron microscopy did not reveal the presence of virus, and etiology has not been determined. Yamashita *et al.* (1990) presented the following evidence suggesting that pigment cell disorders in croaker, *Nibea mitsukurii*, and sea catfish, *Plotosus anguillaris*, may be useful as biomarkers to monitor coastal water pollution for carcinogens: a) known carcinogens induced lesions in both species, b) effluents from a kraft pulp mill located near the contaminated site contained mutagens subsequently identified (4 chloroacetones and 3 alpha dicarbonyl compounds), and c) effluent-induced skin pigment cell hyperplasia in 70 to 100% of laboratory-exposed fish. Field verifications for chromatophoromas and papillomas appeared stronger than that for squamous carcinomas. Additional field studies are needed.

Confounding Issues -- Skin. Viruses may cause epithelial tumors. Additionally, ectoparasites can produce lesions that grossly resemble neoplasms. Papillomas, squamous carcinomas, and chromatophoromas have been induced by carcinogens in the laboratory and have been found in elevated frequencies on benthic feral fishes associated with contaminated sediments. Thus, these lesions represent usable biomarkers, as long as care is taken to distinguish them from skin lesions having a viral or parasitic etiology.

Secondary sex characteristics can also be a confounding effect. Morphometric studies (Zuchelkowski *et al.* 1981, 1986; Schwerdtfeger *et al.* 1979a,b) have shown differences in both the amount of mucus produced and mucous cell morphology between the sexes. Under acute conditions of exposure to acid pH (sulfuric acid addition to aquarium water, final pH 4.0) the volume of brown bullhead *Ictalurus nebulous*, skin occupied by mucous granules increases (Zuchelkowski *et al.* 1986). Immature females respond by increased size (hypertrophy) of mucous cells, while immature males increase both the number of mucous cells (hyperplasia) and their size (hypertrophy). Since males and females differ in mucous cell responses, field studies should be designed to evaluate the sexes separately. Skin reflects secondary sexual changes with breeding behavior. Breeding tubercles develop in the skin of males. Calcium levels in scales of females become reduced as the metal is released for incorporation in vitellogenin (Walker 1987).

Musculoskeletal System

Vertebral deformities (scoliosis, kyphosis, and lordosis) have frequently been reported in wild fish populations (Bengtsson 1974, 1975; Valentine 1975; Van deKamp 1977; Sloof 1982; Baumann and Hamilton 1984). Although commonly referred to as "broken back" syndrome, not all spinal deformities are directly attributable to fracture of vertebral bodies. Two mechanisms have been proposed for development of spinal deformities. The first centers on alteration of structurally critical biological processes involved with the collagenous matrix or mineral content of vertebrae. Causative agents implicated include: hereditary defects, defective embryonic development induced by high water temperature or low dissolved oxygen, radiation, vitamin C or B_{12} deficiency, heavy metals, and xenobiotics (Bengtsson 1974, 1975; Mayer *et al.* 1977; Murai and Andrews 1978).

Vitamin C is an especially critical factor in bone development because of its key role in the production of hydroxyproline, an essential component of the collagenous matrix of bones. Vitamin C deficiency may be caused by several diet factors including: oxidant-dependent depletion, diet, increased utilization, or decreased absorption. Cadmium depletes vitamin C by increasing utilization or decreasing absorption in mullet, *Mugil cephalus* (Thomas *et al.* 1982). Further, di-2-ethyl hexyl phthalate (DEHP) and several organochlorine compounds (toxaphene, Kepone, mirex, Aroclor 1254, and 2,4-DMA) have been shown to alter bone structure by interfering with vitamin C metabolism (Mayer *et al.* 1977; Mehrle and Mayer 1977). Some fish species lack the ability to synthesize vitamin C and may thus be more sensitive to dietary deficiency (Yamamoto *et al.* 1978).

In addition to contributing to decreased vitamin C levels, cadmium may also inhibit the uptake of calcium across the gill (Verbost *et al.* 1987; Reid and McDonald 1988) which in turn may lead to decreased plasma levels and reduced bioavailability for bone deposition (Roch and Maly 1979). Cadmium (Muramoto 1981), lead (Varanasi and Gmur 1978), and zinc (Saiki and Mori 1955; Sauer and Watabe 1984) may also weaken the structural integrity of bone by displacing calcium from normal binding sites.

The second mechanism leading to vertebral deformity is muscular tetany. Muscular tetany has been induced in the laboratory by electrical current (Spencer 1967), acute temperature change (Brungs 1971), parasitic infestation, *Myxosoma cerebralis* (Hoffman *et al.* 1962); heavy metals (zinc, cadmium, lead) (Bengtsson 1974; Bengtsson 1975; Holcombe *et al.* 1976), organochlorine pesticides (toxaphene, chlordecone) (Merhle and Mayer 1975; Couch *et al.* 1977; Stehlik and Merriner 1983), trifluralin (Couch *et al.* 1979), and organophosphate pesticides (parathion and malathion) (Weis and Weis 1989). Organophosphates cause spinal fractures by inhibiting cholinesterase which leads to a buildup of acetylcholine at nerve endings and to muscular tetany. Organochlorines appear to cause both muscular tetany and vitamin C depletion.

Confounding Issues -- Musculoskeletal System. The pleuripotential status of etiology may limit its use. However, nutritional deficiency may be expected to cause defects in other bones and possibly cartilage. Gill deformities with vertebral defects may argue for a more systemic, metabolic etiology. The spinal curvature accompanying acute neuromuscular spasms after exposure of embryos and larvae to some pesticides (Weis and Weis 1989) apparently diminishes when fish are placed in clean water. Therefore, with careful attention to the age of affected individuals, historical observations of gross deformities in the species, monitoring after placement in clean water, attention to reference populations controlled for age and sex, biomechanical tests, and X-ray analysis, the majority of non-toxic etiologies can be ruled out (Hinton *et al.* 1991a).

Nervous System Including Special Sense Organs

An obvious paradox exists between the importance of the central nervous system (CNS; brain and spinal cord) and the special sense organs and the paucity of attention they have received as sites of histopathologic alterations (Hinton *et al.* 1991a). Lipid soluble compounds such as solvents readily gain entry into the CNS. The neural tissue proper is poor in toxicant-metabolizing systems, yet is sensitive to many toxicants because of its inherent high metabolic activity. The major components of the CNS that may be of toxicological interest are neurons, supporting cells (oligodendrocytes, Schwann cells, and astrocytes), and cells of the vascular support system. Any number of these can be the target of a toxicant. The fish brain has essentially the same components as other vertebrates (Ferguson 1989). In contrast to higher organisms, the fish CNS is capable of at least some regeneration (Bernstein 1970).

Little is known about CNS histopathology, either induced by toxic or infectious agents, probably because of the complexity of the system and lack of studies (Ferguson 1989). Data suggest that CNS lesions in fish are both dentifiable and analyzable. Gliosis is a common response to injury of fish CNS (Ferguson 1989); vacuolization of optic lobes has been observed in salmonids after exposure to the pesticide carbaryl (Walsh and Ribelin 1975) and in optic and olfactory cortex of medaka (Hinton and Hawkins, personal communication). The herbicide 2,4-D caused vascular congestion in the brain of bluegills, *Lepomis macrochirus* (Cope *et al.* 1970). Walsh and Ribelin (1975) observed vascular congestion and hyperemia in brains of coho salmon, *Oncorhynchus kisutch*, and lake trout, *Salvelinus namaycush*, exposed to a variety of pesticides, but only 2,4-D exposure was sufficient for diagnostic value. My laboratory has investigated larvae of striped bass, *Morone saxatilis*, acutely exposed to the herbicide molinate. Submeningeal hemorrhage has been shown (unpublished observations).

Although a rather large number of neoplasms arising from various elements in the CNS and affecting a wide variety of species have been reported (Masahito *et al.* in press), a toxicopathic etiology and their use as biomarkers are as yet undetermined. Smith (1984) observed massive hyperplastic lesions originating from the primitive meninx (fish equivalent of the dura mater and arachnoid) of fathead minnows following long-term ammonia exposures.

The neurosensory system in fish includes the olfactory organ, lateral line, inner ear, and eye. Unfortunately, these systems have received little attention. Because the olfactory and lateral line organs act as an interface between the environment and the central nervous system, they would be expected to be excellent monitors of exposure to neurotoxicants. Compromise of organs so important in feeding, homing, position, and prey avoidance could have severe impact at both the individual and population level.

The few studies that have examined olfactory organ and lateral line effects in fish indicate that changes in these organs might provide a window to nervous system effects generally, and also help explain some behavioral abnormalities that often accompany toxic exposures (Hinton *et al.* 1991a). Gardner (1975) examined effects of several chemicals on neurosensory structures in the mummichog, *Fundulus heteroclitus*, and found that copper, mercury, and silver caused degeneration of the anterior lateral line and the olfactory organ, while cadmium and zinc caused no changes in these structures. Methoxychlor caused only lateral line lesions in mummichogs. In the Atlantic silverside, *Menidia menidia*, crude oil exposure resulted in hyperplasia of the olfactory sustentacular epithelium. Behavioral changes in silversides were related to structural damage to the lateral line, cephalic sinuses, olfactory organ, and inner ear. Crude oil and its water soluble fractions caused hyperplasia of non-neural cells of the olfactory organs and necrosis of both neurosensory and sustentacular epithelium of the inland silverside, *Menidia beryllina* (Solangi and Overstreet 1982). In the hogchoker, *Trinectes maculatus*, those substances caused severe necrosis of neurosensory and sustentacular cells of the olfactory mucosa. DiMichele and Taylor (1978) exposed mummichogs to naphthalene and observed damage in neurosensory cells of the taste buds, olfactory organ, and lateral line. Sensitivity of these organs to the few compounds tested, their importance to the well-being of the organism, and their unique anatomical organization, warrant further investigation into their utility as loci for environmental toxicologic alteration.

Only a few studies have examined the inner ear and the eye. The semicircular canals, their membranous and bony labyrinths, and endo- and perilymphs were studied in menhaden, *Brevoortia tyrannus*, from Narragansett Bay, Rhode Island, and from the site of a fish kill involving the same species (Gardner 1975). Although the etiology of the condition was not determined, fish from the latter site showed congestion of both endo- and perilymph and a degeneration of membranous labyrinth, including hemorrhage in connective tissues. A cellular infiltration of both endo- and perilymph was pronounced. A great deal of diversity exists in the morphological organization of fish eyes, but retinal form and function are comparatively conservative (Wilcock and Dukes 1989).

Pathological responses of corneal and lens tissues to chemical and other agents are common but non-specific. Cataract, a common ocular lesion caused by numerous factors, may be an indicator of general environmental deterioration.

The site where new retinal cells proliferate was believed to be the origin for neoplastic lesions in medaka exposed to methylazoxymethanol acetate (Hawkins *et al.* 1986). Hose *et al.* (1984) found microphthalmia, patent optic fissure, depressed sensory retinal mitotic rates, retinal folding, and poor retinal differentiation in rainbow trout exposed as juveniles to BaP. Whereas ocular tumors probably are not good biomarkers because of generally rare occurrence and high compound-specificity, other histopathological effects in retina, as seen by Hose *et al.* (1984), deserve study. Lack of experimental work and omission of a central nervous system and special senses in field studies have led to deficient understanding of the role of these systems in fish toxicologic injury and response. This pressing research need should be addressed.

Cardiovascular System and Spleen

Fish peri-, epi-, myo-, and endocardium show alterations with various infectious diseases (Herman 1975; Ferguson 1989). Reports of toxicologic injury to heart are rare despite the fact that recent evidence (Stegeman *et al.* 1989) indicates the presence and inducibility of cytochrome P450 IA1 in this organ where it is preferentially localized within endothelium. Gardner (1975) reported myocardial necrosis in ventricle of Atlantic silverside, *Menidia menidia*, exposed to crude oil (0.14 mg/l) for seven days.

I included the spleen with the cardiovascular system since it is intimately involved with the storage and breakdown of erythrocytes (Ferguson 1989). In addition to vascular channels, germinal centers, and cells of the reticulo-endothelial system, the spleen is often the site of macrophage aggregates (Roberts 1975). Relative abundance and area of these structures have been suggested as health indicators of wild fish (Wolke *et al.* 1985). Aggregates varied in size and number with such factors as aging, starvation, and toxicity of stressors as well as disease (Brown and George 1985; Wolke *et al.* 1985).

In a review of fish histopathology after exposure to various pesticides (Walsh and Ribelin 1975), alterations in blood vessels were commonly noted. The compounds and lesions were: heptachlor -- enlarged erythrocytes of bluegill sunfish, *Lepomis macrochirus*; phenoxyacetate herbicides -- appearance of PAS+ globular material in liver vessels of *Lepomis macrochirus;* hydrothol 191 -- appearance of spherical purple-red masses in vessels of red ear sunfish, *Lepomis microlophus*, sodium arsenite -- endothelial separation and subendothelial myositis in *Lepomis microchirus*; 2,4-D -- blood vessel congestion in brain of *Lepomis microchirus*.

Primary cardiac neoplasms are extremely rare. However neoplasms of the cardiovascular system in general are fairly common. Neoplastic histotypes include hemangiomas, hemangioendotheliomas, hemangioendotheliosarcomas and hemangiopericytic sarcomas. Recently, hemangiopericytoma has been induced in *Poeciliopsis* sp. (Schultz *et al.* 1985); the sheepshead minnow, *Cyprinodon variegatus* (Couch and Courtney 1987); rivulus, *Rivulus ocellatus marmoratus* (Grizzle and Thiyagarajah 1988); and medaka, *Oryzias latipes* (Bunton 1991).

Careful analysis of the heart and blood vessels in fish exposed to various toxicants is needed. With the presence of cytochrome P450 within endothelium of most visceral organs, the potential exists for bioactivation of potential toxicants, such as the polycyclic aromatic hydrocarbons. Given the common occurrence of infectious disease within the cardiovascular system, the investigator must be careful to include reference or control populations, or both.

Endocrine System

This system is comprised of at least 15 tissues (Groman 1982) the products of which are released into the blood stream where they come into contact with target cells located at varying distances from the endocrine source. Perhaps it is these multiple loci and a preferential treatment of economically important fish species of large body size that has done the most to prevent thorough endocrine evaluation in toxicology. The use of small fish species in toxicity testing (Hoover 1984a) has done much to facilitate evaluation of the various endocrine organs; some that have been studied include the pituitary, thyroid, endocrine pancreas, and interrenal tissue.

Wester and Canton (1986) exposed *Oryzias latipes* to the estrogen mimetic, β-hexachlorocyclohexane (HCH) and produced hypertrophy of the TSH-producing cells of the pituitary. In addition, thyroid follicles appeared depleted of colloid signifying activation. In *Anguilla anguilla* exposed to DDT, degranulation of eosinophil cells occurred (Walsh and Ribelin 1975).

When Couch (1984) exposed *Cyprinodon variegatus* to the herbicide trifluralin, (2,6 dinitro-N,N-dipropyl-4-(trifuoromethyl)benzamine), vertebral dysplasia and focal vertebral lesions were seen in association with enlargement and pathologic alteration, primarily within the adenohypophysis. Histopathology included formation of cyst-like enlargements and vascular engorgement with congestion. Morphometry indicated pituitary enlargement as evidenced by means of maximal widths and lengths.

Reduced thyroid follicular cell height occurred in *Carassius auratus* exposed to endrin (Walsh and Ribelin 1975). Several fish species have been shown to develop thyroid hyperplasia with the resultant appearance of follicles at some distance from the normal location. These protruding masses may distort the architecture of the newly invaded organ (Hoover 1984b). The invasive nature of these masses with their lack of a capsule have forced some investigators to refer to them as invasive thyroid adenocarcinomas.

The pesticide, endrin, apparently caused islet cell hyperplasia in cutthroat trout, *Oncorhynchus clarki* (Walsh and Ribelin 1975). Exposure of guppies, *Poecilia reticulata*, and brown trout, *Salmo trutta*, to DDT resulted in necrosis of interrenal cells (Walsh and Ribelin 1975). Of the 15 tissues purported to be endocrine producing sites, I could find no toxicologic literature related to 11. Additional work with the above tissues and pineal, saccus vasculosus, ultimobranchial body, corpuscle of Stannius, urophysis, chromaffin tissue, thymus, pseudobranchiae, enteroendocrine tissue of the gut, interstitial cells of the gonads, and juxtaglomerular cells of the kidney is long overdue. Spawning-related changes may be encountered. With the multifocal localization inherent in the teleost endocrine system, histopathology would be facilitated and cost reduced by using adults of small fish species as test organisms.

Respiratory System

A variety of insults, including low pH (McDonald 1983), transition metals (Laurén and McDonald 1985), heavy metals (Verbost *et al.* 1987), detergents (Abel and Skidmore 1975), and polycationic agents (Greenwald and Kirschner 1976) affect gill structure (Mallatt 1985). The gill is so sensitive, in fact, that laboratory exposure to any of the agents just listed, at concentrations comparable to those found in the wild, causes sufficient acute ionoregulatory disruption to result in death of fish often without time for formation of histologic lesions. Only at acutely lethal concentrations and low pH were mucous coagulation, necrosis of lamellar epithelial cells, and epithelial lifting observed (Packer and Dunson 1972; Skidmore and Tovell 1972; Daoust *et al.* 1984). Chevalier *et al.* (1985) have reported epithelial lifting and chloride cell degeneration in trout from an acidified lake.

At sublethal concentrations, a variety of toxicants have been shown to induce chloride cell hyperplasia (Laurent *et al.* 1985; Perry and Wood 1985; Avella *et al.* 1987). The proliferation of chloride cells is apparently a compensatory response to ion loss, and chloride cell hyperplasia may be an indicator of adaptation to ionoregulatory stress.

In contrast, hyperplasia of undifferentiated epithelial cells, resulting in clubbing and fusion of lamellae, is a less specific lesion, associated with a wide variety of unrelated insults inducing infection by microorganisms, ectoparasitism, phenols, heat, NH_3, and metals. Mucous cell hyperplasia has been reported in trout exposed to low pH (Daye and Garside 1976), but has not been seen in wild fish from acidified lakes (Chevalier *et al.* 1985). Detection of hyperplastic lesions in absence of inflammation, infection, and parasitism has increased significance and can be quantified using morphometrical analysis (Hughes *et al.* 1979).

A potentially useful finding in the gill is the presence of cytochrome P450IA1 in pillar cells. This enzyme (detected by ethoxyresorufin-O-deethylase) is inducible in both scup, *Stenotomus chrysops*, and rainbow trout, *Oncorhynchus mykiss*, and may be localized in tissue sections with the use of monoclonal antibodies (Miller *et al.* 1989). P450IA1 induction in gill, where it is normally below detection limits, may serve as an indicator of exposure to organic contaminants, such as polycyclic aromatic hydrocarbons (PAH) in the environment.

In summary, the gill is a sensitive indicator of environmental stress, including anthropogenic compounds in the water. However, it may be too sensitive in that fish can be killed with the complete absence of histologic lesions and because a variety of factors can result in the same suite of lesions. A thorough review (Mallat 1985) lists gill structural changes induced by toxicants. The author concluded that irritant-induced alterations in gill morphology are largely non-specific. A listing of lesions (Table 5 of Mallat 1985) indicated heavy metals were more often associated with necrosis, hypertrophy, and mucous secretion.

The major drawback to working with the gill is its extreme sensitivity. A broad base of etiologic factors work together to produce a similar set of alterations. Confounding etiologies result in similar pathology. However, if the investigator wishes to determine the "histologic health index," a microscopic extension of the necropsy approach of Goede and Barton (1990) of a population of fishes residing at a specific site, gill should be included.

Reproductive System

Tolerance to stress may be lower in the reproductive tract than in any other organ system (Gerking 1980). Population decline from reproductive impairment is potentially the most serious biological impact of a toxicant-compromised environment. Donaldson (1990) differentiated reproductive indices into short- and long-term indicators. The former include such histologic endpoints as presence of atretic ova and failure to spermiate or ovulate. Toxicants affecting the reproductive system may interfere at all stages of the life cycle, including fertilization, embryonic development, sex differentiation, oogenesis, or spermatogenesis, final maturation, ovulation, or spermiation and spawning (Donaldson 1990). Effects induced by chemical pollutants on gametes or early life history stages were reviewed (Rosenthal and Alderdice 1976; Donaldson and Scherer 1983; McKim 1985; Weis and Weis 1989, c.f. Tables 1 and 2 in Donaldson 1990).

Oocyte or follicular atresia, a normal occurrence in the ovaries of all fish species, can become pathologic following exposure to xenobiotic compounds

(Johnson *et al.* 1988; Cross and Hose 1988, 1989; Kurugagran and Joy 1988; Lesniak and Ruby 1982; McCormick *et al.* 1989; Stott *et al.* 1981). Oocyte atresia, characterized by degeneration and necrosis of developing ova and subsequent infiltration by macrophages, was demonstrated in a study by McCormick *et al.* (1989). Fathead minnows, *Pimephales promelas*, were exposed to different levels of lowered pH in river water in a controlled field experiment. Morphometric analysis of ovaries from those fish demonstrated reproductive impairment when the ratio of atretic oocyte follicles to total ovarian volume exceeded 20%. This linkage of a histopathologic biomarker (oocyte atresia) to a potential population impact (lowered reproductive success) following a toxic event (lowered environmental pH) should encourage other investigators to test such observations with other species and toxicants.

Response to a wide variety of toxicants and conditions produces relatively few histopathologic lesions (Glaiser 1986). Histopathology of female reproductive tissues has resulted from exposure to several heavy metals, including cadmium in herring (Westernhagen, von *et al.* 1974), selenium in bluegills (Sorenson *et al.* 1984), and arsenic in bluegills (Gilderhaus 1966). In the male, histopathologic lesions include testicular atrophy, which may be either related to age or induced by toxicants; sperm reduction, which also may result from exposure to toxicants, especially those that inhibit mitosis; and inflammation due to bacteria, protozoa, or metazoa. Key lesions in the female include ovarian atresia, failure of ovulation often associated with fibrosis and adhesions, and inflammation following bacterial, protozoan, and metazoan infections.

The testis of the male includes the following cellular components that are of histopathologic concern: Sertoli cells, constituting the blood-testis barrier; Leydig cells, which are interstitial cells involved in male hormone production; and developing germ cells including, in order of differentiation, spermatogonia, spermatocytes, spermatids, and spermatozoa. In some species, secretory cells involved in production of spermatic fluid are part of the testis.

Several toxicants have been shown to damage or impair testicular epithelium or sperm. Mercury damages sperm and decreases their motility probably by interfering with flagellar function (McIntyre 1973; Mottet and Landolt 1987). Cadmium induced testicular damage in goldfish followed intraperitoneal (IP) injection (Tafanelli and Summerfelt 1975), and in brook trout followed long-term exposure to sublethal doses as low as 0.1 ppm (Sangalang and O'Halloran 1972). Sangalang *et al.* (1981) observed testicular abnormalities in cod fed PCB, Aroclor 1254. The lesions included disorganization of lobules and spermatogenic elements, inhibition of spermatogenesis, fibrosis in lobule walls, and fatty necrosis in testes of sexually mature specimens and those undergoing rapid spermatogenic proliferation, but not in specimens in which the testes were sexually mature or regressed. The fact that this study demonstrated specific "biomarker" effects in the male reproductive system of a fish following exposure to a contaminant is of considerable environmental importance and indicates the potential utility of lesions to indicate toxicant exposure. A different type of histological effect, induction of intersexuality as indicated by the appearance of ovarian follicles in testis, occurred in medaka exposed to the β-isomer of lindane, β-hexachlorocyclohexane (Wester and Canton 1986), and in redear sunfish after pond exposure to the herbicide Hydrothol 191 (Eller 1969).

Confounding Issues -- Reproductive System. The varied and complicated nature of breeding strategies in fish include the typical male/female pattern, self-fertilizing hermaphroditism, and gynogenesis, in which the male sperm stimulates but does not contribute to ovulation (Ferguson 1989). Exploitation of this rich diversity in reproductive strategy may yield important dividends, especially if the proper baseline studies are conducted. Because the synchrony of gametogenesis in seasonal breeders reduces the difficulty of distinguishing toxicant effects from normal gametocyte turnover, seasonal rather than continuous breeders offer advantages. Mottet and Landolt (1987) proposed basic studies in reproductive biology, toxicology, and physiology with seasonal breeders. Furthermore, many fish species exhibit sex reversal, which may be either natural or induced by hormonal or environmental factors.

The relationship between the occurrence of neoplasms of reproductive organs and toxic exposure is not clear. Two field surveys identified gonadal neoplasms in Great Lakes fish and suggested that their occurrence was related to pollution. These tumors involved germ cells and stroma in goldfish-carp hybrids (Leatherland and Sonstegard 1978) and tumors of gonadal supporting elements in yellow perch (Budd *et al.* 1975). The occurrence of dysgerminoma and seminoma in medaka might be related to exposure to some toxicants, such as trichloroethylene. Perhaps too little data exist to consider reproductive neoplasms in fish as reliable markers of exposure to anthropogenic chemicals.

Excretory System

The kidneys are organs of critical, though varied, function in the highly diverse fish species. One would assume that the renal tissues would be at major toxicologic risk since they receive large volumes of blood flow from both the renal portal venous system and the renal arteries (Walker 1987). In addition, urine produced collectively or individually through glomerular filtration, tubular reabsorption, or tubular secretion serves as a major route of excretion for metabolites of various xenobiotics to which fish have been exposed (Pritchard and Renfro 1982).

Recent laboratory experiments by Reimschuessel *et al.* (1990) are of interest. They exposed goldfish, *Carassius auratus*, to hexachlorobutadiene, a potent nephrotoxicant. New nephrons developed several weeks following termination of exposure. Stereologic quantification determined the volume percent of the kidney occupied by developing (extremely basophilic) nephrons (greater in treated than in controls). Trump *et al.* (1975) reviewed cellular effects of mercury on epithelial cells of fish kidney tubules. Electron micrographs and correlated ion shift physiologic determinations laid a groundwork for mechanistic studies of effects of toxicants on the nephron. Such experiments should encourage field investigations of nephrotoxicity and repair.

Neoplasms in kidneys of wild fish populations are extremely rare. An early review listed one adenoma and two adenocarcinomas of the kidney (Schlumberger and Lucké 1948). A single renal cell adenocarcinoma has subsequently been described in wild fish (Gardner 1975) and laboratory fish (Hawkins *et al.* 1989). In contrast, embryonal kidney neoplasms (nephroblastomas) are readily induced in the laboratory in rainbow trout and other species by a variety of chemicals, such as MNNG, methylazoxymethanol acetate (MAM-AC), and DMBA (Hawkins *et al.* 1989).

Fewer histopathological studies in kidneys probably contribute to the lack of known toxicologic lesions in this important organ. However, it may also reflect either less exposure to toxic metabolites or greater resistance to their effects. As our interest and understanding in renal toxicity increases, such alterations as tubular necrosis and regeneration, glomerular changes (hyaline degeneration, mesangial lysis and fibrosis, visceral or parietal epithelial necrosis), eosinophilic proteinaceous droplets in proximal tubules, and neoplasia may prove indicative of prior exposure to toxicants. Numerous infectious diseases affect the kidney of teleosts (Ferguson 1989). Lesions may mimic those associated with toxicants.

Digestive System

A derivative of gut endoderm, the swimbladder exists in either a physostomic (with a pneumatic duct opening to the esophagus) or physoclistic (with no pneumatic duct) condition in various fish species. Its major function is to regulate buoyancy and, thus, depth of the fish in the water column. The volume of gas in the organ is controlled through the action of the rete mirabile and gas gland. Few lesions of swimbladders have been reported in fish subjected to toxicants in either laboratory or field situations. Laboratory exposure of rainbow trout to MNNG, MAMA, or DMBA has produced benign papillary adenomas of the epithelial mucosa. However, in experimental studies with rainbow trout exposed as embryos, papillary adenomas of swimbladder nearly always occur simultaneously with gastric adenomas (J.D. Hendricks, personal communication). If the latter are observed, the swimbladder should be carefully examined. One reason swimbladder tumors may not have been reported is that they are easy to overlook, especially if the swimbladder is punctured and deflated. Care should be taken to keep the bladder inflated for examination of tumors that can be seen through the transparent bladder wall.

In a single communication (Black, personal communication) and a limited number of reports (Gardner and Pruell 1987; Gardner *et al.* 1989), fish from heavily polluted freshwater and marine environments may develop noninvasive papillary adenomas or polyps in the mucosa of the glandular stomach and to a lesser degree, in the esophagus, pyloric stomach, and small intestine. I found no publication on laboratory exposures resulting in gastrointestinal tract neoplasms. One chapter in an upcoming Atlas on fish neoplasms will be devoted to the digestive system (C. Dawe, personal communication).

Of the digestive system extramural glands of teleosts, perhaps the exocrine pancreas has received the least attention. Submicroscopic degeneration and acute necrosis of the organ were reported in carp, *Cyprinus carpio*, exposed to dalapon (Walsh and Ribelin 1975).

Spontaneous neoplasms are rare. However, Thiyagarajah and Grizzle (1986) and Fournie *et al.* (1987) exposed rivulus, *Rivulus ocellatus marmoratus*, and guppies, *Poecilia reticulata*, to diethylnitrosamine and to methylazoxymethanol acetate, respectively, and produced neoplasms. Neoplasms produced in rivulus included: adenomas, cystadenomas, and adenocarcinomas. Guppy neoplasms were adenomas, acinar cell carcinomas, and adenocarcinomas. Later Grizzle *et al.* (1988) used embryomicroinjection of a direct acting carcinogen and produced a pancreatic acinar cell carcinoma in a single Gulf killifish, *Fundulus grandis*.

Although the liver is not the only site for significant histologic alterations, it is certainly the primary locus based on current experience (Meyers and Hendricks 1985; Murchelano and Wolke 1985; Wolke *et al.* 1985; Myers et al 1987; Harshbarger and Clark 1990; Kranz and Dethlefsen 1990; Vogelbein *et al.* 1990). There are several reasons for this. First, the liver of teleosts is the major site of the cytochrome P450 -- mediated, mixed-function oxidase system (Stegeman *et al.* 1979). This system inactivates some xenobiotics, while activating others to their toxic forms. Secondly, nutrients derived from gastrointestinal absorption are stored in hepatocytes and released for further catabolism by other tissue (Walton and Cowey 1982; Moon *et al.* 1985). Third, bile synthesized by hepatocytes (Schmidt and Weber 1973; Boyer *et al.* 1976) aids in the digestion of fatty acids and carries conjugated metabolites of toxicants (Gingerich 1982) into the intestine for excretion or enterohepatic recirculation. Fourth, the yolk protein, vitellogenin, destined for incorporation into the ovum, is synthesized entirely within the liver (Vaillant *et al.* 1988). Receptors in the liver must bind the hormone, estradiol, for initiation of the signal to begin synthesis of this essential reproductive component. Given the liver's role in various key functions and its metabolic capacity, the hepatotoxic effect of various toxicants is not surprising. Hepatic toxicologic studies in fish have been reviewed recently (Hinton and Laurén 1990a; Hinton *et al.* 1991a).

Hepatocellular Necrosis and Sequelae. Coagulative necrosis, associated with sudden cessation of blood flow to an organ and damage by toxic agents, represents a meaningful alteration. With coagulative necrosis, shapes of cells and their tissue arrangement are maintained, facilitating recognition of the organ and tissue. Necrotic changes occur after cell death and represent the sum of degradative processes. These changes, useful in determining which cells died and underwent necrosis prior to death of the animal or fixation of its organs, are indications of exposure. Coagulative hepatic necrosis must be distinguished from necrosis due to postmortem change. Here the timely administration of fixative is needed. In those cases where this was not achieved, postmortem change would be reflected in all organs of the individual while coagulative hepatic necrosis would be anticipated to be focal or multifocal and within target organ(s). In addition, the process of coagulative necrosis may release chemotactic factors resulting in a localized inflammatory response. Inflammation spatially related to focal or multifocal necrosis would signify that the necrotic process preceded death of the host.

A substantial amount of information has been accumulated, associating coagulative hepatocellular necrosis with exposure to anthropogenic environmental toxicants in both mammals and fish (Wyllie *et al.* 1980; Meyers and Hendricks 1985; Pitot 1988). As information from correlated biochemical and morphologic studies increases, refinement of this important biomarker may provide additional ways to identify general classes of causative agents. While this biomarker signifies prior exposure to toxic chemicals, we cannot rule out the possibility that natural, plant-derived toxins (Hendricks *et al.* 1981) may be etiologic agents. Currently, hepatocellular coagulative necrosis is a useful marker of anthropogenic toxicant exposure.

Hyperplasia of Regeneration. Following necrosis, surviving cells (presumptive stem cells) undergo hyperplasia, thereby regenerating needed hepatocytes to replace those lost. Regenerated cells are small and basophilic, forming small islands of irregular shape. Regenerative islands may be quantified either by using colchicine to establish the ratio of metaphase to prophase nuclei or using tritiated thymidine followed by autoradiography (Zuchelkowski *et al.* 1981, 1986) or immunohistochemistry, using monoclonal antibodies directed against the synthetic thymidine analogue, bromodeoxyuridine (Droy *et al.* 1988; Miller *et al.* 1986). In the absence of evidence of prior infection, this biomarker has high toxicological significance.

Hyperplasia is indicative of extensive prior necrosis from either toxicant exposure or infectious disease processes. In the absence of evidence for parasitic infestation (e.g., fibrotic tracks or cystic spaces) or prior infection by microorganisms, regenerative hyperplasia is a good biomarker of exposure to toxicants. The presence of specific types of inflammatory cells may also be useful in differentiating infectious from toxicant-derived etiologies. These characteristics, plus the concurrent presence of remaining necrotic hepatocytes, facilitates differentiation of regenerative hyperplasia from basophilic focal and nodular change associated with neoplasia (discussed below).

A good example of the utility of regenerative foci as a histopathologic biomarker was a recent epizootic of mortalities in Atlantic salmon, *Salmo salar*, maintained in sea pen culture in Puget Sound, Washington (Kent *et al.* 1988). Salmon in a sea pen at one site showed normal growth and development, while mortality occurred in a similar stock fed identical rations but maintained in a different location. Gross necropsy, assays for infectious disease, and histopathology indicated that hepatic necrosis was the underlying cause of death. Survivors from the affected pen, held for weeks under standard laboratory conditions, developed regenerative islands of small basophilic hepatocytes with increased mitotic activity. The histopathologic finding of hepatic necrosis followed by compensatory regenerative repair in surviving fish illustrates the usefulness of hepatocellular regeneration as a prolonged indicator of prior necrosis.

Bile Ductular/Ductal Hyperplasia. Recent investigations (Hampton *et al.* 1985, 1988, 1989) of trout have shown that this liver is normally enriched with biliary epithelial cells. However, profiles of bile ductules (cuboidal epithelium) and ducts (columnar epithelium with mucus) are infrequent except near the porta hepatis. With ductular/ductal hyperplasia, profiles of these biliary passageways are numerous and contiguous, with abundant branching and coiling. Cytologic features of hyperplastic epithelial cells are normal. This lesion is of a chronic duration in wild fish from chemically contaminated sites (Murchelano and Wolke 1985; Hayes *et al.* 1990). In western Lake Ontario, proliferative biliary diseases (cholangiohepatitis and cholangiofibrosis) of white suckers, *Catostomus commersoni*, were associated with bile duct neoplasms in polluted harbors (Hayes *et al.* 1990). The proliferative biliary disease was less severe in livers of fish from reference sites. These authors suggested that proliferative bile duct epithelial changes could predispose fish to initiation and promotion of bile duct neoplasia (Hayes *et al.* 1990).

Hepatocytomegaly. Hepatocellular hypertrophy, a type of hepatocytomegaly characterized by organelle hyperplasia and enlarged cellular diameter without nuclear changes, may lead to a net gain in the dry mass of the liver. One way in which hepatocytes undergo hypertrophy is through proliferation of endoplasmic reticulum. EM studies have shown this after exposure of mullet, *Mugil cephalus*, to the polynuclear aromatic hydrocarbon, 3-methylcholanthrene (Schoor and Couch 1979), and in channel catfish, *Ictalurus punctatus*, after subacute exposure to Aroclor 1254 (Klaunig *et al.* 1979). Histologic preparations of liver with this condition show swollen hepatocytes with eosinophilic, hyalinized cytoplasm resembling ground glass. Nuclei typically are unaltered.

Megalocytosis is a second type of hepatocytomegaly and is characterized by marked cellular and nuclear enlargement. Enlarged nuclei often contain false and real inclusions, and multinucleated megalocytes may be seen. A condition involving megalocytosis, termed megalocytic hepatosis, is the most frequently encountered idiopathic lesion in the liver of English sole, *Parophrys vetulus*, from contaminant-laden sites within Puget Sound, Washington (Myers *et al.* 1990). These authors have interpreted megalocytosis as manifestation of chronic toxicity of these sediment contaminants. Megalocytosis was seen in fish from chemically contaminated sites in the Kanawha River of West Virginia (Hinton and Laurén, unpublished observations) and in sea pen cultures of Atlantic Salmon in Puget Sound (Kent *et al.* 1988). Megalocytosis has been produced in the laboratory in trout, *Oncorhynchus mykiss*, exposed to pyrrolizidine (senecio) alkaloids (Hendricks *et al.* 1981) and medaka, *Oryzias latipes*, exposed to diethylnitrosamine (Hinton *et al.* 1988a,b). Megalocytes are probably sublethally injured hepatocytes and are able to survive for months (J. Groff, personal communication; Kent *et al.* 1988). With light microscopy, megalocytes are from 3-5 times larger than hepatocytes and their enlarged nuclei frequently show eosinophilic inclusions.

A third type of hepatocytomegaly arising from marked swelling of perinuclear endoplasmic reticulum cisternae is seen in vacuolated cells of liver (Bodammer and Murchelano 1990). Histologically, affected cells possess a clear cytoplasm, small compact nuclei and are markedly vacuolated. In the initial description, a high prevalence of winter flounder, *Pseudopleuronectes americanus*, from Boston Harbor, greater than 25 cm body length, were said to contain groups of vacuolated cells in acinar and tubular patterns which organizationally and cytochemically seemed more like ductal epithelial cells than hepatic parenchymal cells (Murchelano and Wolke 1985). Moore *et al.* (1989) examined younger specimens from the same harbor and described aspects of the pathogenesis of the lesion. They concluded that the earliest lesion, abnormal vacuolation, was in biliary preductular (Hampton *et al.* 1988) epithelial cells. In older fish, i.e. longer than 30 cm, they reported vacuolated hepatocytes as well as biliary epithelial cells. Bodammer and Murchelano (1990) described ultrastructural features in two affected female flounder between 38 and 46 cm in length. They reported vacuolation was restricted to hepatocytes. Due to the apparent involvement of both cell types, the term "hepatocellular vacuolation" is perhaps more accurate. Hepatocellular vacuolation of winter flounder (Murchelano and Wolke 1985; Gardner *et al.* 1989; Moore *et al.* 1989; Bodammer and Murchelano 1990) and windowpane flounder (*Scophthalmus aquosus*) (Murchelano and

Wolke 1985) may be regarded as a variant of hydropic degeneration. Particularly in winter flounder, the condition is encountered in high prevalence in Boston Harbor, Massachusetts and nearby estuaries where it is highly correlated with cholangiocytic neoplasms, less well with hepatocellular neoplasms (Harshbarger and Clark 1990) and may be seen in livers free of neoplasia. The lesion has also recently been detected in rock sole, *Lepidopsetta bilineata*, and starry flounder, *Platicthys stellatus*, from contaminated sites in Puget Sound, Washington (Stehr *et al.* 1990). In the case of this lesion, the magnitude and unique nature of the cellular alterations, along with the relatively high prevalence of fish affected at contaminated sites point to this as a specific biomarker. Even in the absence of laboratory studies demonstrating induction of similar changes by exposure to toxicants, we recommend its use.

Foci of Cellular Alteration - Staining or Tinctorial Change. An early stage in the stepwise histogenesis of hepatic neoplasia is the formation of foci of cellular alteration. This term includes those foci detected by conventional hematoxylin and eosin preparations (foci of tinctorial or staining alteration) and foci detected by enzyme histochemical procedures (foci of enzyme alteration). Only limited use has been made of enzyme histochemistry in fish carcinogenesis studies (Nakazawa *et al.* 1985; Hinton *et al.* 1988b; Laurén *et al.* 1990); extension from laboratory to field investigations has not been made.

A review of serial progression studies in laboratory exposures of fish to chemical carcinogens generally reveals loss of hepatocyte glycogen and necrosis initially (Scarpelli *et al.* 1963; Stanton 1965; Egami *et al.* 1981; Couch and Courtney 1987; Hinton *et al.* 1988a,b; Laurén *et al.* 1990; Hinton *et al.* 1991b). At some time after this initial toxicity, foci of tinctorially altered hepatocytes appear. When viewed after conventional paraffin processing as above, cells of these foci may be basophilic (Egami *et al.* 1981), basophilic or eosinophilic (Stanton 1965), and basophilic or eosinophilic or clear (Couch and Courtney 1987; Hinton *et al.* 1991b). In addition, a fatty vacuolated focus is described below. Foci are usually spherical to oval and show identical architecture with that of surrounding cells.

Cells of basophilic foci appear to differ from other hepatocytes only in the intensity of their staining, apparently due to glycogen depletion (Scarpelli *et al.* 1963; Laurén *et al.* 1990), reduced cellular volume, and real or apparent increase in granular endoplasmic reticulum and mitochondria. The nucleic acids of ribosomes have affinity for the basic dye, hematoxylin, and tumor cells of trout liver contain increased levels of cytoplasmic RNA (Scarpelli *et al.* 1963).

Cells of eosinophilic foci frequently show variation in size as well as altered tinctorial properties. Hendricks *et al.* (1984) reviewed focal alterations in rainbow trout, *Oncorhynchus mykiss*, after exposure to one of various carcinogens. Eosinophilic foci were invariably small, but component hepatocytes were hypertrophic and contained enlarged and abnormally shaped nuclei. These hepatocytes were apparently reduced in glycogen content. Other, larger, eosinophilic foci contained enlarged hepatocytes with homogeneous, hyalinized cytoplasm. Peripheral regions of these foci were sites of lymphocytic infiltration. In the Hendricks *et al.* (1984) report, basophilic portions of hepatocyte cytoplasm and the hepatocyte nuclei appear to have been rearranged toward the periphery of the cell. Given this appearance, smooth endoplasmic reticulum proliferation may account for the enhanced eosinophilia and the peripheral basophilic zone. In the sheepshead minnow, *Cyprinodon variegatus*, exposed to diethylnitrosamine, Couch and Courtney (1987) found cells of eosinophilic foci were more pleomorphic than those of basophilic foci.

Clear cell foci are collections of cells enriched in glycogen. Hendricks *et al.* (1984) showed glycogen storage "nodules," cells of which resembled glycogen enriched cells. Hinton *et al.* (1991b) studied serial sections through clear cell foci produced in medaka by exposure to diethylnitrosamine. Cells of such foci were strongly PAS-positive for glycogen.

The other type of focal alteration encountered in conventional preparations is that containing vacuolated cells. The round margins of the vacuoles and their positive reactions by fat stains (Hinton *et al.* unpublished observations) distinguish these from the vacuolated cells or "vacuolar foci" of Murchelano and Wolke (1985). Focal fatty vacuolation of hepatocytes is a response associated with exposure of fish to a variety of carcinogenic agents (Hendricks *et al.* 1984), and, with other focal alterations, apparently precedes other changes.

Focal fatty vacuolation should be distinguished from diffuse fatty change. Generalized (diffuse) fatty change is seen after a variety of hepatotoxic insults but this condition is also seen in vitellogenic females, in fish which store abundant lipids (van Bohemen *et al.* 1981) and is influenced by nutritional state (Segner and Möller 1984; Segner and Juario 1986; Segner and Braunbeck 1988). Before diffuse fatty change can be used reliably as a biomarker of chemical exposure, the appearance of the liver in the same species of fish, at the same time of year, under normal conditions must be taken into consideration. Therefore, until additional research on the mechanism of fatty liver in representative teleosts and additional field verification is completed, diffuse fatty change alone is not recommended for inclusion as a present biomarker.

It should be noted that focal fatty change is seen in aging control medaka although prevalence is greater in diethylnitrosamine-treated medaka (Hinton *et al.* 1991b). Field investigations have linked focal alterations to contaminated sites usually where hepatic neoplasms were found, and in satellite lesions in tumor-bearing liver. Kranz and Dethlefsen (1990) reported on a multiyear investigation of liver alterations in the bottom dwelling dab, *Limanda limanda*, of the southern North Sea. Histologically examining only livers with gross alterations, they included lesions of cellular alteration within a group of larger lesions termed "neoplastic changes." They did not refer to the lesions as foci but as nodules of varying size (<1.5 mm, 1.5 to 10 mm, and >10 mm). Smallest nodules were concentric areas of basophilic, eosinophilic, or vacuolated hepatocytes. These small nodules were usually multiple structures within the same liver; some phenotypes of individual nodules were mixed. Nodules and overt neoplasms were found in dab from the more contaminated sites. Feral winter and windowpane flounder from contaminated sites in Boston Harbor contained basophilic and eosinophilic foci (Murchelano and Wolke 1985). Hayes *et al.* (1990) reported on liver lesions in white suckers, *Catostomu commersoni*, from contaminated harbors and reference sites of Lake Ontario. Focal cellular alterations were restricted to basophilic and clear cell phenotypes. They cited rodent carcinogenesis studies, suggesting that exclusion of eosinophilic cytoplasmic differentiation typical of the "resistant" nodule phenotype may signify different initiating and/or promoting xenobiotics in those versus other sites. Baumann *et al.* (1990) reported on lesions of feral brown bullhead, *Ictalurus nebulosus*, from two Lake Erie tributaries. Focal hepatocellular alterations were well differentiated lesions, usually less than 1 mm in diameter. Staining differently than surrounding tissue, these lesions were usually basophilic, occasionally acidophilic (eosinophilic), and occasionally clear staining. Myers *et al.* (1987, 1990) have performed histologic multi-year studies of English sole from contaminated and reference sites within Puget Sound, Washington. Basophilic, eosinophilic, and clear cell foci have been reported. Histologically all are reduced in cytoplasmic iron, and rarely contain other liver structures, such as blood vessels, macrophage aggregates, and pancreatic acini. Vogelbein *et al.* (1990) reported on an epizootic of liver neoplasia in the mummichog, *Fundulus heteroclitus*, from sediment heavily contaminated with polycyclic aromatic hydrocarbons. Focal cellular alterations included basophilic, eosinophilic, and clear cells. Apparently a fourth (fatty vacuolated cell phenotype) was included in the clear cell focus.

Hendricks *et al.* (1984) reviewed results in the rainbow trout model and concluded that basophilic foci were the most important of the early altered cells and regarded certain of these foci as microcarcinomas. Hinton *et al.* (1988b) studied medaka foci with conventional and enzyme histochemical procedures. Diethylnitrosamine exposure was associated with formation of basophilic foci whose enzyme and tinctorial properties were identical to cells of eventual hepatocellular neoplasms. Hinton *et al.* (1991b) showed that the basophilic phenotype was expressed in the majority of neoplasms. Eosinophilic and clear cell phenotypes were apparently restricted to focal lesions in medaka.

Foci of cellular alteration, including basophilic, eosinophilic, clear, fat vacuolated, and enzyme-altered, are associated with exposure of various fish to chemical carcinogens in the laboratory. Their environmental relevance (except enzyme altered foci) has been confirmed in field investigations. Foci constitute meaningful indicators of prior exposure to toxicants.

Hepatic Adenoma. This lesion, in rainbow trout, is thought to represent an enlargement of the basophilic focus previously described (Hendricks *et al.* 1984; Nunez *et al.* 1991). Apparently clear cell and eosinophilic variants also occur in English sole (Myers *et al.* 1987) and in brown bullheads (Baumann *et al.* 1990). Hepatocytes appear essentially normal retaining their normal architecture (Myers *et al.* 1987; Nunez *et al.* 1991). Proliferation is evident, through compression of surrounding hepatocytes in larger adenomas, although mitoses are rare. This lesion may be microscopic in size but often forms a bulge at the surface.

Baumann *et al.* (1990) present a different classification. From hepatocellular focal alteration, they describe a larger, more clearly defined subpopulation of hepatocytes, the "hepatocellular nodule." This intermediate stage bridges hepatocellular focal alterations to frank hepatocellular carcinoma. They prefer not to use the term "adenoma," but accept "hepatoma" as a substitute for hepatocellular nodule. Necropsies performed at 6 or 9 months after initiation of hepatocarcinogenesis in rainbow trout (Nunez *et al.* 1991) and compared to necropsies after 12 or 18 months indicate that the adenoma is a transition lesion. Baumann *et al.* (1990) agree, stating that the bridging lesion may become large enough to bulge at the capsular surface and that it can progress to become a full-blown hepatocellular carcinoma. Agreement exists that early, intermediate, and endpoint stages of the process of hepatocarcinogenesis may be seen in the same organ.

Hepatocellular Carcinoma. Depending on the age of the tumor at necropsy, this lesion can vary from microscopic to very large, often occupying a major portion of the organ. Cells of trout hepatocellular carcinoma are predominantly basophilic (Nunez *et al.* 1991). In the brown bullhead, tumor cells are usually basophilic but may be eosinophilic or rarely clear cell (Baumann *et al.* 1990). The margin is less well defined and invasion into otherwise normal parenchyma is common. In spite of invasiveness, rapid proliferation of this lesion usually causes severe compression of surrounding hepatocytes. The masses of cells are usually solid or trabecular, resembling engorged tubules. Mitotic figures are numerous and can be bizarre. Tumor cells may be pleomorphic with some assuming spindle shapes while others are small, polyhedral in shape, and arranged as tight collections with distinct intercellular space (Hendricks *et al.* 1984; Myers *et al.* 1987; Couch and Courtney 1987; Hinton *et al.* 1988a,b; Vogelbein *et al.* 1990). Infrequently, hepatocellular carcinomas metastasize (Baumann *et al.* 1990). Focal cellular alterations (the proximate lesions), the bridging lesions (adenoma or hepatocellular nodule or hepatoma), and the end-point (hepatocellular carcinoma) are biomarkers of exposure and effect.

Cholangioma. These bile duct tumors are characterized by retention of their ductular architecture, presence of distinct margins resulting in a nodular, well-defined mass. Focal ductal elements may be present as well. Ducts may be cystic with papillary projections, but the lining epithelium is cuboidal to low columnar, simple (i.e., single row), and well differentiated (Hendricks *et al.* 1984; Myers *et al.* 1987; Baumann *et al.* 1990).

Cholangiocarcinoma. This lesion is larger; component cells are pleomorphic; mitotic figures and invasion of surrounding tissues are common. At the margin columns of invading cancer interdigitate with nontumorous liver tissue (Baumann *et al.* 1990). Biliary epithelial cells of tumor can form sheets undifferentiated into ductules (Hendricks *et al.* 1984; Myers *et al.* 1987).

Mixed Hepato – Cholangiocellular Carcinoma. In the rainbow trout, mixed hepato-cholangiocellular carcinomas are seen at an equal or even greater frequency than the hepatocellular carcinoma (Nunez *et al.* 1989; Nunez *et al.* 1991). Review of other reports on hepatic tumors in laboratory and feral fishes reveals similar findings. The close association of hepatocytes and biliary epithelial cells in the tubular teleost liver may account for this duality of components (Hampton *et al.* 1988). Nunez *et al.* (1991) discuss the possible origin of this neoplasm and present evidence to suggest that both cell types are indeed neoplastic and and not just caught up in the expansion of one or the other neoplastic cell type.

Both laboratory experience and field investigations support the use of cholangiocelluar neoplasms as present biomarkers. A partial listing of species that developed cholangiocytic tumors after exposure to various established carcinogens includes trout (Hendricks *et al.* 1984; Nunez *et al.* 1991), estuarine sheepshead minnow (Couch and Courtney 1987), rivulus, *Rivulus marmoratus*, (Koenig and Chasar 1984), danio (Stanton 1965) and guppy, *Lebistes reticulatus*, (Simon and Lapis 1984). Dawe *et al.* (1964) reported cholangiocellular neoplasms in white suckers of a freshwater lake in Maryland. Hayes *et al.* (1990) found similar lesions in the same species in Lake Ontario. Winter flounder on the east coast (Murchelano and Wolke 1985; Harshbarger and Clark 1990) and English sole on the west coast (Myers *et al.* 1987) collected from contaminated sediments have also developed these lesions.

Many carefully studied near-shore marine environments and freshwater lakes are polluted with chemicals as indicated primarily by measurements of certain organics and metals in sediments. For finfish especially, there are correlations between the occurrence and/or prevalence of cancerous diseases of the liver and the degree of chemical contamination in their environment. For each biomarker, laboratory toxicopathic link and environmental relevance have been demonstrated. After reviewing epizootiology of neoplasms of bony fish of North America, Harshbarger and Clark (1990) concluded that hepatocellular neoplasms were strongly correlated with exposure to chemical contaminants. Carcinogenic risk for the host fish is related to life history and perhaps to species sensitivity.

CONCLUSIONS

In summary, a number of conclusions concerning the toxicologic histopathology of fish can be stated: 1) Our knowledge of the microscopic anatomy of teleost organ systems is sufficient to permit detection of toxicant-induced alterations. 2) Differential coverage has been given to the various organ systems with some having received little attention despite their critical roles in health and disease. 3) Some lesions of infectious disease may mimic alterations associated with toxicant exposure. 4) Dose response relationships coupled to morphologic endpoints have characterized only a few investigations, hampering the utility of the information in risk assessment and promotion of remedial action. 5) A battery of histopathologic biomarker lesions exists and both laboratory exposure data and relevant field observations in fish from contaminated sites are available. 6) Toxicologic histopathology of fish is a meaningful tool for environmental health assessment.

ACKNOWLEDGEMENTS

The assistance of Swee Joo Teh, Mark S. Okihiro, Joseph Groff, Yvonne Garrett and other students and staff of the Aquatic Toxicology Group at UC Davis is gratefully acknowledged. The work was supported in part by U.S. Public Health Service grants CA 45131 and ES04699 from the National Cancer Institute and the National Institute of Environmental Health Sciences.

REFERENCES

Aquaculture in the United States: Constraints and Opportunities. 1978. A report of the Committee on Aquaculture, Board on Agriculture and Renewable Resources, Commission on Natural Resources, National Research Council, National Academy of Sciences, Washington, D.C. 123 p.

Abel, P.D., and J.F. Skidmore. 1975. Toxic effects of an anionic detergent on the gills of rainbow trout. *Water Res.* 9:759-765.

Adams, S.M., editor. 1990. *Biological indicators of stress in fish.* American Fisheries Symposium 8, p. 191, Bethesda, Maryland.

Ahne, W., and I. Thomsen. 1985. The existence of three different viral agents in a tumor bearing European eel (*Anguilla anguilla*). *Zentralbl. Veterinaermed. [B]* 32:228-235.

Anderson, D.P. 1990. Immunological indicators: Effects of environmental stress on immune protection and disease outbreaks. *Am. Fish. Soc. Symp.* 8:38-50.

Anthony, A., E.L. Cooper, R.B. Mitchell, W.H. Neff, and C.D. Thierren. 1971. Histochemical and cytophotometric assay of acid stress in freshwater fish. *Water Pollut. Res. Serv. EPA-18050.*

Ashley, L.M. 1975. Comparative Fish Histology. In *Pathology of Fishes*. Madison: University of Wisconsin Press.

Avella, M., A. Masoni, M. Bornancin, and N. Mayer-Gostan. 1987. Gill morphology and sodium influx in the rainbow trout (*Salmo gairdneri*) acclimated to artificial freshwater environments. *J. Exp. Zool.* 241:159-169.

Baumann, P.C., and S.J. Hamilton. 1984. Vertebral abnormalities in white crappies, *Pomoxis annularis* Rafinesque, from Lake Decatur, Illinois, and an investigation of possible causes. *J. Fish. Biol.* 25:25-33.

Baumann, P.C., and D.M. Whittle. 1988. Status of selected organics in the Laurentian Great Lakes: An overview of DDT, PCB's, dioxins, furans, and aromatic hydrocarabons. *Aquat. Toxicol.* 11:241-257.

Baumann, P.C., W.D. Smith, and W.K. Parland. 1987. Tumor frequencies and contaminant concentrations in brown bullheads from an industrialized river and a recreational lake. *Trans. Am. Fish. Soc.* 116:79-86.

Baumann, P.C., J.C. Harshbarger, and K.J. Hartmann. 1990. Relationship between liver tumors and age in brown bullhead populations from two Lake Erie tributaries. *Sci. Total. Environ.* 94:71-87.

Bengtsson, B.E. 1974. Vertebral damage to minnows (*Phoxinus phoxinus*) exposed to zinc. *Oikos* 25:134-139.

Bengtsson, B.E. 1975. Vertebral damage in fish induced by pollutants. In *Sublethal Effects of Toxic Chemicals on Aquatic Animals*, ed. J. H. Koeman and J.J.T.W.A. Strik, pp. 23-30. New York, Amsterdam: Elsevier.

Bernstein, J.J. 1970. I. Anatomy and physiology of the nervous system. In *Fish Physiology, The Nervous System, Circulation, and Respiration*, ed. W. S. Hoar and D. J. Randall, pp. 1-90. New York: Academic Press.

Black, J.J. 1983a. Epidermal hyperplasia and neoplasia in brown bullheads (*Ictalurus nebulosus*) in response to repeated applications of PAH containing extract of polluted river sediment. In *Polynuclear Aromatic Hydrocarbons: Formation, Metabolism, and Measurements*, ed. M. W. Cook and A. J. Dennis, Columbus, Ohio: Battelle Press.

Black, J.J. 1983b. Field and laboratory studies of environmental carcinogenesis in Niagara River fish. *J. Great Lake Res.* 9:326-334.

Black, J.J., H. Fox, P. Black, and F. Bock. 1985. Carcinogenic effects of river sediment extracts in fish and mice. In *Water Chlorination Chemistry, Environmental Impact and Health Effects*, ed. R. L. Jolly, R. J. Bull, W. P. Davis, S. Katz, M. H. Roberts, Jr., and V. A. Jacobs, pp. 415-427. Chelsea, Michigan: Lewis Publishers.

Bodammer, J.E., and R.A. Murchelano. 1990. Cytological study of vacuolated cells and other aberrant hepatocytes in winter flounder from Boston Harbor. *Cancer Res.* 50:6744-6756.

Bolender, R.P. 1978. Correlation of morphometry and stereology with biochemical analysis of cell fractions. *Intl. Rev. Cytol.* 55:247-289.

Bolender, R.P. 1981. Stereology: Applications to pharmacology. *Ann. Rev. Pharmacol. Toxicol.* 21:549-573.

Boyer, J.L., J. Swartz, and N. Smith. 1976. Biliary secretion in elasmobranchs. II. Hepatic uptake and biliary excretion of organic anions. *Am. J. Physiol.* 230:974-981.

Brown, E.R., E. Koch, T.F. Sinclair, R. Spitzer and O. Callaghan. 1979. Water pollution and diseases in fish (an epizootiological survey). *J. Environ. Pathol. Toxicol.* 2:917-925.

Brown, C.L. and C.J. George. 1985. Age-dependent accumulation of macrophage aggregates in the yellow perch, *Perca flavescens* (Mitchill). *J. Fish. Dis.* 8:135-138.

Brungs, W.A. 1971. Chronic effects of constant elevated temperature on the fathead minnow (*Pimephales promelas* Rafinesque). *Trans. Am. Fish. Soc.* 100:659-664.

Bucke, D. 1972. Some histological techniques applicable to fish tissues. *Symp. Zool. Soc. London* 30:153-189.

Budd, J.J., D. Schroder, and K.D. Dukes. 1975. Tumors of the yellow perch. In *Pathology of Fishes*, ed. W. F. Ribelin and G. Migaki, pp. 895-906. Madison: University of Wisconsin Press.

Bunton, T.E. 1991. Ultrastructure of hepatic hemangiopericytoma in the medaka (*Oryzias latipes*). *Exp. Molec. Pathol.* 54:87-98.

Chevalier, G., L. Gauthier, and G. Moreau. 1985. Histopathological and electron microscopic studies of gills of brook trout, *Salvelinus fontinalis*, from acidified lakes. *Can. J. Zool.* 63:2062-2070.

Christensen, N.O. 1980. Diseases and anomalies in fish and invertebrates in Danish littoral regions which might be connected with pollution. *Rapp. P.-v. Rèun. Cons. Int. Explor. Mer* 179:103-109.

Cope, O.B., E.M. Wood, and G.H. Wallen. 1970. Some chronic effects of 2,4-D on the bluegill (*Lepomis macrochirus*). *Trans. Am. Fish. Soc.* 99:1-12.

Couch, J.A. 1984. Histopathology and enlargement of the pituitary of a teleost exposed to the herbicide trifluralin. *J. Fish Dis.* 7:157-163.

Couch, J.A., and L.A. Courtney. 1987. N-nitrosodiethylamine-induced hepatocarcinogenesis in estuarine sheepshead minnow (*Cyprinodon variegatus*): neoplasms and related lesions compared with mammalian lesions. *J. Nat. Cancer Inst.* 79:297-321.

Couch, J.A., J.T. Winstead, and L.R. Goodman. 1977. Kepone-induced scoliosis and its histological consequences in fish. *Science* 197:585-587.

Couch, J.A., J.T. Winstead, D.J. Hansen, and L.R. Goodman. 1979. Vertebral dysplasia in young fish exposed to the herbicide trifluralin. *J. Fish. Dis.* 2:35-42.

Cross, J.N., and J.E. Hose. 1988. Evidence for impaired reproduction in white croaker (*Genyonemus lineatus*) from contaminated areas off southern California. *Mar. Environ. Res.* 24:185-188.

Cross, J.N., and J.E. Hose. 1989. Reproductive impairment in two species of fish from contaminated areas off southern California. *Oceans '89*.

Daoust, P.Y., G. Wobeser, and J.D. Newstead. 1984. Acute pathological effects of inorganic mercury and copper in gills of rainbow trout. *Vet. Pathol.* 21:93-101.

Dawe, C.J., M.F. Stanton, and F.J. Schwartz. 1964. Hepatic neoplasms in native bottom-feeding fish of Deep Creek Lake, Maryland. *Cancer Res.* 24:1194-1201.

Dawe, C.J. 1981. Polyoma tumors in mice and X cell tumors in fish, viewed through telescope and microscope. In *Phyletic Approaches to Cancer*, ed. C. J. Dawe, J.C. Harshbarger, S. Kondo, T. Sugimura, and S. Takayama. pp. 19-49. Tokyo: JPN Sci. Soc. Press.

Daye, P.G., and E.T. Garside. 1976. Histopathologic changes in surficial tissues of brook trout, *Salvelinus fontinalis* (Mitchill), exposed to acute and chronic levels of pH. *Can. J. Zool.* 54:2140-2155.

Dethlefsen, J. 1980. Observations on fish diseases in the German Bight and their possible relation to pollution. *Rapp. P.-v. Rèun Cons. Int. Explor.* 179:110-117.

DiMichele, L., and M.H. Taylor. 1978. Histopathological and physiological responses of *Fundulus heteroclitus* to naphthalene exposure. *J. Fish. Res. Board Can.* 35:1060-1066.

Donaldson, E.M. 1990. Reproductive indices as measures of the effects of environmental stressors in fish. *Am. Fish. Soc. Symp.* 8:109-122.

Donaldson, E.M., and E. Scherer. 1983. Methods to test and assess effects of chemicals on reproduction in fish. In *Methods for Assessing the Effects of Chemicals on Reproductive Functions*. ed V. B. Vouk and P. J. Sheehan, pp 365-404. Sussex, England: Wiley Publishers.

Droy, B.F., M.R. Miller, T. Freeland, and D.E. Hinton. 1988. Immuno-histochemical detection of CCl_4-induced, mitosis-related DNA synthesis in livers of trout and rat. *Aquat. Toxicol.* 13:155-166.

Egami, N., Y. Kyono-Hamaguchi, H. Mitani, and A. Shima. 1981. Characteristics of hepatoma produced by treatment with diethylnitrosamine in the fish, *Oryzias latipes*. In *Phyletic Approaches to Cancer*, ed C. J. Dawe, J.C. Harshbarger, S. Kondo, T. Sugimura, and S. Takoyama. pp. 217-226. Tokyo: JPN Sci. Soc. Press.

Eller, L.L. 1969. Pathology of redear sunfish exposed to Hydrothol 191. *Trans. Am. Fish. Soc.* 98:52-59.

Ellis, A.E., R.J. Roberts, and P. Tytler. 1978. The anatomy and physiology of teleosts. In *Fish Pathology*, ed. R. J. Roberts, pp. 13-54. London: Balliere Tindall.

Ellis, A.E. (ed). 1985. *Fish and Shellfish Pathology*. pp. 412. New York: Academic Press.

Fabacher, D.L., C.J. Schmitt, J.M. Besser, and M.J. Mac. 1988. Chemical characterization and mutagenic properties of polycyclic aromatic compounds in sediment from tributaries of the Great Lakes. *Environ. Toxicol. Chem.* 7:529-543.

Ferguson, H.W. 1989. *Systemic Pathology of Fish*. Ames: Iowa State University Press. 263 pp.

Fournie, J.W., W.E. Hawkins, R.M. Overstreet, and W.W. Walker. 1987. Exocrine pancreatic neoplasms induced by methylazoxymethanol acetate in the guppy *Poecilia reticulata*. *J. Nat. Cancer Inst.* 78:715-725.

Gardner, G.R. 1975. Chemically induced lesions in estuarine or marine teleosts. In *The Pathology of Fishes*, ed. W.C. Ribelin and G. Migaki, pp. 657-694. University of Wisconsin Press.

Gardner, G.R. and R.J. Pruell. 1987. Quincy Bay Study, Boston Harbor: A histopathological and chemical assessment of winter flounder, lobster and soft-shelled clam indigenous to Quincy Bay, Boston Harbor and an *in situ* evaluation of oysters including sediment (surface and cores) chemistry. U.S. EPA Report, Region I, Boston, MA.

Gardner, G.R., R.J. Pruell, and L.C. Folmar. 1989. A comparison of both neoplastic and non-neoplastic disorders in winter flounder (*Pseudopleuronectes americanus*) from eight areas in New England. *Mar. Env. Res.* 28:393-397.

Gerking, S.D. 1980. Fish reproduction and stress. In *Environmental Physiology of Fishes*, ed. Ali Ma, pp. 569-587. New York: Plenum Press.

Gilderhaus, P.A. 1966. Some effects of sublethal concentrations of sodium arsenite on bluegills and the aquatic environment. *Trans. Am. Fish Soc.* 95:289-296.

Gingerich, W.H. 1982. Hepatic toxicology of fishes. In *Aquatic Toxicology*, ed. L. Weber, pp. 55-105, New York: Raven Press.

Glaiser, J. 1986. *Principles of Toxicological Pathology*. London/Philadelphia: Taylor and Francis.

Goede, R.W., and B.A. Barton. 1990. Organismic indices and an autopsy-based assessment as indicators of health and condition of fish. *Am. Fish Soc. Symp.* 8:93-108.

Greenwald, L., and L.B. Kirschner. 1976. The effect of poly-L lysine on gill ion transport and permeability in the rainbow trout. *J. Membic. Biol.* 26:371-383.

Grizzle, J.M., and W.A. Rogers. 1976. *Anatomy and histology of the channel catfish*. Auburn, Alabama: Auburn Printing.

Grizzle, J.M., P. Melius, and D.R. Strength. 1984. Papillomas on fish exposed to chlorinated wastewater effluent. *J. Nat. Cancer Inst.* 73:1133-1142.

Grizzle, J.M., M.R. Putnam, J.W. Fournie, and J.A. Couch. 1988. Microinjection of chemical carcinogens into small fish embryos: exocrine pancreatic neoplasm in *Fundulus grandis* exposed to *N*-methyl-*N*'-nitro-*N*-nitrosoguanidine. *Dis. Aquat. Org.* 5:101-105.

Grizzle, J.M. and A. Thiyagarajah. 1988. Diethylnitrosamine-induced hepatic neoplasms in the fish *Rivulus ocellatus marmoratus*. *Dis. Aquat. Org.* 5:39-50.

Groman, D.B. 1982. *Histology of the Striped Bass*. *Am. Fish. Monogr.* 3, pp. 116, Bethesda, Maryland: American Fisheries Society.

Hampton, J.A., P.A. McCuskey, R.S. McCuskey, and D.E. Hinton. 1985. Functional units in rainbow trout (*Salmo gairdneri*, Richardson) liver. I. Histochemical properties and arrangement of hepatocytes. *Anat. Rec.* 213:166-175.

Hampton, J.A., R.C. Lantz, P.J. Goldblatt, D.J. Laurén, and D.E. Hinton. 1988. Functional units in rainbow trout (*Salmo gairdneri*, Richardson) Liver: II. The biliary system. *Anat. Rec.* 221:619-634.

Hampton, J.A., R.C. Lantz, and D.E. Hinton. 1989. Functional units in rainbow trout (*Salmo gairdneri*, Richardson) Liver: III. Morphometric analysis of parenchyma, stroma, and component cell types. *Am. J. Anat.* 185:58-73.

Harshbarger, J.C. 1984. Pseudoneoplasms in ectothermic animals. *Nat. Cancer. Inst. Monogr.* 65:251-273.

Harshbarger, J.C., and J.B. Clark. 1990. Epizootiology of neoplasms in bony fish of North America. *Sci. Total Environ.* 94:1-32.

Hawkins, W.E., J.W. Fournie, R.M. Overstreet, and W.W. Walker. 1986. Intraocular neoplasms induced by methylazoxymethanol acetate in Japanese medaka (*Oryzias latipes*). *J. Nat. Cancer Inst.* 76:453-465.

Hawkins, W.E., W.W. Walker, J.S. Lytle, T.F. Lytle, and R.M. Overstreet. 1989. Carcinogenic effects of 7,12-dimethylbenz[a]anthracene on the guppy *Poecilia reticulata*. *Aquat. Toxicol.* 15:63-82.

Hayes, M.A., I.R. Smith, T.H. Rushmore, T.L. Crane, C. Thorn, T.E. Kocal, and H.W. Ferguson. 1990. Pathogenesis of skin and liver neoplasms in white suckers from industrially polluted areas in Lake Ontario. *Sci. Total Environ.* 94:105-123.

Hendricks, J.D., R.O. Sinnhuber, M.C. Henderson, and D.R. Buhler. 1981. Liver and kidney pathology in rainbow trout (*Salmo gairdneri*) exposed to dietary pyrrolizidine (Senecio) alkaloids. *Exp. Molec. Pathol.* 35:170-183.

Hendricks, J.D., T.R. Meyers, and D.W. Skelton. 1984. Histological progression of hepatic neoplasia in rainbow trout (*Salmo gairdneri*). *Nat. Cancer Inst. Monogr.* 65:321-336.

Herman, R.L. 1975. Some lesions in the heart of trout. In *The Pathology of Fishes*, ed. W. E. Ribelin and G. Migaki, pp. 331-342. Madison: University of Wisconsin Press.

Hinton, D.E. 1990. Histological Techniques. In *Methods for Fish Biology*, ed. C.B. Schreck and P.B. Moyle, pp. 191-211. Bethesda: American Fisheries Society.

Hinton, D.E., and J.A. Couch. 1984. Pathobiological measures of marine pollution effects. In *Concepts in Marine Pollution Measurements*, ed. H.H. White, Maryland Sea Grant Publ., Univ. of Maryland, College Park, MD. pp. 7-32.

Hinton, D.E., E.R. Walker, C.A. Pinkstaff and E.M. Zuchelkowski. 1984. Morphological survey of teleost organs important in carcinogenesis with attention to fixation. *Nat. Cancer Inst. Monogr.* 65:291-320.

Hinton, D.E., and D.J. Laurén. 1990a. *Liver structural alterations accompanying chronic toxicity in fishes: potential biomarkers of exposure*. Boca Raton, Florida: Lewis Publishers.

Hinton, D.E., and D.J. Laurén. 1990b. Integrative histopathological approaches to detecting effects of environmental stressors on fishes. *Am. Fish. Soc. Symp.* 8:51-66.

Hinton, D.E., D.J. Laurén, S.J. Teh, and C.S. Giam. 1988a. Cellular composition and ultrastructure of hepatic neoplasms induced by diethylnitrosamine in *Oryzias latipes*. *Mar. Environ. Res.* 24:307-310.

Hinton, D.E., J.A. Couch, S.J. Teh, and L.A. Courtney. 1988b. Cytological changes during progression of neoplasia in selected fish species. *Aquat. Toxicol.* 11:77-112.

Hinton, D.E., P.C. Baumann, G.R. Gardner, W.E. Hawkins, J.D. Hendricks, R.A. Murchelano, and M.S. Okihiro. 1991a. Histopathological biomarkers. In *Biomarkers: Biochemical, Physiological, and Histological Markers of Anthropogenic Stress*, ed. R.J. Huggett, R.A. Kimerle, P.M. Mehrle, and H.L. Bergman. Boca Raton: Lewis Publishers. In Press.

Hinton, D.E., S.J. Teh, M.S. Okihiro, J.B. Cooke and L.M. Parker. 1991b. Phenotypically altered hepatocyte populations in diethylnitrosamine-induced medaka liver carcinogenesis: Resistance, growth, and fate. *Mar. Environ. Res.* In Press.

Hodgins, H.O., B.B. McCain, and J.W. Hawkes. 1977. Marine fish and invertebrate diseases, host disease resistance, and pathological effects of petroleum. In *Effects of Petroleum on Arctic and Subarctic Marine Environments and Organisms*, ed. D.C. Malins, pp. 95-173. New York: Academic Press.

Hoffman, C.L., C.E. Dunbar, and A. Bradford. 1962. Whirling disease of trout caused by *Myxosoma cerebralis* in the United States. U.S. Bureau of Sport Fishing, Wildlife Special Scientific Report, Fisheries 427.

Holcombe, G.W., D.A. Benoit, E.N. Leonard, and J.M. McKim. 1976. Long-term effects of lead exposure on three generations of brook trout (*Salvelinus fontinalis*). *J. Fish. Res. Board Can.* 33:1731-1741.

Hoover, K.L. 1984a. Use of small fish species in carcinogenicity testing. Proceedings of a Symposium. Nat. Cancer Inst. Monogr. 65. U.S. Department of health and Human Services. Public Health Service. National Institutes of Health. 409 pp.

Hoover, K.L. 1984b. Hyperplastic thyroid lesions in fish. *Nat. Cancer Inst. Monogr.* 65:275-289.

Hose, J.E., J.B. Hannah, H.W. Puffer, and M.L. Landolt. 1984. Histologic and skeletal abnormalities in benzo(a)pyrene-treated rainbow trout alevins. *Arch. Environ. Contam. Toxicol.* 13:675-684.

Hughes, G.M., Perry. S.F., and V.M. Brown. 1979. A morphometric study of effects of nickel, chromium and cadmium on the secondary lamellae of rainbow trout gills. *Water Res.* 13:665-679.

Jenson, N.J., and J.L. Larsen. 1978. The ulcer syndrome in cod in Danish coastal waters. *Intl. Counc. Expl. Sea* 1978/E:28. 15 p.

Johnson, L.L., E. Casillas, T.K. Collier, B.B. McCain, and U. Varanasi. 1988. Contaminant effects on ovarian development in English sole (*Parophrys vetulus*) from Puget Sound, Washington. *Can. J. Fish. Aquat. Sci.* 45:2133-2146.

Kent, M.L., M.S. Myers, D.E. Hinton, W.D. Eaton, and R.A. Elston. 1988. Suspected toxicopathic hepatic necrosis and megalocytosis in pen-reared Atlantic Salmon *Salmo salar* in Puget Sound, Washington, USA. *Dis. Aquat. Org.* 49:91-100.

Kimura, T., M. Yoshimizu, and M. Tanaka. 1981a. Fish viruses: tumor induction in *Oncorhynchus keta* by the herpes virus. In *Phyletic Approaches to Cancer*, ed. C.J. Dawe, J.C. Harshbarger and S. Kondo, T. Sugimura, and S. Takayama, pp. 59-68. Tokyo, JPN Sci. Soc. Press.

Kimura, T., M. Yoshimizu, M. Tanaka, and H. Sannohoe. 1981b. Studies on a new virus (OMV) from *Oncorhynchus masou* - I. Characteristics and pathogenicity. *Fish. Pathol.* 15:143-147.

Kimura, I., N. Taniguchi, H. Kumai, I. Tomita, N. Kinae, K. Yoshizaki, M. Ito, and T. Ishikawa. 1984. Correlation of epizootiological observations with experimental data: Chemical induction of chromatophoromas in the croaker, *Nibea mitsukurii*. *Nat. Cancer Inst. Monogr.* 65:139-154.

Kimura, I. 1988. Aquatic pollution problems in Japan. *Aquat. Toxicol.* 11:287-301.

Kinae, N., M. Yamashita, I. Tomita, I. Kumura, H. Ishida, H. Kumai, and G. Nakamura. 1990. A possible correlation between environmental chemicals and pigment cell neoplasia in fish. *Sci. Total Environ.* 94:143-153.

Klaunig, J.E., M.M. Lipsky, B.F. Trump, and D.E. Hinton. 1979. Biochemical and ultrastructural changes in teleost liver following subacute exposure to PCB. *J. Environ. Pathol. Toxicol.* 2:953-963.

Koenig, C.C., and M.P. Chasar. 1984. Usefulness of the hermaphroditic marine fish *Rivulus marmoratus*, in carcinogenicity testing. *Nat. Cancer Inst. Monogr.* 65:15-33.

Kranz, H., and V. Dethlefsen. 1990. Liver anomalies in dab (*Limanda limanda*) from the southern North Sea with special consideration given to neoplastic lesions. *Dis. Aquat. Org.* 9:171-185.

Kubota, S.S., T. Miyazaki, and S. Egusa. 1982. *Color atlas of fish histopathology*. Tokyo: Shin-Suisan Shingun-sha.

Kurubagaran, R., and K.P. Joy. 1988. Toxic effects of mercuric chloride, methylmercuric chloride, and Emisan 6 (an organic mercurial fungicide) on ovarian recrudescence in the catfish *Clarias batrachus* L. *Bull. Environ. Contam. Toxicol.* 41:902-909.

Laurén, D.J., and D.G. McDonald. 1985. The effects of copper on branchial ion regulation in the rainbow trout, *Salmo gairdneri* Richardson - Modulation by water hardness and pH. *J. Comp. Physiol.* 155:635-644.

Laurén, D.J., S.J. Teh and D.E. Hinton. 1990. Cytotoxicity phase of diethylnitrosamine-induced hepatic neoplasia in medaka. *Cancer Res.* 50:5504-5514

Laurent, P., H. Hobe and S. Dunel-Erb. 1985. The role of environmental sodium chloride relative to calcium in gill morphology of freshwater salmonid fish. *Cell Tissue Res.* 240:675-692.

Leatherland, J.F., and R. Sonstegard. 1978. Structure of normal testis and testicular tumors of cyprinids from Lake Ontario. *Cancer Res.* 38:3164-3173.

Lesniak, J.A., and S.M. Ruby. 1982. Histological and quantitative effects of sublethal cyanide exposure on oocyte development in rainbow trout. *Arch. Environ. Contam. Toxicol.* 13:101-104.

Lock, R.A.C., and A.P. van Overbeeke. 1981. Effects of mercuric chloride and methylmercuric chloride on mucus secretion in rainbow trout (*Salmo gairdneri*) Richardson. *Comp. Biochem. Physiol.* 69C:67-73.

Loud, A.V., and P. Anversa. 1984. Biology of disease: Morphometric analysis of biologic processes. *Lab. Invest.* 50:250-261.

Lucké, B., and H.G. Schlumberger. 1941. Transplantable epitheliomas of the lip and mouth of catfish. *J. Exp. Med.* 74:397-408.

Malins, D.C., B.B. McCain, D.W. Brown, S-L Chan, M.S. Myers, J.T. Landahl, P.G. Prohaska, A.J. Friedman, L.D. Rhodes, W.D. Gronlund, and H.O. Hodgins. 1984. Chemical pollutants in sediments and diseases in bottom-dwelling fish in Puget Sound, Washington. *Sci. Technol.* 18:705-713.

Malins, D.C., M.M. Krahn, D.W. Brown, L.D. Rhodes, M.S. Myers, B.B. McCain, and S-L Chan. 1985a. Toxic chemicals in marine sediment and biota from Mukilteo, Washington: relationships with hepatic neoplasms and other hepatic lesions in English sole (*Parophrys vetulus*). *J. Nat. Cancer Inst.* 74:487-494.

Malins, D.C., M.M. Krahn, M.S. Myers, L.D. Rhodes, D.W. Brown, C.A. Krone, B.B. McCain, and S-L Chan. 1985b. Toxic chemicals in sediments and biota from creosote-polluted harbor: Relationships with hepatic neoplasms and other hepatic lesions in English sole (*Parophrys vetulus*). *Carcinogen* 6:1463-1469.

Mallatt, J. 1985. Fish gill structural changes induced by toxicants and other irritants: a statistical review. *Can. J. Fish Aquat. Sci.* 42:630-648.

Masahito, P., T. Ishikawa, and H. Sugano. Neural and pigment cell neoplasms. In *Neoplasms and Related Disorders in Fishes*, ed. C.J. Dawe, In Press.

Mayer, F.L., P.M. Mehrle, and R.A. Schoettger. 1977. *Collagen metabolism in fish exposed to organic chemicals.* Corvallis, Oregon: U.S. Environmental Protection Agency, Ecological Research Series, EPA/600/3-77-085.

McAllister, P.E., T. Nagabayashi, and K. Wolf. 1977. Viruses of eels with and without stomatopapillomas. *Ann. N.Y. Acad. Sci.* 298:233-244.

McCain, B.B., D.W. Brown, M.K. Krahn, M.S. Myers, Jr., R.C. Clark, S.L. Chan, and D.C. Malins. 1988. Marine pollution problems, North American West Coast. *Aquat. Toxicol.* 11:143-162.

McCarthy, J.F., and L.R. Shugart, eds. 1990a. In *Biomarkers of Environmental Contamination*, pp. 457. Boca Raton, Florida: Lewis Publishers, CRC Press, Inc.

McCarthy, J.F., and L.R. Shugart. 1990b. Biological markers of environmental contamination. In *Biomarkers of Environmental Contamination*, pp. 3-14. Boca Raton, Florida: Lewis Publishers, CRC Press, Inc.

McCormick, J.H., G.N. Stokes, and R.O. Hermanatz. 1989. Oocyte atresia and reproductive success in fathead minnows (*Pimephales promelas*) exposed to acidified hardwater environments. *Arch. Environ. Contam. Toxicol.* 18:207-214.

McDonald, D.G. 1983. The effects of H^+ upon the gills of freshwater fish. *Can. J. Zool.* 61:691-703.

McIntyre, J.D. 1973. Toxicity of methylmercury for steelhead trout sperm. *Bull. Environ. Contam. Toxicol.* 9:98-99.

McKim, J.M. 1985. Early life stage toxicity tests. In *Fundamentals of Aquatic Toxicology*, eds. G.M. Rand and S.R, Petrocelli, pp. 58-95. Washington: Hemisphere Publishing.

Mehrle, P.M., and F.L. Mayer. 1975. Toxaphene effects on growth and development of brook trout (*Salvelinus fontinalis*). *J. Fish Res. Board Can.* 32:609-613.

Mehrle, P.M., and F.L. Mayer. 1977. Bone development and growth of fish as affected by toxaphene. In *Fate of Pollutants in Air and Water Environments. Part 2*, ed I. H. Suffet, pp. 301-304. New York: Wiley Interscience Publishing.

Meyers, T.R., and J.D. Hendricks. 1985. Histopathology. In *Fundamentals of Aquatic Toxicology*, ed. G. M. Rand and S. R. Petrocelli, pp. 283-331. Washington: Hemisphere.

Miller, T.G., and W.C. Mackay. 1982. Relationship of secreted mucus to copper and acid toxicity in rainbow trout. *Bull. Environ. Contam. Toxicol.* 28:68-74.

Miller, M.R., C. Heyneman, S. Walker, and R.G. Ulrich. 1986. Interaction of monoclonal antibodies directed against bromodeoxyuridine with pyrimidine bases, nucleosides, and DNA. *J. Immunol.* 136:1791-1795.

Miller, M.R., D.E. Hinton, and J.J. Stegeman. 1989. Cytochrome P-450 E induction and localization in gill pillar (endothelial) cells of scup and rainbow trout. *Aquat. Toxicol.* 28:68-74.

Mittal, A.K., and J.S.D. Munshi. 1974. On the regeneration and repair of superficial wounds in the chin of *Rita rita* (Ham.) (Bazridae, Pisces). *Acta Anatom.* 88:424-442.

Moon, T.W., P.J. Walsh, and T.P. Mommsen. 1985. Fish hepatocytes: A model metabolic system. *Can. J. Fish Aquat. Sci.* 42:1772-1782.

Moore, M.J., R. Smolowitz, and J.J. Stegeman. 1989. Cellular alterations preceding neoplasia in *Pseudopleuronectes americanus* from Boston Harbor. *Mar. Environ. Res.* 28:425-429.

Mottet, N.K., and M.L. Landolt. 1987. Advantages of using aquatic animals for biomedical research on reproductive toxicology. *Environ. Health Perspect.* 71:69-75.

Möller, H. 1985. A critical review on the role of pollution as a cause of fish diseases. In *Fish and Shellfish Pathology*. ed. A. E. Ellis, pp. 169-182. New York: Academic Press.

Mulcahy, M.F., and A. O'Leary. 1970. Cell-free transmission of lymphosarcoma in the northern pike *Esox lucius* L. (Pisces: Esocidae). *Experientia* 26:891.

Murai, T., and J. Andrews. 1978. Riboflavin requirements of channel catfish fingerlings. *J. Nutrition* 108:1512-1517.

Muramoto, S. 1981. Vertebral column damage and decrease of calcium concentration in fish exposed experimentally to cadmium. *Environ. Pollut.* 24:125-133.

Murchelano, R.A., and J. Ziskowski. 1979. Some observations on an ulcer disease of red hake, *Urophycis chuss*, from the New York Bight. *Int. Counc. Explor. Sea.* C.M. 1979/E:23, 5 p.

Murchelano, R.A. 1982. Some pollution-associated diseases and abnormalities of marine fish and shellfish: A perspective for the New York Bight. In *Ecological Stress and the New York Bight: Science and Management,*. ed. G. F. Mayer, pp. 327-346. Columbia, SC: Estuar. Res. Fed.

Murchelano, R.A., and R.E. Wolke. 1985. Epizootic carcinoma in the winter flounder *Pseudopleuronectes americanus*. *Science* 228:587-589.

Murty, A.S. 1986. *Toxicity of Pesticides to Fish*, Volume II, pp. 143. Boca Raton, Florida: CRC Press.

Myers, M.S., L.D. Rhodes, and B.B. McCain. 1987. Pathologic anatomy and patterns of occurrence of hepatic neoplasms, putative preneoplastic lesions, and other idiopathic hepatic conditions in English sole (*Parophrys vetulus*) from Puget Sound, Washington. *J. Nat. Cancer Inst.* 78:333-363.

Myers, M.S., J.T. Landahl, M.M. Krahn, L.L. Johnson, and B.B. McCain. 1990. Overview of studies on liver carcinogenesis in English sole from Puget Sound; Evidence for a xenobiotic chemical etiology: pathology and epizootiology. *Sci. Total Environ.* 94:33-50.

Nakazawa, T., S. Hamaguchi, and Y. Kyono-Hamaguchi. 1985. Histochemistry of liver tumors induced by diethylnitrosamine and differential sex susceptibility to carcinogenesis in *Oryzias latipes*. *J. Nat. Cancer Inst.* 75:567-573.

Nunez, O., J.D. Hendricks, D.N. Arbogast, A.T. Fong, B.C. Lee and G.S. Bailey. 1989. Promotion of aflatoxin B_1 hepatocarcinogenesis in rainbow trout by 17 β-estradiol. *Aquat. Toxicol.* 15:289-302.

Nunez, O., J.D. Hendricks, and J.R. Duimstra. 1991. Ultrastructure of hepatocellular neoplasms in aflatoxin B_1 (AFB_1)-initiated rainbow trout (*Oncorhynchus mykiss*). *Toxicol. Pathol.* 19:11-23.

O'Connor, J.M., and R.J. Huggett. 1988. Aquatic pollution problems, North Atlantic coast, including Chesapeake Bay. *Aquat. Toxicol.* 11:163-190.

Okihiro, M.S. 1988. Chromatophoromas in two species of Hawaiian butterflyfish, *Chaetodon multicinctus* and *C. miliaris*. *Vet. Pathol.* 25:422-431.

Okihiro, M.S., J.M. Whipple, J.M. Groff, and D.E. Hinton. 1991. Chromatophoromas and related hyperplastic lesions in Pacific Rockfish (*Sebastes*). *Mar. Env. Res.* In Press.

Packer, R.K., and W.A. Dunson. 1972. Anoxia and sodium loss associated with the death of brook trout and low pH. *Comp. Biochem. Physiol.* 41A:17-26.

Papas, T.S., J.E. Dahlberg, and R.A. Sonstegard. 1976. Type C virus in lymphosarcoma in northern pike (*Esox lucius*). *Nature* 261:506-508.

Papas, T.S., T.W. Pry, M.P. Schafer, and R.A. Sonstegard. 1977. Presence of DNA polymerase in lymphosarcoma in northern pike (*Esox lucius*). *Cancer Res.* 37:3214-3217.

Perry, S.F., and C.M. Wood. 1985. Kinetics of branchial calcium uptake in the rainbow trout: Effects of acclimation to various external calcium levels. *J. Exper. Biol.* 116:411-433.

Pickering, A.D., T.G. Pottinger, and P. Christie. 1982. Recovery of the brown trout, *Salmo trutta* L., from acute handling stress: A time course study. *J. Fish. Biol.* 20:229-244.

Pitot, H.C. 1988. Hepatic neoplasia: Chemical induction. In *The Liver: Biology and Pathology*, ed. I. M. Arias, W. B. Jakoby and H. Popper, pp. 1125-1146. New York: Raven Press.

Pritchard, J.B., and J.L. Renfro. 1982. Interactions of xenobiotics with teleost renal function. In *Aquatic Toxicology*, ed. L.J. Weber, pp. 51-106. New York: Raven Press.

Reid, S.D., and D.G. McDonald. 1988. Effects of cadmium, copper, and low pH on ion fluxes in rainbow trout, *Salmo gairdneri*. *Can. J. Fish. Aquat. Sci.* 45:244-253.

Reide, U.N., and A. Reith. 1980. *Morphometry in Pathology*. New York: Gustav Fisher Verlag.

Reimschuessel, R., R.O. Bennett, E.B. May, and M.M. Lipsky. 1990. Development of newly formed nephrons in the goldfish kidney following hexachlorobutadiene-induced nephrotoxicity. *Toxicol. Pathol.* 18:32-38.

Ribelin, W.E., and G. Migaki, eds. 1975. *The Pathology of Fishes*. Madison: University of Wisconsin Press.

Roberts, R.J. 1975. Melanin-containing cells of teleost fish and their relation to disease. In *The Pathology of Fishes*, ed. W.E. Ribelin and G. Migaki, pp. 399-428. Madison: The University of Wisconsin Press.

Roch, M., and E.J. Maly. 1979. Relationship of cadmium-induced hypocalcemia with mortality in rainbow trout (*Salmo gairdneri*) and the influence of temperature on toxicity. *J. Fish. Res. Board Can.* 36:1297-1303.

Rohr, H.P., M. Oberholzer, G. Bartsch, and M. Keller. 1976. Morphometry in experimental pathology: Methods, baseline data and application. *Intl. Rev. Exper. Pathol.* 15:233-325.

Rosenthal, H., and D.F. Alderdice. 1976. Sublethal effects of environmental stressors, natural and pollutional, on marine eggs and larvae. *J. Fish. Res. Board Can.* 33:2047-2065.

Saiki, M., and T. Mori. 1955. Studies on the distribution of administered zinc in the tissues of fishes. *Bull. JPN Soc. Sci. Fish.* 21:945-949.

Sangalang, G.B., and M.J. O'Halloran. 1972. Cadmium-induced testicular injury and alteration of androgen synthesis in brook trout. *Nature (London)* 240:470-471.

Sangalang, G.B., H.C. Freeman, and R. Crowell. 1981. Testicular abnormalities in cod (*Gadus morhua*) fed Aroclor 1254. *Arch. Environ. Contam. Toxicol.* 10:617-627.

Sauer, G.R., and N. Watabe. 1984. Zinc uptake and its effect on calcification in the scales of the mummichog, *Fundulus heteroclitus*. *Aquat. Toxicol.* 5:51-66.

Scarpelli, D.G., M.H. Grieder, and W.J. Frajola. 1963. Observations on hepatic cell hyperplasia, adenoma, and hepatoma of rainbow trout (*Salmo gairdneri*). *Cancer Res.* 23:848-857.

Schlumberger, H.G., and B. Lucké. 1948. Tumors of fishes, amphibians, and reptiles. *Cancer Res.* 8:657-754.

Schmidt, D.C., and L.J. Weber. 1973. Metabolism and biliary excretion of sulfobromophthalein by rainbow trout (*Salmo gairdneri*). *J. Fish. Res. Board Can.* 30:1301-1308.

Schoor, W. P., and J.A. Couch. 1979. Correlations of mixed-function oxidase activity with ultrastructural changes in the liver of a marine fish. *Cancer Biochem. Biophys.* 4:95-103.

Schultz, M.E. and R.J. Schultz. 1985. Transplatable chemically-induced liver tumors in the vivparous fish Poeciliopsis. *Exp. Molec. Pathol.* 42:320-330.

Schwerdtfeger, W.K. 1979a. Morphometrical studies of the ultrastructure of the epidermis of the guppy, *Poecilia reticulata* Peters, following adaptation to seawater and treatment with prolactin. *Gen. Comp. Endocrinol.* 38:476-483.

Schwerdtfeger, W.K. 1979b. Qualitative and quantitative data on the fine structure of the guppy (*Poecilia reticulata* Peters) epidermis following treatment with thyroxine and testosterone. *Gen. Comp. Endocrinol.* 38:484-490.

Segner, H., and H. Möller. 1984. Electron microscopical investigations on starvation-induced liver pathology in flounders *Platichthys flesus*. *Mar. Ecol. Prog. Ser.* 19:193-196.

Segner, H., and J.V. Juario. 1986. Histological observations on the rearing of milkfish, (*Chanos chanos*), by using different diets. *J. Appl. Ichthyol.* 2:162-173.

Segner, H., and T. Braunbeck. 1988. Hepatocellular adaptation to extreme nutritional conditions in ide, *Leuciscus idus melanotus* L. (Cyprinidae). A morphofunctional analysis. *Fish. Physiol. Biochem.* 5:79-97.

Simon, K., and K. Lapis. 1984. Carcinogenesis studies on guppies. *Nat. Cancer Inst. Monogr.*65:71-81.

Sindermann, C.J. 1979. Pollution-associated diseases and abnormalities of fish and shellfish: A review. *Fish. Bull.* 76:717-749.

Sindermann, C.J. 1990. *Principal Diseases of Marine Fish and Shellfish*, Volume I, 2nd Edition, pp. 521. New York: Academic Press.

Skidmore, J.F., and P.W.A. Tovell. 1972. Toxic effects of zinc sulfate on the gills of rainbow trout. *Water Res.* 6:217-120.

Sloof, W. 1982. Skeletal anomalies in fish from polluted surface waters. *Aquat. Toxicol.* 2:157-173.

Smith, C.E. 1984. Hyperplastic lesions of the primitive meninx of fathead minnows, *Pimephales promelas*, induced by ammonia: species potential for carcinogen testing. *Nat. Cancer Inst. Monogr.* 65:119-125.

Smith, I.R., K.W. Baker, M.A. Hayes, and M.A. Ferguson. 1989. Ultrastructure of malpighian and inflammatory cells in epidermal papillomas of white suckers *Catastomus commersoni*. *Dis. Aquat. Org.* 6:17-26.

Solangi, M.A., and R.M. Overstreet. 1982. Histopathological changes in two estuarine fishes, *Menidia beryllina* (Cope) and *Trinectes maculatus* (Bloch and Schneider), exposed to crude oil and its water soluble fractions. *J. Fish Dis.* 5:13-35.

Sonstegard, R.A. 1977. Environmental carcinogenesis studies in fishes of the Great Lakes of North America. *Ann. N.Y. Acad. Sci.* 298:261-269.

Sorenson, E.M.B., P.M. Cumbie, T.C. Bauer, J.S. Bell, and C.W. Harlan. 1984. Histopathological, hematological, condition factor, and organ weight changes associated with selenium accumulation in fish from Belews Lake, North Carolina. *Arch. Environ. Contam. Toxicol.* 13:153-162.

Sparks, A.K. 1972. *Invertebrate Pathology, Non-Communicable Diseases*. New York: Academic Press.

Spencer, S.L. 1967. Internal injuries of largemouth bass and bluegills caused by electricity. *Prog. Fish-Cult.* 29:168-169.

Stanton, M.F. 1965. Diethylnitrosamine-induced hepatic degeneration and neoplasia in the aquarium fish, *Brachydanio rerio*. *J. Nat. Cancer Inst.* 344:117-130.

Stegeman, J.J., R.L. Binder, and A. Orren. 1979. Hepatic and extrahepatic microsomal electron transport components and mixed-function oxygenases in the marine fish *Stenotomus versicolor*. *Biochem. Pharmacol.* 28:3431-3429.

Stegeman, J.J., M.R. Miller, and D.E. Hinton. 1989. Cytochrome P-450 induction and localization in endothelium of vertebrate heart. *Molec. Pharmacol.* 36:723-729.

Stehlik, L.L., and J.V. Merriner. 1983. Effects of accumulated dietary Kepone on spot (*Leiostomus xanthurus*). *Aquat. Toxicol.* 3:345-358.

Stehr, C. 1990. Ultrastructure of vacuolated cells in the liver of rock sole and winter flounder living in contaminated environments. Proceedings of the XIIth International Congress for Electron Microscopy, pp. 522-523.

Stott, G.G., N.H. McArthur, R. Tarpley, V. Jacobs, and R.F. Sis. 1981. Histopathologic survey of ovaries of fish from petroleum production and control sites in the Gulf of Mexico. *J. Fish. Biol.* 18:261-269.

Tafanelli, R., and R.C. Summerfelt. 1975. Cadmium-induced histopathological changes in goldfish. In *The Pathology of Fishes*. ed. W. E. Ribelin and G. Migaki, pp. 613-645. Madison: University of Wisconsin Press.

Tan, B., P. Melus, and J. Grizzle. 1981. Hepatic enzymes and tumor histopathology of black bullheads with papillomas. In *Polynuclear Aromatic Hydrocarbons: Chemical Analysis and Biological Fate*, ed. M. Cooke and A. J. Dennis, pp. 377-386. Columbus, OH: Battelle Press.

Thiyagarajah, A., and J.M. Grizzle. 1986. Diethylnitrosamine-induced pancreatic neoplasms in the fish *Rivulus ocellatus marmoratus*. *J. Nat. Cancer Inst.* 77:141-147.

Thomas, P., M. Bally, and J.M. Neff. 1982. Ascorbic acid status of mullet *Mugil cephalus* Linn., exposed to cadmium. *J. Fish. Biol.* 20:183-196.

Trump, B.F., R.T. Jones, and S. Sahaphong. 1975. Cellular effects of mercury on fish kidney tubules. In *The Pathology of Fishes*, ed. W. E. Ribelin and G. Migaki, pp. 585-612. Madison: University of Wisconsin Press.

Trump, B.F., E.M. McDowell, and A.U. Arstilia. 1980. Cellular reaction to injury. In *Principles of Pathobiology*, ed. R. B. Hill and M. F. LaVia, pp. 20-111. New York: Oxford University Press.

Vaillant, C., C. Le Guellec, and F. Padkel. 1988. Vitellogenin gene expression in primary culture of male rainbow trout hepatocytes. *Gen. Comp. Endocrinol.* 70:284-290.

Valentine, D.W. 1975. Skeletal anomalies in marine teleosts. In *The Pathology of Fishes*, ed. W. R. Ribelin and G. Migaki, Madison: University of Wisconsin Press.

van Bohemen, CH. G., J.G.D. Lambert, and J. Peute. 1981. Annual changes in plasma levels of yolk proteins in *Salmo gairdneri*. *Gen. Comp. Endocrinol.* 40:319-000.

Van de Kamp, G. 1977. Vertebral deformities of herring around the British Isles and their usefulness for a pollution monitoring programme. Int. Counc. Explor. Sea. Continuous Monitoring in Biological Oceanography 1977-E:5, 10 pp.

Varanasi, U., P.A. Robisch, and D.C. Malins. 1975. Structural alterations in fish epidermal mucus produced by water-borne lead and mercury. *Nature (London)* 258:431-432.

Varanasi, U., and D.J. Gmur. 1978. Influence of water-borne and dietary calcium on uptake and retention of lead by coho salmon (*Oncorhynchus kisutch*). *Toxicol. Appl. Pharmacol.* 46:65-75.

Verbost, P.M., G. Flik, R.A.C. Lock, and S.E. Wendelaar Bonga. 1987. Cadmium inhibition of Ca^{2+} uptake in rainbow trout gills. *Am. J. Physiol.* 253:R216-R221.

Vogelbein, W.K., J.W. Fournie, P.A. Van Veld, and R.J. Huggett. 1990. Hepatic neoplasms in the mummichog *Fundulus heteroclitus* from a creosote-contaminated site. *Cancer Res.* 50:5978-5986.

Walker, Jr., W.F. 1987. *Functional Anatomy of the Vertebrates, An Evolutionary Perspective*, pp. 781. Philadelphia: CBS College Publishing.

Walsh, A.H., and W.E. Ribelin. 1975. The pathology of pesticide poisoning. In *The Pathology of Fishes*. ed. W. E. Ribelin and G. Migaki, pp. 515-577. Madison: University of Wisconsin Press.

Walton, M.J., and C.B. Cowey. 1982. Aspects of intermediary metabolism in salmonid fish. *Comp. Biochem. Physiol.* 73B:59-79.

Weibel, E.R. 1979. *Stereological Methods*, Vol I. New York: Academic Press.

Weibel, E.R. 1980. *Stereological Methods*, Vol. II. New York: Academic Press.

Weis, J.S. and P. Weis. 1989. Effects of environmental pollutants on early fish development. *Aquat. Sci.* 1:45-73.

Wellings, S.R. 1969. Neoplasia and primitive vertebrate phylogeny: echinoderms, prevertebrates, and fishes - a review. *Nat. Cancer Inst. Monogr.* 31:59-128.

Wellings, S.R., B.B. McCain, and B.S. Miller. 1976. Epidermal papillomas in Pleuronectidae of Puget Sound, Washington. *Prog. Exp. Tumor. Res.* 20:55-74.

Wester, P.W., and J.H. Canton. 1986. Histopathological study of *Oryzias latipes* (Medaka) after long-term β-hexachlorocyclohexane exposure. *Aquat. Toxicol.* 9:21-45.

Westernhagen, H. von, H. Rosenthal and K.R. Sperling. 1974. Combined effects of cadmium and salinity on development and survival of herring eggs. *Helgol. Wiss. Meeresunters.* 26:416-433.

Wilcock, B.P., and T.W. Dukes. 1989. The eye. In *Systemic Pathology of Fish*. ed. H. W. Ferguson, pp. 156-173. Ames: Iowa State University Press.

Wolke, R.E., R.A. Murchelano, C.D. Dickstein, and C.J. George. 1985. Preliminary evaluation of the use of macrophage aggregates (MA) as fish health monitors. *Bull. Env. Contam. Toxicol.* 35:222-227.

Wyllie, A.H., J.F.G. Kerr, and A.R. Cumi. 1980. Cell death: The Significance of apoptosis. *Intl. Rev. Cytol.* 68:251-306.

Yamamoto, Y., M. Sato, and S. Ikeda. 1978. Existence of L-gulonolactone oxidase in some teleosts. *Bull. JPN Soc. Sci. Fish* 44:775-779.

Yamashita, M., N. Kinae, I. Kimura, H. Ishida, H. Kumai, and G. Nakamura. 1990. The croaker *Nibea mitsukurii* and the sea catfish *Plotosus anguillaris*: Useful biomarkers of coastal pollution. In *Biomarkers of Environmental Contamination*, ed. J. F. McCarthy and L. R. Shugart, pp. 73-84. Boca Raton, Florida: Lewis Publishers, CRC Press, Inc.

Yasutake, W.T., and J.H. Wales. 1983. Microscopic anatomy of salmonids: an atlas. U.S. Fish and Wildlife Service Resource Publication 150.

Yoshimizu, M., M. Tanaka, and T. Kimura. 1987. *Onchorhynchus masou* virus (OMV): Incidence of tumor development among experimentally infected representative salmonid species. *Fish. Pathol.* 22:7-10.

Zuchelkowski, E.M., R.C. Lantz, and D.E. Hinton. 1981. Effects acid-stress on epidermal mucous cells of the brown bullhead *Ictalurus nebulosus* (Le Sueur): A morphometric study. *Anat. Rec.* 200:33-39.

Zuchelkowski, E.M., R.C. Lantz, and D.E. Hinton. 1986. Skin mucous cell response to acid stress in male and female brown bullhead catfish, *Ictalurus nebulosus* (Le sueur). *Aquat. Toxicol.* 8:139-148.

8

Pathobiology of Selected Marine Mammal Diseases

Romona Haebler
Robert B. Moeller, Jr.

INTRODUCTION

Marine mammals are the subset of true mammals that have evolved unique characteristics for survival in an aquatic environment. The five major groups of marine mammals, cetacea, pinnipeds, mustelids, sirenia, and ursids, did not descend from a common phylogenetic line, and therefore vary greatly in their anatomy, physiology, biochemistry, and habitat requirements. Due to these differences, the incidence and types of their diseases also vary. Marine mammals include:

- Order: Cetacea -- whales and dolphins in two suborders, Odontocetes, (toothed whales and dolphins) and Mysticetes (baleen whales).
- Order: Pinnipedia -- pinnipeds, including 33 species of seals, sea lions, and walruses.
- Order: Sirenia -- only two living species, the dugong and manatee.
- Order: Carnivora -- the family Mustelidae with one marine species, the sea otter, and the family Ursidae with one marine species, the polar bear. (Diseases of polar bears will not be included in this review.)

This chapter reviews the pathobiology, especially gross and microscopic changes, associated with certain diseases of marine mammals. Since a detailed discussion of marine mammal pathobiology is beyond the scope of this document, diseases were selected based on prevalence and significance within feral populations of marine mammals. For further details on marine mammal disease and medicine, see Howard (1983) and Dierauf (1990).

An effort was made to point out the relationships between certain stress factors and disease processes. Stressors that affect marine mammal health frequently act in combination to cause pathological change. For example, viral disease is often complicated by secondary bacterial infection, and parasitic

0-8493-8662-4/93/$0.00 + $.50

disease may weaken the animal and increase susceptibility to other infectious organisms. When investigating a disease, it is critical to identify all relevant stressors that may have contributed to pathological changes. Often the primary stressor that initiates a disease process differs from the ultimate cause of death.

VIRAL DISEASES

Numerous viral agents identified in marine mammals have been implicated in explosive epizootics involving groups of susceptible animals in particular geographic locations, e.g., an outbreak of influenza in seals off Cape Cod (Geraci *et al.* 1982) and distemper in seals of the North Sea (Kennedy *et al.* 1988b).

Virology of marine mammals is a relatively new field of study and few cell lines are established (Smith and Skilling 1979). This is a problem when trying to isolate an agent in the face of a serious disease outbreak. One should not hesitate to try conventional cell lines in attempts to culture a putative pathogenic virus. Some viral agents may pose a threat to both marine and terrestrial mammals and could have human health implications, e.g., vesicular exanthema which affects sea lions, pigs, and man. The mutation of viruses to more virulent strains or from nonpathogens to pathogens is a threat to marine mammals, particularly those in danger of extinction.

Pox Viruses

Seal pox virus was first identified in California sea lions by Wilson in 1969. This disease affects numerous species of captive and wild pinnipeds, including harbor seals, California sea lions and South American sea lions (Wilson 1970; Wilson and Poglayen-Neuwall 1971; Wilson *et al.* 1972a; Wilson *et al.* 1972b; Wilson *et al.* 1972c; Dunn and Spotte 1974).

Seal pox is characterized by the formation of multiple, proliferative cutaneous nodules 2-3 cm in diameter. Nodules are most common over head and neck, but may occur anywhere on the body. Lesions often ulcerate and are slow to heal. Focal alopecia is commonly associated with healed areas. The social behavior of head and neck rubbing, which is common in sea lions and other pinnipeds, may allow cutaneous spread of the virus (Wilson 1970; Wilson and Poglayen-Neuwall 1971).

The histologic lesions of seal pox differ, depending on the species of pinniped affected. In California sea lion and harbor seal, there is marked acanthosis with hyperkeratosis and parakeratosis, causing upward proliferation of the epidermis. Affected cells of the stratum spinosum undergo hydropic degeneration and develop prominent 2-15 μm in diameter eosinophilic intracytoplasmic inclusion bodies. Histologic features of pox viral infections in the South American sea lion include downward proliferation of the epidermis with ulceration and pustule formation resembling molluscum contagiosum. Affected epithelial cells contain single, large intracytoplasmic inclusion bodies (Wilson and Poglayen-Neuwall 1971). Though uncomplicated seal pox is not known to be fatal, the disease may cause high morbidity. Clinical course of the disease is approximately 15 weeks.

Dolphin pox virus causes skin lesions in dolphins commonly referred to as "tattoos" (Geraci *et al.* 1979). Clinically, this disease demonstrates prominent, discrete lines of hyperpigmentation of the epidermis. Lesions are usually flat, but occasionally may be raised or depressed. Various design patterns are described as circles, pinholes, or targets. Lesions occur most commonly on the dorsal body, flippers, dorsal fins, and fluke.

Histologically, lesions consist of hydropic degeneration of deep layers of the stratum intermedium. Irregularly shaped or round intracytoplasmic inclusions are present in cells undergoing ballooning degeneration. This is an unusual pox lesion since it is not proliferative. Inflammatory response is minimal and may be partially responsible for persistence of virus and lesions (Geraci *et al.* 1979). Cause of hyperpigmentation is unknown.

Although pox virus does not appear to be detrimental to cetaceans, development and spread of these lesions appear to coincide with periods of stress and poor health (Geraci *et al.* 1979). The virus persists for long periods (several months or years) and slowly spreads. Animals with the typical tattoo lesion do not appear to develop antibodies to the pox virus. Once the animal begins to develop antibodies to this pox virus, the lesion regresses and affected skin becomes raised and bleached. Later the affected area undergoes necrosis and sloughs. If affected skin is biopsied or scraped, the lesion usually regresses in a zonal pattern around the biopsy site (Smith *et al.* 1983).

Calici Virus

A calici virus identified as San Miguel Sea Lion Virus (SMSLV) was first isolated in 1972 (Smith *et al.* 1973). The disease, which occurs in sea lions and seals, is characterized by the formation of vesicles on the flippers which may rupture and form prominent ulcers. Recently, lesions have also been observed on the lips, nose, chin, and gums. Histologically, early lesions consist of spongiosis of the stratum spongiosum, which later progress to subcorneal vesicle formation. Inclusion bodies are not present in the cytoplasm or nucleus.

SMSLV has been isolated from aborted fetuses of sea lions and fur seals. Its role in the death of fetuses or neonates is unclear. Isolation of the virus from clinically normal northern elephant seal pups has been accomplished (Smith and Skilling 1979). Several serotypes of SMSLV have been isolated from aborting sea lions. Such serotypes are indistinguishable from serotypes of vesicular exanthema virus that cause abortions in swine. The relationship of SMSLV infection and abortions in other marine mammals is suspected but not proven (Smith and Boyt 1990).

SMSLV has been isolated from the opaleye fish (*Girella nigricans*) (Smith and Boyt 1990). This fish is known to develop an active infection with viral replication in the spleen. The fish remains infected for at least 31 days. Clinical disease has not been observed in infected fish (Smith and Boyt 1990). A similar calici virus has been isolated from several Atlantic bottlenose dolphins with vesicular lesions that eroded and left shallow ulcers (Smith *et al.* 1983). SMSLV neutralizing antibodies have been found in 10 species of marine mammals including: California sea lion, Stellar sea lion, northern fur seal, northern elephant seal, Hawaiian monk seal, walrus, bowhead whale, California grey whale, Sei whale, sperm whale, and fin whale. This suggests that infection may be widespread in marine mammals throughout the North Pacific (Smith and Boyt 1990). At least one serotype of SMSLV has been isolated from a human accidentally exposed in a laboratory. This individual developed blisters on hands and feet and had flu-like symptoms (Smith and Boyt 1990). Neutralizing antibodies to marine calici viruses were also identified in humans working with these viruses in the laboratory and field (Smith and Latham 1978).

Infectious organisms that usually affect marine organisms may cross the land:sea interface. SMSLV is thought to be responsible for vesicular exanthema of swine (VESV), a deadly disease that first appeared in 1932 (Barlough *et al.* 1986). In isolated outbreaks of disease in the 1930's, 1940's, and 1950's, an intense eradication effort took place at the cost of $39 million in federal indemnities. Transmission of the disease is now thought to be associated with feeding uncooked garbage of marine origin to swine. Infectious virus can clearly cross the land:sea interface and poses risks to terrestrial mammals.

Adenoviruses

In 1978, an adenovirus was identified as the cause of acute hepatitis in five California sea lions stranded near Los Angeles, California (Britt *et al.* 1979). This was the first report of a lethal viral disease in marine mammals. Serological surveys demonstrated that though large numbers of animals were exposed to this adenovirus, the mortality rate was low with only a few outbreaks reported. (Dierauf *et al.* 1981; Britt and Howard 1983b).

Sea lions affected by this agent may die acutely or peracutely with little or no clinical evidence of infection. Gross signs may include icterus, splenomegaly, mesenteric lymphadenopathy, and liver discoloration. Liver lesions consist of coagulative and lytic necrosis of hepatocytes. Often, necrosis is most severe in the centrilobular region of hepatic lobules. Large intranuclear inclusions are present in hepatocytes and occasionally in Kupffer cells. Inflammatory response is usually minimal, with few macrophages, lymphocytes, and neutrophils at the periphery of the necrotic lesions.

The source of sea lion hepatitis virus remains unknown. Though raw sewage that had been dumped into California coastal waters prior to the first reported outbreak may have been involved in this disease, a clear causative role remains speculative (Britt *et al.* 1979). Adenoviruses are present in many mammalian species. Though there are some morphological differences, adenoviral hepatitis in sea lions is similar to infectious canine hepatitis. Further research is needed to determine whether hepatitis in sea lions is caused by an adenovirus specific to sea lions or whether a terrestrial species may have been introduced into the marine environment through sewage effluent or agricultural run-off.

Adenoviruses have been reported in only two other species of marine mammals: a single Sei whale in Antarctica and one bowhead whale in Alaska (Smith and Skilling 1979). In both cases, the virus was an incidental finding.

Influenza Virus

An influenza virus designated A/Seal/Mass/1/80 was identified as cause of a mass epidemic in harbor seals near Cape Cod, Massachusetts from December 1979 to October 1980 (Geraci *et al.* 1982). This agent resembled fowl plague virus and was thought to have been transmitted by sea birds.

Within a dense population of affected seals, morbidity and mortality were high. The course of the disease was acute to peracute. Seals were weak and in respiratory distress with a bloody or white frothy discharge from mouth and nose. Many developed pulmonary, mediastinal and/or subcutaneous emphysema. Death occurred within hours or days. Histologically, lungs had severe bronchopneumonia with hemorrhage in alveoli and prominent necrosis of bronchioles and bronchi. Bacteria and influenza virus may have interacted synergistically to produce this fatal disease process, as seals experimentally challenged with influenza virus developed only a mild respiratory disease. In affected seals, mycoplasma was consistently isolated from all animals. The role of influenza virus in other incidents of seal disease is unknown. Antibodies to A/Seal/Mass/1/80 were identified in grey seals 500 miles north of Cape Cod.

Herpes Viruses

Several herpes viruses have been isolated from harbor seals and a California sea lion (Borst *et al.* 1986; Kennedy-Stoskopf *et al.* 1986). The herpesvirus affecting harbor seals has been known to cause a serious systemic infection. Affected animals demonstrated an acute pneumonia and hepatitis. The pneumonia was characterized as a diffuse interstitial pneumonia with multifocal fibrinous exudation and emphysema. The liver demonstrated massive necrosis of the hepatic parenchyma with minimal mononuclear cell infiltrates. Intranuclear inclusion bodies were not observed. The virus is characterized as a probable alpha-herpesvirus (Brost *et al.* 1986; Frey *et al.* 1989). Although intranuclear inclusions were not present in affected seals in the European outbreaks, cases of seal herpes virus at the Armed Forces Institute of Pathology have demonstrated prominent intranuclear inclusion bodies in the areas of necrosis.

The nature of this virus and the potential to cause disease in wild and captive populations of seals are unknown. This virus was inoculated in a group of young seals with only minimal upper respiratory signs developing (Horvat *et al.* 1989). Phocine herpes viruses likely act similarly to other mammalian herpes viruses; infections are probably most often fatal in young and seriously stressed or debilitated animals.

A herpesvirus has been observed to cause dermatitis in beluga whales (Martineau *et al.* 1988; Barr *et al.* 1989). Lesions consisted of multiple discrete raised pale grey areas on the skin. Histologically, the superficial epidermis undergoes intracellular edema with necrosis and microvesicle formation. Prominent eosinophilic intranuclear inclusion bodies are present in the infected cells. These lesions have been observed in whales in oceanaria and from the St. Lawrence estuary. Again, effects of stress and possible contamination from industrial waste may play an important role in this disease.

A herpesvirus has recently been identified at the Armed Forces Institute of Pathology and has been implicated as cause for extensive oral lesions in sea otters. Clinically, these lesions consist of variably sized irregular white plaques and/or deep, often bilaterally symmetrical ulcers. In severely affected animals the ulcers tend to coalesce to cover extensive areas of the buccal, labial, gingival, and glossal mucosa. Histologically, the lesions reveal extensive chronic ulcers with associated mixed bacterial colonies and separate foci of epithelial necrosis and intracellular edema. Numerous eosinophilic intranuclear inclusion bodies are observed in degenerating and necrotic cells. Although these lesions may be extensive, they do not appear to adversely affect the animals' ability to eat. Infection may go unnoticed if an oral examination is not performed.

Morbillivirus: Pinniped Distemper Virus

A morbillivirus similar to, yet antigenically distinct from, canine distemper virus has killed thousands of seals in Northern Europe (Kennedy *et al.* 1989). Clinically, affected seals were weak and had respiratory distress with a mucopurulent to serous oculonasal discharge. Many animals had subcutaneous emphysema. At necropsy, these animals had edematous lungs with sharply demarcated areas of red consolidation. Emphysema involving interlobular septa and pleura of caudal lung lobes was present. Congestion and a thick mucopurulent exudate were observed in the upper respiratory tract. Many animals also developed emphysema of the mediastinum, neck, and subcutis. Hydropericardium, hydrothorax, and hepatic congestion were common.

Lung lesions consisted of bronchostitial pneumonia with syncytial cells and type II pneumocyte proliferation. Occasional intracytoplasmic inclusions were present. The brain had a nonsuppurative encephalitis characterized by necrosis of neurons (primarily in cerebral cortex), nonsuppurative perivascular cuffs, and gliosis. Many infected neurons contained intranuclear and intracytoplasmic inclusions. Demyelination of the subependymal white matter was observed. Lymphoid tissue had prominent depletion and necrosis of lymphocytes. Several seals had developed a necrotizing nonsuppurative myocarditis. Intranuclear and intracytoplasmic inclusions were observed in gastric mucosa and transitional epithelium of the urinary bladder and renal pelvis (Kennedy *et al.* 1989).

A morbillivirus causing similar lesions was also observed in harbor porpoises found dead in the Irish Sea during the 1988 outbreak. These animals developed pulmonary and central nervous system lesions similar to those observed in affected seals (Kennedy *et al.* 1988b, 1991).

BACTERIAL DISEASES

Bacterial diseases are a leading cause of death in marine mammals (Howard *et al.* 1983b). Though bacterial infections are most often secondary to other conditions, such as parasitic infections or traumatic injury, certain organisms cause specific disease processes in marine mammals. This discussion focuses primarily on bacteria that have gained notoriety as pathogens in marine mammals.

The incidence of bacterial disease differs among species of marine mammals depending on habitat and behavior. Bacterial species and density vary according to proximity with human populations and discharge of human sewage into nearshore waters. Marine mammals, such as pinnipeds, who haul out on land and engage in physical contact, such as fighting, are at higher risk for bacterial infections than other species. The great whales, for example, who live a more isolated life, are at lower risk.

The immune system of animals plays an important role in protecting them from bacterial agents. Individuals that are immunocompromised, either by stress, viral agents, or possibly immunosuppressive chemical agents, are at a greater risk of becoming infected with pathogenic bacteria. Infected animals can act as carriers and ultimately sources of infection to other marine mammals and man. Most of the bacterial diseases affecting marine mammals can cause disease in terrestrial mammals and man.

Erysipelas

Erysipelothrix rhusiopathiae causes two distinct forms of disease in dolphins: an acute septicemic disease and chronic dermatologic disease (Siebold and Neal 1956; Simpson *et al.* 1958; Geraci *et al.* 1966). The acute septicemic disease usually causes sudden death with no clinical signs or gross lesions (Medway 1980). At necropsy, affected animals may occasionally demonstrate multifocal areas of necrosis and inflammation involving numerous organs. The agent can usually be isolated on culture of lymph nodes, heart blood, kidney, liver, and through animal inoculation of the isolate (Medway 1980).

The dermatologic disease is characterized by dermal infarction that results in sloughing of the epidermis (Simpson *et al.* 1958; Geraci *et al.* 1966). Occasionally, micro-infarcts result in rhomboid areas of cutaneous necrosis. Affected animals are reluctant to move and, in a few cases, erosions of the humeroscapular joint have been found (Medway 1980). As with other animals that develop dermatologic signs associated with erysipelas, the dermal infarction is probably a secondary sequela to a septicemia. If the animals are not treated with antibiotics, they may die.

Erysipelas, a systemic bacterial disease first described in swine in 1885, occurs in many species including birds, rodents, sheep, horses, and man. The agent has been isolated from many species of fish. Ingestion may be the route of infection for cetaceans. The disease has been successfully controlled in captive dolphins by vaccination with either killed or modified live vaccine. Humans who handle diseased or dead animals should take precautions since *E. rhusiopathiae* poses a zoonotic threat.

Leptospirosis

Leptospirosis, primarily caused by *Leptospira pomona*, causes renal disease and abortions in California sea lions and Northern fur seals. The agent, a spirochete, is responsible for zoonotic disease in many species of animals worldwide (Vedros *et al.* 1971; Smith *et al.* 1974a, 1977; Dierauf *et al.* 1985).

California sea lions with renal disease caused by leptospirosis exhibit clinical, gross, and histopathological changes. Animals often are depressed, anorexic, pyrexic, and exhibit various degrees of posterior limb paresis. Other signs include: icterus, oral erosions or ulcers, and excessive thirst. Clinically, most animals have a leukocytosis, increased blood urea nitrogen, and creatinine values which indicate renal disease. At necropsy, the kidneys are swollen.

Histologically, a lymphoplasmacytic, interstitial nephritis is present. The spirochete can be identified with dark field microscopy or Warthin-Starry technique in lumina of renal tubules.

Renal disease due to infection from *L. pomona* is the most prevalent urinary disease in California sea lions and is thought to be endemic to sea lion rookeries in California (Sweeny 1986). An epizootic of leptospirosis in California sea lions occurred in animals from central California to Seattle, Washington, and was most prevalent in young adult males (Dierauf *et al.* 1985).

Leptospira pomona infection may cause reproductive failure in California sea lions and Northern fur seals (Smith *et al.* 1974a,b). In newborns and aborted fetuses, the disease is characterized as hemorrhagic septicemia with subcutaneous hemorrhage and hemorrhage into the anterior chamber of the eye, commonly referred to as "red eye."

Mode of transmission of the agent to pinnipeds is not fully understood: rodents and fish may act as reservoirs for *L. pomona*. Some species of wildlife, such as deer, can be infected with the leptospiral organisms. This disease may have human health significance, especially to inhabitants who participate in the fur seal harvest on St. Paul Island where many people show serologic evidence of exposure to the organism (Medway 1980). Reports indicate that sea lions can continue to shed leptospiras in urine for 154 days (Dierauf *et al.* 1985). Because sea lions frequently inhabit near-shore waters in urban areas, leptospirosis may be considered of public health significance.

Pseudomonas Infection

Various species of *Pseudomonas* have been incriminated in causing bacterial disease in both pinnipeds and cetaceans. *Pseudomonas aeruginosa* has caused bronchopneumonia and multiple large cutaneous ulcers in Atlantic bottlenose dolphins. The bacteria are known to progress deep into the cutaneous tissue, causing serious damage to the animal (Diamond *et al.* 1979; Migaki 1987).

Pseudomonas pseudomallei is a pathogen found in Southeast Asia. The organism is a contaminant in water and is thought to gain entrance into the animal through cutaneous wounds. A septicemia soon follows and may result in death. Animals often die peracutely with no clinical signs. Grossly and histologically, lesions are characterized by multifocal areas of necrosis and inflammation of many organs. This agent is infectious to man and may be transmitted via cuts or abrasions as well as aerosol transmission (Medway 1980).

Edwardsiella and *Salmonella* Infection

Species of *Edwardsiella* and *Salmonella* are common pathogens that occur in both cetaceans and pinnipeds (Howard *et al.* 1983b). These Gram-negative enteric bacteria may cause a severe necrotizing enteritis that may be complicated by septicemia. Occasionally, a severe embolic bronchopneumonia has been associated with these organisms. Clinically, affected animals usually demonstrate severe enteritis with necrosis and/or hemorrhage. Septicemic animals usually die peracutely with no clinical signs (Gilmartin *et al.* 1979; Howard *et al.* 1983b). As with many mammals, potential for the marine mammal to become a carrier of *Salmonella* is of concern. Therefore, stranded animals should be monitored for enteric infection.

Dermatophilosis

A cutaneous disease affecting pinnipeds is caused by the actinomycete *Dermatophilus congolensis* (Frese *et al.* 1972). The disease in South American sea lions is described as a scruffy, pustular, or exudative dermatitis, often involving the entire body. Lesions are elevated above the skin, forming prominent scabs. Histologic examination shows characteristic Gram-positive cocci in parallel rows. Similarities between gross lesions of *D. congolensis* and seal pox requires the diagnostician to carefully evaluate histologic lesions in affected animals (Frese *et al.* 1972; Medway 1980; Migaki and Jones 1983). With infections of *Dermatophilus*, mortality is low but morbidity is high.

Sea Otter Hemorrhagic Gastroenteropathy

The cause of sea otter hemorrhagic gastroenteropathy is unknown (Rausch 1953; Lauckner 1985). Animals that develop this disease are usually severely stressed or debilitated. Isolation of bacteria from affected animals has failed to identify a specific bacterial agent conclusively. Affected sea otters develop bloody diarrhea and are usually depressed, anorectic, and often die. Histologically, there is little damage to the intestinal mucosal epithelium: prominent pooling of blood in the mucosa and submucosa is often observed (Rausch 1953). Pathogenesis of this disease is unclear; some feel that the gastrointestinal tract may act as a shock organ; pooling of blood in the intestinal mucosa and diapedesis of blood into the intestinal lumen may give the gross appearance of hemorrhagic enteritis.

Mycobacteriosis

Several species of *Mycobacterium* have been isolated from seals, *M. bovis, M. fortuitum, M. cheloni*, a California sea lion, *M. smegmatis*, and several manatees, *M. chelonei* and *M. marinum*. These infections have been presented as either nonhealing chronic cutaneous lesions or generalized infections with granulomas in various organs. Numerous acid-fast bacteria are observed in these lesions. Because several species of *Mycobacterium* are found in soil and water, persistent nonhealing cutaneous lesions should be cultured for *Mycobacterium*. These organisms have a zoonotic potential; care should be taken when treating and handling these wounds (Boeuer *et al.* 1976; Howard *et al.* 1983a; Morales 1985; Cousins 1987; Gutter 1987; Wells *et al.* 1990).

MYCOTIC DISEASES

Mycotic disease in marine mammals is often systemic and may contribute to mortality (Sweeney *et al.* 1976; Medway 1980; Migaki and Jones 1983). Many mycotic organisms are opportunistic or secondary invaders and therefore pose a serious threat to animals with compromised immune systems. Predisposing disease risk factors include: malnutrition, preexisting disease, coexisting septicemia, prolonged drug therapy, or immune suppression (Migaki and Jones 1983). Infection occurs due to physical contact with the skin or inhalation of spores (Migaki and Jones 1983). Mycotic infections are usually not considered contagious. Most case reports of mycotic disease in marine mammals involve captive animals. Only a few common mycotic diseases will be discussed. For a more thorough review, see Migaki and Jones (1983).

Lobomycosis

Lobomycosis, a fungal disease caused by *Loboa loboi*, affects the skin of Atlantic bottlenose dolphins (Migaki *et al.* 1971a; Caldwell *et al.* 1975). Lesions are restricted only to skin and may be located anywhere on the body. Gross lesions are multiple, whitish, crusty, and nodular. The organism is thought to enter the skin through abrasions and spread via auto-inoculation (Migaki and Jones 1983). Histologically, there is a histiocytic dermatitis containing round yeast forms (5-10 μm in diameter) that connect to each other and form chains. The epidermis over these areas of inflammation is acanthotic with downward growth of rete pegs. Histiocytes are the most common cell type in granulomas of the dermis. Often histiocytes coalesce to form giant cells. Clinically, the animals are not seriously affected by the growth of the organism. However, as these areas become larger, animals may become debilitated and die usually due to secondary bacterial infections. Treatment of this fungal infection has not been successful. Removal of the affected area has shown positive results.

Lobomycosis has been reported in only man and dolphins. In humans, it occurs primarily in natives living in tropical forests of South and Central America. With the exception of one Guiana dolphin (de Vries and Laarman

1973), all cases of lobomycosis in animals have occurred in *Tursiops truncatus*, the Atlantic bottlenose dolphin. Further, all *Tursiops* affected by this disease originated in Florida. This limited species specificity is not understood.

Candidiasis

Candida albicans is the most common species of *Candida* to cause clinical disease in pinnipeds and cetaceans. Infected animals usually develop cutaneous and/or intestinal infections. Lesions often occur at mucocutaneous junctions, including the blowhole and vagina (Medway 1980), and appear as yellow to white creamy plaques. In internal organs, lesions are discrete white areas due to necrosis and inflammation (Migaki and Jones 1983). Ulcerations of the esophagus are believed to be pathognomonic for candidiasis (Medway 1980). Histologically, large colonies of septate hyphae, pseudohyphae, and blastospores can be identified. This organism is extremely common and may be identified histologically as an incidental finding. When the organism invades healthy tissue, the organism is considered pathogenic (Medway 1980). Candidiasis is most commonly observed in severely stressed animals. Most species of marine mammals appear to be susceptible to the organism; case reports include several species of pinnipeds, small cetaceans, and a killer whale.

Other Mycotic Diseases

Other mycotic diseases reported in marine mammals include: sporotrichosis (Migaki *et al.* 1978), coccidioidomycosis (Reed *et al.* 1976), North American blastomycosis (Williamson *et al.* 1959), cryptococcoses (Migaki 1987), histoplasmosis (Wilson *et al.* 1974), aspergillosis (Sweeney *et al.* 1976; Migaki and Jones 1983; Carrol *et al.* 1986), zygomycosis (Sweeney *et al.* 1976), fusariomycosis (Montali *et al.* 1981), and dermatophytosis (Dilbone 1965; Migaki and Jones 1983). Most of these diseases have been reported in captive marine mammals; incidence in feral populations is not well understood.

PARASITIC DISEASES

Parasitic infection causes serious health problems for many marine mammals. The source of parasitic infection is often through ingestion of contaminated food sources rather than directly from the environment, since many parasites of marine mammals have indirect life cycles (Howard *et al.* 1983a). Animals stressed by predisposing factors, such as preexisting disease, malnutrition, or immune suppression, are especially susceptible to the effects of parasitic infection. Environmental stressors, such as climactic change or pollution, may further weaken animals and increase susceptibility to parasitic effects. As discussed earlier, secondary bacterial infections often complicate or exacerbate parasitic disease.

Protozoan Diseases

Toxoplasma gondii has been observed in several species of marine mammals including: the California sea lion (Migaki *et al.* 1977), harbor seal, Northern fur seal (Holshuh *et al.* 1985), Atlantic bottlenose dolphin (Innskeep *et al.* 1990), spinner dolphin (Migaki *et al.* 1990), and manatee (Buergelt and Bonde 1983).

Histologically, infected animals demonstrate disseminated infection with necrosis of numerous organs containing *Toxoplasma* cysts. Affected organs include heart, brain, liver, lung, lymph node, and stomach. Infection with *T. gondii* has been recorded in many species of mammals and birds. The definitive hosts are both domestic and feral felines. Infection by this organism occurs through ingestion of either tissue cysts, which may be found in certain meats, or oocysts, which are passed in cat feces. Exposure may result from contamination of pinniped haul-out areas with feces of either domestic or feral felines, sewage effluent discharge into nearshore waters, or placental transfer. Manatees are herbivorous animals who live in near-shore waters and are known to drink fresh water from sewage outflows. Though still speculative, exposure in the manatee may have resulted from drinking oocyst-contaminated water from a sewage outflow (Buergelt and Bonde 1983). This mechanism has been documented to account for infections of humans from creek water contaminated by jungle cats (Beneson *et al.* 1982). Though this theory has not been proven, it is important to illustrate potential hazards associated with introduction of mammalian pathogens into the marine environment. Exposure of endangered marine mammals to previously unknown pathogens may pose serious health risks.

Transplacental exposure to *T. gondii* has been postulated as the probable route of infection in a newborn harbor seal (Van Pelt and Dietrich 1973). The pup was captured in Cold Bay, Alaska, when it was approximately 1-h-old and died 23 days later. At necropsy, the liver had multiple foci of necrosis; histologically, numerous cysts of *T. gondii* were identified. Transplacental infection is documented in other mammalian species, including man. Infection of the neonate, especially when concurrent with other stressors (e.g., malnutrition or bacterial infection), may adversely impact neonatal survival.

Sarcocystis species have been reported in many species of pinnipeds and cetaceans including: northern fur seal (Brown 1974), bearded seal (Bishop *et al.* 1979), ringed seal (Migaki and Albert 1980), Atlantic pilot whale, sperm whale, striped porpoise, and common dolphin (Munday *et al.* 1978). These protozoan organisms appear to be incidental findings at necropsy. Zoite-filled cysts were identified in myofibers of skeletal muscle and there was no surrounding inflammation. Little is known about their life cycle.

Ectoparasites

In pinnipeds, sucking lice, members of the order Anoplura, are common external parasites. *Antarcophthirus microchir* are especially common on sea lion pups. Lice cause alopecia in infected animals and severe infestations may lead to debilitation of the animals. In many species of pinnipeds, pediculosis is more severe in young animals than in adults. This may be due to crowded conditions in the rookeries (Howard *et al.* 1983a) or concurrent disease in the young. Lice may act as vectors of disease. Though not yet proven, lice are thought to be the intermediate host of the filariid nematodes of pinnipeds (Conlogne *et al.* 1980).

Both large and small species of cetaceans are known to harbor lice. Though several species of lice have been identified, none are known to be associated with clinical disease.

Demodex zalophi, a mite of the family Demodicidae, causes a mange-like condition in California sea lions. *Demodex zalophi* is similar to *D. canis*, demodectic mange mite of dogs, but is thought to be unique to California sea lions (Nutting and Dailey 1980; Howard *et al.* 1983a; Migaki 1987). Affected animals develop alopecia and thickening of the skin over flippers, ventral body, and genitalia. Lesions are characterized by hyperkeratosis, scaling, and excoriation. Mites live in hair follicles and can be identified by skin scraping.

Endoparasites

Parasites of Pinnipeds

Lung mites are common parasites of the nasal passages, trachea, and bronchioles in seals and sea lions (Kim *et al.* 1980). The most common mites observed are *Orthohalarachne diminuata* and *O. attenuata*. The former inhabits the airways of the lung, and *O. attenuata* are found in the nasopharynx (Kim *et al.* 1980). Mites can be seen as small white specks on the respiratory mucosa. Clinically, lung mite infection can cause copious amounts of mucus in the upper respiratory tract and nasal passages, nasal discharge, dyspnea, and coughing (Kim *et al.* 1980; Sweeney 1974).

Parafilaroides decorus is the most common lungworm in young California sea lions (Howard *et al.* 1983a). The intermediate host for this metastrongyloid nematode is the opaleye fish, *Girella nigricans* (Dailey 1970). After ingestion of the fish and release of the larvae into the gastrointestinal tract, the larvae migrate to the lungs where maturation occurs in alveoli. Females release larvae, which migrate up respiratory airways, are swallowed and discharged in the feces. Histologically, uncomplicated infections result in goblet cell hyperplasia of bronchiolar epithelium where there is mucoid obstruction. A suppurative or granulomatous bronchopneumonia is observed if secondary bacterial infection occurs. Grossly, lungs have a patchy or mottled appearance of red and grey hepatization. Other *Parafilarioides* species have been identified in lungs of other species of pinnipeds (Dailey 1970; Migaki *et al.* 1971; Sweeney 1974).

Otostrongylus circultitus is another large nematode parasite that inhabits primary and secondary bronchi of harbor seals and northern elephant seals (Dailey and Stroud 1978). This parasite causes prominent bronchiectasis, with bronchiectatic abscesses being present. Intact bronchi and bronchioles have marked goblet cell hyperplasia of peribronchiolar glands. Mucus plugs may fill bronchioles. Again a suppurative or granulomatous pneumonia may result if secondary bacterial infection occurs (Sweeney 1974).

Contracecum and *Anisakis* are the most common stomach worms of both pinnipeds and cetaceans (Howard *et al.* 1983a). These parasites cause ulceration of gastric mucosa and submucosa during migration. Occasionally, anisakine parasites may induce nodules in gastric mucosa and submucosa. On histologic examination, parasites are found in ulcerated areas. In toothed whales (Odontocetes), these parasites are found in the first and third chambers of the stomach. Though these parasites are usually considered incidental findings, they have been known to cause hemorrhage and melena in both pinnipeds and cetaceans. Life cycle is not fully understood: a crustacean is thought to be the first intermediate host and a fish the second. The Pacific herring is the second intermediate host for *Anisakis* species (Young and Lowe 1969).

Hookworms are common in sea lions and northern fur seals (Olsen and Lyons 1965; Keys 1965; Brown *et al.* 1974; Sweeney 1974). *Uncinaria lucasi* is the most pathologic hookworm in northern fur seal pups. Neonates become infected during ingestion of milk. These larvae become adults in several weeks and cause severe hemorrhage in the intestines, resulting in anemia. Surviving pups shed parasites after three months and become reinfected with third stage larvae by penetration of skin or ingestion. These larvae migrate to the blubber of the ventral abdomen and remain dormant until the larvae are shed in milk.

Zalophotrema hepaticum is the liver fluke of sea lions. This fluke is found in the common bile duct, intrahepatic bile duct, and gallbladder. Though infection usually causes minimal damage, occasionally these flukes cause cystic cavitation of hepatic parenchyma (Howard *et al.* 1983a).

Acanthocephalans, usually of the genus *Corynosoma*, are common intestinal parasites of sea otters and pinnipeds (Rausch 1953; Howard *et al.* 1983a) These parasites bury their head deep into mucosa and submucosa and are usually found in the lower small intestine or colon. In harbor seals, parasites are usually considered incidental findings, but they burrow through the intestinal wall in sea otters and cause peritonitis (Rausch 1953; Keys 1965; Bishop 1979).

Heartworm disease in pinnipeds is caused by both *Dirofilaria immitis* and *Dipetalonema spirocauda*. The latter is the heartworm most commonly found in feral pinnipeds, primarily harbor seals (Dunn and Wolke 1976; Howard *et al.* 1983a). The life cycle is unknown, but the louse is suspected to be the intermediate host. These parasites are observed in the right ventricle of the heart and pulmonary arteries. Severe infection causes dilatation of the ventricle and myocardial hypertrophy. Microscopically, pulmonary arteries and arterioles have prominent intimal proliferation. Secondary chronic passive congestion occurs in severely affected animals (Perry 1969; Forrester *et al.* 1973; Medway and Wieland 1975; Dunn and Wolke 1976; Eley 1981). *Dirofilaria immitis* is occasionally observed in captive pinnipeds and the disease process is similar to that caused by *D. spirocauda*.

The identification of microfilaria is helpful in separating *D. immitis* from *D. spirocauda*. Microfilaria of *D. immitis* are usually larger than *D. spirocauda*. However, microfilaria from *D. spirocauda* cannot be differentiated morphologically from the common nonpathogenic subcutaneous filariid *Dipetalonema odendhali* (Forrester *et al.* 1973), which is often in subcutaneous tissue, in the intermuscular fascia, beneath the parietal peritoneum, free in the abdominal and thoracic cavities, and in the pericardial sac (Perry 1969).

Parasites of Cetaceans

Nasitrema are flukes common in head sinuses of porpoises, dolphins and many toothed whales (Dailey and Stroud 1978; Howard *et al.* 1983a). The parasite is located in submucosal glands of sinuses and occasionally in the middle ear. Though *Nasitrema* rarely cause problems, these nematodes occasionally migrate to the brain and cause serious central nervous system damage. Numerous strandings have been associated with parasitic migration into brain. The life cycle is unknown (Migaki *et al.* 1971b; Dailey and Stroud 1978).

Stenurus species of nematodes are located in lungs of harbor porpoise and dall porpoise. The subpleural nodules are filled with the parasites. Occasionally this parasite will migrate to the tympanic bullae and air sinuses (Johnston and Ridgway 1969; Migaki *et al.* 1971b; Dailey and Stroud 1978).

The nematode *Halocerus* is a lungworm that affects dolphins and porpoises. Parasites inhabit bronchi; heavy infestations may lead to bronchopneumonia.

Granulomas often form within bronchi, become encapsulated, and calcify. *Halocerus* are thought to have a direct life cycle (Woodard *et al.* 1969; Migaki *et al.* 1971b; Dailey and Stroud 1978; Howard *et al.* 1983a).

Several types of flukes are found in stomachs of cetaceans (Woodward *et al.* 1969). *Pholeter gastrophilus* occurs in the second stomach chamber of dolphins. The parasite burrows deep into submucosa, forming prominent small nodules that can be identified by palpation. The mucosa overlying these parasitic nodules usually remains intact (Migaki *et al.* 1971a). *Braunina cordiformis* is another fluke found in the second chamber of the dolphin stomach. Damage caused to gastric mucosa is minimal (Schryver *et al.* 1967).

Cyclorchis campula is a trematode that primarily inhabits bile and pancreatic ducts. This parasite causes extensive irritation of ducts with hyperplasia of ductal epithelium and fibroplasia around the ducts. This chronic irritation may progress to a chronic fibrosing hepatitis and pancreatitis (Migaki *et al.* 1971b; Migaki *et al.* 1979).

NEOPLASMS

The incidence and pathogenesis of neoplasia in marine mammals are poorly understood (Howard *et al.* 1983c). Until recently, reports of neoplastic disease in marine mammals were so rare that there was speculation that marine mammals may have some type of inherent resistance to the development of neoplasms (Howard *et al.* 1983c). With more thorough necropsies and larger numbers of animals examined, types and prevalences of tumors described in marine mammals have increased. Case reports usually describe only individual animals and often refer to captive animals. Assessment of prevalence in feral populations is logistically difficult, and greater scientific efforts are needed.

Interpretation of tumor prevalence must be done cautiously, since several factors may bias reporting of species or organ system involved. Species representation may be skewed toward small marine mammals, such as seals or dolphins, as opposed to large whales, since: 1) detailed necropsies are more commonly performed in the small animals; and 2) seals and dolphins are more frequently maintained in captivity than large whales. Other species that may be overrepresented because of increased likelihood of examination: 1) live in near-shore waters near urban centers; or 2) were killed in native or commercial hunts. Reporting of organ system involved may be biased when tumors involve tissues that are: 1) easy to examine, such as skin; or 2) have other biological significance that requires examination, such as the reproductive system.

In a detailed review of neoplasms in marine mammals, Howard *et al.* (1983c) report that neoplasms in cetaceans appeared to be rather infrequent and, of those found, most were benign. Only five of 36 tumors (14%) were described as malignant. The genital tract was the site of 12 of 36 tumors (33%); 17 tumors (almost 50%) were classified as fibromas or lipomas.

Howard and coauthors reviewed a total of 58 neoplasms in pinnipeds. They cautioned, however, that the larger number of tumors in pinnipeds compared to cetaceans probably resulted from the greater likelihood of detailed necropsies in these animals. Pinnipeds appear to have a higher incidence of malignant neoplasms, with 34 of the 58 tumors (nearly 60%) classified as malignant compared to 14% in cetaceans. Tumors in the genital tract occurred with similar frequency in pinnipeds (28%) and cetaceans (33%); neoplastic disease of the hematopoietic system involved 10 tumor types (18%).

In a study of California sea lions beached along the southern California coast, Howard *et al.* (1983c) reported 35 tumors of which 25 were malignant. Most malignancies involved the lung, biliary system, and genitourinary system. Carcinomas of the lung and biliary system had not previously been reported in pinnipeds.

In a review, Geraci *et al.* (1987), reported 41 tumors in cetaceans. Of these, 13 (31%) occurred in the skin, 13 (31%) in the gastrointestinal system, and 9 (21%) in the female reproductive system, accounting for 83% of tumors reported. They too cautioned about bias in reporting and predicted that the variety of tumors reported in cetaceans will increase with increased observation.

The belugas in the Saint Lawrence estuary of Canada are relatively isolated animals and, despite complete protection for many years, their population is declining. Necropsies on 15 of these animals by Martineau *et al.* (1988) identified a total of six tumors. The St. Lawrence beluga population lives in an area contaminated with industrial effluent. Martineau *et al.* (1988) also reported that the occurrence of benzo(a)pyrene (BAP) adducts in the brains of three whales in this population coincided with incidence of tumors. They suggested that environmental pollution may be a factor in recent population decline.

Many factors, including heredity, diet, hormonal, viral, chemical, and physical agents, influence the frequency and type of tumors that occur in other species of mammals. Whether these factors play a significant role in the pathogenesis of neoplasia in marine mammals is not well understood. A viral etiology has been proposed for certain papillomas (Lambertsen *et al.* 1987). Known mammalian carcinogens are discharged into near-coastal waters and can be biomagnified as they pass through the food web. Exposure to anthropogenic chemicals may be associated with tumor development in marine mammals (Howard *et al.* 1983c; Martineau *et al.* 1987; Martineau *et al.* 1988). Certain environmental chemical pollutants, such as PCBs, may cause estrogenic hormonal effects. Whether this can be related to tumors of the reproductive system is yet to be determined.

Identification of causal factors associated with neoplasia in marine mammals will be difficult due to problems associated with observing large populations of animals over time and estimating exposure to chemical agents. Progress in this area of research could be enhanced through greater sharing and pooling of scientific information. For example, the literature reflects disagreements among pathologists over the diagnosis and interpretation of particular tumors and confusion over exactly how many different tumor types have been identified (Geraci *et al.* 1987). A centralized registry to archive microscopic slides and biological data from marine mammals with tumors would: 1) standardize tumor nomenclature and diagnosis; 2) centralize data from individual animals to improve understanding of temporal and geographic distribution of tumor incidence.

DISEASE ASSOCIATED WITH ENVIRONMENTAL POLLUTION

Pollution of the marine environment is global. The many types of contaminants, including chemical, physical, and infectious agents, pose serious health risks for marine mammals (Britt and Howard 1983a). Animals may be harmed directly or indirectly through loss of habitat and food resources. Pollution of the marine environment can be divided into two general categories: acute, catastrophic events; intentional, chronic discharge of human, animal, and chemical waste.

Catastrophic Events

Catastrophic or episodic events cause a sudden change in the characteristics of a marine ecosystem. Though oil spills are the most common catastrophic events, illegal dumping or effluent release from industrial sources or spills of toxic chemicals from transport ships also may severely damage a marine ecosystem.

The *Exxon Valdez* oil spill in Alaska in 1989 was the most dramatic, well-publicized acute, catastrophic contamination of the marine environment in U.S. history. The accident occurred in Prince William Sound, and the oil spread into the Gulf of Alaska habitat for many species of marine mammals. Within three weeks, the spill area covered approximately 9,600 square miles of ocean and over 1,000 miles of shoreline. Health effects caused by exposure to oil are poorly understood because a spill of this magnitude had never occurred in an area inhabited by so many species of marine mammals. Studies to determine both acute and long-term effects of exposure to crude oil are currently underway in sea otters, harbor seals, sea lions, and large cetacea.

Marine mammals come into direct physical contact with oil either at sea surface or on contaminated beaches. Physical properties of oil in the marine environment vary, depending on type of oil spilled and amount of weathering that occurred post-spill. Characteristics range from light sheen to thick, tenacious, semi-solid material to tar balls. External fouling is most dramatic in animals with thick pelage, such as sea otters, fur seals, or polar bears. In addition to external exposure, animals, such as the sea otter and polar bear, also ingest oil during grooming attempts to remove oil. Oil has less tendency to adhere to animals with smooth skin, such as whales and dolphins. Sea otters are at greatest risk from exposure via this route, since they depend on thick fur for insulation from cold-water temperatures. Death results from loss of thermoregulation caused by physical destruction of insulatory properties of the fur.

For whales who skim feed, baleen plates may become fouled with floating oil, thus compromising their ability to filter food. This is most serious when the consistency of oil is tarry. In polar regions, oil tends to accumulate at the ice edge or in breathing holes used by many species, such as ringed seals, belugas, walruses, and others, thus making contact with oil unavoidable. Contaminated shorelines pose risks to animals who haul out, including seals, sea lions, and walruses. Oil washed from contaminated shorelines or surface oil that sinks may be ingested by grey whales, walruses, and seals who feed on benthic animals.

Oil is composed of a variety of chemical components, many of which are low molecular weight volatile chemicals, including aromatic and aliphatic hydrocarbons. Many of these compounds, such as benzene and toluene, are known to be toxic in most mammalian systems. These compounds are inhaled when animals breathe at the sea surface. Inhalation of toxic volatiles may cause inflammation of mucous membranes in the respiratory tract, lung congestion, emphysema, or possibly pneumonia (Hansen 1985). Geraci and St. Aubin (1982), report that these volatile chemicals readily enter the blood stream and accumulate in other organs, such as the liver and brain, resulting in hepatic or neurologic damage, or both.

Marine mammals are also exposed to oil when they ingest contaminated food. Exposure may be direct when resources are fouled, e.g., oiled mussels for sea otters or oiled sea grass beds for manatees. Exposure may be indirect when prey organisms bioaccumulate petroleum hydrocarbons in their tissue. As stated above, sea otters and polar bears may ingest oil during grooming. Biological effects of ingested oil are complex. For review, see Geraci and St. Aubin (1990).

Chronic Discharge of Waste

Marine waters are polluted with a variety of contaminants known to damage health of lower marine organisms. Due to logistical, legal, and ethical problems associated with research on protected marine mammals, there are few conclusive studies of health-related effects in these animals. Risk to marine mammal health posed by exposure to anthropogenic substances may be high since many animals are long-lived, feed at the top of the food web, and have high body fat content. Lipophilic substances, e.g., polychlorinated biphenyls and other organic chemicals, are known to bioaccumulate. Several papers review and summarize possible marine mammal health effects related to pollution (Addison *et al.* 1973; Risebrough 1978; Calambokidis 1984; Wageman and Muir 1984; Rejinders 1988).

The scientific community is concerned that pollution may play a significant role in pathological processes affecting marine mammals (Gaskin 1982; Britt and Howard 1983a; Howard *et al.* 1983c; Tanabe 1983; Masse 1986; Martineau 1987). In recent years, many unusual events have increased morbidity and mortality among marine mammals. A mass die-off of bottlenose dolphins on the U.S. east coast in 1987-88 killed approximately 50% of near-shore stock (Scott 1988). A 1988 epidemic among seals in the North Sea killed over 17,000 animals. Though many unusual events might result from natural phenomena, e.g., bacterial or viral infection or biological toxins, pollution may be a contributing stressor.

Marine waters of the U.S., including estuaries, coastal waters, and open ocean, are used extensively for the disposal of various types of wastes. Three major sources of pollutants enter the marine environment: 1) point sources discharge municipal and chemical waste through effluent pipes; 2) non-point sources, i.e., agricultural and urban run-off; 3) wastes transported by ship or barge to designated marine dumping sites and discharged at sea. From these sources, various types of pollutants are introduced into ocean waters, including nutrients, pathogens, toxic chemicals, and toxic metals.

Acute and long-term biological effects of contaminants are extremely difficult to assess. Many varied and complex factors may interact to cause or exacerbate a given biological response. Knowledge of fate and transport of pollutants in the marine environment is necessary to understand how pollutant exposure affects marine mammals.

Chronic pollution has no temporal or spatial boundaries. Depending on tides, currents, and atmospheric transport, pollutants may travel long distances and intermix in complex, time- and season-dependent ways. Man-made chemicals may also enter the food web and bioconcentrate or biomagnify as they pass from one trophic level of the ecosystem to another. The effects of environmental contaminants also influence the ecosystem in indirect ways. For example, increased nutrient loading may produce environmental conditions that result in nuisance algal blooms. Environmental factors may interact synergistically with certain natural phenomena. For example, increases in water temperature can enhance bacterial or viral growth rates; some environmental pollutants are immunosuppressive. A combination of such stressors may increase incidence of natural disease processes observed in marine organisms.

Physiological Effects of Environmental Chemicals

Reijenders (1986) conducted the only controlled exposure study in marine mammals to date and found a causal relationship between naturally occurring levels of pollutants and physiological response in marine mammals. These are the only controlled experimental data available and will be described in detail.

Collapse of the common seal, *Phoca vitulina*, population in the western Wadden Sea, Netherlands, was thought to be due to a sharp decline in pup production. Comparative toxicological studies on heavy metals and organochlorines in tissues from seals from this site and reference sites indicated that only the PCB levels differed significantly. Based on epidemiological and experimental data on the ability of PCBs to interfere with mammalian reproduction, a study was designed to evaluate PCBs as a causal factor of reproductive suppression in seals. Two groups of seals were fed a diet containing different levels of pollutants. The average daily intake during 2 years for Group 1 was 1.5 mg PCBs and 0.4 mg pp'-DDE (fish from Wadden Sea) and 0.22 mg and 0.13 mg for Group 2 (fish from northeast Atlantic). Reproductive success of Group 1 was significantly lower than Group 2. The effect occurred at the stage of the reproductive process near implantation. Findings in seals were similar to experiments involving mink where PCBs impaired reproduction (ovulation, mating, and implantation were followed by early abortion or resorption). No circum-annual reduction in hormone levels was observed.

Studies in mink *Mustela vision*, a fish-eating mammal with comparable reproductive physiology, were conducted to determine if pure PCBs had the same effect on reproduction as PCB-polluted fish (Boer 1984). Results show caused by pure PCB were identical to those of PCB-polluted fish and occurred at a low dose (25 μg/day). Reijenders concluded that organochlorines caused reproductive failure in seals.

SUMMARY

Future research should attempt to improve our understanding of how various stressors interact to produce disease in marine mammals. Medical issues concerning wild populations of animals, especially those that live in the open ocean, are varied. Because our abilities to treat either individuals or populations are limited, efforts must focus on preventative measures to ensure the health of such animals. Stress associated with environmental pollution may be controllable and should be a top priority for research. In marine mammals, environmental pollutants are suspected, but not proven, to contribute to infectious disease by weakening immune systems, suppressing reproductive function, and as a predisposing factor to neoplasia. The study of marine mammal toxicologic pathology is necessary for protection and survival of these unique animals -- the mammals that live in the sea.

REFERENCES

Addison, R.F., S.R. Kerr, J. Dale, and D.E. Sergeant. 1973. Variation of organochlorine residue levels with age in Gulf of St. Lawrence harp seals (*Pagophilus groenlandicus*). *J. Fish. Res. Board Can.* 30:595-600.

Barlough, J.E., E.S. Berry, D.E. Skilling, A.W. Smith, and F.H. Fay. 1986. Antibodies to marine caliciviruses in the Pacific walrus (*Odobenus rosmarus & divergens* Illiger) *J. Wildl. Dis.* 22:165-168.

Barr, B., J.L. Dunn, M.D. Daniel, and A. Banford. 1989. Herpes-like viral dermatitis in a beluga whale (*Delphinapterus leucas*) *J. Wildl. Dis.* 25:608-611.

Beneson M.W., E.T. Takafuji, and S.N. Lennon. 1982. Oocyst-transmitted toxoplasmosis associated with ingestion of contaminated water. *New England J. Med.* 307:66-669.

Bishop, L. 1979. Parasite related lesions in a bearded seal (*Erignathus barbatus*). *J. Wildl. Dis.* 15:285-294.

Boer, M. den. 1984. Annual Report: Rijksinstituut voor Natuurbehur (RIN), ed. T. van Rossum. pp. 77-86. Leersum, the Netherlands: RIN.

Boeuer, W.J., C.O. Thorn, and J.D. Wallach. 1976. *Mycobacterium chelonei* infection in a Natterer manatee. *JAVMA* 169:927-929.

Borst, G.H.A., A.C. Walvoot, P.J.H. Reijnders, J. Van der Kamp, A.D.M.E. Osterhaus. 1986. An outbreak of a Herpesvirus infection in harbor seals (*Phoca vitulina*). *J. Wildl. Dis.* 22:1-6.

Britt, J.O. 1975. Acute viral hepatitis in California sea lions. *JAVMA* 175:921-923.

Britt, J.O., and E.B. Howard. 1983a. Tissue residues of selected environmental contaminants in marine mammals. In *Pathobiology of Marine Mammal Diseases, Vol. II.*,ed. E.B. Howard, pp. 79-94. Boca Raton, Florida: CRC Press.

Britt, J.O., and E.B. Howard. 1983b. Virus diseases. In *Pathobiology of Marine Mammals, Vol. I.*, ed. E.B. Howard, pp. 47-67. Boca Raton, Florida: CRC Press.

Britt, J.O., A.Z. Nagy, and E.B. Howard. 1979. Acute viral hepatitis in California sea lions. *JAVMA* 175:921-923.

Brown, R.J., A.W. Smith, and M.C. Keyes. 1974. *Sarcocystis* in the northern fur seal. *J. Wildl. Dis.* 10:53.

Buergelt, C.D., and R.K. Bonde. 1983. Toxoplasmic meningoencephalitis in a West Indian manatee. *JAVMA* 182(11):1294-1296.

Calambokidis, J., J. Peard, G.H. Steiger, J.C. Cubbage, and R.L. Delong. 1984. Chemical contaminants in marine mammals from Washington state. NOAA Technical Memorandum NOS OMS 6. 167 pp. Springfield, VA: National Technical Information Service.

Caldwell, D.K., M.C. Caldwell, J.C. Woodward, L. Ajello, W. Kaplan, and H.M. McClure. 1975. Lobomycosis as a disease of the Atlantic bottle-nosed dolphin (*Tursiops truncatus* Montagu, 1821). *Am. J. Trop. Med. Hyg.* 24:105-114.

Carrol, J.M., A.M. Jasmin, J.N. Bascom. 1986. Pulmonary aspergillosis of the bottle-nosed dolphin (*Tursiops truncatus*). *Am. J. Vet. Clin. Pathol.* 2:139-140.

Conlogne, G.J., J.A. Ogden, and W.J. Foreyt. 1980. Pediculosis and severe heartworm infection in a harbor seal. *Vet. Med.* 75:1184-1187.

Cousins, D.V. 1987. ELISA for detection of tuberculosis in seals. *Vet. Rec.* 121:305.

de Vries, G.A., and J.J. Laarman. 1973. A case of Lobo's disease in the dolphin, *Sotalia quianensis. Aquat. Mamm.* 1:26-33.

Dailey, M.D. 1970. Transmission of *Parafliaroides decorus* (Nematoda: *Metastrongyloidea*) in the California sea lion (*Zalophus californianus*). *Proc. Helminthol. Soc.* 37:215-222.

Dailey, M.D., and R. Stroud. 1978. Parasites and associated pathology observed in cetaceans stranded along the Oregon coast. *J. Wildl. Dis.* 14:503-511.

Diamond, S.S., D.E. Ewing, and G.A. Cadwell. 1979. Fatal bronchopneumonia and dermatitis caused by *Pseudomonas aeruginosa* in an Atlantic bottle-nosed dolphin. *JAVMA* 175:984-987.

Dierauf, L.A. 1990. *Handbook of Marine Mammal Medicine.* Boca Raton, Florida: CRC Press.

Dierauf, L.A., L.J. Lowenstine, and C. Jerome. 1981. Viral hepatitis (adenovirus) in a California sea lion. *JAVMA* 179:1194-1197.

Dierauf, L.A., D.J. Vandenbroek, J. Roletto, M. Koski, L. Amayo, and L.J. Gage. 1985. An epizootic of leptospirosis in California sea lions. *JAVMA* 187(11):1145-1148.

Dilbone, R.P. 1965. Mycosis in a manatee. *JAVMA* 147:1095.

Dunn, J.L., and S. Spotte. 1974. Some clinical aspects of seal pox in captive Atlantic harbor seals. *J. Zool. Anim. Med.* 5:27-30.

Dunn, J.L., and R.I. Wolke. 1976. *Dipetalonema spirocauda* infection in the Atlantic harbor seal (*Phoca vitulina concolor*). *J. Wildl. Dis.* 12:531-538.

Eley, T.J. 1981. *Dipetalonema spirocauda* in Alaskan marine mammals. *J. Wildl. Dis.* 17:65-67.

Forrester, D.J., R.F. Jackson, J.F. Miller, and C.B. Townsend. 1973. Heartworms in captive California sea lions. *JAVMA* 163:568-570.

Frese, K., A. Weber, and E. Weiss. 1972. Dermatophilose bei Mähnenrobben (*Otaria byronia* Blainvelle) - Diagnose und Differentialdiagnose unter besonderer Berücksichtigung von Pockeninfektionen. In *Erkrankungen der Zootiere*, ed. R. Ippen and H.D. Schroder, pp. 217-220. Berlin: Akademie-Verlag.

Frey, H.R., B. Liess, L. Haes, H. Lehmann, and H.J. Marshall. 1989. Herpesvirus in harbor seals (*Phoca vitulina*): isolation, partial characterization, and distribution. *J. Vet. Med.* 36:699-708.

Gaskin, D.E. 1982. Environmental contaminants and trace elements: their occurence and possible significance in Cetacea. In *The Ecology of Whales and Dolphins*, ed. D.E. Gaskin, pp. 393-433. Portsmouth, N.H.: Heinemann Educational Books.

Geraci, J.R. 1990. *Sea Mammals and Oil: Confronting the risks*. pp. 282. San Diego: Academic Press.

Geraci, J.R., B.D. Hicks, and D.J. St. Aubin. 1979. Dolphin pox: a skin disease of cetaceans. *Can. J. Comp. Med.* 43:399-404.

Geraci, J.R., and D.J. St. Aubin. 1982. *Study of the effects of oil on cetaceans*. 274 pp. Washington, DC: Report to the Bureau of Land Management, U.S. Department of the Interior.

Geraci, J.R., R.M. Sauer, and W. Medway. 1966. Erysipelas in dolphins. *Am. J. Vet. Res.* 27(117):597-606.

Geraci, J.R., D.J. St. Aubin, I.K. Barker, R.G. Webster, V.S. Hinshaw, W.J. Bean, H.L. Ruhnke, J.H. Prescott, G. Early, A.S. Baker, S. Madoff, and R.T. Schooley. 1982. Mass mortality of harbor seals: pneumonia associated with influenza A virus. *Science* 215:1129.

Geraci, J.R., N.C. Palmer, and D.J. St. Aubin. 1987. Tumors in cetaceans: analysis and new findings. *Can. J. Fish. Aquat. Sci.* 44:1289-1300.

Gilmartin, W.G., P.M. Vainik, and V.M. Neill. 1979. Salmonellae in feral pinnipeds off the southern California coast. *J. Wildl. Dis.* 15:511-514.

Gilmartin, W.G., R.L. Delong, A.W. Smith, J.C. Sweeney, B.W. Delappe, R.W. Risebrough, L.A. Griner, M.D. Dailey, and D.B. Peakall. 1976. Premature parturition in the California sea lion. *J. Wildl. Dis.* 12:104-115.

Gutter, A.E. 1987. Generalized mycobacteriosis in a California sea lion (*Zalophus californianus*). *J. Zool. Anim. Med.* 18:118-120.

Hansen, D.J. 1985. The potential effects of oil spills and other chemical pollutants on marine mammals occuring in Alaskan waters. U.S. Dept. of the Interior Minerals Management Service Alaska Outer Continental Shelf Region, Anchorage, Alaska, OCS Rep MMS85-0031, 22 pp.

Holshuh, H.J., A.E. Sherrod, C.R. Taylor, B.F. Andrews, and E.B. Howard. 1985. Toxoplasmosis in a feral northern fur seal. *JAVMA* 187:1229-1230.

Horvat, B., T. Willhaus, H.R. Frey, and B. Liess. 1989. Herpesvirus in harbour seals (*Phoca vitulina*): transmission in homologous host. *J. Vet. Med.* 36:715-718.

Howard, E.B. 1983. *Pathobiology of Marine Mammal Diseases. Vol. I and II.* Boca Raton, Florida: CRC Press.

Howard, E.B., J.O. Britt, and G.K. Matsumoto. 1983a. Parasitic diseases. In *Pathobiology of Marine Mammal Diesease, Vol. II*, ed. E.B. Howard, pp. 96-162. Boca Raton, Florida: CRC Press.

Howard, E.B., J.O. Britt, G.K. Matsumoto, R. Itahara, and C.N. Nagano. 1983b. Bacterial diseases. In *Pathobiology of Marine Mammal Diseases, Vol. I.*, ed. E.B. Howard, pp. 70-117. Boca Raton, Florida: CRC Press.

Howard, E.B., J.O. Britt, and J.G. Simpson. 1983c. Neoplasms in marine mammals. In *Pathobiology of Marine Mammal Diseases, Vol. II.*, ed. E.B. Howard, pp. 96-162. Boca Raton, Florida: CRC Press.

Inskeep, W., C.H. Gardiner, R.K. Harris, J.P. Dubey, and R.T. Goldston. 1990. Toxoplasmosis in Atlantic bottle-nosed dolphins (*Tursiops truncatus*). *J. Wildl. Dis.* 26:(3)377-382.

Johnston, D.G., and S.H. Ridgway, 1969. Parasitism in some marine mammals. *JAVMA* 155:1064-1072.

Kennedy S., J.A. Smyth, P.F. Cush, S.J. McCollough, G.M. Allan, and S. McQuaid. 1988a. Viral distemper now found in porpoises. *Nature* 336:21.

Kennedy S., J.A. Smyth, S.J. McCollough, G.M. Allan, F. McNeilly, S. McQuaid. 1988b. Confirmation of cause of recent seal deaths. *Nature* 336:21.

Kennedy, S., J.A. Smyth, P.F. Cush, P. Duignan, M. Platten, S.J. McCollough, and G.M. Allan. 1989. Histopathologic and immunocytochemical studies of distemper in seals. *Vet. Pathol.* 26:97-103.

Kennedy, S., J.A. Smyth, P.F. Cash, M. McAliskey, S.J. McCullough, and B.K. Rima. 1991. Histopathologic and immunocytochemical studies of distemper in harbor porpoises. *Vet. Pathol.* 28:1-7.

Kennedy-Stoskopf, S., M.K. Eckhans, and J.D. Strandberg. 1986. Isolation of a retrovirus and a herpesvirus from a captive California sea lion. *J. Wildl. Dis.* 22(2):156-164.

Keyes, M.C. 1965. Pathology of the northern fur seal. *JAVMA* 147:1091-1095.

Kim, K.C., V.L. Haas, and M.C. Keyes, M.C. 1980. Populations, microhabitat preference, and effects of infestation of two species of *Orthohalarchne (Halarachnidae: Acarine)* in northern fur seal. *J. Wildl. Dis.* 16:45-51.

Lambertsen, R.H., B.A. Kohn, J.P. Sundberg, and C.D. Buergelt. 1987. Genital papillomatosis in sperm whale bulls. *J. Wildl. Dis.* 23:361-367.

Laukner, G. 1985. Diseases of Mammalia: Carnivora. In *Diseases of Marine Animals Vol. IV, Part 2*, ed. O. Kinne, pp. 645-682. Germany: Biologische Anstalt Helgoland Hamlaurg.

Martineau, D., P. Beland, C. Desjardins, and A. Lagace. 1987. Levels of organochlorine chemicals in tissues of beluga whales (*Delphinapterus leucas*) from the St. Lawrence estuary, Quebec, Canada. *Arch. Envrion. Contam. Toxicol.* 16:137-147.

Martineau, D., A. Lagace, P. Beland, R. Higgins, D. Armstrong, and L.R. Shugart. 1988. Pathology of stranded beluga whales (*Delphinapterus leucas*) from the St. Lawrence estuary, Quebec, Canada. *J. Comp. Path.* 98:287-311.

Masse, R., D. Martineau, L. Tremblay, and P. Beland. 1986. Concentrations and chromatographic profile of DDT metabolites and polychlorobiphenyl (PCB) residues in stranded beluga whales (*Delphinapterus leucas*) from the St. Lawrence estuary, Canada. *Arch. Environ. Contam. Toxicol.* 15:567-579.

Medway, W. 1980. Some bacterial and mycotic diseases of marine mammals. *JAVMA* 117:831-834.

Medway, W., and T.C. Wieland. 1975. *Dirofilaria immitis* infection in a harbor seal. *JAVMA* 167:549-550.

Migaki, G. 1987. Selected dermatoses of marine mammals. *Clin. Dermatopathol.* 5:155-164.164.

Migaki, G., and T.F. Albert. 1980. Sarcosporidiosis in the ringed seal. *JAVMA* 177:917-918.

Migaki, G., and S.R. Jones. 1983. Mycotic diseases in marine mammals. In *Pathobiology of Marine Mammal Diseases, Vol. II*, ed. E.B. Howard, pp. 1-127. Boca Raton, Florida: CRC Press.

Migaki, G., M.G. Valerio, B.A. Irvine, and F.M. Garner. 1971a. Lobo's disease in an Atlantic bottle-nosed dolphin. *JAVMA* 149:578-582.

Migaki, G., D. van Dyke, and R.C. Hubbard. 1971b. Some histopathological lesions caused by helminths in marine mammals. *J. Wildlife Dis.* 7:281-289.

Migaki, G., J.F. Allen, and H.W. Casey. 1977. Toxoplasmosis in a California sea lion (*Zalophus californianus*). *Am. J. Vet. Res.* 38:135-136.

Migaki, G., R.D. Gunnels,and H.W. Casey. 1978a Pulmonary cryptococcus in an Atlantic bottle-nosed dolphin (*Tursiops truncatus*). *Lab. Animal. Sci.* 28:603-606.

Migaki, G., R.I. Font, W. Kaplan, and E.D. Asper. 1978b. Sporotrichosis in a Pacific white-sided dolphin (*Lagenrohynchus obliquidens*). *Lab. Anim. Sci.* 28:603-606.

Migaki, G., M.D. Lagios, E.S. Herald, and R.P. Dempster. 1979. Hepatic trematodiasis in a Ganges River dolphin. *JAVMA* 175:926-929.

Migaki, G., T.R. Sawa, and J.P. Dubey. 1990. Fatal disseminated toxoplasmosis in a spinner dolphin (*Stenella longirostris*). *Vet. Pathol.* 27:463-464.

Montali, R.J., M. Bush, J.D. Strandberg, D.L. Janssen, D.J. Boness, and J.C. Whitla. 1981. Cyclic dermatitis associated with *Fusarium* sp. infection in pinnipeds. *JAVMA* 179:1198.

Morales, P. 1985. Systemic *Mycobacterium marinum* infection in an Amazon manatee. *JAVMA* 187:1230-1231.

Munday, B.L., R.W. Mason, W.F. Hartely, R.J.A. Presidente, and D. Obendorf. 1978. *Sarcocystis* and related organisms in Australian wildlife. I. Survey findings in mammals. *J. Wildl. Dis.* 14:417-433.

Nutting, W.B., and M.D. Dailey. 1980. Demodicosis (*Acari: Demodicidae*) in the California sea lion, *Zalophus californianus*. *J. Med. Entomol.* 17:344-347.

Olsen, O.W., and E.T. Lyons. 1965. Life cycle of *Uncinaria lucasi* (Nematoda: *Ancylostomatidae*) of fur seals, *Callorhinus ursinus* Linn., on the Pribil Islands, Alaska. *J. Parasitol.* 51:689-700.

Perry, M.L. 1969. A new species of *Dipetalonema* from the California sea lion and a report of microfilaria from a Steller sea lion (Nematoda: Filarioidea). *J. Parasitol.* 53:1076-1081.

Rausch, R.L. 1953. Studies on the helminth fauna of Alaska. XIII. Disease in the sea otter, with special reference to helminth parasites. *Ecology* 34:584-604.

Reed, R.E., G. Migaki, and J.A. Cummings. 1976. Coccidioidomycosis in a California sea lion. *J. Wildl. Dis.* 12:372-375.

Reijnders, P.J.H. 1986. Reproductive failure in common seals feeding on fish from polluted coastal waters. *Nature* 324:456-458.

Reijnders, P.J.H. 1988. Ecotoxicological perspectives in marine mammalogy: research principles and goals for a conservation policy. *Mar. Mamm. Sci.* 4:91-102.

Risebrough, R.W. 1978. Pollutants in marine mammals, a literature review and recommendations for research. Report PB-290728. Prepared for the Marine Mammal Commission, Washington, D.C. 64 pp. Springfield, VA: National Technical Infromation Service.

Scott, G.P., D.M. Burn, and L.J. Hansen. 1988. The dolphin dieoff: long-term effects and recovery of the population. Proceedings of the Oceans '88 Conference, Baltimore, Maryland.

Schryver, H.C., W. Medway, and J.F. Williams. 1967. The stomach fluke, *Braunina cordiformis*, in the Atlantic bottle-nosed dolphin. *JAVMA* 151:884-886.

Seibold, H.R., and J.E. Neal. 1956. Erysipelas septicemia in the propoise. *JAVMA* 128:537-539.

Simpson, C.F., F.G. Wood, and F. Young. 1958. Cutaneous lesions on a porpoise with erysipelas. *JAVMA* 133:558-560.

Smith, A.W., and A.B. Latham. 1978. Prevalence of vesicular exanthema of swine antibodies among feral mammals associated with the sourthern California coastal zones. *Am. J. Vet. Res.* 39:291-296.

Smith, A.W., and D.E. Skilling. 1979. Viruses and virus diseases of marine mammals. *JAVMA* 175:918-1001.

Smith, A.W., and P.M. Boyt. 1990. Caliciviruses of ocean origin: a review. *J. Zoo. Wildl. Med.* 21(1):3-23.

Smith, A.W., T.G. Akers, S.H. Madin, and N.A. Vedros. 1973. San Miguel sea lion virus isolation, preliminary characterization and relationship to vesicular exanthema of swine virus. *Nature* 224:108-110.

Smith, A.W., R.J. Brown, D.E. Skilling, and R.L. DeLong. 1974a. *Leptospira pomona* and reproductive failure in California sea lions. *JAVMA* 165:996-998.

Smith, A.W., C.M. Prato, W.G. Gilmartin, R.J. Brown, and M.C. Keyes. 1974b. A preliminary report on potentially pathogenic microbiological agents recently isolated from pinnipeds. *J. Wildl. Dis.* 10:54-58.

Smith, A.W., R.J. Brown, D.E. Skilling, H.L. Bray, and M.C. Keyes. 1977. Naturally-occurring leptospirosis in northern fur seals (*Callorhinus ursinus*). *J. Wildl. Dis.* 13:144-148.

Smith, A.W., N.A. Vedros, T.G. Akers, and W.G. Gilmartin. 1978. Hazards of disease transfer from marine mammals to land mammals: review and recent findings. *JAVMA* 173:1131.

Smith, A.W., T.G. Akers, A.B. Latham, D.E. Skilling, and H.L. Bray. 1979. A new calicivirus isolated from a marine mammal. Brief report. *Arch. Virol.* 61:255-259.

Smith, A.W., D.E. Skilling, and S. Ridgway. 1983a. Calicivirus-induced vesicular disease in cetaceans and probable interspecies transmission. *JAVMA* 183(11):1223-1225.

Smith, A.W., D.E. Skilling, S.H. Ridgway, and C.A. Fenner. 1983b. Regression of cetacean tatto lesions concurrent with conversion of precipitation antibody against a pox virus. *JAVMA* 183(11):1219-1222.

Sweeney, J.C. 1974. Common diseases of pinnipeds. *JAVMA* 165:805-810.

Sweeney, J.C. 1986. Marine mammals (Cetacea, Pinnipedia, and Sirenia). Infectious diseases of body systems. In *Zoo and Wild Animal Medicine*, ed. M.E. Fowler, pp. 777-784. Philadelphia: Saunders.

Sweeney, J.C., G. Migaki, P.M. Vainik, and R.H. Conklin. 1976. Systemic mycosis in marine mammals. *JAVMA* 169:946-948.

Tanabe, S., T. Mori, R. Tatsukawa, and N. Miyazaki. 1983. Global pollution of marine mammals by PCBs, DDTs, and HCHs (BHCS). *Chemosphere* 12(9/10):1269-1275.

Van Pelt, R.W., and R.A. Dietrich. 1973. Staphylococcal infection and toxoplasmosis in a young harbor seal. *J. Wildl. Dis.* 9:258-261.

Vedros, N.A., A.W. Smith, J. Schonewald, G. Migaki, and R.C. Hubbard. 1971. Leptospirosis epizootic among California sea lions. *Science* 172:1250-1251.

Wagemann, R., and D.C.G. Muir. 1984. Concentrations of heavy metals and organochlorines in marine mammals in northern waters: overview and evaluation. Canadian Technical Report of Fisheries and Aquatic Sciences No. 1279. Department of Fisheries and Oceans, Winnipeg, Manitoba. 95 pp.

Wells, S.K., A. Gutter, and Van Meter. 1990. Cutaneous mycobacteriosis in a harbor seal: attempted treatment with hyperbaric oxygen. *J. Zoo Wildl. Med.* 21(1) 73-78.

Williamson, W.M., L.S. Lombard, and R.E. Getty. 1959. North American blastomycosis in a northern sea lion. *JAVMA* 135:513-515.

Wilson, T.M. 1970. Seal pox in a free-living harbor seal, *(Phoca vitulina)*. *Ann. Proc. Am. Assoc. Zoo. Vet.* pp. 125-127.

Wilson, T.M., and I. Poglayen-Neuwall. 1971. Pox in South American sea lions (*Otaria byronia*). *Can. J. Comp. Med.* 35:174-177.

Wilson, T.M., and P.R. Sweeney. 1970. Morphological studies of seal pox-virus. *J. Wildl. Dis.* 6:94-97.

Wilson, T.M., A.D. Boothe, and N.F. Cheville. 1972a. Sealpox field survey. *J. Wildl. Dis.* 8:158-160.

Wilson, T.M., N.F. Cheville, and A.D. Boothe. 1972b. Sealpox questionnaire survey. *J. Wildl. Dis.* 8:155-157.

Wilson, T.M., R.W. Dykes, and K.S. Tsai. 1972c. Pox in young, captive harbor seals. *JAVMA* 161:611-617.

Wilson, T.M., N.F. Cheville, and L. Karstad. 1969. Seal pox. *Bull. Wildl. Dis. Assoc.* 5:412-418.

Wilson, T.M., M. Kierstead, and J.R. Long. 1974. Histoplasmosis in a harp seal. *JAVMA* 165:815-817.

Woodward, J.C., S.G. Zam, D.K. Caldwell, and M.C. Caldwell. 1969. Some parasites of dolphins. *Vet. Pathol.* 6, 257-272.

Young, P.C., and D. Lowe. 1969. Larval nematodes from fish of the subfamily Anisakinae and gastrointestinal lesions in animals. *J. Comp. Pathol.* 79:301-313.

9

Invertebrate Diseases -- An Overview

Albert K. Sparks

INTRODUCTION

Although some of the earliest recognized diseases were described from invertebrates, the concept of disease as an important facet in life and death of invertebrates, at both individual and population levels, is of fairly recent vintage. Subsequent to damage claims and law suits by Louisiana oystermen alleging high mortalities of oysters due to pollution, a group of oil companies contracted with the Texas A&M Research Foundation to investigate the extent and causes of oyster mortalities in Louisiana. The resultant investigation not only revealed, for the first time, that an infectious disease was responsible for massive mortalities of a marine invertebrate species, but also initiated a period of exponential growth in molluscan pathology. Around the same time, the classic description of leukocytic infiltration, phagocytosis, and removal of India Ink particles injected into the American oyster, *Crassostrea virginica*, provided a foundation to understand the inflammatory response in mollusks.

The original description of the progression and histopathology of Dermo (Perkinsiosis) has stood unchallenged without modification for almost 50 years. Of at least equal importance to the elucidation of Perkinsiosis as a major disease and, for that matter, the study of oyster mortalities was the development of the Thioglycolate Method for diagnosis of Dermo. It not only provided a rapid, reliable means for large-scale studies of the epizootiology of Dermo, but also allowed estimation of intensity of infection. More importantly, perhaps, it made meaningful research possible on oyster mortalities by biologists in remote, poorly equipped laboratories.

Using the Thioglycolate Diagnostic Technique supplemented by histological examination when practical, Perkinsiosis (known for many years as Dermo) was soon shown to be the principal cause of oyster mortality in the Gulf of Mexico

0-8493-8662-4/93/$0.00 + $.50

and Chesapeake Bay prior to the onset of MSX (*Haplosporidium nelsoni*) epizootics in 1957. Despite the fact that not all investigators were prepared with equipment or experience for histopathological studies of large numbers of dying oysters that failed to respond to Thioglycolate Culture in Delaware Bay in 1957 and Chesapeake Bay in 1958, microscopical examination revealed numerous multinucleated spherical plasmodia throughout the tissues of dying oysters. Because the taxonomic relationships of the organism were unknown, it was given the acronym "MSX" (Multinucleated Sphere Unknown). In response to the wide geographic distribution and catastrophic lethality of the epizootic, university, state, and federal oyster biologists in New Jersey, Delaware, Maryland, and Virginia began a cooperative, often collaborative, research program. In addition to frequent consultations among investigators, annual "Oyster Mortality Workshops" were hosted by the involved laboratories on an informal rotational basis to which interested, competent outsiders were invited. The workshops began as "true workshops" to which investigators brought numerous slides of sectioned material for collaborative examination and discussion, frequently marked by "spirited," often impassioned, debate on interpretation of precisely what was revealed in a 5 μm slice of oyster immortality.

Exchange of research data and preliminary interpretation of results were shared to a degree unparalleled in invertebrate disease research. This exchange contributed greatly to the relatively rapid understanding of the cause and epizootiology of the disease, but also resulted in confusion and some conflict relative to priority in scientific contributions. The workshops gradually became more formal and more general rather than being almost totally concerned with MSX; indeed, the last was hosted by this author on the West Coast and was also the time and place of the formal birth of the Society for Invertebrate Pathology (the organizing committee of SIP met in conjunction with the Fifth Oyster Mortality Workshop).

Although scattered and infrequent reports of crustacean diseases were published earlier, systematic investigations began after the subdiscipline of oyster pathology was well-established. Unsurprisingly, significant mortalities of the economically important American lobster, *Homarus americanus*, in ponds (holding facilities) provided the impetus for intensive research. The causative organism was originally isolated and named *Gaffkya homari*; in the mid-1960's, results were published of investigations that eventually resulted in Gaffkaemia becoming almost certainly the best understood of any invertebrate disease.

A long-time, concentrated investigation of diseases of the blue crab, *Callinectes sapidus*, and similar research on penaeid shrimp were begun in the early 1970's. Much additional knowledge has been accumulated on the diseases of crustaceans subsequent to pioneering efforts briefly mentioned above, but it has been built on the foundation of the earlier work.

Useful information on diseases of other invertebrate taxa is, for the most part, fragmentary. Research was also initially sparked by catastrophic epizootics of commercially important species, beginning with largely unsuccessful attempts to identify the causative organism of "Wasting Disease" in commercial sponges. More recent investigations have been more productive, primarily because they were undertaken by researchers trained in molluscan and crustacean pathology. Current emphasis on environmental monitoring should provide needed impetus to widespread, both phyletic and geographic, research on the role of noninfectious and infectious disease in the life and death of estuarine invertebrates. Because little is known of the vast majority of invertebrates, it is appropriate to address the question of what we need to know before and while we investigate invertebrate diseases.

INFORMATION NEEDED AT THE ORGANISMIC LEVEL

Normal Life Cycle

It may appear redundant to state that we must know the normal to recognize the abnormal, but substantial details of the embryonic development and normal patterns of growth, maturation, reproduction, and senescence are available for only a minuscule, highly selective taxonomically, percentage of invertebrate species. While it may be amusing to imagine the frustration of an imaginary fish pathologist on the Lewis and Clark Expedition vainly seeking the cause of massive mortalities after spawning of the newly discovered Pacific salmon, most of us who have sailed uncharted waters in marine disease investigations have found ourselves in equally ludicrous situations.

Normal Histology

It is again redundant to note that we must have a thorough knowledge of the normal histology of an animal before we can, with any degree of confidence, assess abnormal modifications of the cellular architecture. Implicit in the above statement is the requirement that events, such as the gonadal cycle, especially dramatic histologically in molluscs, and the molting cycle, common in some form in most invertebrates and especially important in arthropods, be recognized even if not thoroughly understood.

It is easy to ignore, as we examine a stained slide by light microscopy, that we are only privy to a minute slice of what had been a living animal at one precise, but probably random, moment in its life. The cellular architecture is not fixed in time like the stones in an ancient Greek Temple or a strata of sedimentary rock, but is dynamic, constantly changing. As a result of our ignorance, necrobiosis (normal cell death) and routine cell renewal are not often considered in invertebrates, despite constant reminders by our own flaking dandruff and sloughing callouses on feet and hands.

Pseudocoelomate bilateria (rotifers, gastrotrichs, kinorhynchs, priapulids, and nematodes) share the characteristic of eutely or cell constancy and do not undergo necrobiosis or cell renewal. In virtually all multicellular animals, cells die and are replaced normally by a rate and pattern that vary with the animal and organ system. To ignore or be unaware of this natural phenomenon can lead to serious errors in evaluation or even recognition of the disease process.

Inflammation and Wound Repair

When cells are injured or destroyed, an immediate protective response, generally termed inflammation, is initiated in most animals. It seeks to destroy, dilute, or isolate injurious agents and dead or damaged cells. Successful conclusion of the inflammatory process is wound repair, with dead or damaged cells replaced by healthy cells, resulting in complete restoration of original tissue architecture. Less than completely successful wound repair ranges from death of the animal at some stage of the inflammatory or reparative process to the *in situ* isolation of the injurious agent and tissue debris by some form of encapsulation. Incomplete understanding of either inflammation or wound repair can easily lead to misinterpretation of the progression or even nature of a disease.

Death and Postmortem Change

Death in invertebrates is difficult to define and even more difficult to recognize because somatic death, or death of the entire organism, typically occurs only after necrosis, or cell death, of large areas of the organism. Definition of death in vertebrates was formerly more simple (when heartbeat ceased), but recent advances in life-support systems have complicated the phenomenon even among the most advanced vertebrates. Nevertheless, somatic death is a more gradual process; for example, cilia on the gills and mantle of oysters continue to beat for several days subsequent to autolysis of much of the rest of the body.

It is obviously important to distinguish normal post mortem changes in a dying individual from the histopathological progression of a disease or inflammatory response and reparative processes of the diseased organism in response to the disease. Unfortunately, descriptions of the rate and pattern of postmortem changes in invertebrates are limited to the Pacific oyster, *Crassostrea gigas*, and the brown shrimp, *Penaeus aztecus*. Although experienced invertebrate pathologists are able, intuitively or from long experience, to recognize differences between autolysis and the inflammatory response or histopathological effects of injurious agents, the need for systematic investigation of the rate and pattern of normal postmortem changes in a wide spectrum of invertebrate taxa is critical.

INFORMATION NEEDED AT THE POPULATION LEVEL

Normal Mortality

Among the most misinformed, but widely accepted generalities in invertebrate biology are statements regarding "normal mortality." Obviously, mortality for any specific group (year class, brood, etc.) is 100%; what we really want to know is the mortality or conversely the survival rate for each stage of the life cycle or a specific time interval under normal environmental conditions, food availability, and predation. Given the prodigious fecundity of the vast majority of invertebrates with planktonic larval stages, only a minute fraction of 1% reach maturity under the most benign conditions.

Prior to the advent of aquaculture, the role of disease in the early life-history of invertebrates was almost entirely speculative. Artificial culture of especially peneaid shrimp has confirmed earlier suspicions that disease may, indeed, be a major factor in invertebrate larval success, even with the realization that the artificial conditions of culture may exacerbate larval epizootics.

"Normal" Microbial Flora/Parasite Load

Every invertebrate, even the most healthy, is an individual ecosystem-host to a plethora of organisms that range in their relationship from benign to pathogenic. It is important to know the associated flora and fauna of an animal under normal conditions in order to evaluate their importance when investigating suspected causes of increased mortalities. A prime, but not unique, example of erroneous accusation of a benign organism as the cause of invertebrate disease was the assertion that the gregarine, *Nematopsis ostrearum*, was responsible for significant mortalities of *C. virginica* in Louisiana and Virginia. However, it must be noted that even an apparently innocuous "guest" may become a serious pest or opportunistic pathogen under adverse environmental conditions.

Environmental Tolerances

There is a great deal of rhetoric among the interested, but often uninformed, public on the environment and, unfortunately from a semantic point of view, on ecology. I applaud the recent involvement of the lay public in an area of great importance, but it behooves those in research on diseases (infectious and noninfectious) of marine and estuarine invertebrates to be even more rigorous in science: more exacting in experimental design, requiring better data on normal environmental parameters and cyclic deviations, and more data on effects of perturbations on life and health of the resident invertebrates. If we are unable to provide this rigor to specifics needed to link cause and effect, then the estuaries and invertebrates that inhabit the environment will be even more vulnerable than at present.

DISEASE STUDIES: SOME PITFALLS TO AVOID

The above information relative to normal life history, histology, inflammatory process, wound repair mechanisms, and pattern of cellular autolysis are data we would like to have prior to initiation of an investigation of suspected disease (infectious or noninfectious) in any animal. However, in the less-than-perfect world in which we work, especially with almost all invertebrate species, the data base ranges from sparse to nonexistent.

This leaves us with three alternatives, none completely satisfactory: 1) initiate thorough studies of the normal prior to delving into the abnormal; 2) attempt to conduct controlled studies of the normal concurrently with separate, primarily field, investigation of the diseases of the population in question; or 3) plunge headlong and unprepared into a full-scale investigation of alleged mortalities of one or more species usually well after alleged mortality has peaked. The first alternative is utopian and probably historically nonexistent; the second, the best one probably can hope for; the third, the most common and the most frustrating, but the best opportunity to succeed brilliantly or fail miserably.

Adequate funding is seldom available for baseline studies; when it is, we soon become diverted by our discovery of disease in experimental animals or pressure generated by reported catastrophic mortalities in the natural population. Conversely, under panic conditions of real or imagined impending doom of a commercially important population, more money becomes available for research than can be wisely utilized. The result is often chaotic, with unknowledgeable and inexperienced investigators claiming success based on fragmentary or misinterpreted data, much to the chagrin of more experienced workers, some of whom have forgotten they earned their spurs in similar, earlier battles.

Despite the apparent waste of money and the damage to the scientific credibility of the discipline as a whole, the net result of these panic-induced infusions of research funding is positive: some new investigators have new ideas, unfettered by familiarity with possibly erroneous results of prior research in the area; new techniques and new, more sophisticated equipment is usually introduced during these episodes, and, most importantly, these funds support significant numbers of graduate students who not only contribute to solution of the immediate problems, but become a cadre for future disease research. It is to these "new boys in town," and those members of "the old boy network" who need reminding, that the following cautions are addressed.[1]

Noninfectious Disease

There is, in my considered opinion, far more potential for error in assessing disease from noninfectious causation than from diseases of infectious nature. Lacking the presence of a known or potential biotic pathogen on which to "hang our hat", we must be meticulous in our evaluations.

Tissue Damage vs. Normal Processes

In controlled laboratory experiments and field exposures, pathological effects of the xenobiotic agent at the tissue level are of paramount importance. Inherent in this assessment is the necessity to be able to differentiate tissue damage caused by the pathogen from the organism's response to the insult. Some abiotic pathogens are so lethal that they kill too quickly for recognizable tissue damage to occur; others, such as phenolics and strong acids or bases, are so corrosive they cause immediate loss of exposed tissues; some, e.g., formalin, ethyl and methyl alcohol, acetic acid are fixatives that provide better tissue sections than the controls; still others are less toxic and even more confusing with interplay between tissue damage and host response difficult to decipher.

General Principles of Chemical Injury

The effect of a chemical agent depends on three variables: 1) vulnerability of individual tissues, 2) mode of action of agent, 3) concentration of agent. Cells and tissues within an individual animal vary significantly in susceptibility to chemical injury; there is even greater variability among divergent invertebrate taxa. The mode of action determines location of major damage. Some toxins exert their major effect at the portal of entry, thus damaging the external body

---

[1]These terms are not intended to be gender-related, though both were at inception and through usage. The term "Old Boy System" originated, I believe, in the British Civil Service and was adopted in most societies prior to acceptance of women into the scientific establishment (Marie Sklodowska excepted). I believe "New Boy in Town" is an American term, but had male connotations for the same reason. Fortunately, numbers of women have increased in both groups.

surface, anterior gastrointestinal tract, or respiratory surfaces; others are harmless at the portal of entry, but cause major damage in tissues in which they are stored; still others exert their maximum effects where they are excreted. Frequently, toxic chemicals have combined sites of damage: e.g., portal of entry and portal of exit (site of excretion) or storage site and portal of exit.

In histopathological evaluation, it is easy to mistake normal postmortem changes in an animal dying from a highly lethal toxin for antemortem systemic necrosis from more chronic chemical injury. Conversely, absence of tissues eroded by a highly corrosive agent can be interpreted as normal autolytic changes.

Chemically Induced Tumors

In my opinion, one of the most biologically important, and certainly one of the most potentially vulnerable to media and public misinterpretation, water-quality questions is the possible carcinogenic effects of pollution of oceans and especially estuaries. It is germane to briefly review some basic definitions and concepts to insure that we are on firm ground semantically as well as biologically.

By definition, tumor is any swelling or abnormal mass of tissue, but the term is now virtually restricted to neoplasia: new growth of cells independent of normal laws of growth of the organism that persists at the same rate and pattern after termination of the stimulus that initiated accelerated growth. Most suspected neoplasms of invertebrates are first observed in tissue sections of individuals. We not only do not know the stimulus, or when or even if it has terminated, but usually do not know the rate and pattern of the suspected neoplastic proliferation.

There are a number of cellular phenomena associated with chemical, as well as other injuries that may be mistaken for neoplasia. Inflammation, both acute and chronic, often results in tremendous numbers of infiltrating cells in a traumatized area, many of which become markedly, even unrecognizably, modified in appearance. Cellular proliferation during wound repair may produce sufficient mitotic activity to excite even the most conservative investigator. And, even more difficult to distinguish from neoplasia, hyperplastic response to a stimulus may produce masses of cells, dramatically modifying the normal architectural integrity.

Once the neoplastic nature of a "growth" is established, classification as to benign or malignant usually follows; indeed, the criteria for differentiation between the two can serve to eliminate many suspected tumors from either category. Benign tumors are characteristically composed of tissue typical of the tissue of origin; growth is expansive rather than invasive and slow, as demonstrated by rare and normal mitotic figures; they are usually encapsulated; and metastasis does not occur. Malignant tumors, in contrast, are usually atypical of their cellular origin (poorly differentiated); growth is invasive as well as expansive. They are never fully encapsulated; growth may be rapid with many, often abnormal, mitotic figures; and metastasis frequently occurs.

The characteristics listed above are based on vertebrate criteria and subject to future modification in invertebrate pathology, but are useful until their validity is disproved. For example, I noted long ago that there were no authenticated cases of metastasis in any invertebrate tumor; that, in my opinion, is still valid today despite some unsubstantiated, if not irresponsible, statements in the literature.

Infectious Diseases

With epizootics caused by infectious diseases, we can usually find an infectious agent associated with sites of tissue damage and host response. Indeed, we may find several possible pathogenic organisms; then the problem becomes one of sorting out the real culprit from prior, essentially nonpathogenic, symbionts and secondary invaders. Information from baseline studies is especially valuable in such situations. Lacking such data, knowledge of relative pathogenicity of related organisms in other hosts is helpful, although it is not always possible to identify the organism taxonomically.

Accurate pathogen identification is particularly difficult when viruses are suspected. Although viral etiology may be indicated by disease signs and progression and even supported by histopathological evidence at the light microscopical level, confirmation and tentative identification require examination of suitable material with transmission electron microscopy by a competent investigator. However, for complete identification of a virus, much more is required, including isolation, purification, and characterization. Despite the demonstrated importance of viral diseases in invertebrate aquaculture and their apparent role in epizootics of wild populations, only one, *Baculovirus penaei*, has been investigated in sufficient detail to satisfy the above criteria. These procedures, as is true of many others, cannot be adequately pursued until marine invertebrate tissue culture is a reality. The major handicap in invertebrate virology is the lack of molluscan and crustacean cell lines necessary to initiate the above listed research. Although not as obvious, cell lines are also essential for investigation of microsporidan diseases of commercially important invertebrates as well as other microbial pathogens. Indeed, the lack of molluscan and crustacean cell lines is probably the current major roadblock in marine invertebrate disease research.

10

Infectious Diseases of Molluscs

Frank O. Perkins

INTRODUCTION

For this presentation, "'disease' denotes a demonstrable, negative deviation from the normal state ... of a living organism. 'Negative' implies an impairment, quantifiable in terms of reduction in ecological potential (e.g., survival, growth, reproduction, energy procurement, stress endurance, competition)" (Kinne 1980, p. 14). The literature reviewed herein mainly concerns accounts where a disease agent is identified or suspected to exist in the context of survival (i.e., mortalities) of host larvae, juveniles, or adults. Growth, reproduction, energy procurement, stress endurance, and competitiveness are not nearly as well addressed in the literature. Infectious disease agents considered herein are those known to cause significant mortalities and for which more than just taxonomic and morphological information is available. This translates to consideration primarily of the microbes. The reader is referred to excellent reviews such as those of Lauckner (1983); Sparks (1985); Sindermann (1990); and Elston (1990) for more complete descriptions of the relevant literature, particularly of lesser-known and apparently less significant species of disease agents. No attempt is made to provide a comprehensive literature review because of space limitations. The purpose of my presentation is to provide the nonspecialist with an overview of major diseases from the literature with some special attention to recent papers that have not been cited in previous reviews. This is followed by a review of what little is known concerning the effects of pollutants on expression of infectious diseases in molluscs.

Of the diseases affecting invertebrate phyla, those of molluscs have been the most extensively investigated, with the exception of arthropod diseases. The numbers of described species that infect molluscs are difficult to estimate because often only genus or general taxonomic group is noted. Dominant groups thus far described are clearly the Protista and metazoan parasites.

0-8493-8662-4/93/$0.00 + $.50

DESCRIPTIONS OF IMPORTANT TAXONOMIC GROUPS

Viruses

Although few viruses have been implicated in significant mortalities of molluscs, where effort is expended, virions or virus-like particles can be found in molluscs (Farley 1978; Lauckner 1983, Johnson 1984; Sparks 1985; Comps 1988). Thus the paucity of reports is probably due to the research effort that has been applied. As recently as 21 years ago, in a book on diseases of marine fish and shellfish, Sindermann (1970) did not mention any viral diseases of molluscs. By 1990 in his second edition, he noted that about 20 viruses had been reported in marine molluscs. Although the circumstantial evidence is often strong where viruses have been observed in molluscan hosts undergoing mortalities, the necessary proof of a virus being the etiologic agent is lacking in all but possibly the following two cases.

Hill (1976) isolated a small 55-60 nm RNA virus (birnavirus) from the clam, *Tellina tenuis*, then induced it to replicate in the bluegill fish cell line BF-2. This success was followed by 12 more virus isolates from *T. tenuis*, the oysters, *Crassostrea gigas* and *Ostrea edulis;* the clam, *Mercenaria mercenaria*; the limpet, *Patella vulgata*; and the periwinkle, *Littorina littorea*. Two virus strains, from *T. tenuis* and *O. edulis*, were used to infect *O. edulis* (Hill and Alderman 1979). Infections resulted in general tissue edema with the most pronounced effects observed in connective tissue around digestive gland tubules. Mortalities appeared to be associated with infection intensity, with higher virus levels yielding higher mortalities; however, there was uncertainty in that control oysters also exhibited significant mortalities after 30 days. Later studies indicated that birnaviruses possibly "do not infect oysters but are more likely present as contaminants in the same way as human enteroviruses" (Hill *et al.* 1986).

The second example of virus isolation followed by possible transmission of infections is found in the work of Orpandy et al. (1981) who isolated retrovirus-like particles of 120 nm from soft-shell clams, *Mya arenaria*, which had a possible hemic neoplasia (= disseminated sarcoma) associated with unusual clam mortalities. The viruses were purified and healthy clams were inoculated, yielding clams with the proliferative disorder. The virus was isolated from experimentally infected clams. Whether the virus caused the sarcoma is uncertain. Nevertheless, it appears that such disorders can be transmitted as seen from studies involving the common cockle, *Cerastoderma edule* (Twomey and Mulcahy 1988) and the bay mussel, *Mytilus edulis* (Elston *et al.* 1988). The agent of disease induction was not demonstrated, but a virus or viruses are suspected. The reader is referred to Peters (1988) for a review of disseminated sarcomas in bivalve molluscs and attempts to identify an etiologic agent.

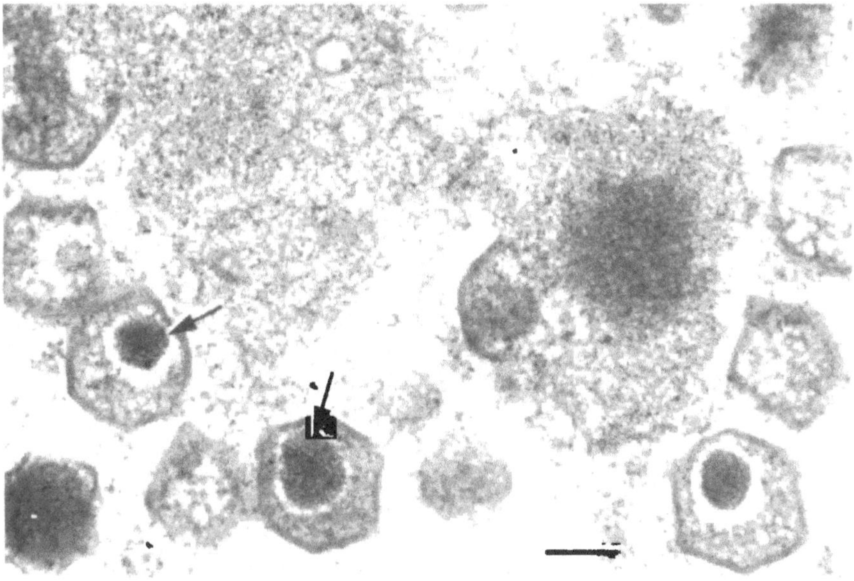

Fig. 10-1. Gill necrosis virus from gill lesion of *C. angulata*. Arrows indicate electron-dense nucleoids of immature virions. Bar = 200 nm. Reprinted with permission of Michel Comps and American Fisheries Society; from Comps (1988).

Other reports of viruses involve mostly ultrastructural observations of viruses associated with sick molluscs and obtained from necrotic tissues. An example is the observation of iridovirus-like particles of 350 nm diameter from gill lesions of the Portuguese oyster *Crassostrea angulata* (Comps *et al.* 1976) (Fig. 10-1). Those oysters along the Atlantic coast of France suffered catastrophic mortalities in the early 1970s. Because transmission of the virus was not accomplished and gill lesions were not induced experimentally, proof that the iridovirus was the causative agent of the mortalities was not obtained.

Evidence that an anthropogenic influence resulted in viral expression and mortalities of a mollusc is found in the work of Farley *et al.* (1972) where *C. virginica* was held at 28-30°C in the coolant discharge from a power plant in Maine. The population experienced marked mortalities, and dying oysters were found to be infected with an intranuclear, herpes-like virus. The virions were 70-90 nm diameter, nonenveloped, and hexagonal. Infected cells were found in large numbers around hemolymph sinuses. Oysters held nearby at 12-18°C and away from the thermally elevated waters experienced 18% mortality as opposed to 52% in the higher temperature waters. Thus, elevation of temperature is implicated in inducing expression of the virus as a disease agent.

In aquaculture hatcheries, viruses have also been implicated in causing diseases of larvae. Oyster velar virus disease (OVVD) is one of the better known. Elston and Wilkinson (1985) described a 228 nm diameter, icosahedral virus enveloped by two bilayered membranous units from *C. gigas* larvae greater than 150 μm shell height. Most virions infect the cytoplasm of the epithelium but also are found in the cytoplasm of the oral and esophageal epithelia. Lesions thus formed are characterized by cellular swelling and detachment.

Mortalities of up to 50% were observed in populations of larvae from a commercial facility where OVVD entities were present. The authors provided protocols for elimination of the disease from a hatchery that essentially consist of: 1) destruction of all stocks shown by appropriate histological and cytological techniques to be infected; 2) careful selection of brood stocks from uninfected estuaries; and 3) sterilization of effluents from the isolated stocks until the hatchery operators are certain that the new stocks are not infected.

In the search for the causative agent of mortalities of the pearl oyster, *Pinctada maxima*, Pass *et al.* (1988) found virus-like inclusions consisting of a large basophilic virogenic-like stroma with forming virons in hypertrophied digestive gland nuclei (Figs. 10-2 and 10-3). Such inclusions have been found in a wide diversity of marine bivalve molluscs (Perkins, unpublished data).

Prokaryotes

A variety of prokaryotes are found to be associated with molluscs; many are suspected to cause diseases of the hosts, but there are few substantiated reports. The microbes include mainly Gram-negative bacteria and some Gram-positive species, as well as Rickettsiae, Chlamydiae, and Mycoplasmas. Cyanobacteria are found to burrow into shells of living molluscs but have not been shown to be disease agents (Lauckner 1983). Since the bacterial flora of molluscs and, in particular, filter-feeding species is extensive, it often has been difficult to determine which bacteria isolated from them are saprobic species and which are obligate or facultative pathogens. Until recently, some workers questioned whether bacteria are primary pathogens of adult bivalves (Lauckner 1983; Tubiash and Otto 1986). A wide range of species of bacteria are easily isolated from sick bivalves and their lesions, but proving that they are the primary agent of disease or just opportunistic species has been difficult.

"Focal necrosis" disease observed in adult *Crassostrea gigas* in Japan and the west coast of the United States is an example of the difficulty experienced by researchers in trying to identify the causative agent of disease of adult bivalves where a bacterium is suspected. Mortalities of oysters were observed in populations where oysters were sexually mature or spawning and water temperature was high (ca. 20°C) (Lipovsky and Chew 1972). As seen in

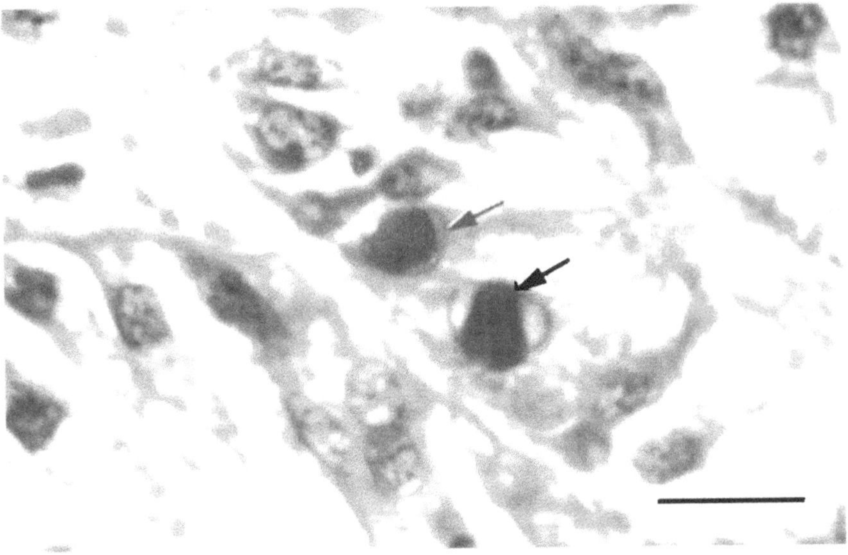

Fig. 10-2. Micrograph of hypertrophied nuclei (arrows) from hepatopancreas epithelium of *Pinctada maxima*. Electron dense inclusions in nuclei are believed to be virogenic stromata, bar = 10 μm.

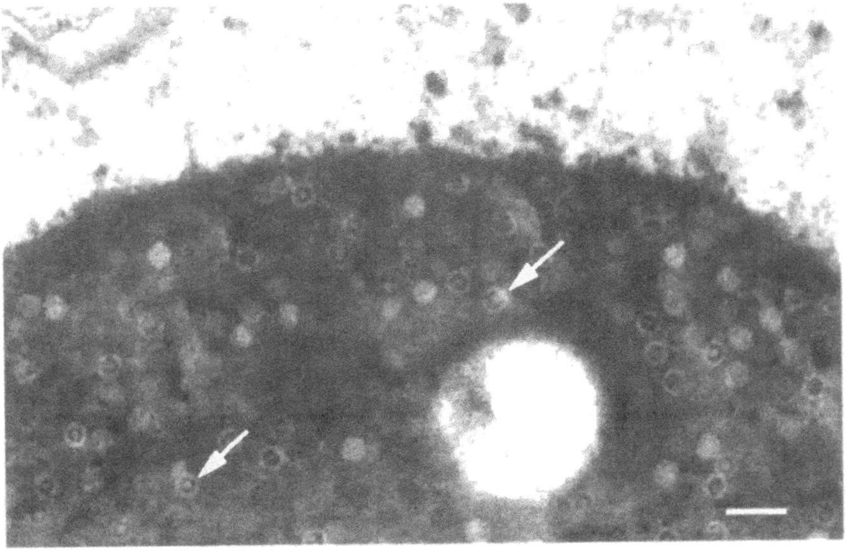

Fig. 10-3. Transmission electron micrograph of nucleus of *P. maxima* with presumptive virogenic stroma in which can be seen presumptive developing virions (arrows), bar = 0.2 μm. (Reproduced with permission of David Pass.)

histological sections, the foci of necrosis ("multiple abscesses" of Imai *et al.* 1968) contain colonies of Gram-positive (very rarely, Gram-negative) bacteria found near the digestive tract. Efforts at isolation and culture of bacteria from oysters believed to have "focal necrosis" disease yielded *Vibrio anguillarum* and *V. alginolyticus* (Grischkowsky and Liston 1974) and *Pseudomonas enalia* (Colwell and Sparks 1967). Recent studies, however, indicate that the causative agent is *Nocardia* sp., an actinomycete-like bacterium which is Gram-positive, acid-fast, beaded and branched, and found in lesions of the connective tissue around the gut and digestive diverticulae (Friedman and Hedrick 1991). Upon gross examination, yellow spots can be seen on the body surface (Elston *et al.* 1987). The disease is now being called nocardiosis (Elston 1990).

Under the artificial conditions of aquaculture hatcheries, bivalve larvae are very susceptible to serious diseases, known as bacillary necrosis or larval vibriosis, caused primarily by *Vibrio anguillarum* and *V. alginolyticus* (Tubiash *et al.* 1970; Tubiash and Otto 1986) or *V. tubiashii* (Hada et al. 1984). Virulence of these *Vibrio* infections is so great as to cause almost 100% mortality in 24 h. Bacteria are opportunistic saprophytes that express themselves as pathogens when larval culture conditions are suboptimal and there is an accumulation of nutrients from sources, such as contaminated and bacterized algal food, dead or dying larvae, or excess bivalve gametes left in culture containers with embryos. When total bacterial concentrations build to greater than 10^5, bivalve mortalities begin and generally progress to catastrophic levels (Castagna and Kurkowski, personal communications). Sindermann and Lightner (1988) stated that, when *Vibrio* spp. concentrations are only 10^2, mortalities can be initiated. Control of the diseases in hatcheries is accomplished by established procedures for maintaining cleanliness of culture containers, avoiding build-up of sick or dying larvae, and maintaining algal food cultures which are not heavily bacterized. Antibiotic treatment is unnecessary when the correct culture protocols are followed. Where antibiotics have been used, the results are often not satisfactory (Lauckner 1983).

Whether comparable diseases occur in natural populations of larvae in estuaries is yet to be demonstrated. Considering the role of nutrient enrichment in hatcheries, a fertile area for future research is to determine whether eutrophication of estuaries leads to elevation of numbers of *Vibrio* spp., which then cause mortalities of bivalve larvae in the estuary. Since the densities of larvae in the natural settings are much lower than those in hatcheries, the concentrations of *Vibrio* spp. needed to induce disease are probably much different from those which cause disease in a hatchery.

Vibrio spp. are also suspected to cause mortalities of the golden-lip pearl oyster, *Pinctada maxima*, as a result of the stress of being transported to a pearl oyster farm at Kuri Bay, Australia, from the Broome region (Eighty Mile

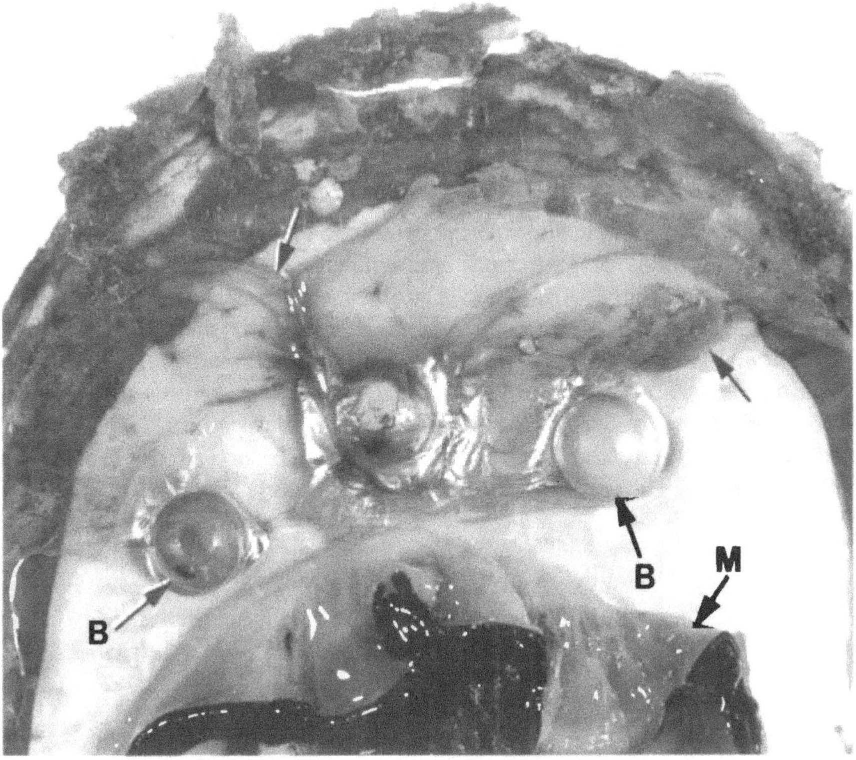

Fig. 10-4. *P. maxima* infected with *Vibrio* sp. which has caused anomalous conchiolin deposits (arrows). Resin bead (B) implants attached to shell for induction of blister pearls; oyster mantle (M) folded back. (Reproduced with permission of David Pass.)

Beach). *Vibrio harveyi* was isolated from the hemolymph of sick oysters and used to experimentally cause the same disease response as seen under natural conditions. Whether it is the only *Vibrio* sp. causing disease or only disease agent is not known. Infected oysters form conchiolin deposits on the nacre that interfere with blister pearl formation even if the oyster does not die (Fig. 10-4). Mortalities as high as 80% have been observed at the farms (Pass *et al.* 1987). It is suspected that the bacterial problem arises as a result of stress on the oyster induced by low winter temperatures of about 18°C, coupled with nutrient enrichment of water in carrier boats used to transport oysters from Australia's Eighty Mile Beach to Kuri Bay. Enrichment probably comes from accumulation of oyster feces and pseudofeces, which stimulate *Vibrio* spp. growth. Thus, measures to lessen mortality were to 1) avoid low temperature shipments, 2) avoid crowding of oysters during shipment, and 3) shorten the time in transit.

Cytophaga-like, gliding bacteria are known to cause pronounced mortalities of juvenile bivalves, including *C. gigas*, *Ostrea edulis*, *M. mercenaria*, *Tapes philippinarum*, *Siliqua patula*, and *Argopecten irradians* (Dungan *et al.* 1989; Elston 1990). The problem is most pronounced in nursery areas where aquacultured bivalves are grown in high concentrations. Whether it occurs under natural conditions is unknown. The bacteria erode and even destroy ligaments of the bivalves, resulting in inability to close the valves, preventing feeding and respiration, and probably allowing opportunistic bacteria to infect the tissues.

Bacterial diseases of gastropods are not as well documented as those of pelecypods (Lauckner 1980). An exception is aeromonasis of the giant African snail, *Achatina fulica*. When severe mortalities of the snail occurred in the 1950's in Sri Lanka, attempts were made to find the causative agent and use it to aid farmers. Sturtevant (1969) identified the agent as *Aeromonas hydrophila* (= *A. liquefaciens* of Mead 1969) following the work of others. Srivastava and Srivastava (1968) attempted to use the bacterium to induce disease in snails with limited success; however, no commercially useful techniques have been developed.

A. hydrophila causes white lesions on the tentacles and on various parts of the head and body. Prevalences of such lesions as high as 94% have been observed in Hawaii coupled with a decline of 85% in the snail populations (Mead 1963). A. *hydrophila* has been shown to be a pathogen of a wide diversity of aquatic vertebrates but no other invertebrates (Khardori and Fainstein 1988). Other bacterial diseases are listed in Table 10-1.

Pelecypod infections caused by the primitive bacteria, Rickettsiae, Chlamydiae, and Mycoplasms are widespread and commonly observed without any known relation to anthropogenic influences (Otto *et al.* 1979; Lauckner 1983; Sparks 1985). Rickettsiae and Chlamydiae usually are found as obligate intracellular parasites, whereas mycoplasms are free-living or intracellular microbes. Because all three groups contain species that cause fatal diseases of humans and mammals, there has been considerable interest in those found in bivalve molluscs consumed by humans. However, no evidence exists to implicate molluscs as vectors of human diseases caused by those prokaryotes. Nevertheless, the possibility exists and requires further examination. There is no conclusive evidence that the primitive bacteria cause diseases in adult bivalve molluscs despite the high concentrations sometimes observed (Meyers 1979; 1981). However, Leibovitz (1989) has demonstrated that *Chlamydia* sp. causes mortalities of larvae and postmetamorphic bay scallops (*Argopecten irradians*), at least under hatchery conditions. Whether the disease is found in other bivalves and under natural conditions remains to be seen.

The Rickettsiae, Chlamydiae, and Mycoplasms generally exist in cytoplasmic vacuoles in the gut and digestive gland epithelia of bivalves where considerable

Table 10-1. Bacterial Diseases

Name of Disease	Causative Agent	Host(s)	Literature Sources
Gastropod acid-fast bacteremia	Unidentified acid-fast bacilli	Snails: *Biomphalaria glabrata* *B. pfeifferi* *Helisoma anceps*	Sparks 1985; Michelson 1961
American oyster vibrio cardiac endema or cardiac *vibriosis*	*Vibrio anguillarum*	Oyster: *Crassostrea virginica*	Sparks 1985
——	*Achromobacter* (?)	Oyster: *Crassostrea gigas*	Sindermann 1970
Vibriosis of juvenile oysters	*Vibrio* sp.	Oysters: *C. virginica*, *Ostrea edulis* Clam: *Mercenaria mercenaria*	Sindermann and Lightner 1988; Elston, Elliot, and Colwell, 1982
——	*Leucothrix mucor*(?)	Clam: *Cardium edule*	Lauckner 1983
Mycelial disease	Unidentified actinomycete (?)	Oysters: *C. virginica* *Ostrea lurida*	Mackin 1962; Lauckner 1983
——	Unidentified actinomycete (?)	Oyster: *C. virginica*	Meyers 1981
Juvenile abalone vibriosis	*Vibrio alginolyticus*	Abalone: *Haliotis rufescens*	Elston and Lockwood 1983
——	Opportunistic Gram-negative bacteria	Squid: *Loligo pealei* *L. plei* *Lolliguncula brevis* *Loligo opalesceni* *Ommastrephes pteropus*	Hulet, *et al.* 1979; Leibovitz, Meyers, and Elston 1977

distension of the host cell (up to 100 μm diameter) and displacement of the host cell organelles occur (Harshbarger *et al.* 1977). In histological sections, the vacuolar inclusions are seen as amorphous basophilic bodies in which individual bacteria can be seen with the aid of oil immersion optics. The ultrastructure of the Rickettsiae consists of two tripartite membranes (the plasmalemma and cell wall) delimiting a rod-shaped, Gram-negative bacterium with typical prokaryote substructure.

As opposed to the bivalves, species in other molluscan taxa have not been shown to harbor the primitive bacteria. However, this probably represents no more than a lack of attention to the matter, because hosts of such bacteria have a wide phylogenetic range.

Eumycota (Fungi)

There are only a few species of Eumycota found to infect and cause disease in molluscs. An example is shell disease of bivalve molluscs best documented for *Ostrea edulis* and found throughout the coastal waters of Western Europe (Alderman 1976). The causative agent is *Ostracoblabe implexa*, tentatively assigned to the Fungi Imperfecti (Deuteromycotina) in that "it has regularly septate hyphae but no known reproductive spores" (Porter 1986, p. 144). The mycelium is straight and thin (2 μm diameter) with globose swellings termed "prochlamydospores." From senescent cultures, new growth on fresh medium occurs from the swellings. The latter are considered to be resistant cells, not reproductive spores.

The disease is expressed by proliferation within the oyster shell and, in advanced stages, penetration of the nacre resulting in irritation of the mantle followed by predominantly conchiolin deposition with some calcification around the sites of penetration. Progression of the infection beyond this stage can result in abnormal thickening of the shell and shell distortion until the mantle cavity is limited. Once the mantle is irritated (it is not invaded), oyster mortalities are associated with the presence of the disease.

A species of the Oomycetes (Mastigomycotina), *Sirolpidium zoophthorum*, infects and kills oyster (*C. virginica*) and clam (*M. mercenaria)* larvae (Vishniac 1955) grown under the artificial conditions of aquaculture. It has been reported from the natural environment (Johnson and Sparrow 1961); however, whether it is significant as a disease agent is not known. Biflagellated heterokont zoospores (i.e., one whiplash flagellum and one flagellum with mastigonemes) of 2x5 μm size are formed that encyst, enlarge to 7-10 μm diameter, then elongate to form a tube of about the same diameter. Throughout the larval tissues, a mycelium develops with thick hyphae and septate constrictions followed sometimes by thallus fragmentation at the septa. In cultures on agar medium, terminal swellings are formed, each with a crosswall at the base. These become zoosporangia from which zoospores escape through a discharge tube. In the larvae, each hyphal segment can metamorphose into a sporangium each with a discharge tube (Davis *et al.* 1954).

Since the original description by Vishniac in 1955, when bivalve culture techniques were in the early stages of being developed, there has been little published, possibly because the disease has not been significant in using more modern techniques of aquaculture. Other reports of fungal diseases in molluscs are listed in Table 10-2.

Table 10-2. Fungal Diseases of Molluscs

Name of pathogen	Host(s)	Literature Sources
Haliphthoros milfordensis	Oyster drill: *Urosalpinx cinerea*	Ganaros 1957; Vishniac 1958
Deuteromycete (?)	Mussel: *Mytilus galloprovincialis*	Vitellaro-Zuccarello 1973
Leptolegnia (or *Leptolegniella*) *marina*	Clam: *Cardium echinatum*	Atkins 1954

Protista

Of all the molluscan diseases, those caused by Protista are by far the most thoroughly studied. This focus is not by chance; the Protista induce the most pronounced mortalities of commercially significant molluscs.

One of the most intensively studied species is *Perkinsus marinus*, an apicomplexan pathogen in the class Perkinsasida (Levine 1988), originally described in 1950 by Mackin, Owen, and Collier as *Dermocystidium marinum*. It parasitizes American oysters (*C. virginica*) from New Jersey to Texas and possibly other species of bivalves in northern Atlantic Ocean waters that can be classified as temperate, subtropical, or tropical. *Perkinsus* spp. have been reported in 67 species of molluscs, all bivalves except four species of gastropods (abalone). They were found in coastal waters of Australia and the Mediterranean Sea, as well as the Atlantic Ocean, Gulf of Mexico, and Caribbean Islands (Perkins 1988; Goggin *et al.* 1989); therefore, the pathogens are most probably found in all the warmer coastal waters of the world. Although four species of *Perkinsus* have been described (Mackin et al. 1950 amended by Levine 1978; Lester and Davis 1981; Azevedo 1989; McGladdery *et al.* 1991), it is uncertain whether there is more than one species in light of the studies of Goggin *et al.* (1989) where infections were generally accomplished from species to species of hosts, using a diversity of *Perkinsus* isolates in which zoospores were the infectious cells. Earlier attempts at cross infections among host species were conducted using minced, infected oyster tissues with little success (Ray 1954; Andrews and Hewatt 1957). The techniques used in Australia should be applied to bivalves elsewhere to see if cross-infections can be induced as readily with other hosts and *Perkinsus* isolates. There will be no lack of material because the large majority of the bivalves examined from the warmer coastal waters are parasitized by *Perkinsus* sp. or spp. (Goggin and Lester 1987; Perkins, unpublished data). As suggested earlier (Perkins 1988), there may be two groups of *Perkinsus* spp., but the work of the Australians raises questions about that suggestion.

In addition to infection studies of Goggin *et al.* (1989), further evidence for the existence of only one *Perkinsus* sp. or few *Perkinsus* spp. lies in the cellular and subcellular studies of Perkins (1976a; 1988), Azevedo (1989), Azevedo *et al.* (1990), and Lester (personal communication). The structures described thus far lead me to wonder if differences are significant enough to warrant designation of more than one (or two?) species.

With the exception of *P. marinus*, little is known about the epizootiology of *Perkinsus* sp. or spp. from the 67 species of molluscs. In the case of the *C. virginica* pathogen, serious mortalities of the host have been documented from the Chesapeake Bay and coastal waters of South Carolina, Florida, and the Gulf of Mexico states. Generally when temperatures and salinities rise above normal, incidences of infection and mortalities increase. In Chesapeake Bay, infections established in spring and early summer result in deaths of hosts in summer and fall, seasons of greatest mortalities regardless of when infections are initiated. Deaths may occur into November and December, with regression of incidences and mortalities during winter and spring (Andrews 1988).

Transmission of infections is direct, from oyster to oyster with meronts or zoospores serving as infective elements when they are filtered from the water. The infective cells apparently penetrate the epithelium of the gill or gut or are carried through it by hemocytes. Multiplication occurs mainly in the connective tissue or between the epithelial cells. Large numbers of *P. marinus* cells may develop and cause extensive tissue damage before death of the host.

The life cycle consists of uninucleate coccoid meronts (2-4 μm), which enlarge to about 10-20 μm and in the process acquire a large eccentric vacuole and a vacuoplast (Fig. 10-5). The latter is formed then disappears as enlargement progresses. Schizogony occurs by successive bipartitioning of the protoplast (karyokinesis followed by cytokinesis then repeated) or possibly progressive cleavage (repeated karyokineses followed by cytokinesis to yield the final number of daughter cells) within the mother cell wall to yield 2-64 meronts, which are released upon rupture of the wall. Biflagellated zoospores may also be formed from large meronts (> ca. 15 μm) that become prezoosporangia when released into sea water. The protoplast divides by successive bipartitioning (Fig. 10-6) then flagella are formed before the last bipartitioning (Fig. 10-7) (Perkins and Menzel 1966). Zoospore formation is best visualized after prezoosporangia are induced to form in fluid thioglycollate culture medium and then placed in sea water. The process that undoubtedly occurs under natural conditions has not been observed in *P. marinus* unless fluid thioglycollate medium is used. It has

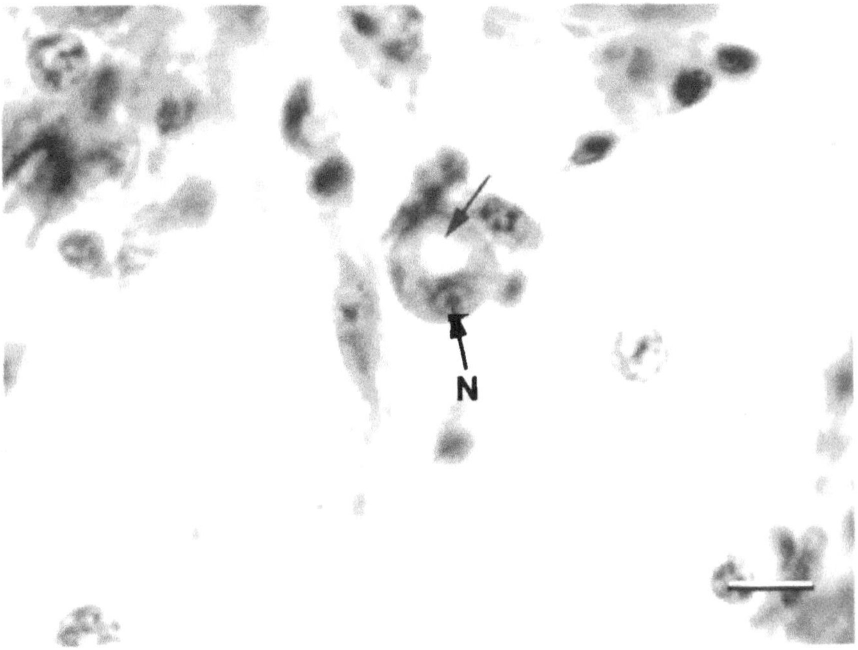

Fig.10-5. Light micrograph of *Perkinsus marinus* meront in connective tissue of *Crassostrea virginica*. Eccentric vacuole (arrow); nucleus with large nucleolus (N), bar = 5 μm.water.

been observed in large *Perkinsus* sp. meronts isolated from *Macoma balthica* tissues without subsequent treatment in the medium (Perkins 1988). Prezoosporangium formation in the culture medium involves marked enlargement of meronts and schizonts from ca. 2-20 μm to 15-100 μm. This enlargement is the basis for an inexpensive rapid diagnostic technique devised by Ray (1952) and used by most workers to detect presence of *Perkinsus* in molluscan tissues. There do not appear to be resistant stages formed, but survival outside of the molluscan hosts may be possible for days as evidenced by the observation of viable meronts in the feces of fish (Hoese 1964) and buccal cavity of a snail (White *et al.* 1987). An alternate host or vector, however, is not necessary to complete the life cycle.

The terms "meront" and "schizont" are used following definitions of Margulis *et al.* (1990). I believe "meront" is more appropriate than "trophozoite" as used by Perkins (1988), because non-flagellated cells of *P. marinus* in bivalve tissue are not motile. Margulis *et al.* (1990) define trophozoites as motile cells and reject more general definition that 1) equates with "vegetative" cells of botanical literature and 2) has been used by protozoologists for decades (Kudo 1966).

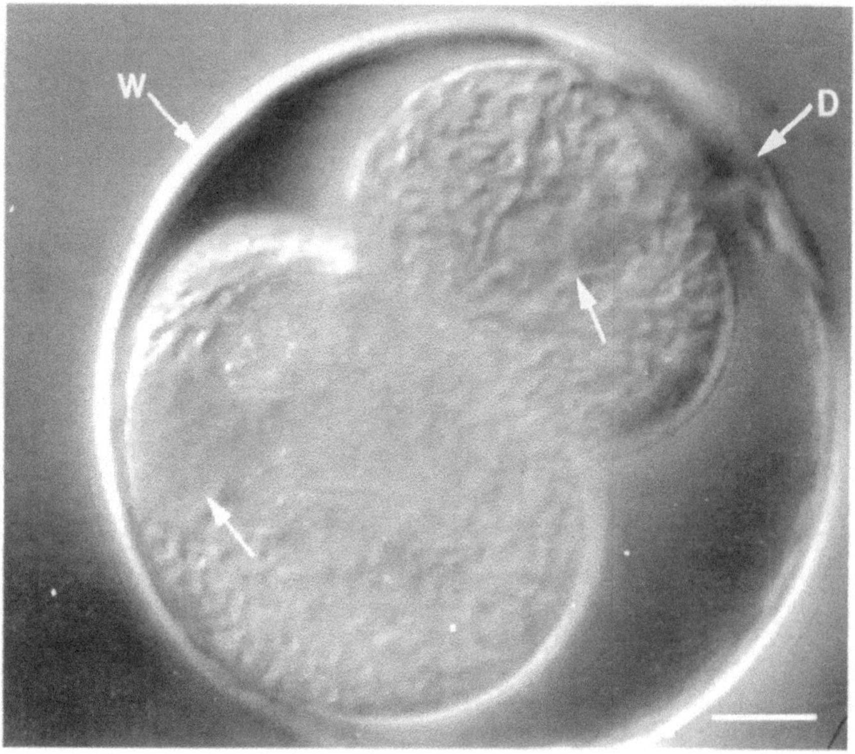

Fig. 10-6. Differential interference micrograph of *P. marinus* zoosporangium with protoplasm undergoing first cleavage. Nuclei (arrows); cell wall (W); discharge pore (D) occluded with plug of wall material, bar = 10 μm.

Another group of molluscan pathogens that has received considerable attention are members of the Haplosporidia (=Acetospora), a phylum of Sporozoa that form spores with various wall surface structures, useful in distinguishing species (Perkins 1990). The molluscan pathogens are members of the genera *Haplosporidium* and *Minchinia* with the greatest information having been generated concerning *H. nelsoni* (commonly known as MSX), which causes significant mortalities of *C. virginica* from Massachusetts to North Carolina with isolated but questionable sightings as far south as Biscayne Bay, Florida, and north to Maine. The origin of *H. nelsoni* remains unknown. It was discovered in 1957 in Delaware Bay when oysters started dying in large numbers. It is not known whether the protist pre-existed in the oysters as a parasite causing negligible mortalities then mutated to become a pathogen or invaded the oyster *de novo* from another host. Obviously the importation of an exotic host species

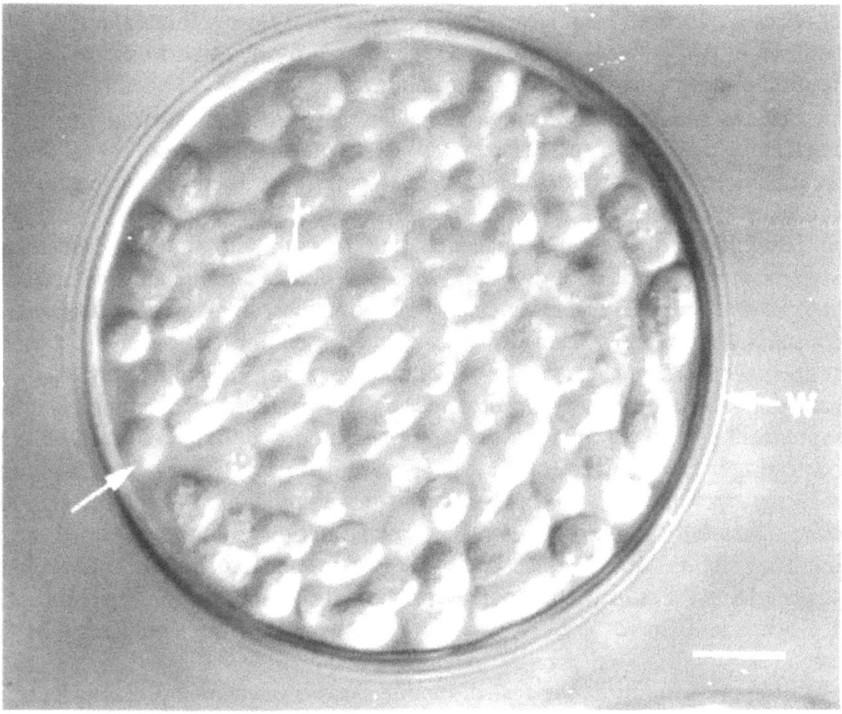

Fig. 10-7. Differential interference micrograph of *P. marinus* zoosporangium prior to final division of the forming zoospores (arrows). Flagella have been formed. Cell wall (W), bar = 10 μm.

into Delaware Bay is a possible source of the problem, but no such importation has been documented. It is known that the distribution of *H. nelsoni* expanded rapidly north and south from Delaware Bay, probably as a result of movement of infected oysters (and an intermediate host?) into the non-endemic waters.

The life cycle of *M. nelsoni* has not been elucidated. Development within the oyster appears to arise from uninucleate or binucleate naked cells first seen intercellularly (unless phagocytized) in or beneath the epithelium of the labial palps or gut. From there they spread throughout the host, most often residing in connective tissue. Nuclei divide and cells enlarge to become plasmodia (i.e., a naked cell with more than one nucleus). Cytokinesis is by multiple irregular fission. The cells have fairly typical eukaryotic organelles with the mitochondria being tubulovesicular. Atypical is the presence of haplosporosomes, generally spheroid, electron dense organelles of 70-250 nm diameter (Perkins 1979). These organelles appear to be characteristic of the Haplosporidia, Paramyxa, and possibly the Myxozoa. Their function is unknown.

Spores are formed from large plasmodia that first form a thin delimiting wall, then one of two possible sporulation sequences are suggested to occur: 1) the plasmodial protoplast subdivides into uninucleate sporoblasts each of which, in turn, cleave internally to form a uninucleate sporoplasm surrounded by an anucleate epispore cytoplasm (Perkins 1971); 2) plasmodial protoplast subdivides into uninucleate sporoblasts which then forms pairs followed by fusion into binucleate sporoblasts. Karyogamy follows, then the cell pinches in halves. The anucleate half envelops the nucleated half to yield the same end result as mechanism 1 (Desportes and Nashed 1983). Whether either sequence occurs is yet to be determined. The mature spore of the molluscan Haplosporidia consists of a uninucleate sporoplasm with a convoluted membranous organelle, termed the spherulosome, at the anterior end and haplosporosomes in addition to other commonly observed eukaryotic organelles. The spore wall consists of a squat, amphora-like cup with an anterior flap of wall material, hinged along one side, and resting over the orifice of the cup (Fig. 10-8). Excystment occurs when the lid swings open and the sporoplasm moves in an amoeboid fashion through the orifice (Azevedo and Corral 1989). Ornamentation is found in the form of strands or ribbons around the spore wall. Perkins and van Banning (1981) have suggested that the substructure of the ornamentation is of taxonomic significance. McGovern and Burreson (1990) have further evaluated the significance of the structures.

After spore formation, the rest of the life cycle is unclear until uninucleate or binucleate cells appear in another host of the same species. No haplosporidian infections have been induced experimentally with infected tissues or spore suspensions (possible exception reported by Barrow 1965). Azevedo and Corral (1989) suggested that sporoplasms, from spores in a given host, may autoinfect that host. Their suggestion was derived from studies of *Minchinia* sp. in the Portuguese clam, *Ruditapes decussatus*. The idea merits further evaluation.

Most of what is known about the epizootiology and general biology of *H. nelsoni* is derived from excellent studies in the Delaware Bay and Chesapeake Bay areas by H.H. Haskin, S.E. Ford, J.D. Andrew, and colleagues (Haskin and Andrews 1988). The parasite is active and causes mortalities of *C. virginica* above 20 °/oo salinity. Below 15 °/oo, infections may occur, but cellular multiplication in the host is curtailed. Below 10 °/oo it does not survive in the host. Thus the geographic distribution of the pathogen is in large part controlled by rainfall and salinity changes in estuaries and their tributaries. This limitation has been fortunate for oyster growers since low salinity sanctuaries have been used to grow oysters. Winter temperatures inhibit activity of MSX and host mortalities are negligible. During summer, mortalities rapidly increase and peak about July-August. For planted oysters moved from low salinity seed beds to higher salinity grow-out areas, the time of planting shifts the time of high

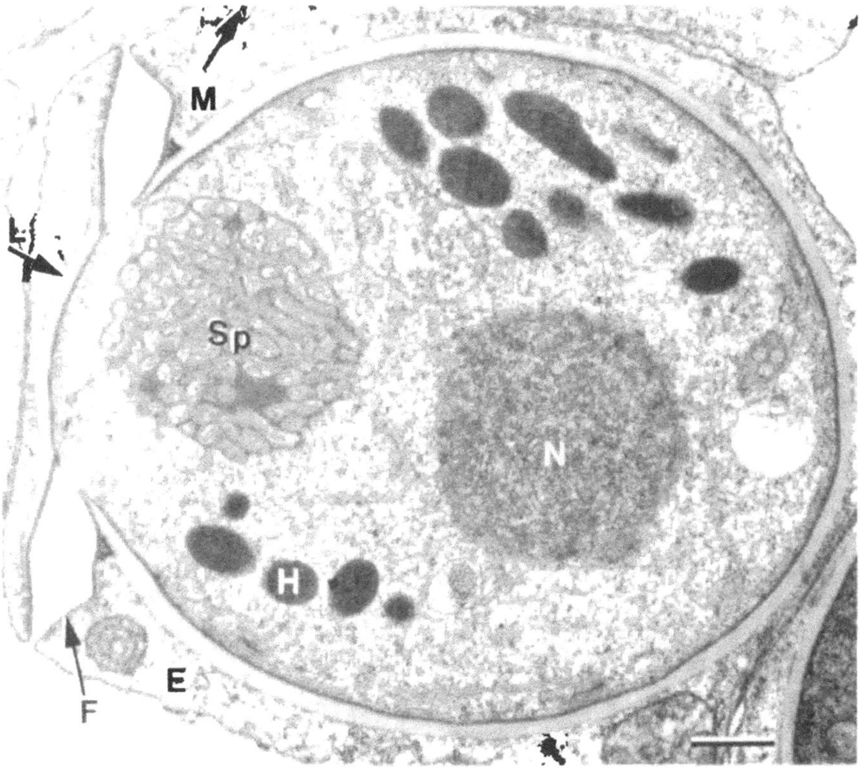

Fig. 10-8. Transmission electron micrograph of immature spore of *Minchinia teredinis*. Epispore higher salinity grow-out areas, the time of planting shifts the time of highcytoplasm (E); spherulosome (Sp); nucleus (N); haplosporosome formative regions (H); microtubule-like structures (M); lid (L) of spore wall; spore wall flange (F), bar = 0.5 μm (with permission of Elizabeth McGovern and Society of Protozoologists; from McGovern and Burreson [1990]).

mortalities. In Chesapeake Bay, August imports of uninfected oysters result in the highest mortalities the following July and August, whereas March imports have high mortalities the next July-August period only 4-5 months later. In Delaware Bay, the same pattern holds except that significant mortalities may occur after only 1 month.

Higher than normal salinities in the Chesapeake Bay and upper Delaware Bay are associated with elevated mortalities from *H. nelsoni* in the endemic areas, whereas, excluding 1985, drought conditions in the lower Delaware Bay do not influence mortalities caused by the pathogen but rather temperatures are important. Very cold winters will lower mortalities in the following summer, leading to the suggestion that cold winters may kill a reservoir host from which infective cells originate. This concept of an intermediate host dominates speculation about the life cycle of *H. nelsoni* and other Haplosporidia (Burreson 1988).

Haplosporidian parasites found in molluscs are listed in Table 10-3. *Bonamia ostreae* is included because haplosporosomes are found in the cytoplasm and internal cleavage as in the Paramyxea does not occur. However, whether it is a haplosporidian is uncertain because spores have not yet been observed. A more complete description is found in the section, entitled *Incertae Sedis.*

Paramyxea

Several bivalve pathogens are found in the phylum (?) Paramyxea, Class Marteiliidea. The best known is Aber disease caused by *Marteilia refringens* in *Ostrea edulis* (European flat oyster). The Paramyxea "are characterized by the formation of propagules (traditionally called spores) consisting of several cells enclosed inside one another that arise by a process of internal cleavage or endogenous budding within a stem cell" (Desportes and Perkins 1990, p. 30). The stem cell is believed to be a rarely seen amoeboid, uninucleate cell that is found between the host cells. It apparently establishes infections of the host and exists for a short time before undergoing one nuclear division then internal cleavage to form an uninucleate cell within an uninucleate cell. The latter is the earliest stage leading to spore formation. Because many cells are observed in the process of sporulation in a host and few stem cells are observed, one assumes that the short-lived stem cells multiply before engaging in sporulation. Otherwise the numerous sporulating cells would each have to arise from a stem cell that had its origin outside the host. However, multiplication of the stem cells has not been observed. An alternate possibility is that the frequently observed single cell-within-a-cell stage is not necessarily committed to sporulation and actually is the stage for proliferation within the host before sporulation. However, how such multiplication might occur is not known.

As sporulation progresses within each of the former stem cells of *M. refringens*, the initially delimited internal cell divides three times to form 8 uninucleate cells (termed secondary cells), each in a vacuole within the former stem cell (Perkins 1976b). Each secondary cell acquires a delimiting wall, then undergoes karyokinesis and internally cleaves to form 3 or 4 uninucleate cells (tertiary cells) each within a vacuole in the cytoplasm of the secondary cell. Each tertiary cell in turn cleaves internally to form spores with 3 uninucleate sporoplasms that are all delimited by a wall (Fig. 10-9). Unlike the results of the previous two cycles of internal cleaving, the 3 uninucleate sporoplasms do not each occupy a separate vacuole. Rather, there is an innermost sporoplasm within an intermediate sporoplasm, which is, in turn, located in an outermost sporoplasm. Haplosporosomes

Table 10-3. Haplosporidian Parasites of Molluscs

Causative Agents	Hosts	Literature Sources
Haplosporidium nelsoni	*C. virginica*	Haskin *et al.* 1966; Couch *et al.* 1966; Perkins 1990
H. costale	*C. virginica*	Andrews and Castagna 1978; Perkins 1969
H. mytilovum	*Mytilus edulis* (mussel)	Field 1922
H. pickfordae	*Heliosoma companulatum; Physa parkeri; P. sayii; Lymnea emarginata; L. stagnalis* (all snails)	Barrow 1965
H. tapetis	*Ruditapes decussatus* (clam)	Vilela 1951; Chagot *et al.* 1987
H. tumefacientis	*Mytilus californianus* (mussel)	Taylor 1966
Haplosporidium sp.	*Crassostrea gigas*	Kern 1976
Haplosporidium sp.	*C. gigas*	Katkansky and Warner 1970
Minchinia armoricana	*Ostrea edulis*	van Banning 1977; Perkins and van Banning 1981
M. chitonis	*Lepidochitona cinereus* (chiton)	Ball 1980
M. dentali	*Dentalium entale* (scaphopod)	Desportes and Nashed 1983
M. teredinis	*Teredo navilis; T. furcifera; T. bartschi* (shipworms)	McGovern and Burreson 1990; Hillman *et al.* 1990
Bonamia ostreae	*Ostrea edulis*	Grizel *et al.* 1988
Bonamia sp.	*Tiostrea lutaria* *Ostrea chilensis*; *O. lurida*	Dinamani et al. 1987 Elston 1990

are present in the outermost sporoplasm (Fig. 10-10). Obviously the original stem cell enlarges markedly during sporulation to accommodate the extraordinary complex of cells. Presumably the tripartite spores serve to infect other hosts, but experimental transmission of the disease among oysters has not been accomplished. As with *Haplosporidium* spp., an intermediate host is suspected. Aber disease is found along the Atlantic coast of Spain, France, and

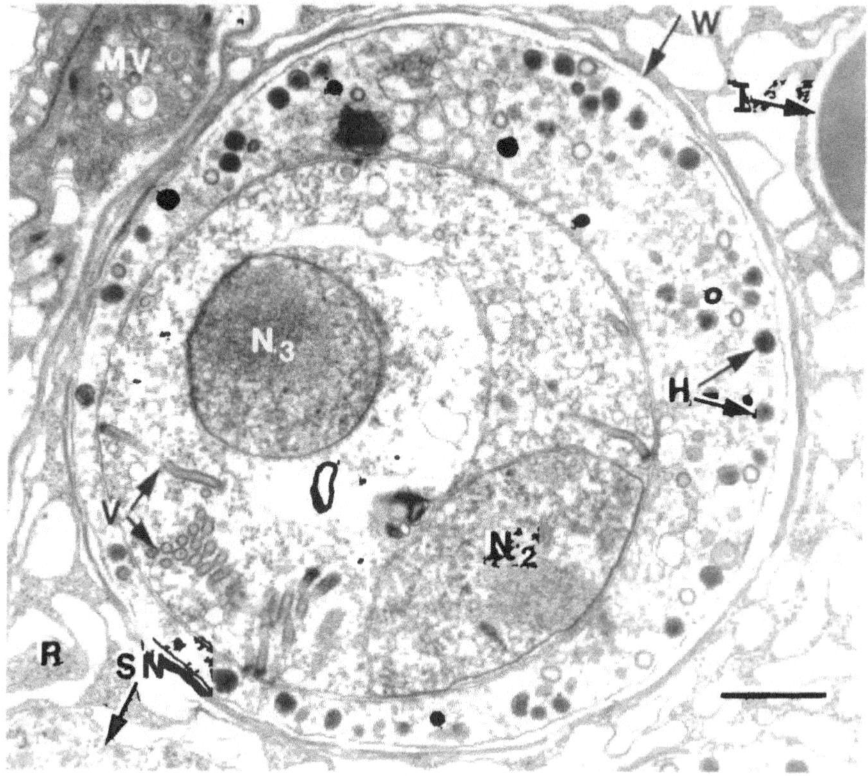

Fig. 10-9. Transmission electron micrograph of *M. sydneyi* immature spore: reticulated cytoplasm (R), inclusion body (I), nucleus (SN) of secondary cell where spores are formed; spore wall (W); haplosporosomes (H) in outermost sporoplasm; nucleus (N_2) and flattened vesicles (V) of intermediate sporoplasm; nucleus (N_3) of innermost sporoplasm; multivesicular body (MV) of stem cell, bar = 0.5 μm (with permission of the *Journal of Parisitology*, from Perkins and Wolfe [1976]).

the Netherlands, and was first recorded in France in 1968 along the northern Brittany coast where it spread northeast and south (Figueras and Montes 1988). The epizootiology is fairly well known, with host mortalities beginning in May, peaking in June-August, and then decreasing in December or January. In early spring, losses are generally negligible. Depending on the location, *M. refringens* is either absent in the host in late winter and early spring or is found in small numbers in only a few hosts. Apparently new infections are established from May to August. Overwintering cells may multiply the following spring to cause mortalities. It appears that at least 17°C is required for establishment of new infections. In some populations of *O. edulis*, infections may be as high as 100%.

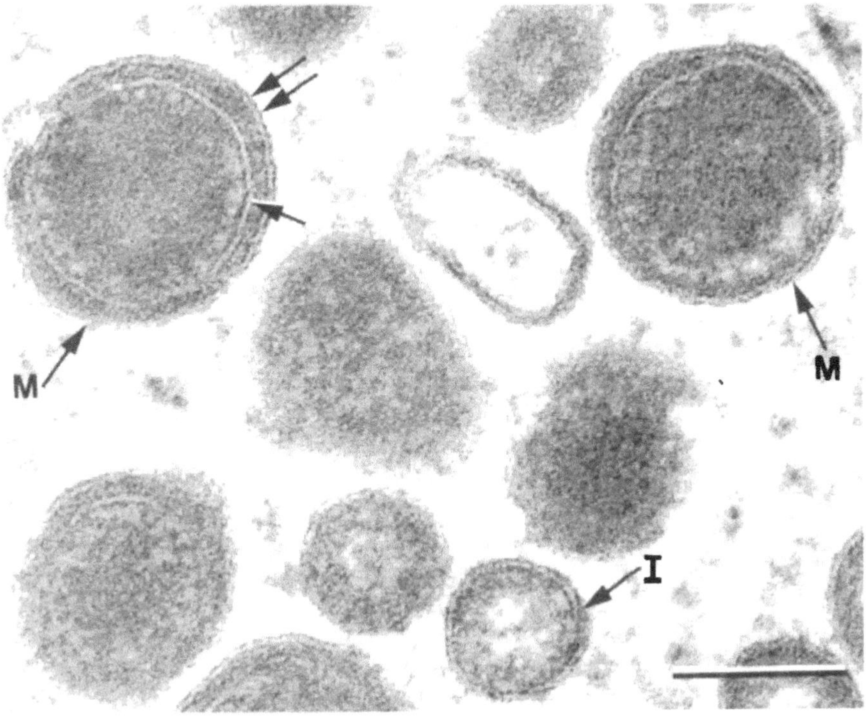

Fig. 10-10. Transmission electron micrograph of mature (M) and incompletely formed (I) haplosporosomes from outermost sporoplasm of *M. sydneyi* spore. Note delimiting unit membrane (double arrow) and internal membrane (arrow), bar = 0.1 μm.

Other species of oysters, *C. angulata* and *C. gigas*, in the region are not reported to be infected by *M. refringens* (except possibly *C. gigas* around Roscoff, France, which were shown to have light infections of early cell stages [Balouet *et al.* 1979]).

During the 1970's, *M. refringens* caused serious mortalities of *O. edulis*, resulting in a French oyster industry almost solely dependent on *O. edulis* before 1968 being transformed to an industry which consisted of 90% *C. gigas* by 1980. After *M. refringens* epizootics became significant, the cultivation of *C. gigas* was necessary to prevent collapse of the French oyster industry. Some recovery of *O. edulis* populations occurred in the late 1970's, but in 1979 a new pathogen (*Bonamia ostreae*) was detected in *O. edulis* on Tudy Island (Brittany coast of France) (Comps et al. 1980). *B. ostreae* proved to be as devastating a pathogen as *M. refringens* and further depressed the *O. edulis* production. Since the appearance of *B. ostreae*, *M. refringens* is considered to be of secondary importance to the *O. edulis* farmers and received little attention by oyster pathologists in the 1980's (see following section).

Incertae Sedis

Bonamia ostreae is a species of unknown phylogenetic affinities, found in hemocytes and extracellularly in tissues of *O. edulis*. In addition, *B. ostreae* or some other unidentified species of *Bonamia* is (are) found in New Zealand oysters (*Tiostrea lutaria*), Chilean oysters (*O. chilensis*), and the Olympic oyster (*O. lurida*) (Elston 1990).

Bonamia spp. are characterized by small 2-6 μm spheroidal basophilic cells generally uninucleate but sometimes binucleate. Rarely multinucleate plasmodia are observed. The ultrastructure is simple with mitochondria, numerous ribosomes, and haplosporosomes. Existence of the latter organelles leads one to think that *Bonamia* spp. are Haplosporidia; however, unless typical spores are observed it will not be possible to assign the species to that taxon. Because internal cleavage has not yet been observed, it is unlikely that *Bonamia* spp. are members of the Paramyxea or Myxozoa, the only other groups that have haplosporosomes.

The range of *B. ostreae* is now along the Atlantic coasts of Spain, France, the Netherlands, England, and Ireland. Although infected oysters have been transferred to the Mediterranean coasts of Spain or France, those areas remain free of the disease (Grizel *et al.* 1988). In an excellent effort at tracking, Elston *et al.* (1986) were able to show that *B. ostreae* in Europe originated in *O. edulis* imported from California.

Transmission (unlike haplosporidian oyster diseases) is from oyster to oyster and may occur throughout the year. Uninfected oysters placed in endemic waters usually become infected in 3-4 months, with highest infection rates occurring in summer. However, low temperatures of 4-5°C do not prevent infections from being established. The influence of salinity is not known. Other bivalves, *C. gigas*, *Mytilus edulis*, *Ruditapes decussatus*, *R. philippinarum* and *Cardium edule*, from *B. ostreae*-endemic areas of France could not be experimentally infected. *Ostrea chilensis* and *O. angasi* were experimentally infected with *B. ostreae* as were *O. edulis* from various geographical locations (Grizel et al. 1988).

Closely related to *Bonamia* spp. are *Mikrocytos mackini*, which causes mortalities of the Pacific oyster, *C. gigas*, in British Columbia (Denman Island Disease), and *M. roughleyi*, which kills the Sydney rock oyster, *Saccostrea commercialis* (Australian Winter Disease; Farley *et al.* 1988). *M. mackini* is ultrastructurally similar to *B. ostreae*, there being differences in positioning of the nucleoli and in the lack (?) of mitochondria. *M. roughleyi* ultrastructure is not known. The authors contend that the two pathogens also differ from *B. ostreae* in that they are found in crassostreid oysters but not in ostreids. Histologically, the former are found in focal abscesses and the latter in generalized infections.

Labyrinthomorpha

Protista in this phylum of uncertain affinities have only in the last 11 years been recognized to be pathogens of molluscs. The organisms are characterized by the formation of ectoplasmic nets from specialized organelles in the cell cortex termed sagenogenetosomes (Figs. 10-11, 10-12). Nets are membrane-bound extensions of the plasmalemma that form a reticulation lacking cell organelles except elements of the endoplasmic reticulum. Nets are used in gliding motility and in nutrient absorption and penetration of the substrate being fed upon. The cells are covered by overlapping disc-like scales formed in a Golgi body (Fig. 10-12). Most of the cells in the life cycle are generally coccoid or spindle-shaped.

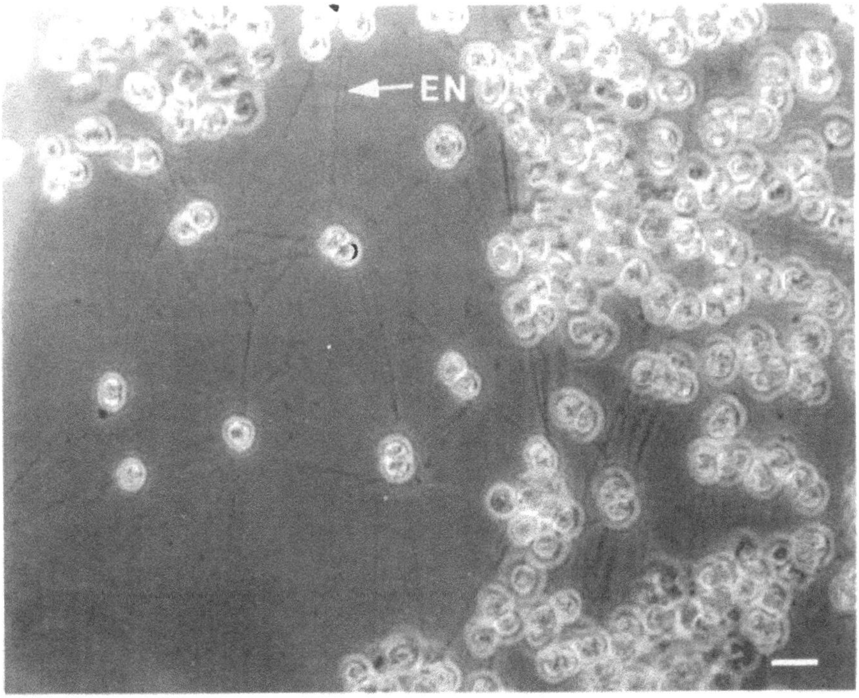

Fig. 10-11. Phase contrast micrograph of *Labyrinthuloides haliotidis* in culture. Ectoplasmic net elements (EN) radiating from trophozoites, bar = 10 µm (with permission of Susan Bower and the *Canadian Journal of Zoology*; from Bower [1987]).

Biflagellated heterokont zoospores are formed with a bilateral array of tubular mastigonemes along the anterior flagellum. Meiosis has been demonstrated in one species (Olive 1975).

The pathogenic species most thoroughly studied is *Labyrinthuloides haliotidis*, which caused mortalities as high as 100% in juvenile populations of the abalone, *Haliotis kamtshatkana* and *H. refuscens*, during aquaculture in British Columbia (Bower 1987). Small (less than ca. 4 mm) abalone younger than 190 days of age (postsetting) were highly susceptible to the disease, with susceptibility dropping markedly in older and larger abalone until infections were rare, even in individuals as small as 4.0 to 10.5 mm. Larger individuals were not infected.

Muscle and nervous tissue of the head and foot are invaded, causing tissue lysis. Other host tissues are generally not invaded. The pathogen is most active

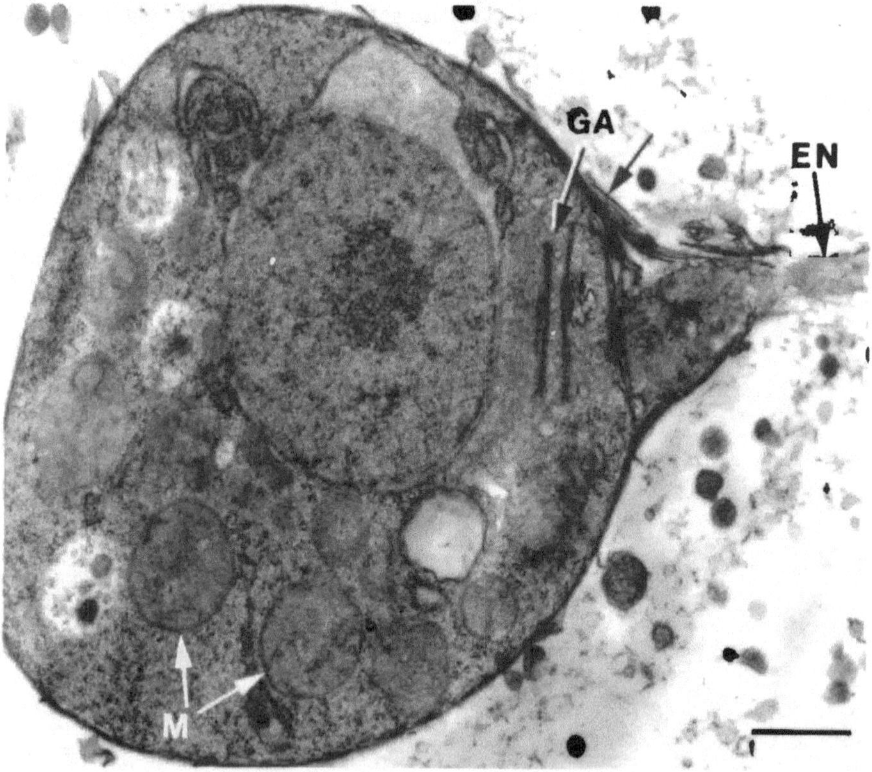

Fig. 10-12. Transmission electron micrograph of *L. haliotidis* trophozoite from infected abalone (*Haliotis kamtschatkana*). Ectoplasmic net element (EN); Golgi apparatus (GA); mitochondria (M); wall plate (arrow), bar = 1 μm (with permission of Susan Bower and the *Canadian Journal of Zoology*; from Bower [1987]).

around 5-10°C. In host tissue, cells of *L. haliotidis* divide by binary fission and form ectoplasmic nets. The latter are involved in host cell destruction as observed in other species of labyrinthomorphids growing on molluscan tissue (Perkins 1973; Jones and O'Dor 1983; McLean and Porter 1987). In culture, *L. haliotidis* forms zoosporangia that release up to 10 biflagellated zoospores.

Bower (1989) found that *L. haliotidis* could be eliminated from hatchery equipment and facilities by rinsing with water containing 25 mg/l of chlorine. She had limited success in treating parasitized abalone when using cycloheximide at 1-2 mg/l for 23 h per day on 5 consecutive days. Other molluscs found to be parasitized by labyrinthomorphids are listed in Table 10-4.

Table 10-4. Labyrinthomorphid Parasites of Molluscs

Causative Agents	Hosts	Literature Sources
Thraustochytrid	Squid (*Illex illecebrosus*)	Jones and O'Dor 1983
Thraustochytrid	Nudibranch (*Tritonia diomedea*)	McLean and Porter 1987
Thraustochytrid	Octopus (*Eledone cirrhosa*)	Polglase 1980

Other Protista

Numerous other Protista are found to be associated with molluscs, most of which appear to be harmless commensals or saprobic microbes. Species include Rhizopoda, Amebomastigota, various zooflagellates, gregarines, coccidians, Microspora, Ciliophora, and Protophyta (unicellular and filamentous algae). The reader is referred to Lauckner (1983), Sparks (1985), and Sindermann (1990) for reviews of the literature on lesser known species of pathogenic Protista and Cyanophyta.

Metazoa

The number of described species of metazoan parasites of molluscs is large as seen in the reviews of Lauckner (1980; 1983), Sparks (1985), Bower and Figueras(1989), and Sindermann (1990). However, with few exceptions, they do not appear to be significant in causing mortalities or reduced reproductive capacity in their molluscan hosts. Therefore, they are not given more detailed attention herein. Some exceptions to the above statement are trematode species in the Bucephalidae and Fellodistomidae families, which may castrate the bivalve host and weaken the adductor muscle(s). Sporocysts and cercaria are involved. Parasitic copepods of the genus *Mytilicola* may weaken the bivalve host, but it is uncertain as to whether the parasite is of significance in causing mortalities (Bower and Figueras 1989).

The most intensively studied metazoan parasites of molluscs are the clinostomatoid and schistomatoid parasites of snails. A diversity of parasite stages may be involved, including miracidia, sporocysts, mother rediae, cercaria, and metacercaria. Whereas morphological anomalies may be induced by the parasites, mortalities are not significant (Cheng 1973).

Examples of hyperparasitism are found in metazoan parasites of molluscs. Examples are Protista that cause death of worms in bivalve molluscs. The haplosporidian, *Urosporidium spisuli*, multiplies then sporulates in the tissues of an immature anisakid nematode, that infects the surf clam, *Spisula solidissima* (Perkins *et al.* 1975). An unidentified haplosporidian infects *Bucephalus* sp. in the oyster, *Crassostrea virginica*, and also causes rupture of the worm in the oyster's tissue (Mackin and Loesch 1954).

POLLUTION AND INFECTIOUS DISEASES

As noted by Sindermann (1983), it is well known that stress changes an animal's resistance to infection. In particular, chemical and thermal stress are known to activate latent infections, stimulate activity of facultative pathogens, and reduce host resistance to obligative or overt pathogens. However, specific information is largely lacking as to possible effects of pollution-induced stress on infectious diseases of molluscs. Some insights have been obtained with respect to *Perkinsus marinus*. Winstead and Couch (1988) exposed oysters (*Crassostrea virginica*) to high concentrations of n-nitrosodiethylamine (DENA) at 20°C, and found that *P. marinus* multiplied to higher levels in exposed oysters as compared to controls and caused higher mortalities. Whether DENA stimulated growth of the pathogen or inhibited host defense mechanisms was not determined. On the other hand, Scott *et al.* (1985) found that chlorine derived from NaOCl at concentrations of 1.00, 0.56, and 0.10 mg/l (chlorine-produced oxidant concentrations of 0.11-0.01 mg/l) did not appear to affect expression of *P. marinus* disease. After 60 days exposure, infection intensities were generally equal in all oysters, with higher mortalities in oysters held at moderate salinities (21-25 ppt) vs. at low salinities (8-10 ppt). The effect of salinity noted in these experiments is in keeping with field and laboratory observations of others where chlorine is not involved. It is not known whether higher concentrations of chlorine would enhance pathogenicity of *P. marinus*.

Chronic exposure of *C. virginica* to low levels ($\leq$ 3.0 ppb) of DDT, toxaphene, and parathion by Lowe *et al.* (1971) resulted in infections with a mycelial fungus that did not invade the control oysters. Lysis of tissues in all major organs resulted with little host cellular response. It appears that the cellular immune system of the experimental oysters was compromised by the organotoxicants, causing only moderate or marginal histopathological and negative growth effects on the host. The study warrants repeating to determine whether the expression of other disease agents in oysters would be affected by the organotoxicants.

Bivalve molluscs living on and in petroleum-contaminated sediments of coastal waters have been studied extensively as a function of the presence of disseminated sarcomas (Sindermann 1979; Peters 1988). The working hypothesis was that petrochemicals and petroleum compounds in the sediments and water stimulated the formation of such neoplasms. However, since disseminated sarcomas have been found in molluscs from a diversity of regions with and without obviously contaminated sediments, it is highly questionable as to whether anthropogenic influences have resulted in chemical environments that induce such disorders. Peters (1988) has provided a detailed review of the extensive literature on this subject, which is mentioned in this chapter on infectious diseases because of the suspicion that viruses are involved as infectious agents which generate the disorders (Elston *et al.* 1988)

ACKNOWLEDGMENTS

I am grateful to Mrs. K.B. Stubblefield and W.W. Jenkins for preparation of figures and Mrs. Janet Walker for typing and editing the manuscript.

REFERENCES

Alderman, D.J. 1976. Fungal diseases of marine animals. In *Recent Advances in Aquatic Mycology*, ed. E.B. Gareth Jones, pp. 223-60. London: Elek Science.

Andrews, J.D. 1988. Epizootiology of the disease caused by the oyster pathogen *Perkinsus marinus* and its effects on the oyster industry. In *Disease Processes in Marine Bivalve Molluses*, ed. W. S. Fisher. *Am. Fish. Soc. Spec. Publ.* 18:47-63.

Andrews, J.D., and M. Castagna. 1978. Epizootiology of *Minchinia costalis* in susceptible oysters in seaside bays of Virginia's eastern shore, 1959-1976. *J. Invert. Pathol.* 32:124-138.

Andrews, J.D., and W.G. Hewatt. 1957. Oyster mortality studies in Virginia. II. The fungus disease caused by *Dermocystidium marinum* in oysters in Chesapeake Bay. *Ecol. Monogr.* 27:1-25.

Atkins, D. 1954. Further notes on a marine member of the Saprolegniaceae, *Leptolegnia marina* n. sp., infecting certain invertebrates. *J. Mar. Biol. Assoc. U.K.* 33:613-625.

Azevedo, C. 1989. Fine structure of *Perkinsus atlanticus* n. sp. (Apicomplexa, Perkinsea) parasite of the clam *Ruditapes decussatus* from Portugal. *J. Parasitol.* 75:627-635.

Azevedo, C., and L. Corral. 1989. Fine structural observations of the natural spore excystment of *Minchinia* sp. (Haplosporida). *Europ. J. Protistol.* 24:168-173.

Azevedo, C., L. Corral, and R. Cachola. 1990. Fine structure of zoosporulation in *Perkinsus atlanticus* (Apicomplexa:Perkinsea). *Parasitology* 100:351-358.

Ball, S.J. 1980. Fine structure of the spores of *Minchinia chitonis* (Lankester, 1885) Labbe, 1896 (Sporozoa:Haplosporida), parasite of the chiton *Lepidochiton cinereus. Parasitology* 81:169-176.

Balouet, G., A. Cahour, and C. Chastel. 1979. Epidemiologie de la maladie de la glande digestive de l'huitre plate: hypothese sur le cycle de *Marteilia refringens. Haliotis* 8:323-326.

Barrow, J.H., Jr. 1965. Observations on *Minchinia pickfordae* (Barrow 1961) found in snails of the Great Lakes region. *Trans. Am. Microscop. Soc.* 84:587-593.

Bower, S.M. 1987. *Labyrinthuloides haliotidis* n. sp. (Protozoa: Labyrinthomorpha), a pathogenic parasite of small juvenile abalone in a British Columbia mariculture facility. *Can. J. Zool.* 65:1996-2007.

Bower, S.M. 1989. Disinfectants and therapeutic agents for controlling *Labyrinthuloides haliotidis* (Protozoa: Labyrinthomorpha), an abalone pathogen. *Aquaculture* 78:207-215.

Bower, S.M., and A.J. Figueras. 1989. Infectious diseases of mussels, especially pertaining to mussel transplantation. *World Aquaculture Rev.* 20(4):89-94.

Bower, S.M., N. McLean, and D.J. Whitaker. 1989. Mechanism of infection by *Labyrinthuloides haliotidis* (Protozoa: Labyrinthomorpha), a parasite of abalone (*Haliotis kamtschatkana*) (Mollusca: Gastropoda). *J. Invert. Pathol.* 53:401-409.

Burreson, E.M. 1988. Use of immunoassays in haplosporidan life cycle studies. In *Disease Processes in Marine Bivalve Molluscs*, ed. W.S. Fisher. *Am. Fish. Soc. Spec. Publ.* 18:298-303.

Chagot, D., E. Bachere, F. Ruano, M. Comps, and H. Grizel. 1987. Ultrastructural study of sporulated instars of a haplosporidian parasitizing the clam *Ruditapes decussatus. Aquaculture* 67:262-263.

Cheng, T.C. 1973. *General Parasitoloty.* pp. 965. New York: Academic Press.

Colwell, R. R., and A.K. Sparks. 1967. Properties of *Pseudomonas enalia*, a marine bacterium pathogenic for the invertebrate *Crassostrea gigas* (Thunberg). *Appl. Microbiol.* 15:980-986.

Comps, M. 1988. Epizootic diseases of oysters associated with viral infections. In *Disease Processes in Marine Bivalve Molluscs*, ed. W. S. Fisher. *Am. Fish. Soc. Spec. Publ.* 18:23-37.

Comps, M., J.-R. Bonami, C. Vago, and A. Campillo. 1976. Une virose de l'huitre portugaise (*Crassostrea angulata* LMK). *C. R. Acad. Sci. Paris*, Ser. D 282:1991-1993.

Comps, M., G. Tigé, and H. Grizel. 1980. Recherches ultrastructurales sur un Protiste parasite de l'huitre plate *Ostrea edulis. C. R. Acad. Sci. Paris*, Ser. D 290:383-384.

Couch, J.A., C.A. Farley, and A. Rosenfield. 1966. Sporulation of *Minchinia nelsoni* (Haplosporida, Haplosporidiidae) in *Crassostrea virginica* (Gmelin). *Science* 153:1529-1531.

Davis, H.C., V.L. Loosanoff, W.H. Weston, and C. Martin. 1954. A fungus disease in clams and oyster larvae. *Science* 120:36-38.

Desportes, I., and N.N. Nashed. 1983. Ultrastructure of sporulation in *Minchinia dentali* (Arvy), an haplosporean parasite of *Dentalium entale* (Scaphopoda, Mollusca); taxonomic implications. *Protistologica* 19:435-460.

Desportes, I., and F.O. Perkins. 1990. Phylum Paramyxea. In *Handbook of Protoctista*, ed. L. Margulis, J.O. Corliss, M.M. Melkonian, and D.J. Chapman, pp. 30-35. Boston: Jones and Bartlett Publ.

Dinamani, P., P.M. Hine, and J.B. Jones. 1987. Occurrence and characteristics of the hemocyte parasite *Bonamia* sp. in the New Zealand dredge oyster *Tiostrea lutaria*. *Dis. Aquat. Org.* 3:37-44.

Dungan, C.F., R.A. Elston, and M. Schiewe. 1989. Evidence for colonization and destruction of hinge ligaments of cultured juvenile Pacific oysters, *Crassostrea gigas*, by Cytophaga-like bacteria. *Appl. Environ. Microbiol.* 55:1128-1135.

Elston, R.A. 1990. *Mollusc diseases*: Guide for the shellfish farmer. Washington Sea Grant Program. Seattle: University of Washington Press.

Elston, R.A., and G.S. Lockwood. 1983. Pathogenesis of vibriosis in cultured juvenile red abalone, *Haliotis rufescens*, Swainson. *J. Fish. Dis.* 6:111-128.

Elston, R.A., and M.T. Wilkinson. 1985. Pathology, management and diagnosis of velar virus disease (OVVD). *Aquaculture* 48:189-210.

Elston, R., C. Farley, and M. Kent. 1986. Occurrence and significance of bonamiasis in European flat oyster *Ostreae edulis* in North America. *Dis. Aquat. Org.* 2:49-54.

Elston, R.A., E.L. Elliot, and R.R. Colwell. 1982. Conchiolin infections and surface coating *Vibrio*: Shell fragility, growth depression and mortalities in cultured oysters and clams, *Crassostrea virginica, Ostrea edulis*, and *Mercenaria mercenaria*. *J. Fish. Dis.* 5:265-284.

Elston, R.A., J.H. Beattie, C. Friedmann, R. Hedrick, and M.L. Kent. 1987. Pathology and significance of fatal inflammatory bacteraemia in the Pacific oyster, Crassostrea gigas Thunberg. *J. Fish. Dis.* 10:121-132.

Elston, R.A., M.L. Kent, and A.S. Drum. 1988. Transmission of hemic neoplasia in the bay mussel, *Mytilus edulis*, using whole cells and cell homogenate. *Dev. Comp. Immunol.* 12:719-27.

Farley, C.A. 1978. Viruses and virus-like lesions in marine mollusks. U.S. Nat. Mar. Fish. Service. *Mar. Fish. Rev.* 40:18-20.

Farley, C.A., W.C. Banfield, G. Kasnic, Jr., and W.S. Foster. 1972. Oyster herpes-type virus. *Science* 178:759-760.

Farley, C.A., P.H. Wolf, and R.A. Elston. 1988. A long-term study of "microcell" disease in oysters with a description on a new genus, *Mikrocytos* (G.N.), and two new species, *Mikrocytos mackini* (Sp. N.) and *Mikrocytos roughleyi* (Sp. N.). *Fish. Bull. (U.S.)* 86:581-593.

Field, I.A. 1922. Biology and economic value of sea mussel, *Mytilus edulis*. *Bull. U.S. Bur. Fish.* 38:127-259.

Figueras, A.J., and J. Montes. 1988. Aber disease of edible oysters caused by *Marteilia refringens*. In *Disease Processes in Marine Bivalve Molluscs*, ed. W. S. Fisher. *Am. Fish. Soc. Spec. Publ. No.* 18:38-46.

Friedman, C.S. and R.P. Hedrick. 1991. Pacific oyster nocardiosis: isolation of the bacterium and induction of laboratory infections. *J. Invert. Pathol.* 57:109-120.

Ganaros, A.E. 1957. Marine fungus infecting eggs and embryos of *Urosalpinx cinerea*. *Science* 125:1194.

Goggin, C.L., and R.J.G. Lester. 1987. Occurrence of *Perkinsus* species (Protozoa, Apicomplexa) in bivalves from the Great Barrier Reef. *Dis. Aquat. Org.* 3:113-117.

Goggin, C.L., K.B. Sewell, and R.J.G. Lester. 1989. Cross-infection experiments with Australian *Perkinsus* species. *Dis. Aquatic Org.* 7:55-59.

Grischkowsky, R.S., and J. Liston. 1974. Bacterial pathogenicity in laboratory-induced mortality of the Pacific oyster *Crassostrea gigas* Thunberg). *Proc. Nat. Shellfish Assoc.* 64:82-91.

Grizel, H., E. Mialhe, D. Chagot, V. Boulo, and E. Bachere. 1988. Bonamiasis: a model study of diseases in marine molluscs. In *Disease Processes in Marine Bivalve Molluscs*, ed. W. S. Fisher. *Am. Fish. Soc. Spec. Publ.* 18:1-4.

Hada, H.S., P.A. West, J.V. Lee, J. Stemmler, and R.R. Colwell. 1984. *Vibrio tubiashii* sp. nov., a pathogen of bivalve molluscs. *Int. J. Syst. Bacteriol.* 34:1-4.

Harshbarger, J.C., S.C. Chang, and S.V. Otto. 1977. Chlamydiae (with phages), mycoplasmas, and rickettsiae in Chesapeake Bay bivalves. *Science* 196:666-668.

Haskin, H.H., and J.D. Andrews. 1988. Uncertainties and speculations about the life cycle of the Eastern oyster pathogen *Haplosporidium nelsoni* (MSX). In *Disease Processes in Marine Bivalve Molluscs*, ed. W.S. Fisher. *Am. Fish. Soc. Spec. Publ.* 18:5-22.

Haskin, H.H., L.A. Stauber, and J.G. Mackin. 1966. *Minchinia nelsoni* n. sp. (Haplosporida, Haplosporidiidae) causative agent of the Delaware Bay oyster epizootic. *Science* 153:1414-1416.

Hill, B.J. 1976. Molluscan viruses: their occurrence, culture, and relationships. *Proc. First Int. Colloq. on Invertebr. Pathol. and IX Annu. Meet. Soc. Invertebr. Pathol.*, pp. 25-29. Queens Univ., Kingston, Ont.

Hill, B.J., and D.J. Alderman. 1979. Observations on the experimental infection of *Ostrea edulis* with two molluscan viruses. *Haliotis* 8:297-299.

Hill, B.J., K. Way, and D.J. Alderman. 1986. IPN-like birnaviruses in oysters:infection or contamination? In *Pathology in Marine Aquaculture*, eds. C. P. Vivares, J.-R Bonami, and E. Jaspers, p. 297. European Aquaculture Society, Spec. Publ. No. 9. Bredene, Belgium. (Abst).

Hillman, R., S.E. Ford, and H.H. Haskin. 1990. *Minchinia teredinis* N. Sp. (Balanosporida, Haplosporidiidae), a parasite of teredinid shipworms. *J. Protozool.* 37:364-368.

Hoese, H.D. 1964. Studies on oyster scavengers and their relation to the fungus *Dermocystidium marinum*. *Proc. Nat. Shellfish Assoc.* 53:161-173.

Hulet, W.H., M.R. Villoch, R.E. Hixon, and R.T. Hanlon. 1979. Fin damage in captured and reared squids. *Lab. Anim. Sci.* 29:528-533.

Imai, T., L. Mori, Y. Sugawara, H. Tamate, J. Oizumi, and O. Itikawa. 1968. Studies on the mass mortality of oysters in Matsushima Bay. VII Pathogenic investigation. *Tohoku J. Agric. Res.* 19:250-265.

Johnson, P.T. 1984. Viral diseases of marine invertebrates. In *Internat. Helgoland Symp. 1983*, eds. O. Kinne and H.-P Bulnheim, Helgol. Meeresunters. 37:65-98.

Johnson, T.W., Jr., and F.K. Sparrow, Jr. 1961. *Fungi in Oceans and Estuaries*. p. 73. New York: Hafner Publishing Company.

Jones, G.M., and R.K. O'Dor. 1983. Ultrastructural observations on a thraustochytrid fungus parasitic on the gills of squid (*Illex illecebrosus* Lesueur). *J. Parasitol.* 69:903-911.

Katkansky, S.C., and R.W. Warner. 1970. Sporulation of a haplosporidian in a Pacific Oyster (*Crassostrea gigas*) in Humbolt Bay, California. *J. Fish. Res. Bd. Can.* 27:1320-1321.

Kern, F.G. 1976. Sporulation of *Minchinia* sp. (Haplosporida, Haplosporidiidae) in the Pacific Oyster *Crassostrea gigas* (Thunberg) from the Republic of Korea. *J. Protozool.* 23:498-500.

Khardori, N., and V. Fainstein. 1988. *Aeromonas* and *Plesiomonas* as etiological agents. *Annu. Rev. Microbiol.* 42:395-419.

Kinne, O. 1980. Diseases of marine animals:general aspects. Chapter 2. In *Diseases of Marine Animals, Vol. I*, ed. O. Kinne, pp. 13-74. New York: John Wiley and Sons.

Kudo, R.R. 1966. *Protozoology*, 5th ed. pp. 1174. Springfield: Charles C. Thomas, Publisher.

Lauckner, G. 1980. Diseases of Mollusca:Gastropoda. Chapter 12. In *Diseases of Marine Animals*, Vol. II, ed. O. Kinne, pp. 311-424. New York: John Wiley and Sons.

Lauckner, G. 1983. Diseases of Mollusca:Bivalvia. Chapter 13. In *Diseases of Marine Animals*, Vol. II, ed. O. Kinne, pp. 477-984. Biologische Anstalt. Helgoland.

Leibovitz, L. 1989. Chlamydiosis: a newly reported serious disease of larval and postmetamorphic bay scallops, *Argopecten irradians* (Lamarck). *J. Fish Dis.* 12:125-136.

Leibovitz, L., T.R. Meyers, and R. Elston. 1977. Necrotic exfoliative dermatitis of captive squid (*Loliga pealei*). *J. Invert. Pathol.* 30:369-376.

Lester, R.J.G., and G.H.G. Davis. 1981. A new *Perkinsus* species (Apicomplexa, Perkinsea) from the abalone *Haliotis ruber*. *J. Invert. Pathol.* 37:181-187.

Levine, N.D. 1978. *Perkinsus* gen. n. and other new taxa in the protozoan phylum Apicomplexa. *J. Parasitol.* 64:549.

Levine, N.D. 1988. *The Protozoan Phylum Apicomplexa*. p. 9. Boca Raton: CRC Press, Inc.

Lipovsky, V.P., and K.K. Chew. 1972. Mortality of Pacific oysters (*Crassostrea gigas*): The influence of temperature and enriched seawater on oyster survival. *Proc. Nat. Shellfish Assoc.* 62:72-82.

Lowe, J.I., P.D. Wilson, A.J. Rick, and A.J. Wilson, Jr. 1971. Chronic exposure of oysters to DDT, toxaphene, and parathion. *Proc. Nat. Shellfish Assoc.* 61:71-79.

Mackin, J.G. 1962. Oyster disease caused by *Dermocystidium marinum* and other microorganisms in Louisiana. *Publs. Inst. Mar. Sci. Univ. Tex.* 7:131-229.

Mackin, J.G., and H. Loesch. 1954. A haplosporidian hyperparasite of oysters. *Proc. Nat. Shellfish Assoc.* 45:182-183.

Mackin, J.G., H.M. Owen, and A. Collier. 1950. Preliminary note on the occurrence of a new protistan parasite, *Dermocystidium marinum* n. sp., in *Crassostrea virginica* (Gmelin). *Science* 111:328-329.

Margulis, L., J.O. Corliss, M. Melkonian, and D.J. Chapman, eds. 1990. Handbook of Protoctista. pp. 769-803. Boston: Jones and Bartlett Publ.

McGladdery, S.E., R.J. Cawthorn, and B.C. Bradford. 1991. *Perkinsus karlssoni* n. sp. (Apicomplexa) in bay scallops *Argopecten irradians. Dis. Aquat. Org.* 10:127-137.

McGovern, E.R., and E.M. Burreson. 1990. Ultrastructure of *Minchinia* sp. spores from shipworms (*Teredo* spp.) in the Western North Atlantic, with a discussion of taxonomy of the Haplosporidiidae. *J. Protozool.* 37:212-218.

McLean, N., and D. Porter. 1987. Lesions produced by a thraustochytrid in *Tritonia diomedes* (Mollusca: Gastropoda: Nudibranchia). *J. Inverteb. Pathol.* 49:223-225.

Mead, A.R. 1963. Disease, decline and predation in the giant snail populations of Hawaii. *Ann. Rept. Amer. Malacol. Union.* p. 22 (Abstract).

Mead, A.R. 1969. *Aeromonas liquefaciens* in the leukodermia syndrome of *Achatina fulica. Malacologia* 9:43 (Abstract).

Meyers, T.R. 1979. Preliminary studies on a chlamydial agent in the digestive diverticular epithelium of hard clams, *Mercenaria mercenaria* (L.) from Great south Bay, New York. *J. Fish Dis.* 2:179-189.

Meyers, T.R. 1981. Endemic diseases of cultured shellfish of Long Island, New York: Adult and juvenile American oysters (*Crassostrea virginica*) and hard clams (*Mercenaria mercenaria*). *Aquaculture* 22:305-330.

Michelson, E.H. 1961. An acid-fast pathogen of freshwater snails. *Am. J. Trop. Med. Hyg.* 10:423-433.

Olive, L.S. 1975. *The Mycetozoans*, pp. 215-241. New York: Academic Press.

Orpandy, J.J., D.W. Chang, A.D. Pronovost, K.R. Cooper, R.S. Brown, and V.J. Yates. 1981. Isolation of a viral agent causing hematopoietic neoplasia in soft shell clam, *Mya arenaria*. *J. Invertebr. Pathol.* 38:45-51.

Otto, S.V., J.C. Harshbarger, and S.C. Chang. 1979. Status of selected unicellular eucaryote pathogens, and prevalence and histopathology of inclusions containing obligate procaryote parasites, in commercial bivalve mollusks from Maryland estuaries. *Haliotis* 8:285-295.

Pass, D.A., R. Dybdahl, and M.M. Mannion. 1987. Investigations into the causes of mortality of the pearl oyster, *Pinctada maxima* (Jamson), in Western Australia. *Aquaculture* 65:149-169.

Pass, D.A., F.O. Perkins, and R. Dybdahl. 1988. Virus-like particles in the digestive gland of the pearl oyster (*Pinctada maxima*). *J. Invert. Pathol.* 51:166-167.

Perkins, F.O. 1969. Electron microscope studies of sporulation in the oyster pathogen, *Minchinia costalis* (Sporozoa:Haplosporida). *J. Parasitol.* 55:897-920.

Perkins, F.O. 1971. Sporulation in the trematode hyperparasite *Urosporidium crescens* De Turk, 1940 (Haplosporida:Haplosporidiidae)--an electron microscope study. *J. Parasitol.* 57:9-23.

Perkins, F.O. 1973. Observations of thraustochytriaceous (Phycomycetes) and labyrinthulid (Rhizopodea) ectoplasmic nets on natural and artificial substrates--an electron microscope study. *Can. J. Bot.* 51:485-91.

Perkins, F.O. 1976a. Zoospores of the oyster pathogen, *Dermocystidium marinum*. I. Fine structure of the conoid and other sporozoan-like organelles. *J. Parasitol.* 62:959-974.

Perkins, F.O. 1976b. Ultrastructure of sporulation in the European flat oyster pathogen, *Marteilia refringens*--taxonomic implications. *J. Protozool.* 23:64-74.

Perkins, F.O. 1979. Cell structure of shellfish pathogens and hyperparasites in the genera *Minchinia*, *Urosporidium*, *Haplosporidium* and *Marteilia*, Taxonomic implications. In *Haplosporidian and Haplosporidian-Like Diseases of Shellfish*, ed. F.O. Perkins. *U.S. Nat. Mar. Fish. Serv. Mar. Fish. Rev.* 41:25-37.

Perkins, F.O. 1988. Structure of protistan parasites found in bivalve molluscs. In *Disease Processes in Marine Bivalve Molluscs*, ed. W.S. Fisher. *Am. Fish. Soc. Spec. Publ.* 18:93-111.

Perkins, F.O. 1990. Haplosporidia. In *Handbook of Protoctista*, ed. L. Margulis, J. O. Corliss, M. Melkonian, and D.J. Chapman, pp. 19-29. Boston: Jones and Bartlett Publishers Corp.

Perkins, F.O., and R.W. Menzel. 1966. Morphological and cultural studies of a motile stage in the life cycle of *Dermocystidium marinum*. *Proc. Nat. Shellfish Assoc.* 56:23-30.

Perkins, F.O., and P.H. Wolf. 1976. Fine structure of *Marteilia sydneyi* sp. n. - haplosporidan pathogen of Australian oysters. *J. Parasitol.* 62:528-538.

Perkins, F.O., and P. van Banning. 1981. Surface ultrastructure of spores in three genera of Balanosporida, particularly in *Minchinia armoricana* van Banning, 1977--the taxonomic significance of spore wall ornamentation in the Balanosporida. *J. Parasitol.* 67:866-874.

Perkins, F.O., D.E. Zwerner, and R.K. Dias. 1975. The hyperparasite, *Urosporidium spisuli* sp. n. (Haplosporea), and its effects on the surf clam industry. *J. Parasitol.* 61:944-949.

Peters, E.C. 1988. Recent investigations on disseminated sarcomas of marine bivalve molluscs. In *Disease Processes in Marine Bivalve Molluscs*, ed. W.S. Fisher.*Am. Fish. Soc. Spec. Publ.* 18:74-92.

Polglase, J.L. 1980. A preliminary report on the thraustochytrid(s) and labyrinthulid(s) associated with a pathological condition in the lesser octopus *Eledone cirrhosa*. *Bot. Mar.* 23:699-706.

Porter, D. 1986. Mycoses of marine organisms:an overview of pathogenic fungi. In *The Biology of Marine Fungi*, ed. S.T. Moss, pp. 141-53. Cambridge: Cambridge University Press.

Ray, S.M. 1952. A culture technique for the diagnosis of infections with *Dermocystidium marinum* Mackin, Owen, and Collier in oysters. *Science* 116:360-361.

Ray, S.M. 1954. Biological studies of *Dermocystidium marinum*. *The Rice Institute Pamphlet. Special Issue, November, 1954.* Houston: The Rice Institute.

Scott, G.I., E.O. Oswald, T.I. Sammons, D.S. Baughman, and D.P. Middaugh. 1985. Interactions of chlorine-produced oxidants, salinity, and a protistan parasite in affecting lethal and sublethal physiological effects in the Eastern or American oyster. In *Water Chlorination: Chemistry, Environmental Impact and Health Effects*, Vol. 5, ed. R.L. Jolley *et al.*, pp. 463-480. Chelsea, Michigan: Lewis Publishers Inc.

Sinderman, C.J. 1970. *Principal diseases of marine fish and shellfish*. pp. 369. New York: Academic Press.

Sindermann, C.J. 1979. Pollution-associated diseases and abnormalities of fish and shellfish: a review. *U.S Fish. Bull.* 76(4):717-749.

Sindermann, C.J. 1983. An examination of some relationships between pollution and disease. *Rapp. P.-v. Reun. Cons. Int. Explor. Mer* 182:37-43.

Sindermann, C.J. 1990. *Principal diseases of marine fish and shellfish*, 2nd ed. Vol. 2. *Diseases of marine shellish.* pp. 521. New York: Academic Press.

Sindermann, C.J., and D.V. Lightner. 1988. *Disease diagnosis and control in North American marine aquaculture*. 2nd. rev. ed. pp. 271-273. New York: Elsevier Science Publ.

Sparks, A.K. 1985. *Synopsis of invertebrate pathology exclusive of insects*. pp. 423. New York: Elsevier Science Publishers.

Srivastava, P.E. and Y.N. Srivastava. 1968. Role of snails' diseases in the biological control of *Achatina fulica*. Bowdich, 1822 in the Andmans. *Veliger* 10:320-321.

Sturtevant, A.B., Jr. 1969. A taxonomic study of the genus *Aeromonas*. M.S. Thesis. 58 pp. University of Alabama, Tuscaloosa, AL.

Taylor, R. 1966. *Haplosporidium tumefacientis* sp. n., the etiologic agent of a disease of the California Sea Mussel, *Mytilus californianus* Conrad. *J. Invert. Pathol.* 8:109-121.

Tubiash, H.S., and S.V. Otto. 1986. Bacterial problems in oysters. A review. In *Pathology in marine aquaculture*, ed. C.P. Vivares, J.-R Bonami, E. Jaspers. European Aquaculture Society. Spec. Publ. 9:233-242. Bredene, Belgium.

Tubiash, H.S., R.R. Colwell, and R. Sakazaki. 1970. Marine vibrios associated with bacillary necrosis, a disease of larval and juvenile bivalve molluscs. *J. Bacteriol.* 103:272-273.

Twomey, E., and M. Mulcahy. 1988. Transmission of a sarcoma in the cockle *Cerastoderma edule* (Bivalvia; Mollusca) using cell transplants. *Devel. Compar. Immunol.* 12:195-200.

van Banning, P. 1977. *Minchinia armoricana* sp. nov. (Haplosporida), a parasite of the European flat oyster, *Ostrea edulis*. *J. Invert. Pathol.* 30:199-206.

Vilela, H. 1951. Sporozoaires parasites de la palourde *Tapes descussatus* L. Rev. Fac. Ciencias, Universidade de Lisboa, Serie 2C 1:379-386.

Vishniac, H. 1955. Morphology and nutrition of a new species of *Sirolpidium*. *Mycologia* 7:633-645.

Vishniac, H. 1958. A new marine phycomycete. *Mycologia* 50:66-79.

Vitellaro-Zuccarello, L. 1973. Ultrastructure of the byssal apparatus of *Mytilus galloprovincialis*. I. Associated fungal hyphae. *Mar. Biol.* 22:225-230.

White, M.E., E.N. Powell, S.M. Ray, and E.A. Wilson. 1987. Host-to-host transmission of *Perkinsus marinus* in oyster (*Crassostrea virginica*) populations by the ectoparasitic snail *Boonea impressa* (Pyramidellidae). *J. Shellfish Res.* 6:1-5.

Winstead, J.T., and J.A. Couch. 1988. Enhancement of protozoan pathogen *Perkinsus marinus* infections in American oysters *Crassostrea virginica* exposed to the chemical carcinogen n-nitrosodiethylamine (DENA). *Dis. Aquat. Org.* 5:205-213.

11

Noninfectious Diseases of Marine Molluscs

Thomas C. Cheng

INTRODUCTION

Within the realm of noninfectious diseases of estuarine and marine animals are those due to 1) nutritional deficiencies, 2) genetically controlled abnormalities, 3) environmental insults, or 4) uncertain causes. Some are undoubtedly functionally interrelated. Among molluscs, relatively little is known about the existence and impact of noninfectious diseases. Even less is known about pathogenic mechanisms underlying such conditions. The intent of this contribution is to review examples of known instances of noninfectious diseases in estuarine and marine molluscs and to speculate on their implications. Several avenues of future research are suggested.

NUTRITIONAL DEFICIENCIES

Little is known about pathological alterations associated with nutritional deficiencies in molluscs. Considerable basic information has been accumulated on carbohydrate metabolism (for reviews, see Goudsmit 1972; de Zwaan 1983; Livingstone and de Zwaan 1983), lipid and sterol components and metabolism (for reviews, see Voogt 1972, 1983), and nitrogen metabolism (for reviews, see Florkin and Bricteux-Gregoire 1972; Bishop *et al.* 1983) in these invertebrates. Nutrition and consequences of malnutrition are important aspects of the pathobiology of marine molluscs that require increased attention. As Cheng (1990a) pointed out, effects of malnutrition and unwholesome diets are varied; however, such effects can be reflected in manifestation of immunodeficiency diseases and elevated pathogenicity of viral, microbial, and parasitic diseases, in addition to toxicologic and teratologic diseases.

0-8493-8662-4/93/$0.00 + $.50

GENETIC ABNORMALITIES

Other than structural abnormalities caused by predators, parasites, and epiphoronts, e.g., the formation of mud blisters on the nacre of oysters by the annelid *Polydora* (Haswell 1886; Leloup 1937; Kavanagh 1940; Lunz 1940; Medcof 1946; and others), little has been reported pertaining to abnormalities of marine molluscs, especially genetically controlled ones. It is known that supernumerary siphons, either functional or nonfunctional, occur on the hard clam, *Mercenaria*, and the soft-shell clam, *Mya arenaria* (Potter and Kuff 1967; Tubiash *et al.* 1968). It remains undetermined as to whether these abnormalities are directly controlled by genes or represent abnormal regeneration subsequent to injury. Similarly, bifurcation of the foot of clams has been reported (Pelseneer 1923; Atkins 1931) as has the bifurcation of the adductor muscle of oysters (Gunter 1957; Pauley and Sayce 1967). These anomalies represent curiosities and are of no serious economic importance.

Two additional structural anomalies that have been reported in marine bivalves deserve brief mention: the so-called "kidney stones" and pearls. Kidney stones, which are calcium phosphate concretions, have been reported in hard clams (*M. mercenaria*) and scallops (*Argopecten irradians*) along the Atlantic coast of the United States (Potts 1967; Doyle *et al.* 1978; Gold *et al.* 1982). These calcified bodies may occlude the entire kidney and thus lead to nephritic dysfunction. It is not known if the development of these structures is genetically controlled or due to some other cause(s). However, as a result of the induced formation of kidney stones in *Donax trunculus*, Mauri and Orlando (1982) suggested that the abnormality was related to pollution or the influence of otherwise abnormal environments. Their suggestion was based on the finding that calcium phosphate concretions were more numerous and larger in specimens of *D. trunculus* from polluted sites than from reference sites.

Nacrezation, or pearl formation, has been considered another type of structural abnormality by some (Lauckner 1983; Sindermann 1990). The extensive literature pertaining to this phenomenon has been reviewed by Cheng and Rifkin (1970) and Lauckner (1983). As pointed out by Cheng and Rifkin (1970), nacrezation actually represents one type of internal defense mechanism in molluscs, although its manifestation is in the form of a multi-layered nacreous tunic surrounding some foreign particle. Nacrezation is mentioned because Richards (1970, 1972) demonstrated that pearl formation is genetically controlled in the gastropod *Biomphalaria glabrata*. Richards (1972) also suggested "that genetics might be involved in pearl formation in some other molluscs."

Malformation of the shells of marine molluscs has been reported periodically (Pelseneer 1920, 1923; Blake 1929; Lauckner 1983; and others). Again, the responsible factors are not known. Relative to such representing phenotypic expressions of responsible genes, Shuster (1966) reported the occurrence of a specimen of *Mercenaria mercenaria* with distorted and asymmetrical valves from Narragansett Bay, Rhode Island: "The Narragansett Bay specimen resembles closely some ancient fossilized genera of non-burrowing molluscs, especially *Exogyra* and *Gryphaea*. Perhaps whatever happened to the present quahaug specimen triggered a *latent genetic mechanism* (italics added) for shell shape that has been dominant in the oyster family for millions of years." Shuster's comment suggests that latent, dominant genes are responsible for what are now considered to be deformed valves that have been activated. This important concept should be borne in mind when considering influence of environmental factors, including anthropogenic pollutants, on marine molluscs.

As reviewed later, the relative frequent occurrence of hemocytic neoplasia in marine bivalves which may be in some yet undetermined way associated with pollutants and/or viruses has been recorded. Frierman and Andrews (1976) reported that they examined over 100 small, laboratory-reared, inbred populations of oysters, *Crassostrea virginica*, among which two exhibited high prevalences of hemocytic neoplasia. They also reported that one of these revealed high mortalities that appeared to be directly related to the high prevalence of the disease. No other diseases occurred. Thus, it would appear that the high susceptibility to hemocyte neoplasia and mortality from it may be genetically regulated.

ENVIRONMENTAL INSULTS

During the past two decades, considerable attention has been directed toward the impact of the environment, especially anthropogenic alterations, on marine animals, including molluscs. From what is known, this appears to be a serious problem that deserves intensive investigation, especially at the molecular and biochemical levels. Molluscan diseases resulting from environmental insults can be broadly categorized into three classes: neoplastic diseases, toxicologic diseases, and diseases of uncertain nature.

Neoplastic Diseases

Prior to a discussion of possible correlations between neoplastic diseases in marine molluscs and exogenous factors, a definition of neoplasms (or neoplasias) appears appropriate. The definition, which is increasingly more difficult to establish, in my opinion, has been most appropriately addressed by Sparks (1985). According to Sparks: "(neoplasms are) growth disturbances characterized by excessive, abnormal, proliferation of cells, independent of normal growth-regulating mechanisms of the animal, and persisting after termination of the stimulus that initiated growth." Neoplasms are either malignant (i.e., progressive growth, which if not checked, spread to distant sites, terminating in death) or benign (i.e., not malignant).

The first neoplastic growth for a marine mollusc was recorded by Ryder (1887) in American oyster, *C. virginica*. Lauckner (1983), Sparks (1985), and to some extent Peters (1988) and Sindermann (1990) have provided detailed reviews of most of these cases. Most earlier reports were of benign tumors. However, since 1969, a number of malignant neoplasms have been reported in marine bivalves, although earlier cases are known. In view of recent reviews, another detailed review is not warranted. Reported cases are tabulated in Tables 11-1, 11-2. In addition, numerous reports of hemocytic neoplasias have been reported in several species of marine bivalves (Table 11-3). Etiology of neoplastic diseases remains unclear, with some claiming that initiating cause rests with environmental pollution. Further, it remains uncertain as to whether all reported cases share the same etiology or represent similar conditions. In fact, there is some evidence that at least two cell types are involved and probably two distinct diseases occur (Moore *et al.* 1991).

Yevich and Barszcz (1976,1977) and Reinisch *et al.* (1983) referred to the condition in *Mya arenaria* as "hematopoietic neoplasia." Farley (1969a), who reported the first recent case, referred to the disease in the American oyster, *C. virginica*, as a "probable neoplastic disease of the hematopoietic system." Farley and Sparks (1970) referred to a comparable disease in the Olympia oyster, *Ostrea lurida*, as "vesicular cell neoplasm" and "hematopoietic neoplasm." Similarly, Harshbarger *et al.* (1979) referred to the disease in *C. virginica* as "hematopoietic neoplasms," and Cosson-Mannevy *et al.* (1984) designated the disease in *Mytilus edulis* as "neoplasm of hematopoietic origin." In another species of marine bivalve, the soft-shell clam, *M. arenaria*, Brown *et al.* (1977), Reinish *et al.* (1984), Farley *et al.* (1986), and Farley (1989) referred to the neoplastic disease as "hematopoietic neoplasms."

In all of the designations mentioned above, the adjective "hematopoietic" was employed. This suggests that the investigators knew that the stem cells from which the proliferative neoplastic cells arose were germinal and/or stem hemocytes. This, of course, is an error since the hematopoietic site(s) in bivalves remains unknown (Cheng 1981). It is recognized that a general, unproven consensus exists that bivalve hemocytes may have their origin, at least in part, from the multiplication and differentiation of Leydig cells (=vesicular connective tissue cells) (Cheng 1981). Thus, if the nature of the disease associated with neoplastic hemocytes is to be understood, we must determine beyond doubt the ontogenesis of bivalve hemocytes. It is noted, however, that Smolowitz *et al.* (1989), by employing an indirect immunoperoxidase technique, demonstrated that

Table 11-1. Reported Neoplasms in Species of Oysters.

Oyster species	Neoplasms	Authority
Crassostrea virginica	Mesenchymal tumor	Ryder 1887
C. virginica	Benign nodular tumor or normal-appearing Leydig cells	Smith 1934
C. virginica	Neoplastic cells in connective tissue and hemolymph	Newman 1972
C. virginica	Lesions in gills	Harshbarger *et al.* 1979
C. virginica	Invasive neoplasms originating from germinal epithelium	Farley 1976
Crassostrea gigas	Benign nodular tumor	Sparks *et al.* 1966a
C. gigas	Rectal, fecal impaction	Sparks *et al.* 1964b
C. gigas	Tumor anterior to adductor muscle	Sparks *et al.* 1969
C. gigas	Tumor attached to mantle	Harshbarger 1976
C. gigas	Mantle growth	Pauley and Sayce 1967
C. gigas	Polypoid growth	Pauley and Sayce 1967
[0]*C. gigas*	Invasive epithelioma	Pauley and Sayce 1972
C. gigas	Myofibroma	Pauley and Sayce 1968
[0]*C. gigas*	Malignant neurofibroma	Pauley *et al.* 1968
[0]*Crassostrea commercialis*	Malignant epitheliomas	Wolf 1969, 1971
C. commercialis	Tumors in pericardial cavity	Dinamani and Wolf 1973
Pinctada margaritifera	Fibro-vascular polyps	Dix 1972

[0] = diagnosed to be malignant.

Table 11-2. Reported Neoplasms in Species of Clams.

Clam species	Neoplasms	Authority
Mya arenaria	Papillary growths around rectum	Butros 1948
M. arenaria	Firm growth on siphon	Pauley and Cheng 1968
M. arenaria	Wrinkled fungiform swelling on siphon	Pauley and Cheng 1968
[0]*M. arenaria*	Gill epithelium hyperplasia	Barry *et al.* 1971
M. arenaria	Hyperplastic lesions in kidney	Barry *et al.* 1971
Tresus nuttali	Papilloma-like growth on siphon	DesVoigne *et al.* 1970
T. nuttali	Polypoid and papillary lesions on foot	Taylor and Smith 1966
Saxidomus giganteus	Large polypoid growth on foot	Pauley 1967
[0]*Mercenaria mercenaria*	Ovarian tumor	Yevich and Barry 1969
Mercenaria spp.	Gonadal neoplasms	Hasselman *et al.* 1987
[0]*Mya arenaria*	Gonadal tumor	Barry and Yevich 1975

[0] = diagnosed to be malignant

monoclonal antibodies (Mab 4A9) developed against neoplastic hemocytes in *M. arenaria* also reacted with a small subpopulation of normal circulating hemocytes as well as a type of connective tissue cell. Therefore, they hypothesized that that connective tissue cell serves as the progenitor of the small subpopulation of normal hemocytes as well as the neoplastic cells.

It needs to be recalled that other investigators who have studied a similar condition have been more conservative in their terminology. Specifically, Mix (1975) and Mix *et al.* (1977), who studied what was apparently the same or a similar disease in *Ostrea lurida*, designated it as "neoplastic disease" and "cellular proliferative disorder." Also, Mix (1983), Elston *et al.* (1988a,b), and Moore et al. (1991) employed the designation "hemic neoplasm" for the disease in *M. edulis*; Green and Alderman (1983) employed the term "proliferating hemocytic condition" for the disease in *M. edulis*; Twomey and Mulcahy (1984) designated a similar condition in the cockle, *Ceratoderma edule*, as a "proliferative disorder of possible hemic origin; "and Poder and Auffret (1986) and Auffret and Poder (1986) referred to the disease in *C. edule* as a "sarcoma, probably of hemocytic origin." Couch (1970) referred to one of the earliest cases as "sarcoma-like."

Table 11-3. Reported Cases of Hemocytic Neoplasia in Marine Bivalves.

Molluscan species	Location	Hemocytic neoplasia	Authority
Mytilus edulis	Yaquina Bay, OR	Neoplastic hemocytes	Farley 1969b
M. edulis	Yaquina Bay, OR	Hemocytic neoplasm	Mix 1983
M. edulis	British Columbia	Hemocytic neoplasm	Cosson-Mannevy *et al.* 1986
M. edulis	British waters	Hemocytic neoplasm	Lowe and Moore 1978; Green and Alderman 1983
M. edulis	Puget Sound, WA	Hemocytic neoplasia	Elston *et al.* 1988a
Ostrea edulis	Spain, Yugoslavia	Hemocytic neoplastic cells	Alderman *et al.* 1977
O. edulis	Northwest French coast	Hemocytic neoplasm	Franc 1975; Balouet *et al.* 1986
Ostrea lurida	Pacific northwest, U.S.	Hemocytic neoplasm	Jones and Sparks 1969; Sparks 1970; Mix *et al.* 1979a
Ostrea chilensis	Chile	Hemocytic neoplasm	Mix and Breese 1980
Crassostrea gigas	Matsushima Bay, Japan	Hemocytic neoplasm	Farley 1969a
C. gigas	Northwest French	None among 8000 oysters	Balouet *et al. 1986*
Crassostrea virginica	Chesapeake Bay	Hemocytic neoplasm	Frierman and Andrews 1976; Harshbarger *et al.* 1979; Farley 1969a; Couch 1969
C. virginica	Chesapeake Bay	Sarcoma with large basophilic blastoid cells	Couch 1969, 1970
Mya arenaria	Brunswick, Maine	Hemocytic neoplasm	Yevich and Barszcz 1976, 1977
M. arenaria	New England, U.S. (10 populations)	This neoplasm is apparently a viral infectious disease	Brown *et al.* 1976, 1977
M. arenaria	New Bedford Harbor, MA	Hemocytic neoplasm	Reinsch *et al.* 1984
M. arenaria	Long Island Sound	Hemocytic neoplasm	Brousseau 1987
M. arenaria	Chesapeake Bay	Hemocytic neoplasm	Farley *et al.* 1986
M. arenaria	Connecticut coast	Neoplastic hemocytes	Brousseau and Baglivo 1990
Cerastoderma edule	Cork Harbor, Ireland	Infectious disease suggested	Twomey and Melcahy 1984
E. edule	Northern Brittany coast, France	Cockles with hemocytic disease	Auffret and Poder 1986; Poder and Auffret 1986

Also, to be even less definitive, Newman (1972), Harshbarger *et al.* (1979), and Farley (1976) intentionally referred to the condition as undifferentiated sarcoma.

This hemocytic disease, or complex of diseases, as pointed out by Farley *et al.* (1986), shares the following characteristics: 1) occurrence of dense cellular infiltration, especially in connective tissue associated with the digestive tract, gonads, mantle, and ctenidia; 2) cellular abnormalities including enlarged, irregular nucleus, multiple nucleoli, increased nuclear:cytoplasmic ratios, and frequent mitotic figures; 3) folded nuclear membrane and deformed mitochondria when examined at the EM level; 4) affected hemocytes that are nonphagocytic when host is parasitized by protozoans; and 5) affected cells that fail to aggregate and adhere to glass *in vitro*.

It is important to emphasize that Auffret and Poder (1986) pointed out that the cause(s) of such assumed hemocytic neoplasias is unknown, although in the case of *M. arenaria*, Cooper and Chang (1982) implicated a retrovirus and in the cockle, *Cerastoderma edule*, and Twomey and Mulcahy (1988) reported successful transmission by cell transplants. In other cases, Farley (1969a), Yevich and Barszcz (1977), Mix and co-workers (1981), and Mix (1982) suggested that oil pollution may be a contributing factor. Although there is no irrefutable evidence to indicate this, the most convincing data supporting this view are those of Mix *et al.* (1981) and Mix (1982) who reported a correlation between tissue levels of polycyclic aromatic hydrocarbons, including benzo(a)pyrene, and neoplasia in mussels, *M. edulis*, from the Oregon coast. Similarly, Reinisch *et al.* (1984) suggested that pollution with polychlorinated biphenyl (PCB) in growing areas may be implicated in enhanced prevalences of hemocytic neoplasia in *M. arenaria*. An analysis of published information leads me to conclude that it remains questionable whether there is a cause-and-effect relationship between neoplasia in marine molluscs and oil pollution.

It is of interest that Couch (1985) has pointed out that absence of an infectious agent or pathognomonic indication does not preclude the possibility that certain lesions that appear to be noninfectious, i.e., an infectious agent is not present, could have been vicariously caused by, or related to, infectious agents, e.g., delayed inflammatory response to some previously present agent. Also, in spite of intense interest in the hemocytic neoplasia in marine molluscs and its possible causes, the actual lethality of the disease, other than the periodic occurrence of mass mortalities, was not ascertained until Brouseau and Baglivo (1990) reported that, as a result of a mark and recapture experiment, 51% of clams, *M. arenaria*, determined to have "high severity neoplasia" died within 6 months as compared 8% death among clams with "low severity neoplasia" and only 3% death among normal clams.

Kent *et al.* (1989) demonstrated that *M. edulis* with hemocytic neoplasia portrayed impaired cell-mediated immunity. As to whether this condition may lead to greater susceptibility to infections and less efficiency in sequestering pollutants remains to be elucidated. The answer to the possible cause-and-effect relationship between environmental pollutants and the underlying mechanisms and hemocytic (and other) neoplasms in marine molluscs must await future research. Nevertheless, I wish to take this opportunity to suggest where the future should be focused.

Of the several lines of proof pertaining to the development of malignant neoplasms in vertebrates, the viral etiology of cancers is the most popularly pursued. Nonetheless, there are those who are searching for DNA sequences in cancer cells that provoke uncontrolled proliferation when introduced into normal cells. Even in the case of known chemical carcinogens, these molecules are the stimulators of uncontrolled mitosis, and hence are mitogens. It is in this role that such chemicals are involved in carcinogenesis. Furthermore, as Ames and Gold (1990) pointed out, mitogenesis increases mutagenesis, and it is the latter genetic event that leads to carcinogenesis. Prior to delving into what I think should be the major approach to understanding the causes of neoplasms in lower animals, including molluscs, a brief review of the current state of knowledge relative to carcinogenesis would appear appropriate.

Cancer cells, by definition, proliferate in defiance of normal controls. Cancer cells retain many of the features of the specific cell type from which they are derived. Most neoplastic cells are believed to originate from a single cell that has undergone somatic mutation, but the progeny of this cell must undergo further change, probably involving several additional mutations, before they become cancerous. In mammals, tumor progression, which usually takes several years, reflects the operation of evolution by mutation and natural selection among somatic cells. One assumes that in molluscs, because of the relatively shorter life spans, the development of malignant neoplasms is a briefer process. The rate of tumor progression in mammals is accelerated by both mutagenic agents (tumor initiators) and nonmutagenic agents (tumor promoters). Both initiators and promoters affect gene expression, stimulate cell proliferation, and alter the balance between mutant and nonmutant cells. Many factors contribute to oncogenesis, some of which are environmental.

Since oncogenesis has a mutational basis and involves natural selection, no two cancers are genetically identical. Nevertheless, all malignant neoplastic cells disrupt the normal restraints on cell proliferation, and for each type of cell there is a finite number of ways by which such disruption can occur. According to the presently favored interpretation, changes in a relatively small set of genes appear responsible for deregulation of cell division in cancer (Klein 1987; and others). Actually, two classes of regulatory genes are involved; those whose products help stimulate cell proliferation and those whose products help inhibit the process (Fig. 11-1). Correspondingly, there are two mutational routes leading to uncontrolled cell proliferation. One results in the stimulation of gene hyperactivity, and the second leads to inactivation of an inhibitory gene (Fig. 11-2).

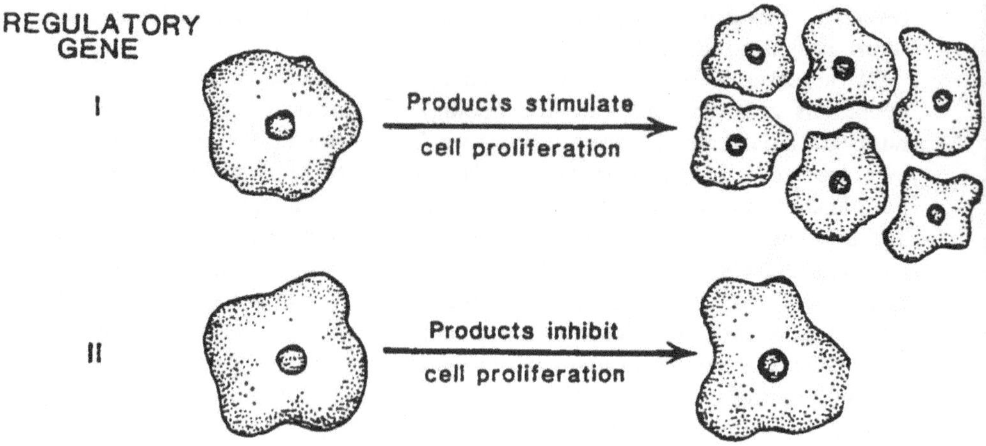

Fig. 11-1. Schematic drawing showing two classes of regulatory genes (I and II) and the effects of their products on cell proliferation.

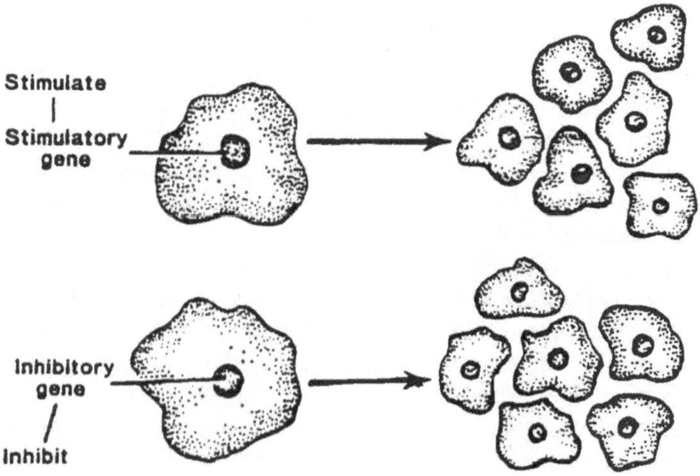

Fig. 11-2. Schematic drawing showing two mutational routes that lead to uncontrolled cell proliferation. One results in the stimulation of gene hyperactivity; the other leads to the inactivation of an inhibitory gene.

In the first, which has a dominant effect, the alteration only needs to involve one allele. The altered gene is known as an oncogene while the normal gene, i.e., one in which neither allele has undergone mutation, is known as a proto-oncogene (see review by Cooper 1990). In the second type, both alleles must be inactivated or deleted to eliminate the inhibition. The inactivated or deleted gene has been designated as a tumor suppressor gene.

In addition to altered genes, there is another type of genetic alteration that can lead to carcinogenesis. Specifically, the system controlling cell division can be subverted by foreign nucleic acids introduced into the cell by a virus. Such a virus, known as a tumor virus, can be either a DNA or RNA virus. A number of tumor viruses have been associated with human cancers. In molluscs, the only known instance of a possible viral etiology of a neoplastic disease is a retrovirus associated with hemocytic neoplasm in *M. arenaria* by Cooper and Chang (1982). This implication, however, remains to be unequivocally confirmed.

Considering that it has only been 13 years since the revelation that mammalian oncogenes are derived from mutation or rearrangement of normal cellular progenitors (for review, see Rose and Trnka 1988), it is amazing how much progress has been made in understanding the molecular basis for manifestations of oncogenes. A great deal, however, remains to be elucidated (for reviews, see Rose and Trnka 1988; Weinberg 1989; Cooper 1990).

It is beyond the scope of my presentation to present a summary sketch of malignant neoplasms of viral etiology. For an abbreviated but lucid account, see Alberts *et al.* (1989). At this point, I wish to share some thoughts relative to nonviral oncogenesis. As stated earlier, there is a strong correlation between carcinogenesis and mutagenesis. This is especially apparent for three classes of agents: 1) tumor viruses, which introduce foreign DNA into the cell (already discussed); 2) chemical carcinogens, which typically cause simple local changes in the nucleotide sequence; 3) ionizing radiation typically cause chromosome breaks and translocation.

Prior to some considerations pertaining to chemical carcinogens, which fall within the domain of this contribution, some general comments relative to the induction of carcinogenesis would appear to be in order. Based on studies on higher animals, primarily mammals, it is generally accepted that a given cancer cannot be caused by a single event or a single cause. Rather, as a rule, malignant neoplasias result from the chance occurrence in one cell the cumulative effects of several independent mutagenic accidents rather than a single mutation (Cairns 1975; Nowell 1976; and others). It thus follows that the cell's environment influences the frequency of inductive events in a variety of ways. It is noted, however, that there are some unusual carcinogenic agents that increase the frequency of critical events to the point where it is almost certain, given a high enough dosage, that at least one cell in the organism will become cancerous. This applies to chemical carcinogens.

Relative to the uptake and accumulation of known carcinogens by aquatic animals, Couch and Harshbarger (1985) and Couch *et al.* (1979) suggested that both vertebrates and invertebrates, including *C. virginica*, may be good indicators of the presence of such compounds and have presented a protocol for testing accumulation. Such studies, although useful, still do not reveal the mechanisms responsible for oncogenesis at the molecular level.

In experimental studies involving mammals, many disparate chemicals have been shown to be similarly carcinogenic. Some chemicals act directly on target cells, while others exert their effect only after they are metabolized to a more effective form. One of the more common metabolic pathways that results in a more reactive form involves a set of intracellular enzymes known as the cytochrome P-450 oxidases. Normally, these enzymes are involved in detoxification of ingested toxins and foreign lipid-soluble materials that are readily excreted. Cytochrome P-450 oxidases cannot perform this function if reacted with such compounds as aflatoxin-2, 3-epoxide. Rather, they are directly converted into carcinogens. Almost all known chemical carcinogens, although structurally quite diverse, are mutagens. Also, they can react with DNA. On the other hand, there are a few chemical carcinogens, the so-called tumor promoters, which are not mutagens. As a result of studies, again involving mammals, it is known that tumor promoters can induce carcinogenesis in cells that were exposed to a tumor initiator, i.e., a true mutagenic compound that initiates vulnerability of a cell towards becoming malignant. The most studied tumor promoters are phorbol esters, e.g., tetradecanoylphorbol acetate (TPA), the artificial activators of protein kinase C. Exposure to phorbol esters (and other tumor promoters) results in cancer only if cells are exposed to a tumor initiator.

This sketch of roles of humoral initiators and promoters emphasizes that similar studies and related mechanistic molecular studies should be conducted on molluscs before any statement, other than an inference, can be made about any environmental pollutant as a cause of neoplasms. It follows the identification of oncogenes and proto-oncogenes must be forthcoming if we are to understand the basis for oncogenesis in aquatic molluscs and other invertebrates.

Since, as stated, oncogenesis, like the evolution of species, is based on mutations and natural selection, with neoplastic cells representing evolving "new species," four major governing parameters should be borne in mind if we are to elucidate the factors that enhance the development of malignancies. We must ascertain 1) the mutation rate, i.e., the probability per gene per unit time that any given cell in the organism will undergo change; 2) the number of cells in the organism; 3) the reproduction rate, i.e., the average number of generations of progeny produced per unit time; and 4) the selective advantage derived by successful mutant cells, i.e., the ratio of the number of surviving progeny cells the mutant cells produce per unit time to the number of surviving progeny cells produced by nonmutant cells. As in the case of individuals in a population, the selective advantage to the progeny of mutant cells depends on both the nature of the mutation and environmental conditions. Additional complications arise if inheritable epigenetic changes occur, either randomly or in reaction to specific stimulators. Thus, the mutation rate per cell is not the only significant variable in oncogenesis.

It should be apparent from the above that experimental oncology among marine organisms, including molluscs, has a long way to go to reach the state of knowledge that presently exists for vertebrates, especially mammals. Although the reporting of histopathologic information is important, it cannot lead to conclusive evidence pertaining to causative mechanisms. Consequently, in view of the fact that the three major stimulators of neoplastic diseases in higher animals are certain viruses, chemical carcinogens, and ionizing radiation, I suggest that increased attention be directed to understanding intricacies associated with these stimulators. One basic tool of common utility to all lines of investigation is cell and tissue culture. Without the ability to establish and maintain cell lines suitable for maintenance of viruses, little progress can be made in elucidating events leading to and during viral oncogenesis. Similarly, the availability of cultured cells would permit definition of intracellular chemical pathways, resulting in development of cancer cells. In the area of ionizing radiation, detailed cytological information relative to aberrant chromosomal behavior could be obtained more readily on a quantitative basis on cells cultured *in vitro*.

If the oncogenetic patterns that have been elucidated in mammalian cells, (e.g., occurrence of oncogenes, initiators, promoters) occur in molluscan systems, their isolation and characterization will be simplified when cells maintained *in vitro* become available. Furthermore, mutation rates and reproduction rates could be ascertained with greater ease by using cells cultured *in vitro*.

In addition to the establishment of cell cultures, biochemical and molecular biological tools essential to reveal sites and nature of oncogenes, proto-oncogenes, various aspects of mutations, etc. must be adopted, sometimes with essential modifications, if progress is to be made. In brief, the causes and intermediate processes leading to histopathologically identifiable neoplasms remain to be explored in nonrefutable ways in molluscs.

Toxicological Diseases

As a result of interest in the possible deleterious effects of toxic pollutants in the aquatic environment, numerous studies have been carried out to determine the toxicity of additives. As to be expected, most of these studies, especially the earlier ones, have involved the determination of LC_{50}s, TLm_{50}s, and LD_{50}s and the lethality of toxicants on finfish. With little attention to actual disease processes, possibly initiated by the toxicants. However, since the late 1960's and early 1970's, there has developed an interest in the sublethal effects as well, but, again, primarily of finfish (Koeman and Strik 1975).

Ecological and economic problems associated with mass mortalities of aquatic animals caused by pollutants are of sufficient interest that an annual publication of geographic areas, possible toxicants, estimated number of fish killed, severity, and duration of critical effect was initiated by the U.S. Public Health Service in 1960. This publication was taken over by the Federal Water Pollution Control Administration in 1965 and is presently issued by the EPA. Further, a handbook compiled by Johnson and Finley (1980)for the U.S. Department of the Interior listed species of U.S. aquatic invertebrates and fishes used in acute toxicity testing and chemical toxicants that have been tested and the results obtained. Similar data on a world-wide basis have been compiled by the Food and Agriculture Organization (FAO) of the United Nations (1978).

In this section, I attempted to present a glimpse of the types of research that pertain to toxicological problems as related to marine molluscs. It is not meant to be a thorough survey of published literature, which is voluminous. Nevertheless, during preparation of this section, I examined numerous publications and found a dearth of basic principles.

Organic Toxicants

A variety of organic compounds are lethal to all animals. As Sparks (1985) has pointed out, such compounds as ethyl and methyl alcohol, metaldehyde, formaldehyde, phenol, cresol, and strong acids and bases fall into this category.

Recently, the FAO (1990) has distributed a list of organochlorines which are toxic in the marine environment. The most significant point in this report is that there does not appear to be a correlation between the molecular weights of the compounds and their potential toxicity. This needs to be reexamined since this conclusion is not supported, at least in part, by quantitative structure-activity relationship data (reviewed later).

Certain chlorinated hydrocarbons, particularly orthodichlorobenzene, have been used alone or in combination with other pesticides around the periphery of or directly on oyster beds to discourage predators (Loosanoff *et al.* 1960; Davis *et al.* 1961); the pesticide is usually mixed with sand or some other inert carrier prior to application. Loosanoff *et al.* (1960) reported that oysters, *C. virginica*, were "unharmed" after 3 h immersion in 100 ppm orthodichlorobenzene but were almost all killed after 24 h. Sublethal effects were not studied.

To provide some degree of historic completeness, it is noted that with the development and wide usage of several organochlorines as pesticides and in industrial applications during and after World War II, this category of organics entered the aquatic environment. Among such additives was dichlorodiphenyl trichloroethane (DDT), a widely used contact insecticide. It is still used but in considerably reduced quantity. According to Waugh *et al.* (1952) and Waugh and Ansell (1956), DDT was employed in Great Britain to control fouling of oyster beds by barnacles until it was discovered that organisms other than barnacles were affected. Butler (1966) reported that oysters, *C. virginica*, exposed to 0.001-0.002 ppm of DDT at 30°C appeared normal in growth and behavior but concentrated the compound in their tissues at levels up to 70,000 times the amount present in the ambient sea water. Also, Butler *et al.* (1962) reported that when young oysters were exposed to concentrations as low as 0.1 ppm, there was essentially total inhibition of growth within 24 h. Even upon exposure to 0.0001 ppm of DDT for 24 h, the growth of young oysters was approximately 20% of that of unexposed oysters. Davis (1961) reported that *C. virginica* larvae exposed to 1 ppm of DDT suffered 100% mortality within 6 days.

Lindane, a rapid-acting insecticide, has also been employed in shellfish predator and pest control (Hanks 1963). The lethality of lindane to marine molluscs varies. Lindsay (1963) reported that the bentnose clam, *Macoma nasuta*, and the soft-shell clam, *Mya arenaria*, associated with oyster beds in the Pacific northwest were killed when exposed to <1.0 ppb; however, the Japanese littleneck clam, *Venerupis* sp., was not.

Among other organochlorines, Butler *et al.* (1962), Butler (1964), and Butler and Springer (1963) reported that aldrin, dieldrin, endrin, toxaphene, chlordane, and heptachlor markedly effected shell growth of young oysters and caused irritative shell movements at levels <1.0 ppm. According to Butler (1966), organophosphates are less toxic to oysters than the chlorinated hydrocarbons.

Sevin, a carbamate insecticide, has been used to control predators and pests on oyster beds in the Pacific northwest. According to Butler *et al.* (1962), the growth of young oysters is inhibited by exposure to 1.0 ppm of Sevin. Shaw and Griffith (1967) reported that after several treatments at this concentration, some mortality occurred among adult oysters, but ceased after 2 weeks.

Aroclor is a polychlorinated biphenyl (PCB) structurally similar to DDT. It and other PCBs are typically industrial pollutants. According to Couch and Nimmo (1974), shell growth of *C. virginica* exposed to 100 ppb of Aroclor 1254 for 96 h was totally inhibited; however, normal growth was resumed after the oysters were replaced in uncontaminated water. Thus, the deleterious effect was reversible.

Among organic herbicides that have been tested against marine molluscs, methylurea compounds vary in their toxicity to eggs and larvae of *Mercenaria mercenaria* in inverse relation to their solubility (Davis 1961). Specifically, fenuron and monuron, the most soluble of those tested, were the least toxic, with no effect on the development of clam eggs and larvae at concentrations up to 5.0

ppm. On the other hand, diuron, which is less soluble, significantly reduced the percentage of eggs developing normally at 5.0 ppm. Neburon, the least soluble, inhibited embryonic development at 2.4 ppm. Somewhat surprisingly, Davis (1961) reported that *M. mercenaria* larvae grew better at low concentrations of the methylurea herbicides than in control cultures; however, monuron was toxic at 1.0 ppm and higher, diuron dramatically reduced the growth rate at 5.0 ppm with occasional high mortalities, and neburon caused 100% mortality at 2.4 ppm.

Another group of herbicides, the 2,4-D formulations, have also been tested to some extent on marine molluscs. Specifically, Butler (1965) reported that at 1.0 ppm, the 2,4-D commercial preparation known as Dow Silvex reduced normal development of oyster embryos by about 80% and eventually killed all hatched larvae. Butler also reported that butoxyethanol ester (2,4-D BE) had little effect on embryonic development at 5.0 ppm; however, at 10.0 ppm, it reduced normal development by 75%. At a concentration of 1.0 ppm, 2,4-D BE killed oyster larvae and growth of juvenile oysters exposed to 3.75 ppm for 96 h decreased by 50%. Rawls (1965), as a result of testing several 2,4-D formulations on caged *C. virginica* and *M. arenaria*, reported only 2,4-D acetamide to be dangerously toxic.

Toxic effects of detergents on clams and oysters have also been tested to some extent. Since 1965, commercial detergents in the U.S. have been almost exclusively of the biodegradable linear alkylate sulfonate class. Hidu (1965) reported that the growth of *C. virginica* and *M. mercenaria* larvae decreased at lower concentrations than necessary to cause significant mortalities and oyster larvae were more sensitive to surfactants than clam larvae. Calabrese and Davis (1967) tested the effects of linear alkylate sulfonate detergents on the life cycle stages of *C. virginica* and *M. mercenaria*. They reported that fertilized eggs of these bivalves were killed at lower concentrations that fully developed veligers. These investigators noted that the linear alkylate sulfonate detergents break down rapidly when subjected to sludge from secondary sewage treatment plants. Consequently, most detergents occur in the aquatic environment as degradation products and consequently are reduced in toxicity.

Quantitative Structure-Activity Relationships (QSAR)

Sufficient information is now available to permit establishment of quantitative relationships between the nature and extent of responses of aquatic invertebrates to the structural or associated physicochemical characteristics of additive organic molecules. Such relationships, designated as quantitative structure-activity relationships (QSARs), have been used to 1) indicate modes of toxic action by groups of compounds, 2) predict the effect of single synthetic compounds with biocidal potential, 3) predict the effects of complex pollutant mixtures, 4) identify compounds with unusual modes of action, 5) compare the responses of invertebrate species to toxicants, and 6) assist in design of combined chemical and biological environmental monitoring programs and interpreting results. The idea behind QSAR originated with Crum *et al.* (1968); however, it has been considerably refined since then (for review and detailed explanation, see Donkin and Widdows 1990). In brief, if the QSAR is a linear one, then an association can be assumed.

Donkin and Widdows (1990), who analyzed known QSARs, pointed out these major attributes: 1) Compounds that closely fit a single QSAR line generally have the same toxic mechanism; 2) The toxicity of untested compounds can be predicted if their physicochemical properties are similar to those of compounds for which QSARs are known; 3) Since QSAR can provide information pertaining to expected toxicity, compounds with observed toxicity that deviate significantly from expectations merit further study as previously unknown modes of toxic action or mechanisms may be revealed; 4) The toxicity of mixtures of compounds which conform to the same QSAR is usually additive; 5) QSARs can define the physicochemical characteristics of organic compounds that are likely to be toxic.

QSARs have been assembled for several marine molluscs. As examples, Geyer *et al.* (1982) applied this relationship to the bioconcentration factor (BCF) of organics in *M. edulis*, Hawker and Connell (1986) applied QSARs for bioconcentration in *M. edulis*, *C. virginica*, and *M. arenaria* and demonstrated that considerable similarities between the bioconcentration patterns in different species of marine bivalves. Specifically, the BCF, which is the ratio at steady state between the concentration of the compound in the mollusc (based on wet weight) and the concentration in the water to which it is exposed, is determined, the $\log_{10}$ of the BCF of the chemicals tested (aromatic hydrocarbons, carbamate, organophosphorus pesticides, and a phthalate plasticizer) correlated linearly with $\log_{10}$ of the n-octanol/water partition coefficient ($\log K_{ow}$).

QSARs can also be employed to ascertain the specific source of toxicants (e.g., from food, directly from surrounding water, etc.). For example, Pruell *et al.* (1986), after exposing *M. edulis* to sediments contaminated with PCBs and aromatic hydrocarbons, calculated the $\log_{10}$ bioaccumulation factor (BAF) of the aromatic hydrocarbons from particulate and total (including water) concentrations in the system. They found that a nonlinear relationship was obtained with K_{ow}, indicating reduced bioavailability of the more hydrophobic molecules associated with the sediment. However, when the $\log_{10}$ BCF was calculated from water concentrations, a straight line relationship with $\log K_{ow}$ was obtained. This closely resembled the regressions for bivalve molluscs (Geyer *et al.* 1982; Hawker and Connell 1986) and hence it was concluded that bioaccumulation into *M. edulis* was predominantly from the aqueous phase.

Inorganic Toxicants

Among inorganics, Sparks (1985) has reviewed what is known about the toxicity of cyanide, mercuric compounds, lead, arsenic, copper, cadmium, zinc, and phosphorus on invertebrates. Briefly, in the case of marine molluscs, Galtsoff (1964) found that oysters, *C. virginica*, take up and accumulate lead. Also, Shuster and Garb (1967) reported increased mortality occurred when *C. virginica* are exposed to lead concentrations of 0.1 and 0.2 ppm for 10 weeks, but not at lower concentrations. Lead-challenged oysters portrayed edematous mantles and aggregates of lead chromate crystals and granules (possibly lead proteinate) in the basal portion of the intestinal epithelium and, less commonly, throughout the connective tissue of the visceral mass, gills, and mantle. Small quantities of these inclusions occurred in the reproductive tissues.

Oysters, *C. gigas*, do concentrate mercury (Sparks 1985). Also adult oysters (both *C. gigas* and *C. virginica*) concentrate copper (Marks 1938; and others), although larvae and spat are highly susceptible to copper toxicity. It is noted, however, that Fujiya (1960) reported necrosis and desquamation of gastric epithelium, accompanied by regressive changes in the digestive diverticula of *C. gigas* adults exposed to 0.1-0.5 ppm copper for 2 weeks. It was also demonstrated that there was a decrease of RNA and polysaccharides in the digestive diverticular cells during exposure.

That marine molluscs are capable of taking up and concentrating a variety of metals and other elements and molecules from the environment is now well known (Romeril 1971; Raymont 1972; Ayling 1974; and numerous others.) The pathobiological implication of this phenomenon is that marine molluscs may not initially portray abnormal manifestations associated with certain metals and other elements; as additives are taken up and concentrated in tissues, concentration eventually reaches the level where sublethal effects and even lethality occur.

Zinc, like cadmium and mercury, alters O_2 consumption and hemolymph osmotic concentration in marine isopods but it has essentially no influence on the respiratory rates of *Mytilus edulis* (Brown and Newell 1972). Additional studies involving other metals at various concentrations are needed before any general statements can be made.

Because chlorine is commonly employed as a bactericide or pesticide, Waugh (1964) tested its toxicity to *Ostrea edulis* larvae. He reported that these larvae are unharmed (not killed) when exposed to 10 ppm, the normal industrial concentration, for 10 minutes. Further, Waugh reported considerable growth of *O. edulis* larvae exposed to 20 ppm of chlorine; however, at concentrations between 50 and 200 ppm, almost all exposed larvae were killed.

Since Waugh (1964) tested the toxicity of chlorine by measuring death rates, it is not possible to evaluate possible sublethal effects. On the other hand,

Galtsoff (1964), who reviewed the physiological effects of chlorination on *C. virginica*, stated that at concentrations of 0.01-0.05 ppm of chlorine, the normal physiological functions of oysters (i.e., contraction of cardiac muscles, water transport by gills, feeding, respiration, discharge of excreta) were interfered with. Furthermore, some tolerance is developed as a result of repeated exposure. Nevertheless, effective pumping ceases at concentrations above 1.0 ppm.

Sublethal Effects

As stated earlier, older studies pertaining to the toxicity of pollutants were almost exclusively devoted to the establishment of LC_{50} and LD_{50} of the additives. Commencing in the late 1960s and early 1970s, increased attention was paid to sublethal effects of pollutants, i.e., physiological deviations from the norm associated with pollutants. These deviations can be considered as diseases and hence a few remarks are included. It is noted, however, that the following does not constitute an exhaustive review of the published works on this topic.

Among contaminating metals, Brown and Newell (1972) and Scott and Major (1972) reported that exposure of *Mytilus* to copper resulted in the reduction of O_2 consumption and Thurberg *et al.* (1974) reported that exposure to silver resulted in increased O_2 consumption by *M. edulis* and *M. arenaria*. They found a reduction of filtration rates and abnormal ciliary activity in *M. arenaria* that had taken up dissolved chromium and in *M. edulis* that have taken up both dissolved and particulate chromium. Further, a decline was reported in O_2 consumption of excised gill tissue from both bivalves, suggesting that chromium interfered with an energy supplying metabolic pathway and this was reflected in the inhibition of ciliary activity.

Sublethal concentrations of certain metals can also effect the number of circulating hemocytes in marine molluscs. For example, Suresh and Mohandas (1990a) reported that although there were no variations in the numbers of hemocytes in the bivalve *Sunetta scripta* exposed to 1, 3, and 5 ppm Cu^{2+} in $30^0/_{00}$ salinity sea water during a 5-day period, the numbers of hemocytes in another bivalve, *Villorita cyprinoides* var. *cochinensis*, exposed to 0.15 and 0.30 ppm Cu^{2+} in $15^0/_{00}$ salinity sea water were significantly lower than in controls for 48 h and in specimens exposed to 0.45 ppm Cu^{2+}, the numbers of hemocytes were consistently reduced, commencing after 24 h. Suresh and Mohandas (1990a) attributed the absence of hemocyte number reduction in *S. scripta* to either low concentration of Cu^{2+}, which was far below the lethal dose, or to low uptake of Cu^{2+} in high salinity, either of which could have resulted in the entry of an insufficient amount of Cu^{2+} to cause cell mortality. To the contrary, in the case of *V. cyprinoides*, the concentration range of Cu^{2+} was close to the LC_{50}. Also, the uptake of Cu^{2+} was greater at the lower salinity.

As George *et al.* (1976) and Coombs (1977) had demonstrated earlier, Suresh and Mohandas (1990a) reported the uptake of Cu^{2+} by hemocytes in the bivalves that they studied. It is known that lead and copper, and perhaps other metals, are sequestered within lysosomes of molluscan granulocytes (Cheng unpublished). What effect this has on the synthesis of lysosomal hydrolases and their subsequent release, which play an important role in humoral immunity in molluscs (Cheng 1978, 1979, 1983, 1986), remains unknown. It is noted that Suresh and Mohandas (1990b) speculated that the reduction in the number of circulating hemocytes in Cu^{2+}-exposed *V. cyprinoides* may compromise cell-mediated immunity in this bivalve. Others also have shown that chemical stress can also depress phagocytosis (Fries and Tripp 1980) in studies with *Mercenaria*.

Suresh and Mohandas (1990b), as a result of exposing the two clam species, *S. scripta* and *V. cyrinoides*, to three sublethal concentrations of copper (1, 3, and 5 ppm in the case of *S. scripta*; 0.15, 0.30, and 0.45 ppm in the case of *V. cyprinoides*), reported that 1) activity levels of serum acid phosphatase varied between the two clam species and were dependent on the concentration of Cu^{2+}; 2) Cu^{2+} caused destabilization of the lysosomal membrane in granulocytes resulting in eventual release of this hydrolase into serum; 3) Cu^{2+} inhibited the activity of acid phosphatase; 4) depending on the exposure period and the concentration of the metal ion, enzyme synthesis could be adversely affected. These sublethal effects would influence the humoral immunity of molluscs since it is known that lysosomal hydrolases released into serum play a defensive role (for reviews, see Cheng 1978, 1979; Cheng and Combes 1990).

Similarly, Cheng (1989, 1990) reported that *in vitro* exposure of oyster granulocytes to 1 ppm Cu^{2+} partially inhibited release of lysozyme into serum; exposure to 1 ppm Cu^{2+} stimulated the intracellular synthesis of β-glucuronidase in granulocytes but partially inhibited the release of this enzyme into serum. Thus, sublethal concentrations of these metals can effect internal defense mechanisms of oysters, and molluscs could become more vulnerable to infectious diseases. It is apparent that sublethal effects of certain toxicants could be translated into influencing parameters that affect infectious diseases.

These examples pertaining to the sublethal effects of metals point out that different metals may exert differing effects. In addition, the concentration of the metal, the salinity of the sea water in which it is suspended, and other factors, some yet to be identified, influence its effect on molluscs. It is also apparent that considerable research needs to be forthcoming before any principles can be established relative to the influence of additives, including metals, on both cell-mediated and humoral immunity in marine molluscs.

Exposure to metals and other types of pollutants can also result in sublethal histopathologic alterations. As an example, as a part of the International Mussel Watch Program, during which bioaccumulation, physiological deviations, and histopathological alterations associated with pollutants were studied (Goldberg 1980); Sunila (1988), searched for histopathologic changes in *M. edulis* exposed to 5 ppm of several metals as well as PCB and DDT. She reported that in mussels exposed to cadmium, copper, lead, and silver, an inflammatory reaction in the gill commenced with enlargement of the post-lateral cells, after which the hemolymph sinuses dilated and granulocytes migrated to the epithelia. On the other hand, Sunila found that exposure to cobalt, iron, or dieldrin had no effect. Copper exposure also resulted in endothelial edema and deformation of abfrontal parts of the filaments. Exposure to lead also caused loss of lateral cilia or sloughing of lateral cells, and exposure to silver also resulted in endothelial cells becoming vacuolated. Organic pollutants (PCB, DDT) were reported to be associated with the loosening of the intercellular matrix in epithelia and shrinkage of the cells.

It is noted that although histopathological changes may reflect exposure to pollutants, it is essential that the triggering mechanism(s) and intermediate steps leading to morphologic manifestations be elucidated. Such information would be extremely useful in explaining why different metals and other categories of pollutants have different histopathologic effects.

Relative to the sublethal effects of organic pollutants, especially petroleum hydrocarbons, Gilfallan *et al.* (1977) reported that very low concentrations (3-4 ppm) of aromatic hydrocarbons in *M. arenaria* tissues derived from the environment can cause large reductions in carbon flux. There are other similar reports of the deleterious but nonlethal effects of crude oils (Ottway 1971; Vernberg and Vernberg 1974; Gilfallan 1975; and others). It needs to be pointed out, however, that in interpreting such results, it should be borne in mind that one can expect fluctuations from the normal levels of physiological functions in marine molluscs when placed under stress. Such stress may be due to pollutants but may also be due to other environmental stressors, e.g., changes in temperature, salinity, and other nonadditive changes (see reviews in Vernberg and Vernberg 1974; and Vernberg *et al.* 1977).

Consequently, in order to be certain that the sublethal effects are indeed due to specific pollutants, the responsible mechanisms must be elucidated. In view of what is known about toxicants, a statement by Sparks (1985) represents an excellent perspective of toxicologic diseases in marine invertebrates, including molluscs: "Despite wide variations in susceptibility to specific toxins within the animal kingdom, there are certain generalizations applicable to all chemical injuries. The effect of a chemical agent depends on the vulnerability of individual tissues, the mode of action of the agent, and the concentration of the chemical. Tissues and cells within an animal vary greatly in their susceptibility to chemical injury, and, obviously, there is even greater variation among the diverse invertebrate groups."

Other Pathologies

Another pathologic condition that may be associated with pollution was reported by Couch (1984, 1985). Specifically, he reported that oysters, *C. virginica*, collected in Pensacola and Escambia Bays in Florida, Mobile Bay in Alabama, and Pascagoula Harbor in Mississippi portrayed a condition designated as digestive gland epithelial atrophy. This histopathologic picture was significantly more prevalent in Pascagoula Harbor than at the other collection sites. Examination of the amounts of organic pollutants emptied into each of the sites revealed that Pascagoula Harbor contained the highest levels of base-neutral, organic pollutants. As a result, although a cause-and-effect relationship was not established, Couch (1985) suggested that digestive gland epithelial atrophy in oysters may serve as an indicator of elevated levels of chemical pollutants.

Although there may be some merit to Couch's (1985) suggestion, the histopathologic picture of atrophied digestive gland tubules may be reflective of other physiologic dysfunctions, e.g., starvation. Further, even if a specific class of pollutants will be found experimentally to be associated with the collapsed cell condition, a series of questions need to be resolved. As examples, how do affected cells become targets of the pollutant? In other words, are recognition molecules present on the cells and how do these function? Is the action of the pollutant extra- or intracellular? If it is intracellular, how is the pollutant carried to the reactive organelles, and if stored, at which sites? What is the biochemical basis for the cytopathologic expression? Of course, many other questions need to be asked and answered before the entire saga is elucidated. Thus, the collaboration between pathobiologists of various shades must necessarily be involved as has been promoted by Couch (1988) and others.

Additional structural modifications have been reported in marine bivalves from polluted waters as opposed to laboratory exposures. Seiler and Morse (1988) reported that reabsorptive and secretory kidney cells in *M. arenaria* can reflect impact of pollution. Specifically, they found that kidney cells of clams from polluted sediments included greater numbers of granules compared to clams from an unpolluted site. Also, they found that clams from polluted areas included greater numbers of granulocytes which, in turn, included larger inclusions. They postulate, probably correctly, that granulocytes remove pollutant particles from clams.

CONCLUSIONS

It is hoped that this sampling of known noninfectious diseases in marine molluscs has revealed that our understanding of such conditions has only reached the *alpha* phase, i.e., the pathologies, both structural and functional, have been identified; however, in most instances causative mechanisms remain unknown. Elucidation of such mechanisms would constitute the *beta* phase. Eventually, the genomic pliability that permits pathologic alterations to become manifest should be defined. That would constitute the *gamma* phase.

ACKNOWLEDGMENT

The original information included herein resulted from research supported by a Grant (NA16FL0408-01) from the National Marine Fisheries Service, U.S. Department of Commerce.

REFERENCES

Alberts, B., D. Bray, J. Lewis, M. Raff, K. Roberts, and J.D. Watson. 1989. *Mol. Biol. of the Cell*, 2nd ed. New York: Garland.

Alderman, D.J., P. Van Banning, and A. Perez-Colomer. 1977. Two European Oyster (*Ostrea edulis*) mortalities associated with an abnormal hemocytic condition. *Aquaculture* 10:335-40.

Ames, B.N., and L.S. Gold. 1990. Too many rodent carcinogens: mitogenesis increases mutagenesis. *Science* 249:970-1.

Appeldoorn, R.S., and J.J. Oprandy. 1980. Tumors in soft-shell clams and the role played by a virus. *Maritimes* 24:4-6.

Atkins, D. 1931. Note on some abnormalities of labial palps and foot of *Mytilus edulis*. *J. Mar. Biol. Assoc. U.K.* 17:545-50.

Auffret, M., and M. Poder. 1986. Sarcomatous lesion in the cockle *Cerastoderma edule*. II. Electron microscopical study. *Aquaculture* 58:9-15.

Ayling, G.H. 1974. Uptake of cadmium, zinc, copper, lead and chromium in the Pacific oyster, *Crassostrea gigas*, grown in the Tamar River, Tasmania. *Water Res.* 8:729-38.

Balouet, G., M. Poder, A. Cahour, and M. Auffret. 1986. Proliferative hemocytic condition in European flat oysters: a 6-year survey. *J. Invert. Pathol.* 48:208-15.

Barry, M.M., and P.P. Yevich. 1975. The ecological, chemical and histopathological evaluation of an oil spill site. III. Histopathological studies. *Mar. Pollut. Bull.* 6:171-73.

Barry, M.M., P.P. Yevich, and N.H. Thayer. 1971. Atypical hyperplasia in the soft-shell clam *Mya arenaria*. *J. Invert. Pathol.* 17:17-27.

Bishop, S.H., L.L. Ellis, and J.M. Burcham. 1983. Amino acid metabolism in molluscs. In *The Mollusca. Vol. 1. Metabolic Biochemistry and Molecular Biomechanics*, ed. K.M. Wilbur and P.W. Hochachka, pp. 244-327. New York: Academic Press.

Blake, J.H. 1929. An abnormal clam. *Nautilus* 38:89-90.

Brouseau, D.J. 1987. Seasonal aspects of sarcomatous neoplasia in *Mya arenaria* (soft-shell clam) from Long Island Sound. *J. Invert. Pathol.* 50:269-76.

Brouseau, D.J., and J.A. Baglivo. 1990. Field and laboratory comparisons of mortality in normal and neoplastic *Mya arenaria*. *J. Invert. Pathol.* 57:59-65.

Brown, B., and R.C. Newell. 1972. The effects of copper and zinc on the metabolism of the mussel *Mytilus edulis*. *Mar. Biol.* 16:108-18.

Brown, R.S., R.E. Wolke, and S.B. Saila. 1976. A preliminary report on neoplasia in feral populations of the soft-shell clam, *Mya arenaria*, prevalence, histopathology and diagnosis. *Proc. Int. Colloq. Invert. Pathol.* 1:151-58.

Brown, R.S., R.E. Wolke, S.B. Saila, and C. Brown. 1977. Prevalence of neoplasia in 10 New England populations of the soft-shell clam (*Mya arenaria*). *Ann. N.Y. Acad. Sci.* 298:522-34.

Butler, C.K. 1965. Field tests of herbicide toxicity to certain estuarine animals. *Chesapeake Sci.* 6:150-61.

Butler, P.A. 1964. Commercial fisheries investigations. *U.S. Fish Wildl. Serv. Circ.* 199:5-28.

Butler, P.A. 1966. Pesticides in the environment and their effects on wildlife. *J. Appl. Ecol.* 3(suppl.):253-59.

Butler, P.A., and P.F. Springer. 1963. Pesticides: A new factor in coastal environments. *Trans. N. Amer. Wildl. Natur. Res. Conf.* 28:378-90.

Butler, P.A., A.J. Wilson, Jr., and A.J. Rick. 1962. Effect of pesticides on oysters. *Proc. Nat. Shellfish Assoc.* 51:23-32.

Butros, J. 1948. A tumor in a fresh-water mussel. *Cancer Res.* 8:270-72.

Cairns, J. 1975. Mutation selection, and the natural history of cancer. *Nature* 255:197-200.

Calabrese, A., and H.C. Davis. 1967. Effects of 'soft' detergents on embryos and larvae of the American oyster (*Crassostrea virginica*). *Proc. Nat. Shellfish Assoc.* 57:11-6.

Capuzzo, J.M., and J.J. Sasner, Jr. 1977. The effect of chromium on filtration rates and metabolic activity of *Mytilus edulis* L. and *Mya arenaria* L. In *Physiological Responses of Marine Biota to Pollutants*, ed. F.G. Vernberg, A. Calabrese, F.P.Thurberg, and W.B. Vernberg, pp. 225-37. New York: Academic Press.

Cheng, T.C. 1978. The role of lysosomal hydrolases in molluscan cellular response to immunologic challenge. *Comp. Pathobiol.* 4:59-71.

Cheng, T.C. 1979. The role of hemocytic hydrolases in the defense of molluscs against parasites. *Haliotis* 8:193-209.

Cheng, T.C. 1981. Bivalves. In *Invertebrate Blood Cells 1: General Aspects. Animals without True Circulatory Systems to Cephalopods*, ed. N.A. Ratcliffe and A.F. Rowley, pp. 233-300. London: Academic Press.

Cheng, T.C. 1983. The role of lysosomes in molluscan inflammation. *Am. Zool.* 23:129-44.

Cheng, T.C. 1986. Specificity and the role of lysosomal hydrolases in molluscan inflammation. *Int. J. Tissue React.* 8:439-45.

Cheng, T.C. 1989. Immunodeficiency diseases in marine mollusks: measurements of some variables. *J. Aquat. Anim. Health* 1:209-16.

Cheng, T.C. 1990a. Introductory remarks to nutritional pathology section. In *Pathology in Marine Science*, ed. F.O. Perkins and T.C. Cheng, pp. 439-40. San Diego: Academic Press.

Cheng, T.C. 1990b. Effects of in vivo exposure of *Crassostrea virginica* to heavy metals on hemocyte viability and activity levels of lysosomal enzymes. In *Pathology in Marine Science*, ed. F.O. Perkins and T.C. Cheng, pp. 513-24. San Diego: Academic Press.

Cheng, T.C., and C. Combes. 1990. Influence of environmental factors on the invasion of molluscs by parasites: with special reference to Europe. In *Biological Invasions in Europe and the Mediterranean Basin*, ed. R. di Castri, A.J. Hansen, and M. Debussche, pp. 307-32. Dordrecht: Kluwer Academic Publishers.

Cheng, T.C., and E. Rifken. 1970. Cellular reactions in marine molluscs in response to helminth parasitism. In *A Symposium on Diseases of Fishes and Shellfishes*, ed. S.F. Snieszko, pp. 443-96. Washington, D.C.: American Fisheries Society, special publ. No. 5.

Coombs, T.L. 1977. Uptake and storage mechanisms of heavy metals in marine organisms. *Proc. Anal. Div. Chem. Soc.* 14:219-21.

Cooper, G.M. 1990. *Oncogenes*. Boston: Jones and Bartlett.

Cooper, K.R., and P.W. Chang. 1982. A review of the evidence supporting a viral agent causing a hematopoietic neoplasm in the soft-shelled clam, *Mya arenaria*. *Proc. Int. Colloq. Invert. Pathol.* 3:271-2.

Cooper, K.R., R.S. Brown, and P.W. Chang. 1982a. Accuracy of blood cytological screening techniques for the diagnosis of a possible hematopoietic neoplasm in the bivalve mollusc, *Mya arenaria*. *J. Invert. Pathol.* 39:281-9.

Cooper, K.R., R.S. Brown, and P.W. Chang. 1982b. The course and mortality of a hematopoietic neoplasm in the soft-shell clam, *Mya arenaria*. *J. Invert. Pathol.* 39:149-57.

Cosson-Mannevy, M.A., C.S. Wong, and W.J. Cretney. 1984. Putative neoplastic disorders in mussels (*Mytilus edulis*) from southern Vancouver Island waters, British Columbia, *J. Invert. Pathol.* 44:151-60.

Couch, J.A. 1969. An unusual lesion in the mantle of the American oyster, *Crassostrea virginica*. *Nat. Cancer Inst. Monogr.* 31:557-62.

Couch, J.A. 1970. Sarcoma-like disease in a single specimen of the American oyster. In *Comparative Leukemia Research 1969* ed., R.M. Dutcher, p. 647. Basel: Karger.

Couch, J.A. 1984. Atrophy of diverticular epithelium as an indicator of environmental irritants in the oyster, *Crassostrea virginica*. *Mar. Environ. Res.* 14:525-26.

Couch, J.A. 1985. Prospective study of infectious and noninfectious diseases in oysters and fishes in three Gulf of Mexico estuaries. *Dis. Aquat. Organ.* 1:59-82.

Couch, J.A. 1988. Role of pathobiology in experimental marine biology and ecology. *J. Exp. Mar. Biol. Ecol.* 118:1-6.

Couch, J.A., and D.R. Nimmo. 1974. Ultrastructural studies of shrimp exposed to the pollutant chemical polychlorinated biphenyl (Aroclor 1254). *Bull. Soc. Pharm. Ecol. Pathol.* 2:17-20.

Couch, J.A., and J.C. Harshbarger. 1985. Effects of carcinogenic agents on aquatic animals: An environmental and experimental overview. *Environ. Carcinogen. Rev.* 3:63-105.

Couch, J.A., L.A. Courtney, J.T. Winstead, and S.S. Foss. 1979. The American oyster (*Crassostrea virginica*) as an indicator of carcinogens in the aquatic environment. In *Animals as Monitors of Environmental Pollutants*, pp. 65-84. Washington, D.C.: National Academy of Sciences.

Crum Brown, A., and T.R. Fraser. 1968. On the connection between chemical constitution and physiological action. I. On the physiological action of the salts of ammonium bases, derived from strychnia, brucia, thebaia, codeia, morphia and nicota. *Trans. Roy. Soc. Edinburgh* 25:151-62.

Davis, H.C. 1961. Effects of some pesticides on eggs and larvae of oysters (*Crassostrea virginica*) and clams (*Venus mercenaria*). *Commer. Fish. Rev.* 23:8-23.

Davis, H.C., V.L. Loosanoff, and C.L. MacKenzie, Jr. 1961. Field tests of a chemical method for the control of marine gastropods. *Bur. Commer. Fish. Biol. Lab., Milford, Conn. Bull.* 25:1-24.

De Zwaan, A. 1983. Carbohydrate catabolism in bivalves. In *The Mollusca Vol. 1. Metabolic Biochemistry and Molecular Biomechanics*, ed. K.M. Wilbur and P.W. Hochachka, pp. 138-75. New York: Academic Press.

DesVoigne, D.M., M.C. Mix, and G.B. Pauley. 1970. A papillomalike growth on the siphon of the horse clam, *Tresus nuttali* . *J. Invert. Pathol.* 15:262-70.

Dinamani, P., and P.H. Wolf. 1973. Multiple tumors in the pericardial cavity of an Australian rock oyster, *Crassostrea commercialis* (Iredale and Roughley). *Int. J. Cancer Res.* 11:293-99.

Dix, T.G. 1972. Two mesenchymal tumors in a pearl oyster, *Pinctada marganitifera*. *J. Invert. Pathol.* 20:317-20.

Donklin, P., and J. Widdows. 1990. Quantitative structure-activity relationships in aquatic invertebrate toxicology. *Rev. Aquat. Sci.* 2:375-98.

Doyle, L.J., N.J. Blake, C.C. Woo, and P.P. Yevich. 1978. Recent biogenic phosphorite: concretions in mollusk kidneys. *Science* 199:1431-33.

Elston, R.A., M.L. Kent, and A.S. Drum 1988a. Progression, lethality, and remission of hemic neoplasia in the bay mussel, *Mytilus edulis*. *Dis. Aquat. Organ.* 4:135-42.

Elston, R.A., M.L. Kent, and A.S. Drum. 1988b. Transmission of hemic neoplasia in the bay mussel, *Mytilus edulis*, using whole cells and cell homogenate. *Dev. Comp. Immunol.* 12:719-29.

Farley, C.A. 1969a. Probable neoplastic disease of the hematopoietic system in oysters, *Crassostrea virginica* and *Crassostrea gigas*. *Nat. Cancer Inst. Monogr.* 31:541-55.

Farley, C.A. 1969b. Sarcomatoid proliferative disease in a wild population of blue mussels (*Mytilus edulis*). *J. Nat. Cancer Inst.* 43:509-16.

Farley, C.A. 1976. Proliferative disorders in bivalve mollusks. *Mar. Fish. Rev.* 38:30-33.

Farley, C.A. 1989. Selected aspects of neoplastic progression in molluscs. In *Progressive Stages of Malignant Growth/Development. Vol. I, Part 5: Comparative Aspects of Tumor Progression*, ed. H.E. Kaiser, pp. 24-31. New York: Nijhoff.

Farley, C.A., and A.K. Sparks. 1970. Proliferative diseases of hemocytes, endothelial cells, and connective tissue cells in mollusks. *Bibl. Haematol.* 36:610-17.

Farley, C.A., S.V. Otto, and C.L. Reinisch. 1986. New occurrence of epizootic sarcoma in Chesapeake Bay soft shell clams (*Mya arenaria*). *Fish. Bull.* 84:851-57.

Florkin, M., and S. Bricteux-Gregoire. 1972. Nitrogen metabolism in mollusks. In *Chemical Zoology Vol. VII. Mollusca*, ed. M. Florkin and B.T. Scheer, pp. 301-48. New York: Academic Press.

Food and Agriculture Organization (FAO) of the United Nations. 1978. *Inventory of Data on Contaminants in Aquatic Organisms*. FAO Fisheries Circ. No. 338, Rev. 1. Rome: FAO.

Food and Agriculture Organization (FAO) of the United Nations. 1990. *Review of Potentially Harmful Substances: Choosing Priority Organochlorines for Marine Hazard Assessment*. Reports and Studies. No. 42, Rome: FAO.

Franc, A. 1975. Hyperplasie hemocytaire et lesions chez l'huitre plate (*Ostrea edulis* L.). *Comp. Rend. Hebd. Seances Acad. Sci.*, Ser. D. 280:495-98.

Frierman, E.M., and J.D. Andrews. 1976. Occurrence of hematopoietic neoplasms in Virginia oysters. *J. Nat. Cancer Inst.* 56:319-24.

Fries, C.R., and M.R. Tripp. 1980. Depression of phagocytosis in *Mercenaria* following chemical stress. *Dev. Comp. Immunol.* 4:233-44.

Fujiya, M. 1960. Studies on the effects of copper dissolved in sea water on oysters. *Bull. JPN Soc. Sci. Fish.* 26:462-68.

Galstoff, P.S. 1964. The American oyster *Crassostrea virginica* Gmelin. *U.S. Fish. Wildl. Serv. Fish. Bull.* 64:1-480.

George, S.G., B.J.S. Pirie, and T.L. Coombs. 1976. The kinetics of accumulation and excretion of ferric hydroxide in *Mytilus edulis* (L.) and distribution in the tissues. *J. Exp. Mar. Biol. Ecol.* 23:71-84.

Geyer, H., P. Shehan, D. Kotzias, D. Freitag, and F. Korte. 1982. Prediction of ecotoxicological behaviour of chemicals: relationship between physicochemical properties and bioaccumulation of organic chemicals in the mussel *Mytilus edulis*. *Chemosphere* 11:1121-33.

Gilfillan, E.S. 1975. Decrease of net carbon flux in two species of mussels caused by extracts of crude oil. *Mar. Biol.* 29:53-58.

Gilfillan, E.S., D.W. Mayo, D.S. Page, D. Donovan, and S. Hanson. 1977. Effects of varying concentrations of petroleum hydrocarbons in sediments on carbon flux in *Mya arenaria*. In *Physiological Responses of Marine Biota to Pollutants*, ed. F.J. Vernberg, A. Calabrese, F.P. Thurberg, and W.B. Vernberg, pp. 299-314. New York: Academic Press.

Gold, K., G. Capriulo, and K. Keeling. 1982. Variability in the calcium phosphate concentration load in the kidney of *Mercenaria mercenaria*. *Mar. Ecol. Prog. Ser.* 10:97-99.

Goldberg, E.D. 1980. *The International Mussel Watch: Report of a Workshop Sponsored by the Environmental Studies Board*, Commission *on Natural Resources*. Washington, DC: National Research Council.

Goudsmit, E.M. 1972. Carbohydrates and carbohydrate metabolism in Mollusca. In *Chemical Zoology Vol. VII Mollusca*, ed. M. Florkin and B.T. Scheer, pp. 219-43. New York: Academic Press.

Green, M., and D.J. Alderman. 1983. Neoplasia in *Mytilus edulis* L. from United Kingdom waters. *Aquaculture* 30:1-10.

Gunter, G.P. 1957. An abnormal Virginia oyster with a bifurcated muscle. *Proc. Nat. Shellfish Assoc.* 48:152-154.

Hanks, R.W. 1963. Chemical control of the green crab, *Carcinus maenas* (L.). *Proc. Nat. Shellfish Assoc.* 52:75-86.

Harshbarger, J.C. 1976. Description of polyps and epidermal papillomas in three bivalve mollusk species. *Mar. Fish. Rev.* 38:25-29.

Harshbarger, J.C., S.V. Otto, and S.C. Chang. 1979. Proliferative disorders in *Crassostrea virginica* and *Mya arenaria* from the Chesapeake Bay and intranuclear virus-like inclusions in *Mya arenaria* with germinomas from a Maine oil spill site. *Haliotis* 8:243-48.

Hasselman, D.M., E.C. Peters, and N.J. Blake. 1987. Neoplasia in the hard shell clam (genus *Mercenaria*). Prog. Abst. 20th Annu. Meet. Soc. Invert. Pathol. p. 4.

Haswell, W.A. 1886. On a destructive parasite of the rock oyster (*Polydora ciliata* and *Polydora polybranchia* n. sp.). *Proc. Linn. Soc. New So. Wales* 10:272-5.

Hawker, D.W., and D.W. Connell. 1986. Bioconcentration of lipophilic compounds by some aquatic organisms. *Ecotoxicol. Environ.* 11:184-92.

Hidu, H. 1965. Effects of synthetic surfactants on the larvae of clams (*M. mercenaria*) and oysters (*C. virginica*). *J. Water Pollut. Control Fed.* 37:262-70.

Johnson, W.W., and M.T. Finley. 1980. *Handbook of Acute Toxicity of Chemicals to Fish and Aquatic Invertebrates*. U.S. Fish. Wildl. Serv. Resour. Publ. 137.

Jones, E.J., and A.K. Sparks. 1969. An unusual histopathological condition in *Ostrea lurida* from Yaquina Bay, Oregon. *Proc. Nat. Shellfish. Assoc.* 59:11.

Kavanaugh, L.D. 1940. Mud blisters in Japanese oysters imported to Louisiana. *La. Conserv.* Autumn 1940:31-34.

Kent, M.L., R.A. Elston, M.T. Wilkinson, and A.S. Drum. 1989. Impaired defense mechanisms in bay mussels, *Mytilus edulis*, with hemic neoplasia. *J. Invert. Pathol.* 53:378-86.

Klein, G. 1987. The approaching era of the tumor suppressor genes. *Science* 238:1539-45.

Koeman, J.H., and J.J.T.W.A. Strik (eds.). 1975. *Sublethal Effects of Toxic Chemicals on Aquatic Animals*. Amsterdam: Elsevier.

Lauckner, G. 1983. Diseases of Mollusca: Bivalvia. In *Diseases of Marine Animals Vol. II. Introduction, Bivalvia to Scaphopoda*, ed. O. Kinne, pp. 476-1038. Hamburg: Biologische Austalt Helgoland.

Leloup, E. 1937. Contributions a l'étude de la faune Belge. VIII. Les dégâts causés par le ver polychete *Polydora ciliata* (Johnston) daus les coquilles des bigorneaux et des huîtres. *Bull. Mus. Hist. Nat. Belg.* 13:104.

Lindsay, C.E. 1963. Pesticide tests in the marine environment in the state of Washington. *Proc. Shellfish Assoc.* 52:87-97.

Livingstone, D.R., and A. de Zwaan. 1983. Carbohydrate metabolism in gastropods. In *The Mollusca Vol. I. Metabolic Biochemistry and Molecular Biomechanics*, ed. K.M. Wilbur and P.W. Hochachka, pp. 177-242. New York: Academic Press.

Loosanoff, V.L., C.L. MacKenzie, Jr., and L.W. Shearer. 1960. Use of chemicals to control shellfish predators. *Science* 131:1522-23.

Lowe, D.M., and M.N. Moore. 1978. Cytology and quantitative cytochemistry of a proliferative atypical hemocytic condition in *Mytilus edulis* (Bivalvia, Mollusca). *J. Nat. Cancer Inst.* 60:1455-59.

Lunz, G.A. 1940. The annelid worm, *Polydora*, as an oyster pest. *Science* 92:310.

Marks, G.H. 1938. The copper content and copper tolerance of some species of mollusks of the southern California Coast. *Biol. Bull.* 75:224-37.

Mauri, M., and E. Orlando. 1982. Experimental study on renal concretions in the wedge shell *Donas trunculus* L. *J. Exp. Mar. Biol. Ecol.* 63:47-57.

Medcof, J.C. 1946. The mud-blister worm, *Polydora*, in Canadian oysters. *J. Fish. Res. Board Can.* 6:498-505.

Mix, M.C. 1975. Proliferative characteristics of atypical cells in native oysters (*Ostrea lurida*) from Yaquina Bay, Oregon. *J. Invert. Pathol.* 26:289-298.

Mix, M.C. 1982. Cellular proliferative disorders in bay mussels (*Mytilus edulis*) from Oregon estuaries. *Proc. Int. Colloq. Invert. Pathol.* 3:266-67.

Mix, M.C. 1983. Haemic neoplasms of bay mussels, *Mytilus edulis* L., from Oregon: occurrence, prevalence, seasonality and histopathological progression. *J. Fish. Dis.* 6:239-48.

Mix, M.C., and W.P. Breese. 1980. A cellular proliferative disorder in oysters (*Ostrea chilensis*) from Chiloe, Chile, South America. *J. Invert. Pathol.* 36:123-24.

Mix, M.C., J.W. Hawkes, and A.K. Sparks. 1979a. Observations on the ultrastructure of large cells associated with putative neoplastic disorders of mussels, *Mytilus edulis*, from Yaquina Bay, region. *J. Invert. Pathol.* 34:41-56.

Mix, M.C., S.R. Trenholm, and S.I. King. 1979b. Banzo[a]pyrene body burdens and the prevalence of proliferative disorders in mussels (*Mytilus edulis*) in Oregon. In *Animals as Monitors of Environmental Pollutants*, pp. 53-64. Washington, DC: National Academy of Sciences.

Mix, M.C., R.L. Schaffer, and S.J. Hemingway. 1981. Polynuclear aromatic hydrocarbons in bay mussels (*Mytilus edulis*) from Oregon. In *Phyletic Approaches to Cancer*, ed. C.J. Dawe, J.C. Harshbarger, S. Kondo, T. Sugimura, and S. Takayama, pp. 167-77. Tokyo: JPN Soc.

Mix, M.C., H.J. Pribble, R.T. Riley, and S.P. Tomasovic. 1977. Neoplastic disease in bivalve mollusks from Oregon estuaries with emphasis on research on proliferative disorders in Yaquina Bay oysters. *Ann. N. Y. Acad. Sci.* 298:356-73.

Moore, J.D., R.A. Elston, A.S. Drum, and M.T. Wilkinson. 1991. Alternate pathogenesis of systemic neoplasia in the bivalve mollusc *Mytilus*. *J. Invert. Pathol.* In Press.

Newman, M.W. 1972. An oyster neoplasm of apparent mesenchymal origin. *J. Nat. Cancer. Inst.* 48:237-43.

Nowell, P.C. 1976. The clonal evolution of tumor cell populations. *Science* 194:23-28.

Oprandy, J.J., P.W. Chang, A.D. Prouvost, K.R. Cooper, R.S. Brown, and V.J. Yates. 1981. Isolation of a viral agent causing hematopoietic neoplasia in the soft-shell clam, *Mya arenaria*. *J. Invert. Pathol.* 38:45-51.

Ottway, S. 1971. The comparative toxicities of crude oils. In *The Ecological Effects of Oil Pollution on Littoral Communities*, ed. E.B. Cowell, pp. 78-87. Barking, U.K.: Applied Science.

Pauley, G.B. 1967. A butter clam (*Saxidomus giganteus*) with a polypoid-tumor on the foot. *J. Invert. Pathol.* 9:577-79.

Pauley, G.B. 1969. A critical review of neoplasia and tumor-like lesions in mollusks. In *Neoplasms and Related Disorders in Invertebrates and lower Vertebrate Animals. Nat. Cancer Inst. Monogr.* 31:509-39.

Pauley, G.B., and T.C. Cheng. 1968. A tumor on the siphons of a soft-clam (*Mya arenaria*). *J. Invert. Pathol.* 11:504-6.

Pauley, G.B., and C.S. Sayce. 1967. Descriptions of some abnormal oysters (*Crassostrea gigas*) from Willapa Bay, Washington. *Northwest Sci.* 41:155-9.

Pauley, G.B., and C.S. Sayce. 1972. An invasive epithelial neoplasm in a Pacific oyster, *Crassostrea gigas*. *J. Nat. Cancer Inst.* 49:897-902.

Pauley, G.B., A.K. Sparks, and C.S. Sayce. 1968. An unusual internal growth associated with multiple watery cysts in a Pacific oyster (*Crassostrea gigas*). *J. Invert. Pathol.* 11:398-405.

Pelseneer, P. 1920. Les variations et leur hérédité chez les mollusques. *Mém. Acad. Belg. Cl. Sci.* (Ser.2) 5:1-826.

Pelsenner, P. 1923. Variations dans les mollusques. *Ann. Soc. Zool. Belg.* 54:68-78.

Peters, E.C. 1988. Recent investigations on the disseminated sarcomas of marine bivalve molluscs. In *Disease Processes in Marine Bivalve Molluscs*, ed. W.S. Fisher, pp. 74-92. Bethesda, Maryland: American Fisheries Society, Special Publication 18.

Poder, M., and M. Auffret. 1986. Sarcomatous lesion in the cockle *Cerastoderma edule*. I. Morphology and population survey in Brittany, France. *Aquaculture* 58:1-8.

Potter, M., and E. Kuff. 1967. A developmental anomaly of the siphon of the soft-shell clam, *Mya arenaria*, from Chesapeake Bay. In *Activities Report of the Registry of Tumors in Lower Animals for the Period September 1, 1966 to March 31, 1967*, ed. J.C. Harshbarger, p. 11. Washington, D.C.: Smithsonian Institution.

Potts, W.T.W. 1967. Excretion in the mollusks. *Biol. Rev.* 41:1-41.

Pruell, R.J., J.J. Lake, W.R. Davis, and J.G. Quinn. 1986. Uptake and depuration of organic contaminants by blue mussels (*Mytilus edulis*) exposed to environmentally contaminated sediment. *Mar. Biol.* 91:497-515.

Rawls, C.K. 1965. Field test of herbicide toxicity to certain estuarine animals. *Chesapeake Sci.* 6:150-61.

Raymont, J.E.G. 1972. Some aspects of pollution in Southampton water. *Proc. Roy. Soc. London*, Ser. B, 180:451-68.

Reinisch, C.L., A.M. Charles, and J. Troutner. 1983. Unique antigens on neoplastic cells of the soft shell clam *Mya arenaria*. *Dev. Comp. Immunol.* 7:33-39.

Reinisch, C.L., A.M. Charles, and A.M. Stone. 1984. Epizootic neoplasia in soft shell clams collected from New Bedford Harbor. *Hazard. Waste* 1:73-81.

Richards, C.S. 1970. Pearl formation by *Biomphalaria globrata*. *J. Invert. Pathol.* 15:459-60.

Richards, C.S. 1972. *Biomphalaria glabrata* genetics: pearl formation. *J. Invert. Pathol.* 20:37-40.

Romeril, M.G. 1971. *Laboratory Note* No. *RD/L/N31/71*. Leatherhead, England: Central Electricity Research Laboratories.

Rose, N.R., and Z. Trnka (eds.). 1988. *Cellular Oncogene Activation*. New York: Marcel Dekker.

Ryder, J.A. 1887. On a tumor in the oyster. *Proc. Acad. Nat. Sci. Philadelphia* 39:25-27.

Scott, D., and C.W. Major. 1972. The effect of copper (II) on survival, respiration, and heart rate in the common blue mussel, *Mytilus edulis*. *Biol. Bull.* 143:679-88.

Seiler, G.R., and M.P. Morse. 1988. Kidney and hemocytes of *Mya arenaria* (Bivalvia): normal and pollution-related ultrastructural morphologies. *J. Invert. Pathol.* 52:201-14.

Shaw, W.N., and G.T. Griffith. 1967. Effects of polystream and drillex on oyster setting in Chesapeake Bay and Chincoteaque Bay. *Proc. Nat. Shellfish. Assoc.* 57:17-23.

Shuster, C.N., Jr. 1966. A uniquely shaped quahog. *Maritimes* 10:14.

Shuster, C.N., Jr., and F.C. Garb. 1967. A note on the histopathological condition of oysters exposed to lead. Abstr. 9th Shellfish Pathology Conference, January 27-28, 1967.

Sindermann, C.J. 1990. *Principal Diseases of Maine Fish and Shellfish*, vol.2, 2nd ed. San Diego: Academic Press.

Smith, G.M. 1934. A mesenchymal tumor in an oyster (*Ostrea cieginica*). *Am. J. Cancer* 22:838-41.

Smolowitz, R.M., D. Miosky, and C.L. Reinisch. 1989. Ontogeny of leukemic cells of the soft shell clam. *J. Invert. Pathol.* 53:41-51.

Sparks, A.K. 1985. *Synopsis of Invertebrate Pathology Exclusive of Insects*. Amsterdam: Elsevier.

Sparks, A.K., G.B. Pauley, R.R. Bates, and C.S. Sayce. 1964a. A mesenchymal tumor in a Pacific oyster, *Crassostrea gigas* (Thurberg). *J. Insect Pathol.* 6:448-52.

Sparks, A.K., G.B. Pauley, R.R. Bates, and C.S. Sayce. 1964b. A tumor-like fecal impaction in a Pacific oyster, (*Crassostrea gigas*). *J. Insect Pathol.* 6:453-56.

Sparks, A.K., G.B. Pauley, and K.K. Chew. 1969. A second mesenchymal tumor from a Pacific oyster (*Crassostrea gigas*). *Proc. Nat. Shellfish. Assoc.* 59:35-39.

Sunila, I. 1988. Acute histological responses of the gill of the mussel, *Mytilus edulis*, to exposure by environmental pollutants. *J. Invert. Pathol.* 52:137-41.

Suresh, K., and A. Mohandas. 1990a. Effect of sublethal concentrations of copper on hemocyte number in bivalves. *J. Invert. Pathol.* 55:325-31.

Suresh, K., and A. Mohandas. 1990b. Number and types of hemocytes in *Sunetta scripta* and *Villorita cyprinoides* var. *eochinensis* (Bivalvia), and leukocytosis subsequent to bacterial challenge. *J. Invert. Pathol.* 55:312-18.

Taylor, R.L., and A.C. Smith. 1966. Polypoid and papillary lesions in the foot of the gaper clam, *Tresus nuttali*. *J. Invert. Pathol.* 8:264-66.

Thurberg, F.P., A. Calabrese, and M.A. Dawson. 1974. Effects of silver on oxygen consumption of bivalves at various salinities. In *Pollution and Physiology of Marine Organisms*, ed. F.J. Vernberg and W.B. Vernberg, pp. 67-78. New York: Academic Press.

Tubiash, H.S., C.N. Shuster, Jr., and J.A. Couch. 1968. Anomalous siphons in two species of bivalve mollusks. *Nautilus* 81:120-25.

Twomey, E., and M.F. Mulcahy. 1984. A proliferative disorder of possible hemic origin in the common cockle, *Cerastoderma edule*. *J. Invert. Pathol.* 44:109-11

Twomey, E., and M.F. Mulcahy. 1988. Transmission of a sarcoma in the cockle *Cerastoderma edule* (Bivalvia:Mollusca) using cell transplants. *Dev. Comp. Immunol.* 12:195-200.

Vernberg, F.J., and W.B. Vernberg (eds.). 1974. *Pollution and Physiology of Marine Organisms*. New York: Academic Press.

Vernberg, F.J., A. Calabrese, F.P. Thurberg, and W.B.Vernberg (eds.). 1977. *Physiological Responses of Marine Biota to Pollutants*. New York: Academic Press.

Voogt, P.A. 1972. Lipid and sterol components and metabolism in Mollusca. In *Chemical Zoology Vol. VII. Mollusca*, ed. M. Florkin and B.T. Scheer, pp. 245-300. New York: Academic Press.

Voogt, P.A. 1983. Lipids: their distribution and metabolism. In *The Mollusca Vol. I. Metabolic Biochemistry and Molecular Biomechanics*, ed. K.M. Wilbur and P.W. Hochachka, pp. 329-70. New York: Academic Press.

Waugh, G.D. 1964. Observations on the effects of chlorine on the larvae of oysters (*Ostrea edulis* L.) and barnacles (*Elminius modestus* Darwin). *Ann. Appl. Biol.* 54:423-40.

Waugh, G.D., and A. Ansell. 1956. The effect on oyster spatfall of controlling barnacle settlement with DDT. *Ann. Appl. Biol.* 44:619-25.

Waugh, G.D., F.B. Hawes, and F. Williams. 1952. Insecticides for preventing barnacle settlement. *Ann. Appl. Biol.* 39:407-15.

Weinberg, R.A. 1989. *Oncogenes and the Molecular Origins of Cancer*. New York: Cold Spring Harbor Laboratory.

Wolf, P.H. 1969. Diseases and parasites in Australian commercial shellfish. *Haliotis* 8:75-83.

Wolf, P.H. 1971. Unusually large tumor in a Sydney rock oyster. *J. Nat. Cancer Inst.* 46:1079-84.

Yevich, P.P., and M.M. Barry. 1969. Ovarian tumors in the quahog *Mercenaria mercenaria*. *J. Invert. Pathol.* 14:266-7.

Yevich, P.P., and C.A. Barszcz. 1976. Gonadal and hematopoietic neoplasms in *Mya arenaria*. *Mar. Fish. Rev.* 38:42-3.

Yevich, P.P., and C.A. Barszcz. 1977. Neoplasia in soft-shell clams (*Mya arenaria*) collected from oil-impacted areas. *Ann. N.Y. Acad. Sci.* 298:409-26.

12

Infectious Diseases of Marine Crustaceans

James E. Stewart

INTRODUCTION

In examining disease among crustaceans, there is merit in looking briefly at the animal group they represent. According to Barnes (1968), the phylum Arthropoda consists of about 800,000 species or 80% of the known animal species. Some 750,000 of these belong to the Insecta, while another 26,000 species form the class Crustacea. The Crustacea are mainly marine, and, as Brock and Lightner (1990) point out, many occupy basic positions in aquatic food chains or are the subject of important commercial fisheries. Currently a few (mainly shrimp and prawns) are being cultured extensively for human food.

Arthropoda comprise the largest proportion of the world's animals; this phylum, as would be expected, has significant impact on all ecosystems. The effects on man alone, directly, and indirectly range from major threats to such mainstays as agriculture, forestry, and human health to its benefits as aids to mineralization of organic products, pollination, and the direct production of food. Oddly, despite size and importance of the group, knowledge recorded for it is not as large or fundamental as one would expect. There has been a heavy emphasis on taxonomy, anatomical descriptions, life cycles, and behavior; less emphasis has been placed on physiology, biochemistry, nutritional requirements, and optimal environmental requirements. This is especially true for Crustacea.

Much of the difficulty in acquiring this knowledge presumably stems from the highly diverse forms of the individual species, their complicated life cycles, and the enormous range of their habitats. Nevertheless, even such central themes as the comparative biochemistry of well-known arthropods or the physiology of the blood for the more common species have not been subjected to the highly organized and comprehensive treatment given to similar topics among the mammals. Even the biochemistry of the bacterium, *Escherichia coli*, is far better known than that of any arthropod.

0-8493-8662-4/93/$0.00 + $.50

Not surprisingly, this general observation also applies in the field of arthropod diseases, and the same lack of fundamental information seriously complicates the study of diseases. Except for some academic studies, the driving forces for investigations of arthropod diseases relate directly to the utility of the species. If the animal is useful to man (e.g., bees and certain large crustaceans), the disease is usually studied to aid in preserving the animal; when the arthropod destroys items of value to man or acts as a vector for human disease (e.g., spruce budworm, gypsy moth, or mosquitoes), it is termed a pest and its diseases are studied with increasing success, largely for controlling the pest.

Obviously, different objectives result in approaches which have elements in common, but foster markedly different investigations. Those interested in elimination of pests view the animals' barriers to transmission and intrinsic or induced resistance factors as obstacles and gauge effectiveness in terms of population control or eradication. Those seeking to preserve a species for its utility must examine and understand these same factors, explore the means whereby such factors can be enhanced, and understand non-disease elements important to the animals' well-being, such as environmental and nutrient requirements.

Although not comparable in volume to mammalian disease studies, a number of studies have been carried out on arthropod diseases, mostly on terrestrial insects, with microbial control as the objective. Because a holistic approach is necessary to study diseases of Crustacea and the animals of most interest are large-bodied, long-lived, and often abundant in culture, it is possible that studies of crustacea will serve as excellent models to gain insights into disease among all arthropods. If diseases are recognized more fully as models, the field will not only be advanced, but the vision of people, such as Steinhaus and his contemporaries who pioneered efforts to draw terrestrial and aquatic invertebrate pathology together, will be vindicated and fulfilled.

The appropriateness of using crustacean diseases as models can be gauged by considering the wide range of agents identified in or isolated from crustaceans or that have been shown to cause infections or debilitating infestations. These include the following: viruses, bacteria and rickettsia, fungi, protozoans, helminths, and parasitic crustaceans. These diseases have been reviewed comprehensively and in depth by Sindermann (1990), whose second edition of "Principal Diseases of Marine Fish and Shellfish" has appeared recently; by Brock and Lightner (1990) in their just completed major review, "Diseases Caused by Microorganisms," and the chapter entitled "Diseases of Crustacea" in Kinne's long awaited Volume III of "Diseases of Marine Animals." Since crustacean diseases have been reviewed recently, the infectious diseases will be treated here in summary form to describe various disease forms and defense mechanisms and illustrate research needs and possible directions.

VIRUSES

The first viral disease of any marine invertebrate was reported by Vago (1966) for the shore crab, *Macropipus depurator*, on the French Mediterranean coast. Since then the list has increased to more than 30. Briefly, viruses have been seen in or have been associated with disease in penaeid shrimp, the blue crab, *Callinectes sapidus*, the shore crab, *Macropipus*, and members of the crab genera *Carcinus* and *Paralithodes* plus the isopod *Portunion* (Sindermann 1990; Brock and Lightner 1990; Johnson 1984a; Couch 1981). Table 12-1, drawn from Brock and Lightner (1990), Sindermann (1990) and Sindermann and Lightner (1988), lists the viral families implicated in infections of or found to be present in marine crustaceans. As Brock and Lightner warn, these classifications are only tentative since, with one exception (*Baculovirus penaei* Couch), the necessary fundamental data are lacking. These same authors have issued a further warning concerning perceived deficiencies, "It is clear, however, that marine crustacean virus taxonomy deserves an intensive research effort."

Occurrence of viruses in marine crustaceans does not mean that a disease will necessarily result; many are latent for large parts of the crustacean life cycle and some have never been associated with any pathological condition. Most viruses can infect more than one host species; shrimp culture has been most affected by viral disease notably through infection with *Baculovirus penaei*. Infections occurring frequently in the larval stages are enhanced by stress induced by overcrowding, abnormal temperatures, and pollutants, among others. Detection occurs through observation of various gross signs and increased rates of mortality followed by examination either by light microscopy (observation of inclusion bodies and distinctive pathological changes often involving hemocytes and hepatopancreatic cells) or electron microscopy (viral arrays and associated ultrastructures). Crustacean tissue culture cell lines and serological techniques for diagnosis and unequivocal rapid identification are not available.

BACTERIA AND RICKETTSIA

Although there are many reports of bacteria in association internally and externally with marine crustaceans, associations have proven to be pathogenic in only a few instances. A number of crustacean infections occur from facultative pathogens as a result of damage to the integument or impairment of resistance produced by adverse conditions. In addition, there is an intestinal microflora, made up largely of Gram-negative bacteria, which as long as it remains *in situ* should not be considered pathogenic.

Table 12-1. Viral Families or Groupings Recorded in Crustaceans

DNA	RNA
Baculoviridae	Picorna-like
Herpes-like	Reo-like
Parvo-like	Rhabdo-like
	Bunya-like
	Birna-like (similar to Reoviruses)

Opinion differs concerning sterility of the hemolymph of healthy crustaceans. Some authors, e.g., Bang (1970) and Johnson (1983), argue that the hemolymph is normally sterile; others, e.g., Colwell *et al.* (1975), and Sizemore *et al.* (1975) report that potential pathogens (Gram-negative bacteria) have been isolated from healthy, recently captured blue crabs, and thus the hemolymph is not normally sterile. An explanation for these apparently contradictory perceptions can perhaps be drawn from the work of Cornick and Stewart (1966) who found that as many as 20% of lobsters, newly captured, had bacteria in their hemolymph: *Micrococci*, pseudomonads, *Brevibacterium*, and *Achromobacter*.

They were non-pathogenic and disappeared over the course of the first few days of captivity at temperatures between 5 and 15°C. Presumably they had entered from the environment through punctures caused during capture and subsequent handling. This initial transient infection was observed repeatedly and should not be considered a disease; lobsters held in captivity under good conditions and routinely examined had sterile hemolymph.

The bacteria observed most frequently in terms of crustacean disease are members of the genus *Vibrio*, *Aerococcus viridans* (var) *homari*, *Leucothrix mucor*, a chlamydia-like organism and several rickettsia-like organisms. They have been shown to affect lobsters (including juveniles), crabs, and shrimp generally in culture or capture situations; little information is available to gauge the impact of disease among populations of crustaceans in the wild.

Vibriosis

Gram-negative bacterial septicemic disease or vibriosis has been observed in captured marine crustaceans or cultured populations exposed to stress where it can result in severe mortalities. Systemic infections or septicemias have been reviewed (Sindermann and Lightner 1988; Sindermann 1990; Brock and Lightner 1990). They point out that although an occasional pseudomonad is involved, infections are usually caused by *Vibrio parahaemolyticus*, *V. alginolyticus*, or the odd *Aeromonas* sp.; these septicemias have been particularly severe among cultured shrimp. Brock and Lightner (1990) state that although there are many reports of vibriosis among marine decapod crustaceans, none has been studied in sufficient detail to be well understood. The general view is that Gram-negative organisms (often not identified) predominate in the marine environment are facultative pathogens, part of ubiquitous indigenous microflora which gives rise to vibriosis when a portal of entry (a wound) is effected in a compromised (stressed) host, and the defense mechanisms cannot cope with the invasion by eliminating bacteria or limiting their growth. Diagnosis relies on history, gross and clinical pathology, and microbiological findings (presence of Gram-negative bacteria in substantial numbers) and should be suspected when captive animals exposed to stress suffer high mortalities. Vibriosis has been observed among juvenile and adult lobsters, crabs (*Carcinus, Callinectes* and *Cancer*), and penaeid shrimp at all stages.

Gaffkemia

Unlike the generalized infection "Gram-negative septicemia," gaffkemia is a disease specific to lobsters and is caused by a single, free-living, bacterial species, the Gram-positive, tetrad forming, coccus *Aerococcus viridans* (var.) *homari*, which has proved troublesome only to lobsters of the genus *Homarus* on both sides of the Atlantic. The bacterium was first isolated by Snieszko and Taylor (1947) from lobsters during a disease episode in a holding unit in Maine; their trials to fulfill Koch's postulates proved that the agent was a true pathogen. Hitchner and Snieszko (1947) suggested the name *Gaffkya homari* which persisted until the current name was published in Bergey's Manual (Buchanan and Gibbons 1974). The original name of the causative agent was the basis for the name, gaffkemia, coined by Roskam (1957). Gaffkemia is the most intensively studied of all crustacean diseases and has been reviewed comprehensively (Sindermann 1970, 1971, 1977, 1990; Sindermann and Lightner 1988; Stewart and Rabin 1970; Stewart 1975, 1980, 1984; and Fisher *et al.* 1978).

The bacterium does not possess invasive properties. It gains access only through wounds and punctures in the lobster's integument, permitting access to hemolymph spaces. Lobsters, although cannibalistic, do not acquire infection through eating moribund or dead infected lobsters; the acidity of the lobster's gastric fluid kills the ingested pathogen in a short period (Stewart *et al.* 1969b). The pathogen conducts activities within the lobster, and its success is explained by its unique capacity to resist or survive the lobster's internal defenses (opsonins, agglutinins, bactericidins, phenoloxidase, and phagocytic capacity) and flourish at expense of the lobster reserves. Only small numbers of a virulent pathogenic strain (no more than 10 bacteria/kg lobster body weight) are necessary to cause an infection; a fatal infection almost invariably results. The times to death are essentially independent of the size of the inoculum causing the infection. The disease is strictly temperature dependent; at 3°C the average time between infection and death was about 180 days ranging downward in time at different intermediate temperatures to 2 days at 20°C. Among lobsters infected with the pathogen and held for 250 days at 1°C, no deaths occurred from gaffkemia. The pathogen, however, was retained in a virulent state; all lobsters given an infective dose 250 days earlier died with a full-blown infection in periods proportionate to deliberate increases in temperature. Thus the pathogen was able to survive in lobsters at 1°C with its virulence intact for a period approximately 2½ times as long as this temperature would be experienced even in the northern part of the lobsters' range -- an important epidemiological factor.

In addition to its capacity for survival in the lobster, *A. viridans* (var.) *homari* is a free living agent which can survive for months in muds in lobster habitats and ponds as well as in slime and debris on the interior surfaces of tanks and plumbing, thereby ensuring its availability for future infections and epidemics. On occasion it is the cause of severe economic losses in the lobster industry. It can be detected and identified readily by standard microbiological procedures of culture and isolation and by hemolymph smears (Gram stains or serological techniques) when infection levels are sufficiently high.

Infections with Chlamydia-like and Rickettsia-like Organisms

Sparks *et al.* (1985) observed an infection among *Cancer magister* (Dungeness crab) populations in Puget Sound, Washington, which, based largely upon morphological criteria, was attributed to a Chlamydia-like organism. A high incidence in the wild and heavy mortalities in the crab pots and holding facilities were observed. A heavy systemic infection occurred in diseased crabs that also exhibited extreme lethargy. Sparks *et al.* (1985) suggested that the organisms may be important in determining the population levels of Dungeness crab.

The first rickettsia-like organism in marine crustaceans was reported by Bonami and Pappalardo (1980) for *Carcinus mediterraneus*. These authors established infections by injecting healthy crabs with hepatopancreatic tissue from infected crabs. Death followed in about two weeks. Infections with rickettsia-like organisms also have been seen in *Paralithodes platypus* (Johnson 1984b), *Penaeus marginata*, *P. merguiensis* (Brock 1988), and *P. monodon* (Anderson *et al.* 1987). Distribution of rickettsia-like organisms, their identity, and their pathogenicity are unknown. As Brock and Lightner (1990) point out, rickettsia-like disease signs are difficult to distinguish since in all cases other possibly pathogenic agents were also present. Morphology of the agents in tissue is used to detect both chlamydia-like and rickettsia-like organisms.

Epibiontic Growth

Various external growths, e.g., mussels, barnacles, and seaweeds, will occur from time to time, depending on circumstance and location. Except for harm to embryonic forms, the infestations are usually of no consequence and are lost upon ecdysis. Colonization of crustacean surfaces by filamentous microepibionts, however, can be a serious problem, probably by interfering with gas exchange across gill and egg membranes. These infestations can occur during any part of the crustacean life cycle as a product of poor water quality. The most important fouling organism is the filamentous bacterium, *Leucothrix mucor*, often observed on marine crustaceans in both nature and culture. Its potential for occurrence is high since it is ubiquitous and has a high capacity for growth in well-aerated waters on simple organic compounds over a wide temperature range; other unidentified epibionts can occur simultaneously. Conditions extant in many recirculating culture systems have all these conditions and thus are particularly suitable to support microepibionts. Additionally, the infestation may contribute to and provide the basis for fungal infestations. This wide infestation of all crustaceans has been reviewed in detail by Fisher and co-workers (1978), Sindermann and Lightner (1988), and Brock and Lightner (1990).

FUNGAL INFECTIONS

On the basis of a tabulation of citations, Unestam (1973) concluded that fungi outnumber bacteria as pathogens of invertebrates. Johnson (1968, 1970), however, asserted that most records lacked definitive evidence to prove that fungal associations were pathogenic in nature. A large number of reports have been limited to registering presence of fungi and did not provide identification, quantitative data, or effects on the host. The most extensive reports of fungal infestations were studies of cultured marine crustaceans reviewed comprehensively by Sindermann and Lightner (1988), Sindermann (1990) and Brock and Lightner (1990). These authors cite Phycomycetes as the predominant fungal pathogens, followed by Ascomycete (*Trichomaris invadens*) and representatives of the Deuteromycetes (*Fusarium* sp.) that have caused significant losses among cultured species and some losses for wild populations. Detection of fungal infections relies on observation of gross pathology, histological examination, and standard mycological isolation and identification procedures.

Fungal infections occur frequently in marine crustaceans, usually resulting from invasions of injured or stressed hosts; eggs and larvae seem to be highly vulnerable. The literature on fungal diseases of economically important marine crustacea is relatively sparse and does not contain any reports of a virulent primary pathogen comparable to the bacterial infection, gaffkemia, or the European crayfish plague caused by the fungus, *Aphanomyces astaci*.

A listing of the main diseases would include the larval mycosis of *Penaeus aztecus* and *P. setiferius* caused by *Lagenidium* sp., especially *L. callinectes*, and *P. duorarum*, and *P. californiensis*, through infection of the juvenile and adult shrimp by *Fusarium solani*; both have caused mortalities. The tanner crab, *Chionoecetes bairdi*, is subject to a condition described as Black Mat Disease caused by *Trichomaris invadens*. Sparks (1982) showed that the infection was not restricted to the external surface, as previously believed, but was invasive, penetrating the thick cuticle and proliferating in underlying epidermal, subepidermal, and deeper tissues. It may have been a lethal infection since heavily infected crabs were prevented from molting.

Callinectes sapidus and other crabs have been affected by sponge disease caused by *Lagenidium callinectes*. This disease causes destruction of up to 25% of the egg masses (sponge) and on occasion larvae (Sindermann 1990; Sindermann and Lightner 1988). In addition, *L. callinectes* infections have been reported for *Cancer magister* larvae (Armstrong *et al*. 1976) and on the eggs of the estuarine shrimp, *Palaemon macrodactylus* (Fisher 1983).

Among lobsters, the first fungal disease was reported by Herrick (1909) for hatchery-reared lobster larvae. The fungus spread from the point of infection, destroying all internal tissues and leaving the body as a chitinous shell packed with mycelia. Lightner and Fontaine (1975) observed 35% mortalities in a 12-month period among immature lobsters (reared in a closed aquarium system), which they attributed to infection with a *Fusarium* sp. The fungus formed a "black spot" on the exoskeleton; sections from the exoskeleton showed hyphae in the cuticle and subcuticular tissues. The animals fed well and behaved normally, but did not survive ecdysis. These authors considered the disease to be very similar to the *Fusarium* infection of the Kuruma prawn, *Penaeus japonicus* named "Black Gill Disease" (Egusa and Ueda 1972).

Infestations by Phycomycetes were reported by Fisher *et al.* (1978) and Nilson *et al.* (1976) for lobsters held in a semiclosed recirculating system maintained at 20°C. The first was apparently caused by *Haliphthoros milfordensis* in postlarval juvenile forms of both European and American lobsters (44% mortalities) by invading under the carapace and through the gill area or at soft flexible joints of the appendages. Destruction of internal tissues and impairment of molting were considered the cause of death. Since the infection seemed to affect juveniles that had lighter integuments, Fisher *et al*. (1978) suggested minor wounds or weak fungal chitinolytic capacity as effecting a portal of entry; it was felt that older larger lobsters with heavier integuments could ward off the infection. The second infection among American lobster larvae was caused by a *Lagenidium* sp. that replaced the internal tissues, causing mortalities of 90% within 72 h of infection. Again the thickness of the integument was believed to be the critical factor confining the infection to the thinly covered egg and larval stages. Sordi (1958) cited in Sindermann (1970) and Unestam (1973) recorded infections in lobster gills by two deuteromycetes, *Didymaria palinuri* and *Ramularia branchiales*, in an aquarium in Italy.

Since many fungal infections occurred in recirculating seawater systems at high and relatively constant temperatures, it is worth noting the characteristics of such systems. They are usually rich in dissolved nutrients, highly oxygenated conditions and operate at temperatures high enough to promote rapid growth and development. These conditions favor growth of the euryhaline fungi and bacterial epibionts and place stress on lobsters or other crustaceans. Thus, in animals stressed by rapid growth, frequent molting, and consequential integumental variation, conditions are optimal for development of infections by opportunistic fungi and epibionts.

PROTOZOANS, HELMINTHS, AND PARASITIC CRUSTACEANS

Other unicellular forms that cause pathological signs and conditions among crustaceans are often grouped under the heading of protozoans, using the most liberal or Protistan criteria for inclusion of life forms in this phylum. Using this broad brush approach and recognizing that protozoans will exist anywhere there is sufficient moisture, it is not surprising that Sparks (1985) concluded that "protozoan parasites are the most common cause of disease in invertebrates."

Couch (1983), Sindermann and Lightner (1988) and Sindermann (1990), have provided reviews of the protozoan infestations of the most serious instances of parasitism. The Microspora lead the list as the most severe (occurring in shrimp and crabs where they are obligately parasitic and frequently cause extensive mortalities), followed by opportunistic but free-living forms, such as the ciliate, *Paranophrys* (shrimp, crabs, and lobster), and the amoebae, *Paramoeba perniciosa* (crabs), that invade injured animals and multiply in the host tissues. Internal parasites, such as the gregarines, are usually benign and cause problems only when conditions favor their multiplication to excessive numbers.

Dinoflagellates have been recorded as pathogens. Newman and Johnson (1976) described *Hematodinium perezi* as a severe pathogen of blue crabs. Other crabs also have been recorded as vulnerable. Meyers *et al*. (1987) and Meyers *et al*. (1990) showed that the Tanner crabs, *Chionoecetes opilio* and *C. bairdi*, are infected by a *Hematodinium*-like dinoflagellate, that invades all host tissues, giving rise to the Bitter Crab Syndrome and to serious economic losses. Parasite prevalence of up to 35% among newly captured crabs was observed, and it was suggested that periods of up to 16 months are usual between infection and death. Meat recovered from affected crabs had a chalky texture and an astringent aftertaste, thereby making the product unacceptable as human food.

Crustaceans are host to a wide variety of helminth parasites (e.g., trematodes, cestodes, nematodes, acanthocephalans, parasitic turbellarians, nemerteans, and leeches). According to Sindermann (1990), their overall effects on shrimp, crabs, and lobsters are minimal except for exceptional massive invasions. Effects on crustaceans of the few that are pathogenic are usually listed as sublethal although serious, i.e., destruction of gonads, nutrient depletion, and structural abnormalities among other effects. Significant host mortalities are rare (Sindermann 1990).

COURSE OF A NON-SPECIFIC INFECTION

Shell Disease Syndrome

In early discussions, the Shell Disease Syndrome was described as a specific disease resulting from infection with either bacteria or fungi, depending largely on whether the author was North American or European. Rosen (1970), however, in a comprehensive review, defined the disease as a single syndrome manifesting itself by "progressive chitinolysis and necrosis of the exoskeleton of aquatic crustaceans"; chitinoclastic microorganisms have been associated with the condition in virtually all cases. Additional reviews have been provided (Sindermann 1970, 1971, 1977, 1989, 1990; Fisher *et al.* 1978, Stewart 1980, 1984; Sindermann and Lightner 1988; Brock and Lightner 1990).

The condition exists in all environments where crustaceans occur and in all crustaceans studied. It was first observed by Hess (1937) among lobsters held in captivity in Nova Scotia; it is usually rare in natural populations. The condition consists of erosion and pitting of the exoskeleton which increases with time and is often associated with blackened areas of the shell resulting from melanization. In Europe, where fungi are often considered the causative agent, it has been called "burn spot" or "burned spot" disease (in German, Brandfleckenkrankheit). It is easily transferred experimentally.

There are several possible explanations for apparently contradictory beliefs regarding causes of shell disease: 1) the syndrome is not the result of invasion by a specific microorganism, but may be caused by different microorganisms acting in concert or in succession differently on different host species; 2) the disease is not microbial in origin, but a result of opportunistic microorganisms exploiting an injury or impaired state of the host; 3) environmental conditions may be precipitating factors.

Despite many conflicting reports regarding its etiology, there are agreements on the general course of its development. The exterior layer of the crustacean shell is a very thin cuticle of proteolipid material (including a high content of polyphenols) called the epicuticle, followed in order by three chitinous layers, the exocuticle (calcified pigmented layer), the calcified endocuticle, and the bottom layer, the non-calcified endocuticle.

The epicuticle is generally impervious to biochemical attack and is an effective barrier to microbial invasion. It is, however, subject to mechanical abrasion; once the epicuticle is breached, the chitinous layers are open to attack by chitinoclastic organisms ubiquitous in the aquatic environment. Thus, the prime conditions are disruption of the epicuticle under circumstances that prevent its ready repair, followed by opportunistic invasion by microorganisms. Factors that can impair restoration of the epicuticle include continual mechanical damage, enzymatic attack, adverse temperatures, inadequate diet, and sustained high pollution levels of particular kinds.

Evidence for each of these exists for different crustaceans. Baird (1950) and Wilder and McLeese (1961) described one form of shell disease emanating from and around wounded areas in lobster claws produced by wooden pegs used to immobilize them. In controlled experiments, Malloy (1978) showed that removal or mechanical breaching of the epicuticle was a prerequisite for development of the lesions in the shells of live *Homarus americanus*. In their studies of crayfish plague caused by the fungus *Aphanomyces astaci*, Nyhlen and Unestam (1975) and Söderhäll (1978) showed that the epicuticle of the crayfish, *Astacus astacus*, was breached apparently partly by an extracellular protease produced by the fungal pathogen. Malloy (1978) showed that lesions developing on abraded portions of the shells of live lobsters after application of chitinoclastic bacteria (*Vibrio/Beneckea*) were markedly influenced by temperature and stage of ecdysis. In extensive exposures at temperatures held constant for the duration of the experiments, Malloy noted that more infections occurred among post-ecdysal lobsters than in those approaching ecdysis; further, more infections were established among lobsters held between 2 and 5°C than at higher temperatures. Lobsters at the lower temperatures were less active metabolically and thus less able to effect cuticular repair. Fisher *et al.* (1976) demonstrated that diet was an important factor in epicuticular repair and an important element in the onset of shell infections. In studies on the culture of juvenile American lobsters (4th to 12th stage), they observed that rapidly molting animals fed an apparently inadequate synthetic diet were virtually without an epicuticle in contrast to those fed brine shrimp. Mortalities among animals exposed to infection during extended synthetic dietary regimes reached 80%, and a chitinolytic, Gram-negative bacterium was recovered from the shell lesions of all affected lobsters. Young and Pearce (1975) and Gopalan and Young (1975) found high incidences of shell lesions in crabs (*Cancer irroratus*), lobsters (*Homarus americanus*), and shrimp (*Crangon septemspinosa*) taken from the New York Bight, an area of considerable pollution. Healthy lobsters exposed for up to six weeks in aerated seawater containing sewage also developed typical lesions and erosions, while controls maintained in clean seawater did not. Affected lobster gills were fouled with granular material and a dark brown coating; gill filaments were often eroded and underlying areas became necrotic. Other examples of similar pollution effects can be found in literature cited above.

Mortalities will occur with Shell Disease Syndrome, presumably as a result of the impairment of gas exchange or by infections resulting from the invasion of pathogens through the weakened portions of the shell. Eroded appendages have not been seen to regenerate, although superficial damage is sloughed on ecdysis without affecting the new shell.

COURSE OF INFECTIONS BY SPECIFIC PATHOGENS

In this review, it may appear that most crustacean infections result from opportunistic agents that exploited adverse situations. Although there is considerable merit in this view, resulting largely from the way the field developed, some infections are known to be highly specific. The fatal bacterial infection of lobsters, gaffkemia caused by *A. viridans* (var.) *homari*, and the crayfish plague (Krebspest) caused in the European crayfish, *Astacus* sp. by the fungus *Aphanomyces astaci*, are explored below to illustrate unique traits of pathogens and the means whereby they interact with the host's resistance factors.

Crustaceans in general are equipped with a battery of defenses that under most circumstances are highly effective in coping with infective agents. These consist of an external sheath or shell described above, where the intact epicuticle is proof against most biochemical assaults. Protection of the shell is bolstered by internal factors, hemolymph coagulation, bactericidins, agglutinins, opsonins, phagocytosis, encapsulation, and melanization (phenoloxidase activity). Some internal factors can be enhanced or induced by immunogens quantitatively and in some instances qualitatively (Stewart and Zwicker 1974).

Gaffkemia

The specific lobster pathogen, *A. viridans* (var.) *homari*, lacks external enzymes and hence has no invasive powers; it passes the lobster's only effective barrier to an active infection, an intact integument, through ruptures caused by crowded conditions or fighting. A single bacterial cell of a virulent strain is probably all that is required to initiate an infection since the bacterium is resistant to all of the internal defense mechanisms.

The course of establishing the infection appears to be consistent with the following sequence (Stewart 1980, 1984). Upon entry to the hemolymph, *A. viridans* (var.) *homari* encounters an inducible bactericidin (Acton *et al.* 1969) present in the hemolymph plasma in an inactive form (Stewart and Zwicker 1972). This bactericidin is adsorbed readily in the inactive or active state by bacteria, including the pathogen. In numbers proportional to the insult, hemocytes immediately congregate at the site of the shell wound or rupture in response to introduced foreign material. A number of hemocytes rupture releasing, in the vicinity, several factors. One is the enzyme that converts the plasma fibrinogen to fibrin to seal the wound with a hard non-retracting clot consisting of fibrin and ruptured hemocytes. Another is an activator for the bactericidin already adsorbed on the bacteria, thereby beginning the killing of susceptible agents; *A. viridans* (var.) *homari*, which adsorbes the bactericidin readily, is not affected by it. An additional hemocyte factor(s) released is the agglutinin/opsonin, which upon adsorption agglutinates most agents or promotes the phagocytosis of susceptible bacteria. Virulent strains of *A. viridans* (var.) *homari* are not agglutinated nor do they adsorb the agglutinins; avirulent strains, however, do adsorb the agglutinins (unpublished observation). The difference between the two is probably related to the capsule: virulent strains develop a large capsular layer, while the avirulent strains possess virtually no capsular layer (Stewart 1984). Phagocytosis of the pathogen occurs, but is ineffective since the pathogen is not killed and apparently grows in the hemocyte (Cornick and Stewart 1968). Following introduction to the hemolymph, pathogen numbers first decline there, concentrating and increasing in the hepatopancreas and heart, a phase which is followed by exponential growth in these tissues and in the hemolymph and skeletal muscle as shown in Fig. 12-1 (Stewart and Arie 1973). At 15°C, the time between infection and death is approximately 14 days, a period in which major physiological changes occur. Early in the infection the number of circulating hemocytes declines dramatically and by the eighth day have virtually disappeared. As shown in Fig. 12-2, this disappearance has a marked negative effect on hemolymph clotting times (Stewart *et al.* 1969a) since the hemocytes carry, among other factors, the enzyme that converts fibrinogen to fibrin to form the hemolymph clot; thus, lobsters in the late stages of the disease will hemorrhage when wounded. As the pathogen develops *in vivo* at the expense of the lobster's glucose and non-protein nitrogen (it cannot use complex carbohydrates, lipids, or proteins directly), the glycogen levels of the hepatopancreas, heart, and skeletal muscle decline. Since lobsters cease to feed

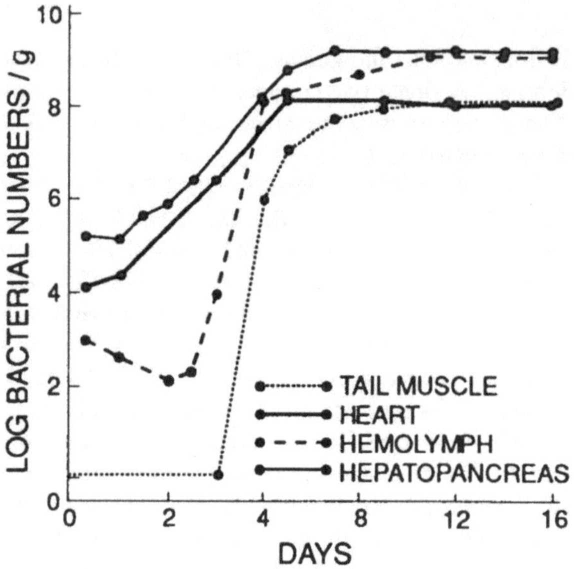

Fig. 12-1. Growth of *A. viridans* (var.) *homari* in tissues of live lobsters held at 15°C. (Redrawn from Stewart and Arie 1973).

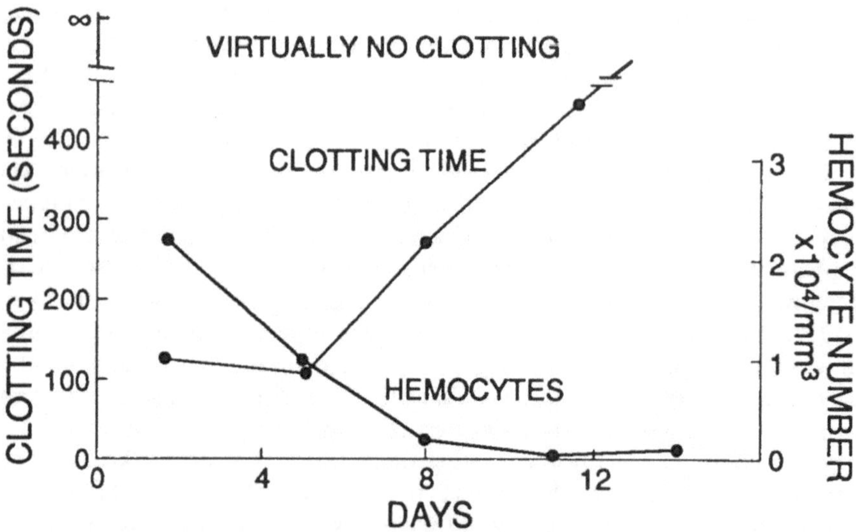

Fig. 12-2. Effect of gaffkemia on hemocyte levels and consequent hemolymph clotting times for lobsters held at 15°C. Values for control lobsters remained essentially constant throughout and equivalent to the initial values in this illustration. (Redrawn from Stewart *et al.* 1969a).

shortly after initiation of the infection, there is no opportunity to restore the glucose (drawn internally from glycogen) and non-protein nitrogen levels from exogenous sources. The pathogen grows as freely in the lobster as it would in a microbiological medium designed specifically to meet its particular needs. Since the lobster's internal defenses are ineffective against it, the rate of growth and abundance of the pathogen in the lobster should determine the time to death (rate of growth and abundance are a function of the available nutrients). This, in fact, occurred in starved lobsters, where the reserve materials were markedly depleted, exhibited smaller numbers of the pathogen in their hemolymph and lived, on average, 2 to 3 times longer following infection than did well fed lobsters that had higher levels of reserve materials (Stewart *et al.* 1972b). Since no toxin is involved (Stewart and Arie 1973), the infection is, in effect, an unequal competition between the lobster and the pathogen for the lobster's reserves.

This is borne out by the fact that in infected moribund lobsters the adenosine triphosphate (ATP) levels of the heart fell to 50% of their initial values and the ATP of the hepatopancreas disappeared entirely (Stewart and Arie 1973). These authors concluded on the basis of these facts that death resulted from a massive dysfunction of the hepatopancreas, the central metabolic organ responsible for absorption of nutrients, synthesis, and conversion.

Rittenburg *et al.* (1979) suggested alternatively that the actual cause of death might be a lack of intracellular oxygen resulting from impairment of oxygen-binding capacity of hemocyanin in infected lobsters. Their measurements, however, were based on the amount of oxygen associated with the hemocyanin rather than its oxygen-binding capacity. Rittenburg *et al.* (1979) showed that hemocyanin in terminal-phase, infected animals had approximately 50% of the oxygen levels associated with hemocyanin in control animals. Hemocyanin itself does not change during the infection (Stewart *et al.* 1969a) and can vary from a high concentration of 70 mg/ml of hemolymph to 8-15 mg/ml in normal (uninfected) lobsters (Stewart *et al.* 1967a,b; Stewart *et al.* 1972a) without apparent ill effects. As noted above, starved animals (i.e., reduced hemocyanin levels) were able to sustain infections with *A. viridans* (var.) *homari* for periods several times longer than fed animals possessing correspondingly higher concentrations of hemocyanin (Stewart *et al.* 1972b). In addition, crustaceans are noted for their capacity to withstand long exposures to low or negligible oxygen pressures (Wolvekamp and Waterman 1960); further, McMahon and Wilkens (1975) showed that the portion of total oxygen in lobsters supplied by hemocyanin varied from a low 25% at ambient external oxygen levels rising proportionately to a maximum 95% at extremely low external oxygen tensions. External oxygen levels in the infection experiments were maintained by aeration of water in which the lobsters were held at levels approaching saturation. Thus, it does not seem likely that a 50% reduction in hemocyanin-bound oxygen would cause death even if hemocyanin was the sole source of oxygen in the animals.

Crayfish Plague

Extensive studies of crayfish plague, a disease of freshwater crustaceans, have produced fundamental insights into a specific infection and the defense mechanisms applicable to all crustaceans. The disease, also called Krebspest, was first observed in Italy in 1860 (Unestam 1973, 1981), is caused by the fungus, *Aphanomyces astaci*, and has gradually eliminated the economically valuable *Astacus* sp. crayfish from most of Europe and parts of Asia. It is believed that the fungus was introduced to highly susceptible European crayfish populations through importation of exotic crayfish species carrying the fungus as normal microflora.

The pathogen, *A. astaci*, produces a swimming zoospore that settles on the cuticle of the crayfish, drops its flagella, forms a cell wall, and germinates. Upon germination, it penetrates the epicuticle (an effective barrier to other pathogens) through a combination of hydrolytic enzymes and mechanical forces (Nyhlen and Unestam 1975; Söderhäll and Unestam 1975, 1979; Söderhäll 1978). When large numbers of zoospores are available, crayfish defenses will be overwhelmed rapidly, and there will be few gross signs. When fewer zoospores are present, the slower infection development causes melanized areas to become apparent in exoskeletons of infected crayfish (reviewed by Alderman and Polglase 1988).

Unestam and Nyhlen (1974) provided an interesting summary of the progress of the fungus through the cuticle of *A. astacus,* where it is melanized, to different degrees at different places in the cuticle. The hyphae growing in the cuticle immediately under the epicuticle were melanized, hyphae in the area intermediate between the epicuticle, and the epidermis covering the soft internal tissues had no melanin associated with them; those hyphae in the area of the epidermis or underlying muscle and hemolymph spaces were quite heavily melanized. Death is caused by the fungal proliferation in the internal tissues of susceptible forms, eg., the *Astacus* sp. The same general series of events were observed for the American crayfish, *Pacifastacus leniusculus*, which is relatively resistant to the fungus. The major difference in *Pacifastacus* was that the hyphae penetrating the cuticle were much more heavily melanized; in fact, they were covered to such an extent that they were almost black. Presumably a higher level of phenoloxidase occurs in *Pacifastacus* sp. and is a factor in providing greater resistance to this fungus. If and when the fungus penetrates internal tissues, the same internal defenses itemized in the previous section come into play; in *Astacus* sp., however, they have not proven adequate to control the fungus, *A. astaci*. Phenoloxidase is regarded as a central feature of the system of defenses not only in crayfish, but in invertebrates generally and deserves to be described more fully.

PHENOLOXIDASE SYSTEM

Söderhäll and colleagues in Sweden, using crayfish as a model, have conducted definitive studies and published widely on the prophenoloxidase activating system and melanization. Much of their work has been summarized in various reviews over the past decade (Söderhäll 1982; Söderhäll and Smith 1986; Söderhäll *et al.* 1988; Johansson and Söderhäll 1989). Söderhäll and Smith (1986) stated: "From the numerous investigations of immunity in arthropods, it is clear that within the hemocoel the recognition of and the response to nonself entities by the hemocytes constitutes the most important step in overcoming systemic infection. However, until recently, very little was known about biochemical events that initiate, control, and regulate cellular activity in these animals. Evidence is now accumulating that the prophenoloxidase-activating (proPO) system -- the enzyme cascade responsible for converting prophenoloxidase into active phenoloxidase during the early stages of melanization -- plays a major role in mediating nonself recognition and host defense in the arthropods."

The prophenoloxidase activating system in crayfish has been formulated over the past decade by Söderhäll and associates (Fig. 12-3). Söderhäll and Smith (1986) have compared this system with the mammalian complement system and feel that the two are analogous with the obvious difference that the proPO system is contained within hemolymph-circulating cells and the complement system is found in mammalian plasma. Its retention in the hemocytes permits tight control of factors which could also harm the host. It offers the additional attraction of detection and recognition of nonself factors (glucans and LPS) from fungi and bacteria accompanied by a rigorously controlled response graded to the actual scale of the insult. It is argued that this comprehensive system within the hemocytes makes them the driving force in the crayfish cellular defenses and that this system or variations on this theme form the central feature of crustacean and arthropod internal defenses. It will be extremely interesting for invertebrate pathologists to see how this picture develops in the near future, what variations occur in the different species, the degree to which it can be manipulated to enhance resistance, and what the relation is to the extremely confusing crustacean agglutinin/lectin capacities. There is no doubt that this proPO system confers on the defenses of the crustaceans, a high degree of flexibility and capacity for proportionate responses.

MICROBIAL CUES + HEMOCYTES

(Beta 1,3 glucans, and/or bacterial lipopolysaccharides [LPS]) + (Granulocytes, Semigranular cells)

results in
RECOGNITION AS NON-SELF
triggering release of proPO enzyme cascade
to plasma and activation as follows:

1) INACTIVE SERINE PROTEASE —— Glucans / LPS ——> ACTIVE SERINE PROTEASE

2) PROPHENOLOXIDASE —— Active Serine Protease ——> PHENOLOXIDASE

1) Production of anntimicrobial quinones and melanin
2) Probable opsonin activity leading to phagocytosis and encapsulation by hyaline and semigranular cells respectively.

Fig. 12-3. Probable crayfish and other crustaceans' internal defense mechanism involving prophenoloxidase (proPO) /phenoloxidase system based largely on reports by Söderhäll and Söderhäll and colleagues (see bibliography).

CONCLUSIONS

The crustaceans are subject to a wide variety of infectious agents. Unless factors, such as adverse environmental conditions including pollution, dietary inadequacies, or extreme infection pressures are imposed, the animals are reasonably well-equipped to ward off infections. If the highly protective continuous chitinous sheath is breached, the extensive battery of internal defenses provides for a flexible and graded response effective against most possible pathogens; parts of this system (phagocytic capacity and bactericidal effects) can be induced to enhance protection. Although much more work needs to be done before internal defense mechanisms are completely understood, major advances have been made -- some of which, e.g., crayfish phenoloxidase system, can serve as models for increasing general understanding of arthropod resistance to disease.

To date many potentially infectious agents have been identified in or associated with crustaceans, though their presence has not always been proven definitively as the cause of specific diseases. More work is needed to ensure which of these are indeed pathogens, not simply adventitious and coincidental occurrences. Not only must the agent be clearly identified but, wherever possible, Koch's Postulates or some reasonable modification should be fulfilled before a disease state is described or a pathogen identified.

Most descriptions of disease have stemmed from culture operations or situations where crustaceans are held captive. Natural populations also suffer from mortalities of which some are caused by disease. In fact, natural population sizes are limited by a combination of food, predation, reproduction, age, and disease. The influence of disease on natural crustacean populations, as for most wild animal species, is essentially unknown, but it can have major impacts on population size and should be understood well enough to be taken into account.

Currently, many studies are impeded by shortcomings in methodology. Viral studies in particular have not progressed as rapidly or as far as desirable because of the lack of suitable tissue culture cell lines to use in diagnostic work or to produce sufficient viral particles for use in developing specific antisera for diagnostic purposes. Serological methods of all sorts would be extremely advantageous, not only for more rapid diagnosis for many agents, but also to simplify field studies needed to assess the presence and possible impacts of disease among natural populations.

Although the question is not treated extensively in this paper, there is no doubt that pollution can play a major role in the development of infections. Two outwardly quite different examples can be used to illustrate this point. The first is the widely described Shell Disease Syndrome discussed earlier in which generalized pollution is listed as a possible contributor to the condition. The second is the definitive experiments on *Baculovirus* infection of *Penaeus duoarum* resulting from exposure to specific agents, Mirex or Aroclor 1254, described by Couch (1974, 1976) and Couch and Courtney (1977). Despite apparent differences in pollutants and infections, both probably result from direct repression of hosts' resistance factors possibly coupled with a stimulation or favoring of opportunistic or latent biological agents. A general predisposition to disease would be expected to result from exposure to pollution; exact nature of patent infection, however, would be governed by the specific host, nature of the pollutant and its concentration, environmental circumstances, and the array of infectious agents (specific and opportunistic) available to take advantage of the situation. Only rarely or under rigidly prescribed conditions would a specific infection be expected to result from exposure to a particular pollutant. Thus, in studies of the influence of pollution on infections (their genesis and subsequent course), emphasis should be placed on the impact of pollutants and accompanying conditions on all aspects of crustacean physiology and should focus intensively on their interaction with the mechanisms crustaceans possess for resistance to disease.

Finally to extend the foregoing, since so many diseases are the consequences of an imbalance in the interactions among the host, pathogen, and the environment, it is essential that we acquire a detailed understanding of the various elements involved. To do this properly, an holistic approach is required in which the biochemistry, physiology, nutrient requirements, behavior, environmental needs, and internal resistance factors of crustacean hosts are studied and understood. This approach must also include an appreciation for how predisposing and precipitating factors, such as pollutants or adverse conditions, can create new diseases or contribute to the worsening of the pathology of traditional diseases.

ACKNOWLEDGEMENTS

I thank Drs. J.A. Brock and D.V. Lightner for their generosity in providing access to a draft of their review entitled "Diseases Caused by Microorganisms" in the chapter "Diseases of Crustacea" in Kinne's "Diseases of Marine Animals" and my wife, Heather, for typing the several drafts of this review.

REFERENCES

Acton, R.T., P.F. Weinheimer, and E.E. Evans. 1969. A bactericidal system in the lobster *Homarus americanus*. *J. Invertebr. Pathol.* 13:463-464.

Alderman, D.J., and J.L. Polglase. 1988. Pathogens, parasites and commensals. In *Freshwater Crayfish: Biology, Management and Exploitation*, ed. D.M. Holdich and R.S. Lowery, pp. 167-212. London: Croom Helm.

Anderson, I.G., M. Shariff, G. Nash and M. Nash. 1987. Mortalities of juvenile shrimp, *Penaeus monodon*, associated with *Penaeus monodon* baculovirus, cytoplasmic reo-like virus, and rickettsial and bacterial infections, from Malaysian brackishwater ponds. *Asian Fish. Sci.* 1:47-64.

Armstrong, D.A., D.V. Buchanan and R.S. Caldwell. 1976. A mycosis caused by *Lagenidium* sp. in laboratory-reared larvae of the Dungeness crab, *Cancer magister*, and possible chemical treatments. *J. Invertebr. Pathol.* 28:329-336.

Bang, F.B. 1970. Disease mechanisms in crustacean and marine arthropods. In *A Symposium on Diseases of Fishes and Shellfishes*, ed. S.F. Snieszko. *Am. Fish. Soc. Spec. Publ.* No. 5:383-404.

Baird, F.T. 1950. Lobster plugs and their effect on the meat of the lobster's claw. *Maine Dept. Sea Shore Fish. Res. Bull.* 2:2-12.

Barnes, R.D. 1968. *Invertebrate Zoology*, 2nd ed. Philadelphia: W.B. Saunders.

Bonami, J.R., and R.Pappalardo. 1980. Rickettsial infection in marine crustacea. *Experientia* 36:180-181.

Brock, J.A. 1988. Rickettsial infection of penaeid shrimp. In *Disease Diagnosis and Control in North American Marine Aquaculture*, 2nd, revised edition, ed. C.J. Sindermann and D.V. Lightner, pp. 38-41. Amsterdam: Elsevier.

Brock, J.A., and D.V. Lightner. 1990. Diseases caused by microorganisms. In *Diseases of Marine Animals. Vol. III Introduction, Cephalopoda, Annelida, Crustacea, Chaetognatha, Echinodermata, Urochordata*, ed. O. Kinne, pp. 245-326. Hamburg: Biologische Anstalt Helgoland.

Buchanan, R.E., and N.E. Gibbons (ed.). 1974. *Bergey's Manual of Determinative Bacteriology*, 7th edition. Baltimore: Williams and Wilkins.

Colwell, R.R., T.C. Wicks, and H.S. Tubiash. 1975. A comparative study of the bacterial flora of the hemolymph of *Callinectes sapidus*. *Mar. Fish. Rev.* 37: 29-33.

Cornick, J.W., and J.E. Stewart. 1966. Microorganisms isolated from the hemolymph of the lobster, *Homarus americanus*. *J. Fish. Res. Board Can.* 23:1451-1454.

Cornick, J.W., and J.E. Stewart. 1968. Interaction of the pathogen *Gaffkya homari* with natural defense mechanisms of *Homarus americanus*. *J. Fish. Res. Board Can.* 25:695-709.

Couch, J.A. 1974. An enzootic nuclear polyhedrosis virus of pink shrimp: Ultrastructure, prevalence, and enhancement. *J. Invertebr. Pathol.* 24:311-331.

Couch, J.A. 1976. Attempts to increase *Baculovirus* prevalence in shrimp by chemical exposure. *Prog. Exp. Tumor Res.* 20: 304-314.

Couch, J.A. 1981. Viral diseases of invertebrates other than insects. In *Pathogenesis of Invertebrate Microbial Diseases*, ed. E. W. Davidson, pp. 127-160. Totowa, NJ: Allanheld, Osmun.

Couch, J.A. 1983. Diseases caused by Protozoa. In *The Biology of Crustacea: Vol. 6 Pathobiology*, ed. A.J. Provenzano, pp. 79-111. New York: Academic Press.

Couch, J.A., and L. Courtney. 1977. Interaction of chemical pollutants and virus in a crustacean: A novel bioassay system. *Ann. N.Y. Acad. Sci.* 298:497-504.

Egusa, S., and T. Ueda. 1972. A *Fusarium* sp. associated with black gill disease of the Kuruma prawn, *Penaeus japonicus* Bate. *Bull. JPN. Soc. Sci. Fish.* 38:1253-1260.

Fisher, W.S. 1983. Eggs of *Palaemon macrodactylus*: III. Infection by the fungus, *Lagenidium callinectes*. *Biol. Bull.* 164:214-226.

Fisher, W.S., T.R. Rosemark, and E.H. Nilson. 1976. The susceptibility of cultured American lobsters to a chitinolytic bacterium. *Proc. World Maricult. Soc.* 7:511-520.

Fisher, W.S., E.H. Nilson, J.F. Steenbergen, and D.V. Lightner. 1978. Microbial diseases of cultured lobsters: A review. *Aquaculture* 14:115-140.

Gopalan, U.K., and J.S. Young. 1975. Incidence of shell disease in shrimp in the New York Bight. *Mar. Pollut. Bull.* 6:149-153.

Herrick, F.H. 1909. Natural history of the American lobster. *Bull. U.S. Bur. Fish.* 29:149-408.

Hess, E. 1937. A shell disease in lobsters (*Homarus americanus*) caused by chitinovorous bacteria. *J. Biol. Board Can.* 3:358-362.

Hitchner, E.R., and S.F. Snieszko. 1947. A study of a microorganism causing a bacterial disease of lobsters. *J. Bacteriol.* 54:48 (abstr.).

Johansson, M.W., and K. Söderhäll. 1989. Cellular immunity in crustaceans and the proPO system. *Parasitol. Today* 5:171-176.

Johnson, P.T. 1983. Diseases caused by viruses, rickettsiae, bacteria and fungi. In *The Biology of Crustacea: Vol. 6 Pathobiology*, ed. A.J. Provenzano, pp. 1-78. New York: Academic Press.

Johnson, P.T. 1984a. Viral diseases of marine invertebrates. *Helgol. Meeresunters.* 37:65-98.

Johnson, P.T. 1984b. A rickettsia of the blue king crab, *Paralithodes platypus*. *J. Invertebr. Pathol.* 44:112-113.

Johnson, T.W. 1968. Saprobic marine fungi. In *The Fungi: Vol. 3*, ed. G.C. Ainsworth and A.S. Sussman, pp. 95-104. New York: Academic Press.

Johnson, T.W. 1970. Fungi in marine crustaceans. In *A Symposium on Diseases of Fishes and Shellfishes*, ed. Stanislas F. Snieszko. *Fish. Soc. Spec. Publ.* 5:405-408.

Lightner, D.V., and C.T. Fontaine. 1975. A mycosis of the American lobster, *Homarus americanus* caused by *Fusarium* sp. *J. Invertebr. Pathol.* 25:239-245.

Malloy, S.C. 1978. Bacteria induced shell disease of lobsters (*Homarus americanus*). *J. Wildl. Dis.* 14:2-10.

McMahon, B.R., and J.L. Wilkens. 1975. Respiratory and circulatory responses to hypoxia in the lobster *Homarus americanus*. *J. Exp. Biol.* 62:637-655.

Meyers, T.R., T.M. Koeneman, C. Botelho, and S. Short. 1987. Bitter crab disease: a fatal dinoflagellate infection and marketing problem for Alaskan Tanner crabs *Chionoecetes bairdi*. *Dis. Aquat. Org.* 3:195-216.

Meyers, T.R., C. Botelho, T.M. Koeneman, S. Short, and K. Imamura. 1990. Distribution of bitter crab dinoflagellate syndrome in southeast Alaskan Tanner crabs *Chionoecetes bairdi*. *Dis. Aquat. Org.* 9:37-43.

Newman, M.W., and C.A. Johnson. 1976. A disease of blue crabs (*Callinectes sapidus*) caused by parasitic dinoflagellates, *Hematodinium* sp. *J. Parasitol.* 6:554-555.

Nilson, E.H., W.S. Fisher, and R.A. Shleser. 1976. A new mycosis of larval lobster (*Homarus americanus*). *J. Invertebr. Pathol.* 27:177-183.

Nyhlen, L., and T. Unestam. 1975. Ultrastructure of the penetration of the crayfish integument by the fungal parasite, *Aphanomyces astaci*, Oomycetes. *J. Invertebr. Pathol.* 26: 353-366.

Rittenburg, J.H., M.L. Gallagher, R.C. Bayer, and D.F. Leavitt. 1979. The effect of *Aerococcus viridans* (var.) *homari* on the oxygen binding capacity of hemocyanin in the American lobster (*Homarus americanus*). *Trans. Am. Fish. Soc.* 108: 172-177.

Rosen, B. 1970. Shell disease of aquatic crustaceans. In *A Symposium on Diseases of Fishes and Shellfishes*, ed. Stanislas F. Snieszko. *Am. Fish. Soc. Spec. Publ.* No. 5:409-415.

Roskam, R.T. 1957. Gaffkemia, a contagious disease in *Homarus vulgaris*. *Int. Counc. Explor. Sea, Shellfish Comm. Rep.* No. 1 pp. 1-4 (mimeogr.)

Sindermann, C.J. 1970. *Principal Diseases of Marine Fish and Shellfish*. New York: Academic Press.

Sindermann, C.J. 1971. Internal defenses of crustacea: A review. *Fish. Bull.* 69: 455-489.

Sindermann, C.J. 1977. *Disease Diagnosis and Control in North American Aquaculture*. Amsterdam: Elsevier.

Sindermann, C.J. 1989. The shell disease syndrome in marine crustaceans. U.S. Department of Commerce NOAA Tech. Memo. NMFS-F/NEC-64:1-43.

Sindermann, C.J. 1990. *Principal Diseases of Marine Fish and Shellfish*, 2nd edition. New York: Academic Press.

Sindermann, C.J., and D.V. Lightner. 1988. *Disease Diagnosis and Control in North American Marine Aquaculture*. Amsterdam: Elsevier.

Sizemore, R.K., R.R. Colwell, H.S. Tubiash, and T.E. Lovelace. 1975. Bacterial flora of the hemolymph of blue crab, *Callinectes sapidus*: numerical taxonomy. *Appl. Microbiol.* 29:393-399.

Snieszko, S.F., and C.C. Taylor. 1947. A bacterial disease of the lobster (*Homarus americanus*). *Science* 105:500.

Söderhäll, K. 1978. Interactions between a parasitic fungus, *Aphanomyces astaci*, Oomycetes, and its crayfish host. II. Studies on the fungal enzymes and on the activation of crayfish prophenoloxidase by fungal components. *Acta Univ. Ups.* 456:1-22.

Söderhäll, K. 1982. Prophenoloxidase activating system and melanization – A recognition mechanism of arthropods? A review. *Dev. Comp. Immunol.* 6:601-611.

Söderhäll, K., and T. Unestam. 1975. Properties of extracellular enzymes from *Aphanomyces astaci* and their relevance in the penetration process of crayfish cuticle. *Physiol. Plant.* 35:140-146.

Söderhäll, K., and T. Unestam. 1979. Activation of serum prophenoloxidase in arthropod immunity. The specificity of cell wall glucan activation by purified fungal glycoproteins of crayfish phenoloxidase. *Can. J. Microbiol.* 25: 406-414.

Söderhäll, K., and V.J. Smith. 1986. Prophenoloxidase - activating cascade as a recognition and defense system in arthropods. In *Hemocytic and Humoral Immunity in Arthropods*, ed. A.P. Gupta, pp. 251-285. New York: John Wiley and Sons.

Söderhäll, K., M.W. Johansson, and V.J. Smith. 1988. Internal defence mechanisms. In *Freshwater Crayfish: Biology, Management and Exploitation*, ed. D.M. Holdich and R.S. Lowery, pp. 213-235. London: Croom Helm.

Sordi, M. 1958. Microsi dei Crostaci decapodi marine. *Riv. Parassitol.* 19:131-137.

Sparks, A.K. 1982. Observations on the histopathology and probable progression of the disease caused by *Trichomaris invadens*, an invasive ascomycete, in the Tanner crab, *Chionoecetes bairdi*. *J. Invertebr. Pathol.* 40:242-254.

Sparks, A.K. 1985. *Synopsis of Invertebrate Pathology Exclusive of Insects*. Amsterdam: Elsevier.

Sparks, A.K., J.F. Morado, and J.W. Hawkes. 1985. A systematic microbial disease in the Dungeness crab, *Cancer magister*, caused by a Chlamydia-like organism. *J. Invertebr. Pathol.* 45:204-217.

Stewart, J.E. 1975. Gaffkemia, the fatal infection of lobsters (genus *Homarus*) caused by *Aerococcus viridans* (var.) *homari*: A review. *Mar. Fish. Rev.* 37:20-24.

Stewart, J.E. 1980. Diseases. In *The Biology and Management of Lobsters, Vol. 1*, ed. J.S. Cobb and B.F. Phillips, pp. 301-342. New York: Academic Press.

Stewart, J.E. 1984. Lobster diseases. *Helgol. Meeresunters.* 37:243-254.

Stewart, J.E., and H. Rabin. 1970. Gaffkemia, a bacterial disease of lobsters (genus *Homarus*). In *A Symposium on Diseases of Fishes and Shellfishes*, ed. S.F. Snieszko. *Am. Fish. Soc. Spec. Publ.* 5:431-439.

Stewart, J.E., and B.M. Zwicker. 1972. Natural and induced bactericidal activities in the hemolymph of the lobster: Products of hemocyte-plasma interaction. *Can. J. Microbiol.* 18:1499-1509.

Stewart, J.E., and B. Arie. 1973. Depletion of glycogen and adenosine triphosphate as major factors in the death of lobsters (*Homarus americanus*) infected with *Gaffkya homari*. *Can. J. Microbiol.* 19:1103-1110.

Stewart, J.E., and B.M. Zwicker. 1974. Comparison of various vaccines for inducing resistance in the lobster *Homarus americanus* to the bacterial infection, gaffkemia. *J. Fish. Res. Board Can.* 31:1887-1892.

Stewart, J.E., J.W. Cornick, and J.R. Dingle. 1967a. An electronic method for counting lobster (*Homarus americanus* Milne Edwards) hemocytes and the influence of diet on hemocyte numbers and hemolymph proteins. *Can. J. Zool.* 45:291-304.

Stewart, J.E., J.W. Cornick, D.M. Foley, M.F. Li, and C.M. Bishop. 1967b. Muscle weight relationship to serum proteins, hemocytes, and hepatopancreas in the lobster, *Homarus americanus*. *J. Fish. Res. Board Can.* 24:2339-2354.

Stewart, J.E., B. Arie, B.M. Zwicker, and J.R. Dingle. 1969a. Gaffkemia, a bacterial disease of the lobster, *Homarus americanus*: Effects of the pathogen, *Gaffkya homari*, on the physiology of the host. *Can. J. Microbiol.* 15:925-932.

Stewart, J.E., A. Dockrill, and J.W. Cornick. 1969b. Effectiveness of the integument and gastric fluid as barriers against transmission of *Gaffkya homari*, to the lobster *Homarus americanus*. *J. Fish. Res. Board Can.* 26:1-14.

Stewart, J.E., G.W. Horner, and B. Arie. 1972a. Effects of temperature, food, and starvation on several physiological parameters of the lobster, *Homarus americanus*. *J. Fish. Res. Board Can.* 29:439-442.

Stewart, J.E., B.M. Zwicker, B. Arie, and G.W. Horner. 1972b. Food and starvation as factors affecting the time to death of the lobster *Homarus americanus* infected with *Gaffkya homari*. *J. Fish. Res. Board Can.* 29:461-464.

Unestam, T. 1973. Fungal diseases of crustaceans. *Rev. Med. Vet. Mycol.* (U.K.). 8: 1-20.

Unestam, T. 1981. Fungal diseases of freshwater and terrestrial crustacea. In *Pathogenesis of Invertebrate Microbial Diseases*, ed. E.W. Davidson, pp. 485-510. Totowa, NJ: Allanheld, Osmun.

Unestam, T., and L. Nyhlen. 1974. Cellular and noncellular recognition of and reactions to fungi in crayfish. *Contemp. Topics Immunobiol.* 4:189-206.

Vago, C. 1966. A virus disease in Crustacea. *Nature* (London). 209:1290.

Wilder, D.G., and D.W. McLeese. 1961. A comparison of three methods of inactivating lobster claws. *J. Fish. Res. Board Can.* 18:367-375.

Wolvekamp, H.P., and T.H. Waterman. 1960. Respiration. In *The Physiology of Crustacea: Vol. I, Metabolism and Growth*, ed. T.H. Waterman, pp. 35-100. New York: Academic Press.

Young, J.S., and J.B. Pearce. 1975. Shell disease in crabs and lobsters from New York Bight. *Mar. Pollut. Bull.* 6:101-105.

13

Noninfectious Diseases of Crustacea With an Emphasis on Cultured Penaeid Shrimp

Donald V. Lightner

INTRODUCTION

A number of years ago, Dr. S.F. Snieszko, an early U.S. pioneer and leader in the fish health field, used three intersecting circles to illustrate the concept of host, pathogen, and environment as interactive factors that determine "disease" (Fig. 13-1). Snieszko's (1973) sphere model illustrates the reality that aquatic species "are not in a vacuum -- we have them in their environments...." All spheres represent variables with no constants. The host can vary (species, strain, age, life stage, nutritional status, and others); the pathogen can vary (in virulence); the environment can vary from ideal (as far as host is concerned) to unsuitable for host. In some extremes, host and environmental interaction are sufficient to produce significant disease, even in absence of a pathogen. Obviously, in other scenarios, marginal environmental conditions favor development of disease from opportunistic pathogens.

Sindermann (1990) listed a number of natural and man-made environmental changes that affect abundance and disease in wild crustacean populations. Among man-made changes were fishing and over-fishing, dredging, toxic chemicals, and abnormal nutrient loads that lead to algal blooms and anoxia. Natural environmental changes, e.g., extreme temperatures or salinities, changes in predator balance, storms, and inadequate food production, affect survival and disease incidence in wild populations.

Interactions among crustacean hosts, their environment, and disease-causing agents determine health and disease status of individual animals and whole populations. Effects of interrelated factors that determine disease are perhaps most easily studied and understood in captive or cultured crustacean populations. Commercial culture of penaeid shrimp has provided both the opportunity and necessity that disease, and factors that affect it, be understood. The rapid

0-8493-8662-4/93/$0.00 + $.50

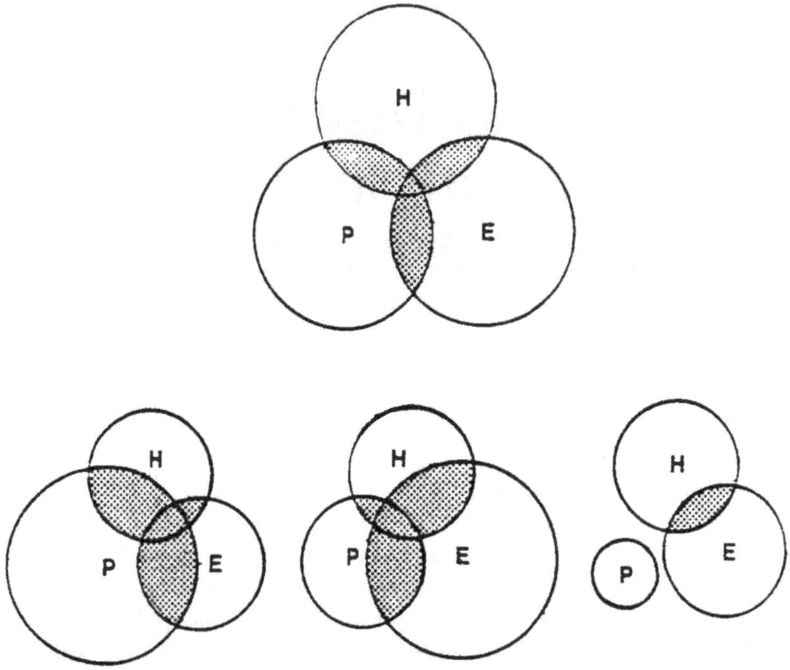

Figure 13-1. Intersecting spheres illustrate the concept of host (H), pathogen (P), and environment (E) as interactive factors which determine "disease." When size of one sphere is increased (by such factors as increased virulence of a pathogen, perturbations of the environment, or nutritional imbalances), incidence or severity of disease increases (as indicated by increased area of overlap of intersecting spheres). Even in absence of a pathogen, environmental factors can cause disease (lower right) (modified from Snieszko 1973).

world-wide development of the shrimp culture industry has been accompanied by the occurrence of numerous significant diseases of noninfectious etiologies, as well as a number of important diseases caused by opportunistic pathogens that invade a host compromised due to environmental perturbations or alterations of intrinsic factors. Among the noninfectious diseases of this important and relatively well-studied group of cultured crustaceans are diseases with etiologies that include toxicants, nutritional deficiencies or imbalances, and environmental extremes. While some noninfectious diseases have simple, straightforward etiologies, others have complex etiologies in which affected animals actually succumb to infections by opportunistic pathogens after initial insult by other predisposing factors that may be environmental, nutritional, or intrinsic.

This overview of noninfectious diseases of crustaceans will largely emphasize diseases of cultured penaeid shrimp with noninfectious etiologies or a mixed etiology in which the environment is an important disease determinant.

NUTRITIONAL DISEASES

While several nutritional disease syndromes are suspected in the penaeid shrimp, only two are well-studied. The best known is the ascorbic acid deficiency syndrome called "black death," and the second is cramped muscle syndrome, a presumed dietary mineral imbalance of cultured penaeid shrimp.

Ascorbic-acid Deficiency Syndrome ("Black Death Disease")

The earliest recognized nutritional disease syndrome of cultured penaeids was originally named "black death" to describe the typical, large black (melanized) lesions that occur in dying shrimp (Lightner *et al.* 1977; Magarelli *et al.* 1979). The disease occurs in penaeids reared in closed systems, aquaria, or flow-through systems in which most or all diet is artificial and without adequate ascorbic acid supplementation. The disease has not been observed in shrimp cultured in ponds, tanks, or raceways where there was primary productivity (growth of algae) (Lightner *et al.* 1979). Shrimp with black death typically display blackened (melanized hemocytic) lesions in tissues with a high collagen content. Such lesions are present in the stomach wall, hindgut wall, gills, and subcuticular tissues at various locations in shrimp, especially at the junction of body and appendage cuticular segments. There is often a terminal bacterial septicemia in shrimp with clinical signs of black death disease; Vibrio spp. and other opportunistic bacteria are typically isolated from the hemolymph of affected shrimp.

Black death disease has not been observed in subadult and adult shrimp and is apparently confined to the juvenile stages of the species. Penaeid shrimp apparently have a limited ability to synthesize the vitamin that meets the nutritional requirements of the older life stages, but not of the more rapidly growing juveniles stages (Magarelli *et al.* 1979).

Table 13-1. Species of Cultured Penaeid Shrimp in Which the Ascorbic Acid Deficiency Syndrome "Black Death" Has Been Observed.

Species	Location	Reference
P. californiensis	Mexico	Lightner 1977, 1988
P. stylirostris	Mexico	Magarelli *et al.* 1979
P. aztecus	Texas	Lightner 1977, 1988
P. japonicus	Japan	Shigueno 1975, Deshimaru and Kuroki 1976
P. japonicus	Tahiti	Lightner 1983

Cramped Muscle Syndrome

Cramped muscle syndrome (CMS) of penaeid shrimp has also been called "cramped tail," "body cramp," and "bent tail." CMS is characterized by an ante-mortem ventral flexure of the abdomen, which is so rigid that it cannot be straightened without tearing the abdominal muscle tissue. The condition typically follows handling, although shrimp have been observed with partially cramped tails in undisturbed ponds (Johnson 1975a). Severe CMS typically results in the rapid death of affected shrimp (Lightner 1988).

Although reported in only seven penaeid species, CMS probably occurs in all penaeids. CMS has been observed in pond-, tank-, and raceway-reared shrimp in North, Central, and South America, Hawaii, Southeast Asia, and the Middle East (Johnson 1990; Lightner 1988; Lightner *et al.* submitted). First reported by Johnson (1975), the cause of CMS has been suggested to be due to physiological or nutritional factors that are enhanced by physical or environmental stressors (Liao *et al.* 1977; Venkataramiah *et al.* 1977; Overstreet, 1978; Johnson 1978 and 1990; Lightner 1988; and Lightner *et al.* submitted).

CMS has been linked experimentally in captive *P. aztecus* exposed to artificial seawater with altered cation content. Highest incidence of CMS (60% of the test group) occurred in shrimp exposed to seawater containing 10% of the normal potassium content of seawater (Venkataramiah *et al.* 1977). Also in the study, CMS was observed, at lower incidences, in shrimp exposed to seawater with 50% normal potassium (20% developed CMS) and with reduced calcium (30%, 10%, and 10% CMS in seawater with 10%, 15%, and 25% normal calcium content, respectively). No CMS was observed in groups of shrimp exposed to seawater in which only magnesium content was varied.

CMS has also been linked to the use of artificial feeds. A diet formulation change (in which crayfish meal was substituted for sun-dried shrimp meal) at Marine Culture Enterprises, a large commercial facility or farm in Hawaii that used super-intensive culture methods (Neal 1973; Salser *et al.* 1978; Wickins 1986), was followed by a severe epizootic of CMS in the juvenile populations of *P. stylirostris*. Subsequent experimental use of feeds, with and without the crayfish meal substitution, confirmed the link between CMS and crayfish meal. The higher content of calcium salts relative to other cations, especially potassium, in crayfish meal when compared to shrimp meal was suggested as the cause of CMS at the affected farm (Lightner *et al.* submitted).

Incidence and severity of CMS in the farm's nursery populations were higher (averaging 44%, with daily incidence in samples ranging from 9 to 84% at peak of the epizootic) than in growout raceway populations of larger shrimp (incidence rates generally below 20%). Daily mortalities were elevated slightly above expected levels in undisturbed juvenile populations, but increased dramatically by as much as 50% in some populations following disturbances, such as transfer from nursery to growout raceways (Lightner *et al.* submitted). Affected shrimp displayed gross signs immediately following handling that ranged from a partial to a complete ventral flexure of the abdomen. In shrimp with complete abdominal flexures, the condition was irreversible; the body musculature rapidly turned white, and death followed within a few hours. In shrimp with a partial abdominal flexure, focal areas of white muscle developed in the abdomen, and often such lesions were resolved within a week.

Histology and TEM of affected muscle in shrimp with acute CMS showed focal to extensive myonecrosis accompanied by moderate inflammatory response and fibrosis of affected areas. Analysis of whole body and muscle cations showed significant alterations of the ratios of Na:K, K:Ca, and Ca:Mg ratios relative to unaffected and wild shrimp values, implicating a neuromuscular ionic imbalance in the pathogenesis of the syndrome. Hence, data from the studies by Venkataramiah *et al.* (1977) and Lightner *et al.* (submitted), suggest that reduced dietary and/or environmental potassium (relative to the cations Ca, Na, and Mg) may be the principal factor in the etiology of CMS in penaeid shrimp.

Selenium Deficiency

Of interest is a recent report linking muscle lesions and appendage loss to a selenium deficiency in cultured *Daphnia magna* (Elendt 1990). While selenium has not been linked to the occurrence of "white muscle" lesions in cultured penaeid shrimp, the association is suspect. Possibly, white muscle lesions are a rarity in cultured penaeid shrimp because selenium is usually present as a mineral supplement (to provide 1 ppm of the mineral) in most shrimp feeds (Akiyama and Dominy 1985).

Chronic Soft-Shell Syndrome

Cultured shrimp with chronic soft-shell syndrome are frequently observed in cultured populations of *P. monodon* in Southeast Asia and in *P. vannamei* in the Americas. It has been most thoroughly studied by researchers working in the Philippines with *P. monodon* (Baticados *et al.* 1986, 1987, 1990; Bautista and Baticados 1990; Baticados and Tendencia 1991). Shrimp with the chronic soft-shell syndrome display thin and persistently soft, rough, and wrinkled shells. They are typically dark colored and heavily colonized by surface-fouling epibionts, such as the colonial peritrich *Zoothamnium sp.* and other epicommensals. Soft-shelled shrimps are lethargic, weak, susceptible to wounding and cannibalism, show poor growth rates, and eventually die. Soft-shelled shrimp are distinguished from normal recently molted shrimp, which have smooth, clean (free of surface-fouling organisms) soft shells that harden within one to two days (Baticados *et al.* 1990).

Histological and histochemical study of soft-shelled *P. monodon* revealed that the exocuticle and endocuticle layers of the exoskeleton of normal hard-shelled shrimp were considerably thicker than those with soft shells. Histochemical demonstration of calcium in hepatopancreas and exoskeleton of normal hard and soft-shelled shrimp differed, with the soft-shelled shrimp showing less intense staining reactions for calcium in the cuticle, but a more intense reaction in the hepatopancreas (Baticados *et al.* 1987). Analysis of these tissues by atomic absorption spectrophotometry showed that calcium and phosphorous were elevated in the hepatopancreas of soft-shelled shrimp, but decreased in the exoskeleton of soft-shelled shrimp relative to normal ones with hard shells (Baticados *et al.* 1986).

Inadequate feed and feeding practices were noted to be closely associated with chronic soft-shell syndrome in the Philippines (Baticados *et al.* 1986). Improper storage of feeds, rancid or low quality feeds, and inadequate feeding in overstocked shrimp culture ponds were linked to the soft-shelled syndrome (Baticados *et al.* 1990). Pond surveys also indicated that occurrence of soft-shelling could be predicted with 98% accuracy under conditions of high soil Ph, low water phosphate, and low organic matter in soil. Of the pond shrimp with the syndrome, 70% had high soil pH (>6), low water phosphate (<1ppm), and low organic matter content (<7%) (Baticados *et al.* 1990).

Subsequent controlled laboratory studies showed that the syndrome could be induced by feeding diets with mineral imbalances of calcium and phosphorous and by exposure to certain pesticides (Baticados *et al.* 1986; Bautista and Baticados 1990; Baticados and Tendencia 1991). Their findings suggest that the soft-shell syndrome is a metabolic disease involving calcium and phosphorous metabolism, but with multiple etiologies. After inducing soft-shell syndrome in laboratory-reared groups of juvenile *P. monodon* with inadequate diets, Bautista and Baticados (1990) fed experimental shrimp eight isocaloric and isonitrogenous diets that varied in calcium and phosphorous content (with Ca:P ratios of 0:0, 0:1, 1:0, 1:0.2, 1:1, 1:2, 0.2:1, and 2:1). After 31 days of feeding, shrimp fed the diet with the 0:0 Ca:P ratio showed no recovery from soft-shell syndrome, while those fed other diets showed recovery rates ranging from 40 to 89%. The optimum feed, which gave the 89% recovery rate, had a Ca:P ratio of 1:1 (Bautista and Baticados 1990).

Certain pesticides also have been linked to chronic soft-shell syndrome in *P. monodon*. Baticados *et al.* (1990) stated that exposure of normal hard-shelled shrimp to very low levels of the pesticides Aquatin (an organostannous compound) and Gusathion A (an organophosphate) for 4 days, or to higher levels of rotenone (10-50 ppm) and saponin (100 ppm), resulted in significant soft-shelling of the exposed shrimp. Laboratory experiments in which shrimp were exposed to Aquatin showed that exposure to as little as 0.0154 ppm of the pesticide for 96 h resulted in soft-shelling in 47-60% of the shrimp (Baticados *et al.* 1986). Likewise, 27-53% of the shrimp exposed to 1.5-150 ppb Gusathion A for 96 h developed shell softening (Baticados and Tendencia 1991).

DISEASES DUE TO PHYSICAL EXTREMES

Gas-Bubble Disease

Gas-bubble disease has been reported to occur in penaeid and caridean shrimp as a result of supersaturation of atmospheric gases and oxygen (Lightner *et al.* 1974; Supplee and Lightner 1976; Parker *et al.* 1976; Brisson 1985; Colt *et al.* 1986). Apparently, shrimp are similar to fish in sensitivity to supersaturation of atmospheric gases in water, with the sources of super saturation also similar. Gas-bubble disease in shrimp due to atmospheric gas (primarily nitrogen because air is 80% nitrogen) has been linked to faulty plumbing and pumps, pump cavitation in seawater wells, and heated effluents from power plants. Levels of nitrogen or atmospheric gas supersaturation required to cause gas-bubble disease in penaeids are not known, but are assumed to be similar to the 118% reported for some fish species by Rucker (1972). To protect fishery resources, the EPA has set a criterion of 110% for gas supersaturation of public waters (Marking 1987), but some salmonid fry are sensitive to levels as low as 104%.

Oxygen-caused gas-bubble disease in penaeid shrimp has also been reported. This condition was reported to occur when dissolved oxygen levels reached or exceeded 250% of normal saturation in seawater of 24 to 26°C and 35 ppt salinity (Supplee and Lightner 1976). Such oxygen levels occasionally occur as a result of dense phytoplankton blooms that are accompanied by calm water conditions and long warm sunny days. Gas-bubble disease due to oxygen supersaturation was not found to necessarily be a lethal condition, if corrective measures (vigorous aeration) were taken to immediately lower the dissolved oxygen content of the culture tank water. Many affected shrimp recovered within a few hours (Supplee and Lightner, 1976). Nitrogen (or atmospheric gas) caused gas-bubble disease, in contrast, is usually lethal to penaeids (Lightner, 1977, 1983, 1985, 1988).

Regardless of the gas that causes the disease, clinical signs are the same. The first sign of gas-bubble disease is a rapid, erratic swimming behavior, soon followed by stupor. Shrimp so affected float helplessly (in all other diseases, dead or dying shrimp sink) near the water surface, with the ventral side of the cephalothorax higher than the abdomen. Fresh preparations of gills or whole tissue examined by microscopy reveals gas bubbles in hemocoel spaces in tissues (Lightner 1977, 1988).

Muscle Necrosis

Muscle necrosis (= spontaneous necrosis or idiopathic muscle necrosis) is the name given to a condition in all species of penaeid and caridean shrimp that is characterized by whitish opaque areas in the striated musculature, especially of the distal abdominal segments (Rigdon and Baxter 1970; Couch 1978; Akiyama *et al.* 1982; Brock 1988; Johnson 1975b, 1978, 1990; Lightner 1988). The condition typically follows periods of severe stress from over-crowding, low dissolved-oxygen levels, sudden temperature or salinity changes, rough handling, etc. It is reversible in its initial stages if stress factors are reduced, but it may be lethal if large areas are affected. When irreversible damage occurs and opportunistic bacteria invade the necrotic areas of the abdomen, the condition is called "tail-rot" by shrimpers and shrimp farmers. This chronic and typically septic form of the disease is readily recognized as the distal portion of the abdomen (or appendages) becomes completely necrotic, turns red, and begins to decompose and slough before the affected individual shrimp actually dies from the condition (Rigdon and Baxter 1970; Johnson 1975, 1977, 1990; Lakshmi *et al.* 1978; Overstreet 1978; Brock 1988; Lightner, 1977, 1983, 1988).

TOXIC DISEASES

Toxigenic algae

A number of algae have been reported to cause (or are suspected to cause) mortalities in cultured penaeid shrimp. Senescent blooms of the diatom *Chaetoceros graeilis* were reported to be toxic to the larval stages of *P. stylirostris* and *P. vannamei* (Simon 1978). It was assumed that toxic substances were released by dead or dying diatom cells.

Dinoflagellate blooms (red tides) have been suspected to have caused serious losses in penaeid shrimp culture in Mexico, but a cause and effect relationship has not been demonstrated. However, a toxicity syndrome called BSX, possibly related to red tides, has been observed in cultured populations of *P. californiensis* and *P. stylirostris* in Mexico (Lightner *et al.* 1980; Lightner 1983, 1988). Shrimp with this syndrome die during molting or after handling; in an affected population, a large percentage of the shrimp have been observed to develop "blunt heads," in which head appendages (antennal scales, antennae, antennal flagella, and eye stalks) are damaged by aberrant behavior of affected shrimp.

Affected shrimp have been observed to repeatedly collide with the culture tank walls for several hours without pausing, thereby eroding the head appendages. Dinoflagellate toxins have been thought to be non-toxic to crustaceans (Sievers 1969), but only short-term toxicity tests were run on shrimp in Sievers' studies. However, during Sievers' tests, the few shrimp that molted also died. That observation, and the probable association of red tides and the BSX syndrome in Mexico, indicate that the importance of red tide toxins to shrimp may be more significant than previously assumed (Lightner 1983, 1988).

Blue-green algae

Blooms of certain filamentous blue-green algae, all belonging to the family Oscillatoriaceae, have been implicated as causing the disease syndrome hemocytic enteritis (HE) primarily in juvenile penaeid and caridean shrimp (Table 13-2). One species of blue-green algae, confirmed by experimentation (McKee 1981) to cause this syndrome, was *Schizothrix calcicola* (Agardh) Gamout (Drouet 1968). That species possesses a potent endotoxin which causes gastroenteritis in man (Keleti *et al.* 1979). Other blue-green algae, *Spirulina subsala* and *Microcoleus lyngbyaceus*, are suspect in the etiology of HE, but neither has been linked experimentally to the syndrome.

While HE is most typical in juveniles of 0.1 to 5 g average weight, it has been observed in 12 to 20 g *P. stylirostris*. HE apparently occurs as the result of algal toxins released in the gut from ingested algae (Lightner 1978, 1983, 1988; Lightner *et al.* 1978). This disease is characterized by necrosis and sloughing of the mucosal epithelium of portions of the shrimp gastrointestinal tract (the midgut and the anterior and posterior midgut caeca) that lack a chitinous lining

Table 13-2. Geographic Locations and Species of Cultured Penaeid and Caridean Shrimp in Which Hemocytic Enteritis Has Been Observed (from Lightner 1985).

Species	Geographic Location
Penaeus duorarum	Florida
P. stylirostris	Mexico, Hawaii, Israel
P. vannamei	Mexico, Hawaii, Texas
P. californiensis	Mexico
P. japonicus	Hawaii
P. monodon	Philippines, Taiwan,Indonesia
Macrobrachium rosenbergii	Philippines, Brazil

(Bell and Lightner 1988), and a consequent marked hemocytic inflammation of affected areas of the gut (Lightner 1988). In some examples, hepatopancreatic atrophy and necrosis are observed typically in shrimp; numerous prominent cytoplasmic eosinophilic autophago lysosomes in hepatopancreatic epithelial cells are present, as are less obvious but common crystalline inclusions in the nucleoli of affected cells (Lightner and Redman 1984).

Death in shrimp with HE may be due to osmotic imbalances or poor absorption of nutrients from the midgut due to destruction of its mucosa; in most instances, death appears to be the result of secondary bacterial septicemias. Species of *Vibrio*, principally *V. alginolyticus*, have been the most commonly isolated from the hemolymph of shrimp with septic HE (Lightner 1978, 1983, 1988; Lightner *et al.* 1978, 1980). Mortality rates in raceway-cultured populations of *P. stylirostris* with HE have reached 85%, but more commonly were less than 20% of affected populations. Runting of shrimp affected by non-lethal HE is a chronic effect of the disease, which results from functional impairment of the midgut, caeca, and hepatopancreas (Lightner 1985).

Aflatoxicosis

Experimental exposure of penaeid shrimp (*P. stylirostris* and *P. vannamei*) to aflatoxin produced prominent inflamed lesions in the hepatopancreas and less prominent, but nevertheless significant, lesions in other tissues, including the lymphoid organ and hematopoietic tissues (Lightner *et al.* 1982; Lightner 1988). Single intramuscular injections of aflatoxin B1 into the tail muscle produced 24- and 96-h median lethal doses of 100.5 and 49.5 mg/kg, respectively. A toxicity curve generated from the data showed no toxicity threshold at the levels tested (25 to 160 mg of aflatoxin per kg shrimp). The mortality response in a feeding study with *P. vannamei* was not dose dependent, but tissue and organ damage were similar to that seen in injected animals (Wiseman *et al.* 1982).

Similar lesions are frequently found in cultured shrimp, suggesting that aflatoxicosis may occur naturally in these animals. Moldy feeds (with *Aspergillis flavis* and *A. parasiticus*) or feed ingredients, which are a common problem in warm and humid regions where shrimp are cultured, may be responsible for some disease syndromes commonly observed in cultured shrimp in which idiopathic hepatopancreatic atrophy, necrosis, and inflammation are a prominent feature (Lightner 1988). Hepatopancreatic lesions almost identical to those observed in aflatoxicosis occur in *P. monodon* with a disease syndrome called "septic hepatopancreatic necrosis" (SHN). Red disease of *P. monodon* (Liao *et al.* 1977; Lightner and Redman 1985; Lightner 1988) is one form of expression of SHN in some moribund shrimp, but not all shrimp with SHN display red pigmentation in the terminal phase. The etiology of SHN (and red disease) is

Table 13-3. Acute Toxicity of Aflatoxin B1 to Various Groups of Animals (from Wiseman *et al.* 1982).

Group	Animal	LD50 (mg/kga or mg/Lb)
Mammals	Rabbit	0.3a
	Hamster	10.2a
Fish	Rainbow trout	0.5a
Crustaceans	Brine shrimp	14.0b
	Copepod (*Cyclops fuscus*)	1.0b
	Penaeid shrimp (*Penaeus stylirostris*)	100.5a

poorly understood; whether the disease has a toxic and/or infectious etiology is not known (Lightner 1988).

Although lesions and mortalities resulted from experimental exposure of juvenile *P. stylirostris* and *P. vannamei* to aflatoxin, shrimp are relatively resistant to aflatoxin; species like trout, rabbits, hamsters, and copepods are more sensitive, by one or two orders of magnitude (Table 13-3). A possible explanation for differences in aflatoxin tolerance may lie in feeding habits. Detritus makes up a significant portion of the diet of penaeid shrimp, thus exposing shrimp continuously to large amounts of bacterial and fungal toxins. In contrast, species with low tolerance to aflatoxin are predatory, omnivorous, or herbivorous, and, hence, may seldom encounter food sources that contain high levels of microbial toxins (Wiseman *et al.* 1982).

Black Gill Disease, Shell Disease, and Other Toxicity Syndromes

Black gills or melanized gills ("burned gills," "black spot disease," etc.) and related melanized cuticular lesions, e.g., shell disease in shrimp and other decapod crustaceans, are a common manifestation of a number of disease syndromes, which are regularly observed in wild and cultured marine and freshwater shrimp. The black or brown pigment present in the gill, appendage, and cuticular lesions is melanin which is formed at the sites of hemocytic inflammation and tissue necrosis (Lightner and Redman 1977). Black gills, shell disease, and related appendage and cuticular lesions are a commonly observed feature of a number of infectious, parasitic, and noninfectious diseases (Cook and Lofton 1973; Rinaldo and Yevich 1974; Johnson 1975b, 1978, 1990; Couch 1977, 1978, 1979; Lightner 1977, 1983, 1985, 1988; Nimmo *et al.* 1977; Overstreet

Table 13-4. Noninfectious Causes of Black (Melanized) Gills.

Heavy metals	Chemical irritants	Other
cadmium copper	potassium permanganate ozone crude oil acids (very low pH seawater) ammonia nitrite chlorine, bromine	ascorbic acid deficiency syndrome gas-bubble disease

1978, 1988; Doughtie and Rao 1983; Doughtie *et al.* 1983; Sparks 1985; Papathanassiou 1985; Brock 1988; Sindermann 1990).

Of special interest in this review are those noninfectious diseases which occur in cultured shrimp and are accompanied by black gills and/or appendage or cuticular lesions (Table 13-4). Among the noninfectious causes of such lesions are certain heavy metals, such as cadmium and copper, that may be encountered in aquaculture settings. Copper-based algicides are frequently used in shrimp culture and their misuse may result in necrosis and melanization of the gills and in cuticular lesions (Williams *et al.* 1982; Williams and Lightner 1988). Similar black gill lesions occur in shrimp exposed to cadmium (Couch 1977, 1978; Nimmo *et al.* 1977) and to a variety of other chemical irritants (Table 13-4) that may be encountered in shrimp culture.

Perhaps the most common cause of black gills and occasional cuticular lesions in cultured shrimp are ammonia and nitrite. Subacute or chronic exposure to as little as 1 ppm unionized ammonia or 10 ppm nitrite may result in gill damage and black gills, and low level mortalities (Chen *et al.* 1990). Such ammonia nitrite concentrations are commonly observed in the pond and tank seawater at shrimp culture facilities that employ intensive or super-intensive methods (Neal 1973; Salser *et al.* 1978; Lightner 1983; Wickins 1986; Chen *et al.* 1989, 1990).

REFERENCES

Akiyama, D., J.A. Brock, and S.R. Haley. 1982. Idiopathic muscle necrosis in cultured freshwater prawn, *Macrobrachium rosenbergii*. VM/SAC, 1119.

Akiyama, D.M., and W.G. Dominy. 1985. Penaeid shrimp nutrition for the commercial feed industry. In *Texas Shrimp Farming Manual*, ed. G.W. Chamberlain, M.G. Haby, and R.J. Miget. Corpus Christi: Texas Agricultural Extension Service.

API. 1986. API Rapid NFT Nonfermenters Identification Codebook. ISBN 2-904243-20-8. API Analytab Products, Sherwood Medical, Plainview, N.Y.

Baticados, M.C.L., R.M. Coloso, and R.C. Duremdez. 1986. Studies on the chronic soft-shell syndrome in the tiger prawn, *Penaeus monodon* Fabricius, from brackishwater ponds. *Aquaculture* 56:271-285.

Baticados, M.C.L., R.M. Coloso, and R.C. Duremdez. 1987. Histology of the chronic soft-shell syndrome in the tiger prawn *Penaeus monodon*. *Dis. Aquat. Org*. 3:13-28.

Baticados, M.C.L., E.R. Cruz-Lacierda, M.C. de la Cruz, R.C. Duremdez-Fernandez, R.Q. Gacutan, C.R. Lavilla-Pitogo, and G.D. Lio-Po. 1990. Diseases of Penaeid Shrimps in the Philippines. Aquaculture Extension Manual No. 16. Aquaculture Department, Southeast Asian Fisheries Development Center (SEAFDEC), Tigbauan, Iloilo, Philippines. 46 p.

Baticados, M.C.L., and E. Tendencia. 1991. Effects of Gusathion A on the survival and shell quality of juvenile *Penaeus monodon*. *Aquaculture* 93:9-19.

Bautista, M.N., and M.C.L. Baticados. 1990. Dietary manipulation to control the chronic soft-shell syndrome in tiger prawn, *Penaeus monodon* Fabricius. In *The Second Asian Fisheries Forum*, ed. R. Hirano and I. Hanyu, pp. 341-344. Manila: Asian Fisheries Society.

Bell, T.A., and D.V. Lightner. 1988. A Handbook of Normal Shrimp Histology. Special Publication No. 1, World Aquaculture Society, Baton Rouge, LA. 114 p.

Brisson, S. 1985. Gas-bubble disease observed in pink shrimps, *Penaeus brasiliensis* and *Penaeus paulensis*. *Aquaculture* 47:97-99.

Brock, J.A. 1983. Diseases (infectious and noninfectious), metazoan parasites, predators, and public health considerations in *Macrobrachium* culture and fisheries. In *Handbook of Mariculture*, Vol. 1, ed. J.P. McVey, pp. 329-370. Boca Raton, Florida: CRC Press.

Brock, J.A. 1988. Diseases and husbandry problems of cultured *Macrobrachium rosenbergii*. In Disease Diagnosis and Control in North American Marine Aquaculture. Ed. C.J. Sindermann and D.V. Lightner, pp. 134-180. Second Edition. Amsterdam: Elsevier Scientific Publishing Co.

Chen, J.C., P.C. Liu, Y.T. Lin, and C.K. Lee. 1989. Highly intensive culture study of tiger prawn *Penaeus monodon* in Taiwan. In *Aquaculture -- a Biotechnology in Progress*, ed. N. de Pauw, E. Jaspers, N. Ackefors, and N. Wilkins, pp. 377-382. Bredene, Belgium: European Aquaculture Society.

Chen, J.C., P.C. Liu, and S.C. Lei. 1990. Toxicities of ammonia and nitrite to *Penaeus monodon* adolescents. *Aquaculture* 89:127-137.

Colt, J., G. Bouck, and L. Fidler. 1986. Review of current literature and research on gas supersaturation and gas bubble trauma. Special Publication Number 1, American Fisheries Society. 52 p.

Cook, D.W., and S.R. Lofton. 1973. Chitinoclastic bacteria associated with shell disease in *Penaeus* shrimp and the blue crab (*Callinectes sapidus*). *J. Wildl. Dis*. 19:154-159.

Couch, J.A. 1977. Ultrastructural study of lesions in gills of a marine shrimp exposed to cadmium. *J. Invertebr. Pathol*. 29:267-288.

Couch, J.A.. 1978. Diseases, parasites, and toxic responses of commercial penaeid shrimps of the Gulf of Mexico and South Atlantic Coasts of North America. *Fish. Bull*. 76:1-44.

Couch, J.A. 1979. Shrimp (Arthropoda: Crustacea: Penaeidae), In *Pollution Ecology of Estuarine Invertebrates*, pp. 235-258. New York: Academic Press.

Deshimaru, O., and K. Kuroki. 1976. Studies on a purified diet for prawn - VII Adequate dietary levels of ascorbic acid and inositol. *Bull. JPN. Soc. Sci. Fish*. 42:571-576.

Doughtie, D.G., and K. Ranga Rao. 1983. Ultrastructural and histological study of degenerative changes leading to black gills in grass shrimp exposed to a dithiocarbamate biocide. *J. Invertebr. Pathol*. 41:33-50.

Doughtie, D.G., P.T. Conklin, and K. Ranga Rao. 1983. Cuticular lesions induced in grass shrimp exposed to hexavalent chromium. *J. Invertebr. Pathol*. 42:249-258.

Drouet, F. 1968. Revision of the classification of the Oscillatoriaceae, Monograph 15 of The Academy of Natural Sciences of Philadelphia, pp. 370. Lancaster, Pennsylvania: Fulton Press.

Elendt, B.-P. 1990. Selenium deficiency in crustacea. *Protoplasma* 154:25-33.

Hunter, B., P.C. Magarelli, Jr., D.V. Lightner, and L.B. Colvin. 1979. Ascorbic acid-dependent collagen formation in penaeid shrimp. *Comp. Biochem. Physiol.* 64B:381-385.

Johnson, S.K. 1975a. Cramped condition in pond-reared shrimp. Leaflet No. FDDL-56, Texas A & M University Fish Disease Diagnostic Laboratory, 2 p.

Johnson, S.K. 1975b. Handbook of Shrimp Diseases, Sea Grant Publ. No. TAMU-SG-75-603, Texas A & M University. 19 p.

Johnson, S.K. 1978. Handbook of Shrimp Diseases, Sea Grant Publ. No. TAMU-SG-75-603 (revised), Texas A & M University 23 p.

Johnson, S.K. 1990. Handbook of Shrimp Diseases, Sea Grant Publ. No. TAMU-SG-90-601, Texas A & M University 25 p.

Keleti, G., J.L. Sykora, E.C. Lippy, and M.A. Shapiro. 1979. Composition and biological properties of lipopolysaccharides isolated from *Schizothrix calcicola* (Ag.) Gamout (Cyanobacteria). *Appl. Environ. Microbiol.* 38:471-477.

Krol, R., W.E. Hawkins, and R.M. Overstreet. In press. Rickettsial and mollicute infections in hepatopancreatic cells of cultured Pacific white shrimp (*Penaeus vannamei*). *J. Invertebr. Pathol.*

Lakshmi, G.J., A. Venkataramiah, and H.D. Howse. 1978. Effect of salinity and temperature changes on spontaneous muscle necrosis in *Penaeus aztecus* Ives. *Aquaculture* 13:35-43.

Liao, I.C., F. Yang, and S. Lou. 1977. Preliminary report on some diseases of cultured prawn and their control methods, Reports on Fish Disease Research (I), JCRR Fisheries Series, Taipei, Taiwan, No. 29:28-33.

Lightner, D.V. 1977. Shrimp diseases. In *Disease Diagnosis and Control in North American Marine Aquaculture Developments in Aquaculture and Fisheries Science*,. Vol. 6, ed. C.J. Sindermann, pp. 10-77. New York: Elsevier Science Publishing Co.

Lightner, D.V. 1978. Possible toxic effects of the marine blue-green alga, *Spirulina subsalsa*, on the blue shrimp, *Penaeus stylirostris*. *J. Invertebr. Pathol.* 32:139-150.

Lightner D.V. 1983. Diseases of cultured penaeid shrimp. In *Handbook of Mariculture, Volume 1. Crustacean Aquaculture*, ed. J.P. McVey, pp. 289-320. Boca Raton, Florida: CRC Press.

Lightner, D.V. 1985. A review of the diseases of cultured penaeid shrimp and prawns with emphasis on recent discoveries and developments. In *Proceedings of the First International Conference on the Culture of Penaeid Prawns/Shrimps*, ed. Y. Taki, J.H. Primavera, Jose A. Llobrera, pp. 79-103. Iloilo, Philippines: Southeast Asian Fisheries Development Center.

Lightner, D.V. 1988. Diseases of Penaeid Shrimp. In *Disease Diagnosis and Control in North American Marine Aquaculture*. Second Edition, ed. C.J. Sindermann and D.V. Lightner,. pp. 8-133. Amsterdam: Elsevier Science Publishers.

Lightner, D.V., and Redman, R.M. 1977. Histochemical demonstration of melanin in cellular inflammatory processes of penaeid shrimp. *J. Invertebr. Pathol.* 30:298-302.

Lightner, D.V., B.R. Salser, and R.S. Wheeler. 1974. Gas-bubble disease in the brown shrimp, (*Penaeus aztecus*). *Aquaculture* 4:81-84.

Lightner, D. V., Fontaine, C.T., and Hanks, K. 1975. Some forms of gill disease in penaeid shrimp. *Proc. World Mariculture Soc.* 6:347-365.

Lightner, D.V., L.B. Colvin, C. Brand, and D.A. Danald. 1977. Black Death, a disease syndrome related to a dietary deficiency of ascorbic acid. *Proc. World Mariculture Soc.* 8:611.

Lightner, D., D. Danald, R. Redman, C. Brand, B. Salser, and J. Reprieto. 1978. Suspected blue-green algal poisoning in the blue shrimp (*Penaeus stylirostris*). *Proc. World Mariculture Soc.* 9:447-458.

Lightner, D.V., B. Hunter, P.C. Magarelli, Jr., and L.B. Colvin. 1979. Ascorbic acid: nutritional requirement and role in wound repair in penaeid shrimp. *Proc. World Mariculture Soc.* 10:513-528.

Lightner, D.V., R.M. Redman, D.A. Danald, R.R. Williams, and L.A. Perez. 1980. Major diseases encountered in controlled environment culture of penaeid shrimp at Puerto Penasco, Sonora, Mexico. Proceedings Third U.S.-Japan Meeting on Aquaculture, Kyoto, Japan. pp. 75-97.

Lightner, D.V., R.M. Redman, M.O. Wiseman, and R.L. Price. 1982. Histopathology of aflatoxicosis in the marine shrimp *Penaeus stylirostris* and *P. vannamei*. *J. Invertebr. Pathol.* 40:279-291.

Lightner, D.V., and R.M. Redman. 1984. Intranucleolar polyhedral crystalline bodies in the hepatopancreas of the blue shrimp *Penaeus stylirostris*. *J. Invertebr. Pathol.* 3. 43:270-273.

Lightner, D.V., and R.M. Redman. 1985. Necrosis of the hepatopancreas in *Penaeus monodon* and *P. stylirostris* with red disease. *J. Fish Dis.* 8:181-188.

Lightner, D.V., R.M. Redman, R.R. Williams, and T.A. Bell. Submitted. Histopathology and clinical chemistry of cramped muscle syndrome in penaeid shrimp and its probable relationship to a dietary mineral imbalance. *J. Fish Dis.*

Magarelli, P.C., Jr., B. Hunter, D.V. Lightner, and L.B. Colvin. 1979. Black Death: an ascorbic acid deficiency disease in penaeid shrimp. *Comp. Biochem. Physiol.* 63A:103-108.

Marking, L.L. 1987. Gas supersaturation in fisheries: causes, concerns, and cures. *U.S. Fish Wildl. Serv. Leafl.* 9. 10 p.

McKee, C. 1981. The Toxic Effect of Five Strains of Blue-green algae on *Penaeus stylirostris* Stimpson, M.S. Thesis, School of Renewable Natural Resources, University of Arizona, Tucson.,

Moriarty, D.J.W. 1976. Quantitative studies on bacteria and algae in the food of the mullet *Mugil cephalus* L. and the prawn *Metapenaeus bennettae* (Racek & Dall). *J. Exp. Mar. Biol. Ecol.* 22:131-143.

Neal, R.A. 1973. Alternatives in aquacultural development: Consideration of extensive versus intensive methods. *J. Fish Res. Bd. Canada* 30: 2218-2222.

Nimmo, D.W.R., D.V. Lightner, and L.H. Bahner. 1977. Effects of cadmium on the shrimps, *Penaeus duorarum*, *Palaemonetes pugio* and *Palaemonetes vulgaris*. In *Physiological Responses of Marine Biota to Pollutants*, ed. F.J. Vernberg, A. Calabrese, F.P. Thurberg, and W.B. Vernberg, pp. 131-183. New York: Academic Press.

Overstreet, R.M. 1978. Marine maladies? Worms, germs, and other symbionts from the Northern Gulf of Mexico. Mississippi-Alabama Sea Grant Consortium. MASGP-78-021. 140 p.

Overstreet, R.M. 1988. Aquatic pollution problems, Southeastern U.S. coasts: histopathological indicators. *Aquatic Toxicol.* 11:213-239.

Papathanassiou, E. 1985. Effects of cadmium ions on the ultrastructure of the gill cells of the brown shrimp *Crangon crangon* (L.)(Decapoda, Caridea). *Crustaceana* 48:6-17.

Parker, N.C., K. Strawn, and T. Kaehler. 1976. Hydrological parameters and gas bubble disease in a mariculture pond and flow-through aquaria receiving heated effluent. Proceedings of Southeastern Association of Game and Fish Commissioners Thirtieth Annual Conference, pp. 179-191.

Rigdon, R.H., and K.N. Baxter. 1970. Spontaneous necroses in muscles of brown shrimp, *Penaeus aztecus* Ives. *Trans. Am. Fish. Soc.* 99:583-587.

Rinaldo, R. G., and Yevich, P. 1974. Black spot gill syndrome of Northern Shrimp *Pandalus borealis*. *J. Invertebr. Pathol.* 24:224-233.

Rucker, R.R. 1972. Gas-bubble disease of salmonids: A critical Review. *U.S. Fish. Wild. Serv. Tech. Pap. No.* 48, 11 p.

Salser, B., L. Mahler, D. Lightner, J. Ure, D. Danald, C. Brand, N. Stamp, D. Moore, and B. Colvin, B. 1978. Controlled environment aquaculture of penaeids. In *Drugs and Food from the Sea. Myth or Reality?*, ed. P.N. Kaul and C.J. Sindermann, pp. 345-355. Norman, Oklahoma: Univ. of Oklahoma Press.

Shigueno, K. 1975. Shrimp Culture in Japan. 153 p. Tokyo: Assoc. Int. Tech. Pro.

Sievers, A.M. 1969. Comparative toxicity of *Gonyaulax monilata* and *Gymnodinium breve* to annelids, crustaceans, molluscs, and a fish. *J. Protozool.* 16:401.

Simon, C. 1978. The culture of the diatom *Chaetoceros gracilis* and its use as a food for penaeid protozoeal larvae. *Aquaculture* 14:105.

Sindermann, C.J. 1990. Principal Diseases of Marine Fish and Shellfish. Volume 2, Second Edition, 516 p. New York: Academic Press.

Snieszko, S.F. 1973. Diseases of fishes and their control in the U.S. In *The Two Lakes Fifth Fishery Management Training Course Report*. pp. 55-66. London: Jansen Services, Dalmeny House.

Sparks, A.K. 1985. Synopsis of Invertebrate Pathology Exclusive of Insects. 423 p. Amsterdam: Elsevier Science Publishers.

Supplee, V.C., and D.V. Lightner. 1976. Gas-bubble disease due to oxygen supersaturation in raceway-reared California brown shrimp. *Prog. Fish Cult.* 38:158-159.

Venkataramiah, A., G.J. Lakshmi, P. Biesiot, J.D. Valleau, G. Gunter. 1977. Studies on the time course of salinity and temperature adaptation in the commercial brown shrimp *Penaeus aztecus* Ives. Gulf Coast Research Laboratory, Ocean Springs, MS, Contract Report H-77-1 for Chief of Engineers, U.S. Army, Washington, DC. 308 p.

Wickins, J.F. 1986. Prawn farming today: opportunities, techniques, and developments. *Outlook Agric.* 15:52-60.

Williams, R.R., J.E. Hose, and D.V. Lightner. 1982. Toxicity and residue studies on cultured shrimp, *Penaeus stylirostris*, treated with the algicide Cutrine-Plus. *Prog. Fish Cult.* 44:196-199.

Williams, R.R., and D.V. Lightner. 1988. Regulatory status of drugs and chemotherapeutants for penaeid aquaculture in the United States. *J. World Aquacult. Soc.* 19: 188-196.

Wiseman, M.O., R.L. Price, D.V. Lightner, and R.R. Williams. 1982. Toxicity of aflatoxin B1 to penaeid shrimp. *Appl. Environ. Microbiol.* 44:1479-1481.

14

Chemically Induced Histopathology in Aquatic Invertebrates

George R. Gardner

INTRODUCTION

Environmental chemicals and pollutants, including metals, aromatic hydrocarbons, substituted aromatic hydrocarbons, dyes, chlorinated hydrocarbons, phenols and substituted phenols, nitrosoamines, crude oil and petroleum, and contaminated sediment cause invertebrate diseases, including neoplasms. Traditional studies of chemical carcinogenesis with molluscan models demonstrated carcinogenicity of coal tar more than 60 years ago. By the 1960's, studies of the tumor-inducing potential for specific PAHs, insecticides, and a dye demonstrated that exposures to environmental chemicals caused neoplasms; co-occurrence of chemicals and neoplastic disease in animals from polluted environments suggested that chemicals could cause neoplasms. Survey information has shown that metals, PAHs, and pesticides dominated studies in the 1960's; nitrosocompounds, PCBs, petroleum, and aflatoxin, were studied in the 1970's; emphasis shifted to studies of complex mixtures of PAHs, PCBs, insecticides, and metals in the 1980's. Such studies of causal relationships contributed to present understanding of consequences associated with chemical contaminants in the environment.

This chapter surveys pathology of invertebrates resulting from exposure to chemicals and environmental hazards without judgment of methodologies, chemicals, or species studied. Reference sources are scientific papers, conference proceedings, monographs, reviews, and synoptic views in existing literature (Hueper 1963; Johnson 1968; Harshbarger and Dawe 1969; Pauley 1969; Krieg 1973; Harshbarger 1977; Bang 1980; Kaiser 1980; Sindermann *et al*. 1980, 1990; Couch 1982; Meyers and Hendricks 1982; Stebbing and Brown 1984 Couch and Harshbarger 1985; Sparks 1985; Mix 1986; Mix 1988; Dawe 1991). The survey includes tumor acquisitions from the Registry of Tumors in Lower Animals (RTLA), National Museum of Natural History, Washington, DC, and the

0-8493-8662-4/93/$0.00 + $.50

Resources of Biomedical and Zoological Specimens published by the Registry of Comparative Pathology, Armed Forces Institute of Pathology (1990). Reports of invertebrate pathology from these sources and the scientific literature are separated into laboratory research and field surveys. Data are summarized by chemical or environmental hazard, and geographic location. An exception is toxicology-related research supported by public health agencies (discussed below). The invertebrate species represented, toxicological methods, toxic effect on target organs, and species specificity are summarized in Tables 14-1, 14-2.

LABORATORY STUDIES OF CHEMICAL EFFECTS

Investigators adapted many toxicology testing methods from classical mammalian single agent/single species studies, this included five different routes of exposure: immersion in solutions, direct application, implants, injection, and dosed feeding. Methods of acute and chronic exposure unique to pollutant hazards include feeding, spiked sediments, microcosm simulation, and *in situ* deployment in the field.

In addition to the experimental and epizootiological approach to chemically-caused disorders, which is the focus of this chapter, it is noteworthy that invertebrates have been used by Public Health Service agencies to determine developmentally hazardous and teratogenic properties of xylene, vinblastine, glycols, and glycolesters in freshwater hydra (Chun *et al.* 1983a; Chun *et al.* 1983b; Johnson 1980; Johnson and Christian 1985; Johnson *et al.* 1982, 1984, 1986), and aromatic and aromatic substituted hydrocarbons and crude oil in marine jellyfish (Spangenberg *et al.* 1980; Spangenberg 1984). Further, a single study analyzed the response to antitumor drugs in fast developing embryos of sand-dollar (Karnofsky and Simmel 1963).

Findings for Specific Toxicants and Some Mixtures

Metals

Silver: Two studies have evaluated chronic exposures of silver to define the toxic characteristics in molluscs. Calabrese *et al.* (1984) and Nelson *et al.* (1983) detailed evidence of argyria or silver deposition and retention in basement membranes of digestive organs and associated connective tissues of mussels and Atlantic slippers. Yevich reported argyrosis in ocean quahogs at dump sites correlated with silver contamination of sediment (P.P. Yevich, unpublished data).

Cadmium: A number of studies have assessed responses to cadmium, including subchronic exposures to platyhelminths (Hall *et al.* 1984, 1986a,b), to fresh-

Table 14-1. Pathological Disorders Produced by Chemical Agents

Agent	Exposure*	Concentration/ Duration	Species	Tissue/Organ	Lesion**	Reference
Metals						
Silver	WC	0,1,5,10 ppb 2,18,21 mo	*M. edulis*	Basement membrane, digestive diverticula	Argyrism	Calabrese *et al.* 1984
	WC	1, 5,10 ppb 24 mo	*C.fornicata*	Basement membrane, digestive diverticula	Argyrism	Nelson *et al.* 1983
Arsenic	WC	5,10,15 ppb 12 mo	*C. virginica*	Circulatory system	Stenosed arteries	Yevich and Zaroogian EPA, Unpubl.
Cadmium	WC	0.001-1 ppm 1 mo	*E. fluviatilis*	Gemmosclere	Malformed; DD	Mysing-gubala and Poirrier 1981
	WC	763 ppb, 15 d	*P. duorarum*	Gill	"Black Gill"	Couch 1977
	WC	5 ppm, 96 h 5-7 d, PE.	*P. duorarum*	Gill	"Black Gill"	Nimmo *et al.* 1977
	WC	0.3,0.5,0.75 ppm	*D. doroto-cephala*	Stem cells in cephalic or pharyngeal region	Reticular cell tumors, DD	Hall *et al.* 1984, 1986 a,b
	WC	20 ppb, 7 wk 2 wk PE	*P. magellanicus*	Digestive diverticula, kidney, reproductive, tentacles	Necrosis; kidney metaplasia	Yevich and Yevich 1987
	WC	0.075,0.15,0.3, 0.5,0.75,1.0 ppm, 30 d	*P. pugio* *P. vulgaris*	Gill, cuticle, appendages	Melanization, "Black Gill" regressive	Lightner 1987 Lightner Pers. Comm.
Chromium	WC	0.5,1,2,4 ppm 28 d	*P. pugio*	Antennal gland, hepatopancreas, gill, midgut	hypertrophy	Doughtie and Rao 1984
	WC	0.5,1,2,4 ppm 28 d	*P. pugio*	Subcuticular epithelium	"Shell disease"	Doughtie and Rao 1984
	WC	6.6 ppm 96 h LC50, 4.5 ppm 8 d LC50	*Mysidopsis sp.*	Hepatopancreas epithelial cells	Rounding and sloughing midgut	Lightner 1983, 1987
Copper	WC	0-1 ppm 8 d	*B. canali-culatum*	Nerve; gills and osphradium	Necrosis; inflammation	Betzer 1975
	WC	20 ppb, 2 wk PE	*P. magellanicus*	Kidney	Metaplasia	Yevich and Yevich 1985
	WC	1 ppm, EC-50 24 h	*Penaeus* sp.	Gills/epipodites	Melanization	Lightner Pers. Comm.
	WC	0,5,100,1000 ppb 48 h	*C. sapidus*	Nerve system	Integrity of dendrites	Bodammer 1979
	WC	1,5,10 ppb 12,18,21 mo	*M. edulis*	Digestive diverticula, ducts	Epithelial erosion	Calabrese *et al.* 1984

*WC = Water Column; PE = Post-exposure; IM = Implant; MC = Microcosm; D = Diet; IS = *In situ;* SP = Sediment particulate; I = Injection; E = Environmental; EA = Epidermal application.
**NOE = No observed effects; DD = Dose-dependent.

Table 14-1 (Continued)

Agent	Exposure	Concentration/ Duration	Species	Tissue/Organ	Lesion	Reference
	WC	0.1-0.5 ppm	Cultured oysters	Digestive diverticula, stomach	Regressive changes, desquamation and necrosis	Fujiya 1960
	WC	250 ppb, 96 h	*M. mercenaria*	Gill	Necrosis and metaplasia	LaRoche *et al.* 1978
	WC	0.02 ppm, 24 h 1 yr PE	*M. edulis*	Gill	Gill deformities	Sunila 1987, 1988
	WC	0.02 ppm, 24 h PE IS Baltic Sea 1 yr	*M. edulis*	Kidney, mantle	watery cysts mantle; cystic kidney 22%	Sunila 1986
Mercury	WC	0.001-1 ppm 1 mo	*E. fluviatilis*	Gemmosclere	Gemmule malformation and growth, DD	Mysing-gubala and Poirrier 1981
	WC	9 ppb-1 d	*N. virens*	Intestine; kidney	Mercury localization in epithelial, peritoneal cells	Jensen and Baatrup 1979
Aromatic Compounds						
B[a]P	WC,IM	WC-24 h	*D. dorotocephala*	Dorsal surface	Neoplastic nodule, Adult-7% Offspring-12%	Foster 1969
	I		*A. australis*	Epithelium	Epithelial neoplasms	Krieg 1970, 1973
	I	B[a]P analogs; 16-20 wk, 0.5 mg	*T. rubescens*	Epithelium	Subcutaneous tumors (33%)	Shimkin *et al.* 1951
	WC/I	Unconfirmed 10-25 wk	*M. mercenaria*	Digestive tubules	NOE	Tripp and Fries 1987
	WC	1-5 ppb	*C. virginica*	Major organs	NOE	Couch *et al.* 1979
3-MC	IM,WC	IM 2 mo, 0.4 ppm, 24 h	*D. dorotocephala*	Dorsal surface	Lethal growth in Adult 7-9%; In offspring papilliform tumors 30%, malformations 32%	Foster 1969
	I	0.02-0.1 ml of 0.5-1% sol. 95-150 d, PE	*A. australis*	Mantle epithelium	Epithelial neoplasms average 3%	Kreig 1970, 1973
	WC	1-5 ppb	*C. virginica*	Mantle	Hemic cell infiltration	Couch *et al.* 1979
Aflatoxin B1	I	2-160 μg/g body wt 25 d	*P. stylirostris*	Hepatopancreas	Necrosis and fibrosis	Lightner *et al.* 1982; Lightner 1987

Table 14-1 (Continued)

Agent	Exposure	Concentration/ Duration	Species	Tissue/Organ	Lesion	Reference
	D	53-300 µg/g feed, 25 d	*P. vannamei*	Mandibular organ	Necrosis	Lightner *et al.* 1982
	WC	0.5-1.0 mg/ml 48 h 14 d PE	*P. fontinales*	Tentacles	Subcellular alteration	Bauer and Tulusan 1973
Azo Compounds						
MAM-Ac	WC	2.5 ppm, 4d, 1 yr PE	*M. edulis*	Digestive diverticula	NOE	Rasmussen 1986
Dyes						
Scharlach Red	EA	Unconfirmed conc. 6-10 d	*D. gonocephala* *C. alpina*	Dorsal surface	Abnormal lethal growths	An der Lan 1962
Direct Blue-15	MC	100 ppb, 40 d	*N. anulata*	Major organ	NOE	Perez *et al.* 1987
Chlorinated Hydrocarbons						
PCBs	WC	0.7 ppb, 30 d	*P. duorarum*	Hepatopancreas	NOE; >viral prevalence	Couch and Courtney 1977
	WC	3.5 ppb, 30 d	*P. duorarum*	Hepatopancreas	Ultrastructural alterations	Couch and Nimmo 1977
	WC	3 ppb, ≥ 30 d	*P. duorarum*	Hepatopancreas	crystalloids in nuclei	Nimmo *et al.* 1975
	WC	5 ppb, 24 wk 12 wk PE	*C. virginica*	Digestive diverticula connective tissue	Inflamation/ regressive	Lowe *et al.* 1972
	WC	5 ppb, 6 mo	*C. virginica*	Connective tissue	Leukocyte infiltration	Nimmo *et al.* 1975
Hexachloro-benzene	WC/I	Unconfirmed conc 10-25 wk	*M. mercenaria*	Digestive tubules	NOE	Tripp and Fries 1987
Pesticides						
Dieldrin	D	0,10,100 ppm calculated dose 2-3X wk/13 wk	*M. balthica*	Major organ systems	NOE	Farley 1977
	WC	0.1,0.01,0.001 ppm 43 d	*C. virginica*	Digestive diverticula, stomach and mantle	NOE	Emanuelsen *et al.* 1978
DDT	WC	1,5,10 ppm >14 d	*D. gonocephala*	Dorsal surface	Abnormal lethal growths by 14 d	An der Lan 1962
DDT, parathion, toxaphene	WC	3 ppb as single and mix, 2 yr	*C. virginica*	Kidney, gill, digestive tubules, visceral ganglion	Necrosis, tubule dimensional changes	Lowe *et al.* 1971
DDT parathion	WC	DDT 0.05 mg/l, 7 d; parathion 0.5 mg/l, 17 h	*C. affinis*	Nerve cells	Nissl substance lesions	Kayser *et al.* 1962
Sevin	WC	5 ,10 ppm; >14 d	*C. alpina*	Dorsal Surface	Proliferative lesions	An der Lan 1962

Table 14-1 (Continued)

Agent	Exposure	Concentration/ Duration	Species	Tissue/Organ	Lesion	Reference
	WC	15,20,25,30 ppm 96 h	*M. nasuta*	Gill mantle, siphon, branchial gland	Necrosis, DD	Armstrong and Millemann 1974
Phenols and Substituted Phenols						
PCP	WC/I	10 and 25 wk	*M. mercenaria*	Digestive tubules	NOE	Tripp and Fries 1987
Nitroso Compounds						
MNU	I	0.25,0.5,1.0, 4.0 mg, IX 12 d 0.5 mg 4X in 24 wk	*M. edulis*	Digestive diverticula	Necrosis, inflammation, ducts, tubules, collagen disease	Rasmussen *et al.* 1985
DPN	I	0.25,0.5,1.0 2.0 mg, 1 inj/ mussel - 12 d 0.5 mg, 4 wkly for 17 wk	*M. edulis*	Digestive diverticula	Necrosis after 12 d; collagen scarring and granulocytomas after 17 wk	Rasmussen *et al.* 1985
DEN	WC	200 ppm, 64 d exp./71 d. avg. latency of tumor induction 400 ppm, 61 d avg.	*U. pictorum*	Digestive diverticula, hemopoietic	Tumors 18% 200 ppm; tumors 68%, 400 ppm	Khudoley and Syrenko 1978
	WC	600 ppm, 28 d	*C. virginica*	Major organs	Parasite enhancement	Winstead and Couch 1988
	WC	130 ppm, 5 wk	*C. virginica*	Hemopoietic	Hemic tumors in 1/65 possible	Couch *et al.* 1987
DMN	WC	10,100,200 ppm 6 mo	*P. clarkii*	Antennal gland, hepatopancreas	Glandular degeneration; hyperplasia	Harshbarger *et al.* 1971
	WC	200 ppm 51 d, 85 d avg. latency	*U. pictorum*	Digestive diverticula, kidney, hemopoietic	Induced neoplasms, 8%	Khudoley and Syrenko1978
	WC	200 ppm 152 d, 81 d avg. latency			Induced neoplasms, 27%	Khudoley and Syrenko 1978
	WC	100 ppm, alternate d for 14 d	*M. edulis*	Digestive diverticula, blood vessels, connective tissue	NOE	Rasmussen 1982
	I	0.1,0.2,0.4, 0.8 mg, 1 I 15 d	*M. edulis*	Digestive diverticula, connective tissue, gonad	Inflammation, necrosis	Rasmussen *et al.* 1988
	I	0.2 mg-30 wk 8 I	*M. edulis*	Digestive diverticula	Collagenous scarring, granulocytomas	Rasmussen *et al.* 1983
MNNG	I	0.063,0.125, 0.25,0.5 mg, 15 d	*M. edulis*	Digestive diverticula	Necrosis, collagen disease	Rasmussen *et al.* 1983
	I	0.125 mg, 4x-28 wk			Encapsulation by collagen scar	

Table 14-1 (Continued)

Agent	Exposure	Concentration/ Duration	Species	Tissue/Organ	Lesion	Reference
Nitrilotri-acetic Acid	WC	5500 ppm TL50, 168 h	*N. virens*	Major organs	NOE	Eisler *et al.* 1972
	WC	1800 ppm TL50, 168 h	*P. vulgaris*	Digestive diverticula, kidney, major tissue/ organs	Diverticular, renal glandular necrosis	Eisler *et al.* 1972
	WC	1875 ppm TL50, 168 h	*P. longicarpus*	Major organs	NOE	Eisler *et al.* 1972
	WC	>10,000 ppm TL50, 168 h	*M. mercenaria*	Major organs	NOE	Eisler et al. 1972
Petroleum						
Empire Mix Saudi Arabian, Nigerian	IS	4 ppm, 9 mo 1 crude oil in 3 estuarine ponds	*C. virginica*	Digestive diverticula mantle, food groove, body wall	Altered diverticula; hyaline degeneration	Barszcz *et al.* 1978
Crude Oil	WC	Crude oil untreated; crude/dispersant sustained 0.5,5, 55 ppm 1 yr	*M. truncata*	Gill, digestive diverticula	Necrosis, granulocytomas, 3 cases neoplasia	Neff *et al.* 1987
	WC	Crude oil untreated; crude/dispersant sustained 0.5, 5, 55 ppm 1 yr	*M. calcarea*	Gill, digestive diverticula	Vacuolization digestive diverticular epithelium; 1 case hemopoietic tumor	Neff *et al.* 1987
Louisiana Crude Oil	WC	Long-term	*C. virginica*	Ciliated epithelium	Exposure affects ciliary action	Galtsoff *et al.* 1935
Waste Motor Oil	WC	20 ppm, up to 60 d	*C. virginica*	Circulatory, GI tract, mantle	Circulatory stenosis; melanization	Gardner *et al.* 1975; Gardner 1978
		20 ppm, up to 60 d	*A. irradians*	Gill and kidney	Gill and plicate impregnated with debris; renal epithelium displaced	Gardner *et al.* 1975
#2 Heating Oil	WC	0.1,0.5 ppm 12 wk	*M. areolata*	Reproductive tissue, zooxanthellae, muscle	degeneration zooxanthellae, atrophy of cells and muscle	Peters *et al.* 1981; Peters 1987
	WC	0.01, 1.0 ppm 2 wk	*C. sapidus*	Hepatopancreas, muscle, gill	NOE	Melzian 1982
	WC	1 ppm, > 30 d	*H. americanus*	Major organs	NOE	Gardner 1978
#2 Heating oil, JP-4 jet fuel	WC	250 ppb; 309 d	*M. arenaria*	Digestive diverticula gonad	Alterations of digestive, epithelia cells; impaired reproduction	Gardner *et al.* 1991b
Diesel and Copper	MC	Cu 25 ppb, oil, 25 ppb 25d	*M. edulis*	Digestive diverticula	Atrophy > basophilic cells	Lowe 1988; Lowe and Clarke 1989

Table 14-1 (Continued)

Agent	Exposure	Concentration/ Duration	Species	Tissue/Organ	Lesion	Reference
Sediment						
Black Rock Harbor	SP	20 mg/l, 11 d	*A. poculata*	Mucous secretory cells	Hyperplasia 100%	Gardner and Yevich 1988
(BRH) Bridge-port, CT	S	50%, 100%, 42 d	*N. incisa*	Cuticle and setae	Macrophages, epithelial change	Yevich *et al.* 1987
BRH	SP	12%; 10 d	*A. abdita*	Mucous cells/tegmental gland, muscle, gills of adults	Pathology mucous cells tegmental gland muscle, gills	Yevich *et al.* 1986
BRH	SP	20 mg/l, 33 d	*H. americanus*	Major organs	NOE	Gardner and Yevich 1988
	SP	20 mg/l BRH sediment 30 d;60 d;30 d 60 d PE; 60 d 30 d PE	*C. virginica*	Kidney, gastro-intestinal gonad, gill, heart, neural	Tumors after 30 d, average rate 14%	Gardner and Yevich 1986, Gardner *et al.* 1987, 1991a
BRH	SP	20 mg/l, 5-7 d 20 mg/l, 28 d	*M. edulis*	Digestive ducts, heart	Necrosis (100%) Myxoma in 2 of 4 animals	Gardner and Yevich 1988
BRH	SP,E	10 mg/l, 30 mg/l, 28 d 1-7 mo (Long Island Sound disposal site)	*M. edulis*	Gill, kidney, intestinal, reproductive	Gills altered in (80%); ovaries (20%); intestinal and renal (7%) NOE in deployed mussels	Yevich *et al.* 1987
BRH	SP	50 mg/l, 100 mg/l, 10 d	*M. edulis*	Reproductive, heart	Pathology of reproductive 25% (50 mg/l) and 87% (100 mg/l); cardiac tumors 23% (100 mg/L)	Yevich *et al.* 1986
BRH	SP	20 mg/l, 15 d	*M. arenaria*	Kidney, red gland, digestive tubules, ganglia	Inflammation of gill and siphon (100%)	Gardner and Yevich 1988
BRH	SP	20 mg/l, 11 d	*A. irradians*	Digestive tubules, kidney, connective tissue	Inflammation, necrosis digestive tubules kidney (100%)	Gardner and Yevich 1988
Miscellaneous						
Phorbol-ester	WC	0.01 ppm added to cadmium sol. 30 d	*D. dorotecephala*	Stem cells	Reticuloma 17.2% after phorbol-ester; 50% in phorbol and cadmium	Hall, Morita and Best 1986 a,b

Table 14-2. Pathological Disorders at Different Contaminated Sites.

Pollutant	Source	Site	Species	Tissue/Organ	Lesion	Reference
Northeast Atlantic						
Petroleum impacted	E	New England 17 sites	*M. arenaria*	Whole animal; index of 5 conditions	Hypoplasia, hyperplasia, lipofusin, neoplasia, hemic proliferation	Walker *et al.* 1981
Petroleum impacted	E	New England 10 sites/survey	*M. arenaria*	Hemopoietic	Hemic neoplasia	Brown *et al.* 1977
Petroleum/ organic pollutants	E	Searsport, ME	*M. arenaria*	Gill, kidney	Hyperplasia of gill and kidney	Barry *et al.* 1971
#2 Heating Oil	E	Searsport, ME	*M. arenaria*	Hemopoietic, gonadal	Hemic neoplasms; germimonas average rate	Barry and Yevich 1975; Gilfillan *et al.* 1977; Harshbarger *et al.* 1979
Urban unknown	E	Narragansett Bay, RI	*M. mercenaria* *A. irradians*	Kidney	Nephrocalcinosis	Doyle *et al.* 1978
Urban unknown	E	Narragansett Bay, RI	*M. mercenaria*	Kidney	Nephrocalcinosis	Barry and Yevich 1969;Rheinberger *et al.* 1979;Seiler and Morse 1988
Urban unknown	E	Narragansett Bay, RI	*M. mercenaria*	Ovary	Ovarian tumors	Yevich and Barry 1969; Barry and Yevich 1972
Organic solvents	E	Military landfill RI	*M. arenaria*	Gill, kidney, hemopoietic	Gill/kidney hyperplasia hemic neoplasms	Gardner *et al.* 1990
Urban sewage	E	Quincy Bay Boston MA	*H. americanus*	All major organs	NOE	Gardner and Pruell 1988
	E	Quincy Bay	*M. arenaria*	Gill, kidney gonad	Gill inflammation 80%, hyperplasia 59%; kidney hyperplasia 72%	Gardner and Pruell 1988
	IS	Quincy Bay	*C. virginica*	Kidney, GI tract	Renal cell tumors, enteric adenomas	Gardner and Pruell 1988
PCBs, inorganics	E	New Bedford MA	*M. arenaria*	Pericardial gland, kidney hemocytes	Increased granulation	Seiler and Morse 1988
PCBs, inorganics	E	New Bedford MA	*G. demissa*	Byssus gland	Teratocarcinoma	Gardner *et al.* 1990
Unknown	E	Bay of Fundy Canada	*H. pyriformis*	Tunic	Supernumerary siphons	Anderson *et al.* 1977
Sediment	E	BRH, Central Long Isalnd Sound (CLIS)	*C. americanus*	Organs	Necrosis, loss of mucous secretory cells, ptychocysts	Peters and Yevich 1989
Sediment	E	BRH 30 d. *in situ*	*C. virginica*	Gill water tube	Tumor (7%)	Gardner and Yevich 1988; Gardner *et al.* 1991a

Table 14-2. Pathological Disorders at Different Contaminated Sites.

Pollutant	Source	Site	Species	Tissue/Organ	Lesion	Reference
BRH dredge materials	E	CLIS	*M. edulis*	Major organs	NOE	Yevich *et al.* 1987
BRH dredge materials	E	CLIS 36 d	*C. virginica*	Gill water tube	Water tube tumor (7%)	Gardner *et al.* 1987, 1991a
Herbicides	E	Searsport, ME	*M. arenaria*	Gonad, hemoplastic	Germinomas, hemopoietic neoplasia	Gardner *et al.* 1991b
Herbicides 2, 4 -D, 2,4,5-T, Tordon 101	E	Forestry, Dennysville, ME	*M. arenaria*	Gonad, siphon	Germinomas 35% teratoid siphons 7%	Gardner *et al.* 1991b
2, 4 -D, 2,4,5-T, Tordon 101	E	Rogue Bluff, ME Blueberry agrochemicals	*M. arenaria*	Gonad	Germinomas 3%	Gardner *et al.* 1991b
Mid-Atlantic						
Farm pesticides	E	Chesapeake Bay, MD	*M. balthica*	Gill	Carcinoma rate 2-10%	Christensen *et al.* 1974
Urban pollution	E	Raritan Bay, NJ	*M. mercenaria* *M. edulis*	Organs	Significant histochemical reaction	Farley 1988 b
			C. virginica	Hemopoietic	Inflammatory responses	Farley 1988a
Polluted environ-ments	E	NOAA, 11 sites	*M. edulis*	Connective tissue, digestive gland, plicate, gonad	Inflammation, mucous cell hyperplasia, ciliate infection	NOAA
South Atlantic and Gulf Coast						
Unknown pollutant	E	Davis Bayou, MS	*P. aztecus* *P. setiferus*	Muscle of 6th abdominal segment	Hamartoma	Overstreet and Van Devender 1978
Complex organic	E	3 Northern Gulf Coast estuaries	*C. virginica*	Digestive gland, hemopoietic	Non-specific inflammation; epithelial atrophy; hemic neoplasms	Couch 1978, 1984, 1985
Complex organic	E	Pascagoula harbor, MS	*C. virginica*	Digestive gland	Atrophy 100%,	Couch 1984
Complex organic	E	Indian River, FL	*M. mercenaria*	Gonad	Gonadal neoplasia	Hesselman *et al.* 1987, 1988
Northwest Pacific						
Copper	E	Howe Sound Vancouver, B.C.	*M. carlottensis*	Digestive diverticula kidney, gut	Inflammation necrosis, vacuolation	Bright 1987
B[a]P (Tissue)	E	Yaguina Bay, OR	*M. edulis*	Major organs	Hemic disorders	Mix *et al.* 1979

Table 14-2. Pathological Disorders at Different Contaminated Sites.

Pollutant	Source	Site	Species	Tissue/Organ	Lesion	Reference
	E	Puget Sound	*C. capitata*	Major organs	NOE	Malins *et al.* 1980
	E	Puget Sound	*C. capitata*	Major organs	NOE	Malins *et al.* 1980
	E	Puget Sound	*C. capitata*	Major organs	NOE	Malins *et al.* 1980
	E	Puget Sound	*P. pinnata*	Major organs	NOE	Malins *et al.* 1980
Complex sediment	E	Puget Sound	*C. alaskensis*	Bladder, gill, hepato-pancreas, antennal gland	Necrosis, melanized nodules/granulomas in gill	Malins *et al.* 1980, 1982
Complex sediment	E	Puget Sound	*Cancer* sp.	Gill, hepatopancreas, bladder, midgut	Necrosis, epithelial metaplasia, melanized nodules	Malins *et al.* 1980, 1982
Complex sediment	E	Puget Sound	*Pandalus* sp. *Cancer* sp.	Gill, hepatopancreas, antennal gland, bladder	Necrosis, melanized nodules/granulomas in gill	Malins *et al.*1980, 1982; Olson and Myers 1986
Complex sediment	E	Sinclair Inlet, Puget Sound	*A. castrenis*	Digestive tubules	Necrosis 21%	Malins *et al.* 1980 Olson and Myers 1986
		Sinclair Inlet, Puget Sound	*M. carlottenis*	Digestive tubules	Necrosis 100 %; NOE in clams at other Puget Sound Locations	Malins *et al.* 1980 Olson and Myers 1986
Europe						
Petroleum	E	*Amoco Cadiz* Spill, France	*C. edule*	Connective tissue	Sarcomas in 3 wild populations; at *Amoco Cadiz* spill and reference sites	Auffret and Poder 1986; Poder and Auffret 1986
Light Arabian and Iranian crude	E	*Amoco Cadiz* Spill	*C. gigas* *O. edulis*	Digestive diverticula tissue, gills, gonad	Destructive necrosis of digestive glands 7 yr after wreck of *Amoco Cadiz*; gonadal 2 yr.	Berthou *et al.* 1987; Neff 1979
Copper	E	Finland Industrial Pollution	*M. edulis*	Digestive diverticula, kidney	Ulcers, hemorrhage of digestive glands, cystic kidney	Sunila 1988
Industrial pollution	WC/E	Baltic Sea near discharges	*M. edulis*	Gill	Deformities in gill of mussels deployed similar to observations in laboratory studies	Sunila 1987
Metals	E	Langesund fiord, Norway PAH, PCB	*M. edulis*	Digestive diverticula	Modified structure, changes in lipid levels	Lowe 1988

water sponges (Mysing-gubala and Poirrer 1981), and acute and chronic exposure to crustaceans (Couch 1977, 1978; Nimmo *et al.* 1977; Lightner 1987). In planaria cadmium acted as a carcinogen. In evaluations of chemical carcinogenesis with phorbol ester 12-O-tetradecanoylphorbol-13-acetate (TPA) added to cadmium-containing solutions, it acted as a tumor initiator. Studies of sponge gemmule growth demonstrated a positive correlation with alterations and increasing concentrations of cadmium. In crustaceans, the response to cadmium included gill melanization, a condition known as "Black Gill Disease."

Chromium: Doughtie and Rao (1984) found hexavalent chromium had a preference for antennal glands, hepatopancreas, gill, and midgut of shrimp when given in solution. Doughtie and co-workers (1983) observed 50% of surviving shrimp with gross characteristics of "shell disease." Using similar exposure methods, Lightner (1983, 1987) determined chromium (hexavalent) caused rounding and sloughing of midgut and hepatopancreas epithelial cells.

Copper: Nine studies detailed results from acute and chronic copper exposures to arthropods and bivalve and gastropod molluscs. With few exceptions, it generally affected a high frequency of lesions in respiratory organs. Gill alterations resulted from exposure in >50% of the studies, followed in order by kidney (38%), digestive diverticula glands and neural elements (25%), and mantle, ctenidia, and epipodites (13%). Short-term tests defined exogenous melanization on gill and epipodites of shrimp (D.V. Lightner, Arizona State University, Tucson, personal communication), neural alterations in blue crabs (Bodammer 1979), and gill necrosis in channeled whelk (Betzer and Yevich 1975) and quahogs (LaRoche *et al.* 1973). In addition, long-term tests methods have shown respiratory organ deformations from exposure to 200 ppb copper for 24 h and a 1-year period of post-exposure of mussels (Sunila, 1987; 1988). Sunila (1988) correlated similar deformations in natural populations of Baltic Sea mussels with metal contamination emanating from iron and steel factory discharges. Studies of nephrotoxic change indicated the combination of controlled exposure and 4 months of *in situ* deployment in the Baltic Sea led to watery cysts; condition frequency increased to 22% after 1 year (Sunila 1986, 1987). In studies of toxic responses in mussels and bay scallops, cellular alterations of digestive diverticula dominated (Calabrese *et al.* 1984; Yevich and Yevich 1985).

Mercury: Jensen and Baatrup (1988) demonstrated persistence from exposures to mercury in marine polychaetes, with an autometallographic silver enhancement method to confirm localization in intestine, nephridia, epidermis, and cuticle. Studies of freshwater sponge gemmules have shown a relationship between mercury concentration and growth alterations (see section on cadmium) (Mysing-gubala and Poirrier 1981).

Aromatic Hydrocarbons

Methylcholanthrene (MCA) and benzopyrene (B[a]P) stimulate papilliform tumors (30%), other malformations (32%) in planaria (Foster 1969), and papillomas, benign and malignant adenopapillomas, adenomas, and an adenocarcinoma in apple snails (3%) (Krieg 1970, 1973) when administered to epidermal surfaces. These compounds are ineffective carcinogens in oysters when administered in water column solutions (Couch *et al.* 1979), but B[a]P appears to be a coeffective tumor inducer when given in sediment (Gardner *et al.*, in press). In toxicological testing for the National Toxicology Program, Spangenberg (1984) determined B[a]P and nine other PAHs cause jellyfish segmentation or strobilation delays, somatic mutations, and teratogenic malformations.

In studies of B[a]P carcinogenesis, Shimkin and co-workers (1951) demonstrated that barnacles attached to creosote-treated wooden pilings in a Corona Del Mar, California, marina contained benzopyrene in tissues. By extracting B[a]P and accompanying analogs from these crustaceans and injecting them under epidermal surfaces of mice, they produced subcutaneous tumors in 4 of 12 animals receiving 0.5 mg injections of material after 16, 17, 19, and 20 weeks, and 2 of 12 mice receiving 0.25 mg after 17 and 19 weeks. They also reported the presence of 3,4-benzpyrene in tissues of the goose barnacle, *Mitella polymerus*, collected from pier pilings. This approach to study pathways of PAH carcinogens in the food web could prove useful for investigations of seafood contamination and human risk.

Substituted Aromatic Hydrocarbons

Findings of studies on developmental and structural changes affected by substituted aromatic compounds included aniline in coelenterates (Spangenberg 1984), aflatoxin B1 in crustaceans (Lightner *et al.* 1982; Lightner 1987), and gastropods (Bauer and Tulusan 1973). Results of studies of aflatoxin B1 included findings of necrosis and fibrosis of shrimp hepatopancreas when administered by injection, and necrosis of mandibular organ when the compound was given a diet. Bauer and Tulusan showed aflatoxin produced subcellular alterations (segregation and zoning of nucleolar components) in snail tentacle epithelium characteristic of structural lesions observed in cell cultures and hepatic tissue of rat. Four studies in shrimp and oysters showed changes from chronic exposures of aroclor 1254. These studies showed that the compound augments shrimp-baculovirus interactions (Couch and Nimmo 1974; Couch and Courtney 1977), stimulated increased endoplasmic reticulum, free ribosomes, membrane whorls, and nuclear degeneration (Lowe *et al.* 1972), and also crystalloids in nuclei of hepatopancreatic cells (Nimmo *et al.* 1975). The compound in oysters also induced leucocytic infiltration, and histological alterations of digestive diverticula and associated connective tissue. A single report discusses morphological changes in quahog digestive diverticula from exposures to hexachlorobenzene (Tripp and Fries 1987). The chlorinated compound failed to elicit structural lesions in these and other tissues.

Seven studies summarized toxic and carcinogenic properties of insecticides in helminths, crustaceans, and molluscs. Studies by An der Lan (1962) and Foster (1969) have shown that DDT and Sevin in planaria lead to abnormal, lethal, epidermal growths when administered in solution to helminths. Kayser and co-workers (1962) determined that DDT and parathion in crayfish caused neurological changes, including swelling, chromatolysis, and agglutination of nissl substance around cell nuclei; Lowe *et al.* (1971) indicated that DDT, parathion, and toxaphene as a mixture in young oysters leads to kidney, visceral ganglion, gill, digestive tubule, and connective tissue lesions. In other studies, the insecticide Sevin in bent-nosed clams produced lesions in gill, mantle, siphon, and suprabranchial glands (Armstrong and Millemann 1974), while dieldrin in oysters, when exposed in a static water system (Emanuelsen *et al.* 1987), or in baltic macoma clams, by route of calculated doses (Farley 1977) was ineffective.

Dyes

There are two studies of carcinogenic dyes, tumor induction studies in planaria have shown that Scharlach Red dye produced abnormal lethal growths much like those fostered by DDT and Sevin when applied to flatworms (two species) (An der Lan 1962). In contrast, results of microcosm simulation studies in nutclams indicated the aromatic azo compound "Direct Blue-15" was ineffective in these bivalve molluscs (Perez *et al.* 1987).

Phenols and Substituted Phenols

Toxicological testing of phenol and cresol in studies with jellyfish (Spangenberg 1984) and pentachlorophenol in relation to morphological changes in quahog digestive epithelial cells complete the list of studies of these compounds (Tripp and Fries 1987). Phenol and cresol caused segmentation delays, phenol also caused mutations and teratogenic changes, and pentachlorophenol was ineffective.

Nitroso Compounds

Seventeen studies detailed toxic and carcinogenic effects of nitrosoamines, including N-nitrosodimethylamine (DMNA), N-nitrosodiethyl-amine (DENA),

N-methyl-n-nitro-n-nitrosoguanidine (MNNG), N-methyl-nitrosurea (MNU), N-nitrosodipropylamine (NDPA), and nitrilotriacetic acid (NTA). Invertebrate models studied included freshwater crayfish (Harshbarger *et al.* 1971), freshwater mussels (Khudoley and Syrenko 1978), blue mussels (Rasmussen 1982, 1986; Rasmussen *et al.* 1983a,b, 1985), and eastern oysters (Couch *et al.* 1987, Winstead and Couch 1988). In addition, Eisler *et al.* (1972) tested effects of acute exposures to several marine macroinvertebrate and fish species.

The characterization of DMNA in crayfish by Harshbarger and co-workers represents the first systematic measure of nitrosoamine effects in aquatic invertebrates. Exposure to DMNA (10-200 ppm) leads to antennal gland degeneration and hepatopancreas tubular cell hyperplasia when given in a water medium. Later studies have shown tumor-inducing action of DMNA and DENA in freshwater mussels (Khudoley and Syrenko 1978), but not in blue mussels (Rasmussen 1982; Rasmussen *et al.* 1983) or oysters (Couch *et al.* 1987). Carcinogenesis studies of DENA and DMNA (solutions of 200-400 ppm) in freshwater mussels have shown that exposure leads to digestive epithelial blast cell, renal, and hemopoietic neoplasms. Extension studies have tested carcinogenicity of DENA, DMNA, MAM-Ac, MNNG, MNU, and NDPA in blue mussels by route intramuscular injection, and DENA and MAM-Ac by water column solution and found non-neoplastic action was stimulated in connective tissue and in circulatory, respiratory, and digestive systems (Rasmussen 1982, 1983a, 1983b, 1985). Rasmussen concluded that failure of DMNA and other nitroso compounds to show tumor-inducing action achieved in fresh water mussels by Khudoley and Syrenko probably to be related to degree of exposure, duration of study, and concentration of compound.

Crude Oil and Petroleum

Influence of crude oil and petroleum exposures to marine invertebrates has been well-studied in laboratory and field investigations. Galtsoff *et al.* (1935) determined Louisiana crude oil exposures to oysters caused a corresponding ciliary dysfunction. Since then, the results of laboratory studies have shown that home-heating oil (No.2 fuel oil) and diesel fuel affect histopathology of stony coral (Peters *et al.* 1981; Peters 1987), softshell clams (Gardner *et al.* 1991b), and mussels (Lowe 1988; Lowe and Clarke 1989), but has little influence on blue crabs (Melzian 1982). Histopathological studies show that No.2 fuel oil in coral promoted loss of zooxanthellae, atrophy of muscle and mucous epithelia, and impaired reproductive potential, and altered digestive diverticular epithelia, and impaired gonadal development in softshell clams. Findings from studies of diesel fuel and copper in mussels have shown cellular changes of digestive epithelial similar to those observed in other bivalve molluscs. Investigations of waste motor oil in oysters and bay scallop indicated the metal contaminated oil influenced lesion development in branchial efferent veins and digestive diverticula of oysters, and gill and kidney of bay scallop, and mantle surfaces of both species (Gardner *et al.* 1975).

Two studies analyzed the effects of controlled spills of crude oil in temperate and arctic regions. Empire Mix, Saudi Arabian, and Nigerian crude oils in different estuarine ponds stimulated action in oyster digestive diverticula and reproductive organs (Barszcz *et al.* 1978). Crude oil in a Canadian arctic ecosystem was related to vacuolation in softshell clam digestive epithelium (Neff *et al.* 1987), who reported neoplastic changes in truncate softshell clams (4 cases) exposed to Lagomedio crude oil and Corexit 9527, and chalky macoma (1 case) exposed to the untreated crude oil. These studies of petroleum toxicity provide the only evidence of petroleum stimulated carcinogenicity in bivalve molluscs.

Chemically contaminated Cedar Creek, Black Rock Harbor (BRH), Bridgeport, Connecticut, sediment caused several conditions in polychaete worms, amphipods, and mussels when given as particulate suspensions, and as a solid (Yevich *et al.* 1986, 1987) including defects in marine worm parapodia and cutaneous tissue, in amphipod gill, tegmental gland, and muscle, and in mussel heart (cardiac myxomas 23%) and gonads. Several studies of BRH sediment carcinogenicity have proven that carcinogens present in Cedar Creek sediment are capable of stimulating a significant excess of adenomas, myxomas, papillomas, and blastomas when administered as particulates of filterable size (Gardner and Yevich 1986; Gardner *et al.* 1987, 1991a). Results of sediment carcinogenicity studies in oysters deployed in Cedar Creek have shown that these molluscs develop gill filament and water tube tumors. Information from the studies shows that sediment has high levels of polychlorinated biphenyls (PCBs), polycyclic aromatic hydrocarbons (PAHs), polycyclic aromatic ketones (PAKs) and polycyclic aromatic quinones (PAQs), chlorinated hydrocarbons and carbazoles, and nickel, lead, cadmium, chromium, iron, copper, zinc, manganese, mercury, and possibly carcinogenic aromatic amines (Gardner *et al.* 1987). The uptake of these chemical compounds from filterable sediment particles by mussels and oysters stabilizes after about 7 days, and tumors develop within 30 days when administered by particulate concentrations of <20 mg/l. Most tumors are benign, although malignant variants occur.

FIELD STUDIES OF CHEMICAL EFFECTS

The modern survey studies of invertebrate diseases in chemically degraded aquatic environments consist of examinations of indigenous species 88%, and deployment of healthy organisms into degraded environments 12% of the time.

Summary information by area below shows that these diseases correspond to environmental concentrations of PAHs, PCBs, insecticides, metals, petroleum, sewage, and other forms of contamination in nearly all coastal regions of the United States. Chemical analyses of sedentary invertebrate organism tissues generally show that animals collected in nearshore and urban estuaries have elevated concentrations of these same xenobiotics.

United States and Canada

Maine: Studies of molluscs environmentally exposed to heating oil and jet fuel spilled into Long Cove, Searsport, Maine, led to the first report of hemopoietic and gonadal neoplasms in softshell clams (Barry and Yevich 1975). The presence of these hemic neoplasms at a petroleum-impacted site urged investigations within other oil-impacted areas in the Northeast Atlantic region (Barry and Yevich 1975; Yevich and Barszcz 1976, 1977; Brown *et al.* 1977; Gilfillian *et al.* 1977; Gardner *et al.* 1991b; Harshbarger *et al.* 1979; Peters 1988; Walker *et al.* 1981). These studies fail to support petroleum as a significant factor in the etiology of these hemic disorders. Recent data have shown the prevalence of germinomas consistently averages 22% at Searsport, 3% at Machiasport, and 35% at Dennysville, Maine (Gardner *et al.* 1991b). The tumors are mostly benign, although aggressive variations occur; they develop at most ages without favoring one sex over another. These studies bolster the idea that phenoxyacetic acid herbicides could be involved in this disease of softshell clams (See carcinogenesis, field survey section). These studies also described pericardial mesotheliomas and teratoid siphon anomalies in Dennysville clams. Studies of homologous siphon atypia in tunicates of Deer Island located approximately 10 miles east of Dennysville (Anderson *et al.* 1977) have not determined the causal agent.

Narragansett Bay, Rhode Island: Invertebrate diseases in Narragansett Bay from impacts of pollution are being studied with some regularity by nearby academic, state and federal regulatory agencies. These studies of Bay pollution determined that contamination extends as a gradient from Gaspee Point near Providence to Block Island Sound. In early studies, Yevich and Barry (1969) and Barry and Yevich (1972) reported the first molluscan tumor in 3% of Rose Island hardshell clams. Rose Island occurs in the moderate to lower level of the pollution gradient, and bay pollution was not associated with the gonadal tumor at the time of study. In studies of molluscan metabolism, Rheinberger *et al.* (1979) and Doyle *et al.* (1978) determined that amorphous formations of calcium phosphate (nephrocalcinosis) in quahogs and bay scallops was correlatable to the pollution gradient. The appearance and increase of gill inflammation, gill hyperplasia, kidney hyperplasia, and rickettsiosis in softshell clams between 1960-1990 has also been associated with the pollution gradient (Gardner *et al*. 1990).

Central Long Island Sound (CLIS), New York: The U.S. Army Corps of Engineers conducted dredging operations in Cedar Creek, Bridgeport, Connecticut, in the mid-1980's accompanied by studies of fate and effects of sediment associated chemicals released to the environment from dredging operations and disposal in CLIS. A number of reports published between 1985-1987 detail these activities (see bibliography). Stebbing and Brown (1984) encouraged ecotoxicological testing of hydroids, anemones, and corals. Peters and Yevich (1989) studied the burrowing anemone-like anthozoan, *Ceriantheopsis americanus*, at the CLIS site. They studied animals from the area at pre-disposal, and examined anemones at 3 stations after disposal. The initial excellent condition of specimens deteriorated after disposal activities began. The studies indicate accumulations of cellular debris and necrosis in all areas of the body. Ultimately these changes disrupted burrow, or tube-formation.

Massachusetts: Persistence of high environmental levels of PCBs and heavy metals in New Bedford Harbor (NBH) was sufficient for U.S. EPA regulators to designate the area as a "Superfund Site." Two studies of co-occurrence of high levels of environmental contaminants and disease have shown histological evidence of disease processes in bivalve molluscs. In a survey of ribbed mussels, Gardner (1990) found a teratocarcinoma in 1 of 30 animals (RTLA #4976). Seiler and Morse (1988) correlated qualitative and quantitative differences of granulocytes in softshell clams with harbor pollution. Studies of winter flounder in different areas of New England correlate prevalence of liver neoplasms with environmental levels of contaminants and NBH winter flounder have one of the highest prevalences (Gardner *et al*. 1989). Findings suggest that the area offers opportunity for studies on responses of fish and invertebrates as indicators of environmental contamination and consequent risks to ebbing fisheries resources.

Concern about sewage contamination in Quincy Bay led to a Congressionally mandated study of Quincy Bay, Boston Harbor, in relation to the Nut Island Sewage Treatment Plant. Gardner and Pruell (1988) evaluated pathology and chemistry showing presence of PCBs, PAHs, and metals including carcinogens in sediment surface and cores, and in tissues of lobster, softshell clams, fish, and oysters deployed *in situ*. Study results have shown Quincy Bay pollution did not affect structural disturbances in lobster, supporting previous evidence suggesting the forms of environmental pollution in the northeast do not often cause internal diseases in these animals. There still could be functional risks to lobster, because data analysis of tissues shows the highest levels of chlorinated hydrocarbons (in lobster hepatopancreas 113 $\mu g/g$ dry wt.), and the highest PAH concentrations in the species studied. Tissue analysis has shown PAH levels in clams intermediate between lobster and fish, and pathological examinations revealed the remaining populations, one at "the Moons" and the other "Moon Island," had been decimated by gill inflammation (80%), gill hyperplasia (59%), kidney hyperplasia (72%), and rickettsiosis (52%), and that gamete maturation favoring male over female probably caused spawning asynchrony. Analysis of PAHs, PCBs, and metals in healthy oysters deployed *in situ* at four different locations in Quincy Bay near the Nut Island Sewage Treatment plant probably contributed to a 6% rate of renal cell and enteric adenomas in these animals, 10% of these oysters became infected by papovavirus, or "Ovacystis Disease" during deployment. The public health risks these contaminant levels and viral infections may pose are not known. The study prompted the U.S. EPA and State of Massachusetts to issue a health advisory for contaminants in fish and shellfish from Quincy Bay. Estimated cancer risks and recommendations derived from the Quincy Bay Study data have been discussed (Cooper *et al.* 1991).

New York Bight: In studies of mid Atlantic region, and Great Bay and Raritan Bays, New Jersey, Farley (1988a,b) found affected bivalve mollusc connective tissues, digestive glands, gills, plicates and gonads and suggested the lesions were "probably caused by copper." In other studies in the Chesapeake Bay area, Christensen and co-workers (1974) cited a possible association between agrochemicals and gill carcinomas in *Macoma spp*., and Harshbarger and co-workers (1979) described the occurrence of proliferative disorders in Chesapeake Bay oysters and softshell clams.

Northern Gulf Coast estuaries: Two studies revealed a relationship between pollution in the South Atlantic Gulf Coast and diseases of shrimp and oysters. Field studies in Mississippi estuaries have shown white and brown shrimp at the most polluted sites have hamartomas (Overstreet and Van Devender 1978). Studies of organochlorine, phosphate, and carbamate pesticides, napthalenes, and mercury pollution in Gulf shrimp and oysters by Couch (1982, 1984, 1985) indicated, histological background variation was mostly within normal ranges. Atrophy of digestive glands in oysters was correlated with "organic pollutant chemicals" from least to the most polluted Gulf sites, and a proliferative blood disease was considered to be a possible neoplastic development.

Indian River, Florida: Basic studies on reproductive dynamics of Indian River, Florida, hard clam (*Mercenaria* spp.) revealed invasive gonadal neoplasms (Hesselman *et al.* 1988, 1989; Hesselman and Arnold 1990). These monomorphic, basophilic germ cell tumors averaged 11.6% of the population, while peak prevalence reached as high as 60%. Histological studies following disease progression failed to show a relationship between tumor development and factors of seasonal variation. These studies have shown that tumors affected females more often than males.

Puget Sound, Washington: Malins and co-workers selected several annelids, crustaceans, and molluscs to identify and define areas of chemical contamination in Puget Sound (Malins *et al.* 1980, 1982; Olsen and Myers, U.S. Marine Fisheries Service, unpublished). Elevated levels of PCBs, chlorinated pesticides, other chlorinated organic compounds and metals in Puget Sound sediment have been correlated with concentrations in tissues marine organisms. Histological evaluations of marine worms, shrimp, crabs, and clams primarily detailed 6 lesion types in shrimp, and crabs that may have been related to chemical pollution of Puget Sound sediment. Tubular metaplasia, vesiculate hepatopancreatic cells, and necrosis of hepatopancreas and bladder were among the observed effects. Melanized nodules in gill stems of one crab species, and necrosis of digestive tubular epithelium of clams were included. The outcome of these studies in Puget Sound was consistent with, and supported other surveys in the U.S. that have shown relationships between chemical contamination of sediment and disease processes.

British Columbia, Vancouver: Bright (1987) selected the bivalve mollusc, *Macoma carlottensis*, to study effects of copper mining in Howe Sound, British Columbia. He found pathology in digestive diverticula, ctenidia, kidney, and ovary of these indigenous clams. Further, the observed frequency of these conditions were directly correlatable to copper mine tailings as a function of distance from a point source.

Oregon: In studies of Yaquina Bay mussels Mix and coworkers (1979) found measurable levels of benzopyrene in tissues, and histological evidence of proliferative disorders. The occurrence of B[a]P body burdens correlated to hemic disorders in nearly 10% of all Yaquina Bay mussels at the time of survey. A similar relationship existed for winter flounder liver neoplasms and B[a]P in sediment (Gardner *et al.* 1989).

Arctic

Two studies of molluscan pathology related to petroleum exposures in arctic marine environments (Hodgins *et al.* 1977; Neff *et al.* 1987). These studies have shown petroleum in two species of softshell clams causes tissue alterations similar to those observed in a number of studies of petroleum in molluscs at temperate climates. These investigators associated the influence of crude oil with hemic neoplasms.

Europe

In the Baltic Sea, Finland, Sunila (1987, 1988) developed information on pollution-related pathology in Baltic Sea mussels at 23 polluted and unpolluted sites. Mussels located near iron and steel factory point-source discharges to the sea had gill deformation, ulceration of digestive lining epithelium, and cystic cavities in kidneys. Sunila correlated the influence of factory pollution on mussel lesions, and acute copper exposure (24 h to 0.2 ppm) followed by post-exposure monitoring. These studies by Sunila on copper corroborated findings by Farley (1988b) who attributed gill alteration to pollution, particularly copper. In Brittany, France, a number of studies of the *Amoco Cadiz* oil spill of 1978 in molluscs, including flat and Pacific oysters, razor clams, and cockles, have shown vacuolar lesions in digestive epithelial linings, gonadal atrophy in oysters, and sarcomas in cockles (Auffret and Poder 1986; Berthou *et al.* 1987; Neff and Haensly 1982; Poder and Auffret 1986; and Yevich, unpublished data). By following lesion progression, Berthou and co-workers found total destruction of digestive epithelial mucosa. Atrophy of tubular epithelium in gonads appeared after the first and second post-spill year. The destructive necrosis of gonadal organs in these oysters led to lost reproductive capability after one year. They determined that oysters were at risk because gonadal pathology persisted for more than five years after the spill event. In their studies, Auffret and Poder found sarcomatous lesions in cockles in two different populations near the site of *Amoco Cadiz* spillage and reference area. Cockles at the reference site had the highest tumor prevalence (46%); cause of the condition remains unknown.

SUMMARY

Laboratory Experimental Studies

Harmful effects of environmental chemicals are most commonly observed at portals of chemical entry, concentration, and excretion including organs of respiration, metabolism, and excretion. Lesions are found less frequently in organs of reproduction, and neurological control. The experimental evidence has shown toxic induced lesions found at higher phylogenetic levels are found also in invertebrates (Sparks 1985; Couch 1978). These structural and functional modifications include invasiveness and relocation of the cells of resulting neoplasms.

Organ and Species Specificity and Multiple Effects

Studies of crustacean gill and epipodite melanism resulting from exposures of copper and cadmium indicate environmental contaminants may be associated with carapace pigmentation in wild crustaceans. Investigations of pathologic pigmentation indicate the condition in shrimp and crabs of Puget Sound (Malins

1982) and euphasiids at Deep Water Dumpsite 106 in the North Atlantic Ocean (MacLean *et al.* 1981; Sindermann *et al.* 1989) correspond with pollution. Research has not detected the fundamental mechanisms/processes leading to melanotic pigmentation of the carapace.

Toxicologic characterization of arsenic and silver poisoning in organs of molluscs compared well with results observed in mammals. Deposit of silver (argyrism) in enteric tissue resulting from chronic exposure and from ocean dumping activities involving metals showed a high degree of specificity. Silver specificity for site of action in molluscs makes argyria a potential marker for assessment of ocean disposal sites, ocean outfalls, and other pollution gradients containing silver.

Research activities indicate invertebrate sensitivity to toxic effects of copper, silver, PCBs, nitroso compounds, crankcase oil, home heating oil, jet fuel, and contaminated sediment usually included multiple alterations of atrophy, vacuolation, necrosis, and ulceration of diverticular surfaces. This is a generalization that may have some exceptions, but appears sound in molluscs from areas of Raritan Bay, New Jersey, polluted by heavy metals, in Howe Sound, British Columbia, polluted by mine tailings, in Gulf of Mexico estuaries polluted by organics, and at locations of catastrophic petroleum spills. These distinctive responses may be a useful starting point for effective application of wider investigative techniques.

Carcinogenesis

Detailed experimental studies of carcinogenesis have been accomplished in platyhelminthes, arthropods, and molluscs. Lesions produced by carcinogenic compounds suggest that invertebrate cellular growth disorders are similar to those seen at higher phylogenetic levels. In addition, studies have described enzyme systems that transform xenobiotics and initiate neoplastic diseases in molluscs and crustaceans (Lee 1981; Anderson 1985; Stegeman and Lech 1991).

PAHs and turbellarians, cirripedia and gastropod molluscs: In early studies of PAH carcinogenicity pioneer investigators used intact turbellarians and gastropods as model systems of tumor induction and transmission to study rates and manner of growth in established tumors. Studies of chemical carcinogenesis in apple snails by Krieg (1970) resulted in the first chemically induced molluscan neoplasms to prove transplantability (Krieg 1970, 1973). The assessment of interactions of cadmium and TPA in flatworms (Hall *et al.* 1984, 1986a,b) demonstrated the principle of cocarcinogenesis.

Characterization of effects in organs of oysters resulting from chronic exposures of PAHs have shown positive alterations in rate of uptake, metabolism, and correlative pathology (Couch *et al.* 1979). Couch's view that this relationship of susceptibility in oysters could serve as a model indicator of chemical carcinogens in the aquatic environment has been supported by later investigations (Pittinger *et al.* 1985, 1987; Gardner and Pruell 1988; Gardner 1991a). In contrast to molluscs, studies of crustaceans demonstrate that they accumulate potent carcinogens in tissues and organs, but these compounds fail to act as inducers of neoplasms (Gardner and Yevich 1988; Gardner and Pruell 1988). The carcinogenic potency of accumulations of these compounds in crustaceans and molluscs are unquestionably demonstrated for mammals and fish. Studies of neoplasms in rodents injected with B[a]P homologs extracted from barnacles (Shimkin *et al.* 1951) and tumors in winter flounder from a diet of chemically contaminated mussels (Gardner *et al.* 1991a) revealed the potential human health risk from consumption of contaminated crustaceans and molluscs.

The role of certain potent carcinogens, e.g., nitrosamines, in inducing neoplasms in invertebrates, e.g., bivalve molluscs, in nature, is unknown. Analytical methods now available indicate amines are mostly undetectable in the environment of contaminated harbors like Puget Sound (Tetra Tech 1986); therefore, significance of their risk to wild invertebrates remains uncertain.

Compounds bound to sediment are involved in tumor formation in molluscs and fish. Several investigations have shown that these compounds can cause tumors in bivalve molluscs, particularly oysters. Experimentally, Gardner and coworkers (in press) determined that selected aromatic amines, chlorinated hydrocarbons, and heavy metals at 1X and 10X the level found in polluted environments, like BRH, can cause renal and enteric adenomas in oysters exposed in sediment as a filterable particulate for 30 days. The oyster model allows us to ask relevant questions about the carcinogenic process and extend laboratory studies to environmental risk assessment.

Field Survey Studies

Organ and Species Specificity

Specific lesion types correlatable with pollution gradients include: crystalloid formation in hepatopancreas of shrimp exposed to organic pollution and PCBs; deformed gills in mussels exposed to copper; nephrocalcinosis in bivalve molluscs exposed to metals and sewage; alterations in circulating hemocytes of molluscs from polluted embayments; and cardiac neoplasms in bivalve molluscs exposed to substances at ocean dumpsites. Disordered cardiac structure in bivalve molluscs from sites contaminated by herbicides, sewage, and metals has occurred in the Hardscrabble River, Maine (Gardner 1991b), the Philadelphia ocean dumpsite, and the Atlantic Ocean Deepwater Site 106 (K. Eller, University of Connecticut, Avery Point, and P.P. Yevich, unpublished data), thus suggesting that chemicals may be involved in the neoplastic disease process. The induction of mesotheliomas and myxomas in bivalve molluscs by xenobiotics contained in Black Rock Harbor sediment provided strong, clear-cut evidence that environmental pollution is involved in neoplastic disease. These findings could be of practical value in studies of heart structural and functional changes produced by the disease, as well as assessments of pollution.

Investigations have shown patterns of identifiable diseases targeting the gill, digestive glands, and kidney of molluscs at high frequencies in environments contaminated by aromatic hydrocarbons, chlorinated hydrocarbons, insecticides, and metals. Certain characteristic forms of structural modification are apparent. Studies of petroleum-induced pathogenesis show that lesions typically occur in digestive and reproductive organs, for example. The same chemical agents that cause tumor formation may foster increased susceptibility to parasitism and pathogens (Couch 1978; Gardner and Pruell 1987; Möller 1987; Gardner *et al.* 1990); thus, studies of these interactions may provide an integrated holistic approach to studies of ecological risk.

Carcinogenesis

Studies of several shellfish and fish species from the Great Lakes, Puget Sound, Yaquina Bay, Boston Harbor, New England coastal waters, and other areas led to early speculation that PAHs, especially B[a]P, are important in carcinogenesis in aquatic organisms. Mussels studied by Mix and co-workers (1979) showed correlative evidence between tissue levels of B[a]P and a proliferative disorder in mussels. The development of *in situ* data linking rapid uptake of aromatic hydrocarbons, mutagenic activity, and tumor formation in oysters to environmental contamination (Pittinger *et al.* 1985; Pittinger *et al.* 1987; Gardner and Pruell 1987; Gardner *et al.* 1991a) lends support to the above relationship of B[a]P in mussels. A similar relationship has been established between sediment levels and winter flounder liver neoplasms (Gardner *et al.* 1989).

Agrochemicals and high prevalences of neoplasms, including germinomas in molluscs, carcinomas, melanomas, and papillomas in bullhead fish, and cutaneous fibroblastic neoplasms of green sea and loggerhead turtles, co-occur at sites of known or suspected contamination, including Searsport, Machiasport, and Dennysville, Maine (Gardner *et al.* 1991b), Indian River, Florida (Hesselman *et al.* 1988), and lakes of Florida citrus groves (Harshbarger and Clark 1990; Jacobsen *et al.* 1989; Harshbarger, this volume). These areas are known to have received repetitive applications of phenoxyacetic acid herbicides and more than 50 other pesticides, herbicides, and fungicides. Co-occurrence of high environmental exposures and tumor-bearing animals are thought to be related by the action of watershed runoff. Experimental studies showing carcinogenicity of DDT and Sevin in planarians (An der Lan 1962; Foster 1969), and epizootiological studies linking phenoxyacetic acid herbicide defoliating agents (i.e., Agent Orange) to an excess of seminomas in military dogs in Vietnam between 1967 and 1972 (Hayes *et al.* 1990) strengthen the above evidence.

The U.S. EPA developed an historic data base for spatial and temporal patterns of U.S. cancer mortality rates and trends. The rate of mortality from cancer of the ovary, fallopian tube, and broad ligament for females from Washington County, Maine, and Indian River and Brevard Counties, Florida, was insignificant in 1950-1969, but exceeded the national average by a significant margin in the 1970's (Riggan *et al.* 1987). These demographic data show that boundaries of excessive mortality rates in humans from gonadal cancer and neoplastic conditions in molluscs co-occur in Maine and Florida. This does not imply that high concentrations of insecticides may be involved in these human health problems, or that data extrapolations can be made from epizootiological study of clams to predict human risk. It does imply that lower animals may provide clues for the human situation.

CONCLUSION

The broad occurrence of carcinogens in the world's aquatic ecosystems (Kraybill 1977) and contamination of aquatic seafood (Dawe 1991) have encouraged a public commitment to protect sensitive environmental systems and safeguard human health. These concerns have been underscored by presence of recognizable neoplasms in lower animals, particularly molluscs from most regions of the U.S. This survey has shown that carcinogens, mutagens, and genotoxic agents in the water, sediment, or tissues, solely or in combination, may cause neoplasms in vertebrates and invertebrates either directly or through the food web. In contrast, high concentrations of PAHs, PCBs, insecticides, and metals are not known to cause neoplasms in arthropods, but, as in the case of barnacles, tissue carcinogens in these animals may be potent and capable of causing tumors in higher trophic forms. Thus, the scientific character of invertebrate pathology has increased our understanding of chemicals representing a risk for aquatic ecosystems. Further, this discipline can be useful in an agenda for scientific progress. For example, studies of shellfish neoplasia have presented insights for experimentalists from other fields (i.e., genetics, genetic alteration of cellular oncogenes, DNA adducts, transfection assays, and others) (Jackim *et al.* 1990; Hinton *et al.* 1992). Ultimately, invertebrate models may be used to understand, predict, and quantify ecologic and human risks that define safe and unsafe environments.

REFERENCES

An der Lan, H. 1962. Histopathologische auswirkungen von insektiziden (DDT und Sevin) bei wirbellosen und ihre cancerogene beurteilung. *Mikroskopie* 17:85-112.

Anderson, R.S. 1985. Metabolism of a model environmental carcinogen by bivalve molluscs. *Mar. Environ. Res.* 17:137-140.

Anderson, R.S., L.A. Jordan, and J.C. Harshbarger. 1977. Tunic abnormalities of the urochordate *Halocynthia pyriformis*. *J. Invert. Pathol.* 30:160-168.

Armstrong, D.A., and R.E. Millemann. 1974. Pathology of acute poisoning with the insecticide Sevin in the bent-nosed clam, *Macoma nasuta*. *J. Invert. Pathol.* 24:201-212.

Auffret, M., and M. Poder. 1986. Sarcomatous lesion in the cockle *Cerastoderma edule*. II. Electron microscopical study. *Aquaculture* 58:9-15.

Bang, F.B.. 1980. Monitoring pathological changes as they occur in estuaries and in the ocean in order to measure pollution (with special reference to invertebrates). *Rapp. P.-v. Reun. Cons. Int. Explor. Mer* 179:118-124.

Barry, M.M., and P.P. Yevich. 1972. Incidence of gonadal cancer in the quahog *Mercenaria mercenaria*. *Oncology* 26:87-96.

Barry, M.M., and P.P. Yevich. 1975. The ecological, chemical and histopathological evaluation of an oil spill site. Part III. Histopathological studies. *Mar. Pollut. Bull.* 6:171-173.

Barry, M.M., P.P. Yevich, and N.H. Thayer. 1971. Atypical hyperplasia in the softshell clam *Mya arenaria*. *J. Invert. Pathol.* 17:17-27.

Barszcz C.A., P.P. Yevich, L.R. Brown, J.D. Yarbrough, and C.D. Minchew. 1978. Chronic effects of three crude oils on oysters suspended in estuarine ponds. *J. Environ. Toxicol.* 1:879-896.

Bauer, L., and A.H. Tulusan. 1973. Subcellular alterations produced by the carcinogen Aflatoxin B1 in tentacle epithelium of the mollusk, *Physa fontinales* L. *Z. Krebsforsch.* 79(3):185-192.

Berthou, F., G. Balouet, G. Bodennec, and M. Marchand. 1987. The occurrence of hydrocarbons and histopathological abnormalities in oysters for seven years following the wreck of the *Amoco Cadiz* in Brittany (France). *Mar. Environ. Res.* 23:103-133.

Betzer, S.B., and P.P. Yevich. 1975. Copper toxicity in *Busycon canaliculatum* L. *Biol. Bull.* 148:16-25.

Bodammer, J.E. 1979. Preliminary observations on the cytopathological effects of copper sulphate on the chemoreceptors of *Callinectes sapidus*. In *Marine Pollution: Functional Responses*. ed. W.B. Vernberg, A. Calabrese, F.P. Thurberg, and F.J. Vernberg, pp. 223-236. New York: Academic Press.

Bright, D.A. 1987. A case study of histopathology in *Macoma carlottensis* (Bivalvia, Tellinidae) related to mine-tailings discharge and review of pollution induced invertebrate pathology, shades of selye? Presented at annual meeting of Natl. Shellfisheries Assoc. August 9-13, Halifax, NS, Canada.

Brown, R.S., R.E. Wolke, S.B. Saila, and C.W. Brown. 1977. Prevalence of neoplasia in 10 New England populations of the softshell clam (*Mya arenaria*). *Ann. N.Y. Acad. Sci.* 298:522-534.

Calabrese, A., J.R. MacInnes, D.A. Nelson, R.A. Greig, and P.P. Yevich. 1984. Effects of long-term exposure to silver or copper on growth, bioaccumulation and histopathology in the blue mussel, *Mytilus edulis*. *Mar. Environ. Res.* 11:253-274.

Christensen, D.J., C.A. Farley, and F.G. Kern. 1974. Epizootic neoplasms in the clam *Macoma balthica* L. from Chesapeake Bay. *J. Nat. Cancer Inst.* 52(6):1739-1947.

Chun, Y.H., E.M. Johnson, and Bradley E.G. Gabel. 1983a. Relationship of developmental stage to effects of vinblastine on the artificial "embryo" of hydra. *Teratology* 27:95-100.

Chun, Y.H., E.M. Johnson, Bradley E.G. Cabel, and Albert S.A. Cadogan. 1983b. Effects of vinblastine sulfate on the growth and histologic development of reaggregated hydra. *Teratology* 27:89-94.

Cooper, C.B., M.E. Doyle, and K. Kipp. 1991. Risks of consumption of contaminated seafood: the Quincy Bay case study. *Environ. Health Perspect.* 90:133-140.

Couch, J.A. 1977. Ultrastructural study of lesions in gills of a marine shrimp exposed to cadmium. *J. Invert. Pathol.* 29(3):267-288.

Couch, J.A. 1978. Diseases, parasites and toxic responses of commercial penaeid shrimps of the Gulf of Mexico and south Atlantic coasts of North America. *Fish. Bull.* 76(1):1-44.

Couch, J.A. 1985. Prospective study of infectious and noninfectious diseases in oysters and fishes in three Gulf of Mexico estuaries. *Dis. Aquat. Org.* 1:59-82.

Couch, J.A. 1982. Aquatic animals as indicators of environmental exposure. *J. Environ. Sci. Health* 17:473-476.

Couch, J.A. 1984. Atrophy of diverticular epithelium as an indicator of environmental irritants in the oyster, *Crassostrea virginica*. *Mar. Environ. Res.* 14:525-526.

Couch, J.A., and D.R. Nimmo. 1974. Ultrastructural studies of shrimp exposed to the pollutant chemical, polychlorinated biphenyl (aroclor 1254). *Bull. Soc. Pharmacol. Environ. Pathol.* 11(2):17-20.

Couch, J.A., and L. Courtney. 1977. Interaction of chemical pollutants and virus in a crustacean: a novel bioassay system. *Ann. N.Y. Acad. Sci.* 298: 497-504.

Couch, J.A., and J.C. Harshbarger. 1985. Effects of carcinogenic agents on aquatic animals: an experimental overview. *Environ. Carcinog. Rev.* 3:63-105.

Couch, J.A., J.T. Winstead, and L.A. Courtney. 1987. Development of bivalve mollusc models for carcinogenesis studies. Paper read at 20th Ann. Mtg. of Soc. Invert. Pathol., Gainesville, FL.

Couch, J.A., L.A. Courtney, J.T. Winstead, and S.S. Foss. 1979. The American oyster (*Crassostrea virginica*) as an indicator of carcinogens in the aquatic environment. In *Animals as Monitors of Environmental Pollutants*., ed. S.W. Nielsen, G. Migaki, and D.G. Scarpelli pp. 65-83. Washington, DC: National Academy of Science.

Dawe, C.J. 1991. Introduction: focus and objectives of symposium on chemically contaminated aquatic food resources and human cancer risk. *J. Environ. Health Perspect.* 90:3.

Doughtie, D.G., and K.R. Rao. 1984. Histopathological and ultrastructural changes in the antennal gland, midgut, hepatopancreas and gill of grass shrimp following exposure to hexavalent chromium. *J. Invert. Pathol.* 43:89-108.

Doughtie, D.G., P.J. Conklin, and K.R. Rao. 1983. Cuticular lesions induced in grass shrimp exposed to hexavalent chromium. *J. Invert. Pathol.* 42:249-258.

Doyle, L.J., N.J. Blake, C.C. Woo, and P.P. Yevich. 1978. Recent biogenic phosphorite: concentrations in mollusk kidneys. *Science* 199:1431-1433.

Eisler, R.E., G.R. Gardner, R.J. Henekey, G. LaRoche, D.F. Walsh, and P.P. Yevich. 1972. Acute toxicology of sodium nitrilotriacetic acid (NTA) and NTA-containing detergents to marine organisms. *Water Res.* 6:1009-1027.

Emanuelsen, M., J.L. Lincer, and E. Rifkin. 1978. The residue uptake and histology of American oysters (*Crassostrea virginica* Gemlin) exposed to dieldrin. *Bull. Environ. Contam. Toxicol.*, 19(1):121-129.

Farley, A.C. 1977. Neoplasms in estuarine mollusks and approaches to ascertain causes. *Ann. N.Y. Acad. Sci.* 298:225-232.

Farley, A.C. 1988a. A computerized coding system of organs, tissues, lesions and parasites of bivalve mollusks and its application in pollution monitoring with *Mytilus edulis*. *Mar. Environ. Res.* 24:243-249.

Farley, A.C. 1988b. Histochemistry as a tool for examining possible pathologic cause-and-effect relationships between heavy metal and inflammatory lesions in oysters, *Crassostrea virginica*. *Mar. Environ. Res.* 24:271-275.

Foster, J.A. 1969. Malformations and lethal growths in planaria treated with carcinogens. *Nat. Cancer Inst. Monogr.* 31:683-691.

Fujiya, M. 1960. Studies on the effects of copper dissolved in sea water on oyster. *Bull. JPN. Soc. Sci. Fish*. 26:462-467.

Galtsoff, P.S., H.F. Prytherch, R.O. Smith, and V. Koehring. 1935. Effects of crude oil pollution on oysters in Louisiana waters. *Bull. U.S. Bur. Fish*. 48(18):143-210.

Gardner, G.R., and R.J. Pruell. 1988. Quincy Bay Study, Boston Harbor: A histopathological and chemical assessment of winter flounder, lobster and softshell clams indigenous to Quincy Bay, Boston Harbor and an *in situ* evaluation of oysters including sediment (surface and cores) chemistry. U.S. EPA.

Gardner, G.R., and P.P. Yevich. 1988. Comparative histopathological effects of chemically contaminated sediment on marine organisms. *Mar. Environ. Res*. 24:311-316.

Gardner, G.R., and P.P. Yevich. 1986. Renal carcinoma in American oysters exposed to contaminated estuarine sediment. Presented at IV International Colloquium of Invertebrate Pathology, Veldhoven, Netherlands.

Gardner, G.R., R.J. Pruell, and A.R. Malcolm. Chemical induction of tumors in oysters by a mixture of aromatic and chlorinated hydrocarbons, amines and metals. *Mar. Environ. Res*. in press.

Gardner, G.R., R.J. Pruell, and L.C. Folmar. 1989. A comparison of both neoplastic and non-neoplastic disorders in winter flounder (*Pseudopleuronectes americanus*) from eight areas in New England. *Mar. Environ. Res*. 28: 393-397.

Gardner, G.R., P.P. Yevich, and P.F. Rogerson. 1975. Morphological anomalies in adult oyster, scallop, and Atlantic silversides exposed to waste motor oil. In *Proceedings of Conference on Prevention and Control of Oil Pollution*, ed. R.W. Scott, pp. 473-477, 1801 K Street N.W, Washington, DC: American Petroleum Institute.

Gardner, G.R., P.P. Yevich, J.C. Harshbarger, and A.R. Malcolm. 1991a. Carcinogenicity of Black Rock Harbor, Bridgeport, CT, sediment to the eastern oyster and trophic transfer of Black Rock Harbor carcinogens from the blue mussel to the winter flounder. *J. Environ. Health Perspect*. 90: 53-66.

Gardner, G.R., P.P. Yevich, J. Hurst, P. Thayer, S. Benyi, J.C. Harshbarger, and R.J. Pruell. 1991b. Germinomas and teratoid siphon anomalies in softshell clams, *Mya arenaria*, environmentally exposed to herbicides. *J. Environ. Health Perspect*. 90:43-51.

Gardner, G.R., T. Gleason, W. Munns, G. Pesch, D. Phelps, and P.P. Yevich. 1990. Pathology of some bivalve molluscs from Allen Harbor and Narragansett Bay, RI, and Boston and New Bedford Harbor, MA, at V International Colloquium of Invertebrate Pathology, August 20-24. Adelaide, Australia.

Gardner, G.R., P.P. Yevich, A.R. Malcolm, and R.J. Pruell. 1987. Carcinogenic effects of Black Rock Harbor sediment on American oysters and winter flounder. Final Report to National Cancer Institute, U.S. EPA, Narragansett, RI.

Gilfillan,E.S., S.A. Hanson, D.S. Page, D. Mayo, J. Cooley, J. Chelfant, T. Archambeault, and J.C. Harshbarger. 1977. Final Report to the Department of Environmental Protection, Augusta, Maine.

Hall, F.L., M. Morita, and J.B. Best. 1984. Tumorigenic response of planarians to mammalian carcinogens. Presented at Amer. Soc. Zool., Denver, Colorado, December 27-30.

Hall, F.L., M. Morita, and J.B. Best. 1986a. Neoplastic transformation in the planarian: I. Cocarcinogenesis and histopathology. *J. Exper. Zool*. 240:211-227.

Hall, F.L., M. Morita, and J.B. Best. 1986b. Neoplastic transformation in the planarian: II. Ultrastructure of malignant reticuloma. *J. Exper. Zool*. 240:229-244.

Harshbarger, J.C. 1977. Role of the Registry of Tumors in Lower Animals in the study of environmental carcinogenesis in aquatic animals. *Ann. N.Y. Acad. Sci*. 298:280-289.

Harshbarger, J.C., and C.J. Dawe. 1969. Symposium on neoplasms and related disorders of invertebrate and lower invertebrate animals. *Nat. Cancer Inst. Monogr. 31*.

Harshbarger, J.C., and J. Clark. 1990. Epizootiology of neoplasms in bony fish of North America. *Science Total Environ.*, 94:1-32.

Harshbarger, J.C., Otto, S.V., and S.C. Chang. 1979. Proliferative disorders in *Crassostrea virginica* and *Mya arenaria* from the Chesapeake Bay and intranuclear virus-like inclusions in *Mya arenaria* with germinomas from a Maine oil spill site. *Haliotis* 8:243-248.

Harshbarger, J.C., G.E. Cantwell, and M.F. Stanton. 1971. Effects of N-Nitroso-dimethylamine on the crayfish, *Procambarus clarkii*. Presented at IV International Colloquium on Insect Pathology, p. 425-430. Soc. Invert. Pathol., College Park, MD.

Hayes, H.M., Tarone, R.E., Casey, H.W. and D.L. Huxsoll. 1990. Excess of seminomas observed in Vietnam service U.S. military working dogs. *J. Nat. Cancer Inst.* 82(12):1042-1046.

Hueper, W.C. 1963. Environmental carcinogenesis in man and animals. *Ann. N.Y. Acad. Sci.* 108: 963-1031.

Hesselman, D.M., and W.S. Arnold. 1990. Distribution and prevalence of gonadal neoplasms within the Indian River clam (*Mercenaria* spp.) population. Presented at National Shellfisheries Association, Williamsburg, Virginia, April 1-5.

Hesselman, D.M., N.J. Blake, and E.C. Peters. 1988. Gonadal neoplasms in hard shell clams *mercenaria* spp., from the Indian River, Florida: Occurrence, prevalence, and histopathology. *J. Invert. Pathol.*, 52:436-446.

Hesselman, D.M., B.J. Barber, and N.J. Blake. 1989. The reproductive cycle of adult hard clams, *Mercenaria* spp. in The Indian River Lagoon, Florida. *J. Shellfish Res.*, 8(1):43-49.

Hinton, D., P. Baumann, G.R. Gardner, W. Hawkins, J. Hendricks, R. Murchelano, and M. Okihiro. In Press. Histopathological biomarkers. In *The existing and potential value of biomarkers in evaluating exposure and environmental effects of toxic chemicals*, ed. R.J. Huggett. Washington, DC: Society of Environmental Toxicology and Chemistry.

Hodgins, H.O., B.B. McCain, and J.W. Hawkes. 1977. Marine fish and invertebrate diseases, host disease resistance, and pathological effects of petroleum. In *Effects of petroleum on arctic and subarctic marine environments and organisms. II. Biological effects.*, ed. D.C. Malins, pp. 95-173. New York: Academic Press.

International Agency for Cancer Research. Monographs on the evaluation of the carcinogenic risk of chemicals to humans, Vols. 1-46, Supplements 1-7. World Health Organization, Lyon, France, 1972-1989.

Jackim, E., G.G. Pesch, A.R. Malcolm, and G.R. Gardner. 1990. Application of biomarkers to predict responses of organisms exposed to contaminated marine sediments. In *Carcinogenic, Mutagenic, and Teratogenic Marine Pollutants: Impact on Human Health and the Environment, Vol. 5*, ed. Grandjean, pp. 165-174. The Woodlands, Texas: Portfolio Publishing Co.

Jacobsen E.R. J.L. Manfell, J.P. Sundberg, L. Hajjar, M.E. Reichmann, L.M Ehrhart, M. Walsh, and F. Murru. 1989. Cutaneous fibropapillomas of Green Turtles. *J. Comp. Pathol.* 101:39-52.

Jensen P.K., and E. Baatrup. 1988. Histochemical demonstration of mercury in the intestine, nephridia and epidermis of the marine polychaete *Nereis virens* exposed to inorganic mercury. *Mar. Biol.* 97: 533-540.

Johnson, E.M. 1980. A subvertebrate system for rapid determination of potential teratogenic hazards. *J. Environ. Pathol. Toxicol.* 4:153-156.

Johnson, E.M., and M.S. Christian. 1985. The hydra assay for detecting and ranking developmental hazards. *Concepts Toxicol.* 3:107-113.

Johnson, E.M., R.M. Gorman, B.E.G. Gabel, and M.E. George. 1982. The *Hydra attenuata* system for detection of teratogenic hazards. *Teratog. Carcinog. Mutagen.* 2:263-276.

Johnson, E.M., B.E.G. Gabel, and J. Larson. 1984. Developmental toxicity and structure/activity correlates of glycols and glycol ethers. *Environ. Health Perspect.* 57:135-139.

Johnson, E.M., B.E.G. Gabel, M.S. Christian, and E. Sica. 1986. The Developmental Toxicity of xylene and xylene isomers in the hydra assay. *Toxicol. Appl. Pharmacol.* 82:323-328.

Johnson, P.T. 1968. An Annotated Bibliography of Pathology in Invertebrates Other Than Insects. Minnesota: Burgess Publishing Company.

Kaiser, H.E. 1980. Species-specific Potential of Invertebrates for Toxicological Research. Maryland: University Park Press.

Karnofsky, D.A., and E.B. Simmel. 1963. Effects of growth-inhibiting chemicals on the sand-dollar embryo, *Echinarachnius parma*. *Prog. Exp. Tumor Res.* 3:254-295.

Kayser, H., D. Ludemann, and H. Neumann. 1962. Veranderungen an nervenzellen insektizidvergiftung bei Fischen und Krebsen. *Ztschr. Angew. Zool.* 49: 135-148.

Khudoley, V.V., and O.A. Syrenko. 1978. Tumor induction by n-nitroso compounds in bivalve mollusks *Unio pictorum*. *Cancer Lett.* 4:349-354.

Kraybill, H.F. 1977. Global distribution of carcinogenic pollutants in water. *Ann. N.Y. Acad. Sci.* 298:80-89.

Krieg, K. 1970. *Ampullarius australis* d'Orbigny (Molluscs, Gastropoda) as an experimental animal in oncological research. A contribution to the study of carcinogenesis in invertebrates. *Neoplasma* 19:41-49.

Krieg, K. 1973. Invertebrates in Tumor Research. Dresden: Verlag Theodor Steinkopff.

LaRoche, G., G.R. Gardner, R.E. Eisler, E.H. Jackim, P.P. Yevich, and G.E. Zaroogian. 1973. Analysis of toxic responses in marine poikilotherms. In *Bioassay Techniques and Environmental Chemistry*. ed. G.E. Glass. pp. 199-216. Michigan: Ann Arbor Science.

Lee, R. 1981. Mixed function oxygenases (MFO) in marine invertebrates. *Mar. Biol. Lett.* 2: 87-105.

Lightner, D.V. Diseases of cultured penaeid shrimp. 1983. In *CRC Handbook of Mariculture, Vol 1. Crustacean Aquaculture*. ed. J.P. McVey, pp. 289-320. Boca Raton, FL: CRC Press, Inc.

Lightner, D.V. 1987. Examples of lesions in the crustacea due to the toxic effects of organic and inorganic toxins. Presented at the Soc. Invert. Pathol., Gainsville, FL, July 20-24, 1987.

Lightner, D.V., R.M. Redman, R.L. Price, and M.O Wiseman. 1982. Histopathology of aflatoxicosis in the marine shrimp *Penaeus stylirostris* and *P. vannamei*. *J. Invert. Pathol.* 40:279-291.

Lowe, D.M. 1988. Alterations in cellular structure of *Mytilus edulis* resulting from exposure to environmental contaminants under field and experimental conditions. *Mar. Ecol. Prog. Ser.* 46:91-100.

Lowe, D.M., and K.R. Clarke. 1989. Contaminant-induced changes in the structure of the digestive epithelium of *Mytilus edulis*. *Aquatic Toxicol.* 15: 345-358.

Lowe, J.I., P.D. Wilson, A.J. Rick, and A.J. Wilson, Jr. 1971. Chronic exposure of oysters to DDT, Toxaphene and Parathion. *Proc. Nat. Shellfish Assoc.* 61:71-79.

Lowe, J.I., P.R. Parish, J.M. Patrick, Jr. and J. Forester. 1972. Effects of the polychlorinated biphenyl aroclor 1254 on the American oyster *Crassostrea virginica*. *Mar. Biol.* 17:209-214.

MacLean, S.A., Farley, C.A., Newman, M.W., and A. Rosenfield. 1981. Gross and microscopic observations on some biota from deep water dumpsite 106. In *Ocean Dumping of Industrial Wastes*, ed. B.H. Ketchum, D.R. Kester, and P.K. Park, pp. 421-437. New York: Plenum Publ.

Malins, D.C., B.B. McCain, D.W. Brown, A.K. Sparks, and H.O. Hodgins. 1980. Chemical contaminants and biological abnormalities in central and southern Puget Sound. NOAA Technical Memorandum, OMPA-2, NMFS, Seattle, WA.

Malins, D.C. 1982. Alterations in the cellular and subcelluar structure of marine teleosts and invertebrates exposed to petroleum in the laboratory and field: A critical review. *Can. J. Fish. Aquat. Sci.* 39:877-889.

Malins, D.C., B.B. McCain, D.W. Brown, A.K. Sparks, H.O. Hodgins, and S. Chan. 1982. Chemical contaminants and abnormalities in fish and invertebrates from Puget Sound. NOAA Technical Memorandum, OMPA-19, NMFS, Seattle, WA.

Melzian, B.D. 1982. Acute toxicity, histopathology, and bioconcentration-retention studies with No.2 fuel oil and the blue crab, *Callinectes sapidus* Rathbun. Ph.D. Dissertation, University of Rhode Island.

Meyers, T.R., and J.D. Hendricks. 1982. Summary of tissue lesions in aquatic animals induced by controlled exposures to environmental contaminants, chemotherapeutic agents, and potential carcinogens. *Mar. Fish Rev*. 44(12):1-17.

Mix, M.C. 1986. Cancerous diseases in aquatic animals and their association with environmental pollutants: a critical literature review. *Mar. Environ. Res*. 20:1-141.

Mix, M.C. 1988. Shellfish diseases in relation to toxic chemicals. *Aquat. Toxicol*. 11:29-42.

Mix, M.C., Trenholm, S.R., and K.I. King. 1979. Benzo[a]pyrene body burdens and the prevalence of proliferative disorders in mussels (*Mytilus edulis*) in Oregon. In *Animals as Monitors of Environmental Pollutants*, ed. S.W. Nielsen, G. Migaki, and D.G. Scarpelli, pp. 52-64. Washington, DC: National Academy of Science.

Möller, H. 1987. Pollution and parasitism in the aquatic environment. *Int. J. Parasitol*. 17(2): 353-361.

Mysing-gubala, M., and M.A. Poirrier. 1981. The effects of cadmium and mercury on gemmule formation and gemmosclere morphology in *Ephydatia fluviatilis* (Porifera: Spongillidae). *Hydrobiologia* 76:145-148.

Neff, J.M. 1979. Polycyclic Aromatic Hydrocarbons in the Aquatic Environment: Sources, Fates and Biological Effects. Barking, Essex: Applied Science Publishers Ltd.

Neff, J.M., and J.W. Anderson. 1981. Responses of Marine Animals to Petroleum and Specific Petroleum Hydrocarbons. London: Applied Science Publishers Ltd.

Neff, J.M., and W.E. Haensly. 1982. Long-term impact of the *Amoco Cadiz* crude oil spill on oysters *Crassostrea gigas* and plaice *Pleuronectes platessa* from Aber Benoit and Aber Wrac'h, Brittany, France. I. Oyster histopathology II. Petroleum contamination and biochemical indices of stress in oysters and plaice. In: *Ecological Study of the* Amoco Cadiz *Oil Spill*. pp. 259-328. Boulder, CO.: NOAA/CNEXO Joint Scientific Comm.

Neff, J.M., R.E Hillman, R.S. Carr, R.L. Buhl, and J.I. Lahey. 1987. Histopathologic and biochemical responses in arctic marine bivalve molluscs exposed experimentally to oil. *Arctic* 40(1):220-229.

Nelson, D.A., R.A. Calabrese, R.A. Greig, P.P. Yevich, and S. Chang. 1983. Long-term silver effects on the marine gastropod, *Crepidula fornicata*. *Mar. Ecol. Prog. Ser*. 12:155-165.

Nesnow, S., M. Argus, H. Bergman, K. Chu, C. Frith, T. Helmes, R. McGaughy, V. Ray, T.J. Slaga, R. Tennant, and E. Weisberger. 1987. Chemical carcinogens. A review and analysis of the literature of selected chemicals and the establishment of the Gene-Tox carcinogen data base. *Mutat. Res*. 185:1-195.

Nimmo, D.R., D.V. Lightner, and L.H. Bahner. 1977. Effects of cadmium on the shrimps, *Penaeus duorarum*, *Palaemonetes pugio* and *Palaemonetes vulgaris*. In: *Physiological Responses of Marine Biota to Pollutants*, ed. J.F. Vernberg, A.J. Calabrese, F.P. Thurnberg, and W.B. Vernberg, p. 131-184. New York: Academic Press.

Nimmo, D.R., D.J. Hansen, J.A. Couch, N.R. Cooley, P.R. Parrish, and J.I. Lowe. 1975. Toxicity of aroclor 1254 and its physiological activity in several organisms. *Arch. Environ. Contam. Toxicol*. 3(1):22-39.

Overstreet, R.M., and T. Van Devender. 1978. Implication of an environmentally induced hamartoma in commercial shrimps. *J. Invert. Pathol*. 31:234-238.

Pauley G.B. 1969. A critical review of neoplasia and tumor-like lesions in mollusks. *Nat. Cancer Inst. Monogr*. 31:509-531.

Perez, K.T., E.W. Davey, G.E. Morrison, A.W. Soper, N.F. Lackie, D.W. Winslow, R. Blasco, R. Johnson, and S. Marino. 1987. Environmental assessment of a substituted benzidine based azo dye, direct blue-15, in experimental microcosms. EPA, Narragansett, RI, Contribution No. 882.

Peters, E.C. 1988. Recent investigations of the disseminated sarcomas of marine bivalve molluscs. *Am. Fish Soc. Spec. Pub*. 18:74-92.

Peters, E.C. 1987. Pathological effects of chemicals on coelenterates. Presented at the Society for Invert. Pathol. Annu. Meeting, Gainesville, FL, July 20-24, 1987.

Peters, E.C., and P.P. Yevich. 1989. Histopathology of *Ceriantheopsis americanus* (Cnidaria: Ceriantharia) exposed to Black Rock Harbor dredge spoils in Long Island Sound. *Dis. Aquat. Org.* 7:137-148.

Peters, E.C., P.A. Meyers, P.P. Yevich, and N.J. Blake. 1981. Bioaccumulation and histopathological effects of oil on a stony coral. *Mar. Pollut. Bull.* 12(10):333-339.

Pittinger, C.A., A.L. Buikema, Jr. and J.O. Falkingham, III. 1987. *In situ* variations in oyster mutagenicity and tissue concentrations of polycyclic aromatic hydrocarbons. *Environ. Toxicol. Chem.* 6:51-60.

Pittinger, C.A., A.L. Buikema, Jr., S.G. Horner, and R.W. Young. 1985. Variation in tissue burdens of polycyclic aromatic hydrocarbons in indigenous and relocated oysters. *Environ. Toxicol. Chem.* 4:379-387.

Poder, M., and M. Auffret. 1986. Sarcomatous lesion in the cockle *Cerastoderma edule* I. Morphology and population survey in Brittany, France. *Aquaculture* 58:1-8.

Resources of Biomedical and Zoological Specimens. 1990. Second edition, Registry of Comparative Pathology. Washington, DC: Armed Forces Institute of Pathology.

Rasmussen, L. 1982. Light microscopical studies of the acute toxic effects of n-nitrosodimethylamine on the marine mussel, *Mytilus edulis*. *J. Invert. Pathol.* 39:66-80.

Rasmussen, L.P.D. 1986. Cellular reactions in molluscs with special reference to chemical carcinogens and tumors in natural populations of bivalve molluscs. IV International Colloquium on Invertebrate Pathology, Veldhoven, Netherlands.

Rasmussen, L.P.D., E. Hage, and O. Karlog. 1983a. Light and electron microscopic studies of the acute and chronic toxic effects of n-nitroso compounds on the marine mussel, *Mytilus edulis* (L.) I. N-nitrosoethylamine. *Aquat. Toxicol.* 3:285-299.

Rasmussen, L.P.D., E. Hage, and O. Karlog. 1983b. Light and electron microscopic studies of the acute and chronic toxic effects of n-nitroso compounds on the marine mussel, *Mytilus edulis* (L.) II. N-methyl-n-nitro-n-nitrosoguanidine. *Aquat. Toxicol.* 3:301-311.

Rasmussen, L.P.D., E. Hage, and O. Karlog. 1985. Light and electron microscopic studies of the acute and long-term effects of N-nitrosodipropylamine and N-methylnitrosourea on the marine mussel *Mytilus edulis*. *Mar. Biol.* 85:55-65.

Riggan, W.B., J.P. Creason, W.C. Nelson, K.G. Manton, M.A. Woodbury, E. Stallard, A.C. Pellom, and J.C. Beaubier. 1987. U.S. Cancer Mortality Rates and Trends, 1950-1979. Volume IV: Maps. EPA/600/1-83/015e, U.S. EPA, Health Effects Research Laboratory, Research Triangle Park, NC.

Rheinberger, R., G.L. Hoffman, and P.P. Yevich. 1979. The kidney of the quahog (*Mercenaria mercenaria*) as a pollution indicator. In *Animals as Monitors of Environmental Pollutants*, ed. S.V. Nielsen, G. Migaki, and D.G. Scarpelli, pp. 199-129. Washington, DC: Nat. Acad. Sci.

Seiler G.R., and P.M. Morse. 1988. Kidney and hemocytes of *Mya arenaria* (*Bivalvia*): Normal and pollution-related ultrastructural morphologies. *J. Invert. Pathol.* 52:201-214.

Shimkin, M.B., B.K. Koe, and L. Zechmeister. 1951. An instance of the occurrence of carcinogenic substances in certain barnacles. *Science* 113:650-651.

Sindermann, C.J. 1990. Principal Diseases of Marine Fish and Shellfish. San Diego: Academic Press.

Sindermann, C.J., F. Csulak, T.K. Sawyer, R.A. Bullis, D.W. Engel, B.T. Estrella, E.J. Noga, J.B. Pearce, J.C. Rugg, R. Runyon, J.A. Tiedemann, and R.R. Young. 1989. Shell diseases of the New York Bight. NOAA Tech Memo. NMFS-F/NEC-74.

Sindermann, C.J., F.B. Bang, N.O. Christensen, V. Dethlefsen, J.C. Harshbarger, J.R. Mitchell, and M.F. Mulcahy. 1980. The role and value of pathobiology in pollution effects monitoring programs. *Rapp. P.-v. Reun. Cons. Int. Explor. Mer.* 179:135-151.

Spangenberg, D.B. 1984. Use of the *Aurelia* metamorphosis test system to detect subtle effects of selected hydrocarbons and petroleum oil. *Mar. Environ. Res.* 14:281-303.

Spangenberg, D.B., K. Ives, and K. Patten. 1980. Strobilation aberrations in *Aurelia* induced by petroleum-related aniline and phenol. In *Developmental and Cellular Biology of Coelenterates*, ed. Tardent and Tardent, pp. 263-269. Amsterdam: Elsevier.

Sparks A.K. 1985. *Synopsis of Invertebrate Pathology*. Amsterdam: Elsevier.

Stebbing, A.R.D., and B.E. Brown. 1984. Vol I; Marine ecological tests with coelenterates. In *Ecological testing for the marine environment*, ed. G. Persoone, E. Jaspers, and C. Claus., 798 p. Bredene, Belgium: State Univ. Ghent and Inst. Mar. Sci. Res.

Stegeman, J.J,. and J.J. Lech. 1991. Cytochrome P-450 monooxygenase systems in aquatic species: carcinogen metabolism and biomarkers for carcinogen and pollutant exposure. *J. Environm. Health Perspect.*, 90: 101-109.

Sunila, I. 1988. Pollution-related histopathological changes in the mussel *Mytilus edulis* L. in the Baltic Sea. *Mar. Environ. Res.* 24:277-280.

Sunila, I. 1987. Histopathological effects of environmental pollutants on the common mussel, *Mytilus edulis* L. (Baltic Sea), and their application in marine monitoring. Ph.D. Dissertation, Univ. of Helsinki.

Sunila, I. 1986. Cystic kidneys in copper-exposed mussels, *Mytilus edulis* L. IV International Colloquiem of Invertebrate Pathology, Soc. Invert. Pathol., ed. R.A. Samson, J.M. Vlak, and D. Peters. Veldhoven, Netherlands.

Tetra Tech, Inc. 1986. User's Manual for the Pollutant of Concern Matrix. Final Report to U.S. EPA, Region X.

Tripp, M.R., and C.R. Fries. 1987. Histopatholgoy and Histochemistry. In *Pollutant Studies in Marine Animals*, ed. C.S. Giam and L.E. Ray, pp. 111-154, Boca Raton, FL: CRC Press.

Walker, H.A., E. Lorda, and S.B. Saila. 1981. A comparison of the incidence of five pathological conditions in softshell clams, *Mya arenaria*, from environments with various pollution histories. *Mar. Environ. Res.* 5:109-123.

Winstead, J.T., and J.A. Couch. 1988. Enhancement of protozoan pathogen *Perkinsus marinus* infections in American oysters *Crassostrea virginica* exposed to the chemical carcinogen n-nitrosodiethylamine (DENA). *Dis. Aquat. Org.* 5:205-213.

Yevich, P.P., and M.M. Barry. 1969. Ovarian tumors in the quahog, *Mercenaria mercenaria*. *J. Invert. Pathol.* 14(2):266-267.

Yevich P.P., and C.A. Barszcz. 1976. Gonadal and hematopoietic neoplasms in *Mya arenaria*. *Mar. Fish. Rev.* 38(10):42-43.

Yevich P.P., and C.A. Barszcz. 1977. Neoplasia in softshell clams (*Mya arenaria*) collected from oil impacted sites. *Ann N.Y. Acad. Sci.* 298:4,09-426.

Yevich, C.A., and P.P. Yevich. 1985. Histopathological effects of cadmium and copper on the sea scallop, *Placopecten magellanicus*. In *Marine Pollution Physiology, Recent Advances*, ed. F.J. Vernberg, F.P. Vernberg, A. Calabrese, W.B. Vernbérg, pp. 187-198, Belle W. Baruch Lib. Mar. Sci. No. 13.

Yevich P.P., C. Yevich, and G. Pesch. 1987. Effects of Black Rock Harbor dredged material on the histopathology of the blue mussel *Mytilus edulis* and polychaete worm *Nephtys incisa* after laboratory and field exposures. U.S. EPA Tech. Rpt. D-87-8, Narragansett, RI.

Yevich P.P., C.A. Yevich, K.J. Scott, M. Redmond, D. Black, P.S. Schauer, and C.E. Pesch. 1986. Histopathological effects of Black Rock Harbor dredged material on marine organisms. U.S. EPA Tech. Rpt. D-86-1, Narragansett, RI.

15

Diseases of Other Invertebrate Phyla: Porifera, Cnidaria, Ctenophora, Annelida, Echinodermata

Esther C. Peters

INTRODUCTION

Five phyla (Porifera, Cnidaria, Ctenophora, Annelida, Echinodermata) examined in this paper represent some of the most diverse and numerous species found in the water column and benthos of estuarine and coastal marine environments. These animals range in size from microscopic (worms) to tens of cubic meters (stony corals). Despite our lack of understanding about their life histories and interrelationships, their importance should not be underestimated.

Besides their role in complex food webs as predators and prey (see Paine 1966; Connell 1975; Hoppe 1988), members of these groups may also physically modify their environment in a variety of ways. For example, the stony corals produce calcium carbonate exoskeletons that contribute to construction of extensive reefs. The reefs form and protect land masses in tropical and subtropical seas around the world (Goreau *et al.* 1979) and provide complex habitats to support fisheries and tourism industries (Stevenson and Marshall 1974; Robinson 1979). At the same time, sponges and sea urchins dissolve this carbonate (e.g., Hein and Risk 1975; Glynn *et al.* 1978). Some species of polychaetes also build hard reefs or dense mats while constructing tubes. Other polychaetes are well-known for bioturbation activities in soft-bottom habitats, altering geochemical cycles (e.g., Aller 1978), as well as survival and trophic relationships of associated organisms (Wilson 1981).

Although few animals in these groups are used as food for humans, many of support commercial interests. Even though synthetic sponges have replaced much of the market (Vicente 1989), several species of demosponges are still harvested. Some polychaete worms are used as fish bait (Creaser and Clifford

0-8493-8662-4/93/$0.00 + $.50

1986; Olive and Cadnam 1990). One of the most active fields of recent research has been the extraction of bioactive substances, such as antitumor, antiviral, or cytotoxic agents from a variety of sponges, cnidarians, and echinoderms (among other marine groups) (Munro *et al.* 1987). Medically important prostaglandins were obtained from soft corals before being commercially synthesized (Bayer and Weinheimer 1974). New sunscreens are being developed from reef organisms (Dunlap and Chalker 1986; W.C. Dunlap, Australian Institute of Marine Science, Townsville, Queensland, Australia, personal communication).

Within these phyla, many symbiotic associations are known, including a variety of microorganisms in interspecific associations on and within tissues of other organisms. These relationships can be classified along a continuum, from mutualistic symbioses beneficial to both microorganism and host, to parasitic symbioses where the microorganism derives a nutritional benefit from the host. A parasite that impairs vital functions of its host can cause disease and death of the host, and is then known as a pathogen (Ahmadjian and Paracer 1986). The association the microorganism has with its host can be difficult to determine, since viruses, bacteria, or other microorganisms can be present on or within organisms in the absence of clinical signs of disease (e.g., Colwell and Liston 1962; Unkles 1977; Gotto 1979; Johnson 1984; Paul *et al.* 1986). Mutualistic symbioses of chemoautotrophic bacteria that enable their hosts to live in potentially toxic environments or to subsist on nutritionally limited diets have been discovered in representatives of these phyla (Guerinot and Patriquin 1981; Giere and Langheld 1987). In other cases, a true parasite only harms a host that is stressed by some other biotic or abiotic factor. An abiotic disease may be complicated by secondary infections from microorganisms that are not normally pathogenic.

For these phyla, little is also known about the regulation of symbiont populations by the host, how the susceptibility or relative resistance of the host can change, depending on the size of the population of microorganisms present, effectiveness of changes in host defense mechanisms, and how these factors vary with the stage of development of disease (Rohde 1982). There is often a high degree of host specificity related to susceptibility, but mechanisms by which a pathogen causes disease in a host appear varied (for reviews on invertebrate pathology and disease processes see Kinne 1980; Sparks 1985).

As in commercially important shellfish, infectious diseases caused by microorganisms, parasites, and non-infectious (nutritional, environmental, or genetic) disorders have been reported in each of these phyla. However, many reports in the literature involve descriptions of "parasites" where the true nature of the association with the host has not been experimentally determined or of mass mortalities where the etiologic agent has not been identified. For these phyla, diseases have usually been discovered by scientists in other fields, such as ecology, physiology, or taxonomy. These diseases have usually not been investigated by multidisciplinary pathobiology teams at major marine laboratories, as has been done with fish and shellfish disease research supported by government laboratories and the seafood industries. Often, a disease is not recognized until a major segment of the population has been affected and there are few survivors to study.

Thus, while the ecological impact of the mass mortality of a particular species may have been investigated, or lesions caused by a "parasite" to an individual host may have been superficially characterized, the etiology and pathogenesis of most diseases in these phyla should be examined. Moreover, some disease problems observed in these "forgotten" phyla may have devastating consequences for the condition of the habitat and associated species, including commercially important human food organisms. This paper focuses on several recent disease problems in these five phyla that warrant attention of multidisciplinary teams to examine relationships among environmental conditions, pathogens, and their hosts.

Phylum Porifera

Over 5000 known species of sponges constitute a significant portion of the biomass in some environments, such as coral reefs. Only a few species, members of the genera *Spongia* and *Hippospongia*, are commercially harvested in the northern Mediterranean, northern Caribbean, and eastern Gulf of Mexico (Vicente 1989). Until recently, only diseases affecting commercial sponge populations have received much attention.

Early Disease Reports

The first indication that sponges could become diseased were observations from the late 1800's of the destruction of sponges in the Indian Ocean and eastern Gulf of Mexico, possibly by a fungus (see Lauckner 1980a). In this century, Allemand-Martin (1906, cited in Lauckner 1980a) discovered moribund specimens of *Hippospongia equina* from the Mediterranean coast of Tunisia that were covered by a white, greyish, or greenish liquid containing bacteria. Some sponges placed in clean running seawater recovered. Further work suggested that insufficient water exchange in the spongocoel advanced the disease and that the disease was usually found only at shallower depths (Allemand-Martin 1914, cited in Lauckner 1980a).

Extensive mortalities occurred in species of commercial sponges (*Spongia* and *Hippospongia*) beginning in late 1938 in the Bahamas (Galstoff *et al.* 1939). The mortalities spread rapidly throughout the Caribbean, reaching the commercial sponge beds off the west coast of Florida by February and March 1939, and moved to British Honduras in the summer of 1939 (Galstoff 1942). Although some sponges recovered, commercial fisheries were eradicated (Smith 1941). On close examination, healthy-appearing sponges were rotting internally even before gross signs were apparent. The earliest external lesion consisted of a "bald patch" that formed on the surface of affected animals, followed by "rotting" of tissue beneath the patch with complete loss of architectural integrity, killing the specimen within a week of the initial appearance of the diseased patch. The spongin fibers were not affected. The lesions always contained long slender aseptate filaments absent from the normal tissue. The filaments were believed to be a fungus and were placed in the genus *Spongiophaga* (Galstoff 1942; Johnson and Sparrow 1961, cited in Sparks 1985). Galstoff (1942; reviewed in Sparks 1985) was able to transmit the disease by "contact" between diseased and healthy sponges. However, injection of dissociated cells from diseased sponge did not cause the disease. The only environmental change noted was a slight increase in salinity in lagoon waters due to drought; sponge beds in areas with maximum salinities were not as severely damaged. Non-commercial sponges were not affected, nor were other organisms. Occasionally, some sponges recovered when rotten tissue was physically separated by a callous similar in structure to the lining of the main oscula.

This epizootic was believed to have been a fungal infection that was transmitted by coastal counter currents to the various sponge beds (Smith 1941; Sparks 1985). There have been no further reports of this "sponge-wasting disease" in the Caribbean or of fungal infections (Sparks 1985; Vicente 1989).

Although the fungus had been consistently found in all diseased sponges during this epizootic, Dosse (1940) suspected that the fungus might be a secondary invader and that bacteria were the primary pathogen. He examined bacterial flora in diseased *Hippospongia communis* var. *meandriniformis* from the Bahamas and compared it to healthy *Spongia officinalis* var. *lacinulosa* and *Cacospongia cavernosa* from the Mediterranean Sea near Naples. He found that when the Mediterranean species were exposed to warmer water temperatures in the laboratory, histological examinations revealed changes in the bacterial flora within the mesohyl comparable to that seen in diseased *H. communis*. However, he also observed the fungal filaments at the border of dead and living tissue in sponges from the Bahamas.

Lauckner (1980a) noted that more recent studies have confirmed presence of probable mutualistic bacteria in healthy sponges and that determination of the causal agent(s) of disease in sponges may be obscured by secondary infestations from seawater populations of microorganisms. Many species of sponges harbor various bacterial symbionts within the intercellular matrix (mesohyl) of their skeletons (Sarà and Vacelet 1973; see review by Lauckner 1980a). These associations often involve mutualistic unicellular or multicellular phototropic cyanobacteria that may provide nutrition for their sponge hosts (Wilkinson 1980; Lauckner 1980a; see also Gaino and Pronzato 1987; Rutzler 1990), and other bacteria that use host metabolic wastes (Wilkinson 1978a,b). The composition of the microflora varies among sponge species (Wilkinson *et al.* 1981).

However, for many bacteria found in symbiotic associations with sponges, the type of relationship with their host and what they provide to their host are unclear. In phototropic associations, host sponges may be able to control populations of symbionts by archeocyte phagocytosis to maximize benefits to the sponge (Wilkinson 1978b). Many, but not all, sponges also can produce antimicrobial compounds to control pathogenic microorganisms (Jakowska and Nigrelli 1960; Burkholder and Ruetzler 1969; Bakus *et al.* 1990), as found by Gaino and Pronzato (1987). Laboratory studies, in support of Dosse's earlier study, by Bertrand and Vacelet (1971) and Gaino and Pronzato (1987) noted that sponges with degenerating mesohyl have altered microflora populations and tissue-penetrating bacteria normally absent from host tissues.

Mediterranean Epizootic

Another epizootic occurred among commercial sponge species in the Mediterranean beginning in 1986. The disease affected *Spongia officinalis* and *Hippospongia communis* off Tunisia, Cyprus, and Turkey, severely curtailing the sponge industry in countries that were suffering from overexploitation of the sponge beds (Ben Mustapha and Vacelet 1991; Vacelet 1991). This disease was also observed off Greece, Egypt, and Italy (J. Vacelet, Station Marine d'Endoume, Marseille, France, personal communication).

In contrast to the lack of damage to spongin fibers during the Caribbean epizootic, Gaino and Pronzato (1989) found differences in the structure of the spongin fibers between healthy and diseased specimens of *Spongia officinalis* collected from the coast of Portofino, Italy. Up to 60% of the specimens at this locality were diseased in the spring of 1987. Grossly, the sponge body was either partially or totally reduced to a mesh of brittle fibers (Fig. 15-1). Microscopically, the fibers contained canaliculi opening to the outside. Bacteria were found within spaces in the canaliculi as well as the outside of spongin fibers. They also noted that some affected sponges isolated damaged areas from living tissue by formation of a colorless "callus" and recovered. Similar bacteria within canaliculi (Fig. 15-2) were found in diseased *Hippospongia communis* from Tunisia and in two non-commercial, but similarly affected, sponges of the genus *Ircinia* from Cyprus and Turkey (J. Vacelet, personal communication). Although the cause of the mortalities has not been verified experimentally, Gaino and Pronzato (1989) proposed that fiber-invading bacteria were the most probable

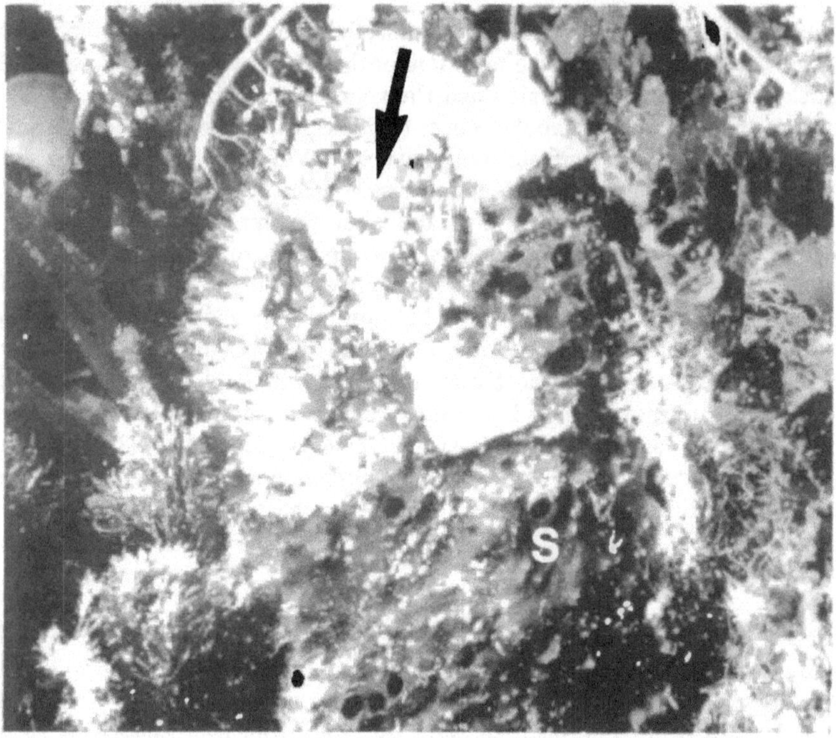

Fig. 15-1. Mediterranean sponge epizootic. Gross lesions on commercial sponge, *Hippospongia communis*, seen during epizootic off Tunisia in September 1989. Note brittle meshwork of spongin fibers appearing where tissue has sloughed off (arrow), with living sponge still present below (S). (Photo credit J. Vacelet, Station Marine d'Endoume, Marseille, France)

agents. They further suggested that because these bacteria morphologically resembled the symbionts found in healthy sponges under certain physiological conditions, sponge tissues may be unable to control the proliferation of the normal microflora.

J. Vacelet (personal communication) noted that there has been no demonstrated correlation of the disease with pollution, but earliest observations of the disease followed an unusual period of still and warm water in Tunisia. Sponges living at depths greater than 40 to 60 m and sponges in colder waters of the western Mediterranean or the Sea of Marmara have been less affected. Thus, it appears that high temperatures may have increased virulence of the disease. However, experiments to examine this relationship have not been performed.

Hummel *et al.* (1988) studied effects of increased temperature and reduced water flow rates on bacterial invasion of sponges, and demonstrated that two

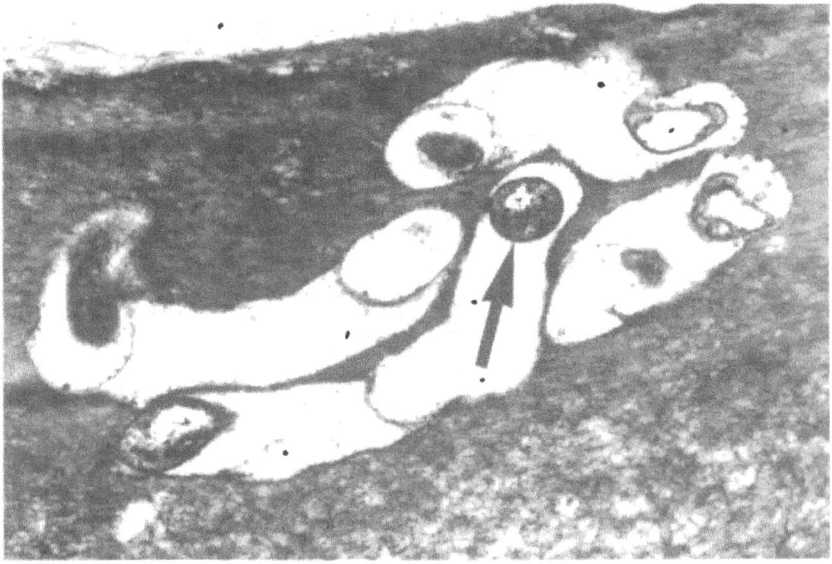

Fig. 15-2. Photomicrograph of bacteria (arrow) in canaliculi of spongin fiber found in diseased sponges (in Fig. 15-1) during the Mediterranean epizootic. Magnification X 18,000. (Photo credit M. F. Gallisian and J. Vacelet, Station Marine d'Endoume, Marseille, France)

bacterial isolates found in marine sediments and waters caused disease in inoculated crumb-of-bread sponge, *Halichondria panicea*. This sponge has low antimicrobial activity. However, these bacteria had not previously been reported to be infectious to sponges in the wild. Vicente (1989) analyzed species richness patterns of commercial sponges in the greater Caribbean region and proposed that these sponges had evolved under slightly cooler climatic conditions. As seawater temperatures appear to have risen in the period from 1900 to 1950, total or partial extinctions of these species occurred in the Antilles and elsewhere, either as a direct or indirect result of the change in environmental conditions.

Other Observations

Rützler (1988) described a disease apparently caused by the inability of the mangrove demosponge, *Geodia papyracea*, to control the quantities of the cyanobacterial symbionts within its tissues. The cyanobacteria multiplied faster than the sponge archeocytes could remove the excess, resulting in the destruction of the host sponge tissue, possibly by the secretion of toxic substances from the cyanobacteria. As in other observations of diseased sponges, this species was able to produce collagen-like spongin barriers and slough off the decaying tissue. Globular pseudogemmules of spongin microfibrils formed around the cyanobacteria trapped inside the archeocytes and were expelled when the decomposing sponge cortex broke apart (see also Connes 1967). *G. papyracea* occurs in other localities without the cyanobacterium. Rützler (1988) speculated that this might be a newly evolving mutualistic relationship in which the host's control mechanisms were developing (Jeon 1987), but acknowledged that it is not known if this cyanobacterium contributes any nutritional benefit to the sponge or is only a pathogen in a very slowly developing disease. There has been no research on adverse changes in environmental conditions that may be related to development of this disease.

Rützler (1990) reported parasitic filamentous algae (the green alga, *Ostreobium* cf. *constrictum*, and the red alga, *Acrochaetium spongicolum*) boring into spongin fibers of the Caribbean demosponge *Mycale laxissima* and other sponges. Other micro- and macro-organisms live in sponges, although their trophic relationships with sponges are little understood (see review by Lauckner 1980a). While some organisms may alter or damage sponge tissue, none are reported to have caused epizootic diseases in sponge populations, perhaps because sponges possess a variety of cellular defense mechanisms (Connes *et al.* 1971).

There is one intriguing report by Vacelet and Gallissian (1978) on the presence of intranuclear adenovirus-like particles found in abnormal giant cells in 1 out of over 100 specimens of the sponge, *Verongia cavernicola*, near Marseille, France, and smaller cytoplasmic inclusion particles, resembling a picornavirus (Johnson 1984). However, we do not know if viruses could be responsible for causing epizootic disease in these organisms. Although there have been a few isolated reports of structural anomalies (sometimes associated with symbionts) in sponges, we do not know of any cases of neoplasia in these organisms (Lauckner 1980a; Sparks 1985). Another potential sign of the effects of environmental stress or disease on sponges occurred during the last decade, as sponges containing photosynthetic symbionts were observed to "bleach" in reef environments along with cnidarians (Vicente 1990), probably in response to warmer than normal water temperatures (see next section). Clearly, many questions remain on diseases caused by microorganisms in sponges, on sponge host-defense mechanisms for control of symbiotic microorganisms as well as foreign pathogens, and on the influence of environmental conditions.

Phylum Cnidaria

The cnidarians also constitute a large and very diverse group of marine and estuarine organisms that inhabit both the benthos and water column and are further distinguished by variations in life history phases. Although most cnidarians appear to have short life spans, some Anthozoa apparently have indeterminate life spans (Brock 1984), and the scleractinian or stony corals may survive for centuries (Buddemeir and Kinzie 1976). As in the Porifera, members of this phylum are known to have associations with a variety of micro- and macro-organisms (see review by Lauckner 1980b). But the influence of these associates on the host, and whether they may cause significant disease in host populations, are still poorly understood in many cases, particularly for medusoid members of the classes Hydrozoa and Schyphozoa (Sparks 1985).

Numerous parasites have been reported in members of the class Anthozoa (see Phillips 1973; Gotto 1979; Lauckner 1980b; Stock 1988), but there have been few reports of their effects on most of these organisms. However, because scleractinian corals are such large and dominant members of reef communities around the world, and because of their demise in many areas, a number of recent studies have been conducted on their pathogens and abiotic diseases.

Black Band Disease

Black Line, or Black Band Disease (BBD), was reported in faviid corals first from Bermuda (Garrett and Ducklow 1975) and later in the Caribbean, Indo-Pacific, and Red Sea (Antonius 1985a). BBD is characterized by a thin black mat of fine filaments separating bare coral skeleton from living coral tissue. The disease usually starts at a site of physical injury and moves a few millimeters per day across the coral surface (Antonius 1981b, 1985b; Rützler *et al.* 1983). The causal agent has been identified as a cyanobacterium, *Phormidium corallyticum* (Rützler and Santavy 1983; Rützler *et al.* 1983; Taylor 1983; Antonius 1985b), and is believed to cause necrosis of the tissue by a toxic exudate (Fig. 15-3). The ability of the cyanobacterium to infect coral tissues varies with the species and, under experimental conditions, with the method of infection (Antonius 1988). A fungus was found in BBD-infected colonies of Venezuelan corals (Ramos-Flores 1983), and fungal infections have been reported in necrotic tissues of scleractinians and octocorals (Glynn *et al.* 1989) and bleached scleractinians and hydrocorals (see below, Jaap 1985; Te Strake *et al.* 1988), but the role of fungi as primary pathogens of corals or secondary invaders has not been investigated.

The appearance of BBD on a particular reef may be influenced by environmental conditions (Dustan 1977; Antonius 1985b). Water eutrophication in aquaria enhanced virulence of BBD in experimental infections (Antonius 1981a). Taylor (1983) also noted that development of infection was dependent on abnormal physiological stress or trauma that lowered coral resistance and released potential substrates for the cyanobacteria. Antonius (1988) confirmed

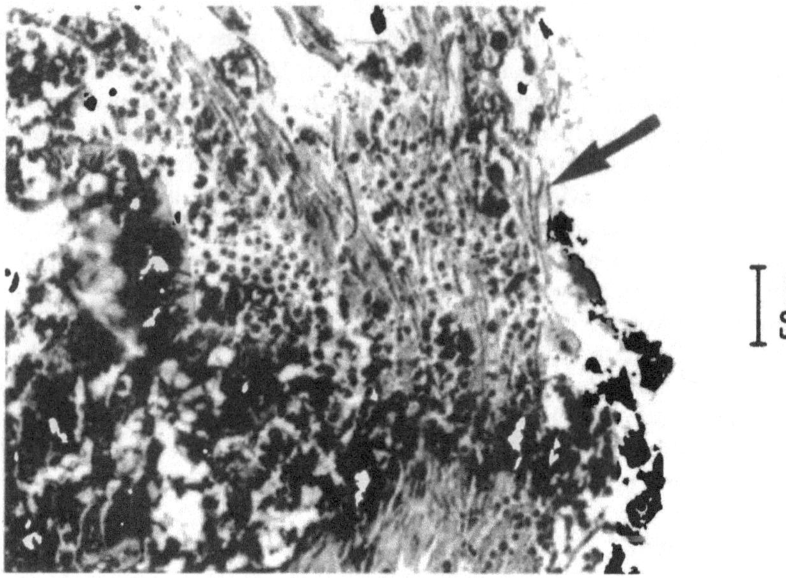

Fig. 15-3. Photomicrograph of *Phormidium corallyticum* cyanobacterial filaments (arrow) in necrotic tissue of brain coral afflicted with black band disease. Taylor's Modification of Brown and Brenn's stain for Gram-positive and Gram-negative bacteria, from Registry of Tumors in Lower Animals (RTLA), accession number 5111, contributed by E. C. Peters. Scale bar = 50 microns. (Photo credit E. C. Peters)

that infections of normally BBD-resistant species had occurred in polluted environments. Although BBD is usually found only on widely scattered coral heads (Garrett and Ducklow 1975; Edmunds 1991), perhaps through chance encounters with *P. corallyticum* trichromes, an epizootic affecting most of the faviid corals within the Looe Key National Marine Sanctuary (Fig. 15-4) began in 1986. This incident prompted further studies on etiology of this disease, treatment of affected corals, and the role of abiotic factors in the development of the epizootic at this site (B.D. Causey, National Oceanic and Atmospheric Administration [NOAA], Looe Key National Marine Sanctuary, Florida, personal communication). More recently, black band disease outbreaks have been sighted off Belize, the Cayman Islands, St. Vincent, and the Bahamas (J.H. Hudson, NOAA, Key Largo National Marine Sanctuary, Florida, personal communication).

Cyanobacterial infections of soft corals, *Pseudopterogorgia acerosa* and more rarely on *P. americana*, were found by Feingold (1988) off Sand Key in the northern Florida Keys. The disease agent here was identified as *Phormidium corallyticum*, which spreads along the axis and removes living coral tissue

Fig. 15-4. Divers examining black band disease outbreak on a large colony of the scleractinian coral *Montastraea annularis* in the Looe Key National Marine Sanctuary, Florida in 1986. (Photo credit J. H. Hudson, National Oceanic and Atmospheric Administration (NOAA), Key Largo National Marine Sanctuary, Florida)

(coenenchyme), up to 5 mm per day. This infection caused mortalities of several gorgonian colonies, but the disease was reduced or disappeared as water temperratures dropped below 20°C. Other infected colonies appeared to recover, although reinfections could occur. Appearance of the disease at this site could not be attributed to pollution or other abiotic factors. Cooler temperatures also appeared to moderate damage by BBD in stony corals (Dustan 1977; B.D. Causey, personal communication). Sea fans and gorgonians are also dying due to rapid growth of mats of cyanobacterium *Lyngbya* at a 100 x 75 m site north of North Key Largo Dry Rocks in the Florida Keys (J.H. Hudson, personal communication). Guzmán and Cortés (1984) theorized that black band disease was responsible for the disappearance of *Gorgonia flabellum* tissues off Costa Rica, because only axial skeletons of sea fans remained at the site after the epizootic. The agent was never identified. Sea fans have also mysteriously disappeared from reefs off Colombia, and a variety of gorgonians experienced mortalities off Trinidad (J. Garzón-Ferreira, INVEMAR, Santa Marta, Colombia, personal communication; Laydoo 1983).

Antonius (1981a,b; 1985a,b) noted that the appearance of bare white skeleton, signaling the start of sloughing of coral tissue at the rate of a few millimeters per day, was often a natural starting point for an attack of BBD microorganisms. Loss of tissue could occur either at the base of the colony or adjacent to abandoned holes of boring organisms on the coral surface. Antonius could not find any microorganism that might have removed the tissue prior to infection by *P. corallyticum*, and proposed that tissue sloughing was caused by a toxic shock to coral tissues in contact with unidentified species of algae or other epibenthos, or resulting from toxic interactions of interspecific coral aggressions.

White Band Disease

Recent studies indicate that corals can show similar signs of disease but may be afflicted with different pathogens or abiotic diseases. The terms "white band disease" (= WBD), "white plague," or "white death" have been applied to the sloughing of tissue in all species of corals. However, rapid loss of tissue may actually represent the end (or beginning) stages of different diseases. Tissue sloughing was first recognized and termed WBD in Caribbean acroporid corals (all three species of elkhorn and staghorn corals) in the mid-1970's (Gladfelter *et al.* 1977; Gladfelter 1982), and has since been reported throughout the region (Rogers 1985). These disease signs also occur in nine species of acroporids in the Red Sea and the Philippines (Antonius 1985a). Tissue sloughs off the exoskeleton from the base of the branches and proceeds to the tips (Fig. 15-5). This condition is distinguishable from predator damage.

Because of the importance of the acroporids as major reef framework builders, the presence of this disease on reefs has caused concern. At most sites, only a few colonies of *Acropora* spp. are affected. However, the *A. palmata* at Tague Bay and nearby Buck Island National Monument, St. Croix, U.S. Virgin Islands, have been decimated by this disease during the last decade. Extensive losses of *A. cervicornis* have also been reported in the Caribbean, possibly due to this disease and mediated by physical damage from hurricanes (Bak and Criens 1981; Knowlton *et al.* 1981; van Duyl 1983).

Except for this latter susceptibility due to physical damage, there has not been any correlation of diseased corals with adverse anthropogenic or natural environmental factors (Gladfelter 1982; Peters 1984). Peters and co-workers (1983) discovered unusual Gram-negative rod-shaped bacteria living in colonies or aggregates ("ovoid basophilic bodies") in the calicoblast epidermis lining the gastrovascular canals of diseased *Acropora* spp. from Tague Bay and Bonaire, Netherlands Antilles (Figs. 15-6). Bacterial aggregates were found not only near the periphery of sloughing tissues, but throughout the branch. Most apparently

Fig. 15-5. Elkhorn coral, *Acropora palmata*, afflicted with white band disease (bare white skeleton appearing at base of branches) at Grecian Rocks, Key Largo National Marine Sanctuary. (Photo credit J. C. Halas, NOAA, Key Largo National Marine Sanctuary, Florida)

healthy *A. palmata* and *A. cervicornis* sampled at Tague Bay in August 1981 contained these aggregates, with more bacteria per unit area of tissue in colonies showing signs of the disease than in apparently healthy ones (Peters 1984). Such bacteria were not found in healthy acroporid corals. Approximately 95% of the *A. palmata* colonies on the Tague Bay forereef had died by October 1986. Whether this microorganism is actually responsible for the loss of living coral tissue or is a secondary pathogen is still unknown. Interestingly, at the turn of the century, Duerden (1902) illustrated ovoid colonies of unidentified microorganisms similar to those found in WBD in *A. cervicornis* from Jamaica.

Sloughing of tissues from the base of the coral colony has also been observed in non-acroporid corals (e.g., Antonius 1985a; Dustan 1977). Peters (1984) examined this condition in several acroporid, poritid, and faviid corals collected from field sites in the Caribbean and from the Coral Reef Microcosm, Smithsonian Institution, Washington, DC. Although tissues on these colonies appeared superficially normal at the time of collection, microscopically the tissues exhibited varying stages of degeneration and necrosis, with no observable microorganisms present. Because the microcosm corals had experienced higher nutrient levels and sediment resuspension not normally found on reefs where they were collected, Peters (1984) proposed the term "Stress-Related-Necrosis" (SRN) for cases characterized by sloughing degenerating tissues in the absence of obvious pathogens to distinguish the lesions from WBD. A virus or bacterium may still be found to be associated with SRN. Dustan (1977) reported that inoculations of apparently healthy colonies of *Mycetophyllia ferox*, *M. lamarckiana*, and *Colpophyllia natans* with tissue from "plague-infected" colonies of *M. ferox* and *C. natans* caused "plague" in the healthy colonies, but further work has not been undertaken. SRN may be a reversible condition. Antonius (1981b, 1985b) noted that it was rarely lethal for the whole colony in faviids. Gladfelter *et al.* (1977) also noted an occasional reversal of tissue sloughing in *A. palmata*, suggesting that these colonies may not have been infected with the bacteria or may have recovered from the infection.

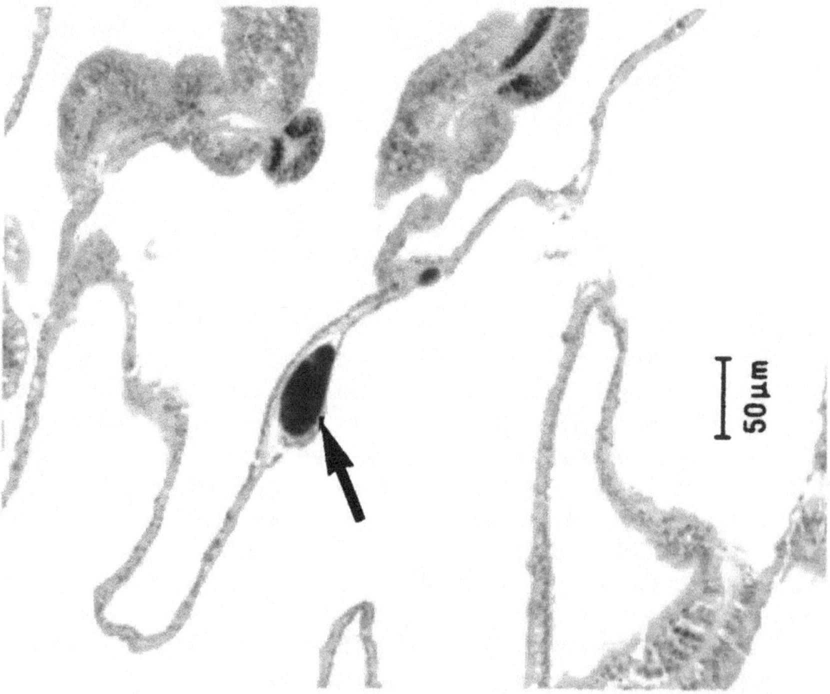

Fig. 15-6. White band disease. Ovoid basophilic bodies in calicoblast epidermis of *Acropora palmata*. Hematoxylin and eosin, RTLA 5070, collected by T. van't Hof. Scale bar = 50 microns. (Photo credit E. C. Peters)

Bleaching

Another gross indication of disease or environmental stress in corals and many other cnidarians is the loss of brownish pigments from tissues. These organisms have a mutualistic symbiotic relationship with single-cell dinoflagellate algae, known as zooxanthellae, that live in gastrodermal tissues. Zooxanthellae utilize cnidarian metabolic waste products and gases and produce nutrients by photosynthesis that are then taken up by coral tissues. Deposition of the calcium carbonate exoskeleton in scleractinians is enhanced by the presence of these algae (Goreau *et al.* 1979). Loss of zooxanthellae or algal pigments is termed bleaching, and has been documented in various field and laboratory experiments during periods of localized environmental stress, including high and low temperature extremes, light and salinity fluctuations, sedimentation, chemical pollution, and starvation (see reviews by Brown and Howard 1985; Williams and Bunkley-Williams 1990). However, we do not understand mechanisms by which densities of zooxanthellae in the gastrodermis are controlled, or how various environmental stresses may lead to bleaching in such organisms (Steen and Muscatine 1987).

Slight loss of color by tissues has been observed in acroporid corals afflicted with WBD (E. C. Peters, unpublished). Upton and Peters (1986) also noted that protozoa (Apicomplexa) living in coral tissues caused localized adverse host tissue reactions characterized by patchy bleaching. Isolated cnidarian bleachings with unknown etiologies have occurred on Caribbean and Indo-Pacific reefs (Williams and Bunkley-Williams 1990). Widespread partial or complete bleaching of many species of coral reef cnidarians (including stony corals, fire corals, gorgonians, sea anemones, and zoanthids) began to be reported in the early 1980's in the Indo-Pacific and Caribbean (Coffroth *et al.* 1990), increasing in frequency and occurrence worldwide in the past decade. Williams and Bunkley-Williams (1990) reviewed the literature detailing these bleaching events, focusing on the global tropical bleachings of 1987-88 and alterations in reef habitats.

A variety of species were affected, but severity of bleaching differed among species, with depth and locality (Fig. 15-7). Although most bleached cnidarians regained their zooxanthellae within a few weeks (Jaap 1979, 1985; Hayes and Bush 1990), many species suffered partial or complete mortalities (e.g., Glynn 1983; Lasker *et al.* 1984; Glynn *et al.* 1985; Harriott 1985; Lang *et al.* 1988; Williams and Bunkley-Williams 1990). Longer-term effects included reduced productivity and respiration rates, reduced skeletal growth rates, changes in tissue biochemistry, loss of gametogenesis, and tissue atrophy (Goreau and Macfarlane 1990; Savina 1990; Szmant and Gassman 1990). Although some workers speculated that biotic disease of either the zooxanthellae or coral was

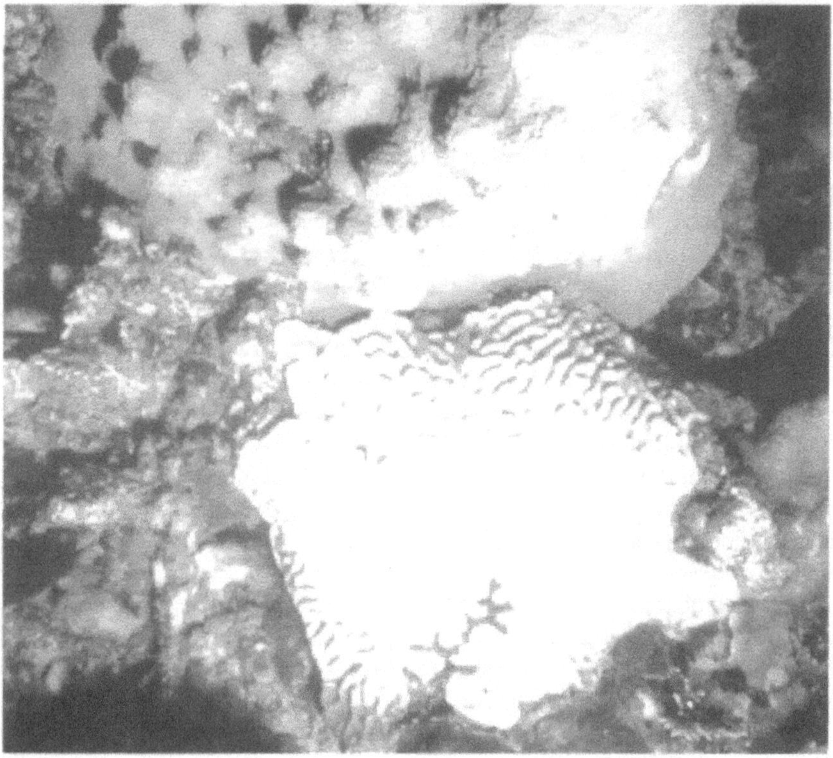

Fig. 15-7. Bleaching of brain coral, *Diploria strigosa*, during 1988 event off Puerto Rico, note *Montastraea annularis* with normal pigmentation above completely bleached coral. (Photo credit L. Bunkley-Williams, University of Puerto Rico, Lajas, Puerto Rico)

the cause of the zooxanthellae expulsion, widespread occurrence of bleaching in so many different animals (see Coffroth *et al.* 1990; Vicente 1990), and mortalities in non-zooxanthellate and non-cnidarian organisms on reefs suggested that environmental factors might be responsible.

Further research narrowed the causal agent to elevated temperatures documented at most reef sites. Possible conflicting stresses were identified as reduced water circulation, hypersaline water masses due to increased evaporation, increased solar ultraviolet irradiation, sedimentation, or turbidity, and pollution (Williams and Bunkley-Williams 1990). Most data support elevated temperatures as the cause, with recovery of zooxanthellae in coral tissues occurring as seawater temperatures dropped (Brown and Suharsono 1990; Cook *et al.* 1990; Glynn and D'Croz 1990; Jokiel and Coles 1990).

However, other studies indicate that although temperature increases of only one to a few °C over normal summer seawater regimes can cause expulsion of zooxanthellae, elevated temperatures also reduce the amounts of protective UV-absorbing compounds in tissues. Increased solar UV irradiation can occur during periods of calm water and can reduce the photosynthetic pigment content of algae, a condition seen in some affected corals (Hoegh-Guldberg and Smith 1989; Szmant and Gassman 1990). Thus, the animal/algal association may be damaged by the toxicity of active oxygen species to the zooxanthellae (Lesser *et al.* 1990).

Despite our lack of understanding of complex processes observed during bleaching, the importance of these organisms as indicators of global environmental change or disease has prompted attention from news media, Congressional hearings (Hollings 1988, 1990), and plans for additional monitoring and research. Another bleaching event began in the Caribbean and off Hawaii and Okinawa as seawater temperatures increased abnormally during summer in 1990. More extensive mortalities are expected due to weakening of reef organisms by previous episodes (Langreth 1990), with changes in structure and function of coral reef communities (Goenega *et al.* 1989; Edmunds 1991).

Other mortalities of corals have been attributed to changes in environmental conditions. Antonius (1977) described a rapid (within minutes) sloughing of tissues termed "Shut-Down-Reaction" (SDR) in heat- or cold-acclimated laboratory specimens that were under attack by black band disease microorganisms or predators. He also observed that cases of SDR occurred in stressed and physically wounded corals in the field and proposed that mass mortalities of corals at Hens and Chickens Reef off Plantation Key, Florida, in 1969-1970 were due to SDR acting on winter cold-stressed corals there. However, there have been no further reports on this phenomenon. Although Antonius (1977) reported that the SDR was easily spread from one coral colony to another, no pathogens were isolated.

Smith (1975) hypothesized that coral mortalities due to a 1971 "red tide" off the west coast of Florida were caused by bacterial and fungal infections, oxygen depletion, and hydrogen sulfide poisoning, rather than algal toxins. Bacterial growth increases with increased production of coral mucus and could cause death by sulfide poisoning, oxygen depletion, and attack of coral tissue (e.g., Ducklow and Mitchell 1979; Paul *et al.* 1986). Hodgson (1990) demonstrated that tetracycline-sensitive bacteria associated with the coral mucus produced during sedimentation stress are involved in tissue necrosis. Guzmán *et al.* (1990) reported mass mortality of reef corals (as well as fishes and other invertebrates) during severe dinoflagellate blooms of *Cochlodinium catenatum* and *Gonyaulax monilata*. He also observed production of copious amounts of mucus, but suggested that smothering by mucus, toxic compounds, and oxygen depletion by high densities of phytoplankton and decomposition of dead organisms were responsible for the mortalities.

Tumors

Several types of tumors have been found in cnidarians, although the etiology of theses lesions and influence of environmental conditions on their origins are still unknown. Again, the reef organisms have been the most studied. The gorgonian corals, *Pseudoplexaura* spp., have conspicuous coenenchymal swellings caused by a filamentous chlorophyte, *Entocladia endozoica*. Algal tumors have been found at various reefs along the Florida Keys (Goldberg and Makemson 1981; Goldberg *et al.* 1984), in *Pseudoplexaura porosa, P. flagellosa, P. wagenaari*, and *Plexaura flexuosa* in Santa Marta-Tayrona National Park, Caribbean coast of Colombia (Botero 1990), and in *Pseudopterogorgia* sp. off Costa Rica (Guzmán and Cortés 1984). There has been no correlation of the appearance of these tumors with known environmental stresses. Goldberg *et al.* (1984) noted that there were no pathological changes associated with the polyp layer surrounding tumors nor interference with reproduction. The number of scleroblasts and granular amoebocytes increased and then algal filaments were encapsulated by mesoglea.

Another type of algal tumor was examined in the sea fan *Gorgonia ventalina* by Morse *et al.* (1977, 1981). In this disease, filamentous algae produce a hyperplasia of the coral mesenchyme with secretion of abnormal gorgonin tubules that encapsulate but do not kill algae. These algae were analyzed biochemically and tentatively identified as of the order Siphonales, either a member of the genus *Ostreobium* or a close relative. This algal genus is a well-known endolithic associate of scleractinian corals. Authors noted that the distribution of tumors on sea fans varied with location in Bonaire, Netherlands Antilles, and Maqueripe Bay, Trinidad. The only sites where they occurred were in close proximity to petroleum tanker traffic lanes and loading depots, suggesting that exposure to oil hydrocarbons, possibly in conjunction with other environmental factors, such as predation, might influence tumorigenesis in the algal infection. However, further studies were not done, and recent sightings of these lesions in sea fans from relatively pristine reefs off Costa Rica and Puerto Rico do not support the association with pollution (Guzmán and Cortés 1984; E.H. Williams, Jr., University of Puerto Rico, personal communication). Although Morse and colleagues described these lesions as neoplastic, there has been no evidence to suggest that excessive gorgonin deposition would continue if algal filaments were removed from the axis. Such continued cellular proliferation and gorgonin deposition after the stimulus has ceased would be a requirement for the designation of a neoplastic disease.

The calcium carbonate exoskeleton of scleractinian corals is frequently altered by animal associates (e.g., Scott 1987; Zibrowius and Grygier 1985), but few of these organisms produce adverse effects in their hosts. Conversely, metacercaria of the digenetic trematode *Plagioporus* sp. feed on host tissues and then encyst in conspicuous elevated nodules on the surface of two species of scleractinian corals, *Porites compressa* and *P. lobata*, from Kanehoe Bay, Oahu, Hawaii (Cheng and Wong 1974). The cyst wall is secreted by the parasite, produces distortions of the gastrovascular cavity and cellular alterations within tentacles of the polyps, and disrupts normal calcium carbonate deposition by the calicoblast epidermis. If gastrodermal tissues of the host are ruptured, partial encapsulation of the metacerariae by coral cells occurs. Skeletal growth in parasitized corals can be reduced by almost 50%, but Aeby (1991) noted that reduced growth occurred even when the corallivorous butterflyfish, *Chaetodon multicinctus*, that feeds on parasitized polyps was excluded from feeding by caging. However, uncaged corals fed on by fish exhibited reduced levels of parasites, as infected polyps that were removed during fish feeding were replaced by healthy ones.

Cases of true neoplasia exist in scleractinians. Peters and co-workers (1986) reviewed reports of anomalous skeletal formations and identified calicoblastic epitheliomas in acroporid corals from western Atlantic and Indo-Pacific sites. These tumors are characterized by the appearance of protuberant whitened calcified nodules on normal polyps. Histopathological examinations of soft tissues in these skeletal masses revealed proliferation of gastrovascular canals and associated calicoblastic epidermis, with degeneration of normal polyp structures and loss of zooxanthellae. As the tumor enlarges, mucous secretory cells in the epidermis covering the lesion disappear, and the tissue becomes ulcerated and invaded by filamentous algae. There is also reduced skeletal accretion in surrounding branches of affected colonies (Bak 1983). No pathogens or parasites have been associated with these skeletal anomalies. The authors noted that genetic and environmental factors may affect distribution of the tumors, but cases of calicoblastic epitheliomas have been found only in isolated clusters of acroporid corals. Several other types of tumorous skeletal anomalies, possibly representing neoplastic diseases, have been found occasionally in corals, but the nature of tissue alterations has not been confirmed by histological examination (Peters *et al.* 1986).

In summary, cnidarians are now known to be afflicted by a variety of diseases, and there is sufficient evidence to suggest that environmental factors are important in mediating the virulence of some of these conditions. More research needs to be done to examine mechanisms of pathogenesis and to clarify etiologies. Although planktonic medusoid forms of cnidarians may be difficult to survey in most estuarine and coastal marine environments, the larger colonial benthic members of the class Anthozoa may prove to be important sentinel organisms for monitoring effects of environmental stress and disease.

Phylum Ctenophora

In contrast to our knowledge of diseases in the phylum Cnidaria, little is known about the health of the combjellies, partly because they are quite difficult to collect intact and to maintain in aquaria. However, some progress has been made, especially on the roles of xenogenous organisms found in association with these delicate animals.

Parasitosis or Paratenesis?

Ctenophores are known to be hosts for a variety of protozoan and metazoan organisms, some of which may be parasites of these animals. However, in many associations, ctenophores are suspected of being paratenic hosts, in which the larval trematodes, larval tetraphyllidean cestodes, and nematodes are merely

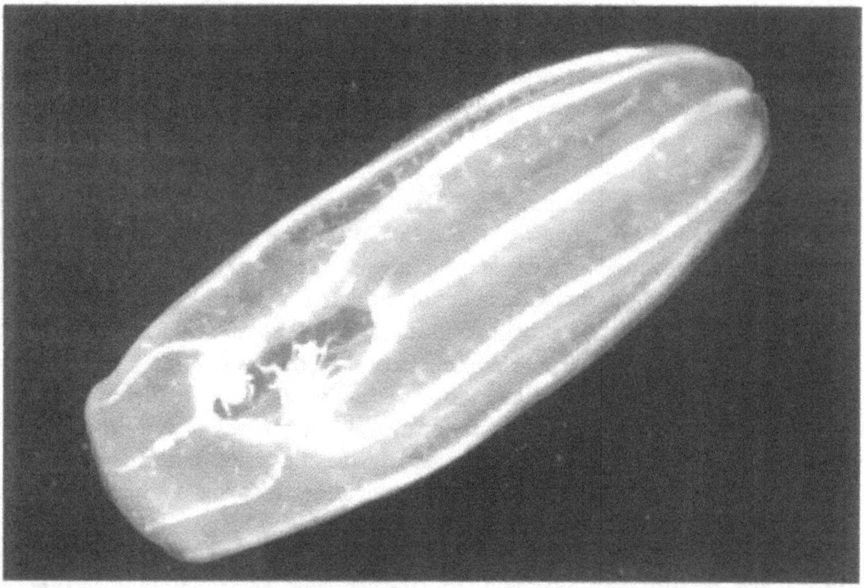

Fig. 15-8. Ctenophore associates. Two *Hyperia galba* on the ctenophore *Beroe cucumis*. After boring a hole into their host, the amphipods feed on the food that the host captures (usually other ctenophores), apparently without further damaging the host. (Photo credit G. R. Harbison, Woods Hole Oceanographic Institute, Massachusetts)

carried in their tissues and only undergo further development in their final host, usually crustaceans or fish that feed on ctenophores.

Cercaria can also migrate from the combjelly and penetrate epithelial tissues of the final host to gain access. The ctenophores obtain parasites by feeding on small crustaceans, then pass them when ctenophores are preyed upon by fish. Lauckner (1980c) reviewed these associations and noted that pathologic effects of the worms on the ctenophores were not well-studied, although some authors had suggested reduced growth rates or reduced longevity for infected individuals.

Another controversial association is that of hyperiid amphipods, *Hyperoche medusarum* and *H. mediterranea*, with ctenophores, *Pleurobrachia bachei* (Lauckner 1980c), *Lampetia pancerina* (Hoogenboom and Hennen 1985), *Beroe ovata* and *Mnemiopsis leidyi* (Cahoon *et al.* 1986), as well as other medusae

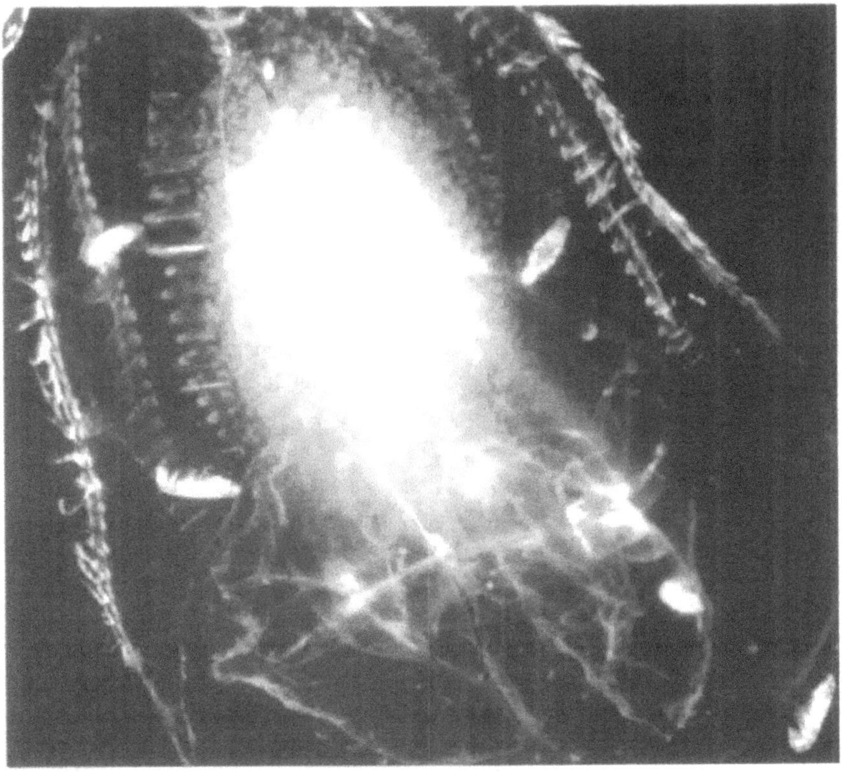

Fig. 15-9. Juvenile *Streetsia* sp. amphipods on the ctenophore *Leucothea* sp. (Photo credit G.R. Harbison)

(Laval 1980). *Hyperia galba* has been found in *P. pileus* and *B. cucumis*. Bowman *et al.* (1963) noted that *H. galba* (Fig. 15-8) feeds on the host ctenophore tissue in addition to other particles. However, various authors have ascribed different feeding modes and locations of attachment in different age classes of *P. bachei*. Lauckner (1980c) and Laval (1980) proposed that *H. medusarum* was a facultative endoparasite in ctenophores during early development and switched to an ectoparasitic or free-living phase, preferring larval fish for food at a later developmental stage. Other amphipods (e.g., *Streetsia* spp.) may heavily infect and consume whole ctenophores held in the laboratory (G.R. Harbison, Woods Hole Oceanographic Institute, Massachusetts, personal communication; Fig. 15-9).

Little is known of their effects at the population level. Yip (1984) examined the parasite load of *Pleurobrachia pileus* in an effort to determine whether these fish parasites contributed to observed fluctuations in numbers of ctenophores in Galway Bay, Ireland. Metacercaria of the trematode *Opechona* (*Pharyngora*) *bacillaris* was the most common of five parasites, especially in summer. These metacercariae were usually found attached by suckers to the external wall of the posterior pharynx or tentacular sheath of the ctenophore, but also occurred in other locations. In only two cases, nematodes pierced the epidermis or were embedded in the mesoglea, although didymozoids were often embedded. The author did not observe any damage to hosts by these parasites, but recorded a sharp decline in ctenophores either during or after peak infections in late spring and early summer over three years. He concluded that parasitism (weakening the hosts by using ctenophore resources) as well as food shortages and predation (as discussed in Greve's 1971 study) were responsible for reductions in ctenophores in late summer. There is still much to be learned about these associations and their effects on ctenophores.

Other Observations

Microbial diseases have not been reported in the combjellies, except under conditions of starvation and crowding stress in the laboratory (Lauckner 1980c). The only report of an abiotic disease in these animals was by Greve (1971), who correlated low water temperatures (3 °C) with degeneration of ciliary plates in up to 100% of *Pleurobrachia pileus* in the German Bight. No tumors have been reported in the Ctenophora (Lauckner 1980c; Sparks 1985).

These investigations examined ctenophores from north temperate waters. Our knowledge of the parasites or pathogens of combjellies in tropical marine and estuarine environments is limited, but there have been observations of combjelly associates with *Ocyropsis maculata* (Venus girdle) and hemiurids in the

Caribbean (G.R. Harbison, Woods Hole Oceanographic Institute, Massachusetts, personal communication). Because of their importance as predators and prey in food webs of estuarine and nearshore marine areas, and as intermediate hosts for parasites of crustaceans and fish, further examinations of health and susceptibility of ctenophores to biotic and abiotic diseases are in order.

Phylum Annelida

Diseases of freshwater and terrestrial annelid worm groups have received much attention in comparison to the paucity of studies on infectious and non-infectious diseases of marine and estuarine polychaetes, oligochaetes, and leeches. Polychaete worms are among the most species-rich benthic groups and are important components of community trophic structure and energy budgets (Fauchald and Jumars 1979). Polychaetes far outnumber other marine and estuarine members of the Annelida, but oligochaetes and leeches, most common in terrestrial and freshwater habitats, have also been recognized as important inhabitants of coastal waters. Most leeches are parasitic but some are free-living. Oligochaetes are of subordinate importance both numerically and ecologically in marine habitats, but they increase in abundance and ecological relevance in estuarine areas where marine competitors are excluded. The biology and ecology of marine and estuarine oligochaeta have recently been reviewed by Giere and Pfannkuche (1982), who noted that their role in nearshore ecosystems is still largely speculative. Interestingly, these worms are highly resistant to pollution, remaining as the sole surviving fauna. Bioturbation by oligochaetes can effectively deplete pollutants from sediments (Giere and Pfannkuche 1982).

Although population oscillations have been observed in the field and laboratory for some polychaete species, such as *Capitella capitata*, other species are known to form stable populations over time (Chesney and Tenore 1985; Reise 1979). Similarly, oligochaete populations fluctuate seasonally, with peaks in the spring, or spring and autumn, although tubificid standing stocks appear to be maintained throughout the year (Giere and Pfannkuche 1982). The role of diseases in altering their population size and trophic relationships is still unknown. Recently, a mass mortality (up to 99%) of lugworms, the polychaete *Arenicola marina*, occurred on the coast of South Wales, Great Britain (Olive and Cadnam 1990). Other invertebrates were also affected, but two species of polychaetes remained healthy. In conjunction with other factors, the distribution of the mortalities suggested that a bloom of the dinoflagellate *Gyrodinium aureolum* (= *Gymnodinum* cf. *nagasakiense*) was responsible. The authors suspected that earlier reports of lugworm mortalities could also be due to toxic phytoplankton blooms.

Polychaete Pathogens and Parasites

In the polychaetes, the only known viral disease is a DNA virus with characteristics of the Iridoviridae that infects maturing gametocytes in male *Nereis diversicolor* off Boulogne, France. This virus, named *Nereis* Iridescent virus, or NIV, prevents formation of spermatids; thus, the host is sterile. Injections of virus from infected tissue caused 100% infection of normal male worms (Devauchelle and Durchon 1973, cited in Johnson 1984).

Diseases caused by bacteria are virtually unknown in these animals, possibly because they may possess well-developed humoral and cellular defense mechanisms (Anderson 1980; Anderson and Chain 1982; Chain and Anderson 1983). Although Anderson and colleagues reported bactericidal activity in the coelomic fluid of *Glycera dibranchiata*, Dales and Dixon (1980) did not find such antibacterial activity in *Neoamphitrite figulus*, *Arenicola marina*, or *Nereis diversicolor*. Coelomic fluid from *G. dibranchiata* has a normal and highly variable background level of naturally occurring bacteria. Roch *et al.* (1990) also found wide variation in hemolytic activity of eight species of homogenized polychaete worms from seven families, further underscoring the extreme range of susceptibilities present in these annelids. Exposure to xenobiotics impaired the ability of amoebocytes of *G. dibranchiata* to recognize foreign material (Anderson 1987), although other parameters were not significantly affected and no disease appeared. This phenomenon was interpreted as similar to other marine invertebrates that exhibited increased viral, bacterial, and parasitic susceptibilities during exposure to pollutants (e.g., Couch and Courtney 1977; Winstead and Couch 1988).

Recently, P.J.W. Olive (unpublished, cited in Olive and Cadnam 1990) reported a vibriosis affecting the nervous system of *Nereis virens*. Bacterial cells of *Vibrio* cf. *parahaemolyticus* were isolated from lesions evident on the sense organs, and behavioral abnormalities occurred, with the polychaetes eventually dying. Douglass and Jones (1991) found a large Gram-negative bacterium infecting white cauliflower-like lesions in the cuticular epithelium of the spionid *Polydora nuchalis* with another morphological type of bacterium present in the epithelium, and other possibly pathogenic bacteria in *Scolelepsis maculata*.

A number of interesting microorganisms have been found in the polychaetes, but we have no information on their pathologic effects on the hosts (Seibert 1973; Sparks 1985). Carton (1967) examined the tissue response of the sabellid worm, *Spirographis spallanzani*, to infestation by copepods of the genus *Sabelliphilus*. *S. sarsi* had little effect on the body wall of its natural partner, but settlement on atypical host species of *Spirographis* resulted in dramatic host defense reactions that destroyed most copepods. Porchet-Henneré and M'Berri (1987) noted that *Nereis diversicolor* hosts coccidians, as well as gregarines within the coelomic cavity, and observed that in the field the worms "seem able to endure the presence of numerous parasites very well." In experiments, they examined the effects of infection with an orthonectid mesozoan, yeast, and coccidian. *N. diversicolor* resisted bacterial infections, and foreign organisms were encapsulated without damage to host worms.

One of the most studied associations in freshwater and terrestrial annelids has been the presence of gregarines (Apicomplexa: Sporozoea) in the gut and coelomic cavity of the worms. Kudo (1966) noted that in aquatic environments, some gregarines are known to cycle between crustacean and molluscan hosts. Gymnospores voided in the feces of crustaceans enter molluscan hosts through the gills, mantle, or digestive system, where they become paired and fuse. The zygotes develop into naked or encapsulated sporozoites within phagocytes of the molluscan host. When crustaceans feed on these molluscs, the gregarines develop into cephaline gregarines (trophozoites) in the stomach and midgut of the specific crustacean host.

Nematopsis spp. and *Cephalolobus* spp. trophozoites occur in the anterior midgut cecum and midgut of many species of commercially harvested and cultured penaeid shrimp (Sindermann 1990). An infection of *Nematopsis* sp.

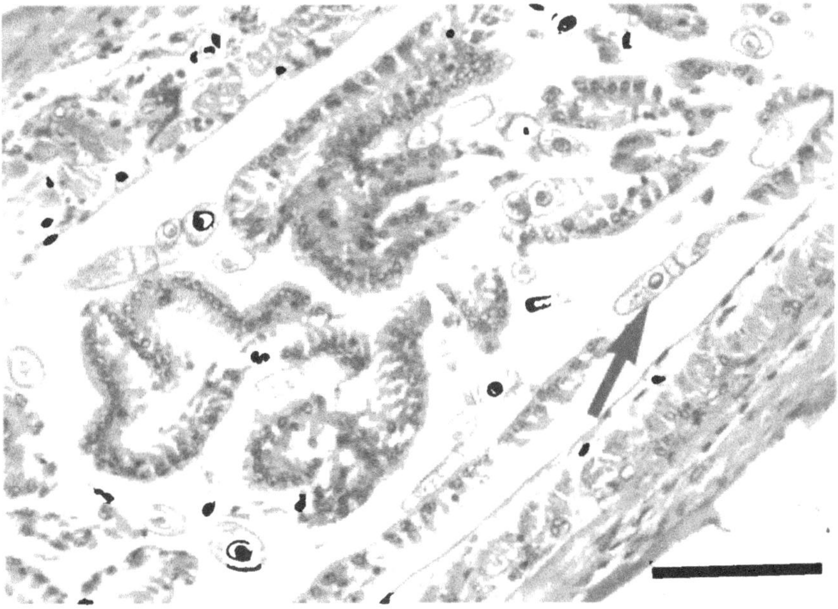

Fig. 15-10. Polychaetes as alternate hosts. Gregarine trophozoites (arrow) in hyperplastic midgut of white shrimp, *Penaeus vannamei*, from Ecuador. Hematoxylin and eosin. Scale bar = 50 microns. (Photo credit D. V. Lightner, University of Arizona, Tucson)

in *Penaeus vannamei* from shrimp ponds in Ecuador has recently been investigated by Lightner (D.V. Lightner, University of Arizona, Tucson, Arizona, personal communication). The epimerite of the gregarine trophozoite stage attaches to the intestinal mucosa of the shrimp, either between or on cells, causing necrosis and sloughing of the epithelium. The one-cell stage induces hyperplasia of the midgut epithelial cells with villous formation (Fig. 15-10). Sporozoites may also be found within the shrimp, with sexual stages in the hindgut only. Although the parasite infection itself is probably not fatal, severe midgut lesions may reduce growth and provide routes of entry for opportunistic pathogens (like *Vibrio* sp.) that can kill postlarvae and juvenile shrimp. To control this parasite, Lightner and colleagues searched for alternate molluscan host(s) for the sporozoite stage, but found none. Instead, sporozoites were discovered in polychaete worms, *Polydora cirrosa*, living in mud on the bottom of ponds (Fig. 15-11).

Infection studies were performed in an isolated lab in Ecuador by feeding *P. vannamei* postlarvae with clams, mussels, mud, and polychaetes from ponds

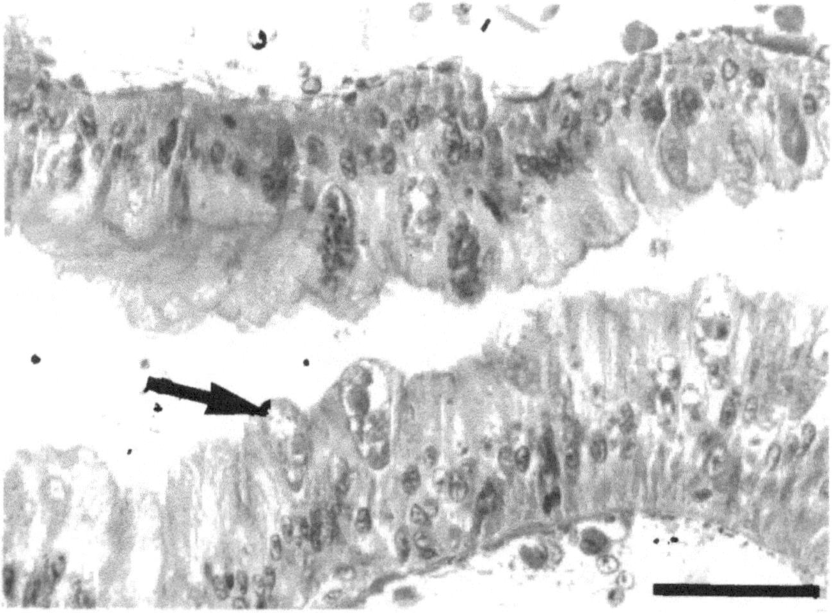

Fig. 15-11. Unidentified endoparasites (arrow) in intestinal epithelium of the spionid polychaete *Polydora cirrosa* associated with gregarine outbreaks in shrimp ponds in Ecuador. Hematoxylin and eosin. Scale bar = 25 microns. (Photo credit D. V. Lightner)

having the gregarines (D.V. Lightner, personal communication). Only the shrimp fed worms developed gregarine infections (100% incidence, and many severe cases). Ecuadorean shrimp farmers control the gregarines in affected ponds with a feed impregnated with an undisclosed anticoccidian drug borrowed from the poultry industry.

Besides serving as an alternate host for the *Nematopsis* sp. parasite of shrimp, this polychaete also appears to have its own gregarine parasite (D.V. Lightner, personal communication). Gregarine parasites representing both all life stages and only the trophozoite stage have been recorded from other polychaetes (e.g., Shrevel 1963; Desportes and Théodoridès 1986; Landers and Gunderson 1986; K. Fauchald, Department of Invertebrate Zoology, National Museum of Natural History, Washington, D.C., personal communication).

However, these have been only descriptive taxonomic studies, with no mention of pathologic effects on their hosts, or whether alternate hosts may figure in the life cycle of these protozoans. Douglass and Jones (1991) found at least 26 species of parasites (including 13 gregarines, 16 apicomplexans, 4 microsporidia, 3 ciliates, 1 mesozoan, and bacteria) in a survey of 9 spionid polychaetes. Although they did not discuss the effects of these protozoans and mesozoans in the paper, Douglass noted that the gregarine *Selenidium* sp. infecting *Boccardia proboscidea* and the orthonectid mesozoan were clearly pathogenic in laboratory culture, but their effects on wild populations of polychaetes were still unknown (T.G. Douglass, California State University, Long Beach, California, personal communication).

As for metazoan parasites, Køie (1985) has identified polychaetes as the second intermediate hosts for the metacercarial stage of the digenetic trematode, *Lepidapedon elongatum*. The final host for this trematode is the Atlantic cod, *Gadus morhua* or *G. ogac*. Brown and Prezant (1986) discovered metacercariae of the family Echinostomatidae in the polychaete, *Scolopos fragilis*, during a study of the trophic ecology of this burrowing deposit feeding worm at Cape Henlopen tidal flat near Lewes, Delaware. Polychaetes previously had been only infrequently identified as intermediate hosts of trematodes. The encysted metacercariae were found year-round, and often up to 100% of the worms were infected, with highest degree of parasitism in the summer months. Cercariae of *Himasthla quisetensis* taken from the snail *Ilyanassa obsoleta* readily infected the polychaetes, suggesting they were the same species. The authors further noted that the worms may be foraged upon by avian predators at low tide, and stated that "the observations reported here are significant because they suggest that annelids play a far greater role as intermediate hosts for some trematodes than has previously been realized." Other parasites of marine molluscs, crustaceans, and fish, including commercially important species, may require an alternate host that occurs in this poorly studied phylum.

Hirudinea and Oligochaeta Associates

We know virtually nothing about diseases in marine leeches. *Calliobdella vivida* (Burreson and Zwerner 1982), a common parasite of the hogchoker *Trinectes maculatus* and other species of Atlantic and Gulf Coast fish of North America, is a vector for blood parasites of *Trypanoplasma bullocki*, transmitting this protozoan while feeding on the fish (Burreson 1982). Infective stages of these trypanoplasms migrate anteriorly from the crop and accumulate in the proboscis sheath of the leech, while others remain throughout the crop and postceca, but not the intestine. The flagellates apparently do not harm the leech. However, mass mortalities of yearling summer flounder, *Paralichthys dentatus*, from Chesapeake Bay, Virginia, and Pamlico Sound, North Carolina, in early 1978 and the York River, Virginia, in 1981, were linked to infections in fish weakened by cold water temperatures. Burreson and Zwerner (1984) documented the role of temperature in mortalities from experimental infections.

Marine oligochaetes are common hosts for protozoan parasites in the orders Astomata (Ciliata: Holotricha) and Gregarinida (Sporozoa: Telosporidia) (see Giere and Pfannkuche 1982). The ciliates have been found in the gut of enchytraeids, tubificids, and naidids, where they absorb dissolved nutrients without apparent pathogenic effects on the host, even in cases of massive infections. Host-specificity does not seem to be very strict. One report by Dahl (1960) noted an unusually thick layer of chloragogenic cells on the outside of the gut in association with a heavy ciliate infestation of *Paranais litoralis*. He suggested this was the reason for the population minimum in August, although similar population decreases have been reported in late summer without the worms being infested (Giere and Pfannkuche 1982).

Gregarinid sporozoa of the family Monocystidae, genus *Monocystis*, occur in the coelomic cavity and seminal vesicles of oligochaetes. Young trophozoites cause hypertrophy of the worms' chloragocytes; however, even heavy infestations do not seem to cause histological abnormalities or sterilization effects. Distribution of parasitized animals is often patchy. A few other parasites are reported in marine oligochaetes, but their effects on hosts and populations have not been described. So, whether these associations cause significant pathosis or contribute to observed population fluctuations remains to be further investigated.

The oligochaetes also provide nutritious food for demersal young fish, sometimes selectively by species, as well as food for macrofauna: polychaetes, turbellarians, small crustaceans, such as shore crabs and shrimp, gastropod molluscs, and water birds. Their importance as prey items suggests that they may also play a role as intermediate hosts for parasites of these organisms. The only report of such an association in marine or estuarine waters is that of Lichtenfels and Stroup (1985), who found 1 in 1767 oligochaetes (*Limnodrilus* sp.) in the Chesapeake Bay contained a third stage nematode of the genus *Eustrongylides*, a parasite of the benthic mummichog fish *Fundulus heteroclitus*. The authors had observed increased prevalences of these nematodes in fish collected from warmer waters at a power plant, and suspected that the abundant oligochaetes in sediments there harbored the larval stages of the nematodes.

The freshwater tubificid oligochaete, *Tubifex*, has been implicated as the alternate host for the myxosporean agent of "whirling disease" (*Myxosoma cerebralis*) that causes serious damage to young species of salmonid fish in hatcheries. Observing that freshly isolated spores from fish did not cause the disease in exposed fish, Markiw and Wolf (1983) investigated the role of tubificids in the development of the disease, and found that only fish fed worms from the sediments of affected hatcheries developed the disease. Tubificids sectioned histologically revealed spores (Fig. 15-12) that showed focal reactivity

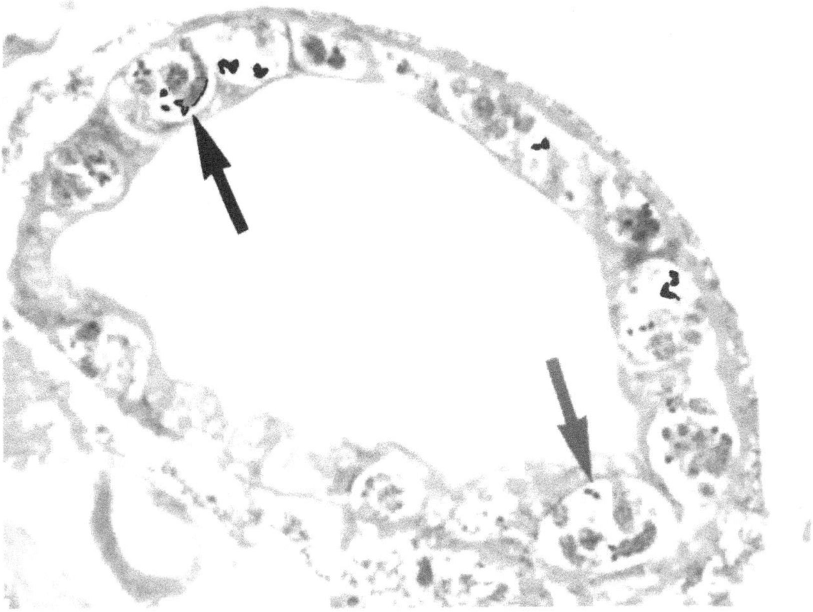

Fig. 15-12. Salmonid whirling disease. Histological cross-sections of tubificid oligochaete *Tubifex tubifex* experimentally infected with actinosporean *Triactinomyxon* 130 days after exposure to *Myxobolus cerebralis* spores. Note several cysts in the gut wall of tubificid (arrows); each cyst contains eight actinosporeans. The three polar capsules (arrows) of the upper body or epispore are intensely stained and the numerous internal sporozoites (sporoplasms) are stained lightly. May-Grunwald Giemsa. Magnification X 350 (Photo credit M. E. Markiw, U. S. Fish and Wildlife Service, National Fish Health Research Laboratory, Kearneysville, West Virginia)

to *Myxosoma cerebralis* fluorescent antibodies. Wolf and Markiw (1984) identified the worm spores as belonging to the actinosporean *Triactinomyxon gyrosalmo*. Similar actinosporeans had long been noted as parasites of annelids.

Corliss (1985) reviewed morphology of the spore stage found in worms, and noted that it was very different from spores found in fish. He further clarified taxonomic problems associated with this discovery, stating that it was the first discovery of a single species with stages apparently belonging to two widely separated taxons. He suggested the term *Triactinomyxon* as a collective-group name, and pointed out that the correct name for the parasite was *Myxobolus cerebralis*. Wolf and co-workers (1986) described the effects of the cysts on worms, noting that cysts were found in the gut of segment 17 and posteriad in the annelids, but not anterior, and that infected worms often appeared pale, displaying an opaque outer layer, with generalized anterior swelling.

More recently, Hamilton and Canning (1987) refuted this hypothesis, finding no hatching or development of the spores in *Tubifex* into actinosporeans. They also noted the absence of *T. gyrosalmo* (they considered it identical to *T. dubium*) in three of four sites where whirling disease had been diagnosed. They noted that if there were direct relationships between the Myxosporea and other invertebrates, then it appears that too few actinosporeans have been found to support this determination.

However, Kent and co-workers (1990) noted that more recent studies have confirmed the results of Wolf and Markiw (1984) with the same and other myxosporeans. These authors are examining the life cycle of *M. articus*, a parasite infecting the brains of sockeye salmon *Oncorhynchus nerka*. Lumbriculid worms, *Eclipidrilus* sp., were found in sediments from Sproat Lake, Vancouver Island, where almost 100% of the sockeye are infected with *M. articus*. All fish fed these worms in the laboratory became infected, while fish fed fresh spores obtained from infected fish did not become infected.

Although whirling disease occurs only in freshwater fish populations, it may have unknown impacts on seawater fauna, and indicates the importance of looking for alternate life cycle stages of other fish parasites in worms. Additionally, too little is known about the influence of abiotic factors on the diseases of marine and estuarine annelids, and their relationship to population fluctuations or host dynamics. Interestingly, no neoplasms have been found in marine annelids (Dales 1983).

Phylum Echinodermata

Echinoderms often form large and conspicuous assemblages in benthic habitats, and are often top predators in their communities or controlling agents of seagrass or kelp beds. As a result, they have received much attention from ecologists, physiologists, biochemists, and even invertebrate pathobiologists. Jangoux (1987a,b,c,d) has extensively reviewed the literature on diseases and host defense mechanisms in the echinoderms. Some of the most interesting and important early studies on invertebrate immunity were performed by the late F. B. Bang, using starfish or sea stars. For example, Bang and Lemma (1962) discovered that bacteremia occurred naturally in field-collected starfish, and could be induced by injection of bacterial isolates into autotomized individuals placed in stagnant water. However, these infections usually resolved within three days. Further work indicated that the coelomic fluid and associated cells of this starfish and other echinoids provide a variety of defense mechanisms to control bacterial infections in the coelomic cavity (Hetzel 1965; Johnson and Beeson 1966; Millott 1969), and eliminate foreign microorganisms or tissue (Ghiradella 1965; Johnson and Chapman 1970a,b; Bang 1982; Coulon and Jangoux 1987). Other bacteria normally populate the gut or other organs (e.g., Unkles 1977). Jangoux (1987d) observed that members of the Class Asteroidea possessed more efficient defense mechanisms than those of the Echinoidea, and that most biotic agents were well-tolerated (see below).

Turton and Wardlaw (1987) discovered pathogenic marine yeasts that caused mortalities of *Echinus escuelentus* when injected into the coelomic cavity. The yeast may have acted by depressing host defense mechanisms, and secondary bacterial infections may have been responsible for the mortalities. Naturally occurring yeast infections are unknown in these organisms. Birkeland and Lucas (1990) reviewed studies on a bacterial disease of *Acanthaster planci*. Although the starfish normally contained an unusual undescribed bacterium in the body wall, mucus secretions, and pyloric caeca, under conditions of stress (held in aquaria) these bacteria became facultative pathogens. Other secondary infections by *Vibrios* and other bacteria also occurred.

Assorted Associations and Tumors

Many different types of symbiotic associations with other organisms are known in the echinoderms, but again, most reports lack details on the nature of these relationships and pathologic effects on their hosts or on host populations (see Jangoux, above, and Sparks 1985). Parasitic organisms include protozoans, flatworms, snails, barnacles, and copepods (Powers 1933, 1935; Gotto 1979; Lambert 1981; Cannon 1986; Cannon and Jennings 1988; Birkeland and Lucas 1990). Although most parasites do not seem to harm echinoderm hosts, Jangoux (1987c) noted that ectoparasitic crabs could kill their hosts. Emson and co-workers (1985) reported severe damage to and possible death of brittlestars heavily parasitized by copepods. One copepod associated with the basketstar, *Astrophyton muricatum*, is normally ectoparasitic or occasionally found in its stomach. Williams and Wolfe-Waters (1990) reported that 1 of 75 infected basketstars (49.3% of total examined) contained 49 of these copepods, *Doridocola astrophyticus*, in its stomach. This animal appeared emaciated and lacked gonads. An endoparasitic gastropod, *Thyonicola americana*, can interfere with suspension feeding activities of its host holothurian, *Eupentacta quinquesmita* (Byrne 1985).

Five species of gregarines co-occur in the burrowing *Echinocardium cordatum*. Coulon and Jangoux (1987) reported that the trophozoite stages found in the gut were harmful to the host, but coelomocytes reacted against those free-stages that invaded the coelomic cavity, killing the parasites, especially those paired in syzygy. Gregarine gametocysts that developed were not attacked by coelomocytes and did not harm the echinoid. Some remained within the host until its death, possibly for up to 10 years, at which time they would be released to the sediments, starting the cycle again. Hagen (1987) reported the association of large-scale destruction of kelp beds on the coast of northern Norway by urchins, *Strongylocentrotus droebachiensis*, with infections by adult stages of a large ovoviviparous nematode endoparasite, *Echinomermella matsi*. The parasite was found in the perivisceral coelom of 6 to 56% of urchins in the overgrazed area. The author further noted that hatching of juvenile nematodes was probably lethal for the host, and urchin mortality increased in areas with high parasitism. *Echinomermella grayi* has been found in the coelom of *Echinus escuelentus* in England, but not in Norway.

Tumorous lesions of unknown etiologies have been reported in 7 out of 95 specimens of the brittlestar, *Ophiocomina nigra*, from Plymouth, England (Fontaine 1969) and in specimens of the sand dollar, *Echinarachnius parma*, from the north Pacific and Atlantic Oceans (E.C. Peters, R. Mooi. and K.L. Price, unpublished). The brittlestar lesions (up to 3 per animal) had pigmented noncalcified masses of mitotically active epidermal melanocytes, representing several stages of development. Fontaine (1969) thought that these were of minor pathological significance to the individual in terms of health and survival in a natural population, so perhaps they were not analogous to melanomas, but rather were analogous to pigmented junctional nevi. However, there has been no further research on this condition. The sand dollar lesions (cases now archived in the RTLA) are calcified, irregularly shaped, often verrucose, protuberances on the aboral surface of the test, and appear to be a neoplastic disease originating in the sclerocytes. Birkeland and Lucas (1990) illustrated an unusual condition of the spines in *Acanthaster planci*, found on three or four specimens near Vila, E'fate Island, Vanuatu, in 1983. Each starfish had from 2 to 12 small protuberances from the aboral surface, each protuberance covered with its own spines. These spines were smaller than the normal adjacent epidermal spines and "had apparently developed with the lump" (p. 142). The authors further noted "the lumps may be malignant neoplasms or they may be responses to endoparasites; they have not been otherwise reported." Because no histological studies were performed on these lesions, the suggested diagnosis of malignant neoplasm is unwarranted.

Parasitic Castration

Four disease conditions found in sea stars, sea cucumbers, and sea urchins have received much attention in recent years. The first is parasitic castration. Pearse (1982) noted that parasitic nematodes in the gonads of sea urchins block gametogenesis, but did not elaborate. Whitfield and Emson (1988) observed a 44% reduction in brooding capacity in the ophiuroid, *Amphipholis squamata*, in intertidal rock pools in South Devon, UK, due to the endoparasitic copepod, *Parachordeumium amphiurae*. They compared this association with brood parasitism, in which the brood parasite interpolates its offspring into the protective and feeding mechanisms utilized by the host for its own young. This parasitism did not cause stunted embryological development, just reduced numbers of brooded development stages, and possibly contributed to host mortality (density-dependent, parasite-induced host mortality).

Although holothurians are common in shallow-water environments, little is known about parasite infestations that affect their reproductive capacities. However, Tyler and Billett (1987) found that testes of deep-sea elasipodid holothurians (families Psychropotidae, *Psychropotes longicauda* and *Benthodytes sordida*) from the North Atlantic were infected with protozoan parasites. In the former, parasitic castration occurred in at least 2 of the testes in 24 out of 27 males. In the latter, 80% of the males had a low level of sporozoan infestation. The nature of the parasite was not further identified.

The parasitic ciliate *Orchitophyra stellarum* was discovered in the testes of the asteroid *Asterias rubens* in the North Atlantic near France and described by Cépède (1907a,b). Later, it was reported in gonads of male *Asterias forbesi* from Long Island Sound by Piatt (Anonymous 1935). It has occasionally been found in females (see Jangoux 1987a; Bouland and Jangoux 1988). Galstoff and Loosanoff (1939) considered it a natural control of starfish populations in Long Island Sound. The ciliate, infecting from 1-20% of males, depending on locality, and 1% of females, destroyed gonads and caused partial or complete castration. In *A. rubens*, Vevers (1951) found total castration of 28% of male starfish sampled. Bang (1982) investigated an indirect effect of the infestation, absence of *in vitro* clumping of coelomocytes. More recently, Bouland and Jangoux (1988) found a low prevalence of the ciliate in specimens of *A. rubens* collected from the coast of the Netherlands and described partial destruction of the germinal epithelium with further damage by host phagocytes. Although

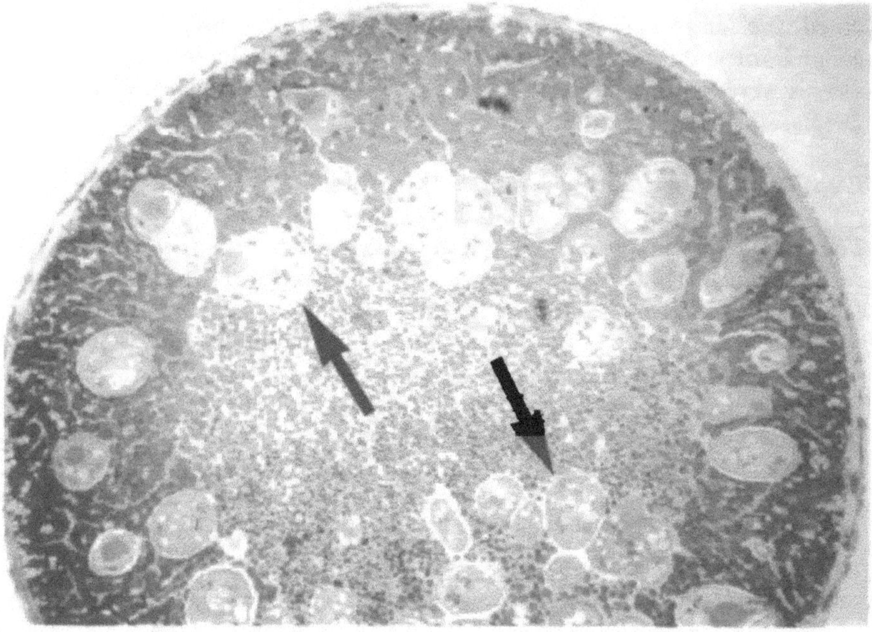

Fig. 15-13. Parasitic castration in sea stars. One micron-thin section of plastic-embedded gonad from *Asterias forbesi* with *Orchitophyra stellarum* infection (arrows). Magnification approximately X 200. (Photo credit C. W. Walker, University of New Hampshire, Durham)

phagocytes normally occurred in post-spawn animals, these cells differed from the phagocytes found in hemal and coelomic systems.

Wicklow *et al.* (submitted) further described the *O. stellarum* organism, derived from infected *A. vulgaris* (Fig. 15-13) from New England waters (= *A. rubens*, M.S. Downey, National Museum of Natural History, Washington, D.C., personal communication). They noted that endoparasites were only present during the differentiative phase of spermatogenesis (February-May) and in severe infections, spermatozoa were completely removed from all 10 testes. Although this ciliate does not appear to cause other diseases in sea stars of the genus *Asterias*, and infected individuals usually recover (Bang 1982), Leighton and co-workers (1991) found the same ciliate infesting the sea star, *Pisaster ochraceus*, at two sites on the southern coast of British Columbia, Canada. In this case, not only were male starfish effectively castrated by the ciliate, but mortalities of the male starfish also occurred. They proposed that the ciliate has been recently introduced into the Pacific Ocean, resulting in the more virulent nature of this parasite in *P. ochraceous*, as compared with Atlantic and Mediterranean hosts. Because this starfish is a top predator and the single dominant starfish at many localities on the Pacific coast of North America, the mortalities seen as the result of the introduction of this endoparasite could signal drastic changes in the diversity of inter-tidal habitats there (see also Paine 1966). Whether the ciliate may infect other sea stars in the Pacific is also unknown (Waldichuk 1990).

Bald Sea Urchin Disease

Sea urchins are also powerful modifiers of marine and estuarine habitats, particularly noted for their effects on benthic macroalgae and other competitors for space on hard substrata (e.g., Breen and Mann 1976; Carpenter 1981). Thus, occurrences of epizootic diseases in these organisms were first noticed by ecologists and physiologists. Pearse *et al.* (1977) reported that *Strongylocentrotus franciscanus* off southern California suffered progressive loss of spines and tube feet, with discrete lesions of epidermal necrosis. Similar signs occurred during mortalities of *Paracentrotus lividus* in the Mediterranean (Hobaus *et al.* 1981). Effects of these urchin mortalities on the species diversity and growth of epiphytes on seagrasses were documented by Pearse and Hines (1979) and Bourdouresque and co-workers (1981). The characteristic lesion is a central necrotic region of disorganized tissue surrounded by a ring of swollen tissue containing the bactericidal red spherule cells and vibratile cells. The necrotic area is greenish-colored, completely denuded, and free of spines.

Gilles and Pearse (1986) examined epidermal lesions in populations of *S. purpuratus* in Carmel and Monterey Bays, California. They were able to initiate similar lesions in laboratory experiments with 2 of 14 bacteria isolated from field lesions, *Vibrio anguillarum* and *Aeromonas salmonicida*. These marine bacteria, commonly found in sediments and seawater, are known pathogens. Lesions were only formed if animals were injured; lesions often disappeared, and no mortalities were seen. Maes and Jangoux (1984) termed the lesions "bald sea urchin disease," noting its occurrence in 10 species in temperate areas and in Europe. They determined that several microorganisms invaded the lesions, but also isolated a single strain of bacterium that again produced lesions when applied to slightly injured healthy urchins. They proposed that this bacterium might be insensitive to antibacterial substances naturally produced by echinoids (Maes *et al.* 1986), but also observed that most lesions healed spontaneously. Thus, the role of pathogenic bacteria in causing mass mortality of California urchins remains in question.

Green Sea Urchin Amoebiasis

During autumn months of 1980-83, coincident with unusually warm water temperatures, mass mortalities of the green urchin, *Strongylocentrotus droebachiensis,* were reported from the Atlantic coast of Nova Scotia, resulting in the loss of over 99% of the urchin biomass in these habitats, and altering species diversity and biomass of the flora and fauna (Scheibling 1984; Scheibling and Raymond 1990). The disease was characterized by progressive loss of oral, tube foot, and spine function. Internally, the urchins displayed lesions in the body wall and associated tissues (water vascular system, nerves, spine bases), and changes in the composition of the coelomic fluid, with infiltration of amoebocytes to all tissues, resulting in reduced red and white spherule cells and incomplete clotting capabilities. Early work by Li and co-workers (1981) tentatively identified a *Labyrinthomyxa*-like organism in gut and gonad tissues of affected urchins, but Jones and co-workers (1984,1985) isolated a pathogenic amoeba from diseased urchins. This organism, *Paramoeba invadens*, was sparsely distributed in the body wall, water vascular system, nerves, and gut in individuals in intermediate or late disease stages and did not elicit a specific response (phagocytosis) by infiltrating coelomocytes (Jones and Scheibling 1985).

Scheibling and Stephenson (1984) were able to transmit the disease among urchins in the laboratory, suggesting that a water-borne infective agent was involved, and discussed the possibility of secondary bacterial or fungal infections responsible for mortalities. Further experiments by Jellett and co-workers (1988) correlated the increase of disease with increasing water temperature (16°C, not usually seen in north temperate locations). They also observed reductions in the concentration of coelomocytes, especially red, white, and vibratile cells, but not phagocytes. This reduction in coelomocytes is similar to that found in bacterial infections. The mechanism by which the amoeba caused the disease was apparently not due to the production of toxins. Because levels of protein in the coelomic fluid were elevated, they proposed that the amoebae either stimulated urchin autolysis or produced degradative enzymes.

Roberts-Regan and co-workers (1988) reviewed the known characteristics of the disease and noted that epidermal lesions, like those of "bald sea urchin disease," had occurred in low incidence during and after mortalities in *S. droebachiensis* off Nova Scotia. Experiments supported earlier evidence that the lesions resulted from proliferation of bacteria in injured areas, although these authors found the bacteria to be normal test microflora. The urchins always recovered. They suggested that "bald sea urchin disease" may actually have various other causal agents, and encouraged more research to distinguish acute and chronic diseases in these organisms.

There were no further mortalities in *Strongylocentrotus droebachiensis* off Nova Scotia until September 1990 (J.F. Jellett, Bedford Institute of Oceanography, Dartmouth, Nova Scotia, and R.E. Scheibling, Dalhousie University, Halifax, Nova Scotia, personal communication). Populations around Halifax and St. Margaret's Bay were heavily impacted, and unusually warm water temperatures were noted. However, the pathogen of this epizootic was never identified. Other mass mortalities of echinoderms have been linked to abnormally warm seawater temperatures (Dugan *et al.* 1982; Tsuchiya *et al.* 1987), but it is not known what, if any, pathogens were involved in these episodes.

Caribbean Sea Urchin Mass Mortalities

In contrast to the well-studied Canadian mortalities in *S. droebachiensis*, the causal agent of the most extensive epizootic ever reported for marine invertebrates, the mass mortalities of the long-spined sea urchin, *Diadema antillarum*, in the Caribbean and western Atlantic Ocean, have never been identified. Urchin mortalities were first reported on reefs off Panama in January 1983, and subsequently from other sites around the Caribbean and Bermuda until early 1984 (Lessios 1988), reaching Tobago in January-March 1984 (R.S. Laydoo, Institute of Marine Affairs, Republic of Trinidad and Tobago, personal communication). Populations were reduced by 85-100% (Bak *et al.* 1984; Lessios *et al.* 1984a,b; Hunte *et al.* 1986). Only adult urchins were affected, rarely juveniles, and some populations remained healthy at a few isolated sites. This was the only species of echinoid affected.

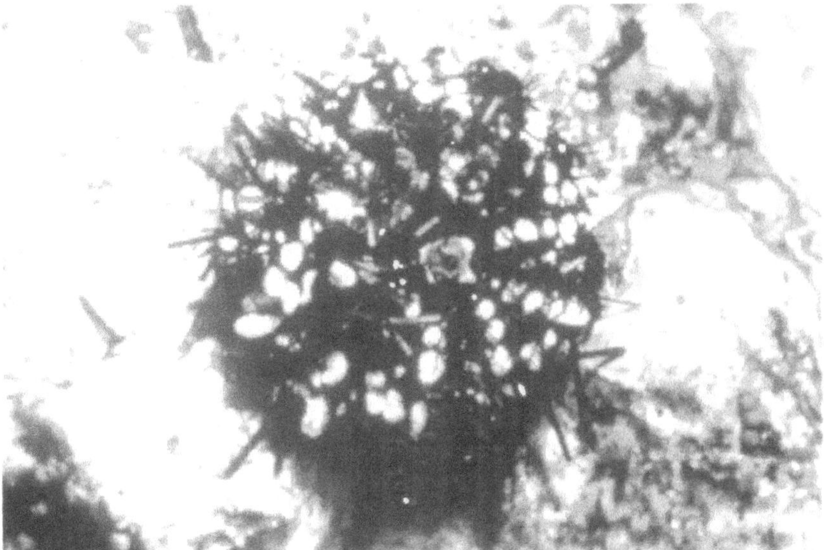

Fig. 15-14. Long-spined sea urchin mortality. Diseased *Diadema antillarum*, showing loss of epidermis and spines during 1983 epizootic off Trinidad. (Photo credit, Institute of Marine Affairs, Republic of Trinidad and Tobago)

The first sign of the disease was accumulation of sediment on the spines and sloughing of the spine epidermis. This was followed by: loss of pigment in the skin covering the spine muscles, peristome, and anal cone; breaking and loss of the spines exposing tubercules and areoles on the test; and inability to cling to the substratum as tube feet became flaccid and no longer fully retracted. Finally, large patches of skin and spines sloughed off and the test disintegrated (Fig. 15-14). Diseased urchins died within four days to six weeks, depending on locality (Bak *et al.* 1984; Lessios *et al.* 1984b). Normally hidden during the day, some diseased urchins moved to the reef during daylight and were attacked by fishes that normally do not prey on healthy *D. antillarum*. However, some urchins apparently survived the disease. These urchins were found after mortalities, exhibiting regenerating distal ends of spines and healed skin lesions, similar to the appearance of lesions in "bald sea urchin disease."

The pattern of mortalities followed major Caribbean currents from west to east, with a few exceptions (Lessios 1988), reminiscent of that seen during the sponge disease epizootic of the late 1930's. The wide geographical spread and species-specificity suggested a water-borne pathogen as the most likely causal agent (Lessios *et al.* 1984a). No abnormal fluctuations in tidal levels, sea surface temperatures, rainfall, or salinity were noted.

A few samples of moribund urchins were collected during the 1983 die-off and fixed for histopathological examination, but were in poor condition and could not be examined (R.E. Scheibling, personal communication). Bauer and Agerter (1987) isolated three strains of Gram-positive anaerobic spore-forming rods identified as *Clostridium perfringens* and *C. sordelli* from gonads of two individuals of *D. antillarum* that died showing similar signs of disease in September 1983 in a flow-through seawater aquarium at the University of Miami, Florida (J.C. Bauer, Rosenstiel School of Marine and Atmospheric Science, University of Miami, Florida, personal communication). Subsequent exposure of healthy *D. antillarum* to cultures of these bacteria caused death in 10 h to 6 days, depending on water temperature. Unexposed controls suffered no mortalities. The bacteria were reisolated from tissues of experimentally infected urchins, but the authors were unable to detect them in sediments or healthy field-collected urchins during July-August 1984 or January-February 1985.

Another die-off of *D. antillarum* was observed on reefs off St. Croix, U.S. Virgin Islands, in September-October 1985. Urchins representing four progressive stages of the disease, similar to those lesions seen during the first mortalities, were fixed and processed for examination by light microscopy. While most tissues of the apparently healthy animals were in good condition (Fig. 15-15), Gram-positive micrococci (0.5 - 0.8 microns in diameter) were evident in the mucoid cells of the glandular crypts in the esophagus and in connective tissue and muscle bundles of the peristome spines (Fig. 15-16), and ampullae. As the

Fig. 15-15. Normal spine muscle of *Diadema antillarum* from Coral Reef Microcosm, National Museum of Natural History. Leaver *et al.* procedure for Gram-positive and Gram-negative bacteria. Scale bar = 50 microns. (Photo credit E. C. Peters)

disease progressed, more bacteria and atrophy and necrosis of all tissues were evident, but no micrococci were found in the gonads (E.C. Peters, R.C. Carpenter, and R.E. Scheibling, unpublished). Unfortunately, no samples were available for microbial isolations. Because bacteria were also found in "apparently healthy" animals, it is possible that urchins had already been infected or that some bacteria occurred naturally (Bauer and Agerter 1987).

Because the mortalities began off Panama, there have been speculations that the pathogen was introduced from ships or other human activities (Bak *et al.* 1984; Lessios 1988). Bauer and Agerter (1987) noted that *Clostridium perfringens* has been associated with water pollution; however, there is insufficient evidence to confirm that this pathogen was responsible for mortalities. Williams and Bunkley-Williams (1987) proposed that a virus was the primary pathogen, but no virus infection is yet known in the echinoderms (Jangoux 1987a). Interestingly, specimens of *D. antillarum* maintained in aquaria receiving ambient seawater on the west coast of Panama (eastern Pacific Ocean) were unaffected by the disease, so the origin of the pathogen remains a mystery (Lessios *et al.* 1984b).

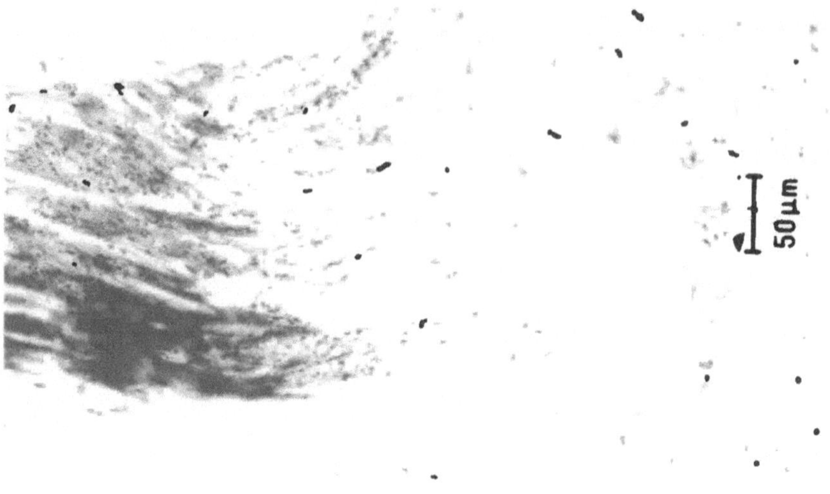

Fig. 15-16. Spine muscle from diseased *Diadema antillarum* collected during secondary die-off from St. Croix in 1985. Note accumulation of Gram-positive staining bacteria and atrophy and degeneration of muscle fibers. Leaver *et al.* procedure for Gram-positive and Gram-negative bacteria. Scale bar = 50 microns. (Photo credit E. C. Peters)

Although Hughes and co-workers (1987) reported no recovery of urchin populations off Jamaica after three years, recruitment of juveniles occurred in other areas (Hunte *et al.* 1986; Williams *et al.* 1986; Lessios 1988). Thick mats of fleshy and filamentous macroalgae now cover some reefs, with declines in coral species, crustose coralline algal cover, and clionid sponges (see Carpenter 1990a). The herbivorous fish have not been able to remove all algae that the urchins formerly grazed (de Ruyter van Stevenick and Bak 1986; Hughes *et al.* 1987; Carpenter 1990b). Further mortalities of *D. antillarum* and *Tripneustes ventricosus* have occurred at several localities following the first episode (e.g., Panama/November 1985: Lessios 1988; St. Croix/October 1985: Carpenter 1990a) and more recently (Jamaica, Cayman Islands/spring-summer 1990, E.H. Williams, Jr., personal communication), as well as localized mortalities of the urchins *Astropyga magnifica* and *Eucidaris tribuloides* off the northwest coast of Puerto Rico in the winter of 1984-1985 (Williams *et al.* 1986). A team of microbiologists, a histopathologist, and a virologist assembled to investigate the *Diadema* disease were finally able to obtain samples of diseased urchins from a mass mortality that occurred off the Florida Keys in April 1991 and are attempting to identify the pathogen involved. Much remains to be learned about the susceptibility of this species and other urchins to pathogenic microorganisms and other factors that may influence these disease outbreaks.

CONCLUSIONS

This review has discussed infectious and non-infectious diseases in five invertebrate phyla that are major contributors to the structure and function of shallow estuarine and coastal marine ecosystems around the world, and has touched on the serious repercussions such diseases may have, not only on the individual organisms, but at the population level as well. Regrettably, too little is known about the etiologies and pathogenesis of diseases in these organisms, including recent epizootics (Table 15-1), particularly the effects at different stages of development.

It is apparent that temperature stress may be an important factor in the development of diseases that affect large populations (Jellett *et al.* 1888; Williams and Bunkley-Williams 1990). The effects of predicted global warming may prove catastrophic for many groups of organisms (Hollings 1988, 1990). Williams and Bunkley-Williams (1990) have noted that major marine ecological disturbances are increasing around the world. Whether these mass mortalities and outbreaks of disease are the result of large-scale natural environmental processes, such as the El Nino-Southern Oscillation of 1982-1983 (Arntz and Tarazona 1990; Coffroth *et al.* 1990), and/or the result of anthropogenic activities remains to be determined.

As noted by Overstreet (1988), pollutant- or toxicant-modified communities, species, individuals, and lesions must be distinguished from those that may be affected either directly by natural environmental factors, or those that have indirect effects, such as on prey organisms or on vectors of parasites and pathogens (Möller 1987). Physical injuries to organisms by anchors, divers, ship hulls, and nets, as well as by predators, can also facilitate the spread of disease. Introductions of new pathogens into new environments may occur not only by transportation and dumping of bilgewater, but also in the shipment and culture of organisms without adequate quarantine controls to protect the natural inhabitants near water outfalls.

The role of these phyla as vectors of diseases of fish, molluscs, or crustaceans also needs more research (Jangoux 1987d). Some pathogens may also prove to be valuable as biological control agents (Miller 1985), should they be needed, but many questions remain about the optimal conditions and effects on populations. Furthermore, those disease outbreaks examined thus far suggest that there are genetically determined patterns of recovery and resistance to pathogens or abiotic diseases that warrant more studies (e.g., May 1983; May and Anderson 1983; Anderson 1986; Vicente 1989; Williams and Bunkley-Williams 1990).

Although there are few reports of neoplasms and neoplasm-like lesions in these phyla, further study may reveal more of these conditions, or uncover

Table 15-1. Important Recent Epizootics Recognized in Five Phyla. (Note lack of information on the etiology of most of these diseases [adapted from Williams and Bunkley-Williams 1990].)

Species or Group	Condition	Location	Dates	Causal Agent	Reference
PHYLUM PORIFERA					
Commercial sponges: *Spongia, Hippospongia,* also *Ircinia, Sarcotragus* spp.	mass mortalities	eastern Mediterranean: Tunisia, Greece, Cyprus, Turkey, Egypt, Italy	August 1986 to present	spongin-boring bacterium, temperature?	Reiswig 1988; Gaino and Pronzato 1989
various species	mortalities	Caribbean Sea	Summer 1987	?	Vicente 1989
Chondrilla nucula	partial mortalities	Puerto Rico	-	filamentous algal overgrowth	Vicente 1989
PHYLUM CNIDARIA					
Acropora palmata, A. cervicornis	white band disease, mass mortalities	western Atlantic, Caribbean Sea	1977 to present	bacterium or environmental degradation	Gladfelter *et al.* 1977 Bak and Criens 1981 Knowlton *et al.* 1981 Rogers 1985
Acropora palmata, Porites porites	coral mortalities	Anegada, British Virgin Islands	1979 to present	?	Brown 1987
Gorgonia spp.	sea fan mass mortalities	Trinidad Costa Rica, Panama	1981 1983 to present	? black band disease or abiotic factors?	Laydoo 1983; Guzman and Cortes 1984
		Colombia	1986-1988	?	J. Garzón-Ferreira, pers. comm.
Montastraea annularis, Diploria strigosa, other scleractinians	black band disease epizootic	Looe Key National Marine Sanctuary, Florida	August 1987 to present	cyanobacterium, environmental degradation?	B.D. Causey, pers. comm.

Table 15-1. (Continued)

Species or Group	Condition	Location	Dates	Causal Agent	Reference
Pseudopterogorgia acerosa and *P. americana*	black band disease	Florida Keys	mid-1980's	cyanobacterium	Feingold 1988
numerous species	bleaching, mortalities	worldwide	1980's to present	seawater temperature increase	Williams and Bunkley-Williams 1987
PHYLUM CTENOPHORA					
No epizootics or mass mortalities reported. Seasonal population fluctuations known to occur.					
PHYLUM ANNELIDA					
Arenicola marina	morbidities and mortalities	England	September 1990	dinoflagellate bloom?	Olive and Cadnam 1990
PHYLUM ECHINODERMATA					
Heliaster kubiniji, *Othilia tenuispina*, *Pisaster* spp., *Patiria miniata*, *Stichopus parvimensis*	mass mortalities	Gulf of California	Summer 1978, 1979	high temperatures?	Dugan *et al.* 1982
Strongylocentrotus droebachiensis	mass mortalities, epidermal loss not marked	Nova Scotia	Autumn months 1980 to 1983	*Paramoeba invadens*	Jones *et al.* 1984
		Newfoundland	?	?	Hooper 1980, cited in Jones *et al.* 1985
		Nova Scotia	September 1990	?	
Strongylocentrotus franciscanus	spine, tube foot loss, epidermal necrosis	California	?	bacteria?	Pearse *et al.* 1977

Table 15-1. (Continued)

Species or Group	Condition	Location	Dates	Causal Agent	Reference
Paracentrotus lividus	as above, mass mortalities	Mediterranean	?	?	Hobaus *et al.* 1981
diadematid urchins	mass mortalities	Hawaiian Islands	1981	?	Birkeland and Eldredge 1988
Diadema antillarum	mass mortalities	wider Caribbean	1983-1984	?	Lessios *et al.* 1984 Lessios 1988
	secondary die-off	U.S. Virgin Islands Panama	October 1985 November 1985	? ?	Lessios 1988
	mortalities	Jamaica, Cayman Islands	April 1990 to present	?	E.H. Williams, Jr., pers. comm.
	mortalities	Florida Keys	April 1991	?	G.S. Sullivan, J.H. Hudson, pers. comm.
Astropyga magnifica, *Eucidaris tribuloides*	mortalities	Puerto Rico	December 1984 January 1985	? ?	Williams *et al.* 1986
Echinometra mathaei	mass mortalities	Okinawa	June 1986	temperature	Tsuchiya *et al.* 1987
Pisaster ochraceous	mortalities	British Columbia	1986/1987	ciliate, *Orchitophyra stellarum*	Leighton *et al.* 1991
starfish	mortalities	British Columbia	1990	?	Anonymous 1990
Echinocardium cordatum	mass mortalities	England	September 1990	dinoflagellate bloom?	Olive and Cadnam 1990
starfish and other organisms	mass mortalities	White Sea	1990	?	Jones 1990

medically important reasons why they are absent from certain species (Munro *et al.* 1987). Additional members of these phyla may also prove valuable in the development of biomedical models for disease research (Marine Life Resources Workshop 1989). Viruses may prove to be common etiologic agents of disease in these organisms, but cell cultures will be needed to explore this possibility.

Couch (1988) acknowledged that there are many opportunities for cooperative investigations among pathobiologists and scientists from other backgrounds to determine the natural biological responses and processes of disease and dysfunction on populations and species of estuarine and marine organisms. The field of invertebrate pathobiology remains open for major contributions on the diseases of the "forgotten phyla" in estuarine and coastal marine waters.

ACKNOWLEDGMENTS

I thank H.B. McCarty for critical review of the manuscript. I am most grateful to G.S. Aeby, J.C. Bauer, T.E. Bowman, E.M. Burreson, B.D. Causey, T.G. Douglass, P.J. Edmunds, F. Fauchald, J. Garzón-Ferreira, J.C. Halas, G.R. Harbison, J.H. Hudson, J.F. Jellett, M.L. Kent, R. Laydoo, H.A. Lessios, D.V. Lightner, M.E. Markiw, J. Vacelet, R.E. Scheibling, M.J. Smith, C.W. Walker, C.M. Wahle, L.A. Ward, L. Bunkley-Williams, and E.H. Williams, Jr., for helpful discussions and for sharing their research. J.K. Liebeler assisted in the literature search. A National Institutes of Health, National Research Service Award Training Fellowship (Award Number F32 CA08427-03 from the National Cancer Institute) in the Registry of Tumors in Lower Animals, Smithsonian Institution, provided support during manuscript preparation.

REFERENCES

Aeby, G.S. 1991. Behavioral and ecological relationships of a parasite and its hosts within a coral reef system. *Pacific Science* 45(3):263-269.

Ahmadjian, V., and S. Paracer. 1986. *Symbiosis: An Introduction to Biological Associations.* Hanover, New Hampshire: University Press of New England.

Allemand-Martin, A. 1906. Étude de physiologie appliquée sur la spongiculture sur de côtes de Tunisie. Thesis, University Lyon.

Allemand-Martin, A. 1914. Contribution à l'étude de la culture des éponges. *C. R. Ass. Advent. Sci. Tunis* 1914:375-377.

Aller, R.C. 1978. The effects of animal-sediment interactions on geochemical processes near the sediment-water interface. In *Estuarine Interactions*, ed. M.L. Wiley, pp. 157-72. New York: Academic Press.

Anderson, R.M. 1986. Genetic variability in resistance to parasitic invasion: population implications for invertebrate host species. *Symposia the Zoological Society of London* 56:239-274.

Anderson, R.S. 1980. Hemolysins and hemagglutinins in the coelomic fluid of polychaete annelid, *Glycera dibranchiata*. *Biol. Bull.* 159:259-268.

Anderson, R.S. 1987. Immunocompetence in invertebrates. In *Pollutant Studies in Marine Animals*, ed. C. S. Giam and L. E. Ray, pp. 93-110. Boca Raton, FL: CRC Press, Inc.

Anderson, R.S., and B.M. Chain. 1982. Antibacterial activity in the coelomic fluid of a marine annelid, *Glycera dibranchiata*. *J. Invert. Pathol.* 40:320-326.

Anonymous. 1935. Discovers important parasite of starfish. *U.S. Bur. Fish. Serv. Bull.* 247:3-4.

Anonymous. 1990. Mysterious starfish mortality near Powell River, British Columbia. *Mar. Pollut. Bull.* 21(10):462.

Antonius, A. 1977. Coral mortality in reefs: a problem for science and management. Proc. *Third Int. Coral Reef Symp.* 2:3-6.

Antonius, A. 1981a. Coral reef pathology: a review. *Proc. Fourth Int. Coral Reef Symp.* 2:3-6.

Antonius, A. 1981b. The "band" diseases in coral reefs. *Proc. Fourth Int. Coral Reef Symp.* 2:6-14.

Antonius, A. 1985a. Coral diseases in the Indo-Pacific: a first record. *P.S.Z.N.I.: Mar. Ecol.* 6:197-218.

Antonius, A. 1985b. Black band infection experiments on hexacorals and octocorals. *Proc. Fifth Int. Coral Reef Cong.* 6:155-160.

Antonius, A. 1988. Black band disease behavior on Red Sea reef corals. *Proc. Sixth Int. Coral Reef Symp.* 3:145-150.

Arntz, W.E., and J. Tarazona. 1990. Effects of El Niño 1982-83 on benthos, fish and fisheries off the South American Pacific coast. In *Global Ecological Consequences of the 1982-83 El Niño-Southern Oscillation*, ed. P.W. Glynn, pp. 328-58. Amsterdam: Elsevier.

Bak, R.P.M. 1983. Neoplasia, regeneration, and growth in the reef-building coral *Acropora palmata*. *Mar. Biol.* 77:221-227.

Bak, R.P.M., and S.R. Criens. 1981. Survival after fragmentation of colonies of *Madracis mirabilis*, *Acropora palmata* and *A. cervicornis* (Scleractinia) and the subsequent impact of a coral disease. *Proc. Fourth Int. Coral Reef Symp.* 2:221-228.

Bak, R.P.M., M.J.E. Carpay, and E. D. de Ruyter van Steveninck. 1984. Densities of the sea urchin *Diadema antillarum* before and after mass mortalities on the coral reefs of Curacao. *Mar. Ecol. Prog. Ser.* 17:105-108.

Bakus, G.J., B. Schulte, S. Jhu, M. Wright, G. Green, and P. Gomez. 1990. Antibiosis and antifouling in marine sponges: laboratory versus field studies. In *New Perspectives in Sponge Biology*, ed. K. Rützler, pp. 102-8. Washington, DC: Smithsonian Institution Press.

Bang, F.B. 1982. Disease processes in seastars: a Metchnikovian challenge. *Biol. Bull.* 162:135-148.

Bang, F.B., and A. Lemma. 1962. Bacterial infection and reaction to injury in some echinoderms. *J. Insect Pathol.* 4:401-414.

Bauer, J.C., and C.J. Agerter. 1987. Isolation of bacteria pathogenic for the sea urchin *Diadema antillarum* (Echinodermata: Echinoidea). *Bull. Mar. Sci.* 40:161-165.

Bayer, F.M., and A.J. Weinheimer (eds.). 1974. *Prostaglandins from Plexaura homomalla: Ecology, Utilization and Conservation of a Major Medical Marine Resource: A Symposium*. Coral Gables, Florida: University of Miami Press.

Ben Mustapha, K., and Jean Vacelet. 1991. Etat actuel des fonds spongifères de Tunisie. In *Les Especes Marines à Protéger en Méditerranée (Marine Species in Need of Protection in the Mediterranean)*, Colloque International, Carry-le-Rouet, France, November 1989, ed. C.-F. Boudouresque, M. Avon and V. Gravez, pp. 43-46. Marseille, France: GIS Posidonies.

Bertrand, C., and Jean Vacelet. 1971. L'association entre éponges cornées et bactéries. *C. R. Séance Acad. Paris* 273:638-641.

Birkeland, C., and L.G. Eldredge. 1988. Large-scale phenomena. *Coral Reef Newsl.* 19:57.

Birkeland, C., and J.S. Lucas. 1990. *Acanthaster planci: Major Management Problems of Coral Reefs*. Boca Raton, FL: CRC Press.

Botero, L. 1990. Observations on the size, predators and tumor-like outgrowths of gorgonian octocoral colonies in Santa Marta, Caribbean coast of Colombia. *Northeast Gulf Sci.* 11:1-10.

Bouland, C., and M. Jangoux. 1988. Infestation of *Asterias rubens* (Echinodermata) by the ciliate *Orchitophyra stellarum*: effect on gonads and host reaction. *Dis. Aquat. Organ.* 5:239-242.

Bourdouresque, C.F., H. Nedelec, and S.A. Sheppard. 1981. The decline of a population of the sea urchin *Paracentrotus lividus* in the Bay of Port-Cros (Var, France). *Rapp. P. -v. Reun. Cons. Int. Explor. Mer* 27:223-224.

Bowman, T.E., C.D. Meyers, and S.D. Hicks. 1963. Notes on the associations between hyperiid amphipods and medusae in Chesapeake and Narragansett Bays and the Niantic River. *Chesapeake Sci.* 4(3):141-146.

Breen, P.A., and K.H. Mann. 1976. Changing lobster abundance and the destruction of kelp beds by sea urchins. *Mar. Biol.* 34:137-142.

Brock, M.A. 1984. Senescence in *Campanularia flexuosa* and other cnidarians. In *Invertebrate Models in Aging Research*, ed. D.H. Mitchell and T.E. Johnson, pp. 15-44. Boca Raton, FL: CRC Press.

Brown, B., and R.S. Prezant. 1986. Occurrence of a digenetic trematode in a polychaete. *J. Invert. Pathol.* 48:239-241.

Brown, B.E. 1987. Worldwide death of corals - natural cyclical events or man-made pollution? *Mar. Pollut. Bull.* 18(1):9-13.

Brown, B.E., and L.S. Howard. 1985. Assessing the effects of "stress" on reef corals. *Adv. Mar. Biol.* 22:1-63.

Brown, B.E., and Suharsono. 1990. Damage and recovery of coral reefs affected by El Niño related seawater warming in the Thousand Islands, Indonesia. *Coral Reefs* 8:163-170.

Buddemeir, R.W., and R.A. Kinzie. 1976. Coral growth. *Oceanog. Mar. Biol. Ann. Rev.* 14:183-225.

Burkholder, P.R., and K. Ruetzler. 1969. Antimicrobial activity of some marine sponges. *Nature* (London) 222:983-984.

Burreson, E.M. 1982. The life cycle of *Trypanoplasma bullocki* (Zoomastigophorea: Kinetoplastida). *J. Protozool.* 29(1):72-77.

Burreson, E.M., and D.E. Zwerner. 1982. The role of host biology, vector biology, and temperature in the distribution of *Trypanoplasma bullocki* infections in the lower Chesapeake Bay. *J. Parasitol.* 68(2):306-313.

Burreson, E.M., and D. E. Zwerner. 1984. Juvenile summer flounder, *Paralichthys dentatus*, mortalities in the western Atlantic Ocean caused by the hemoflagellate *Trypanoplasma bullocki*: evidence from field and experimental studies. *Helgol. Meeresunter.* 37:343-352.

Byrne, M. 1985. The life history of the gastropod *Thyonicola americana* Tikasingh, endoparasitic in a seasonally eviscerating holothurian host. *Ophelia* 24(2):91-101.

Cahoon, L.B., C.R. Tronzo, and J.C. Howe. 1986. Notes on the occurrence of *Hyperoche medusarum* (Kroyer) (Amphipoda, Hyperiidae) with Ctenophora off North Carolina, U.S.A. *Crustaceana* 51(1):95-97.

Cannon, L.R.G. 1986. The Pterastericolidae: parasitic turbellarians from starfish. In *Parasite Lives: Papers on Parasites, Their Hosts and Their Associations*, ed. M. Cremin, C. Dobson and D.E. Moorhouse, pp. 15-32. St. Lucia/London/New York: University of Queensland Press.

Cannon, L.R.G., and J.B. Jennings. 1988. *Monocystella epibatis* n. sp., a new aseptate gregarine hyperparasite of rhabdocoel turbellarians parasitic in the Crown of Thorns Starfish, *Acanthaster planci* Linnaeus, from the Great Barrier Reef. *Arch. Protistenkd.* 136:267-272.

Carpenter, R.C. 1981. Grazing by *Diadema antillarum* (Philippi)[sic] and its effects on the benthic algal community. *J. Mar. Res.* 39:749-765.

Carpenter, R.C. 1990a. Mass mortality of *Diadema antillarum*. I. Long-term effects on sea urchin population dynamics and coral reef algal communities. *Mar. Biol.* 104:67-77.

Carpenter, R.C. 1990b. Mass mortality of *Diadema antillarum*. II. Effects on population densities and grazing intensity of parrotfishes and surgeonfishes. *Mar. Biol.* 104:79-86.

Carton, Y. 1967. Spécificité parasitaire de *Sabelliphilus sarsi*, parasite de *Spirographis spallanzani*. 2. Réactions histologiques de hôtes lors des contaminations expérimentales. *Arch. Zool. Exper. Générale* 109(1):123-144.

Cépède, C. 1907a. La castration parasitaire des étoiles de mer mâles par un nouvel infusoire astome: *Orchitophyra stellarum* n.g., n. sp. *C.R. Hebd. Séances Acad. Sci. Paris* 145:1305-1306.

Cépède, C. 1907b. Sur un nouvel infusoire astome, parasite des testicules des étoiles de mer. Considérations générales sur les Astomata. *C. R. Assoc. Fr. Avanc. Sci.* 36:258.

Chain, B.M., and R.S. Anderson. 1983. A bactericidal and cytotoxic factor in the coelomic fluid of the polychaete, *Glycera dibranchiata*. *Develop. Compar. Immunolo.* 7:625-628.

Cheng, T.C., and A.K.L. Wong. 1974. Chemical, histochemical, and histopathological studies on corals, *Porites* spp., parasitized by trematode metacercariae. *J. Invert. Pathol.* 23:303-317.

Chesney, E.J., Jr., and K.R. Tenore. 1985. Oscillations of laboratory populations of the polychaete *Capitella capitata* (Type I): their cause and implications for natural populations. *Mar. Ecol. Prog. Ser.* 20:289-296.

Coffroth, M.A., H.R. Lasker, and J.K. Oliver. 1990. Coral mortality outside of the eastern Pacific during 1982-1983: relationship to El Niño. In *Global Ecological Consequences of the 1982-83 El Niño-Southern Oscillation*, ed. Peter W. Glynn, pp. 141-82. Amsterdam: Elsevier.

Colwell, R.R., and J. Liston. 1962. The natural bacterial flora of certain marine invertebrates. *J Insect Pathol.* 4:23-33.

Cook, C.B., A. Logan, J. Ward, B. Luckhurst, and C.J. Berg, Jr. 1990. Elevated temperatures and bleaching on a high latitude coral reef: the 1988 Bermuda event. *Coral Reefs* 9:45-49.

Connell, J.H. 1975. Some mechanisms producing structure in natural communities. In *Ecology and Evolution of Communities*, ed. Martin L. Cody and Jared M. Diamond, pp. 460-90. Cambridge, Massachusetts: Belknap Press of Harvard University Press.

Connes, R. 1967. Réactions de défense de l'éponge *Tethya lyncurium* Lamarck, vis-à-vis de micro-organismes et de l'amphipode *Leucothoe spinicarpa* Abildg. *Vie Milieu* (Series A) 18:281-289.

Connes, R., J. Paris, and J. Sube. 1971. Réactions tissulaires de quelques démosponges vis-a-vis de leurs commensaux et parasites. *Le Naturaliste Canada* 98:923-935.

Corliss, J.O. 1985. Consideration of taxonomic-nomenclatural problems posed by report of myxosporidians with a two-host life cycle. *J. Protozool.* 32(4):589-591.

Couch, J.A. 1988. Role of pathobiology in experimental biology and marine ecology. *J. Exper. Mar. Biol. Ecol.* 118:1-6.

Couch, J.A., and L. Courtney. 1977. Interaction of chemical pollutants and virus in a crustacean: a novel bioassay system. *Annals of the New York Academy of Science* 298:497-504.

Creaser, E.P., and D.A. Clifford. 1986. The size frequency and abundance of subtidal bloodworms (*Glycera dibranchiata* Ehlers) in Montsweag Bay, Woolwich-Wiscasset, Maine. *Estuaries* 9:200-207.

Coulon, P., and M. Jangoux. 1987. Gregarine species (Apicomplexa) parasitic in the burrowing echinoid *Echinocardium cordatum*: occurrence and host reaction. *Dis. Aquat. Organ.* 2:135-145.

Dahl, I.O. 1960. The oligochaete fauna of of 3 Danish brackish water areas. *Meddeleser Fra Fiskfriog Havundersogelser N.S.* 2:1-20.

Dales, R.P. 1983. Observations of granulomata in the polychaetous annelid *Nereis diversicolor*. *J. Invert. Patho.* 42:288-291.

Dales, R.P., and L.J.R. Dixon. 1980. Responses of polychaete annelids to bacterial infection. *Comp. Biochem. Physiol.* 67A:391-396.

de Ruyter van Steveninck, E.D., and R.P.M. Bak. 1986. Changes in abundance of coral-reef bottom components related to mass mortality of the sea-urchin *Diadema antillarum*. *Mar. Ecol. Prog. Ser.* 34:87-94.

Desportes, I., and J. Théodoridès. 1986. *Cygnicollum lankesteri* n.sp., grégarine (Apicomplexa, Lecudinidae) parasite des annelides polychètes *Laetmonice hystrix* et *L. producta*; particularites de l'appareil de fixation et implications taxonomiques. *Protistologica* 22:47-60.

Devauchelle, G., and M. Durchon. 1973. Sur la présence d'un virus, du type Iridovirus, dans les cellules males de *Nereis diversicolor* (O.F. Muller). *C. R. Acad. Sci.* Ser. D Paris 277:463-466.

Dosse, G. 1940. Bakterien und Pilzbefunde Sowie pathologische und Faulnisvorgange in Meeres- und Susswasserschwammen. *Zeitschrift fur Parasitenkunde* 11:321-356.

Douglass, T.G., and I. Jones. 1991. Parasites of California spionid polychaetes. *Bull. Mar. Sci.* 48:308-317.

Ducklow, H.W., and R. Mitchell. 1979. Bacterial populations and adaptations in the mucus layers on living corals. *Limnol. Oceanogr.* 24: 715-725.

Duerden, J.E. 1902. *West Indian Madreporarian Polyps.* Washington, DC: Seventh Memoir, National Academy of Sciences 8:399-599.

Dugan, M.L., T.E. Miller, and D.A. Thomson. 1982. Catastrophic decline of a top carnivore in the Gulf of California rocky intertidal zone. *Science* 216:989-991.

Dunlap, W.C., and B.E. Chalker. 1986. Identification and quantitation of near-UV absorbing compounds (S-320) in a hermatypic scleractinian. *Coral Reefs* 5:155-159.

Dustan, P. 1977. Vitality of reef coral populations off Key Largo: recruitment and mortality. *Environ. Geol.* 2:51-58.

Edmunds, P.J. 1991. The extent and effect of black band disease on a Caribbean reef. *Coral Reefs* 10:161-165.

Emson, R.H., P.V. Mladenov, and I.C. Wilkie. 1985. Studies of the biology of the West Indian copepod *Ophiopsyllus reductus* (Siphonostomatoida: Cancerillidae) parasitic upon the brittlestar *Ophiocomella ophiactoides.* *J. Nat. Hist.* 19:151-171.

Fauchald, K., and P. Jumars. 1979. The diet of worms: a study of polychaete feeding guilds. *Oceanogr. Mar. Biol. Annu. Rev.* 17:193-284.

Feingold, J.S. 1988. Ecological studies of the cyanobacterial infection on the Caribbean sea plume *Psuedopterogorgia acerosa* (Coelenterata: Octocorallia). *Proceedings of the Sixth International Coral Reef Symposium* 3:157-162.

Fontaine, A.R. 1969. Pigmented tumor-like lesions in an ophiuroid echinoderm. *Nat. Cancer Inst. Monogr.* 31:255-261.

Gaino, E., and R. Pronzato. 1987. Ultrastructural observations of the reaction of *Chondrilla nucula* (Porifera, Demospongiae) to bacterial invasion during degenerative processes. *Cah. Biol. Mar.* 28:37-46.

Gaino, E., and R. Pronzato. 1989. Ultrastructural evidence of bacterial damage to *Spongia officinalis* fibres (Porifera, Demospongiae). *Dis. Aquat. Organ.* 6:67-74.

Galstoff, P.S. 1942. Wasting disease causing mortality of sponges in the West Indies and Gulf of Mexico. *Proc. Eighth Am. Sci. Congr.* 3:411-421.

Galstoff, P.S., and V.L. Loosanoff. 1939. Natural history and method of controlling the starfish (*Asterias forbesi* Desor). *Bull. U.S. Bur. Fish.* 49(31):75-132.

Galstoff, P.S., H.H. Brown, C.L. Smith, and F.G.W. Smith. 1939. Sponge mortality in the Bahamas. *Nature* 143:807-808.

Garrett, P., and H. Ducklow. 1975. Coral diseases in Bermuda. *Nature* 253:349-350.

Ghiradella, H.T. 1965. The reaction of two starfishes, *Patiria miniata* and *Asterias forbesi*, to foreign tissue in the coelom. *Biol. Bull.* 128:77-89.

Giere, O., and C. Langheld. 1987. Structural organization, transfer and biological fate of endosymbiotic bacteria in gutless oligochaetes. *Mar. Biol.* 93:641-650.

Giere, O., and O. Pfannkuche. 1982. Biology and ecology of marine oligochaeta, a review. *Oceanogr. Mar. Biol. Annu. Rev.* 20:173-308.

Gilles, K.W., and J.S. Pearse. 1986. Disease in sea urchins *Strongylocentrotus purpuratus*: experimental infection and bacterial virulence. *Dis. Aquat. Org.* 1:105-114.

Gladfelter, W.B. 1982. White band disease in *Acropora palmata*: implications for the structure and growth of shallow reefs. *Bull. Mar. Sci.* 32:639-643.

Gladfelter, W.B., E.H. Gladfelter, R.K. Monahan, J.C. Ogden, and R.F. Dill. 1977. Coral Destruction. In *Environmental Studies of Buck Island Reef National Monument*, U.S. National Park Service Report, pp. XI-1-7.

Glynn, P.W. 1983. Extensive "bleaching" and death of reef corals on the Pacific coast of Panama. *Environ. Conserv.* 11:133-146.

Glynn, P.W., and L. D'Croz. 1990. Experimental evidence for high temperature stress as the cause of El Niño-coincident coral mortality. *Coral Reefs* 8:181-191.

Glynn, P.W., G.M. Wellington, and C. Birkeland. 1978. Coral reef growth in the Galapagos: limitation by sea urchins. *Science* 203:47-49.

Glynn, P.W., E.C. Peters, and L. Muscatine. 1985. Coral tissue microstructure and necrosis: relation to catastrophic coral mortality in Panama. *Dis. Aquat. Organ.* 1:29-37.

Glynn, P.W., A.M. Szmant, E.F. Corcoran, and S.V. Cofer-Shabica. 1989. Condition of coral reef cnidarians from the Northern Florida Reef Tract: Pesticides, heavy metals, and histopathological examination. *Mar. Pollut. Bull.* 20:568-576.

Goenega, C., V.P. Vicente, and R.A. Armstrong. 1989. Bleaching induced mortalities in reef corals from La Parguera, Puerto Rico: a precursor of change in the community structure of coral reefs? *Caribb. J. Sci.* 25(1-2):59-65.

Goldberg, W.M., and J.C. Makemson. 1981. Description of a tumorous condition in a gorgonian coral associated with a filamentous green alga. *Proceedings of the Fourth International Coral Reef Symposium* 2:685-697.

Goldberg, W.M., J.C. Makemson, and S.B. Colley. 1984. *Entocladia endozoica* sp. nov., a pathogenic chlorophyte: structure, life history, physiology and effect on its coral host. *Biol. Bull.* 166:368-383.

Goreau, T.F., N.I. Goreau, and T.J. Goreau. 1979. Corals and coral reefs. *Sci. Am.* 241(2):124-136.

Goreau, T.J., and A.H. Macfarlane. 1990. Reduced growth rate of *Montastrea annularis* following the 1987-1988 coral-bleaching event. *Coral Reefs* 8:211-215.

Gotto, R.V. 1979. The association of copepods with marine invertebrates. *Adv. Mar. Biol.* 16:1-109.

Greve, W. 1971. Okologische Untersuchungen an *Pleurobrachia pileus*. I. Freilanduntersuchungen. *Helgol. wiss. Meeresunters.* 22:303-325.

Guerinot, M.L., and D.G. Patriquin. 1981. The association of N_2-fixing bacteria with sea urchins. *Mar. Biol.* 62:197-207.

Guzmán, H.M., and J. Cortés. 1984. Mortand de *Gorgonia flabellum* Linnaeus (Octocorallina: Gorgoniidae) en la costa Caribe de Costa Rica. *Revistas Biologia Tropica* 32(2):305-308.

Guzmán, H.M., J. Cortés, P.W. Glynn, and R.H. Richmond. 1990. Coral mortality associated with dinoflagellate blooms in the eastern Pacific (Costa Rica and Panama). *Mar. Ecol. Progr. Ser.* 60:299-303.

Hagen, N.T. 1987. Sea urchin outbreaks and nematode epizootics inVestfjorden, northern Norway. *Sarsia* 72:213-229.

Hamilton, A.J., and E.U. Canning. 1987. Studies on the proposed role of *Tubifex tubifex* (Muller) as an intermediate host in the life cycle of *Myxosoma cerebralis* (Hofer, 1903). *J. Fish Dis.* 10:145-151.

Harriott, V.J. 1985. Mortality rates of scleractinian corals before and during a mass bleaching event. *Mar. Ecol. Progr. Ser.* 21:81-88.

Hayes, R.L., and P.G. Bush. 1990. Microscopic observations of recovery in the reef building scleractinian coral, *Montastrea annularis*, after bleaching on a Cayman reef. *Coral Reefs* 8:203-209.

Hein, F.J., and M.J. Risk. 1975. Bioerosion of coral heads: inner patch reefs, Florida reef tract. *Bull. Mar. Sci.* 25:133-138.

Hetzel, H.R. 1965. Studies on holothurian coelomocytes. II. The origin of coelomocytes and the formation of brown bodies. *Biolog. Bull.* 128:102-111.

Hobaus, E., L. Fenaux, and M. Hignette. 1981. Premières observations sur les lésions provoquees par une maladies affectanct le test des oursin en Méditerranée occidentale. *Rapp. P.-v. Reun. Cons. Int. Explor. Mer Méditerranee* 27:221-222.

Hoegh-Guldberg, O., and G.J. Smith. 1989. The effect of sudden changes in temperature, light and salinity on the population density and export of zooxanthellae from the reef corals *Stylophora pistillata* Esper and *Seriatopora hystrix* Dana. *J. Exper. Mar. Biol. Ecol.* 129:279-303.

Hodgson, G. 1990. Tetracycline reduces sedimentation damage to corals. *Mar. Biol.* 104:493-496.

Hollings, E.F. 1988. *Bleaching of Coral Reefs in the Caribbean*. Oral and Written Testimony before the Commerce, Justice, State, Judiciary, and related agencies, Appropriations Subcommittee, U.S. Senate, 10 November 1987, pp. 1-142. Washington, DC: U.S. Government Printing Office.

Hollings, E.F. 1990. *Coral Bleaching*. Testimony to the Senate Committee on Commerce, Science, and Transportation, National Ocean Policy Study, U.S. Senate, October 1990.

Hoogenboom, J., and J. Hennen. 1985. Étude sur les parasites du macrozooplancton gélatineux dans la rade de Villefranche-sur-Mer (France), avec description des stades de dévelloppement de *Hyperoche mediterranea* Senna (Amphipoda, Hyperiidae). *Crustaceana* 49:233-243.

Hooper, R. 1980. Observations on algal-grazer interactions in Newfoundland and Labrador. In *Proceedings of the Workshop on the Relationships between Sea Urchin Grazing and Commercial Plant/Animal Harvesting*, ed. J.D. Pringle, G.J. Sharp, and J. F. Caddy, pp. 120-124. Halifax, Nova Scotia, Canada: Canadian Technical Report, Fisheries and Aquatic Sciences, No. 954.

Hoppe, W.F. 1988. Growth, regeneration and predation in three species of large coral reef sponges. *Mar. Ecol. Prog. Ser.* 50:117-125.

Hughes, T.P., D.C. Reed, and M.J. Boyle. 1987. Herbivory on coral reefs: community structure following mass mortalities of sea urchins. *J. Exper. Mar. Biol. Ecol.* 113:39-59.

Hummell, H., A.B.J. Sepers, L. de Wolf, and F.W. Melissen. 1988. Bacterial growth on the marine sponge *Halichondria panicea* induced by reduced water-flow rate. *Mar. Ecol. Prog. Ser.* 2:195-198.

Hunte, W., I. Côté, and T. Tomascik. 1986. On the dynamics of the mass mortality of *Diadema antillarum* in Barbados. *Coral Reefs* 4:135-140.

Jaap, W.C. 1979. Observations on zooxanthellae expulsion at Middle Sambo reef, Florida Keys, USA. *Bull. Mar. Sci.* 29(3):414-422.

Jaap, W.C. 1985. An epidemic zooxanthellae expulsion during 1983 in the lower Florida Keys coral reefs: hyperthermic etiology. *Proc. Fifth Int. Coral Reef Cong.* 6:143-148.

Jakowska, S., and R.F. Nigrelli. 1960. Antimicrobial substances from sponges. *Ann. NY Acad. Sci.* 90:913-916.

Jangoux, M. 1987a. Diseases of Echinodermata. I. Agents microorganisms and protistans. *Dis. Aquat. Organ.* 2:147-162.

Jangoux, M. 1987b. Diseases of Echinodermata. II. Agents metazoans(Mesozoa to Bryozoa). *Dis. Aquat. Organ.* 2:205-234.

Jangoux, M. 1987c. Diseases of Echinodermata. III. Agents metazoans(Annelida to Pisces). *Dis. Aquat. Organ.* 3:59-83.

Jangoux, M. 1987d. Diseases of Echinodermata. IV. Structural abnormalities and general considerations on biotic diseases. *Dis. Aquat. Organ.* 3:221-229.

Jellett, J.F., A.C. Wardlaw, and R.E. Scheibling. 1988. Experimental infection of the echinoid *Strongylocentrotus droebachiensis* with *Paramoeba invadens*: quantitative changes in the coelomic fluid. *Dis. Aquat. Organ.* 4:149-157.

Jeon, K.W. 1987. Change of cellular "pathogens" into required cell components. *Ann. NY Acad. Sci.* 503:359-371.

Johnson, P.T. 1984. Viral diseases of marine invertebrates. *Helgol. Meeresunters.* 2 37:65-98.

Johnson, P.T., and R.J. Beeson. 1966. *In vitro* studies on *Patiria mininata* (Brandt) coelomocytes, with remarks on revolving cysts. *Life Sciences* 5:1641-1666.

Johnson, P.T., and F.A. Chapman. 1970a. Abnormal epithelial growth in sea urchin spines (*Strongylocentrotus franciscanus*). *J. Invert. Pathol.* 16:116-122.

Johnson, P.T., and F.A. Chapman. 1970b. Infection with diatoms and other microorganisms in sea urchin spines (*Strongylocentrotus franciscanus*). *J. Invert. Pathol.* 16:268-276.

Johnson, T.W., and F.K. Sparrow. 1961. Fungi and marine animals. In *Fungi in Oceans and Estuaries*, ed. J. Cramer, pp. 62-92. New York: Hafner Publishing Company.

Jokiel, P.L., and S.L. Coles. 1990. Response of Hawaiian and other Indo Pacific reef corals to elevated temperature. *Coral Reefs* 8:155-162.

Jones, G.M., and R.E. Scheibling. 1985. *Paramoeba* sp. (Amoebida, Paramoebidae) as the possible causal agent of sea urchin mass mortality in Nova Scotia. *J. Parasitol.* 71(5):559-565.

Jones, G.M., A.J. Hebda, R.E. Scheibling, and R.J. Miller. 1985. Histopathology of the disease causing mass mortality of sea urchins (*Strongylocentrotus droebachiensis*) in Nova Scotia. *J. Invert. Pathol.* 45:260-271.

Jones, G.M., R.E. Scheibling, A.J. Hebda, and R.J. Miller. 1984. Amoebae in tissues of diseased echinoids (*Strongylocentrotus droebachiensis*) in Nova Scotia. In *Echinodermata: Proc. Fifth Int. Echinoderm Conf.*, ed. B.F. Keegan and B.D.S. O'Connor, pp. 269-73. Rotterdam, The Netherlands: A.A. Balkema.

Jones, P. 1990. Mass White Sea mortalities. *Mar. Pollut. Bull.* 21(1):502.

Kent, M.L., D.J. Whitaker, and L. Margolis. 1990. Experimental transmission of the myxosporean *Myxobolus arcticus* to sockeye salmon using an aquatic oligochaete, *Eclipidrilus* sp. (Lumbriculidae). *Fish Health Sec./Am. Fish. Soc. Newsl.* 18(4):4-5.

Kinne, Otto. (ed.). 1980. *Diseases of Marine Animals. I. General Aspects, Protozoa to Gastropoda.* New York: John Wiley & Sons.

Køie, M. 1985. On the morphology and life-history of *Lepidapedon elongatum* (Lebour, 1908) Nicoll, 1910 (Trematoda, Lepocreadiidae). *Ophelia* 24(3):135-153.

Knowlton, N., J.C. Lang, M.C. Rooney, and P. Clifford. 1981. When hurricanes kill corals: evidence for delayed mortality in Jamaican staghorns. *Nature* 294:251-252.

Kudo, R.R. 1966. *Protozoology*, 5th edition. Springfield, IL: Charles C. Thomas,

Lambert, P. 1981. *The Sea Stars of British Columbia.* British Columbia Provincial Museum, Handbook 39:20-1.

Landers, S.C., and J. Gunderson. 1986. *Pterospora schizosoma*, a new species of aseptate gregarine from the coelom of *Axiothella rubrocincta* (Polychaeta, Maldanidae). *J. Protozool.* 33(2):297-300.

Lang, J.C., R.I. Wicklund, and R.F. Dill. 1988. Depth- and habitat-related bleaching of zooxanthellate reef organisms near Lee Stocking Island, Exuma Cays, Bahamas. *Proc. Sixth Int. Coral Reef Symp.* 3:269-274.

Langreth, R.N. 1990. Bleached reefs: is a warm-water cycle stripping corals of their life-blood? *Science News* 138:364-365.

Lasker, H.R., E.C. Peters, and M.A. Coffroth. 1984. Bleaching of reef coelenterates in the San Blas Islands, Panama. *Coral Reefs* 3:183-190.

Lauckner, G. 1980a. Diseases of Porifera. In *Diseases of Marine Animals. I. General Aspects, Protozoa to Gastropoda*, ed. Otto Kinne, pp. 139-65. New York: John Wiley & Sons.

Lauckner, G. 1980b. Diseases of Cnidaria. In *Diseases of Marine Animals. I. General Aspects, Protozoa to Gastropoda*, ed. Otto Kinne, pp. 167-237. New York: John Wiley & Sons.

Lauckner, G. 1980c. Diseases of Ctenophora. In *Diseases of Marine Animals. I. General Aspects, Protozoa to Gastropoda*, ed. Otto Kinne, pp. 239-253. New York: John Wiley & Sons.

Laval, P. 1980. Hyperiid amphipods as crustacean parasitoids associated with gelatinous zooplankton. *Oceanogr. Mar. Biol. Annu. Rev.* 18:11-56.

Laydoo, R. 1983. Recent mass mortality of gorgonians in Trinidad. Paper read at 18th Annual Meeting of the Association of Island Marine Laboratories of the Caribbean, 13-17 August 1984, at Institute of Marine Affairs, St. James, Trinidad.

Leighton, B.J., J.D.G. Boom, C. Bouland, E.B. Hartwick, and M.J. Smith. 1991. Castration and mortality in *Pisaster ochraceous* (Brandt) parasitized by *Orchitophyra stellarum* Cepede (1907) (Ciliata). *Dis. Aquat. Organ.* 10:71-73.

Lesser, M.P., W.R. Stochaj, D.W. Tapley, and J.M. Shick. 1990. Bleaching in coral reef anthozoans: effects of irradiance, ultraviolet radiation, and temperature on the activities of protective enzymes against reactive oxygen. *Coral Reefs* 8:225-232.

Lessios, H.A. 1988. Mass mortality of *Diadema antillarum* in the Caribbean: what have we learned? *Annu. Rev. Ecolog. System.* 19:371-393.

Lessios, H.A., D.R. Robertson, and J.D. Cubit. 1984a. Spread of *Diadema* mass mortality through the Caribbean. *Science* 226:335-337.

Lessios, H.A., J.D. Cubit, D.R. Robertson, M.J. Shulman, M.R. Parker, S.D. Garrity, and S.C. Levings. 1984b. Mass mortality of *Diadema antillarum* on the Caribbean coast of Panama. *Coral Reefs* 3:173-182.

Li, M.F., J.W. Cornick, and R.J. Miller. 1982. Studies of recent mortalities of the sea urchin (*Strongylocentrotus droebachiensis*) in Nova Scotia. *Rapp. P.-v. Con. Int. Expl. Mer* 1982:46.

Lichtenfels, J.R., and C.F. Stroup. 1985. *Eustrongylides* sp. (Nematoda: Dioctophymatoidea): first report of an invertebrate host (Oligochaeta: Tubificidae) in North America. *Proc. Helminthological Soc. Washington* 52(2):320-323.

Maes, P., and M. Jangoux. 1984. The bald-sea-urchin disease: a bacterial infection. In *Echinodermata: Proceedings of the Fifth International Echinoderm Conference*, ed. B. F. Keegan and B. D. S. O'Connor, pp. 313-4. Rotterdam, The Netherlands: A. A. Balkema.

Maes, P., M. Jangoux, and L. Fenaux. 1986. La maladie de l'"oursin-chave": ultrastructure de lésions et charactérisation de leur pigmentation. *Ann. Inst. Océanograph. Paris* 62:37-45.

Marine Life Resources Workshop. 1989. Marine models in biomedical research. *Biolog. Bull.* 176:337-348.

Markiw, M.E., and K. Wolf. 1983. *Myxosoma cerebralis* (Myxozoa: Myxosporea) etiologic agent of salmonid whirling disease requires tubificid worm (Annelida: Oligochaeta) in its life cycle. *J. Protozool.* 30(3):561-564.

May, R.M. 1983. Parasitic infections as regulators of animal populations. *Am. Sci.* 71:36-45.

May, R.M., and R.M. Anderson. 1983. Epidemiology and genetics in the coevolution of parasites and hosts. *Proc. Royal Soc. London B* 219:281-313.

Miller, R.J. 1985. Sea urchin pathogen: a possible tool for biological control. *Mar. Ecol. Prog. Ser.* 21:169-174.

Millott, N. 1969. Injury and the axial organ of echinoids. *Experientia* 25/7:756-757.

Möller, H. 1987. Pollution and parasitism in the aquatic environment. *Int. J. Parasitol.* 17(2):353-361.

Morse, D.E., A.N.C. Morse, and H. Duncan. 1977. Algal "tumors" in the Caribbean sea fan, *Gorgonia ventalina*. *Proc. Third Int. Coral Reef Symp.* 1:623-629.

Morse, D.E., A. Morse, H. Duncan, and R.K. Trench. 1981. Algal tumors in the Caribbean octocorallian, *Gorgonia ventalina*: II. Biochemical characterization of the algae, and first epidemiological observations. *Bull. Mar. Sci.* 31(2):399-409.

Munro, M.H.G., R.T. Luibrand, and J.W. Blunt. 1987. The search for antiviral and anticancer compounds from marine organisms. In *Bioorganic Marine Chemistry, Volume I*, ed. P.J. Scheuer, pp. 93-176. Berlin/Heidelberg: Springer-Verlag.

Olive, P.J.W., and P.S. Cadnam. 1990. Mass mortalities of the lugworm on the South Wales coast: a consequence of algal bloom? *Mar. Pollut. Bull.* 21(11):542-545.

Overstreet, R.M. 1988. Aquatic pollution problems, Southeastern U.S. coasts: histopathological indicators. *Aquat. Toxicol.* 11:213-239.

Paine, R.T. 1966. Food web complexity and species diversity. *Am. Naturalist* 100:65-75.

Paul, J.H., M.F. DeFlaun, and W.H. Jeffrey. 1986. Elevated levels of microbial activity in the coral surface microlayer. *Mar. Ecol. Prog. Ser.* 33:29-40.

Pearse, J.S. 1982. Gametogenesis, aboral ring, axial complex, tube feet of an echinoid (abstract). In *Echinoderms: Proceedings of the International Conference, Tampa Bay*, ed. J.M. Lawrence, p 479. Rotterdam, the Netherlands: A.A. Balkema.

Pearse, J.S., and A.H. Hines. 1979. Expansion of a Central California kelp forest following the mass mortality of sea urchins. *Mar. Biol.* 51:83-91.

Pearse, J.S., D.P. Costa, M.B. Yellin, and C.R. Agegian. 1977. Localized mass mortality of red sea urchin, *Strongylocentrotus franciscanus*, near Santa Cruz, California. *U.S. Dept. Comm. Fish. Bull.* 75:645-648.

Peters, E.C. 1984. A survey of cellular reactions to environmental stress and disease in Caribbean scleractinian corals. *Helgol. Meeresunters.* 37:113-137.

Peters, E.C., J.J. Oprandy, and P.P. Yevich. 1983. Possible causal agent of "white band disease" in Caribbean acroporid corals. *J. Invert. Pathol.* 41:394-396.

Peters, E.C., J.C. Halas, and H.B. McCarty. 1986. Calicoblastic neoplasms in *Acropora palmata*, with a review of reports on anomalies of growth and form in corals. *J. Nat. Cancer Inst.* 76(5):895-912.

Phillips, P.J. 1973. Protozoan symbionts from the anemone *Bunodosoma cavernata* from Galveston Island, Texas. *Gulf Res. Rep.* 4(2):186-190.

Porchet-Henneré, E., and M. M'Berri. 1987. Cellular reactions of the polychaete annelid *Nereis diversicolor* against coelomic parasites. *J. Invert. Pathol.* 50:58-66.

Powers, P.B.A. 1933. Studies on the ciliates from sea urchins. I. *Biolog. Bull.* 65:106-21.

Powers, P.B.A. 1935. Studies on the ciliates of sea-urchins: a general survey of the infestations occurring in Tortugas echinoids. *Papers from the Tortugas Laboratory* 29:293-326.

Ramos-Flores, T. 1983. Lower marine fungus associated with black line disease in star corals (*Montastrea annularis*, E. & S.). *Biolog. Bull.* 165:429-435.

Reise, K. 1979. Spatial configurations generated by motile benthic polychaetes. *Helgol. Wiss. Meeresunter.* 32:55-72.

Reiswig, H.M. 1988. Natural commercial sponges. *Int. Sponge Newsl.* 2:1.

Robinson, H. 1979. A Geography of Tourism. Estover, Plymouth, England:MacDonald and Evans.

Roch, P., A. Giangrande, and C. Canicatti. 1990. Comparison of hemolytic activity in eight species of polychaetes. *Mar. Biol.* 107:199-203.

Rogers, C.S. 1985. Degradation of Caribbean and western Atlantic coral reefs and decline of associated fisheries. *Proc. Fifth Int. Coral Reef Congr.* 6:491-6.

Rohde, K. 1982. *Ecology of Marine Parasites*. St. Lucia, Australia: Univ. Queensland Press.

Roberts-Regan, D.L., R.E. Scheibling, and J.F. Jellett. 1988. Natural and experimentally induced lesions of the body wall of the sea urchin *Strongylocentrotus droebachiensis*. *Dis. Aquat. Organ.* 5:51-62.

Rützler, K. 1988. Mangrove sponge disease induced by cyanobacterial symbionts: failure of a primitive immune system? *Dis. Aquat. Organ.* 5:143-149.

Rützler, K. 1990. Associations between Caribbean sponges and photosynthetic organisms. In *New Perspectives in Sponge Biology*, ed. K. Rützler, pp. 455-66. Washington, DC: Smithsonian Institution Press.

Rützler, K., and D. Santavy. 1983. The black band disease of Atlantic reef corals. I. Description of a cyanophyte pathogen. *P.S.Z.N.I.: Mar. Ecol.* 4:301-319.

Rützler, K., D.L. Santavy, and A. Antonius. 1983. The black band disease of Atlantic reef corals. III. Distribution, ecology, and development. *P.S.Z.N.I.: Marine Ecology* 4:329-358.

Sarà, M., and J. Vacelet. 1973. Écologie des Démosponges. In *Traite deZoologie. Tome III, Fasicule 1*, ed. Pierre-P. Grassé, pp. 462-576. Paris: Masson.

Savina, L.A. 1990. Physiological changes in *Montastrea annularis* during recovery from induced bleaching with high temperature and light. Annual Meeting of the American Society of Zoologists *et al.*, December 27-30 1990, San Antonio, Texas.

Scheibling, R.E. 1984. Echinoids, epizootics and ecological stability in the rocky subtidal off Nova Scotia, Canada. *Helgol. Meeresunters.* 37:233-242.

Scheibling, R.E., and R.L. Stephenson. 1984. Mass mortality of *Strongylocentrotus droebachiensis* (Echinodermata: Echinoidea) off Nova Scotia, Canada. *Mar. Biol.* 78:153-164.

Scheibling, R.E., and B.G. Raymond. 1990. Community dynamics on a subtidal cobble bed following mass mortalities of sea urchins. *Mar. Ecol. Prog. Ser.* 63:127-145.

Scott, P.J.B. 1987. Associations between corals and macro-infaunal invertebrates in Jamaica, with a list of Caribbean and Atlantic coral associates. *Bull. Mar. Sci.* 40:271-286.

Shrevel, J. 1963. Grégarines nouvelles de Nereidae et Eunicidae (Annélides polychetes). *C.R. Soc. Biol.* 157:814-816.

Siebert, A.E. 1973. A description of *Haplozoon axiothellae* n. sp. an endosymbiont of the polychaete *Axiothella rubrocincta*. *J. Phycol.* 9:185-190.

Sindermann, C.J. 1990. *Principle Diseases of Marine Fish and Shellfish. Volume 2. Diseases of Marine Shellfish.* San Diego, California: Academic Press.

Smith, F.G. 1941. Sponge disease in British Honduras, and its transmission by water currents. *Ecology* 22:415-421.

Smith, G.B. 1975. The 1971 red tide and its impact on certain reef communities in the mid-eastern Gulf of Mexico. *Environ. Lett.* 9: 141-152.

Sparks, A.K. 1985. *Synopsis of Invertebrate Pathology Exclusive of Insects.* Amsterdam/Oxford/New York: Elsevier Science Publishers B.V.

Steen, R.G., and L. Muscatine. 1987. Low temperature evokes rapid exocytosis of symbiotic algae by a sea anemone. *Biolog. Bull.* 172:246-263.

Stevenson, D.K., and N. Marshall. 1974. Generalizations on the fisheries potential of coral reefs and adjacent shallow-water environments. *Proc. Second Int. Coral Reef Symp.* 1:147-156.

Stock, J.H. 1988. Copepods associated with reef corals: a comparison between the Atlantic and the Pacific. *Hydrobiologia* 167/168:545-547.

Szmant, A.M., and N.J. Gassman. 1990. The effects of prolonged "bleaching" on the tissue biomass and reproduction of the reef coral *Montastrea annularis*. *Coral Reefs* 8:217-224.

Taylor, D.L. 1983. The black band disease of Atlantic reef corals. II. Isolation, cultivation, and growth of *Phormidium corallyticum*. *P.S.Z.N.I.: Mar. Ecol.* 4:321-328.

Te Strake, D., W.C. Jaap, E. Truby, and R. Reese. 1988. Fungal elements in *Millepora complanata* Lamarck, 1816 (Cnidaria: Hydrozoa) after mass expulsion of zooxanthellae. *la. Scientist* 51:184-188.

Tsuchiya, M., K. Yanagiya, and M. Nishihara. 1987. Mass mortality of the sea urchin *Echinometra mathaei* (Blainville) caused by high water temperature on the reef flats in Okinawa, Japan. *Galaxea* 6:375-385.

Turton, G.C., and A.C. Wardlaw. 1987. Pathogenicity of the marine yeasts *Metschnikowia zobelli* and *Rhodotorula rubra* for the sea urchin *Echinus escuelentus*. *Aquaculture* 67:199-202.

Tyler, P.A., and D.S.M. Billett. 1987. The reproductive ecology of elasipodid holothurians from the N. E. Atlantic. *Biol. Oceanogr.* 5:273-296.

Unkles, S.E. 1977. Bacterial flora of the sea urchin *Echinus esculentus*. *Applied Environ. Microbiol.* 34(4):347-350.

Upton, S.J., and E.C. Peters. 1986. A new and unusual species of coccidium (Apicomplexa: Agammococcidorida) from Caribbean scleractinian corals. *J. Invert. Pathol.* 47:184-193.

Vacelet, J. 1991. Statut des éponges commerciales en Méditerranée. In *Les Épecès Marines a Protéger en Méditerranée (Marine Species in Need of Protection in the Mediterranean)*, Colloque International, Carry-le-Rouet, France, Novembre 1989, ed. Charles-Francois Bourdesque, Michel Avon, and Vincent Gravez, pp. 35-42. GIS Posidonies Marseille, France.

Vacelet, J., and M. -F. Gallissian. 1978. Viruslike particles in cells of the sponge *Verongia cavernicola* (Demospongia, Dictyoceratida) and accompanying tissue changes. *J. Invert. Pathol.* 31:246-254.

van Duyl, F.C. 1983. The distribution of *Acropora palmata* and *Acropora cervicornis* along the coasts of Curacao and Bonaire, Netherlands Antilles. Paper read at Interdisciplinary Studies in Coral Reef Research, Annual Meeting of the International Society for Reef Studies, at Leiden, The Netherlands, December 1983.

Vevers, H.G. 1951. The biology of *Asterias rubens*. II. Parasitization of the gonads by *Orchitophyra stellarum* Cepede. *J. Mar. Biolog. Assoc. UK* 29:619-624.

Vicente, V.P. 1989. Regional commercial sponge extinctions in the West Indies: are recent climatic changes responsible? *Mar. Ecol. Prog. Ser.* 10:179-191.

Vicente, V.P. 1990. Response of sponges with autotrophic endosymbionts during the coral-bleaching episode in Puerto Rico. *Coral Reefs* 8:199-202.

Waldichuk, M. 1990. Mysterious starfish deaths in coastal British Columbia. *Mar. Pollut. Bull.* 21(5):222.

Whitfield, P.J., and R.H. Emson. 1988. *Parachordeumium amphiurae*: a cuckoo copepod? *Hydrobiologia* 167/168:523-531.

Wicklow, B.J., A.C. Borror, and C.W. Walker. Submitted. Endocytosis of the spermatogenic cells of *Asterias vulgaris* (Echinodermata) by the gonadal endosymbiont, *Orchitophyra stellarum* (Ciliophora). *Biolog. Bull.*

Wilkinson, C.R. 1978a. Microbial associations in sponges. I. Ecology, physiology and microbial populations of coral reef sponges. *Mar. Biol.* 49:161-167.

Wilkinson, C.R. 1978b. Microbial associations in sponges. III. Ultrastructure of the *in situ* associations in coral reef sponges. *Mar. Biol.* 49:177-185.

Wilkinson, C.R. 1980. Cyanobacteria symbiotic in marine sponges. In *Endocytobiology: Endosymbiosis and Cell Biology*, ed. W. Schwemmler and H. E. A. Schenk, pp. 553-563. Berlin: Walter de Gruyter.

Wilkinson, C.R., M. Nowak, B. Austin, and R.R. Colwell. 1981. Specificity of bacterial symbionts in Mediterranean and Great Barrier Reef sponges. *Microb. Ecol.* 7:13-21.

Williams, E.H., Jr., and L. Bunkley-Williams. 1990. The world-wide coral reef bleaching cycle and related sources of coral mortality. *Atoll Res. Bull.* 335:1-71.

Williams, E.H., Jr., and L.Bunkley-Williams. 1987. Caribbean marine mass mortalities: a problem with a solution. *Oceans* 30(4):69-75.

Williams, E.H., Jr. and T.J. Wolfe-Waters. 1990. An abnormal incidence of the commensal copepod, *Doridocola astrophyticus* Humes, associated with injury of its host, the basketstar, *Astrophyton muricatum* (Lamarck). *Crustaceana* 59(3):302.

Williams, L.B., E.H. Williams, Jr., and A.G. Bunkley, Jr. 1986. Isolated mass mortalities of the sea urchins *Astropyga magnifica* and *Eucidaris tribuloides* in Puerto Rico. *Bull. Mar. Sci.* 38:391-393.

Wilson, W.H., Jr. 1981. Sediment-mediated interactions in a densely populated infaunal assemblage: the effects of the polychaete *Abarenicola pacifica*. *J. Mar. Res.* 39(4):735-748.

Winstead, J.T., and J.A. Couch. 1988. Enhancement of protozoan pathogen *Perkinsus marinus* infections in American oysters *Crassostrea virginica* exposed to the chemical carcinogen n-nitrosodiethylamine (DENA). *Dis. Aquat. Organ.* 5:205-213.

Wolf, K., and M.E. Markiw. 1984. Biology contravenes taxonomy in the Myxozoa: new discoveries show alternation of invertebrate and vertebrate hosts. *Science* 225:1449-1452.

Wolf, K., M.E. Markiw, and J.K. Hiltunen. 1986. Salmonid whirling disease: *Tubifex tubifex* (Müller) identified as the essential oligochaete in the protozoan life cycle. *J. Fish Dis.* 9(1):79-81.

Yip, S. Y. 1984. Parasites of *Pleurobrachia pileus* Muller, 1776 (Ctenophora), from Galway Bay, western Ireland. *J. Plankton Res.* 6(1):107-121.

Zibrowius, H., and M. J. Grygier. 1985. Diversity and range of scleractinian coral hosts of Ascothoracida (Crustacea: Maxillipoda). *Ann. Inst. Oceanogr.* 61(2):115-138.

16

Interactions of Pollutants and Disease in Marine Fish and Shellfish

Carl J. Sindermann

INTRODUCTION

Tentative indications of possible pollution-associated disease conditions in marine fish and shellfish began to appear in the scientific literature in the 1960's and early 1970's. As examples, Young (1964) reported a variety of gross pathological changes in fish taken near southern California's giant ocean outfalls. Young and Pearce (1975) found black gills and shell erosion in crustaceans from the New York Bight. Mahoney and coworkers (1973) reported fin erosion in 22 species of fish from the New York Bight apex. During the same period, Rosenthal and his associates began what was to become a long-term study of effects of pollutants on fish larvae (Kinne and Rosenthal 1967; Rosenthal and Stelzer 1970; Rosenthal 1971; Rosenthal and Sperling 1974; von Westernhagen *et al.* 1974).

In the late 1970's and throughout the 1980's, evidence for an association of numerous disease conditions in marine fish and shellfish with degraded habitats increased dramatically. Chronologically information accumulated about skeletal anomalies in fish (Valentine 1975; Matsusato 1978, 1986), ulcerations in cod (Christensen 1981), epidermal papillomas in several species of flatfish (Stich *et al.* 1977; Dethlefsen 1984; Dethlefsen and Lang 1988), and liver tumors in flatfish (Malins *et al.* 1984, 1988; Murchelano and Wolke 1985; Vethaak 1987). Much available information about current understanding of effects of pollutants on marine animals was summarized admirably in a symposium on *Toxic Chemicals and Aquatic Life: Research and Management* in Seattle in 1986 and published in Volume 11 of *Aquatic Toxicology* (1988).

The evidence, though substantial, for an association of polluted habitats and fish and shellfish diseases, should still be considered as largely "circumstantial" or "inferential," since direct cause and effect relationships have not been demonstrated to the satisfaction of all observers. As recently as 1985, a work-

0-8493-8662-4/93/$0.00 + $.50

shop convened by the Working Group on Pathology of the International Council for the Exploration of the Sea (ICES) concluded that "an established link between diseases and pollution does not exist" and that "the most likely general effect of pollution is a non-specific lowering of resistance, resulting in appearance of variable disease signs rather than the specific induction of an identifiable syndrome" (Anonymous 1985).

This assessment by ICES workshop participants seems unnecessarily conservative and negative, in view of the information summarized in this paper. Whereas it is certainly true that much of the evidence is circumstantial and more data are needed, it seems nonetheless that some measure of association between certain fish and shellfish diseases and pollution has been demonstrated. Studies employing a combination of field observations and experimental exposures to contaminants could do much to further reduce uncertainties. The body of evidence that indicates a statistical relationship between degraded estuarine/coastal environments and certain disease conditions in fish and shellfish has grown to a point where other observers have concluded that a relationship *does* exist, and that what remains to be done is to acquire additional quantitative data and to augment supporting experimental information -- especially that concerning the physiological/biochemical bases for observed pathology (Passino 1984). Additionally, much more difficult studies of effects of pollution-associated diseases on abundance of resource species are still necessary. Results of analyses of disease-caused effects on survival are important to resource managers, even though such investigations constitute only a subset of the scientific examination of pollution/disease relationships.

The objectives of this paper are: 1) to examine the present understanding of several specific disease conditions that may be associated with habitat degradation -- including the ecological/physiological/biochemical mechanisms involved; 2) to assess validity of associations that have been proposed; and 3) to suggest avenues of research that may prove to be productive in the 1990's.

DISEASE CONDITIONS ASSOCIATED WITH HABITAT DEGRADATION

Most disease conditions considered here are traceable to tissue damage and metabolic disturbances induced by toxic chemicals, or (in the case of infectious diseases) to increased infection pressure from expanded populations of facultative pathogens, combined with suppression of internal and external defense mechanisms of stressed animals.

In the search for a sensible categorization of the various kinds of disease conditions to be discussed, one obvious approach would be to look at the animals at risk from the viewpoint of vulnerability to toxic levels of environmental contaminants, and to group the observed effects in this way. If this is done, areas of vulnerability that are apparent immediately include: eggs and larvae, integument, filtering/detoxifying systems, cell and tissue metabolic pathways, and neurosensory systems. The following sections are based on this structure with the recognition that it is artificial, and that some disease conditions may be consequences of multiple effects of pollutants through multiple pathways. Also, listing these five categories does not imply prioritization, since little agreement would be found. Some authorities would consider effects on metabolic pathways as all-important, while others would argue for primacy of effects on internal defenses. Still others might argue for inclusion of effects on respiratory, circulatory, and digestive/assimilative systems. Categories chosen are simply convenient and somewhat inclusive groupings of alternative ways to dissect and discuss the pollution/disease relationship explored in earlier papers (Sindermann 1972, 1976, 1977, 1979, 1983, 1984, 1988a) (Fig. 16-1).

Effects of Pollutants on Early Life History Stages

Abundance of year classes of commercial fish and shellfish can be influenced, sometimes drastically, by environmental events before and after spawning or during larval development. The possible role of contaminants in affecting survival is difficult to determine, except in experimental situations -- but some important insights have been gained about abnormalities and mortalities in early life history stages (eggs, embryos, and larvae).

During the past two decades, two separate lines of investigation of pollutant effects on eggs and larvae have developed. The first is concerned with genetic and cytologic abnormalities in developing eggs; the second with abnormal larval development (with subsequent mortality). These investigations are obviously aspects of a common theme, and a wedding of disciplines is now occurring. Pioneering studies of cytotoxic and mutagenic effects of contaminants on marine fish eggs and larvae have been reported by Longwell (1976, 1977, 1978), Longwell and Hughes (1980), Longwell *et al.* (1983, 1984), Chang and Longwell (1984), and Hughes *et al.* (1986). Genetic disarray, represented morphologically by high prevalences of chromosomal anomalies in developing eggs and larvae, characterized samples from polluted waters. Examination of samples of developing embryos of Atlantic mackerel from the New York Bight plankton disclosed cytological abnormalities (in the form of disruption of the mitotic apparatus) and chromosomal abnormalities (ranging from stickiness and chromosome bridges to complete pulverization). Statistical correlations of high prevalences of chromosomal anomalies with degree of environmental contamination provided evidence for possible impacts on estuarine/coastal populations.

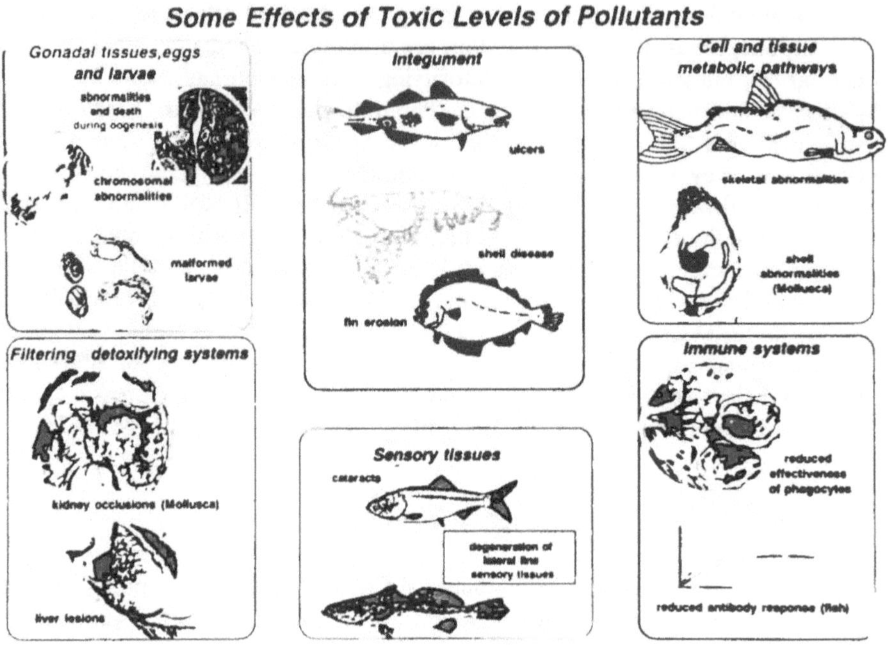

Fig. 16-1. Examples of effects of pollutants on marine fish and shellfish.

Counterpart studies of gross effects of pollutants on developing fish larvae by Rosenthal and colleagues began in 1967 (Kinne and Rosenthal 1967; Rosenthal and Stelzer 1970; Rosenthal 1971; von Westernhagen *et al.* 1974, 1975; Rosenthal and Alderdice 1976; Hansen *et al.* 1985; Rosenthal *et al.* 1986) and have been continued by Dethlefsen and von Westernhagen (Dethlefsen 1980, 1984, 1985, 1988; Dethlefsen *et al.* 1987; von Westernhagen 1988; von Westernhagen *et al.* 1981, 1987, 1988). Among the many significant findings reported in this excellent series of papers are: 1) exposure of maturing females to low concentrations of contaminants -- especially those which are bioaccumulated -- can affect gonad tissue, with effects expressed in the next generation; 2) life cycle stages most vulnerable to contaminants are maturing females, early embryos, early hatched larvae, and larvae at transition from yolk sac to feeding; 3) a wide range of morphological, behavioral, and physiological abnormalities in larvae result from exposure to contaminants, in rough proportion to the environmental level of the particular contaminant; 4) common morphological abnormalities include malformed lower jaw, eye deformities, anomalies in vertebral column, and reduced size at hatching; 5) common physiological abnormalities include reduced heart rate, reduced swimming ability, disturbance in equilibrium, and reduced feeding; 6) early developmental stages showed the highest malformation rates.

Convergence of the two methodologies -- cytogenetic and teratogenic -- is now taking place, augmented by experimental studies of contaminant-exposed spawning fish and by chemical analyses of tissue and environmental samples (e.g., Dethlefsen *et al.* 1985, 1986; von Westernhagen *et al.* 1987, 1988). Chromosomal anomalies reported by Longwell and colleagues suggest that some gross abnormalities in larvae observed by Rosenthal and his colleagues may have a genetic base, even though the kinds of derangements seen in the genetic material have not yet been correlated directly with the kinds of morphological and physiological disturbances seen. However, statistical associations of cytogenetic and teratogenic anomalies have been made with degraded environments. The unified approach, beginning with chromosomal damage in the ovaries of parent females and extending to gross larval abnormalities (Fig. 16-2), has great potential in bioindicator work and implications in population dynamics of estuarine/coastal species in polluted habitats (von Westernhagen *et al.* 1981, 1988; Sindermann 1984).

Pollutants may act as "developmental toxicants," exerting their effects, particularly on embryos, through several metabolic pathways (Weis and Weis 1987). Malformations of embryos of Baltic herring, *Clupea harengus membras*, were found to be abundant and in great variety following exposure to crude oil and chemical dispersants (Linden 1976). In a study of pelagic fish embryos from the North Sea, malformations were found with high frequency in several species (dab, *Limanda limanda*; flounder, *Platichthys flesus*; whiting, *Merlangius merlangus*; cod, *Gadus morhua*; and plaice, *Pleuronectes platessa*) in regions with known high contamination levels (Dethlefsen *et al.* 1986).

Abnormalities induced by pollutants are only part of the spectrum of pollution effects; it is also important to examine *mortalities* that may result from exposure of pre-spawning females to contaminated habitats. Some evidence exists. Chlorinated hydrocarbons can contribute substantially to larval mortality by accumulation in tissues and transfer to eggs from the parent females. In one example of larval mortality that seems related to environmental contamination, evidence was found that mortalities of larval winter flounder, *Pseudopleuronectes americanus*, in a Massachusetts estuary could be related to pesticide pollution (Smith 1973). Adult females concentrated DDT, DDE, and heptachlor epoxide in their ovaries as spawning approached, mortality of post-yolk-sac larvae approached 100%.

Other earlier studies have provided evidence that high tissue concentration of chlorinated hydrocarbons in spawning adults can cause mortalities in developing eggs. Reproductive failure of a sea trout population in Texas was attributed

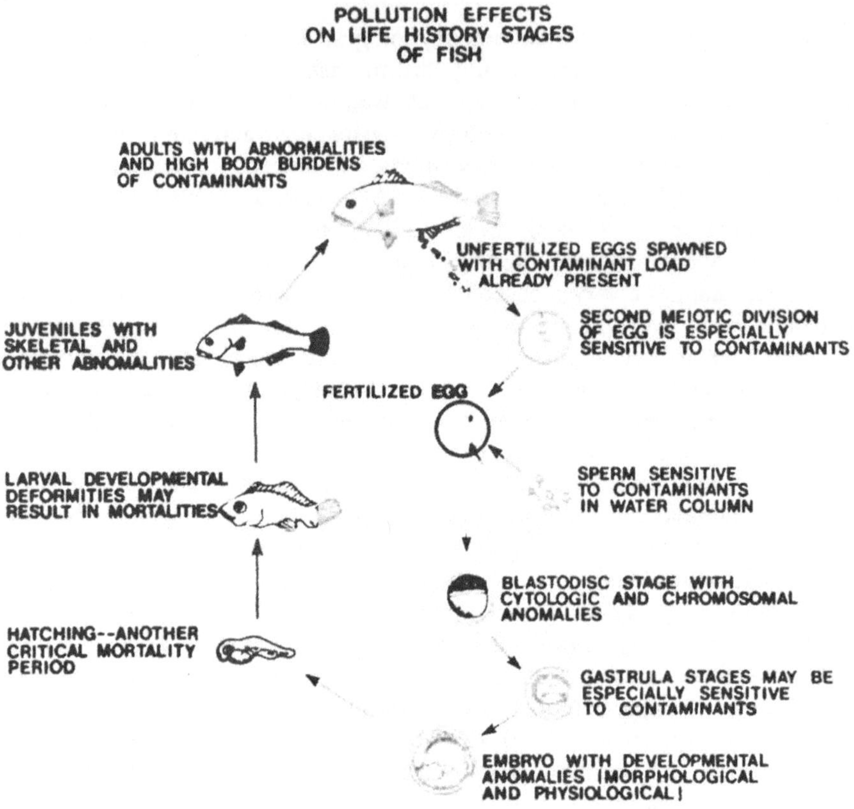

Fig. 16-2. Stages in the life history of fish where pollutant effects have been observed.

to this phenomenon (Butler *et al.* 1972). The sea trout population inhabited an estuary contaminated heavily with DDT, and DDT concentration in the ovaries reached a peak of 8 ppm prior to spawning, compared to less than 0.5 ppm in sea trout from less contaminated estuaries. Spawning seemed normal, but eggs failed to develop.

Several recent studies have indicated that high levels of PCBs in gonads of spawning fish (Baltic flounder, *Platichthys flesus*; Baltic herring, *Clupea harengus*; striped bass, *Morone saxatilis*; and North Sea whiting, *Merlangius merlangus*) can result in low (< 50%) viable hatch of eggs, due to abnormal development and high embryo mortality (von Westernhagen *et al.* 1981, 1988; Hansen *et al.* 1985; Westin *et al.* 1985; Cameron *et al.* 1986). On the California coast, white croakers, *Genyonemus lineatus*, in contaminated waters near Los Angeles had high body burdens of chlorinated hydrocarbons, greater early oocyte atresia, lower fecundity, and lower fertilization rates than fish from reference areas (Cross and Hose 1988). Other effects of environmental pollutants on fecundity have been reported. As an example, a study of ovarian development in plaice, *Pleuronectes platessa*, captured on the French coast in 1979 and 1980, near the site of the *Amoco Cadiz* oil spill in 1978, revealed reduced concentrations of developing follicles and no mature follicles (Stott *et al.* 1983).

Some experimental evidence indicates that specific pollutants may have effects on early life stages of fish. Studies by Weis *et al.* (1981) of killifish, *Fundulus heteroclitus*, disclosed that some females from unpolluted coastal areas produced eggs which were much more resistant to methyl mercury than eggs from other females (as measured by percentages of developmental anomalies which followed exposure). When a population from a heavily polluted coastal area was examined, a much higher percentage of the females produced "resistant" eggs. However, subsequent studies have indicated that even though embryos from polluted areas were more resistant to methyl mercury toxicity, adults seemed less tolerant as determined by mortality and rate of fin regeneration (Weis *et al.* 1982). A recent review (Weis and Weis 1987) summarized an excellent series of papers by the authors and their associates, published from 1974-1986, on experimental studies of pollutant-induced developmental abnormalities in killifish and other species. Included were considerations of teratogenic effects, optic and cardiac malformations, as well as skeletal defects.

Effects of contaminants also included hormonal imbalances, which may affect ovulation or spawning (Struhsaker 1977; Wedemeyer *et al.* 1984). Additionally, hatchability of eggs and viability of larvae may be impaired, either by high tissue contaminant levels in parent females or by toxic levels of contaminants in the spawning environment (Ernst and Neff 1977; Smith and Cameron 1979; Hansen *et al.* 1985).

Of course the basic question that must be asked is, "Can defective embryonic development and high embryo mortality due to pollution affect recruitment?" Observations relative to this question have been proposed by von Westernhagen and colleagues (1988). According to their reasoning, total mortality during the embryonic stage of marine fish has been estimated to be high -- 95 to 99% for species such as Baltic cod and plaice. At this mortality level, decreases in survival due to embryo abnormalities at observed levels from 22 to 33% would be too small to detect in unexploited populations, but, in overexploited populations in which spawning stocks have been reduced severely, the added impact of abnormal embryonic development and high embryo mortality might result in reduced recruitment. Authors also might point out that, in the North Sea, highest prevalences of embryo malformations occurred in highly polluted areas off the mouths of the Rhine and Elbe rivers and near the dumping zone for titanium dioxide wastes.

Effects of Pollutants on the Integument

The integument of aquatic animals is in continuous contact with environmental contaminants, so it is logical to assume that abnormalities could be consequences of such an intimate relationship. This assumption has been supported by results of studies of several disease conditions -- particularly fin erosion and ulcerations in fish and shell disease in crustaceans. Some of the clearest and statistically most defensible associations of pollution and disease are signaled by the presence of integumentary lesions. Microbial pathogens are often, but not always, implicated.

Fin Erosion

Probably the best known but still the least understood disease of fish from polluted waters is a nonspecific condition known as "fin rot" or "fin erosion," a syndrome which seems clearly associated with degraded estuarine or coastal environments -- to the extent that it has been proposed as an index of pollutant-induced disease (O'Connor *et al.* 1987). Fin erosion has been reported from degraded estuarine/coastal areas in many parts of the world -- from the New York Bight (Mahoney *et al.* 1973; Murchelano 1975, 1982; Ziskowski and Murchelano 1975), California coast (Young 1964; Southern California Coastal Water Research Project 1973; Mearns and Sherwood 1974; Cross 1985), Puget Sound (Wellings *et al.* 1976), Biscayne Bay and Escambia Bay in Florida (Couch 1975; Sindermann *et al.* 1978), the Gulf of Mexico (Overstreet and Howse 1977), the Irish Sea (Perkins *et al.* 1972), the coast of Japan (Nakai *et al.* 1973), the Swedish coast (Lindesjöö and Thulin 1990), and elsewhere.

The fin erosion syndrome seems to occur in at least two types: in bottom fish, where damage to fins seems site-specific and related to direct contact with contaminated sediments, and in pelagic nearshore species, characterized by more generalized erosion, but with predominant involvement of the caudal fin.

Some species are either more resistant to fin erosion or are exposed differentially to toxic substances in water or sediments. A study by Wellings *et al.* (1976) in a heavily polluted arm of Puget Sound (the Duwamish River), in which over 6,000 fish of 29 species were examined, disclosed fin erosion only in starry flounder, *Platichthys stellatus*, and English sole, *Parophrys vetulus*. The authors briefly described observations of liver pathology in starry flounder from the area where fin erosion was common. Histopathology included increased fat deposition in hepatic cells, fibrosis, and vascular distention. Subsequent studies (Pierce *et al.* 1978, 1980) disclosed that all starry flounders from the Duwamish estuary with fin erosion also had severe liver lesions. A correlation of liver pathology and fin erosion was found also in Dover sole, *Microstomus pacificus*, from the California coast. Fin erosion, with prevalences up to 33%, was reported from over 30 species of marine fish in polluted sites in southern California (Cross 1985; Malins *et al.* 1987). It seems quite likely that the "fin erosion" syndrome in fish may include participation of some or all of the following: 1) chemical stressors, possibly acting on mucous and/or epithelial cells; 2) physiological responses of fish to prolonged stress (particularly changes in peripheral blood circulation); 3) suppression of immune responses; 4) marginal dissolved oxygen concentrations, possibly enhanced by a sulfide-rich environment; and 5) secondary bacterial invasion in at least some instances.

Ulcerations

Vibrio and other bacteria have been implicated in a number of reports of ulcerations in fish -- in fact, next to fin erosion, ulcerations with bacterial etiology are probably the commonest abnormalities in fish from polluted waters. Ulcers may be integumentary or penetrating; where bacterial isolations have been made from ulcerated tissue, *Vibrio anguillarum* has been by far the most predominant organism, with pseudomonads and aeromonads in lesser abundance. It seems reasonable that many infections that produce grossly visible ulcerations in fish are bacterial (although viruses and fungi have been implicated in a few instances) and often are due to pathogens of the genera *Vibrio*, *Pseudomonas*, or *Aeromonas* (Lamolet *et al.* 1976). Ulceration often begins with scale loss or formation of small papules, followed by sloughing of the skin, exposing underlying muscles that also may be destroyed. Bacterial ulcers may have rough or raised irregular margins and often are hemorrhagic. Ulcers may or may not be associated with fin erosion.

Epizootic ulcerative syndromes have been reported with increasing frequency in fish from many parts of the world, including the east coast of the U.S. (Sindermann 1988b). Primary etiology is uncertain for many outbreaks, although viruses, bacteria, fungi, and other pathogens have been proposed in specific geographic locations. Environmental stress, often a consequence of pollution, has also been implicated in at least some reported epizootics.

The ulcerative lesions do not constitute a single disease entity, since their characteristics may differ in different host species and areas (Fig. 16-3). Such lesions can be considered as generalized responses of fish to infection and/or abnormal environmental conditions. Types of ulcerations have been described, some with several developmental stages (Christensen 1981), and mortalities have been observed in some outbreaks.

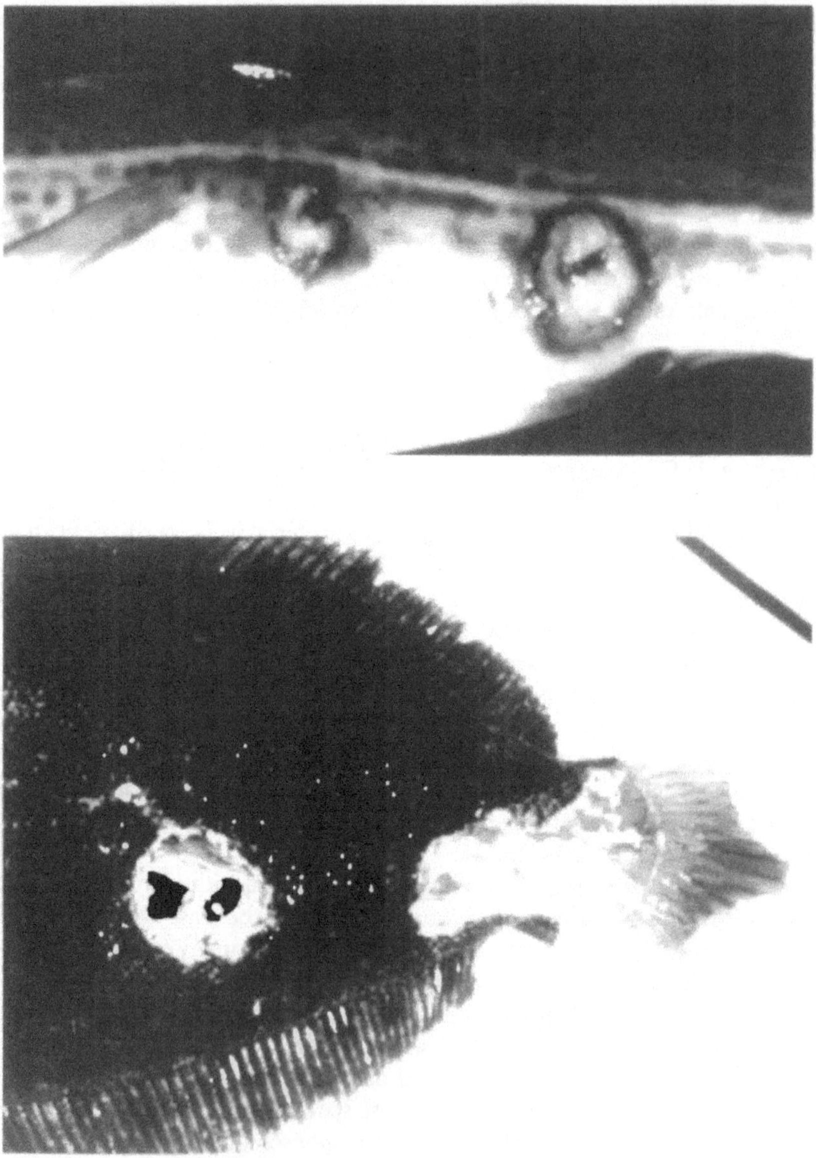

Fig. 16-3. Examples of ulcerations in fish. Advanced lesions in cod from Danish coastal waters (top) and lesion in winter flounder resulting from experimental *Vibrio* infections (bottom). (Photographs courtesy of Dr. I. Dalsgaard and Dr. R. Robohm)

Shell Disease in Crustaceans

A disease condition in Crustacea commonly referred to as "shell disease" or "exoskeletal disease" or "shell erosion" has been associated with badly degraded estuarine and coastal waters. This abnormality can be considered in some ways as the invertebrate analogue of fin erosion in fish. Shell disease has been observed in many crustacean species and under many conditions, both natural and artificial (Sindermann 1989) (Fig. 16-4). Actual shell erosion seems to involve activity of chitinoclastic microorganisms, with subsequent secondary infection of underlying tissue by facultative pathogens. Initial preparation of the exoskeletal substrate by mechanical, chemical, or microbial action probably is significant; thus, high bacterial populations and the presence of contaminant chemicals in polluted environments, as well as extensive detrital and epibiotic fouling of gills, could combine to make shell disease a common phenomenon and a significant mortality factor in crustaceans inhabiting degraded environments.

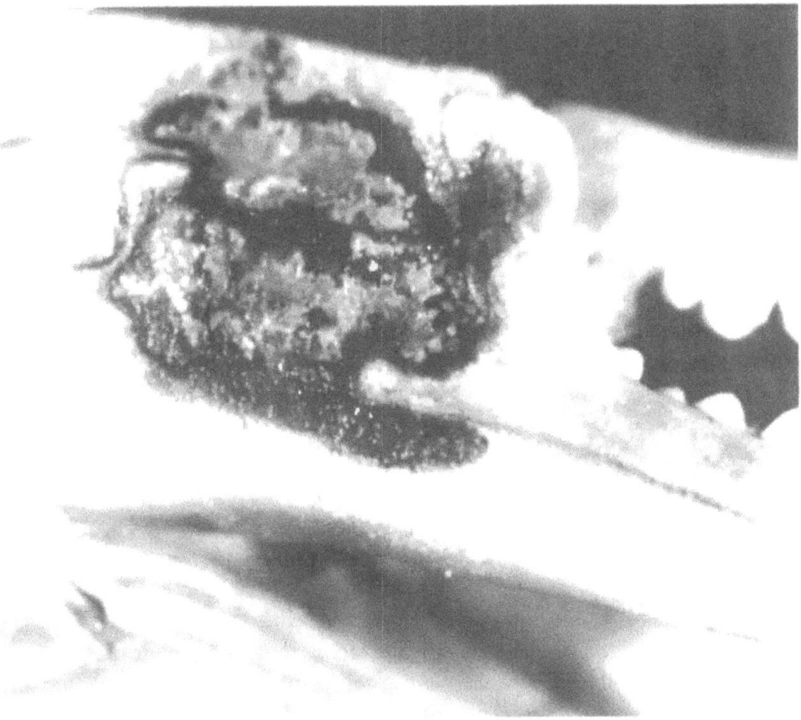

Fig. 16-4. Gross signs of shell disease in the claw of a blue crab, *Callinectes sapidus*. (Photograph courtesy of Dr. R. M. Overstreet)

A study of shell disease in lobsters from Massachusetts waters disclosed highest prevalences in samples from the most polluted sites -- particularly Boston Harbor and Buzzards Bay (Estrella 1984). Mortalities were not observed, but population impacts were considered likely.

A detailed review of shell disease in crustaceans of commercial importance (Anonymous 1989) resulted in the following conclusions about relationships with pollution:

> Shell disease may occur with higher prevalence and greater severity in polluted areas than in those not degraded by man's activities. The balance between metabolic processes associated with new shell formation and infection by microbes capable of utilizing chitin may be disturbed by environmental changes affecting normal shell formation or favoring growth of chitin-utilizing microbes. Such disturbances may be consequences of pollution.
>
> Evidence exists for an association of shell disease with habitat degradation. Prevalences have been found to be high in crustaceans from polluted sites; prevalences show trends similar to the black gill syndrome, which also has a statistical association with extent of pollution. Experimental exposures of crustaceans to contaminated sediments, heavy metals, biocides, petroleum, and petroleum derivatives can result in the appearance of the black gill syndrome, often accompanied by shell disease.

The physiology of crustaceans -- especially that related to hormonal control of molting -- has been elucidated by many studies during the past half-century. The role of pollutants in altering metabolic pathways involved is of course an area of concern. As an example, abnormal production of the steroid-molting hormones may inhibit cuticular synthesis, whereas hormonal insufficiency may delay or prevent molting, thereby affecting growth and survival. These metabolic anomalies may enhance effects of shell disease, especially when accompanied by diminution of other cellular and humoral mechanisms of internal defense. The significance of a prophenoloxidase-activating system in crustacean responses to infection has been pointed out in a series of papers by Söderhäll (1982, 1983, 1986) and associates (1982, 1983, 1984, 1985, 1986); the functioning of this system in the presence of pollutants can be important to host survival.

The appearance of shell erosion therefore may be the consequence of a disturbed balance between processes of chitin maintenance and repair and the activities of chitinoclastic microorganisms -- this disturbance created by either natural or man-made environmental changes. Critical to understanding the relationship are environmental, genetic, and immunological factors which may either promote repair or, conversely, enhance exoskeletal degradation. Also essential to full understanding of the shell disease syndrome are further experimental studies, particularly those concerned with identification of specific microorganisms capable of pathogenesis, experimental manipulation of predisposing environmental variables, and the immunologic responses of hosts to cuticular disruption.

Effects of Pollutants on Filtering/Detoxifying Systems

Chemical examinations of fish and shellfish tissues from grossly contaminated habitats frequently disclose high levels of heavy metals, chlorinated hydrocarbons, and polycyclic aromatic hydrocarbons in critical metabolic sites (livers of fish, hepatopancreas of crustaceans). Pathological changes have also been seen in filtering organs, such as the gills of fish and the kidneys of bivalve molluscs. Several disease conditions in such sites have been studied, and relationships with polluted habitats noted.

Liver Tumors

Probably the best evidence for a relationship of tumors and coastal/estuarine pollution can be found in several studies of hepatocellular neoplasms in fish. Hepatomas in Atlantic hagfish, *Myxine glutinosa*, were studied in the early 1970's by Fange *et al.* (1975) and Falkmer *et al.* (1976, 1977). Hagfish from a polluted Swedish estuary had prevalences of neoplastic livers that were five times higher than those sampled in the open sea, and a possible association with PCB contamination in the estuary was suggested. Soon afterwards, Smith *et al.* (1979) found that 25% of the livers of Atlantic tomcod, *Microgadus tomcod*, from the polluted Hudson River estuary contained neoplastic nodules and hepatocellular carcinomas, with highest prevalences in older fish. Others suggested a possible association of hepatomas with elevated PCB levels in the Hudson River and in livers of some specimens (Klauda *et al.* 1981).

In the late 1970's and the 1980's, the association of progressively severe liver pathology and several types of hepatic neoplasms with badly degraded estuarine/coastal waters became more evident. Several reports demonstrated this relationship. In the North Atlantic, winter flounder, *Pseudopleuronectes americanus*, from several degraded areas on the U.S. east coast (New Haven Harbor, upper Narragansett Bay, Boston Harbor) were reported to have prevalences of 3.4 to 7.5% tumors classified as hepatocarcinomas or cholangiocarcinomas (Murchelano and Wolke 1985) (Fig. 16-5). In the North Pacific, prevalences of hepatomas as high as 16% were found in English sole, *Parophrys vetulus*, from the polluted Duwamish River near Seattle, Washington, a river known to contain high levels of PCBs and many other hydrocarbons (McCain *et al.* 1977, 1982; Malins *et al.* 1983). In subsequent studies, liver neoplasms were also found (in lesser numbers) in rock sole, *Lepidopsetta bilineata*; starry flounder, *Platichthys stellatus*; and Pacific staghorn sculpin, *Leptocottus armatus*. The neoplasms were found in samples from several polluted sites in the Puget Sound area (Commencement Bay, Eliot Bay, Everett Harbor, and Mukilteo Harbor) (Malins *et al.* 1984, 1988). The latter study included a detailed analysis of contaminants in tissues and sediments. Positive correlations were obtained between neoplasm prevalence in bottom-dwelling fish and levels of "certain individual groups of sediment-associated chemicals" (aromatic hydrocarbons, chlorinated hydrocarbons, and heavy metals). As part of the same study, Krahn *et al.* (1986) reported significant positive correlations among prevalences of neoplasms and other lesions and the concentrations of

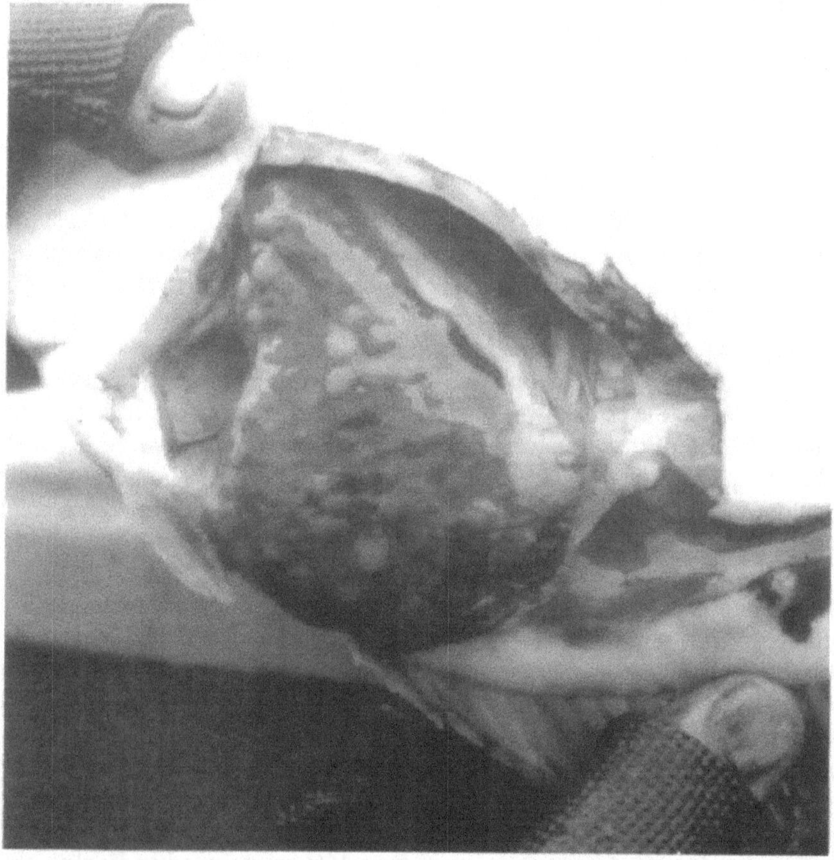

Fig. 16-5. Gross lesions in liver of winter flounder, *Pseudopleuronectes americanus*. (Courtesy of Dr. R.A. Murchelano)

metabolites of aromatic hydrocarbons in bile. Additionally, organic-free radicals, possibly derived from polycyclic aromatic hydrocarbons (PAHs), were found in significantly higher concentrations in liver microsomes of fish with lesions than in normal (Malins *et al.* 1983). A clear cause and effect relationship was not claimed by authors of these studies in the Pacific Northwest. Limitations included inability to identify all sediment-associated contaminants and uncertainty about the synergistic/antagonistic interactions among classes of chemicals. Further southward on the Pacific coast, low prevalences of liver lesions, including neoplasms, were seen in white croaker, *Genyonemus lineatus*, from three polluted sites in the Los Angeles area (Malins *et al.* 1987).

Liver lesions described as tumors have also been reported recently from flounder, *Platichthys flesus*, and dab, *Limanda limanda*, from Dutch coastal waters (Vethaak 1987). Prevalences of gross lesions in fish older than three or four years were locally as high as 40%, and were higher in samples from polluted areas than those from less polluted waters. In other studies, livers of ruffe, *Gymnocephalus cernua*, from the Elbe estuary in Germany were found to have various lesions, including neoplastic nodules. Prevalences in larger sexually mature fish were as high as 32% (Kranz and Peters 1985; Peters *et al.* 1987).

What seems to be emerging from a number of studies of different species in different parts of the world, is a sequence of histopathological changes in livers, including pre-neoplastic changes in liver parenchyma cells (Pierce *et al.* 1978; Köhler and Hölzel 1980; Bucke and Feist 1984; Hendricks *et al.* 1984; Kranz and Peters 1985; Becker *et al.* 1987; Myers *et al.* 1987). As an example, four kinds of hepatic lesions have been recognized in studies of flatfish tumors in the Pacific Northwest: megalocytic hepatosis, nuclear pleomorphisms, foci of cellular alteration, and neoplasms. According to Myers *et al.* (1987), these lesions appear to be related sequentially in the histogenesis of neoplasia in English sole. The progression of pathological changes in livers seems roughly correlated with the extent of estuarine degradation and the length of residence of fish in the estuary. Also, high prevalences of macrophage aggregates (proposed pollution indicators) have been seen in histological material from polluted waters.

In addition to field observations and subsequent correlations of liver tumors and pollution, there is some experimental evidence for induction of liver neoplasms in marine/euryhaline fish by known carcinogens of higher vertebrates (summarized by Ishikawa and Takayama 1979; Hendricks 1982; Couch and Harshbarger 1985; Kimura and Ando 1986; Overstreet 1988). The euryhaline Japanese medaka, *Oryzias latipes*, has been found to be sensitive to carcinogens (methylazoxymethanol acetate and diethylnitrosamine) in several studies. Liver, ocular, and other tumors developed after brief exposure of early life history stages (Ishikawa *et al.* 1975; Aoki and Matsudaira 1977, 1981, 1984; Egami *et al.* 1981; Kimura *et al.* 1984, 1986; Hawkins *et al.* 1986). Medaka juveniles exposed to benzo[a]pyrene (B[a]P) developed what appeared to be sequential lesions (foci of cellular alteration, adenomas, and hepatocellular carcinomas) (Hawkins *et al.* 1988b). Another euryhaline species, the sheepshead minnow, *Cyprinodon variegatus*, developed hepatic and other neoplasms after exposure to methylazoxymethanol acetate and diethylnitrosamine (Hawkins *et al.* 1985a, b, 1988a; Couch and Courtney 1985, 1987). An aggressive, possibly pre-neoplastic lesion, labeled "spongiosis hepatis," was induced in sheepshead minnows by experimental exposure to N-nitrosodiethylamine (Couch 1991). Additionally, the euryhaline species, *Rivulus marmoratus*, developed liver tumors after exposure to diethylnitrosamine (Koenig and Chasar 1984). Hepatic neoplasms were also induced in rainbow trout, *Oncorhynchus mykiss*, by microinjection of embryos and dietary exposure or intraperitoneal injection of juveniles with B[a]P (Black *et al.* 1985; Hendricks *et al.* 1985).

Information about excretory system concretions of calcium phosphate in bivalves is of recent origin (Doyle *et al.* 1978; Gold *et al.* 1982). Clams, *Mercenaria mercenaria*, and scallops, *Argopecten irradians*, from the U.S. east coast develop such concretions, which may occlude the entire kidney. A relationship with polluted or abnormal environments has been proposed, based on the successful experimental induction of the deposits in the wedge shell, *Donax trunculus* (Mauri and Orlando 1982), and the finding that concretions in clams from polluted sites were more numerous and larger than in clams from reference sites.

Kidney Abnormalities in Molluscs

Kidney concretions are apparently distinct from renal plugs described earlier as part of a stress syndrome in clams, *Mercenaria mercenaria*, from hydrocarbon-polluted sites in Narragansett Bay (Jeffries 1972). That condition was characterized by a black tar-like mass which collected in the renal sac in quantities that appeared to plug tubules and interfere with kidney function. Masses of black hemocytes also were seen in mantle and kidney tissues.

A different effect of pollution on the molluscan kidney was reported by Gardner *et al.* (1988), who found that oysters, *Crassostrea virginica*, exposed experimentally to contaminated sediments from a Connecticut estuary developed renal neoplasms.

Effects of Pollutants on Cell and Tissue Metabolic Pathways

This category of effects has a significant degree of overlap with the previous one on filtration/detoxification systems, but is concerned more with consequences of abnormal levels of chemicals on the biochemistry of cells and tissues, as reflected by such abnormalities as some skeletal deformities in fish or anomalies in shell deposition in molluscs, in the presence of high levels of contaminants, such as tributyltin. Such abnormalities occur when filtration/ detoxification systems fail to cope with the influx of particular toxic chemicals, which may spill over into general circulation and affect cellular metabolic pathways. Examples of some gross effects are skeletal deformities in fish and shell abnormalities in bivalve molluscs.

Skeletal Deformities in Fish

Skeletal anomalies, particularly those of the spinal column, are commonly observed in fish and are the subject of extensive literature. Such anomalies may be genetic, resulting from mutations or recombinations; epigenetic, acquired during embryonic development; or postembryonic, acquired during larval or postlarval development (Hickey 1972). Evidence exists for a hereditary basis for some skeletal anomalies, but other evidence points to effects of environmental factors, such as, temperature, salinity, dissolved oxygen, radiation, dietary deficiencies, and toxic chemicals.

Several reports from Japan refer to high and increasing occurrences of skeletal anomalies in fish. Komada (1974) and Ueki and Sugiyama (1976) observed increasing numbers of malformed sweetfish or ayu, *Plecoglossus altivelis*, in rivers and culture farms. Skeletal abnormalities in mullet and eight other species from the Inland Sea of Japan were reported by Matsusato (1973). In a recent study, Matsusato (Dr. T. Matsusato, National Institute of Aquaculture, Tamaki-cho, Mie-ken, Japan 519-04 (personal communication, 1983) found increasing prevalences of two skeletal syndromes (bent spines and fused vertebrae) in an estuary (Kurose River) contaminated upstream with chlorinated hydrocarbons and organophosphates. His sample consisted of 28,000 fish of 68 species. Matsusato (1986) also summarized all reports of skeletal anomalies (especially spinal fractures) in wild fish of Japan, concluding that occurrences were nationwide, with highest prevalences in agricultural rather than industrial areas, possibly because of pesticide contamination of habitats.

Deformed fin rays and associated skeletal abnormalities have been observed in winter flounders from the highly polluted waters of the New York Bight (Ziskowski *et al.* 1980). Observations on skeletal anomalies and related developmental defects have been summarized (Sindermann *et al.* 1978). Related anomalies, in the form of disruption in normal scale patterns and even scale reversal, have been noted in samples from polluted waters of Biscayne Bay, FL (W. Kandrashoff, Port Salermo, FL, unpublished data). Presence and frequency of such scale pattern anomalies may, with skeletal abnormalities, be good indicators of the extent of environmental degradation (Sindermann *et al.* 1980).

Experimental evidence exists for induction of skeletal abnormalities by exposure to environmental contaminants. Couch *et al.* (1977) reported severe scoliosis and associated pathology in the sheepshead minnow, *Cyprinodon variegatus*, exposed to the organochloride pesticide Kepone®. Authors concluded that scoliosis was a secondary effect of Kepone toxicity, with the nervous system or calcium metabolism as primary target. Couch *et al.* (1979) also found that trifluralin induced extensive osseous hyperplasia in vertebrae of sheepshead minnows when life history stages from zygote to 28-day juveniles were exposed to 25-50 ppb trifluralin. Centra of vertebrae, thickened by active osteoblasts and fibroblasts, increased up to 10-30 times normal dimensions -- a striking sublethal effect. Bengtsson and Larsson (1986) found increased incidence of vertebral defects in sculpin exposed to effluents containing heavy metals. Disturbed plasma ionic balance was observed in deformed fish, and decreased collagen synthesis was proposed as a possible explanation for the abnormalities (Bengtsson *et al.* 1988).

Shell Abnormalities in Bivalve Molluscs

Pacific oysters, *Crassostrea gigas*, were introduced to the coast of France in the late 1960's and to the British east coast in the early 1970's. By mid-1970, reduced growth and malformed shells were apparent, and the cause was found to be organotin compounds in antifouling paints used on boats. Abnormally thickened shells in exposed oyster populations had an open laminar structure, a formation described as "chambering," sometimes with a proteinaceous gel secreted between layers. In a French study, abnormalities in shell calcification reached 90% during 1980-1982 in the Bay of Arcachon, but fell to 40% in 1983-1985, after use of tributyltin in antifouling paint was banned (Alzieu *et al.* 1982, 1986). Severity of deformations and tissue levels of tin also decreased during the same period, confirming chemical etiology of the condition. Extreme sensitivity of oysters to tributyltin is apparent from findings that concentrations of only 50 ng/l could induce shell malformations (Alzieu *et al.* 1986). In a British study, *C. gigas* spat cultured in presence of low environmentally feasible levels of the contaminant grew less well than controls and developed pronounced thickening and chambering of the upper shell valve (Waldock and Thain 1983). Higher levels of exposure resulted in increasing percentages of mortalities. [It seems relevant that in other studies, experimental exposure of crabs, *Uca pugilator*, to tributyltin retarded limb regeneration, delayed ecdysis, and produced deformities in regenerated appendages (Weis *et al.* 1987). Decreased survival and reduced growth of lobster and crab larvae also followed exposure to tributyltin (Laughlin and French 1980).]

Effects of Pollutants on Neurosensory Systems

It might be anticipated that neurosensory systems of fish, especially receptors, would be affected by exposure to polluted environments, and some evidence for this exists. Vulnerability to chemical toxicants has been demonstrated experimentally in olfactory and lateral line cells of marine teleosts. Gardner and LaRoche (1973) and Gardner (1975), exposed the killifish, *Fundulus heteroclitus*, and the silverside, *Menidia menidia*, to copper, silver, and mercury, causing severe neural pathology in the form of necrosis in cellular elements of the anterior lateral line system and severe degenerative changes in olfactory organs, especially necrosis of olfactory epithelium. Hara (1970, 1972) observed that olfactory responses of salmon were inhibited by exposure to copper and mercury.

As Gardner (1975) pointed out, effects of neurotoxic chemicals can be potentially lethal in polluted habitats. Loss of the olfactory sense could interfere with spawning migrations of anadromous species. Feeding behavior may be severely disrupted by impairment of lateral line and olfactory sensory capabilities. Additionally, toxic effects of pollutants on sensory organ systems, in Gardner's words, "...are significant even if they do not cause permanent neurological damages, for a temporary disability that prevents an organism from relating to a viable environment for only moments can be disastrous." This cogent observation can be extended to effects on larval survival. In one example, larval winter flounder, *Pseudopleuronectes americanus*, and haddock, *Melanogrammus aeglefinus*, exposed to copper developed moderate to severe olfactory lesions which could interfere with feeding and other behavior (Bodammer 1981).

Petroleum hydrocarbons also have been shown to damage the olfactory epithelium of fish. Variable pathological changes were observed in the silverside (Gardner 1975). Exposure to crude oil and its fractions caused hyperplasia, metaplasia, and dilation and congestion of blood vessels. Hawkes (1980) observed degenerative changes in the chemosensory cilia of the olfactory organs of larval sand sole, *Psettichthys melanostictus*, exposed to the water-soluble fraction of crude oil. According to the author, "the degree of structural alteration observed indicated severe damage to the receptor organelles."

Damage to fish eye lens tissue can also occur from exposure to petroleum hydrocarbons. Long-term (6-month) exposure of cunner, *Tautogolabrus adspersus*, to an oil slick, and a long-term (8-month) diet of crude oil resulted in grossly observable lens opacity in rainbow trout, *Oncorhynchus mykiss* (Hawkes 1977; Payne *et al.* 1978). Hydration of lens fiber cells of trout, thought to indicate early stages in cataract formation, were reported by Hawkes (1980) in trout exposed to petroleum.

In field studies of other species, Bender *et al.* (1988) observed cataracts in fish from heavily contaminated areas of the Elizabeth River in Virginia. Prevalences in the most contaminated zones were 21% in weakfish, *Cynoscion regalis*; 18% in croaker, *Micropogonias undulatus*; and 10% in spot, *Leiostomus xanthurus*. As an indication of the severity of contamination in those zones, fin erosion was seen in 30% of toadfish, *Opsanus tau*, and 11% of hogchokers, *Trinectes maculatus*. The abnormalities were reported to increase in frequency at stations whose sediments contained high levels of PAHs (Huggett *et al.* 1987), and cataracts were induced by experimental exposure of spot to PAH-contaminated Elizabeth River sediments (Hargis and Zwerner 1989).

Neural damage in the brain and retina of embryos of surf smelts, *Hypomesus pretiosus*, resulted from experimental exposure to crude oil fractions (Hawkes and Stehr 1982). Necrotic neurons were seen in the developing forebrain and the neuronal layer of the retina of 27-day embryos after repeated exposures. Damage to retinal receptor cells took the form of cytoplasmic vacuolation and lysis of mitochondria. Mitochondrial disruption in brain cells of larval Pacific herring, *Clupea harengus pallasi*, also resulted from exposure to crude oil (Cameron and Smith 1980).

As pointed out by Hawkes and Stehr (1982), neural tissues of some fish seem to be particularly affected by certain petroleum hydrocarbons; and fish brain tissue can bioaccumulate relatively high levels. As supporting evidence, they cite the work of Roubal *et al.* (1977) and Collier *et al.* (1980), reporting the sequestering of benzene, naphthalene, methylnaphthalene, and anthracene in the brains of rainbow trout and coho salmon, *Oncorhynchus kisutch*.

Optic and forebrain malformations have been reported in numerous studies of developing fish embryos and larvae. The extent and nature of abnormalities have been summarized in an excellent recent review by Weis and Weis (1987). The range of anomalies includes disorganized retinas, abnormal pigment distribution, invasive blood sinuses, protruding lenses, microphthalmia, anophthalmia, and cyclopia, after exposures to pollutants, such as crude oil and its fractions, oil dispersants, insecticides, and methyl mercury. The conclusions of Weis and Weis are highly relevant here: 1) pollutants can disrupt metabolic processes at critical developmental stages, causing morphological abnormalities; 2) pollutants are relatively nonspecific as teratogens, so that many different classes may elicit the same developmental responses; and 3) environmental conditions can be critical, so that interactions of pollutants with natural environmental factors (and with other pollutants) must be examined.

CONCLUSIONS

Based on published information, and recognizing the need for greater depth in many investigations, it is possible to propose some general conclusions. Disease is a significant limiting factor in estuarine/coastal populations of fish and shellfish; its effects may be enhanced by stresses resulting from abnormal environmental conditions, including (but not limited to) increases in population of facultative pathogens and toxic chemical concentrations that may affect metabolism.

The critical role of environmental stressors, particularly toxic pollutants, in disease is becoming more and more obvious, and the pathways leading to pathology are receiving more attention. Most disease conditions considered in this paper are indicators of stress. In fish, fin erosion, ulcerations, and decreased resistance to facultative pathogens are the indicators. In crustaceans, shell disease and black gills are signs; and in bivalve molluscs, poor growth and condition, mantle recession, and kidney concretions may give clues. This list of stress-induced abnormalities, which are consequences of disturbed metabolism, may be augmented by other conditions, such as, liver neoplasia in fish and shell abnormalities in bivalves, that seem related more directly to effects of toxic levels of contaminants in the immediate environment, but are of course still indicators of disturbed metabolism.

Effects of contaminants on living organisms are fundamentally biochemical, resulting in malfunctions in cellular metabolic processes and eventually in morphological/physiological changes that become apparent at higher levels of organization and are identified as "disease" (Fig. 16-6).

During the past two decades, important new findings have added significantly to information about pollution and disease. Examples are: 1) Tumors and other liver lesions have been reported from flatfish sampled in grossly polluted estuarine locations in the U.S. and in Europe. 2) Genetic abnormalities in developing fish embryos and morphological abnormalities in larvae seem related in several studies to the extent of chemical pollution. 3) Shell disease and associated "black gill disease" of crustaceans have been reported with higher prevalences and greater severity in polluted habitats. 4) Ulcerations in fish have been reported with increasing frequency in many parts of the world, and an association with environmental stressors has been hypothesized frequently.

Because complex mixtures of chemical contaminants occur in badly degraded waters, specific pathologies in fish and shellfish cannot often be associated with specific contaminants in a cause and effect relationship. This is, however, feasible in experimental populations, and some disease conditions, such as fin erosion in fish have been induced by laboratory exposures to contaminants. The problem has been stated precisely by Dethlefsen (1988): "Given the present knowledge of ecosystems interactions it is unrealistic to expect that marine or aquatic science will be able in the near future to produce results that unequivocally document a connection between specific dysfunctions and specific pollutants. Such causal relationships can be demonstrated only for substances

polluted areas. Pollution-associated disease conditions have been recognized and studied in estuarine/coastal environments that humans have helped to create -- the New York Bight apex, mouths of Elbe and Rhine rivers, the Duwamish River in Puget Sound, Tokyo Harbor, and the Houston ship canal, to mention a few. Decreases in contamination of these areas at some point should be reflected in a reduction in prevalence and severity of disease conditions described in this paper.

REFERENCES

Alzieu, C., M. Héral, Y. Thibaud, M.J. Dardignac, and M. Feuillet. 1982. Influence des peintures antisalissures a base d'organostanniques sur la calcification de la coquille de l'huitre *Crassostrea gigas*. *Rev. Trav. Inst. Peches Marit.* 44:301-349.

Alzieu, C., J. Sairjuan, J.P. Deltreil, and M. Borel. 1986. Tin contamination in Arcachon Bay: effects on oyster shell anomalies. *Mar. Pollut. Bull.* 17:494-498.

Anonymous. 1985. Report of the study group on biological effects techniques. *Int. Counc. Explor. Sea Doc. C.M.*1985/E:48, 3 pp.

Anonymous. 1989. Shell disease of crustaceans in the New York Bight. New York: U.S. Environmental Protection Agency (Region 2) and National Oceanic and Atmospheric Administration (NMFS and NOS), *Rep. Work. Group Shell Dis.* 87 pp.

Aoki, K., and H. Matsudaira. 1977. Induction of hepatic tumors in teleost (*Oryzias latipes*) after treatment with methylazoxymethanol acetate: brief communication. *J. Nat. Cancer Inst.* 59:1747-1749.

Aoki, K., and H. Matsudaira. 1981. Factors influencing tumorigenesis in the liver after treatment with methylazoxymethanol acetate in a teleost, *Oryzias latipes*, In *Phyletic Approaches to Cancer*, ed. C.J. Dawe, J.C. Harshbarger, S. Kondo, T. Sugimura, and S. Takayama. pp. 202-216. Tokyo: JPN. Sci. Soc. Press.

Aoki, K., and H. Matsudaira. 1984. Factors influencing methylazoxymethanol acetate initiation of liver tumors in *Oryzias latipes*: carcinogen dosage and time of exposure. *Nat. Cancer Inst. Monogr.* 65:345-351.

Becker, D.S., T.C. Ginn, M. Landolt, and D.B. Powell. 1987. Hepatic lesions in English sole (*Parophrys vetulus*) from Commencement Bay, Washington (USA). *Mar. Environ. Res.* 23:153-173.

Bender, M.E., W.J. Hargis, Jr., R.J. Huggett, and M.H. Roberts, Jr. 1988. Effects of polynuclear aromatic hydrocarbons on fishes and shellfish: an overview of research in Virginia. *Environ. Res.* 24:237-241.

Bengtsson, Å., B.E. Bengtsson, and G. Lithner. 1988. Vertebral defects in fourhorn sculpin, *Myoxocephalus quadricornis* L., exposed to heavy metal pollution in the Gulf of Bothnia. *J. Fish Biol.* 33:517-529.

Bengtsson, B.E., and Å. Larsson. 1986. Vertebral deformities and physiological effects in fourhorn sculpin (*Myoxocephalus quadricornis*) after long-term exposure to a simulated heavy metal-containing effluent. *Aquat. Toxicol.* 9:215-229.

Black, J.J., A.E. MacCubbin, and M. Schiffert. 1985. A reliable, efficient, microinjection apparatus and methodology for the *in vivo* exposure of rainbow trout and salmon embryos to chemical carcinogens. *J. Nat. Cancer Inst.* 75:1123-1128.

Bodammer, J.E. 1981. The cytopathological effect of copper on the olfactory organs of larval fish (*Pseudopleuronectes americanus* and *Melanogrammus aeglefinus*). *Int. Counc. Explor. Sea Doc. C.M.*1981/E:46, 11 pp.

Bucke, D., and S.W. Feist. 1984. Histological changes in livers of dab *Limanda limanda* L. *Int. Counc. Explor. Sea Doc.* C.M.1984/F:8, 6 pp.

Butler, P.A., R. Childress, and A.J. Wilson. 1972. The association of DDT residues with losses in marine productivity, In *Marine Pollution and Sea Life*, ed. M. Ruivo. pp. 262-266. London: Fish. News Ltd.

Cameron, J.A., and R.L. Smith. 1980. Ultrastructural effects of crude oil in early life stages of Pacific herring. *Trans. Am. Fish. Soc.* 109:224-228.

Cameron, P., H. Von Westernhagen, V. Dethlefsen, and D. Janssen. 1986. Chlorinated hydrocarbons in North Sea whiting (*Merlangius merlangus*) and effects on reproduction. *Int. Counc. Explor. Sea Doc.* C.M.1986/E:25, 10 pp.

Chang, S., and A.C. Longwell. 1984. Examining statistical associations of malformations, cytopathology and cytogenetic abnormality of Atlantic mackerel embryos with indicator levels of environmental contaminants in the New York Bight. *Int. Counc. Explor. Sea Doc.* C.M.1984/E:11, 9 pp.

Christensen, N.O. 1981. The use of fish diseases in marine pollution effects monitoring programmes. *Bull. Eur. Assoc. Fish Pathol.* 1:7-9.

Collier, T.K., M.M. Krahn, and D.C. Malins. 1980. The disposition of naphthalene and its metabolites in the brain of rainbow trout (*Salmo gairdneri*). *Environ. Res.* 23:35-41.

Couch, J.A. 1975. Histopathologic effects of pesticides and related chemicals on the livers of fishes, In *The Pathology of Fishes*, ed. W.E. Ribelin and G. Migaki. pp. 559-584. Madison, WI: Univ. Wisconsin Press.

Couch, J.A. 1991. Spongiosis hepatis: chemical induction, pathogenesis, and possible neoplastic fate in a teleost fish model. *Toxicol. Pathol.* 19:237-250.

Couch, J.A., and L.A. Courtney. 1985. Attempts to abbreviate time to endpoint in fish hepatocarcinogenesis assays, In *Water Chlorination, Environmental Impact and Health Effects*, ed. R. Jolley, W. Brungs, and R. Cummings. pp. 156-173. Ann Arbor, MI: Ann Arbor Science.

Couch, J.A., and L.A. Courtney. 1987. N-nitrosodiethylamine-induced hepatocarcinogenesis in estuarine sheepshead minnow (*Cyprinodon variegatus*): neoplasms and related lesions compared with mammalian lesions. *J. Nat. Cancer Inst.* 79:297-321.

Couch, J.A., and J.C. Harshbarger. 1985. Effects of carcinogenic agents on aquatic animals: an environmental and experimental overview. *Environ. Carcinog. Rev.* 3:63-105.

Couch, J.A., J.T. Winstead, and L.R. Goodman. 1977. Kepone-induced scoliosis and its histological consequences in fish. *Science* 197:585-587.

Couch, J.A., J.T. Winstead, D.J. Hansen, and L.R. Goodman. 1979. Vertebral dysplasia in young fish exposed to the herbicide trifluralin. *J. Fish Dis.* 2:35-42.

Cross, J.N. 1985. Fin erosion among fishes collected near a southern California municipal wastewater outfall (1971-82). *Fish. Bull.* 83:195-206.

Cross, J.N., and J.E. Hose. 1988. Evidence for impaired reproduction in white croaker (*Genyonemus lineatus*) from contaminated areas off southern California. *Mar. Environ. Res.* 24:185-188.

Dethlefsen, V. 1980. Observations on fish diseases in the German Bight and their possible relation to pollution. *Rapp. P.-v. Réun. Cons. Int. Explor. Mer* 179:110-117.

Dethlefsen, V. 1984. Diseases in North Sea fishes. *Helgol. Meeresunters.* 37:353-374.

Dethlefsen, V. 1985. Krankheiten von Nordseefischen als Ausdruck der Gewässerbelastung. *Abh. Naturwiss. Ver. Bremen* 40:233-252.

Dethlefsen, V. 1988. Status report on aquatic pollution problems in Europe. *Aquat. Toxicol.* 11:259-286.

Dethlefsen, V., and T. Lang. 1988. Skeletal deformities of cod (*Gadus morhua* L.) in the German Bight and the southern North Sea. *Int. Counc. Explor. Sea Doc.* C.M.1988/E:24, 37 pp.

Dethlefsen, V., P. Cameron, and H. Von Westernhagen. 1985. Untersuchungen über die Häufigkeit von Missbildungen in Fischembryonen der südlichen Nordsee. *Inf. Fischwiss.* 32:22-27.

Dethlefsen, V., P. Cameron, A. Berg, and H. Von Westernhagen. 1986. Malformations of embryos of spring spawning fishes in the southern North Sea. *Int. Counc. Explor. Sea Doc. C.M.*1986/E:21, 13 pp.

Dethlefsen, V., P. Cameron, H. Von Westernhagen, and D. Janssen. 1987. Morphologische und chromosomale Untersuchungen an Fischembryonen der südlichen Nordsee in Zusammenhang mit der Organochlor- kontamination der Elterntiere. *Veröff. Inst. Küsten-Binnenfisch.* 97:1-57.

Doyle, L.J., N.J. Blake, C.C. Woo, and P.P. Yevich. 1978. Recent biogenic phosphorite: concretions in mollusk kidneys. *Science* 199:1431-1433.

Egami, N., Y. Kyono-Hamaguchi, H. Mitani, and A. Shima. 1981. Characteristics of hepatoma produced by treatment with diethylnitrosamine in the fish, *Oryzias latipes*. In *Phyletic Approaches to Cancer*, ed. C.J. Dawe, J.C. Harshbarger, S. Kondo, T. Sugimura, and S. Takayama. pp. 217-226. Tokyo: JPN. Sci. Press.

Ernst, V.V., and J.M. Neff. 1977. The effects of the water-soluble fractions of No. 2 fuel oil on the early development of the estuarine fish, *Fundulus grandis* Baird and Girard. *Environ. Pollut.* 14:25-35.

Estrella, B.T. 1984. Black gill and shell disease in American lobster (*Homarus americanus*) as indicators of pollution in Massachusetts Bay and Buzzards Bay, Massachusetts. Massachusetts Department of Fish and Wildlife Recreation Vehicle Division, Marine Fishery Publication, No. 14049-19-125-5-85-C.R., 17 pp.

Falkmer, S., S. O. Emdin, Y. Östberg, A. Mattisson, M.L. Johansson Sjobeck, and R. Fänge. 1976. Tumor pathology of the hagfish, *Myxine glutinosa*, and the river lamprey, *Lampetra fluviatilis*. A light-microscopical study with particular reference to the occurrence of primary liver carcinoma, islet-cell tumors, and epidermoid cysts of the skin. *Progr. Exp. Tumor Res.* 20:217-250.

Falkmer, S., S. Marklund, P.E. Mattsson, and C. Rappe. 1977. Hepatomas and other neoplasms in the Atlantic hagfish (*Myxine glutinosa*): a histopathologic and chemical study. *Ann. N.Y. Acad. Sci.* 298:342-355.

Fänge, R., C. Rappe, and S. Falkmer. 1975. Liver tumors in the Atlantic hagfish (*Myxine glutinosa* L.), In *Sublethal Effects of Toxic Chemicals on Aquatic Animals*, ed. J.H. Koeman and J.J.T.W.A. Strik. pp. 103-109. Amsterdam: Elsevier.

Gardner, G. R. 1975. Chemically induced lesions in estuarine or marine teleosts, In *The Pathology of Fishes*, ed. W.E. Ribelin and G. Migaki. pp. 57-691. Madison, WI: Univ. Wisconsin Press.

Gardner, G.R., and G. LaRoche. 1973. Copper induced lesions in estuarine teleosts. *J. Fish. Res. Board Can.* 30:363-368.

Gardner, G.R., P.P. Yevich, and J.C. Harshbarger. 1988. Neoplastic disorders in American oysters (*Crassostrea virginica*) exposed to contaminated sediment in the laboratory and in the field. Abstract. San Diego, CA: *Proc. Soc. Invertebr. Pathol.*

Gold, K., G. Capriulo, and K. Keeling. 1982. Variability in the calcium phosphate concretion load in the kidney of *Mercenaria mercenaria*. *Mar. Ecol. Prog. Ser.* 10:97-99.

Hansen, P.-D., H. Von Westernhagen, and H. Rosenthal. 1985. Chlorinated hydrocarbons and hatching success in Baltic herring spring spawners. *Mar. Environ. Res.* 15:59-76.

Hara, T.J. 1970. An electrophysiological basis for olfactory discrimination in homing salmon: a review. *J. Fish. Res. Board Can.* 27:565-586.

Hara, T.J. 1972. Electrical responses of the olfactory bulb of Pacific salmon *Oncorhynchus nerka* and *Oncorhynchus kisutch*. *J. Fish. Res. Board Can.* 29:1351-1355.

Hargis, W.J., and D.E. Zwerner. 1989. Some effects of sediment-borne contaminants on development and cytomorphology of teleost eye-lens epithelial cells and their derivatives. *Mar. Environ. Res.* 28:399-405.

Hawkes, J.W. 1977. The effects of petroleum hydrocarbon exposure on the structure of fish tissues, In *Fate and Effects of Petroleum Hydrocarbons in Marine Ecosystems and Organisms*, ed. D.A. Wolfe. pp. 115-128. New York: Pergamon Press.

Hawkes, J.W. 1980. The effects of xenobiotics on fish tissues: morphological studies. *Fed. Proc.* 39: 3230-3236.

Hawkes, J.W., and C.M. Stehr. 1982. Cytopathology of the brain and retina of embryonic surf smelt (*Hypomesus pretiosus*) exposed to crude oil. *Environ. Res.* 27:164-178.

Hawkins, W.E., R.M. Overstreet, J.W. Fournie, and W.W. Walker. 1985a. Development of aquarium fish models for environmental carcinogenesis: tumor induction in seven species. *J. Appl. Toxicol.* 5:261-264.

Hawkins, W.E., R.M. Overstreet, W.W. Walker, and C.S. Manning. 1985b. Tumor induction in several small fish species by classical carcinogens and related compounds, In *Water Chlorination, Environmental Impact and Health Effects*, ed. R. Jolley, W. Brungs, and R. Cummings. pp. 429-438. Ann Arbor, MI: Ann Arbor Science.

Hawkins, W.E., J.W. Fournie, R.M. Overstreet, and W.W. Walker. 1986. Intraocular neoplasms induced by methylazoxymethanol acetate in Japanese medaka (*Oryzias latipes*). *J. Nat. Cancer Inst.* 76:453-465.

Hawkins, W.E., R.M. Overstreet, and W.W. Walker. 1988a. Carcinogenicity tests with small fish species. *Aquat. Toxicol.* 11:113-128.

Hawkins, W.E., W.W. Walker, R.M. Overstreet, T.F. Lytle, and J.S. Lytle. 1988b. Dose-related carcinogenic effects of water-borne benzo[a]pyrene on livers of two small fish species. *Ecotoxicol. Environ. Saf.* 16:219-231.

Hendricks, J.D. 1982. Chemical carcinogens in fish, In *Aquatic Toxicology*, ed. L.J. Weber. pp. 149-211. New York: Raven Press.

Hendricks, J.D., T.R. Meyers, and D.W. Shelton. 1984. Histological progression of hepatic neoplasia in rainbow trout (*Salmo gairdneri*). *Nat. Cancer Inst. Monogr.* 65:321-336.

Hendricks, J.D., T.R. Meyers, D.W. Shelton, J.L. Casteel, and G.S. Bailey. 1985. Hepatocarcinogenicity of benzo[a]pyrene to rainbow trout by dietary exposure and intraperitoneal injection. *J. Nat. Cancer Inst.* 74:839-851.

Hickey, C.R., Jr. 1972. Common abnormalities in fishes, their causes and effects. *N.Y. Ocean Sci. Lab. Tech. Rep.* 0013, 20 pp.

Huggett, R.J., M.E. Bender, and M.A. Unger. 1987. Polynuclear aromatic hydrocarbons in the Elizabeth River, Virginia, In *Fate and Effects of Sediment Bound Chemicals in Aquatic Systems*, ed. K.L. Dickson, A.W. Maki, and W. Brungs. pp. 327-341. New York: Pergamon Press.

Hughes, J.B., D.A. Nelson, D.M. Perry, J.E. Miller, G.R. Sennefelder, and J.J. Periera. 1986. Reproductive success of the winter flounder (*Pseudopleuronectes americanus*) in Long Island Sound. *Int. Counc. Explor. Sea Doc. C.M.*1986/E:10, 11 pp.

Ishikawa, T., and S. Takayama. 1979. Importance of hepatic neoplasms in lower vertebrate animals as a tool in cancer research. *J. Toxicol. Environ. Health* 5:537-550.

Ishikawa, T., T. Shimanine, and S. Takayama. 1975. Histological and electron microscopy observations on diethylnitrosamine-induced hepatomas in small aquarium fish (*Oryzias latipes*). *J. Nat. Cancer Inst.* 55:906-916.

Jeffries, H.P. 1972. A stress syndrome in the hard clam, *Mercenaria mercenaria*. *J. Invertebr. Pathol.* 20:242-251.

Kimura, I., and M. Ando. 1986. Fish neoplasms and environmental chemicals. *J. Hyg. Chem.* 32:317-334. (In Japanese)

Kimura, I., N. Taniguchi, H. Kumai, I. Tomita, N. Kinae, K. Yoshizaki, M. Ito, and T. Ishikawa. 1984. Correlation of epizootiological observations with experimental data: chemical induction of chromatophoromas in the croaker, *Nibea mitsukurii*. *Nat. Cancer Inst. Monogr.* 65:139-154.

Kimura, I., M. Ando, Y. Wakamatsu, K. Ozato, N. Kinae, and J.C. Harshbarger. 1986. Chemical induction of pigment cell neoplasm and hyperplasia in fishes, In *Structure and Function of Melanin, Vol. 3*, ed. K. Jimbow. pp. 82-94. Sapporo: Fuji-Shoin Co.

Kinne, O., and H. Rosenthal. 1967. Effects of sulfuric water pollutants on fertilization, embryonic development and larvae of the herring, *Clupea harengus*. *Mar. Biol.* (Berl.) 1:65-83.

Klauda, R., T. Peck, and G. Rice. 1981. Accumulation of polychlorinated biphenyls in the Atlantic tomcod (*Microgadus tomcod*) collected from the Hudson River estuary in New York. *Bull. Environ. Contam. Toxicol.* 27:829-835.

Koenig, C.C., and M.P. Chasar. 1984. Usefulness of the hermaphroditic marine fish, *Rivulus marmoratus*, in carcinogenicity testing. *Nat. Cancer Inst. Monogr.* 65:15-33.

Köhler, A., and F. Hölzel. 1980. Investigation on health conditions of flounder and smelt in the Elbe estuary. *Helgol. Meeresunters.* 33:401-414.

Komada, N. 1974. Studies on abnormality of bones in anomalous "ayu," *Plecoglossus altivelis*. *Gyobo Kenkyu* 8:127-135. (In Japanese, Engl. summary)

Krahn, M.M., L.D. Rhodes, M.S. Myers, I.K. Moore, W.D. MacLeod, Jr., and D.C. Malins. 1986. Associations between metabolites of aromatic compounds in bile and the occurrence of hepatic lesions in English sole (*Parophrys vetulus*) from Puget Sound, Washington. *Arch. Environ. Contam. Toxicol.* 15:61-67.

Kranz, H., and N. Peters. 1985. Pathological conditions in the liver of ruffe, *Gymnocephalus cernua* (L.) from the Elbe estuary. *J. Fish Dis.* 8:13-24.

Lamolet, L., C. Chepeau, and B. Rea. 1976. Etude sur les pollutions le long du littoral francais de sud de la mer du nord. Dunkerque, France: *Inst. Sci. Tech. Peches Marit.* 81 pp.

Laughlin, R.B., and W.J. French. 1980. Comparative study of the acute toxicity of a homologous series of trialkyltins to larval shore crabs, *Hemigrapsus nudus*, and lobster, *Homarus americanus*. *Bull. Environ. Contam. Toxicol.* 25:802-809.

Linden, O. 1976. The influence of crude oil and mixtures of crude oil/dispersants on the ontogenetic development of the Baltic herring, *Clupea harengus membras* L. *Ambio* 5:136-140.

Lindesjöö, E., and J. Thulin. 1990. Fin erosion of perch *Perca fluviatilis* and ruffe *Gymnocephalus cernua* in a pulp mill effluent area. *Dis. Aquat. Org.* 8:119-126.

Longwell, A.C. 1976. Chromosome mutagenesis in developing mackerel eggs sampled from the New York Bight. U.S. Dept. Commerce, *NOAA Tech. Memo. No.* ERL MESA-7, 61 pp.

Longwell, A.C. 1977. Genetic effects, In *The Argo Merchant Oil Spill and the Fishery Resources of Nantucket Shoals and Georges Bank: A Summary of Assessment Activities and Preliminary Results*. pp. 43-47. U.S. Dept. Commerce, Nat. Mar. Fish. Serv., Narragansett Lab. Rep. No. 77-10.

Longwell, A.C. 1978. Field and laboratory measurements of stress response at the chromosome and cell levels in planktonic fish eggs and the oil problem, In *In the Wake of the Argo Merchant*. pp. 116-125. Kingston, RI: Univ. Rhode Island, Center for Ocean Management Studies.

Longwell, A.C., and J.B. Hughes. 1980. Cytologic, cytogenetic, and developmental state of Atlantic mackerel eggs from sea surface waters of the New York Bight, and prospects for biological effects monitoring with ichthyoplankton. *Rapp. P.-v. Réun. Cons. Int. Explor. Mer* 179:275-291.

Longwell, A.C., D. Perry, J.B. Hughes, and A. Herbert. 1983. Frequencies of micronuclei in mature and immature erythrocytes of fish as an estimate of chromosome mutation rates - results of field surveys on windowpane flounder, winter flounder and Atlantic mackerel. *Int. Counc. Explor. Sea Doc. C.M.*1983/E:55.

Longwell, A.C., D. Perry, J.B. Hughes, and A. Herbert. 1984. Embryological, cytopathological and cytogenetic analysis of '74, '77 and '78 planktonic Atlantic mackerel eggs in the New York Bight. *Int. Counc. Explor. Sea Doc. C.M.*1984/E:13.

Mahoney, J.B., F.H. Midlige, and D.G. Deuel. 1973. A fin rot disease of marine and euryhaline fishes in the New York Bight. *Trans. Am. Fish. Soc.* 102:596-605.

Malins, D.C., M.S. Myers, and W.T. Roubal. 1983. Organic free radicals associated with idiopathic liver lesions of English sole (*Parophrys vetulus*) from polluted marine environments. *Environ. Sci. Technol.* 17:679-685.

Malins, D.C., B.B. McCain, D.W. Brown, S.-L. Chan, M.S. Myers, J.T. Landahl, P.G. Prohaska, A.J. Friedman, L.D. Rhodes, D.G. Burrows, W.D. Gronlund, and H.O. Hodgins. 1984. Chemical pollutants in sediments and diseases of bottom-dwelling fish in Puget Sound, Washington. *Environ. Sci. Technol.* 18:705-713.

Malins, D.C., B.B. McCain, D.W. Brown, M.S. Myers, M.M. Krahn, and S.-L. Chan. 1987. Toxic chemicals, including aromatic and chlorinated hydrocarbons and their derivatives, and liver lesions in white croaker (*Genyonemus lineatus*) from the vicinity of Los Angeles. *Environ. Sci. Technol.* 21:765-770.

Malins, D.C., B.B. McCain, J.T. Landahl, M.S. Myers, M.M. Krahn, D.W. Brown, S.-L. Chan, and W.T. Roubal. 1988. Neoplastic and other diseases in fish in relation to toxic chemicals: an overview. *Aquatic Toxicol.* 11:43-67.

Matsusato, T. 1973. On the skeletal abnormalities in marine fishes. I. The abnormal marine fishes collected along the coast of Hiroshima. *Bull. Nansei Reg. Fish. Res. Lab.* 6:17-68. (In Japanese; Engl. summary)

Matsusato, T. 1978. Skeletal abnormalities of marine fishes found in natural waters, and their value as an environmental index. *Bull. Nansei Reg. Fish. Res. Lab.* 53:183-252. (In Japanese)

Matsusato, T. 1986. Studies on the skeletal anomaly of fishes. *Bull. Nat. Inst. Aquacult.* 10:57-179. (In Japanese)

Mauri, M., and E. Orlando. 1982. Experimental study on renal concretions in the wedge shell *Donax trunculus* L. *J. Exp. Mar. Biol. Ecol.* 63:47-57.

McCain, B.B., K.V. Pierce, S.R. Wellings, and B.S. Miller. 1977. Hepatomas in marine fish from an urban estuary. *Bull. Environ. Contam. Toxicol.* 18:1-2.

McCain, B.B., M. Myers, U. Varanasi, D.W. Brown, L.D. Rhodes, W.D. Gronlund, D.G. Elliot, W.S. Palsson, H.O. Hodgins, and D.C. Malins. 1982. Pathology of two species of flatfish from urban estuaries in Puget Sound. *U.S. Environmental Protection Agency*, EPA-600/7-82-001, 100 pp.

Mearns, A.J., and M.J. Sherwood. 1974. Environmental aspects of fin erosion and tumors in southern California Dover sole. *Trans. Am. Fish. Soc.* 103:799-810.

Mix, M.S. 1986. Cancerous diseases in aquatic animals and their association with environmental pollutants: a critical literature review. *Mar. Environ. Res.* 20:1-136.

Murchelano, R.A. 1975. The histopathology of fin rot disease in winter flounder from the New York Bight. *J. Wildl. Dis.* 11:263-268.

Murchelano, R.A. 1982. Some pollution-associated diseases and abnormalities of marine fish and shellfish: a perspective for the New York Bight, In *Ecological Stress and the New York Bight: Science and Management*, ed. G.F. Mayer. pp. 327-346. Columbia, SC: Estuarine Research Federation.

Murchelano, R.A., and R.E. Wolke. 1985. Epizootic carcinoma in the winter flounder, *Pseudopleuronectes americanus*. *Science* 228:587-589.

Myers, M.S., L.D. Rhodes, and B.B. McCain. 1987. Pathologic anatomy and patterns of occurrence of hepatic neoplasms, putative preneoplastic lesions, and other idiopathic conditions in English sole (*Parophrys vetulus*) from Puget Sound, Washington. *J. Nat. Cancer Inst.* 78:333-363.

Nakai, Z., M. Kosaka, S. Kudoh, A. Nagai, F. Hayashida, T. Kubota, M. Ogura, T. Mizushima, and I. Uotani. 1973. Summary report on marine biological studies of Suruga Bay accomplished by Tokai University 1964-72. *J. Fac. Mar. Sci. Technol. Tokai Univ.* 7:63-117.

O'Connor, J.S., J.J. Ziskowski, and R.A. Murchelano. 1987. Index of pollutant-induced fish and shellfish disease. U.S. Dept. Commerce, *NOAA Spec. Rep.*, 29 pp.

Overstreet, R.M. 1988. Aquatic pollution problems, southeastern U.S. coasts: histopathological indicators. *Aquat. Toxicol.* 11:213-239.

Overstreet, R.M., and H.D. Howse. 1977. Some parasites and diseases of estuarine fishes in polluted habitats of Mississippi. *Ann. N.Y. Acad. Sci.* 298:427-462.

Passino, D.R.M. 1984. Biochemical indicators of stress in fishes: an overview, In *Contaminant Effects on Fisheries*, ed. V.W. Cairns, P.V. Hodson, and J.O. Nriagu. pp. 37-50. New York: John Wiley & Sons.

Payne, J.F., J.W. Kiceniuk, W.R. Squires, and G.L. Fletcher. 1978. Pathological changes in a marine fish after a 6-month exposure to petroleum. *J. Fish. Res. Board Can.* 35:665-667.

Perkins, E.J., J.R.S. Gilchrist, and O.J. Abbott. 1972. Incidence of epidermal lesions in fish of the north-east Irish Sea area, 1971. *Nature* (Lond.) 238:101-103.

Peters, N., A. Köhler, and H. Kranz. 1987. Liver pathology in fishes from the lower Elbe as a consequence of pollution. *Dis. Aquat. Org.* 2:87-97.

Pierce, K.V., B.B. McCain, and S.R. Wellings. 1978. The pathology of hepatomas and other liver abnormalities in English sole (*Parophrys vetulus*) from the Duwamish River estuary, Seattle, Washington. *J. Nat. Cancer Inst.* 60:1445.

Pierce, K.V., B.B. McCain, and S.R. Wellings. 1980. Histopathology of abnormal livers and other organs of starry flounder *Platichthys stellatus* (Pallas) from the estuary of the Duwamish River, Seattle, Washington, U.S.A. *J. Fish Dis.* 3:81-91.

Rosenthal, H. 1971. Wirkungen von "Rotschlamm" auf Embryonen und Larven des Herings *Clupea harengus*. *Helgol. Meeresunters.* 22:366-376.

Rosenthal, H., and R. Stelzer. 1970. Wirkungen von 2,4- und 2,5-dinitrophenol auf die Embryonalentwicklung des Herings *Clupea harengus*. *Mar. Biol.* (Berl.) 5:325-336.

Rosenthal, H., and K.-R. Sperling. 1974. Effects of cadmium on development and survival of herring eggs, In *The Early Life History of Fish*, ed. J.H.S. Blaxter. pp. 383-396. Berlin: Springer-Verlag.

Rosenthal, H., and D.F. Alderdice. 1976. Sublethal effects of environmental stressors, natural and pollutional, on marine fish eggs and larvae. *J. Fish. Res. Board Can.* 33:2047-2065.

Rosenthal, H., M. McInerney-Northcott, C.J. Musial, J.F. Uthe, and J.D. Castell. 1986. Viable hatch and organochlorine contaminant levels in gonads of fall spawning Atlantic herring from Grand Manan, Bay of Fundy, Canada. *Int. Counc. Explor. Sea Doc. C.M.*1986/E:26, 26 pp.

Roubal, W.T., T.K. Collier, and D.C. Malins. 1977. Accumulation and metabolism of carbon-14 labelled benzene, naphthalene, and anthracene by young coho salmon (*Oncorhynchus kisutch*). *Arch. Environ. Contam. Toxicol.* 5:513-529.

Sindermann, C.J. 1972. Some biological indicators of marine environmental degradation. *J. Wash. Acad. Sci.* 62:184-189.

Sindermann, C.J. 1976. Effects of coastal pollution on fish and fisheries—with particular reference to the Middle Atlantic Bight, In *Middle Atlantic Continental Shelf and the New York Bight*, ed. M.G. Gross. pp. 281-301. Lawrence, KS: *Am. Soc. Limnol. Oceanogr., Spec. Symp. 2.*

Sindermann, C.J. 1977. Recent data on possible associations of coastal/estuarine pollution with fish and shellfish diseases. *Int. Counc. Explor. Sea Doc. C.M.*1977/E:14, 33 pp.

Sindermann, C.J. 1979. Pollution-associated diseases and abnormalities of fish and shellfish: a review. *Fish. Bull.* 76:717-749.

Sindermann, C.J. 1983. An examination of some relationships between pollution and disease. *Rapp. P.-v. Réun. Cons. Int. Explor. Mer* 182:37-43.

Sindermann, C.J. 1984. Fish and environmental impacts. *Arch. FischWiss.* 35:125-160.

Sindermann, C.J. 1988a. Biological indicators and biological effects of estuarine/coastal pollution. *Water Resour. Bull.* 24:931-939.

Sindermann, C.J. 1988b. Epizootic ulcerative syndromes in coastal/estuarine fish. U.S. Dept. Commerce, *Nat. Mar. Fish. Serv., NOAA Tech. Memo.* NMFS-F/NEC-54, 37 pp.

Sindermann, C.J. 1989. The shell disease syndrome in marine crustaceans. U.S. Dept. Commerce, *Nat. Mar. Fish. Serv., NOAA Tech. Memo.* NMFS-F/NEC-64, 43 pp.

Sindermann, C.J., J.J. Ziskowski, and V.T. Anderson, Jr. 1978. A guide for the recognition of some disease conditions and abnormalities in marine fish. U.S. Dept. Commerce. *Nat. Mar. Fish. Serv., Tech. Ser. Rep.* 14, 65 pp.

Sindermann, C.J., F.E. Bang, N.O. Christensen, V. Dethlefsen, J.C. Harshbarger, J.R. Mitchell, and M.F. Mulcahy. 1980. The role and value of pathobiology in pollution effects monitoring programs. *Rapp. P.-v. Réun. Cons. Int. Explor. Mer* 179:135-151.

Smith, C.E., T.H. Peck, R.J. Klauda, and J.B. McLaren. 1979. Hepatomas in Atlantic tomcod *Microgadus tomcod* (Walbaum) collected in the Hudson River estuary in New York. *J. Fish Dis.* 2:313-319.

Smith, R.L., and J.A. Cameron. 1979. Effect of water soluble fraction of Prudhoe Bay crude oil on embryonic development of Pacific herring. *Trans. Am. Fish. Soc.* 108:70-75.

Smith, R.M. 1973. Pesticide residues as a possible factor in larval winter flounder mortality, In *Proceedings of a Workshop on Egg, Larval, and Juvenile Stages of Fish in Atlantic Coast Estuaries*, ed. A.L. Pacheco. pp. 173-180. Sandy Hook, NJ: U.S. Dept. Commerce, *Nat. Mar. Fish. Serv., NOAA-NMFS Tech. Publ.* No. 1.

Söderhäll, K. 1982. Prophenoloxidase activating system and melanization — a recognition mechanism of arthropods? A review. *Dev. Comp. Immunol.* 6:601-611.

Söderhäll, K. 1983. 1,3-glucan enhancement of protease activity in crayfish haemocyte lysate. *Comp. Biochem. Physiol. B* 74:221-224.

Söderhäll, K. 1986. The cellular immune system in crustaceans, In *Fundamental and Applied Aspects of Invertebrate Pathology*, ed. R.A. Samson, J.M. Vlak, and D. Peters. pp. 417-420. Wageningen, The Netherlands: Found. Fourth Int. Colloq. Invertebr. Pathol.

Söderhäll, K., and R. Ajaxon. 1982. Effect of quinones and melanin on mycelial growth of *Aphanomyces* spp. and extracellular protease of *Aphanomyces astaci* parasite on crayfish. *J. Invertebr. Pathol.* 39:105-109.

Söderhäll, K., and V.J. Smith. 1983. Separation of the haemocyte populations of *Carcinus maenas* and other marine decapods and prophenoloxidase distribution. *Dev. Comp. Immunol.* 7:229-239.

Söderhäll, K., and L. Häll. 1984. Lipopolysaccharide induced activation of prophenoloxidase activating system in crayfish haemocyte lysate. *Biochim. Biophys. Acta* 797:99-104.

Söderhäll, K., and V.J. Smith. 1986. The prophenoloxidase activating system: the biochemistry of its activation and role in arthropod cellular immunity with special reference to crustaceans, In *Immunity in Invertebrates*, ed. M. Brehelin. pp. 208-223. Berlin: Springer-Verlag.

Söderhäll, K., A. Wingren, M.W. Johansson, and K. Bertheussen. 1985. The cytotoxic reaction of hemocytes from the freshwater crayfish, *Astacus astacus*. *Cell. Immunol.* 94:326-330.

Southern California Costal Water Research Project. 1973. The ecology of the southern California bight: implications for water quality management. El Segundo, CA: *South. Calif. Coast. Water Res. Proj.*, Ref. No. SCCWRP TR 104.

Stich, H.F., A.B. Acton, K. Oishi, F. Yamazaki, T. Harada, T. Hibino, and H.G. Moser. 1977. Systematic collaborative studies on neoplasms in marine animals as related to the environment. *Ann. N.Y. Acad. Sci.* 298:374-388.

Stott, G.G., W.E. Haensly, J.M. Neff, and J.R. Sharp. 1983. Histopathologic survey of ovaries of plaice, *Pleuronectes platessa* L., from Aber Wrac'h and Aber Benoit, Brittany, France: long-term effects of the *Amoco Cadiz* crude oil spill. *J. Fish Dis.* 6:429-437.

Struhsaker, J.W. 1977. Effects of benzene (a toxic component of petroleum) on spawning Pacific herring, *Clupea harengeus pallasi*. *Fish. Bull.* 75:43-49.

Ueki, N., and T. Sugiyama. 1976. On abnormality of bones in artificially produced ayu, *Plecoglossus altivelis* (preliminary note). *Bull. Fish. Exp. Sta. Okayama Pref.* 50:360-362. (In Japanese)

Valentine, D.W. 1975. Skeletal anomalies in marine teleosts, In *The Pathology of Fishes*, ed. W.E. Ribelin and G. Migaki. pp. 695-718. Madison, WI: Univ. Wisconsin Press.

Vethaak, A.D. 1987. Fish diseases, signals for a diseased environment? In *Proceedings of the 2nd North Sea Seminar, Vol. 2*, ed. G. Peet. pp. 41-61. Rotterdam, Amsterdam: Werkgroep Nordzee.

Waldock, M.J., and J.E. Thain. 1983. Shell thickening in *Crassostrea gigas*: organotin antifouling or sediment induced? *Mar. Pollut. Bull.* 14:411-415.

Wedemeyer, G.A., D.J. McLeay, and C.P. Goodyear. 1984. Assessing the tolerance of fish and fish populations to environmental stress: the problems and methods of monitoring, In *Contaminant Effects on Fisheries*, ed. V.W. Cairns, P.V. Hodson, and J.D. Nriagu. pp. 163-195. New York: John Wiley & Sons.

Weis, J.S., and P. Weis. 1987. Pollutants as developmental toxicants in aquatic organisms. *Environ. Health Persp.* 71:77-85.

Weis, J.S., P. Weis, M. Heber, and S. Vaidya. 1981. Methylmercury tolerance of killifish (*Fundulus heteroclitus*) embryos from a polluted vs non-polluted environment. *Mar. Biol.* 65:283-287.

Weis, J.S., P. Weis, and M. Heber. 1982. Variation in response to methylmercury by killifish (*Fundulus heteroclitus*) embryos, In *Aquatic Toxicology and Hazard Assessment*, ed. J.G. Pearson, R. Foster, and W.F. Bishop. pp. 108-119. Washington, DC: Fifth Conf. Am. Soc. Test. Materials.

Weis, J.S., J. Gottlieb, and J. Kwiatkowski. 1987. Tributyltin retards regeneration and produces deformities of limbs in the fiddler crab, *Uca pugilator*. *Arch. Environ. Contam. Toxicol.* 16:321-326.

Wellings, S.R., C.E. Alpers, B.B. McCain, and B.S. Miller. 1976. Fin erosion disease of starry flounder (*Platichthys stellatus*) and English sole (*Parophrys vetulus*) in the estuary of the Duwamish River, Seattle, Washington. *J. Fish. Res. Board Can.* 33:2577-2586.

Westernhagen, H. Von. 1988. Sub-lethal effects of pollutants on fish eggs and larvae, In *Fish Physiology*, ed. W.S. Hoar and D.J. Randall. pp. 253-346. New York: Academic Press.

Westernhagen, H. Von, H. Rosenthal, and K.-R. Sperling. 1974. Combined effects of cadmium and salinity on development and survival of herring eggs. *Helgol. Meeresunters.* 26:416-433.

Westernhagen, H. Von, V. Dethlefsen, and H. Rosenthal. 1975. Combined effects of cadmium and salinity on development and survival of garpike eggs. *Helgol. Meeresunters.* 27:268-282.

Westernhagen, H. Von, H. Rosenthal, V. Dethlefsen, W. Ernst, U. Harms, and P.-D. Hansen. 1981. Bioaccumulating substances and reproductive success in Baltic flounder *Platichthys flesus*. *Aquat. Toxicol.* 1:85-99.

Westernhagen, H. Von, K.-R. Sperling, D. Janssen, V. Dethlefsen, P. Cameron, R. Kocan, M. Landolt, G. Fürstenberg, and K. Kremling. 1987. Anthropogenic contaminants and reproduction in marine fish. *Ber. Biol. Anst. Helgol.* 3:1-70.

Westernhagen, H. Von, V. Dethlefsen, P. Cameron, J. Berg, and G. Fürstenberg. 1988. Developmental defects in pelagic fish embryos from the western Baltic. *Helgol. Meeresunters.* 42:13-36.

Westin, D.T., C.E. Olney, and B.A. Rogers. 1985. Effects of parental and dietary organochlorines on survival and body burdens of striped bass larvae. *Trans. Am. Fish. Soc.* 114:125-136.

Young, J.S., and J.B. Pearce. 1975. Shell disease in crabs and lobsters from New York Bight. *Mar. Pollut. Bull.* 6:101-105.

Young, P.H. 1964. Some effects of sewer effluent on marine life. *Calif. Fish Game* 50:33-41.

Ziskowski, J.J., and R.A. Murchelano. 1975. Fin erosion in winter flounder. *Mar. Pollut. Bull.* 6:26-28.

Ziskowski, J.J., V.T. Anderson, and R.A. Murchelano. 1980. A bent fin ray condition in winter flounder (*Pseudopleuronectes americanus*) from Sandy Hook and Raritan Bays, New Jersey, and Lower Bay, New York. *Copeia* 1980:895-899.

17

Modulation of Nonspecific Immunity by Environmental Stressors

Robert S. Anderson

INTRODUCTION

Exposure of marine animals to sublethal concentrations of xenobiotics may alter homeostatic physiological mechanisms that may be expressed immediately or in the future. In the case of immunomodulation, changes in the organism's ability to detect and/or respond to foreign or "nonself" material, such as pathogens, neoplastic cells, parasites, may threaten its health and survival. Reports of increased incidence of infectious disease in fish taken from polluted environments support this concept. This chapter describes some mechanisms of immunological compromise that may be affected by environmental stressors.

Fish appear to possess lymphocyte subpopulations and immunoglobulins analogous to other vertebrates and thus show immunological specificity and memory. These phenomena are difficult to demonstrate in shellfish; however, a well-developed nonspecific, phagocyte-based defensive capability comprises the first line of defense in shellfish and finfish. Phagocytic cells not only recognize, engulf, and destroy microorganisms but also can be involved in cellular encapsulation, production of immunoregulatory molecules, participation as accessory cells in lymphocyte responses, and the processing and presentation of antigens. Polymorphonuclear cells and macrophages of fish and macrophage-like blood cells of bivalves are thought to play important roles in internal defense systems, regardless of the relative sophistication of the host's adaptive immune capabilities. Methodology available to study leukocyte function is extensive and continually developing, enabling more precise and quantitative data. Therefore, I have concentrated on our current understanding of effects of environmental factors on phagocytes of bivalve molluscs and fish and to indicate some possible

0-8493-8662-4/93/$0.00 + $.50

directions for future work. This is not intended to be a complete literature review; representative studies are cited to describe current areas of research.

BIVALVE STUDIES

Total and Differential Hemocyte Counts

Exposure of clams (*Mercenaria*) to >10 ppb phenol produced hemocyte lysis (Fries and Tripp 1980), as evidenced by the presence of free nuclei in the hemolymph and virtual elimination of fibrocytes that normally comprise about 5% of circulating hemocytes. They also found that granulocytes declined from ~80% to ~30% in a dose-dependent fashion with increasing phenol concentrations to 100 ppb; this effect was observed to at least 50 ppm. Ultrastructural studies showed that phenol exposure produced various morphological changes in granulocytes, whereas hyalinocytes retained their usual morphology. Apparent selective cytotoxicity of chemicals for certain hemocyte classes and their ability to affect important functions like phagocytosis clearly indicate the need for additional studies. For example, Balouet and Poder (1985) found elevated hemocyte counts in bivalves collected from the site of a major oil spill. Anderson (1981a), on the other hand, could find no significant changes in total or differential hemocyte counts in *Mercenaria* after exposure to sublethal concentrations of pentachlorophenol or hexachlorobenzene. Many important details concerning dose, route of exposure, and timing need to be addressed before the question of pollutant-specific effects can be resolved.

Mussels, *Mytilus edulis*, of uniform size were exposed to several concentrations of Prudhoe Bay crude oil emulsion and their tissues and hemocytes examined at intervals for 2 months (McCormick-Ray 1987). The study was conducted during the mussels' most metabolically active period in the summer. After 4-5 weeks in 740 μg/oil, mussels showed a reduced total hemocyte count due to a marked reduction in granulocytes. After the same period in 390 μg/l oil, reductions in total hemocyte count and granulocyte count were less impressive. After 2-month exposure to these oil concentrations, the total hemocyte and granulocyte counts recovered to normal or slightly above normal levels; however, significant reduction in numbers of small agranulocytes was seen. The agranulocytes typically make up only ~10% of the blood cells and are less phagocytically active than the granulocytes; therefore, it is possible that fluctuations of the granulocyte population are more important to the defensive capacity of this bivalve. However, after 2-month high-oil exposure, phagocytic response of hemocytes was reduced (not correlated with the relative numbers of granulocytes present).

Despite complexities of these data, it is possible that differences in circulating cell type profiles may indicate stress and have implications for monitoring.

Other investigators have focused on effects of heavy metal exposure on bivalve hemocyte numbers. As in the case of above-mentioned organics, responses were variable in the few available studies. Ruddell and Rains (1975) reported increased numbers of granulocytes in Cu^{2+}-exposed oysters. Pickwell and Steinert (1984) found increased levels of granulocytes, but a concurrent decline in numbers of circulating macrophages in mussels after Cu^{2+} exposure. However, exposure of mussels to a comparatively high dose of tributyltin did not produce this effect (Pickwell and Steinert 1988). Total and differential hemocyte counts in oysters, *Crassostrea virginica*, were determined after 1 day-2 week exposure to 1 ppm Cu^{2+} or Cd^{2+} (Cheng 1988a), a concentration shown not to influence phagocytosis *in vitro* (Cheng and Sullivan 1984); Cu^{2+} exposure had no effect on total cell count, hemopoiesis, or recruitment. The differential counts were affected by both metal treatments and, ultimately, Cu^{2+} reduced and Cd^{2+} increased the percentage of hyalinocytes. Since granulocytes are thought to be more actively phagocytic than hyalinocytes (Foley and Cheng 1975) and the primary source of bactericidal hydrolases in hemolymph (Cheng *et al.* 1975), Cd^{2+} could be considered immunosuppressive despite its stimulatory effect on total hemocyte numbers.

Suresh and Mohandas (1990a) quantified the circulating hemocytes in two clam species after exposure to sublethal concentrations of copper. In a marine species, *Sunetta scripta*, 1-5 ppm Cu^{2+} exposure had no effect on hemocyte numbers; but in the more estuarine clam, *Villorita cyprinoides*, 0.15-0.45 ppm Cu^{2+} generally depressed total hemocyte count. Differing results reflect trends in Cu^{2+} LD50 values in the two species and the influence of salinity on Cu^{2+} uptake rates. However, actual Cu^{2+} uptake rates by the animals and/or hemocytes were not measured. It is difficult to say whether observed decreases in hemocyte numbers resulted from cytotoxicity, diapedesis of metal-laden hemocytes out of bivalves, or lack of recruitment. However, at least theoretically, a reduction of circulating hemocytes could adversely affect immunocompetency in bivalve molluscs.

Molluscan hemocytes are also known to play an important role in detoxification of pollutants, such as heavy metals. Metals accumulate in large concretions (membrane-limited lipofuscin granules or tertiary lysosomes) in kidney epithelial cells (e.g., Rheinberger *et al.* 1979; George *et al.* 1982; George 1983). Seiler and Morse (1988) also observed that kidney cells in clams, *Mya arenaria*, collected from polluted sediments contained more of these granules than seen in animals from a comparatively pristine site. Further, numbers of circulating granulocytes were significantly elevated and bore larger inclusions in clams from the polluted site. The presence of particle-laden granulocytes in kidney blood spaces suggests that they may be an integral part of the pericardial gland-kidney excretory process. It is likely that hemocytes provide the means whereby pollutants are initially transported to the kidney. Pinocytosis of soluble factors from bivalve serum was first shown by Feng (1965). The ability of blood cells to take up particulate material from the hemolymph was demonstrated by Cuenot (1914) and Stauber (1950), and has been repeatedly confirmed. In addition to the postulated mechanism cited above where hemocytes actively remove particulate and dissolved foreign material from blood spaces and deliver these wastes to the kidney or other epithelial cells for eventual excretion, there is evidence that the cells themselves may cross the epithelium to release their contents outside the organism (e.g., Stauber 1950). These interesting phenomena will not be discussed here as the focus of this paper is immunotoxicology. However, it is clear that the study of effects of pollutant stress on total numbers and differential counts of bivalve hemocytes may provide information not only related to immunological status but also to other key physiological processes.

Interpretations of the effects of experimental manipulations as indicated by changed hemocyte counts in hemolymph samples must be made with caution. It is possible that considerable numbers of hemocytes may not be found in the systemic circulation at any particular time due to infiltration of tissues during inflammatory responses, variations in blood flow and strength of heartbeat, and involvement in diapedesis. Workers in this field also recognize the high variability in cell counts among individual bivalves of the same species, as well as the apparent variability in cell counts in repeat samples from the same individual. Analysis of changes in differential counts may provide more reliable information, but this presupposes consistent involvement, under experimental and control conditions, of the various cell types in tissue infiltration, avidity of attachment, diapedesis, and other functions.

Phagocytic Activity

Exposure of clams (*Mercenaria)* to >1 ppb phenol produced damage to gill and digestive tract epithelia, which may increase susceptibility to microbial infection and disease (Fries and Tripp 1976). Subsequently, it was shown that exposure to 10 ppb phenol substantially reduced numbers of phagocytic blood cells and the average number of yeast cells ingested per hemocyte (Fries and Tripp 1980). *Mercenaria* cells exposed to hexachlorobenzene (HCB) or pentachlorophenol (PCP) did not show significantly reduced phagocytosis of yeast; in fact, the opposite trend was observed after PCP exposure (Anderson 1981a). However, the specific phagocytic target particle used may be important in endocytosis studies. For example, reduced uptake and killing of bacteria by hemocytes from

Mercenaria were seen after exposure to either PCP or HCB (Anderson 1988a). More recently, hemocytes from oysters inhabiting a PAH-polluted environment also showed a reduced phagocytic capacity (Sami *et al.* 1990).

Effects of heavy metals on *in vitro* phagocytosis of polystyrene latex spheres by *C. virginica* hemocytes were measured by Cheng and Sullivan (1984). They found that exposure to 0.5-5.0 ppm Hg^{2+} suppressed phagocytosis; however, these concentrations were also reported to produce dose-dependent cytotoxicity. Certain metals did not affect phagocytosis at <5 ppm (Mn^{2+}, Cd^{2+}, Pb^{2+}, and Zn^{2+}), while others caused apparent stimulation of latex uptake at various concentrations (Co^{2+}, Cr^{3+}, Cu^{2+}, Fe^{3+}, Hg^{2+} at 0.1 ppm, and Sn^{2+}). Cheng (1988b) extended these studies by examining the effects of *in vivo* exposure of *C. virginica* to 1 ppm Cd^{2+} or Cu^{2+} on hemocytic phagocytosis. *In vitro*, this concentration had no effect on latex uptake (Cheng and Sullivan 1984). However, *in vivo*, the metal exposure of the hemocytes (i.e., the actual hemolymph level of metals) was unknown and test particles used for exposure studies were bacteria, *E. coli*, rather than latex beads. Although Cu^{2+} was reported to inhibit phagocytosis (the percentage of granulocytes ingesting bacteria) and Cd^{2+} stimulated phagocytosis, the same metal concentrations had an opposite effect on the mean number of bacteria taken up per granulocyte (Ca^{2+} stimulated, Cd^{2+} inhibited). If the cells destroy bacteria predominately by intracellular mechanisms, one may assume that total numbers taken up are more important in resistance to infection than the percentages of cells ingesting bacteria; therefore, the above results suggest that Cd^{2+} may mediate reduced defensive capacity.

Bivalve molluscs are subject to several putative neoplastic disorders, the most studied of which is hemic proliferative disease as it occurs in the soft shell clam, *Mya arenaria* (Brown *et al.* 1977; Cooper *et al.* 1982a, Cooper *et al.* 1982b; Farley *et al.* 1986) and the bay mussel, *Mytilus edulis* (Farley 1969; Cosson-Mannevy et al. 1984; Mix 1983; Elston *et al.* 1988). In both species the disease can be progressive, invasive, and ultimately fatal. In sharp contrast to normal hemocytes, transformed or neoplastic cells have enlarged nuclei, minimal cytoplasm, and reduced ability to spread on glass (Elston *et al.* 1988). A comparable condition characterized by proliferation of abnormal hemocytes has also been described in cockles and oysters of several species (Farley 1968; Alderman *et al.*1977; Twomey and Mulcahy 1984).

Various manifestations of immunological compromise have been associated with neoplasia in higher animals. It is likely that neoplastic transformation is not an uncommon event and that these cells are usually recognized as foreign and eliminated before they can exert any harmful effects. In some cases, defects in immunological recognition or effector mechanisms may allow the transformed cells to proliferate unchecked. Kent *et al.* (1989) described impaired defense mechanisms in *M. edulis* with hemic neoplasia. In mussels with advanced disease, the ability to clear injected bacteria (*Cytophaga* sp.) was significantly impaired. However, this defect seems to be an expression of the inability of transformed hemocytes to phagocytize foreign particles, rather than an impairment of the phagocytic capacity of those morphologically normal hemocytes still present in the clams. The observations on phagocytic behavior were performed with heat-killed yeast cells *in vitro*, not *Cytophaga*, but probably can be extrapolated to bacteria. Serum components apparently do not influence the phagocytic capabilities of normal or transformed hemocytes. Substitution of normal plasma for plasma from affected mussels did not increase phagocytosis by neoplastic hemocytes, and plasma from neoplastic animals did not inhibit phagocytosis by normal hemocytes. In this instance, there is little to suggest that the development and/or progression of this disease resulted from reduced immunosurveillance since the untransformed hemocytes continue to function normally, at least with regard to phagocytic activity. However, the phagocytic ability of the neoplastic cells seems to be severely limited and, since these cells can make up 90% of the hemocytes in heavily infected mussels, their presence can lead to serious impairment of defense mechanisms. This may account for the observations of bacterial septicemia reported in mussels with terminal hemic neoplasia (Kent *et al.* 1989).

Hemocytic and Serum Lysosomal Enzymes

The biological role of serum lysozyme in bivalve molluscs is commonly thought to be involved with internal defense mechanisms by virtue of its antimicrobial activity. Early studies showed that serum lysozyme levels decreased in MSX-infected *C. virginica* but were elevated in *Bucephalus*-infected oysters (Feng and Canzonier 1970). They also measured higher levels in winter than in summer. This seasonal fluctuation of lysozyme, while subject to considerable individual variation, has recently been confirmed by Chu and LaPeyre (1989), but they could find no correlation between this hemolymph enzyme and parasitism of oysters by *Perkinsus marinus.*

Increased levels of serum lysozyme were induced in mussels exposed to relatively high (100-700 ppb) Cu^{2+} concentrations; these values returned to normal after removal of the toxicant (Pickwell and Steinert 1984). However, an increment in serum lysozyme was not seen in bivalves exposed to 0.7 ppb tributyltin (Pickwell and Steinert 1988). *In vitro* studies with hemolymph from *Mya arenaria* showed lysozyme to be significantly inhibited by the presence of 5 μM zinc acetate or 0.6 μM lead acetate (Cheng and Rodrick 1974). Anderson *et al.* (1981b) found low and highly variable amounts of lysozyme in cell-free hemolymph of *Mercenaria*; however, the hemocytes contained a more substantial amount. The effect of pentachlorophenol or hexachlorobenzene exposure of the clams on intracellular lysozymes was measured. At comparatively low tissue burdens of the pollutants, intrahemocytic lysozyme was significantly induced, but this effect was lost as the PCP or HCB concentrations increased. It was proposed that the increment in activity in the low exposure groups might reflect chemically-induced macrophage activation, as can be produced by these pollutants in other aquatic animals (Anderson *et al.* 1984).

Release of the lysosomal marker enzyme acid phosphatase from hemolymph cells into the serum of several clam species was studied after exposure to sublethal concentrations of copper by Suresh and Mohandas (1990a). Lysosomes have the ability to compartmentalize and accumulate various organic contaminants and metals, thereby serving as a detoxification system until their storage capacity is exceeded. The lysosomal membrane can destabilize as a result of pollutant stress releasing hydrolases into the cells and ultimately into the extracellular fluids. Hemocytes are probably the primary source of serum-borne hydrolases in molluscs (Cheng and Rodrick 1980), thus levels of those enzymes could provide a means to quantify sublethal effects of pollutant exposure (Cheng 1983). The results of this study suggested that alterations in serum acid phosphatase levels produced by copper exposure varied from species to species. In some cases the enzyme was elevated (possibly by lysosomal destabilization or hypersecretion), and in other, the level was reduced (either by reduced synthesis or direct inactivation by copper ions). Relationships of dose and duration of exposure to these changes were not obvious.

Reactive Oxygen Intermediates

Reactive oxygen intermediates (ROI), such as superoxide anion, hydrogen peroxide (H_2O_2), singlet oxygen, the hydroxyl radical, etc., produced by bivalve hemocytes, probably participate in protective antimicrobial mechanisms. ROI have been more extensively studied in vertebrates and will be discussed in greater detail in the section on fish phagocytes. Nakamura and coworkers (1985) quantified H_2O_2 production by resting scallop hemocytes and reported increased activity after stimulation with Concanavalin A or when ingesting several species of bacteria. The hemocytes of gastropods (Dikkeboom *et al.* 1987, 1988) and several oyster species were shown to produce chemiluminescence (CL) *in vitro*, an indication of ROI production. Several implications of ROI in bivalve disease resistance were made recently using the CL method. For example, *Bonamia ostreae*, an important pathogen in *Ostrea edulis*, elicited no CL response upon phagocytosis by hemocytes of *O. edulis* or *C. gigas* (Hervio *et al.* 1989). This suggests that the parasite may escape normal cytotoxic mechanisms by inhibiting ROI production or by producing ROI scavengers. Similarly, *Pecten maximus* hemocytes show no CL response upon ingesting viable rickettsia-like pathogens, but have a vigorous CL response to killed rickettsia or other foreign particles (LeGall *et al.* 1989).

The use of CL to quantify ROI modulation by xenobiotic exposure of bivalve hemocytes is a very recent development. The CL response of *C. virginica* blood cells either exposed *in vitro* or collected from oysters after various periods of exposure to selected heavy metals, pesticides or other organic compounds, was measured by Larson *et al.* (1989). They found that Cu^{2+} in both *in vivo* and *in vitro* studies was the most effective agent tested with regard to ability to depress the CL response, although most of the compounds seemed to suppress CL at high exposure levels. Certain compounds, such as cadmium, aluminum, zinc, dieldrin and naphthalene, apparently caused increased CL at low levels, but this effort was usually reversed at higher concentrations.

Fisher *et al.* (1990) found that *in vitro* tributyltin (TBT) treatment of hemocytes from *C. virginica* or *C. gigas* produced dose-dependent suppression of CL and cell locomotion, but had little effect on their ability to assume an ameboid shape. The effective TBT concentrations (40-400 ppb) were higher than what could be expected in environmental samples; nonetheless, hemocytes of field-exposed oysters are probably exposed to high TBT levels as a result of the oyster's ability to bioaccumulate this compound.

Humoral Factors

Since the main topic of this chapter is xenobiotic-induced modulation of hemocyte function, the discussion of humoral factors will be very brief. This is appropriate because there is no evidence currently available to indicate their susceptibility to pollutant exposure, with the possible exception of lysozyme and acid phosphatase (discussed elsewhere in this chapter).

Naturally occurring and inducible nonlysozyme bactericidal factors can be demonstrated in sera from many invertebrate species, including abalone (Cushing *et al.* 1971), but have not been found in *Crassostrea virginica* (Weinheimer *et al.* 1969) or *Mercenaria* (Anderson, unpublished). Bachere *et al.* (1989) described a factor in oyster hemolymph with enzymatic activity related to the serine esterases that neutralized T3 coliphage. Earlier, Feng (1966) was unable to show neutralization of *Staphylococcus aureus* phage 80 *in vitro*, using hemolymph from untreated or phage-injected *C. virginica*.

In *Mytilus edulis*, serum hemolysin and hemagglutinin were synthesized and secreted by circulating hemocytes (Wittke and Renwrantz 1984; Leippe and Renwrantz 1988). Similar hemolytic factors have been reported in both freshwater clams (Yoshino and Tuan 1985) and *Mercenaria*, in which the hemolysin could be induced by injection of erythrocytes (Anderson 1981b).

There is no evidence to suggest more than a superficial analogy between this factor and vertebrate complement.

Bivalve serum may also contain bacterial agglutinins. In *Mercenaria*, a natural agglutinin for an unidentified Gram-negative, rod-shaped marine bacterium (designated RS-005) was described by Arimoto and Tripp (1977). This lectin was shown to opsonize this strain of bacteria; therefore, it could be involved in both the physical removal of bacteria from the hemolymph by promoting clumping and in facilitation of cellular recognition and ingestion of the bacteria. Circulating agglutinins were shown to opsonize several bacterial species in phagocytosis experiments using *C. gigas* hemocytes (Hardy *et al.* 1977). An agglutinin against marine *Flavobacterium* was also reported in *Mercenaria* serum by Anderson (1981a).

Literature on molluscan hemagglutinins is too vast to be summarized here in depth. Serum agglutinins against various mammalian red blood cells have been described in almost every bivalve studied and are often implicated as immunological recognition factors (as reviewed by Cheng *et al.* 1984; Olafsen 1988). These lectins react with saccharide moieties on hemocytes and foreign substances. In addition to defensive roles, they may also have mitogenic properties (Hardy *et al.* 1978). Anderson (1981a) reported that in *Mercenaria* exposed to PCP and HCB, no significant alterations in hemolysin, hemagglutinin, or bacterial agglutinin activities could be detected. Apparently few or no other studies of experimental or environmental modulation by xenobiotic exposure of bivalves have been published. It is curious that, despite their ubiquitous nature and putative role in host defense mechanisms, little work has been performed on the effect of pollutant exposure on hemagglutinin activity.

Cell Surface Receptors

Lectins have been postulated to serve as primitive recognition molecules ever since the observation by Tripp (1966) that serum agglutinin(s) facilitated phagocytosis of erythrocytes by oyster hemocytes. The presence of serum and cell-bound lectins (opsonins) in oysters and other bivalves has been confirmed by many investigators (e.g., Hardy *et al.* 1977; Vasta and Marchalonis 1982; Renwrantz and Stahmer 1983; Vasta *et al.* 1984). Although serum lectins may facilitate ingestion of certain phagocytic target particles by oyster hemocytes, their participation in the process is not absolutely required and may be of limited significance in other bivalves such as *Mercenaria* (Tripp 1989). Studies of effects of pollutant exposure on the expression, distribution, and mobility of hemocyte receptors for naturally occurring lectins have yet to be carried out.

The interaction of various plant lectins with bivalve hemocyte membranes has also been studied. Concanavalin A (Con A) binding has received considerable attention particularly regarding receptor redistribution and/or lectin-mediated cell adherence or phagocytosis (e.g., Renwrantz and Cheng 1977; Yoshino *et al.* 1979; Cheng *et al.* 1980; Gebbinck 1980; Schoenberg and Cheng 1980, 1981; Yoshino 1981a, 1981b). Con A-receptor complexes were cleared rapidly from the cell surface by endocytosis and accumulated in a single intracellular aggregate associated with a cap-like structure.

Concanavalin A receptors on oyster hemocytes mediate the binding of this lectin and may play a role in cellular recognition mechanisms. Flow cytometric analysis of FITC-conjugated Con A binding showed fewer receptors on cells from oysters collected from a polycyclic aromatic hydrocarbon-polluted site than from a less contaminated river (Sami *et al.* 1990). Phagocytic function was also impaired in the oysters from the PAH-impacted area. Both phagocytic impairment and reduced expression of Con A receptors were reversible if oysters were placed in "clean" water; both effects could be produced in previously unexposed animals when placed on PAH-containing sediments.

Temperature and Salinity Effects on Hemocyte Activity

In addition to the possible influence of xenobiotics on host defense capacities of bivalves and other marine invertebrates, other environmental conditions have been shown to affect these processes. Oyster hemocytes respond to elevated temperatures by increased pinocytosis (Feng 1965), chemotaxis and phagocytosis (Feng and Feng 1974; Foley and Cheng 1975). Similar observations of increased cell mobility and foreign particle binding by oyster hemocytes at higher temperatures were made by Fisher and Tamplin (1988). In addition to determining rates of locomotion, Fisher and his coworkers measured a novel hemocyte activity assay termed TTS, or time to hemocyte spreading, defined as the time required for ameboid hemocytes to appear along the entire periphery of the cell aggregates that form *in vitro* after hemolymph withdrawal. Changes in salinity can affect both rate of locomotion and TTS (Fisher and Newell 1986). The alterations induced in these cellular parameters by environmental factors, such as acute salinity change seem to differ, depending on the habitat of the oysters (estuarine vs. oceanic) according to Fisher *et al.* (1989). These authors suggest that, in addition to hemocyte effects directly influenced by temperature and salinity, there is also an annual cycle in the defensive potential of hemocytes with a low in summer when parasitism by major oyster pathogens is at its peak.

Effects of Chemicals on Resistance to Infection

During attempts to identify immunological mechanisms susceptible to manipulation by sublethal pollutant exposure, Anderson *et al.* (1981) found that such treatment impaired the ability of *Mercenaria* to eliminate injected bacteria. They studied the response of the clams to experimental infection with a marine *Flavobacterium* sp. after exposure to the model marine pollutants hexachlorobenzene (HCB) or pentachlorophenol (PCP) in recirculated and flow-through systems. The bacterial dose that was injected could be virtually 100% cleared from the hemolymph of unexposed clams during the 4-h experiment. The average percent clearance of the bacteria was significantly reduced in clams held in recirculated systems with tissue burdens of ~900 ppb PCP or ~1800 ppb HCB; the response was less marked in clams in the flow-through tanks with tissue burdens of ~400 ppb PCP or ~80 ppb HCB. Pollutant-mediated impaired bacterial clearance appeared to show dose dependency in that the effect correlated directly with tissue concentration of the pollutants achieved at the time of assay. Xenobiotic exposure resulted not only in reduced clearance rates in clams still capable of eliminating the bacteria, but also produced certain individuals in every experimental group who seemed totally unable to control the infection, as evidenced by viable bacterial counts greater or equal to the counts immediately after injection. This inability to clear *Flavobacterium* was never seen among controls that consistently cleared 90-97% of the injected dose. Subsequent studies showed that hemolymph withdrawn from pollutant-exposed animals was deficient in bactericidal activity, -- a result of impaired antimicrobial activity of the hemocytes that occurred independent of serum factors (Anderson 1988a).

Examples of xenobiotic-induced immunosuppression in aquatic invertebrates are not abundant. However, in shrimp, *Penaeus duorarum*, the prevalence and intensity of *Baculovirus* infection was enhanced by exposure to polychlorinated biphenyls, the insecticide Mirex, or by stress associated with crowding (Couch and Nimmo 1974; Couch 1976; Couch and Courtney 1977).

A more recent example of apparently decreased resistance to disease in a bivalve after *in vivo* exposure to xenobiotics has been described. *Perkinsus marinus* is an epizootic parasite of the American oyster and has caused mass mortalities in Gulf Coast and Atlantic estuaries (Lauckner 1983; Sparks 1985). Infections with this parasite were reported to be enhanced in oysters exposed to relatively high concentrations of a chemical carcinogen, n-nitrosodiethylamine (DENA), for periods of time from 17-28 days (Winstead and Couch 1988). Most oysters in their study had light to moderate *Perkinsus* infections prior to exposure; of 105 oysters exposed to 600 ppb DENA, 40% died within 14 days and 86% of survivors showed heavy infections. In contrast, none of the controls died or showed histological evidence of heavy infections. Heavy infection with this parasite is usually seen as massive invasion of the gut epithelium and typically triggers a significant hemocytic response. This hemocytic infiltration in response to heavy *Perkinsus* infection was much reduced or absent in the exposed oysters. Thus, it appears that, in some instances, chemical stressors can suppress the oyster's natural cellular defense capabilities. Aside from accelerating progression of the disease from the typical period of months (from light infection to death) to less than 3 weeks, DENA exposure seemed to promote invasion and destruction of atypical target areas, such as the nonciliated digestive tubules. Also, the parasite usually was dormant in the gut epithelia for months at temperatures >20°C; however, in DENA-exposed oysters, *Perkinsus* reproduced rapidly and caused pathogenic damage and death at 20°C. The details of the mechanisms underlying these effects are complex and probably involve several aspects of the host-parasite relationship, but data suggest immunosuppression, as seen histologically and in decreased resistance to infection.

Overview

The field of immunotoxicology as applied to bivalve molluscs and other aquatic invertebrates is largely unexplored, particularly when compared to the current understanding of immunotoxicological effects and mechanisms in higher animals. This is surprising in light of our heightened awareness of the dangers of environmental pollution and the economic importance of many aquatic species. As indicated in several reviews on the subject (Anderson 1981a, 1987, 1988b), there appears to be mounting evidence that exposure of these animals to sublethal concentrations of several classes of pollutants can produce modulation of hemocyte activities thought to be important in host defense mechanisms.

Studies to date suggest that hemocytes are more directly involved in invertebrate "immune" responses than are soluble hemolymph factors. Indeed, it is easy to quantify both humoral and cellular components of bivalve hemolymph with putative protective activities; yet only cell-related activities seem to be responsive to chemical stressors. However, such a statement is really intended to inspire more research in the humoral field. Clearly, we need to know much more about normal physiological roles of molecules, such as mammalian erythrocyte agglutinins and lysins, to appreciate them as other than laboratory phenomena. Particularly important areas for research will concern interactions between the hemocytes and serum molecules as related to immune mechanisms, as well as investigations of possible cell regulatory molecules. Methodologies to define phagocyte and serum factor activities are under constant development by researchers who use higher animal models to study immunomodulation. Many of these techniques can be adapted readily for use with bivalve hemocytes and have the potential of yielding more quantitative, more reproducible, and less subjective data than those previously generated by classical studies of serum factor titers and labor-intensive visual enumeration of parameters, such as phagocytic indices. Collaborations between specialists in the defensive capabilities of invertebrates and their counterparts in mammalian immunology and clinical medicine will likely advance our understanding of comparative immunology and immunotoxicology most rapidly.

The quest to produce immunomodulation by chemical stressors under controlled, laboratory conditions and to gather information on possible underlying mechanisms will generate data of great academic interest. However, it will also be essential to determine the "real world" significance of these phenomena with regard to health and survival of natural populations of aquatic organisms exposed to xenobiotics. Application of research findings to practical details of animal management and environmental regulation will be a major challenge. Nevertheless, other significant advances in the fields of biomarker development and immunotoxicological modeling, using alternative animal species, are also likely products of future work.

STUDIES WITH FISH PHAGOCYTES

The effects of stress on the resistance of fish to infectious diseases were first clearly described by Wedemeyer (1970) and Snieszko (1974). The concept that chronic exposure of fish to sublethal levels of toxic chemicals predisposes them to disease through immunosuppression continues to be widely accepted (Iwama 1977; Hetrick *et al.* 1979; Knittel 1981; Sindermann, this volume). The cellular and humoral mechanisms underlying disease resistance in fish and warm-blooded vertebrates are similar; therefore, it is easier to extrapolate immunotoxicological information from studies of laboratory rodents to fish than to aquatic invertebrates. In all animals, certain nonspecific defense mechanisms are used to protect against pathogens and other nonself material that may gain access to the internal milieu. Phagocytosis or encapsulation and subsequent destruction of invading agents by various blood cells can take place independent of the involvement of lymphocytes and/or their products. However fish, like mammals, have well-developed lymphoid systems and a comparable level of immunological sophistication involving immunoglobulins, complement, and other factors. In short, unlike invertebrates, fish show typical aspects of immunological specificity and memory, the characteristics of adaptive immunity. The major immunological dissimilarity between fish and mammals is the strong connection between environmental temperature and immunological vigor in fish.

Fish peripheral blood appears to contain various cell types analogous to T and B lymphocytes, monocytes, and neutrophils as described in mammals. For example in channel catfish, lymphocyte subclasses can be identified by the use of monoclonal antibodies (mAb) against catfish immunoglobulin and cell surface antigens (Sizemore *et al.* 1984; Miller *et al.* 1987). Monocytes can be isolated by adherence to fibronectin or depleted from peripheral blood by Sephadex columns

(Sizemore *et al.* 1984; Clem *et al.* 1985). Catfish neutrophils can be differentiated by the use of a specific mAb developed recently by Bly *et al.* (1990). Cytofluorographic analysis is being increasingly used to quantify stress-induced alterations in blood cell subpopulation profiles and to assess cell trafficking between various tissues. Since this paper deals primarily with modulation of phagocytic cell systems, the large body of information available on chemical suppression of fish immunoglobulins and lymphocyte functions will not be discussed. This information is available in review papers (Zeeman and Brindley 1981; Anderson *et al.* 1984; Anderson 1990).

Primary Stress Responses in Fish

In mammals, various stressors can produce marked alterations in hormone balance by activation of the hypothalamus-hypophysis- adrenal system in an attempt to return the organism to the original homeostatic conditions (Selye 1936, 1976). This adaptive response is marked by mobilization of energy reserves and is accompanied by many secondary effects, including decreased resistance to disease. The importance of the hypothalamic-pituitary-interrenal (HPI) axis in fish responses to various stressors has been reviewed by Donaldson (1981) and will not be treated in detail here. Clearly, external stressors (i.e., environmental insults, handling) and internal stressors (i.e., disease state, starvation) can produce primary responses (increased corticosteroid and catecholamine production), which can lead to various secondary responses (immunological effects and other physiological effects). The indices of HPI axis activity most often measured are plasma cortisol, interrenal histopathology, and interrenal ascorbic acid depletion.

There is evidence for HPI axis response in salmonids exposed to disease-related stress, particularly fungal infections by *Saprolegnia* sp., as well as bacterial infections (see Donaldson 1981). The primary responses of the HPI axis may also be used to measure stress responses in fish to environmental contaminants. It appears that in both disease and pollutant stress, plasma cortisol levels may increase only transiently if the stress is mild and/or chronic, indicating adaptation. There may be a graded plasma cortisol incremental response dependent on stress intensity, and finally a failure to adapt to higher pollutant doses resulting in a sustained elevated level (Donaldson and Dye 1975).

In addition to interrenal responses to stress, fish show activation of adrenergic function, which may make differentiation of the secondary effects of corticosteroids and catecholamines difficult. In many fish, elevation of plasma adrenalin and/or noradrenalin occurs within minutes of the onset of stress, may persist for hours after removal of the stressor, and may vary with type of stressor and species of fish (Mazeaud *et al.* 1977). The effects of catecholamine release on physiological parameters in fish has been reviewed by Mazeaud and Mazeaud (1981). Generally, increased circulation via increased cardiac output is observed, resulting in augmented gill blood flow rates with concomitant gas exchange. Certain osmoregulatory processes are also disrupted by stress in fish. For example, adrenalin increases gill permeability to water (Pic *et al.* 1974), which helps to explain the observation that stressed teleosts in fresh water tend to gain weight via water imbibition, whereas they lose weight in sea water via water loss. Many of the metabolic effects caused by different stressors can be emulated by injection of catecholamines, most notably increased blood glucose levels, increased blood lactic acid, and glycogen breakdown in liver and muscles.

The primary stress responses cited above will receive only superficial treatment in order to present more detailed information on direct effects of environmental stressors on immunological competence. Immunomodulation in aquatic species may be recognized as a significant secondary stress response, but its probable link to classical primary stress effects (at least in fish) should not be ignored. Indeed, work in this laboratory and elsewhere indicates that corticosteriods can directly inhibit fish leukocyte functions.

White Blood Cell Populations

Many stressors will produce typical changes in the numbers of white blood cells (leukocytes) in fish. Generally, these responses include increased abundance of neutrophilic granulocytes (PMN) and decreased numbers of lymphocytes. This is in line with stress-induced atrophy of lymphatic tissue, decreased lymphocyte numbers in hemopoietic centers, and lymphocyte lysis seen in mammals. Stress-related lymphocytopenia and neutrocytophilia in fish have been reported by Weinreb (1958), Slicher *et al.* (1962), Pickford *et al.* (1971), McLeay (1973), and Murad and Houston (1988).

Peters and Schwarzer (1985) have studied the effects of handling and social stress on the blood cell composition of hemopoietic organs (spleen and head kidneys) of rainbow trout. Social conflict was produced by housing a dominant fish with a subordinate partner. The most pronounced cellular changes included a reduction of hemoblasts and lymphocytes, enhanced red blood cell degradation, and an increase in the number of macrophage-like cells. Macrophage-like cells were thought to be hypertrophic leukocytes that take on hetero- and autophagocytic capabilities. Presumably, the observed structural and quantitative changes in the leukocytes reflect a functional activation of these tissues that might provide greater protection against microbes. Peters and Schwarzer (1985) found that social stress produced greater effects on leukocyte parameters than handling and that response could be elicited by aggressive behavior without actual physical interaction. Subordinate trout have elevated serum catecholamines, cortisol, lactate, and glucose and are more susceptible to infection with *Aeromonas hydrophila* (Peters *et al.* 1988). In addition to changes in blood cell morphology and population profiles produced by social confrontation, the pronephric leukocytes of subordinate trout show decreased functional responses, such as nonspecific cytotoxicity and proliferative responses to mitogens (Ellsaesser and Clem 1986; Ghoneum *et al.* 1988). The serum of submissive fish was shown to contain factors capable of suppressing proliferative responses to phytohemagglutinin (PHA) and Con A; the presence of immunoinhibitory substance(s) in laboratory-housed killifish, *Fundulus*, serum had been also described by Miller and Tripp (1982). Evidence for endogenous opioid-mediated suppression of nonspecific cytotoxicity and mitogenic activity in subordinate rainbow trout has been presented by Faisal *et al.* (1989). Immunomodulation in fish by social stress, handling, and laboratory housing is under careful investigation in many laboratories, and the resultant literature is too extensive to be included in this short review on the effects of xenobiotics. However, it is likely that certain basic mechanisms of immunosuppression may be similar. Alterations in blood cell profiles are not always observed in fish exposed to environmental chemicals at concentrations that produce other forms of immunomodulation. For example, rainbow trout exposed to Cd levels that altered cellular responses and antibody titers showed no changes in the normal differential leukocyte counts or in the relative proportions of phagocytic cell types (Thuvander 1989). Reductions in the numbers of blood lymphocytes do not always accompany chemical stress. Flounder and perch exposed to Cd in the laboratory and from a polluted river had elevated numbers of lymphocytes (Johansson-Sjobeck and Larson 1978; Sjobeck *et al.* 1984). In earlier studies, Gardner and Yevich (1970) found that Cd exposure reduced lymphocyte and thrombocyte numbers and increased granulocytes in *Fundulus*. On the other hand, Srivastava and Mishra (1979) recorded increased total leukocyte, thrombocyte and lymphocyte counts in *Colisa fasciatus* exposed to Cd.

Phagocytosis

Weeks and Warinner (1984) quantified phagocytosis of formalin treated *E. coli* by kidney phagocytes from fish collected from a relatively nonpolluted river and a river polluted with polycylic aromatic hydrocarbons (PAH). Phagocytosis was markedly inhibited in fish from the polluted site; preliminary evidence suggested that a similar effect could be produced in control fish exposed to contaminated sediments. The phagocytic activity of cells from the chemically stressed fish returned to normal after the fish were held in clean water for 2-3 weeks. Similar results were recorded using spectrophotometry to quantify phagocytosis of congo red-stained yeast cells (Seeley *et al.* 1990). Seeley and Weeks (in press) used this technique to study phagocytosis by macrophages from oyster toadfish collected along a PAH gradient in the Elizabeth River. Results suggested some degree of chemically induced immunosuppression and were comparable to direct microscopic observations of the cells.

Chemiluminescence

A pronounced respiratory burst is associated with blood cell phagocytic activity; this increase in oxygen uptake by the cells is accompanied by the generation of reactive oxygen intermediates (ROI). Cytotoxic ROIs, such as superoxide anion hydrogen peroxide, hydroxyl radical, and singlet oxygen, are involved in protective antimicrobial activities of the phagocytes, and can be easily detected by measuring luminol-enhanced chemiluminescence (CL) with a luminometer or scintillation counter adapted for single photon counting.

This technique gained acceptance as an indicator of ROI production by fish phagocytes after its initial description by Scott and Klesius (1981) in channel catfish. The chemiluminescent response of phagocytes from the pronephros of striped bass, *Morone saxatilis*, was measured after exposure to four pathogenic bacteria by Stave *et al.* (1983). When the effects of various bacteria:phagocytic ratios were studied, it became clear that the shape of the resultant CL curves, the times of peak activity after phagocytosis, and the peak values could all be altered by different bacterial doses. The response was also more strongly induced by *Vibrio anguillarum* than by several other pathogens, all presented at identical levels. The magnitude and kinetics of the CL response seemed to be a function of bacterial concentration and species. Striped bass phagocytes also showed a dose-dependent CL response to phorbol myristate acetate (PMA) and zymosan (Stave *et al.*, 1984). Opsonization of zymosan by incubation with homologous serum enhanced CL by bass phagocytes, as was the case for catfish phagocytes (Scott and Klesius 1981). Opsonization of living bacteria, *Aeromonas hydrophila*, also enhanced CL by bass pronephros cells; however, bacteria killed by heat-, formalin-, or UV-treatment showed markedly reduced CL when incubated with phagocytes.

In an interesting study, the effect of incubation temperature on the CL response of rainbow trout peripheral neutrophils was compared to that of human neutrophils (Sohnle and Chusid 1983). The peak response of both species was stimulated to an approximately equal degree by either *Staphylococcus aureus* or PMA. However, maximal CL occurred at 4-15°C in fish and 23-37°C in human cells. In both species the time of peak CL was retarded at lower incubation temperatures, possibly due to decreased fluidity of the cell membrane at atypical low temperatures. Elevated temperatures are known to enhance host defense systems in ectothermic animals; the results of Sohnle and Chusid (1983) suggest that initiation of neutrophil ROI production occurs more rapidly at higher than ambient temperatures in trout, although the peak response was not increased. The effects of opsonization and temperature on CL by peripheral blood phagocytes for channel catfish exposed to bacteria, *Edwardsiella ictaluri*, was further investigated by Scott *et al.* (1985). Peak light emission was about 15:1, bacteria:cell ratio, using bacteria opsonized with ~10 μg/ml immune serum. Response was lower with bacteria treated with nonimmune, antigen-absorbed, or heat-inactivated serum. Increased temperature (20-30°C) was accompanied by increased cell-mediated bactericidal activity and peak CL response. As expected, CL was sensitive to presence of superoxide dismutase (SOD) or sodium azide in the medium. The authors suggested that complement, in addition to specific antibody, played a role in opsonization because CL was reduced if bacteria were coated with zymosan- or EDTA-treated immune serum prior to presentation to phagocytes. Noting lack of agreement between the findings of Sohnle and Chusid (1983) and Scott *et al.* (1985) with regard to the effect of temperature on peak CL, Angelidis *et al.* (1988) carried out additional studies on sea bass head kidney phagocytes. Their results were similar to those of Sohnle and Chusid (1983). Opsonized zymosan particle stimulation elicited greater CL from phagocytes at lower temperatures (5-20°C) than higher (25-40°C), but CL peak values occurred more rapidly at higher temperatures.

Clearly the total CL response and its kinetics should be thoroughly understood in any fish model under study before attempting to generalize on particular experimental findings. Existing evidence suggests that, in addition to individual variation, the CL response of a given sample of fish phagocytes may depend on a variety of factors including the species of fish, the anatomical source of the cells, the type of stimulating agent, opsonization of particulate stimuli, cell:particle ratio, incubation temperature, and media components. However, CL remains a very useful means of evaluating phagocytosis and ROI production by phagocytes and shows promise for immunotoxicologic studies.

In fact, preliminary studies suggest that fish phagocytes show significantly reduced CL activity either as a result of *in vitro* exposure to xenobiotics or when obtained from fish exposed to pollutants *in vivo*. Elsasser *et al.* (1986) studied CL response of rainbow trout pronephros phagocytes to *Staphyloccocus aureus* after exposure to various metals. When metals were added immediately before the bacterial stimulus, CL response was increased by Cd but decreased by Al or Cu. Metal levels used were 0.1 times the lowest concentration of the metal found to be cytotoxic to the trout cells. Metal exposure for 1 h produced essentially the same effects as those following immediate exposure. After 24 h exposure, both Al and particularly Cu were immunosuppressive; Cd-exposed cells no longer showed increased CL, but the degree of reduction was slight and variable. If CL response was initiated by phagocytosis of *Staphyloccocus aureus* prior to addition of metals, added Cu was strongly suppressive and added Cd had a stimulatory effect. This direct inhibition of CL by Cu may partially explain observations that Cu exposure can predispose fish to infection when challenged with viral (Hetrick *et al.* 1979) or bacterial (Knittel 1981) pathogens. Warinner *et al.* (1988) examined the response of kidney phagocytes from normal and pollutant-exposed fish. In initial studies of the response to various stimulatory agents, typical interspecies differences were observed with regard to peak CL response and peak time. However, the zymosan-elicited CL response in spot, *Leiostomus xanthurus*, collected from the PAH-polluted Elizabeth River was negligible compared to response spot from the York River (a more pristine site). This same suppression of phagocyte CL could be induced under laboratory conditions by exposing spot to contaminated Elizabeth River sediments.

The effects of *in vitro* exposure to tributyltin (TBT) on CL of phagocytes from kidneys of three species of estuarine fish were studied by Wishkovsky *et al.* (1989). The cells were incubated in the presence of TBT for up to 18 h prior to CL determination. If the CL reaction was initiated by the addition of a phagocytic stimulant (zymosan) immediately after TBT addition, 400 μg/l was generally required to produce significant inhibition. The same was seen after 18 h exposure; however, some evidence of dose-dependency appeared in these data. Whereas the effective TBT concentration used in these *in vitro* studies was at least two orders of magnitude greater than what might be considered an environmental concentration, it is possible that fish cells might be exposed to comparable TBT levels *in vivo* as a result of bioconcentration.

Current work in this laboratory is focused on the modulation of CL response by Japanese medaka head kidney phagocytes by *in vitro* exposure to environmental contaminants and other chemicals (Anderson and Brubacher, in press). Incubation of phagocytes with sublethal concentrations of several xenobiotics for 1-20 h significantly inhibited the magnitude of the peak CL response and the total CL response (the area under the curve), but did not affect the time of peak CL. For example, significant reduction of CL activity followed exposure to 75 μg/ml lead or 15 μg/ml sodium pentachlorophenate. These data show that certain chemicals can work directly on phagocytes to reduce their immunological capacity. However, we showed that hydrocortisone *in vitro* also markedly reduced CL at concentrations of 20 μg/ml and greater. Similar inhibition by bacteria- or PMA-stimulated striped bass phagocytes treated with hydrocortisone was reported by Stave and Roberson (1985). This provides a possible link between a primary stress response (the release of corticosteroids) and a manifestation of immunosuppression (reduced CL) of phagocytes, the primary defensive cells by which fish resist infection by invading pathogens.

Other Effects of Xenobiotics on Phagocytes

Macrophages from fish collected from the PAH-contaminated Elizabeth River were reported to have altered ability to migrate toward bacterial antigens (Weeks *et al.* 1986). The study was conducted in standard Boyden Chambers where macrophages responded to a formalin-treated *E. coli* suspension as the chemoattractant. This manifestation of immunotoxicity, like others produced by the polluted water, was reversed by holding fish in clean water for several weeks.

As noted elsewhere, natural killer cell activity in rainbow trout is suppressed by social confrontation (Ghoneum *et al.* 1988); apparently a similar response can be induced by xenobiotic exposure. Faisal *et al.* (in press) report reduced natural cytotoxic cell (NCC) activity of anterior kidney and splenic leukocytes from *Fundulus* collected from a heavily PAH-contaminated region of the Elizabeth River. The defect may involve reduced ability of the leukocytes to recognize and bind the target cells. These fish have an abnormally high incidence of hepatic neoplasia, which is probably exacerbated by reduced immunosurveillance as indicated by impaired NCC activity. There are several other recent examples of reduced NCC activity in tumor-bearing fish (Schmale and McKinney 1987; Faisal 1989).

CONCLUSION

This chapter has devoted greater attention to modulation of bivalve leukocyte functions than to fish phagocytes. This was done intentionally because of a general tendency of immunobiologists to ignore invertebrate studies or to dismiss them as irrelevant to higher animals. Also, the numbers of thorough, recent reviews of fish immunotoxicology far exceed those dealing with similar phenomena in bivalves. Hopefully, enough information has been presented to convince skeptics that further work on bivalve hemocytes is warranted and that these cells have merit as putative models useful to our understanding of chemically induced dysfunctions of phagocytes at many phyletic levels.

Macrophages of fish and macrophage-like cells of invertebrates are an integral part of the host immune system. In addition to their roles as phagocytic, antimicrobial, and tumoricidal cells, they participate in antigen presentation and cytokine production, and (in conjunction with lymphocytes) are intimately involved in initiation and regulation of adaptive immunity in fish and higher animals. Modulation of capabilities of these cells should produce pronounced effects on immunocompetency and susceptibility to disease. Macrophage function assays are often relatively simple, rapid, inexpensive, and useful for identification and evaluation of potentially immunosuppressive chemicals. The presence of such xenobiotics in the aquatic environment could have insidious effects, producing few or no direct lethal effects but predisposing valuable species to atypically severe responses to pathogens or opportunistic microorganisms. The potential exists for use of macrophage impairment assays as biomarkers of environmental stress induced by single agents or complex mixtures typically encountered in the field.

Advances in basic and applied research on phagocytic effector cells require precise quantitation of their various immunological parameters. The classical methods, such as those based on direct microscopic observation of the cells, were labor-intensive and somewhat subjective. The more modern assays, often developed in the course of biomedical research, do not suffer from these limitations. Almost without exception, these newer procedures can be easily modified for use with either bivalve or fish phagocytes. Notable examples include: the use of labelled lectins and monoclonal antibodies as cell surface receptor probes, the use of flow cytometry to differentiate hemocyte subpopulations and to quantify ingestion of various particulates, and the use of chemiluminescence and colorimetric analysis in kinetic studies of the production of reactive oxygen intermediates. These and other methods could be used to gain new insight regarding comparative immunology and immunotoxicology. Mammalian immunotoxicology is one of the most rapidly developing subdisciplines of toxicology; it is hoped that aquatic immunotoxicology will soon undergo a similar expansion in line with current advances in methodology.

REFERENCES

Alderman, D.J., Van Banning, P., and Perez-Colomer, A. 1977. Two European oyster (*Ostrea edulis*) mortalities associated with an abnormal haemocytic condition. *Aquaculture* 10:335-340.

Anderson, D.P. 1990. Immunological indicators: effects of environmental stress on immune protection and disease outbreaks. In *Biological Indicators of Stress in Fish*, ed. S.M. Adams, pp. 38-50, American Fisheries Symposium 8, Bethesda, MD.

Anderson, D.P., W.B. van Muiswinkel, and B.S. Roberson. 1984. Effects of chemically induced immune modulation on infectious diseases of fish. In *Chemical Regulation of Immunity in Veterinary Medicine*, pp. 187-21. New York: Alan R. Liss, Inc.

Anderson, R.S. 1981a. Effects of carcinogenic and noncarcinogenic environmental pollutants on immunological functions in a marine invertebrate. In *Phyletic Approaches to Cancer*, ed. C.J. Dawe, J.C. Harshbarger, S. Kondo, T. Sagimura, and S. Takayama, pp. 319-331. Tokyo: JPN Scientific Society Press.

Anderson, R.S. 1981b. Inducible hemolytic activity in *Mercenaria mercenaria*. *Dev. Comp. Immunol.* 5:575-585.

Anderson, R.S. 1987. Immunocompetence in vertebrates. In *Pollutant Studies In Marine Animals*, ed. C.S. Giam and L.E. Ray, pp. 93-110. Boca Raton: CC Press, Inc.

Anderson, R.S. 1988a. Effects of pollutant exposure on bactericidal activity of *Mercenaria mercenaria* hemolymph. Extended Abstracts, Division of Environmental Chemistry, American Chemical Society National Meeting. 28:248-250.

Anderson, R.S. 1988b. Effects of anthropogenic agents on bivalve cellular and humoral defense mechanisms. In *Disease Processes in Marine Bivalve Molluscs*, American Fisheries Society, ed. W.S. Fisher, pp. 18:238-242. Special Publication.

Anderson, R.S., and L.L. Brubacher. (in press). Immunotoxicology in the medaka: Chemical modulation of macrophage chemiluminescence. Proceedings of the Third Annual Cancer Research Workshop. U.S. Army Biomedical Research and Development Laboratory, Frederick, MD.

Anderson, R.S., C.S. Giam, L.E. Ray, and M.R. Tripp. 1981. Effects of environmental pollutants on immunological competency of the clam *Mercenaria mercenaria*: Impaired bacterial clearance. *Aquat. Toxicol.* 1:187-145.

Anderson, R.S., C.S. Giam, and L.E. Ray. 1984. Effects of hexachlorobenzene and pentachlorophenol on cellular and humoral immune parameters in *Glycera dibranchiata*. *Mar. Environ. Res.* 14:317-326.

Angelidis, P., F. Baudin-Laurencin, and P. Youinou. 1988. Effects of temperature on chemiluminescence of phagocytes from sea bass, *Dicentrarchus labrax* L. *J. Fish Dis.* 11:281-288.

Arimoto, R., and M.R. Tripp. 1977. Characterization of a bacterial agglutinin in the hemolymph of the hard clam, *Mercenaria mercenaria*. *J. Invertebr. Pathol.* 30:406-413.

Bachere, E., D. Hervio, E. Mialhe, and H. Grizel. 1989. Evidence for a neutralizing activity against T3 coliphage in the *Crassostrea gigas* oyster hemolymph. *Dev. Comp. Immunol.* 13:384.

Balouet, G., and M. Poder. 1985. Reaction of blood cells in *Ostrea edulis* and *Crassosrea gigas*: A nonspecific response of differential cells. *Comp. Pathobiol.* 8:97-108.

Bly, J.E., N.W. Miller, and L.W. Clem. 1990. A monoclonal antibody specific for neutrophils in normal and stressed channel catfish. *Dev. Comp. Immunol.* 14:211-221.

Brown, R.S., R.E. Wolke, S.B. Saila, and C. Brown. 1977. Prevalence of neoplasia in ten New England populations of the soft-shelled clam (*Mya arenaria*). *Ann. N.Y. Acad. Sci.* 298:522-534.

Cheng, T.C. 1983. The role of lysosomes in molluscan inflammation. *Am. Zool.* 23:129-144.

Cheng, T.C. 1988a. *In vivo* effects of heavy metals on cellular defense mechanisms of *Crassostrea virginica*: Total and differential cell counts. *J. Invertebr. Pathol.* 51:207-214.

Cheng, T.C. 1988b. *In vivo* effects of heavy metals on cellular defense mechanisms of *Crassostera virginica*: Phagocytic and endocytotic indices. *J. Invertebr. Pathol.* 51:215-220.

Cheng, T.C., and G.E. Rodrick. 1974. Identification and characterization of lysozyme from the hemolymph of the soft-shelled clam, *Mya arenaria*. *Biol. Bull.* 147:311-320.

Cheng, T.C., and Rodrick, G.E. 1980. Nonhemocyte sources of certain lysosomal enzymes in *Biomphalaria glabrata* (Mollusca: Pulmonata). *J. Invertebr. Pathol.* 35:107-108.

Cheng, T.C., and J.T. Sullivan. 1984. Effects of heavy metals on phagocytosis by molluscan hemocytes. *Mar. Environ. Res.* 14:305-315.

Cheng, T.C., Rodrick, G.E., Foley, D.A., and S.A. Koehler. 1975. Release of lysozyme from hemolymph cells of *Mercenaria mercenaria* during phagocytosis. *J. Invertebr. Pathol.* 25:261-265.

Cheng, T.C., J.W. Haung, H. Karadogan, L.R. Renwrantz, and T.P. Yoshino. 1980. Separation of oyster hemocytes by density gradient centrifugation and identification of their surface receptors. *J. Invertebr. Pathol.* 36:35-40.

Cheng, T.C., J.J. Marchalonis, and G.R. Vasta. 1984. Role of molluscan lectins in recognition processes. *Prog. Clin. Biol. Res.* 157:1-15.

Chu, F-L. E., and J.F. LaPeyre. 1989. Effect of environmental factors and parasitism on hemolymph lysozyme and protein of American oysters (*Crassostrea virginica*). *J. Invertebr. Pathol.* 54:224-232.

Clem, L.W., R.C. Sizemore, C.F. Ellsaesser, and N.W. Miller. 1985. Monocytes as accessory cells in fish immune responses. *Dev. Comp. Immunol.* 9:803-809.

Cooper, K.R., R.S. Brown, and P.W. Chang. 1982a. The course and mortality of a hematopoietic neoplasm in the soft-shell clam, *Mya arenaria*. *J. Invertebr. Pathol.* 39:149-157.

Cooper, K.R., R.S. Brown, and P.W. Chang. 1982b. Accuracy of blood cytological screening techniques for the diagnosis of a possible hematopoietic neoplasm in the bivalve mollusc, *Mya arenaria*. *J. Invertebr. Pathol.* 39:281-289.

Cosson-Mannevy, M.A., C.S. Wong, and W.J. Cretney. 1984. Putative neoplastic disorders in mussels (*Mytilus edulis*) from southern Vancouver Island waters, British Columbia. *J. Invertebr. Pathol.* 44:151-160.

Couch, J.A. 1976. Attempts to increase *Baculovirus* prevalence in shrimp by chemical exposure. *Prog. Exp. Tumor Res.* 20:304-314.

Couch, J.A., and L. Courtney. 1977. Interaction of chemical pollutants and virus in a crustacean: A novel bioassay system. *Ann. N.Y. Acad. Sci.* 298:497-504.

Couch, J.A. and D.R. Nimmo. 1974. Detection of interactions between natural pathogens and pollutants in aquatic animals. In *Proceedings of the Gulf Coast Regional Symposium on Diseases of Aquatic Animals*, ed. R.L. Amborski, M.A. Hood, and R.R. Miller, pp. 261-268. Baton Rouge: Louisiana State University Press.

Cuenot, L. 1914. Les organes phagocytaires des Mollusques. *Arch. Zool. Exp.* Gen. 54:268-305.

Cushing, J.E., E.E. Evans, and M.L. Evans. 1971. Induced bactericidal responses of abalones. *J. Invertebr. Pathol.* 17:446-448.

Dikkeboom, R., J.M.G.H. Tijnagel, E.C. Mulder, and W.P.W. van der Knaap. 1987. Hemocytes of the pond snail *Lymnaea stagnalis* generate reactive forms of oxygen. *J. Invertebr. Pathol.* 49:321-331.

Dikkeboom, R., W.P.W. van der Knaap, W. van den Bovenkamp, J.M.G.H. Tijnagel, and C.J. Bayne. 1988. The production of toxic oxygen metabolities by hemocytes of different snail species. *Dev. Comp. Immunol.* 12:509-520.

Donaldson, E.M. 1981. The pituitary-interrenal axis as an indicator of stress in fish. In *Stress and Fish*, ed A.D. Pickering, pp. 11-47. New York: Academic Press.

Donaldson, E.M., and H.M. Dye. 1975. Corticosteroid concentrations in sockeye salmon (*Oncorhynchus nerka*) exposed to low concentrations of copper. *J. Fish. Res. Board Can.* 32:533-539.

Ellsaesser, C.F., and L.W. Clem. 1986. Haematological and immunological changes in channel catfish stressed by handling and transport. *J. Fish Biol.* 28:511-521.

Elsasser, M.S., B.S. Roberson, and F.M. Hetrick. 1986. Effects of metals on the chemiluminescent response of rainbow trout (*Salmo gairdneri*) phagocytes. *Vet. Immunol. Immunopathol.* 12:243-250.

Elston, R.A., M.L. Kent, and A.S. Drum. 1988. Progression, lethality and remission of hemic neoplasia in the bay mussel, *Mytilus edulis*. *Dis. Aquat. Org.* 4:135-142.

Faisal, M. 1989. Modulation of cell mediated immune responses by lymphocystis disease in the gilthead seabream Sparus aurata. In *Viruses of Lower Vertebrates*, W. Ahne and E. Kurstak, pp. 487-496. Berlin: Springer Verlag.

Faisal, M., F. Chiappelli, I.I. Ahmed, E.L. Cooper, and H. Weiner. 1989. Social confrontation "Stress" in aggressive fish is associated with an endogenous opioid-mediated suppression of proliferative response to mitogens and nonspecific cytotoxicity. *Brain Behavior and Immunol.* 3:223-233.

Faisal, M., B.A. Weeks, W.K. Vogelbein, and R.J. Huggett. (in press). Evidence of aberration of the natural cytotoxic cell activity in *Fundulus heteroclitus* (Pisces: Cyprinodontidae) from the Elizabeth River, Virginia. *Vet. Immunol. Immunopath.*

Farley, C.A. 1968. Probable neoplastic disease of the hematopoietic system in oysters, *Crassostrea virginica* and *Crassostrea gigas*. In *Neoplasms and Related Disorders of Invertebrate and Lower Vertebrate Animals, Nat. Cancer Inst. Monogr.* 31:541-555.

Farley, C.A. 1969. Sarcomatoid proliferative disease in a wild population of blue mussels (*Mytilus edulis*). *J. Nat. Cancer Inst.* 43:590-516.

Farley, C.A., S.V. Otto, and C.L. Reinisch. 1986. New occurrence of epizootic sarcoma in Chesapeake Bay soft shell clams, *Mya arenaria*. *Fish. Bull.* 84:851-857.

Feng, S.Y. 1965. Pinocytosis of proteins by oyster leucocytes. *Biol. Bull.* 129-95-105.

Feng, J.S. 1966. The fate of a virus, *Staphylococcus aureus* Phage 80, injected into the oyster, *Crassostrea virginica*. *J. Invertebr. Pathol.* 8:496-504.

Feng, S.Y., and W.J.Canzonier. 1970. Humoral responses in the American oyster (*Crassostrea virginica*) infected with *Bucephalus* sp. and *Minchinia nelsoni*. *Am. Fish. Soc. Spec. Pub.* 5:497-510.

Feng, S.Y., and J.S. Feng. 1974. The effect of temperature on cellular reactions of *Crassostrea virginica* to the injection of avian erythocytes. *J. Invertebr. Pathol.* 23:22-37.

Fisher, W.S., and R.I. Newell. 1986. Salinity effects on the activity of granular hemocytes of American oysters, *Crassostrea virginica*. *Biol. Bull.* 170:122-134.

Fisher, W.S., and M. Tamplin. 1988. Environmental influence on activities and foreign particle binding by hemocytes of American oysters, *Crassostrea virginica*. *Can. J. Fish. Aquat. Sci.* 45:1309-1315.

Fisher, W.S., M.M. Chintala, and M.A. Moline. 1989. Annual variations of estuarine and oceanic oyster *Crassostrea virginica* Gmelin hemocyte capacity. *J. Exp. Mar. Biol. Ecol.* 127:105-120.

Fisher, W.S., A. Wishkovsky, and F-L.E. Chu. 1990. Effects of tributyltin on defense-related activities of oyster hemocytes. *Arch. Environ. Contam. Toxicol.* 19:354-360.

Foley, D.A., and T.C. Cheng. 1975. A quantitative study of phagocytosis by hemolymph cells of the pelecypods *Crassostrea virginica* and *Mercenaria mercenaria*. *J. Invertebr. Pathol.* 25:189-197.

Fries, C.R., and M.R. Tripp. 1976. Effect of phenol on clams. *Mar. Fish. Rev.* 38:10-11.

Fries, C.R., and M.R. Tripp. 1980. Depression of phagocytosis in *Mercenaria* following chemical stress. *Dev. Comp. Immunol.* 4:233-244.

Gardner, G.R., and P.P. Yevich. 1970. Histological and hematological responses of an estuarine teleost to cadmium. *J. Fish. Res. Board. Can.* 27:2185-2196.

Gebbinck, J. 1980. Some observations on concanavalin A receptors on free cells of marine invertebrates. *Dev. Comp. Immunol.* 4:33.

George, S.G. 1983. Heavy metal detoxication in the mussel *Mytilus eduils* composition of Cd-containing kidney granules (tertiary lysosomes). *Comp. Biochem. Physiol.* C76:53-57.

George, S.G., T.L. Coombs, and B.J.S. Pirie. 1982. Characterization of metal-containing granules from the kidney of the common mussel, *Mytilus edulis*. *Biochem. Biophys. Acta* 716:61-71.

Ghoneum, M., M. Faisal, G. Peters, I.I. Ahmed, and E.L. Cooper. 1988. Suppression of natural cytotoxic cell activity by social aggressiveness in *Tilapia*. *Dev. Comp. Immunol.* 12:595-602.

Hardy, S.W., T.C. Fletcher, and J.A. Olafsen. 1977. Aspects of cellular and humoral defense mechanisms in the Pacific oyster, *Crassostrea gigas*. In *Developmental Immunobiology*, pp. 59-66, ed. J.B. Solomn, and J.D. Horton. Amsterdam: Elsevier/North-Holland Biomedical Press.

Hardy, S.W., A.W. Thomson, and T.C. Fletcher. 1978. Effect of haemolymph and agglutinins from the Pacific oyster (*Crassostrea gigas*) on cultured human and rabbit lymphocytes. *Comp. Biochem. Physiol.* 60A:473-477.

Hervio, D., E. Bachere, E. Mialhe, and H. Grizel. 1989. Chemiluminescent responses of *Ostrea edulis* and *Crassostrea gigas* hemocytes to *Bonamia ostreae* (Ascetopora). *Dev. Comp. Immunol.* 13:449.

Hetrick, F.M., M.D. Knittel, and J.L. Fryer. 1979. Increased susceptibility of rainbow trout to infectious hematopoietic necrosis virus after exposure to copper. *Appl. Environ. Microbiol.* 37:198-201.

Iwama, G. 1977. Some aspects of the interrelationship of bacterial kidney disease infection and sodium pentachlorophenate exposure in juvenile chinook salmon (*Oncorhynchus tshawytscha*). M.Sc. Thesis, University of British Columbia, Vancouver, B.C. 106 pp.

Johansson-Sjobeck, M.-L., and A. Larsson. 1978. The effect of cadmium on the hematology and on the activity of delta- aminolevulinic acid dehydratase (ALA-D) in blood and hematopoietic tissues of the flounder, *Pleuronectes flesus L. Environ. Res.* 17:191-204.

Kent, M.L., R.A. Elston, M.T. Wilkinson, and A.S. Drum. 1989. Impaired defense mechanisms in bay mussels, *Mytilus edulis*, with hemic neoplasia. *J. Invertebr. Pathol.* 53:378-386.

Knittel, M.D. 1981. Susceptibility of steelhead trout *Salmo gairdneri* Richardson to redmouth infection *Yersenia ruckeri* following exposure to copper. *J. Fish Dis.* 4:33-40.

Larson, K.G., B.S. Roberson, and F.M. Hetrick. 1989. Effect of environmental pollutants on the chemiluminescence of hemocytes from the American oyster *Crassostrea virginica*. *Dis. Aquat. Org.* 6:131-136.

Lauckner, G. 1983. Diseases of Mollusca: Bivalvia. In *Diseases of Marine Animals*, Vol. 2., ed. O. Kinne, pp. 477-961. Hamburg: Biologische Anstalt Helgoland.

LeGall, G., E. Bachere, E. Mialhe, and H. Grizel. 1989. Zymosan and specific-rickettsia activation of oxygen free radical production in *Pecten maximus hemocytes*. *Dev. Comp. Immunol.* 13:448.

Leippe, M., and L. Renwrantz. 1988. Release of cytotoxic and agglutinating molecules by *Mytilus* hemocytes. *Dev. Comp. Immunol.* 12:297-308.

Mazeaud, M.M., and F. Mazeaud. 1981. Adrenergic responses to stress in fish. In *Stress and Fish*, ed. Pickering A.D., pp. 49-75. New York: Academic Press.

Mazeaud, M.M., F. Mazeaud, and E.M. Donaldson. 1977. Primary and secondary effects of stress in fish: Some new data with a general review. *Trans. Am. Fish. Soc.* 106:201-212.

McCormick-Ray, M.G. 1987. Hemocytes of *Mytilus edulis* affected by Prudhoe Bay crude oil emulsion. *Mar. Environ. Res.* 22:107-122.

McLeay, D.J. 1973. Effects of cortisol and dexamethasone on the pituitary-interrenal axis and abundance of white blood cell types in juvenile Coho salmon, *Oncorhynchus kisutch. Gen. Comp. Endocrinol.* 21:441-450.

Miller, N.W., and M.R. Tripp. 1982. An immunoinhibitory substance in the serum of laboratory held killifish, *Fundulus heteroclitus L. J. Fish Biol.* 20:309-316.

Miller, N.W., J.E. Bly, F. van Ginkel, C.F. Ellsaesser, and L.W. Clem. 1987. Phylogeny of lymphocyte heterogeneity: Identification and seperation of functionally distinct subpopulations of channel catfish lymphocytes with monoclonal antibodies. *Dev. Comp. Immunol.* 11:739-747.

Mix, M.C. 1983. Haemic neoplasms of bay mussels, *Mytilus edulis L.*, from Oregon: Occurrence, prevalence, seasonality and histopathological progression. *J. Fish. Dis.* 6:239-248.

Murad, A., and A.H. Houston. 1988. Leucocytes and leucopoietic capacity in goldfish, *Carassius auratus*, exposed to sublethal levels of cadmium. *Aquat. Toxicol.* 13:141-154.

Nakamura, M., K. Mori, S. Inooka, and T. Nomura. 1985. *In vitro* production of hydrogen peroxide by the amoebocytes of the scallop, *Patinopectin yessoensis* (Jay). *Dev. Comp. Immunol.* 9:407-417.

Olafsen, J.A. 1988. Role of lectins in invertebrate humoral defense. *Am. Fish. Soc. Spec. Publ.* 18:189-205.

Peters, G., and R. Schwarzer. 1985. Changes in hemopoietic tissue of rainbow trout under influence of stress. *Dis. Aquat. Org.* 1:1-10.

Peters, G., M. Faisal, T. Lang, and I.I. Ahmed. 1988. Stress caused by social interaction and its effect on the susceptibility to *Aeromonas hydrophila* infection in the rainbow trout, *Salmo gairdneri* Rich. *Dis. Aquat. Org.* 4:369-380.

Pic, P., N. Mayer-Gostan, and J. Maetz. 1974. Branchial effects of epinephrine in the seawater-adapted mullet. I. Water permeability. *Am. J. Physiol.* 226:698-702.

Pickford, G.E., A.K. Szivastava, A.M. Slicher, and P.K.T. Pang. 1971. The stress response in the adundance of circulating leukocytes in the killifish, *Fundulus heteroclitus*. *J. Exper. Zool.* 177:109-117.

Pickwell, G.V., and S.A. Steinert. 1984. Serum biochemical and cellular responses to experimental cupric ion challenge in mussels. *Mar. Environ. Res.* 14:245-265.

Pickwell, G.V., and S.A. Steinert. 1988. Accumulation and effects of organotin compounds in oysters and mussels: Correlation with serum, biochemical and cytological factors and tissue burdens. *Mar. Environ. Res.* 24:215-218.

Renwrantz, L.R., and T.C. Cheng. 1977. Identification of agglutinin receptors on hemocytes of *Helix pomatia. J. Invertebr. Pathol.* 29:88-96.

Renwrantz, L., and A. Stahmer. 1983. Opsonizing properties of an isolated hemolymph lectin and demonstration of lectin-like recognition molecules at the surface of hemocytes from *Mytilus edulis. J. Comp. Physiol. B.* 149:535-546.

Rheinberger, R., G.L. Hoffman, and P.P. Yevich. 1979. The kidney of the quahog (*Mercenaria mercenaria*) as a pollution indicator. In *Animals as Monitors of Environmental Pollutants*, pp. 119-129. Washington: National Academy of Science.

Ruddell, C.L., and D.W. Rains. 1975. The relationship between zinc, copper and the basophils of two crassostreid oysters, *C. gigas* and *C. virginica. Comp. Biochem. Physiol.* A,51:585-591.

Sami, S., M. Faisal, and R.J. Huggett. 1990. Influence of PAH-exposure on the expression of Concanavalin A receptors and phagocytic activities of the American oyster (*Crassostrea virginica*) hemocytes. Abstract in *Proc. Chesapeake Research Consortium*, Baltimore, MD.

Schmale, M.C., and E.C. McKinney. 1987. Immune responses in the bicolor damsel fish, *Pomacentrus partitus*, and their potential role in the development of neurogenic tumors. *J. Fish Biol.* 31:161A-166A.

Schoenberg, D.A., and T.C. Cheng. 1980. Lectin-binding specificities of two strains of *Biomphalaria glabrata* as demonstrated by microhemadsorption assays. *Dev. Comp. Immunol.* 4:617.

Schoenberg, D.A., and T.C. Cheng. 1981. Lectin-binding specificities of *Bulinus truncatus* hemocytes as demonstrated by microhemadsorption. *Dev. Comp. Immunol.* 5:145.

Scott, A.L., and P.H. Klesius. 1981. Chemiluminescence: a novel analysis of phagocytosis in fish. In *Developments in Biological Standardization*, ed. D.P. Anderson, W. Hennessen, volume 49, pp. 245-256. Basel: S. Krager.

Scott, A.L., W.A. Rogers, and P.H. Klesius. 1985. Chemiluminescence by peripheral blood phagocytes from channel catfish: Function of opsonin and temperature. *Dev. Comp. Immunol.* 9:241-250.

Seeley, K.R., and B.A. Weeks. (in press). Altered macrophage phagocytic activity in oyster toadfish (*Opsanus tau*) from a highly polluted subestuary.

Seeley, K.R., P.D. Gillespie, and B.A. Weeks. 1990. A simple technique for the rapid spectrophotometric determination of phagocytosis by fish macrophages. *Mar. Environ. Res.* 30:37-41.

Seiler, G.R., and M.P. Morse. 1988. Kidney and hemocytes of *Mya arenaria* (Bivalva): Normal and pollution-related ultrastructural morphologies. *J. Invertebr. Pathol.* 52:201-214.

Selye, H. 1936. A syndrome produced by diverse nocuous agents. *Nature* (London) 138:32.

Selye, H. 1976. *Stress in Health and Disease*. Boston: Butterworth.

Sizemore, R.C., N.W. Miller, M.A. Cuchens, C.J. Lobb, and L.W. Clem. 1984. Phylogeny of lymphocyte heterogeneity: The cellular requirements for in vitro mitogenic responses of channel catfish leukocytes. *J. Immunol.* 133:2920-2924.

Sjobeck, M.-L., C. Haux, A. Larson, and G. Lithner. 1984. Biochemical and hematological studies on perch, *Perca fluviatilis*, from the cadmium-contaminated river Eman. *Ecotox. Environ. Safety* 8:303-312.

Slicher, A.M., G.E. Pickford, and J.N. Ball. 1962. Effect of ACTH and cold-shock on white cell counts of fishes. *Anat. Rec.* 142:327.

Snieszko, S.F. 1974. The effects of environmental stress on outbreaks of infectious diseases of fishes. *J. Fish Biol.* 6:197-208.

Sohnle, P.G., and M.J. Chusid. 1983. The effect of temperature on the chemiluminescence response to neutrophils from rainbow trout and man. *J. Comp. Pathol.* 93:493-497.

Sparks, A.K. 1985. Protozoan diseases. In: *Synopsis of Invertebrate Pathology: Exclusive of Insects*, pp. 239-311. New York: Elsevier Science Publishers.

Srivastava, A.K., and S. Mishra. 1979. Blood dyscrasia in a teleost fish *Colisa fasciatus*, associated with cadmium poisoning. *J. Comp. Pathol.* 89:1-5.

Stauber, L.A. 1950. The fate of India ink injected intracardially into the oyster, *Ostrea virginica* Gmelin. *Biol. Bull.* 98:227-241.

Stave, J.W., and B.S. Roberson. 1985. Hydrocortisone suppresses the chemiluminescent response of striped bass phagocytes. *Dev. Comp. Immunol.* 9:77-84.

Stave, J.W., B.S. Roberson, and F.M. Hetrick. 1983. Chemiluminescence of phagocytic cells isolated from the pronephros of striped bass. *Devel. Comp. Immunol.* 7:269-276.

Stave, J.W., B.S. Roberson, and F. Hetrick. 1984. Factors affecting the chemiluminescent response of fish phagocytes. *J. Fish Biol.* 25:197-206.

Stave, J.W., B.S. Roberson, and F.M. Hetrick. 1985. Chemiluminescent responses of striped bass, *Morone saxatilis* (Walbaum), phagocytes to *Vibrio* spp. *J. Fish Dis.* 8:479-483.

Stave, J.W., T.M. Cook, and B.S. Roberson. 1987. Chemiluminescent response of the striped bass, *Morone saxatilis* (Walbaum), phagocytes to strains of *Yersinia ruckeri*. *J. Fish Dis.* 10:1-10.

Suresh, K., and A. Mohandas. 1990a. Effect of sublethal concentrations of copper on hemocyte number in bivalves. *J. Invertebr. Pathol.* 55:325-331.

Suresh, K., and A. Mohandas. 1990b. Hemolymph acid phosphatase activity pattern in copper-stressed bivalves. *J. Invertebr. Pathol.* 55:118-125.

Thuvander, A. 1989. Cadmium exposure of rainbow trout, *Salmo gairdneri* Richardson: effects on immune functions. *J. Fish Biol.* 35:521-529.

Tripp, M.R. 1966. Hemagglutinin in the blood of the oyster. *J. Invertebr. Pathol.* 8:478-484.

Tripp, M.R. 1989. Factors affecting phagocytosis by clam hemocytes. *Devel. Comp. Immunol.* 13:408-409.

Twomey, E., and M.F. Mulcahy. 1984. A proliferative disorder of possible hemic origin in the common cockle, *Cerastoderma edule*. *J. Invertebr. Pathol.* 44:109-111.

Vasta, G.R., and J.J. Marchalonis. 1982. Serum and hemocyte lectins from the oyster *Crassostrea virginica*. *Am. Zool.* 22:926.

Vasta, G.R., T.C. Cheng, and J.J. Marchalonis. 1984. A lectin on the hemocyte membrane of the oyster (*Crassostrea virginica*). *Cell. Immunol.* 88:475.

Warinner, J.E., E.S. Mathews, and B.A. Weeks. 1988. Preliminary investigations of the chemiluminescent response in normal and pollutant-exposed fish. *Mar. Environ. Res.* 24:281-284.

Wedemeyer, G. 1970. The role of stress in disease resistance of fishes. In A Symposium on diseases of Fishes and Shellfishes, ed. Snieszko, S.F. *Am. Fish. Spec. Publ.* 5:30-35.

Weeks, B.A., and J.E. Warinner. 1984. Effects of toxic chemicals on macrophage phagocytosis in two estuarine fishes. *Mar. Environ. Res.* 14:327-335.

Weeks, B.A., and J.E. Warinner. 1986. Functional evaluation of macrophages in fish from a polluted estuary. *Vet. Immunol. Immunopathol.* 12:313-320.

Weeks, B.A., J.E. Warinner, P.L. Mason, and D.S. McGinnis. 1986. Influence of toxic chemicals on the chemotactic response of fish macrophages. *J. Fish Biol.* 28:653-658.

Weinreb, E.C. 1958. Studies on the histology and histopathology of the rainbow trout (*Salmo gairdneri irideus*). I. Hematology under normal and experimental conditions of inflammation. *Zoologica* 43:145-155.

Weinheimer, P.F., R.T. Acton, and E.E. Evans. 1969. Attempt to induce a bactericidal response in the oyster. *J. Bacteriol.* 97:462-463.

Winstead, J.T., and J.A. Couch. 1988. Enhancement of protozoan pathogen *Perkinsus marinus* infections in American oysters *Crassostrea virginica* exposed to the chemical carcinogen n-nitrosodiethylamine (DENA). *Dis. Aquat. Org.* 5:205-213.

Wishkovsky, A., B.S. Roberson, and F.M. Hetrick. 1987. *In vitro* suppression of the phagocytic responses of fish macrophages by tetracyclines. *J. Fish. Biol.* 31(Supp. A):61-65.

Wishkovsky, A., E.S. Mathews, and B.A. Weeks. 1989. Effect of tributyltin on the chemiluminescent response of phagocytes from three species of estuarine fish. *Arch. Environ. Contam. Toxicol.* 18:826-831.

Wittke, M., and L. Renwrantz. 1984. Quantification of cytotoxic hemocytes of *Mytilus edulis* using a cytotoxicity assay in agar. *J. Invertebr. Pathol.* 43:248-253.

Yoshino, T.P. 1981a. Comparison of concanvalin A-reactive determinants on hemocytes of two *Biomphalaria glabrata* snail stocks: receptor binding and redistribution. *Dev. Comp. Immunol.* 5:229.

Yoshino, T.P. 1981b. Concanavalin A-induced receptor redistribution on *Biomphalaria glabrata* hemocytes: Characterization of capping and patching responses. *J. Invertebr. Pathol.* 38:102-112.

Yoshino, T.P. 1982. Lectin-induced modulation of snail hemocyte surface determinants: clearance of Con A-receptor complexes. *Dev. Comp. Immunol.* 6:451-461.

Yoshino, T.P., and T-L. Tuan. 1985. Soluble mediators of cytolytic activity in hemocytes of the Asian clam, *Corbicula fluminea*. *Dev. Comp. Immunol.* 9:515-522.

Yoshino, T.P., L.R. Renwrantz, and T.C. Cheng. 1979. Binding and redistribution of surface membrane receptors for concanavalin A on oyster hemocytes. *J. Exp. Zool.* 207:439.

Zeeman, M.G., and W.A. Brindley. 1981. Effects of toxic agents upon fish immune systems: a review. In *Immunologic Considerations in Toxicology*, vol 2, ed. R.P. Sharma, pp. 1-60. Boca Raton: CRC Press.

18

Observations on the State of Marine Disease Studies

John A. Couch

"Wer immer strebend sich bemüht, den können wir erlösen" Goethe, *Faust*, Part II

PROGRESS IN UNDERSTANDING MARINE DISEASE PROBLEMS

In the past 30 years the science of aquatic animal pathobiology has evolved in several distinct ways (Couch 1988). In this chapter, I would like to discuss several aspects of this evolution -- the good, the bad, the beautiful, and the plain. I will attempt to place in perspective a view of the present and future of aquatic, and more specifically marine/estuarine, animal pathobiology. My underlying theme will be the unique capability of pathobiology as a multi-disciplinary science, to attack and deal with the problems of linking cause(s) and effect(s) in animal health.

Starting in a positive sense, we should contemplate some of the successes of the last decades. Prior to the 1950-1960 period, many fish and shellfish diseases that we consider commonplace today were unknown or were recognized as new problems. One of the earliest major marine diseases that attracted considerable attention in the U.S. was the *Perkinsus marinus* (= *Dermocystidium marinum*) epizootic in the American oyster, *Crassostrea virginica*. This protozoan disease, first recognized in the early 1950's, became a major natural factor that oyster producers and biologists could not ignore. Since then, due to Drs. Mackin, Ray, Perkins, Andrews, and a host of other workers, we have learned considerably more about this major epizootic killer of oysters. We know that *Perkinsus marinus* kills oysters mainly during the late summer and early fall during peaks

Contribution No. 780, Environmental Research Laboratory, Gulf Breeze, FL

0-8493-8662-4/93/$0.00 + $.50

of high water temperature and salinity; we know that *Perkinsus* is probably an apicomplexan protozoan because of ultrastructural studies not available when it was first discovered (1950's). We know that *Perkinsus*-like parasites are found as epizootic pathogens in a variety of mollusks around the world, including abalone and giant clams on the Great Barrier Reef of Australia (Alder and Braley 1988; Gogin, Sewell, and Lester 1990; Perkins, this volume). Finally, we know that this pathogen can kill oysters without the nefarious help of mankind!

Another devastating shellfish pathogen complex, the haplosporidians, particularly *Haplosporidium nelsoni* and *H. costale*, has provided challenge for the last 35 years to many of the researchers who participated in this meeting. MSX or *H. nelsoni*, may have killed more oysters on the Atlantic coast of the U.S. than any other single factor (except human predation) that we know about during the last three to four decades. What have we learned about these pathogens? We could use the history of MSX as a paradigm of the difficulties, frustrations, complexities, and minor triumphs of a major marine disease problem that affected civilians and scientists alike. The first success related to MSX was its discovery by Drs. Haskin, Stauber, and Mackin and their association of the multinucleated sphere-unknown with internal pathological changes in oysters that were dying by the thousands. The Rutgers group, under Dr. Haskin's leadership, and more recently Dr. Susan Ford's research; the VIMS group under Drs. Andrews, Hargis, and Perkins; and the Oxford NOAA group under Drs. Sindermann, Rosenfield, and Farley over the last three decades have contributed to our present understanding of the epizootiology, life cycle, pathogenesis, and taxonomic status of *H. nelsoni*. Investigators at both Rutgers and VIMS have used the only rational approach to date to attempt to overcome MSX as a severe ecological factor limiting oyster production. They have worked for some time on developing stocks of oysters with innate resistance to the pathogen. A recent publication by Littlewood and Ford (1990) reports that contemporary physiological studies of MSX-infected oysters demonstrate that the "ecological fitness" of these animals is compromised by the stress of response of the host to the parasitic infection. Therefore, investigators are beginning to seriously consider the ecological roles of epizootic, infectious agents in marine animal populations. The role of interaction of anthropogenic induced stress (e.g., pollution or water quality decline) with the end results of enhancing or worsening parasitic-induced stress caused by natural pathogens in estuarine species is being considered more real and probable than ever before. There will be further discussion of this point.

Apart from successes associated with major epizootic disease studies, we should note the many smaller, but significant advances contributed over the last 30 years by individual investigators around the country in their identification, characterization, and biological investigation of new and diverse infectious and non-infectious agents or syndromes in marine invertebrates, fishes, reptiles, and mammals. Though we may consider this initial *discovery* and *characterization* of new agent and pathogen groups in the context of "alpha" pathology, we should remember that these discoveries constitute the primal, very significant first series of steps that establish our knowledge base in the science of marine diseases. Most of the preceding chapters have covered many of those "newly" discovered agents. I will not elaborate in detail, but will point out a few exemplary cases. These cases consist of agents or syndromes that were unknown, or whose natures were little understood 20 to 30 years ago. (Of course, most of these agents and their related diseases were always there, working on their hosts, in the presence of our ignorance -- not a new, but still a humbling realization!)

Several investigators have begun to build informational bases on the actual impacts of some of those agents on their hosts in laboratory, natural, and mariculture situations. For example, some of the penaeid shrimp baculoviruses described in the '70's and '80's are now known to kill their hosts in early metamorphosis but not when the host is adult or mature. The fish viruses may, under the right conditions, kill adults and early stages.

We have learned that some pollutant agents, such as heavy metals, may have specific target tissues, such as gill cells in crustacea and lateral line tissue in fishes. Finally, in terms of new beginnings and initial success stories, we have opened the important, extensive area of impact of pollutants as direct lesion producers or indirect effectors of disease in both marine invertebrates and vertebrates. One of the best examples of the dramatic advances of information and knowledge, in this regard, is the estuarine/marine fish carcinogenesis/indicator/assay studies underway for about 15 years around the country (Dawe and Couch 1984; Couch and Harshbarger 1985). Teleost fishes are representative vertebrate animals; therefore, we have been able to use them effectively in these studies as both field indicators of carcinogens and as models for study of specific neoplasms that occur also in mammals (Couch 1990, 1991).

Because neoplasia, the end point (effect) of carcinogenesis, is evaluated most dependably by histopathology, this area of work is exemplary of how pathobiology can play a central, discriminating role in marine/freshwater environmental issues (Harshbarger, this volume). Not only is good diagnostic histopathology necessary for identification of the neoplasm, but it is also often necessary, at least indirectly, for the resolutions of different causes because of the indicator significance of some different kinds of induced lesions, i.e., hepatic carcinoma, (Harshbarger, this volume; Couch and Harshbarger 1985).

Unfortunately, accumulation of quantity of information is not always followed by understanding of processes or phenomena being studied. This has been true in past and recent ecological and pathobiological studies of estuarine and marine diseases.

Table 18-1 reveals that viruses, bacteria, fungi, protozoa, and toxic chemicals are among the more recently recognized pathogenic agents in marine animals.

Table 18-1. Selected Examples of Recent Marine Mortalities and Causes.

Species Harmed	Extent of Harm and Disease	Probable Cause(s)	Locale(s)
Fish	Massive Kills	Chlorine, Kepone	James River, VA
Fish	Epizootic or Incidental cancers	Environmental Pollutants	Puget Sund; Boston Harbor; Elizabeth River, VA
Fish	Mass Mortalities - Ulcerative mycoses	Fungi, bacteria IF*	U.S. Coasts
Oysters	50% to 100% Mortality	Epizootic protozoan parasites	Atlantic Gulf Coasts
Clams	Leukemic-like disease	Unknown	Atlantic Coast; West Coast
Shrimps	Mortality in shrimp populations	Viruses plus IF*	Atlantic Coast; West Coast
Blue Crabs	Seasonal and periodic outbreaks	Infectious Protozoa	Atlantic Gulf Coasts
Tanner or Snow Crabs	Chronic but lethal disease up to 100% mortalities	Infectious Potozoa	Alaskan Coast
Corals	Bleaching of corals and mass mortalities	Unknown; Climate? Disease?	Caribbean Sea; Florida Keys
Coral Reef, Sea Urchin	95-99% mortalities; Significant alteration of coral reef ecosystems for years	Possible pathogenic bacterium and environmental factors.	Caribbean Sea
Marine Turtles	Mass mortalities	Unknown	South Atlantic Coast
Dolphins	Significant mortalities	Unknown	Atlantic Coast; Gulf Coast?

* IF = interactive factors (See Fig. 18-7).

SIGNIFICANT, UNSOLVED DISEASE PROBLEMS

There are more unsolved marine disease problems, than solved. A few important examples will be used to point out the nature of *kinds* of such problems.

Starting with the general and going to more specific points, we should all agree that a paramount and varied problem is our limited ability to quickly resolve the roles of infectious agents and non-infectious agents in disease etiology. This challenge extends across the spectrum of pathobiology and is particularly relevant to aquatic animal diseases where a plethora of infectious agents may cause syndromes, some of which may have shared features, or signs also characteristic of, or exacerbated by, non-infectious agents (Hinton, this volume). Solution of this problem may be simple, such as the conclusive identification of a pathogen like MSX that is strongly associated with mass mortalities of oysters beyond much doubt, or difficult because the situation may involve a toxicant, several microbial pathogens, the host's altered immune system, and changing environmental factors, such as salinity, temperature, dissolved oxygen (Couch and Nimmo 1974; Plumb, Noga, Anderson, this volume).

We have to start thinking of diseases in marine animals more as complex response processes or syndromes rather than isolated clinical case studies, if we are to systematically resolve causes. A perfect example of this challenge is represented in the research on fungi as opportunistic pathogens in marine fishes as reported by Noga, this volume. Ulcerative mycoses (UM) probably are multifactorial diseases reflecting in their signs and progressions, several environmental influences, fungal invasion, bacterial complications, and extreme chronic inflammation. Preliminary investigations (1970's) of the EPA Laboratory at Gulf Breeze on mullet, croakers, and menhaden afflicted with UM and fin rot from the east coast of Florida illustrate this. It was concluded that osmoregulatory stress of estuarine fishes entering freshwater marshes and streams, in very cold weather, led to oomycete fungal invasions of microscopic dermal lesions or abrasions, followed by necrosis and inflammation. We do not know the role in nature, if any, of water quality (pollutants) on UM's course in menhaden and other species, although we have induced massive fin rot and ulceration in certain marine fishes with PCB's in the laboratory (Couch and Nimmo 1974).

In the absence of any apparent or demonstrable *infectious* pathogen, but with a tissue or cellular lesion in hand, one must consider a plethora of *noninfectious* possibilities. Such a phenomenon is exemplified by digestive diverticula epithelial atrophy (or metaplastic changes) in oysters, which may result from any of several categories of stress including starvation, noxious ambient water, seasonal reproductive cycle stress, or a specific irritant pollutant chemicals (Couch 1985). The contribution resides in being able to resolve and discern the primary causative factor in a given situation.

Another major area that needs attention now is the toxicological pathology of marine species. We are slowly building an information base concerning general and specific cellular and tissue pathological responses to a broad range of pollutants and carcinogens (Meyers and Hendricks 1982; Hinton and Couch 1984; Patton and Couch 1984; Couch 1988; Hinton, Gardner, Sindermann, this volume). We must continue to use both invertebrate and fish models to identify and characterize lesions that result from, or that are associated with, toxicant exposure. It is important to focus on the cell(s) and tissue(s) of origin of such lesions induced experimentally in our models in order to identify and understand developmental or progressive stages in chemically-induced disorders (Fig. 18-1) (Couch 1990; Couch and Courtney 1987; Hinton *et al.* 1988). We can then think and speak more accurately on significance and identification of these or similar lesions in wild marine animals, and make possible the identification of target cells and tissues for specific toxicants, heavy metals, or carcinogens.

We are beginning to understand the similarity and differences in cellular and tissue responses to certain toxicants in both aquatic vertebrates and mammals

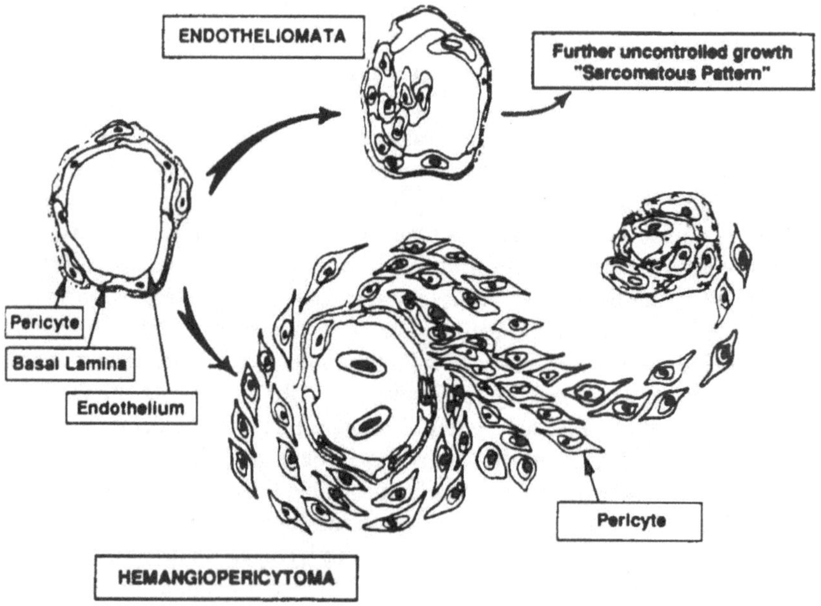

Fig. 18-1. Two tumor types, each arising from proximal but different cell types in capillaries of a fish exposed to a chemical carcinogen; endothelial cells give rise to endotheliomas and pericytes on outer surface of endothelium give rise to hemangiopericytoma.

(Couch and Courtney 1987; Couch 1990). This is well illustrated in the results of carcinogenesis studies with small fish species. The quantitative nature or peculiarities of certain endpoints (e.g., liver cancer in fish vs. mammals) need not detract from their illustration of a type of identical qualitative response potential through a range of the phyletic spectrum (Dawe and Couch 1984; Couch and Harshbarger 1985).

Another unresolved issue of great importance is the large gap in our knowledge of infectious diseases (in their own right) in marine animals; e.g., lack of understanding of life cycles of several key protozoan pathogens of shellfishes and fishes. Chief among these is the haplosporidian, MSX (Fig. 18-2). We do not understand the life-cycle of any haplosporidian even though there are epizootic species in several invertebrate host species. For example, what is the true infectious stage of MSX in oysters? Is there an intermediate host or vector? What happens to the spores? Without this knowledge, there is little possibility of a rational approach to mechanical, chemical, or even genetic control of the pathogen in cultured or wild oysters. Will stages in the life cycle of the pathogen be enhanced by, or weakened by, water quality factors (pollutants)? Investigators need to apply new biomarkers, tagging, or gene probes (i.e., isolation of identifying DNA sequences) to attack this problem.

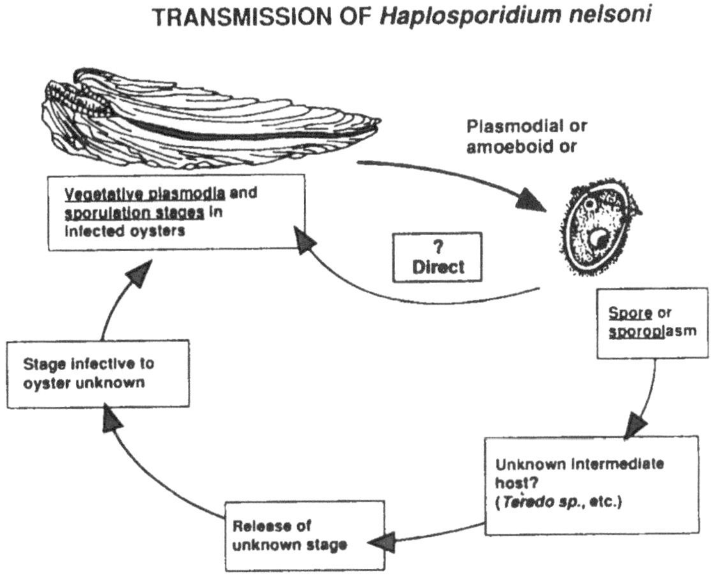

Fig. 18-2. Illustration demonstrating incomplete knowledge of life-cycle and mode of transmission of MSX (*H. nelsoni*), one of the most devastating pathogens in oysters.

Tables 18-2 and 18-3 list some of the key pathologic problems in marine/estuarine species that provide general or specific challenges for research.

Table 18-2. Infectious Disorders

Diseases or Agents	Some Key Challenges[1]
Viruses	a) Roles of newly discovered viruses in mortalities and role of mariculture in their spread. b) Interactions of stressors, host/virus in disease. c) Roles in neoplasia. d) Viral syndromes. e) Use of molecular techniques to solve above.
Rickettsia, Mycoplasma, Chlamydia Bacteria	a) Do they cause disease or are they commensals in such hosts as mollusks, fishes, and marine mammals? b) Carried by marine species to humans? c) Prokaryote disease – agent-specific syndromes. d) Use of molecular probes to follow prokaryotes in hosts.
Ulcerative mycoses	a) Factors that initiate – role of water quality. b) Complete syndrome description. c) Specific etiologic agent identification.
Protozoa	a) Pathogenic vs. commensal role plus enhancing factors. b) Experimental clarification and demonstration of complete life – cycles, hosts, and ranges. c) Fish gill diseases and pseudoneoplasms. d) Syndromes of distinct protozoan diseases. e) Use of molecular probes to identify.
Metazoa	a) Roles in mass mortalities in concert with other agents. b) Symbionts, mutualists, and commensals that may serve as bioindicators of water quality while utilizing their hosts as microhabitats. c) Use of molecular techniques to demonstrate pathogenic mechanisms of metazoa in hosts.

[1] An almost universal challenge to be added to each group below (Tables 18-2 and 18-3) is the identification and elucidation of the roles of *predisposing factors* that affect or modulate agents or diseases listed.

Table 18-3. Non-infectious Disorders

Diseases or Agents	Some Key Challenges
Metals and their cellular effects	a) Mechanisms of action and cellular targets. b) As cofactors in concert with other disease agents. c) Pathognomonic signs and syndromes.
Organic chemicals	a) Roles as etiologic agent in carcinogenesis in fishes and invertebrates. b) Cellular, tissue, organ or system specificities in lesions or disease origin. In invertebrates, fishes, and marine mammals - toxicological pathogenesis. c) Descriptions of toxic syndromes.
Inflammatory, cellular proliferative, and plasias (dys-,hyper-, meta-, neo-.)	a) Roles of interactions, immune response, immunomodulation, and cells and tissues of origin in pathogenesis. b) Are these lesions bioindicative of exposures to chemical or biological agents? c) Roles of these processes in various specific disease syndromes, or as distinct syndromes themselves, (e.g., individual neoplastic diseases).
Genetic disorders	a) Influence of environmental factors on genetic diseases in populations. b) Susceptibility of genetically predisposed populations or individuals to pathogenic agents. c) Chromosomal behavior under influence of toxicants (chromosomal anomalies). d) Abnormal gene function as a pathogenic mechanism. e) Development of molecular techniques to evaluate target loci of genotoxic agents.

FUTURE POSSIBILITIES -- A NEED FOR A SYSTEMATIC, RATIONAL APPROACH TO LINK CAUSE AND EFFECT

For anyone called upon to investigate a mass mortality of marine organisms (shellfish, fish, birds, mammals), there is the memory of how difficult or impossible it may have been to assign or even indict a causative agent or etiology. I am not referring to such obvious events as massive oil spills or release of pesticides in abnormally high quantity, but rather to those frustratingly more subtle cases wherein a multitude of possibilities faced you at the onset. Administrators, the public, and even fellow scientists, not involved, are not very sympathetic at these times because they all want a quick and easy answer. Unfortunately, there are usually few quick and easy answers.

Perhaps, though, we do not have to look upon disease and mass mortality in the sea as some huge lottery; instead we can try rationally to reduce the uncertainty of causes and risk assessments. In order to discuss these possibilities one needs to examine, conceptually, a frame-work of principles and theory of how marine disease studies now may be approached, and the expectations for these approaches.

The Comparative Approach

The first principle considered is that of the intrinsic, comparative nature of the pathobiology of marine/estuarine species. This comparative nature becomes obvious when one considers that marine/estuarine pathobiologists may be concerned with species from all invertebrate and vertebrate phyla and classes, either directly or indirectly (e.g., as victim, host, parasite, or pathogen). One may specialize in certain diseases and perhaps limit one's research to a few or even one species of host; but to be in a position to contribute to the broad science of pathobiology, one must keep a comparative mind-set. If, on the other hand, one is chiefly concerned with the health and disease of marine/estuarine organisms widely and generally, a comparative approach and view are a necessity and must continually underlie the total work and investigation.

This comparative mentality may permit researchers to draw from diverse examples to clarify specific disease enigmas involving similar agents causing disease in diverse species (e.g., bacterial infections in invertebrates vs. fishes vs mammals; or hepatocarcinogenesis in many teleost fishes vs. similar processes in invertebrates and mammals where much less or more work, historically, has been done). Simply put, what is learned in one case may be useful, at least partially, in solving other problems.

Complexity of Pathogenesis in Marine Species

As an extension of dealing with many different invertebrate and vertebrate species (the comparative factor), the innate complexity of many different potential responses of these diverse species to disease-causing factors, further challenges the investigator. An example of this complexity, (Fig. 18-3), which is concerned with only one important exemplary area of diseases in both inverte brate and vertebrate marine species, is that of pathogenesis of neoplasia. The marine pathobiologist studying cancer must be able to resolve confounding associated responses or concurrent lesions or entities, such as necrosis, inflammation, fatty change, cirrhosis, hyperplasia, regeneration, wound repair, and unicellular, amoeboid parasites using subject as host, from the possible endpoint of true neoplasia in both invertebrates with their unique cellular systems and vertebrates with their varied cells and tissues from fishes to mammals.

The real value of understanding the complexities of pathogenesis caused by various agents in diverse organisms is the ability to identify those stages (lesions) in pathogenesis that may be useful as valid intermediate endpoints (see excellent

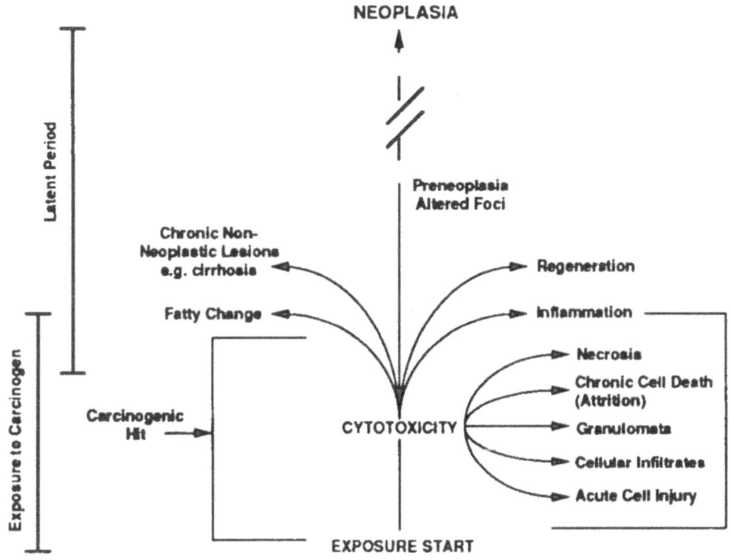

Fig. 18-3. Complexities involving varied and multiple phenomena associated with exposure to carcinogenic agent (either chemical or biological) and carcinogenesis.

discussion on intermediate endpoints in cancer research by Schatzkin *et al.* 1990),biomarkers, indicators, or sentinel markers of exposure and risks of/to agents in the environment (Couch 1985; Couch 1988; Hinton and Couch 1984; Couch and Harshbarger 1985; Hargis and Zwerner 1988; Vogelbein et al. 1990; Harshbarger 1991).

Research Disciplines and Endpoints in Study of Marine Diseases

It is centrally important to remember that ultimate lesions, at the cellular, tissue, organ and organismic levels (Fig. 18-9), may be integrative or summing indicators (signs) of the total biochemical, physiologic, and morphologic impacts of the causative agents. The quest is to find when the level of designative or definitive impact (e.g., a biomarker or bioindicative point) is reflected in a given stage (or lesion) or series of stages of pathogenesis. This can be induced only from the weight of evidence resulting from detailed descriptive and experimental studies of lesion induction, origin, development, progression, and fate (i.e., pathogenesis). Integrative pathobiology is an area where the diagnostic histopathologist, experimental pathologist, experimental and molecular biologist -- each using different, but overlapping approaches -- must work together to synthesize theses of causes and progression of disease.

Unfortunately our knowledge is still in great need of expansion in order to increase our understanding of modes of action and cellular injury for various agents as reflected in Fig. 18-4 (in regard to my earlier use of the paradigm relating to neoplasia as a complex, multifactorial disease).

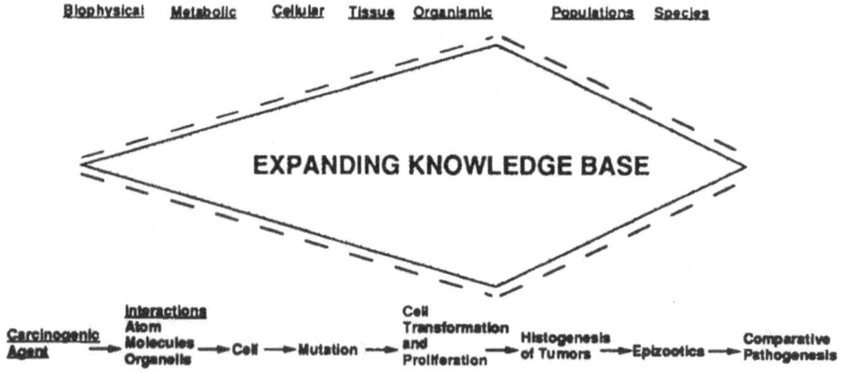

Fig. 18-4. More information at the intermediate cellular, histopathological, and organismic levels exists for marine species' diseases (in this example, cancer) than at the subcellular and population/species levels of biological organization.

For marine diseases, we tend to have more data and information about the grossly observable and intermediate level of effects in our subjects, particularly at the tissue, organ, system, organismic levels of organization (Fig. 18-9), than at the agent (mode of action and structure-activity), molecular, and sub-cellular levels (Figs. 18-4 and 18-9). Other areas, in which quality information and understanding is weak, are at the population and comparative species response potential levels for both infectious and non-infectious agents. We need to balance our studies of effects across the spectrum of biological organization, from the molecular to the population/species levels to better understand critical points of action of disease agents. Marine pathobiology must be an eclectic science, simply because of the complexity of the phenomena it investigates.

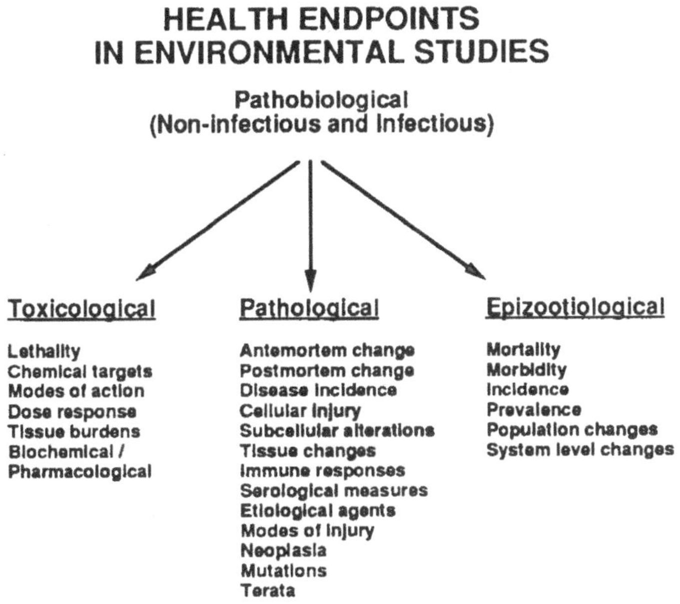

Fig. 18-5. Measurable and observational endpoints used to gauge health/vulnerability of estuarine/marine species. These endpoints can be, and have been, used with many species of invertebrates, fishes, and other marine species. Most of the information in our knowledge bases has been gained through studies utilizing these endpoints. To be added to these lists are some of more recent advances in molecular biology, such as DNA/RNA hybridization (blot) methods for pathogen identification and genetic characterization; immunohistochemistry for aid in determining cell types involved in disease processes; and genetic *markers* for following disease agents in the complex marine environment.

Pathobiology, may be the life science that most demands knowledge of, and use of information, from all levels of biological organization and many disciplines. This is true because the pathobiologist must know or learn and understand normal structure and function as well as malstructure and dysfunction in biological systems (Sparks, this volume). Few, if any, other practitioners of science face broader requirements of specific knowledge.

Fig. 18-5 lists major conventional health (or disease) endpoints that may be identified, estimated, or measured with the tools provided within the combined disciplines of toxicology, pathology, and epizootiology. Attempts must be made to systematically integrate these measures and observations for estuarine/marine species, so that *syndromes* of response may be identified, understood, and used to clarify consequences and for application to predict impacts and risks for both infectious and non-infectious agents. This is the rational, alternate, or additional approach, in contrast and in comparison to presently belabored studies or attempts at isolated, clinical measurements of disease. Both experimental and field studies may be conducted with the intention of establishing and demarcating syndromes that identify specific disease responses in marine species.

As an extension of the disciplinary interaction view, marine pathobiologists must bring themselves up to current understanding in the use of molecular techniques (or enlist, as suggested previously, the help of molecular biologists) to have better and more incisive tools with which to attack problems that have not yielded to more traditional histopathological or microbiological methods (e.g., use of DNA hybridization probes to identify certain stages of infectious pathogens in different marine hosts to add to our understanding of life cycles and relatedness of the pathogens and to understand where they may occur in the ecosystem when not causing overt disease in specific hosts, see Fig. 18-2).

Rational Approaches to Understanding Cause and Effect Relationships

Traditionally the study of diseases in estuarine/marine species has been conducted under the sponsorship and needs of agencies and institutions concerned with ecological problems (usually as a small subset of those needs). Also, in the conventional sense, the word *health* has usually denoted human, or at least mammalian concerns (Fig. 18-6). In the last 20 years, however, there has been a growing recognition, particularly in regard to population effects and comparative species dysfunction, that many endpoints used to measure both individual and population conditions may arise from, and be useful in pursuing, both health and ecological studies. This has been particularly true in regard to the impacts of anthropogenic factors (pollution, environmental stressors) on coastal marine species. Several components, for example, of an ecosystem may be directly or indirectly altered by a single disease process in one or several

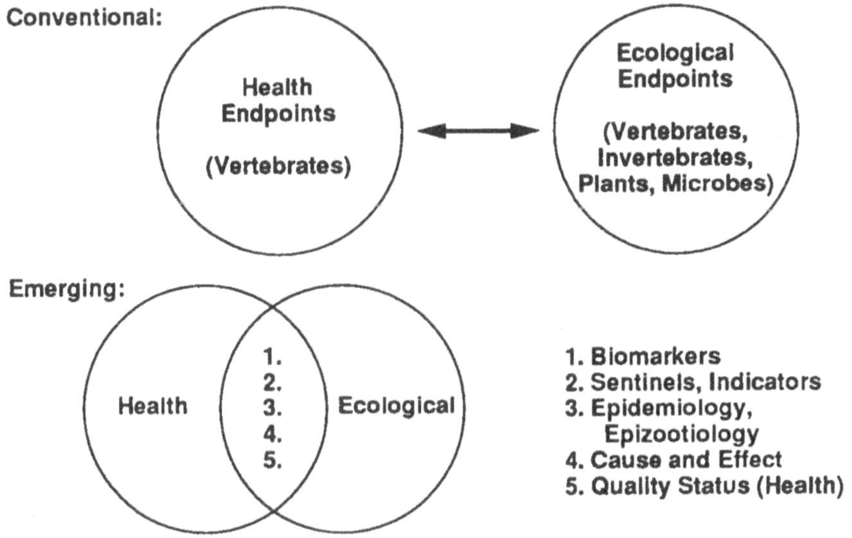

Fig. 18-6. Represented here is the conventional, but perhaps artificial, view of separation of health endpoints and ecological endpoints used in efforts to evaluate mammalian versus other forms of life status. Also represented is an emerging view that realizes there may be overlap in use of both so-called health and ecological measures to establish holistic indicators of individual, population, species and ecosystem status.

species. Prey species populations reduced by chronic disorders induced by a pollutant may have vicarious effects on higher trophic forms. Likewise, infectious epizootics may impact more than the species infected by causing mass mortalities that alter or reduce the normal contribution of that species to the ecosystem (e.g., eel grass mortalities in estuaries worldwide that may be caused by pathogenic slime molds [*Labyrinthula*] can affect all those species that directly or indirectly depend upon the estuarine, eel grass habitat; [Muehlstein 1989]).

There is, therefore, an emerging view that would combine the best and the more interdependent endpoints from the health and ecological disciplines to arrive at biomarker, sentinel, and indicator measurements and evaluations that provide insights into cause and effect relationships and which explain quality or condition status in populations and ecosystem as a result of the cause and effect process (Fig. 18-6). Indeed, as both marine pathobiology and marine ecology mature as sciences, at least in many of their approaches, their disciplinary borders may become blurred in regard to investigating the *health* of ecosystems. That is, a competent marine pathobiologist should become a fair marine biologist, through both training and experience, and a good marine ecologist should become knowledgeable about diseases and their processes in marine/ estuarine species and populations. With the best possible combinations of health and ecological endpoints to be sought and estimated in estuarine/ marine disease situations (Figs. 18-5 and 18-6), a logical, consistent approach is needed. Fig. 18-7 outlines one possible approach to getting at, or understanding, etiologies in marine disease crises, such as massive or significant, unexpected mortalities of plants, invertebrates, fishes, birds, or mammals. In a hypothetical crisis, assuming there is no known syndrome of cardinal signs readily detected that would quickly establish etiology (e.g. finding massive infections of *Perkinsus* in dying oysters which exhibited the other classical "Dermo" disease signs), one probably must use a process of elimination or subtraction of possible causative or contributing factors in order to arrive at the real cause as illustrated in Fig. 18-7. There can be no substitute for experience and broad, but specific knowledge of hosts, disease syndromes, disease processes, and probable etiological agents in optimizing the approach of elimination or subtraction of possible factors. This is a particularly well-made point when one considers the breadth of the categories of physico-chemical factors, toxins, and toxicants, and the vast and diverse groups of potential infectious agents. The initial approach (direction) that an investigative team takes (in reference to Fig. 18-7) in identifying the possible causative agent may not be as important as is its *a priori* knowledge about each possible component in the different groups.

The most complicated situation is encountered when there is interaction among two or more of the factors in the initiation and promotion of the disease process (Fig. 18-7). Thus, one must expect such complications in almost every epizootic or disease process found at work in natural or cultured estuarine/marine populations. Every preceding chapter on specific diseases in this volume attests to that, and the reader is referred particularly to the separate chapters by Kent and Fournie, Overstreet, Lightner, Sindermann, and Anderson for specific instances of interactions among initiating and promoting factors of disease in marine species.

Use of Data and Results to Predict Health or Quality-state of Biota

The competent practice of quality science usually results in understanding that ideally should lend greater predictive capacity to that science, the new science of chaos theory, not withstanding (Gleick 1987). It has been said that the true contribution of science is in its capacity to take detail, generalize, and then predict consistently and validly. In the evolving fields of pathobiology of aquatic species, much new information and data are being generated annually. How should the investigators attempt to utilize varied and growing data to add to their predictive capacity concerning marine diseases (e.g., Fig. 18-4)?

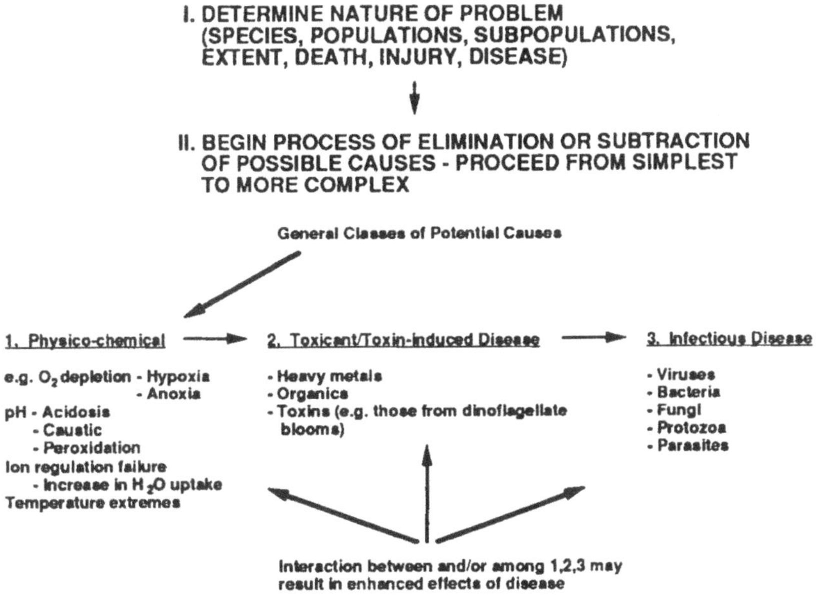

Fig. 18-7. One approach to arriving at cause of major mortalities or marine disease outbreaks would be to use the process of elimination or substraction of possible factors that could be causative through a rigorous collection of physical/chemical evidence, and determination of presence or absence of toxins, toxicants, or infectious agents.

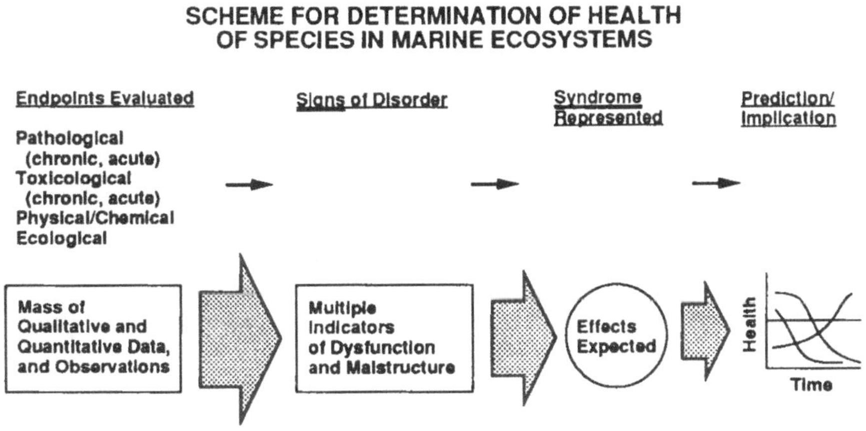

Fig. 18-8. An idealized scheme for handling data and qualitative observations of significance in establishing syndromes that identify specific disease processes.

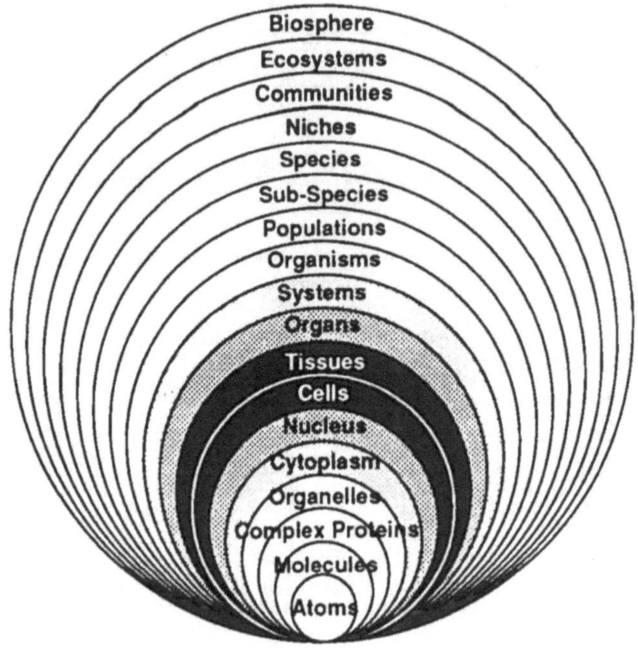

Fig. 18-9. Diagram conceptually represents selected levels of biological organization. The relatively darkened intermediate levels from organism down to the organelle represent biological levels at which we are making most of our significant observations of pathobiological effects, in estuarine/marine systems, species, and populations.

Fig. 18-8 relies heavily on the paradigm established over many decades in biomedicine, but also includes ecological endpoints; it illustrates one conceptual approach to the rational, systematic utilization of large data bases to predict both the state of and the future health of organisms and their populations. The many endpoint measurements (quantitative) or observations (qualitative) that constitute the essential raw information of our knowledge base must be assigned to their proper roles in malstructure and dysfunction as signs of larger disorders. From those groups of related or interdependent signs will emerge processes that constitute syndromes which are critical points at which a process as a definitive (identifiable) entity emerges. If these syndromes, as end results of multifactorial processes, can be characterized and related in reverse fashion to causes by observation, association, rigorous experimentation, and field validation, then one may predict a highly probable outcome of the process (i.e., a clinical course, within the individual, population, or even species and ecosystem of the process), and its eventual impact.

An excellent example of this patient, intensive approach, utilizing pathobiological endpoints to identify a response syndrome to mixed toxicant exposure is that of Braunbeck *et al.* (1990), Burkhardt-Holm *et al.* (1990), Braunbeck and Völk (1991), and Spazer *et al.* (1991), all of the University of Heidelberg. With painstaking thoroughness these authors utilized histologic ultrastructural, and biochemical methods to demonstrate a complex syndrome in Rhine river eels resulting from exposure to the chemical spills in Basle in 1986.

Unfortunately, we are merely in the process of establishing the data sets with only a relatively few marine species, although our understanding of a few select diseases may be far advanced over other diseases that we encounter. Therefore, it is incumbent upon each marine disease researcher to be thinking of the best possible ways to assimilate and integrate their findings into bases for theories and explanatory paradigms that will eventually lead to their use in testing hypotheses, establishing cause and effect relationships, evaluating status of health of marine species, and finally in predicting impacts of disease processes on estuarine/marine populations.

REFERENCES

Alder, J., and R.D. Braley. 1988. Mass mortalities of giant clams on the *Great Barrier Reef. In Giant Clams in Asia and the Pacific*, ed. J.W. Copland, and J.S. Lucas. Canberra: ACIAR.

Braunbeck, T., P. Burkhardt-Holm, and V. Storch. 1990. Liver pathology in eels (*Anguilla anguilla* L.) from the Rhine raiver exposed to the chemical spill at Basle in November, 1986. *Limnologie Aktuell* 1:371-392.

Braunbeck, T., and Alfred Völkl. 1991. Induction of biotransformation in the liver of eel (*Anguilla anguillla* L.) by sublethal exposure to dinitro-o-cresol: and ultrastructural and biochemical study. *Ecotox. Environ. Safety* 21:109-127.

Burkhardt-Holm, P., T. Braunbeck, and V. Storch. 1990. Auswirkung der beim Sandoz-unfall in November 1986 in den Rhine gelangten chemikalien auf die ultrastruktur des Darmes von Aalen. *Limnologie Aktuell* 1:393-404.

Couch, J.A. 1985. Prospective study of infectious and non-infectious diseases in oysters and fishes in three Gulf of Mexico estuaries. *Dis. Aquat. Org.* 1:59-82.

Couch, J.A. 1988. Guest Editorial. Role of pathobiology in experimental marine biology and ecology. *J. Exp. Mar. Biol. Ecol.* 118:1-6.

Couch, J.A. 1990. Pericyte of a teleost fish: ultrastructure, position, and role in neoplasia as revealed by a fish model. *Anat. Rec.* 228:7-14.

Couch, J.A. 1991. Spongiosis hepatis: chemical induction, pathogenesis, and possible neoplastic fate in a teleost fish model. *Toxicol. Pathol.* 19(3):237-250.

Couch, J.A., and D.W. Nimmo. 1974. Detection of interactions between natural pathogens and pollutants in aquatic animals. In *Proceedings Gulf Coast Regional Symposium on Diseases of Aquatic Animals*, pp. 262-268. Louisiana State University Center for Wetland Resources, Baton Rouge, LA.

Couch, J.A., and J. Harshbarger. 1985. Effects of carcinogenic agents on aquatic animals: an experimental and environmental overview. *Environ. Carcinog. Rev.* 3:63-105.

Couch, J.A., and L. Courtney. 1987. N-nitrosodiethylamine-induced hepatocarcinogenesis in estuarine sheepshead minnow (*Cyprinodon variegatus*): neoplasms and related lesions compared with mammalian lesions. *J. Nat. Cancer Inst.* 79:297-321.

Dawe, C.J., and J.A. Couch. 1984. Debate: mouse vs minnow: The future of fish in carcinogenicity testing. *Nat. Cancer Inst. Monogr.* 65:223-235.

Gleick, J. 1987. *Chaos, Making a New Science.* 352 pp. New York: Penguin Books

Goggin, C.L., K.B. Sewell, and R.J.G. Lester. 1990. Tolerances of *Perkinsus* spp. (Protozoa, Apicomplexa) to temperature, chlorine, and salinity. *J. Shellfish Res.* 9:145-148.

Hargis, W.J., and D.E. Zwerner. 1988. Effects of certain contaminants on eyes of several estuarine fishes. *Mar. Environ. Res.* 24:265-270.

Hinton, D.E., and J.A. Couch. 1984. Pathobiological measures of marine pollution effects. In *Concepts in Marine Pollution Measurements*, ed. H.H White, pp. 7-32. Maryland Sea Grant Publ., Univ. of Maryland, College Park, MD.

Hinton, D.E., J.A. Couch, S.J. Teh, and L.S. Courtney. 1988. Cytological changes in progression of neoplasia in selected fish species. *Aquat. Toxicol.* 11:77-112.

Littlewood, D.T.J., and S.B. Ford. 1990. Physiological responses to acute temperature elevation in oysters, *Crassostrea virginica* (Gmelin, 1971), parasitized by *Haplosporidium nelsoni* (MSX)(Haskin, Stauber, and Mackin, 1966). *J. Shelf Res.* 9:159-164.

Meyers, T.R., and J.D. Hendricks. 1982. A summary of tissue lesions in aquatic animals induced by controlled exposures to environmental contaminants, chemo-therapeutic agents, and potential carcinogens. *Mar. Fish. Rev.* 44(12):1-17.

Muehlstein, L.K. 1989. Perspectives on the wasting disease of eelgrass *Zostera marina. Dis. Aquat. Org.* 7(3):211-221.

Patton, J.S., and J.A. Couch. 1984. Can tissue anomalies that occur in marine fish implicate specific pollutant chemicals? In *Concepts in Marine Pollution Measurements*, ed. H.H. White, pp. 511-538. Maryland Sea Grant Publ., Univ. of Maryland, College Park, MD.

Schatzkin, A., L.S. Freedman, M.H. Shiffman, and S.M. Dawsey. 1990. Validation of intermediate endpoints in cancer research. *J. Nat. Cancer Inst.* 82(22):1746-1752.

Spazier, E., V. Storch, and T. Brunbeck (in press). Cytopathology of spleen in eel (*Anguilla anguilla* L.) exposed to the chemical spill in the Rhine river in November, 1986. *Dis. Aquat. Organ.*

Vogelbein, W.K., J.W. Fournie, P.A. VanVeld, and R.J. Huggett. 1990. Hepatic neoplasms in the Mummichog Fundulus heterolclitus from a creosote-contaminated site. *Cancer Res.* 50:5978-5986.

INDEX

A

A/Seal/Mass/1/80 virus, 221
Abalone, *Labyrinthuloides haliotidis* in, 278–279
Aber disease, 272
  distribution of, 273–275
AC, see Anthropogenic contamination
Acanthocephalans
  with agricultural and industrial toxicants, 146–147
  in pinnipeds, 232
Acid phosphatase, release of from hemolymph cells, 489
Acid water, 147
Acropora palmata, with white band disease, 404–406
Acroporids, white band disease in, 404–405
Adenoma, hepatic, 202
Adenoviruses
  of Atlantic cod, 43
  of marine mammals, 220–221
  in plaque epidermal neoplasms, 162
Aerococcus viridans homari, 323–324
  course of disease caused by, 331–334
  in crustaceans, 322
  entry of through ruptured integument, 330–331
  growth of in lobster tissues, 332
Aeromonas infections, 60–61
  hydrophila, 61, 262
  salmonicida, 60–61, 427
  ulceration with, 459
Aeromonasis, 262
Aflatoxicosis, in penaeid shrimp, 352–353
Aflatoxin, immunosuppressive properties of, 4
Aflatoxin B1
  acute toxicity of, 353
  necrosis and fibrosis with, 371
Agglutinin, in bivalve serum, 491
Agrochemicals
  carcinogenic effect of, 382–383
  toxic effect of and parasitic diseases, 146–147
Aldrin, effects of on marine molluscs, 303
Algae
  causing disease in shrimp, 350–351
  pathological effects of ichthyotoxins of, 5–6
  toxins of, 5
Algal blooms, 5–6
Algal tumor, 410
2 Alpha-dicarbonyl compounds, 166–167
Alpha pathology, 513
Ammonia, as cause of black gills, 354
Amoco Cadiz oil spill, 379
Amoebiasis, green sea urchin, 427–429
Amyloodinium ocellatum, 116
  infection of skin and gills with, 12
  life cycle of, 116–118
  mass fish mortalities from, 117–118
  temperature and salinity effects on, 117–118
Anemia, 38–39
Anisakine nematodes, 7
Anisakis, in pinnipeds, 231
Annelids
  diseases of, 415–422
  gregarines associated with, 417
  important recent epizootics in, 435
Antarcophthirus microchir, 230
Anthozoa, 401
Anthropogenic contamination
  bioindicators of, 1
  diseases and acute fish kills with, 3–4
  diseases caused by, 2–5
  efforts to reduce impacts of, 473–474
  immunosuppression with, 4–5
  reproductive problems associated with, 5
  sentinels for, 11–12
Anthropogenic environmental factors, 135–147
Anthropogenic waste, 236–237
Aphanomyces, 105, 106
  astaci, crayfish plague with, 334
  challenge of menhaden with, 98
  growth and sporulation of, 92–93
  in ulcerative mycosis, 92
Apicomplexan protozoan, 121–124
Aquaculture systems, 9–13
Aquatin, causing chronic soft-shell syndrome, 348–349
Arctic marine environments, petroleum pollution in, 378
Argyrism, 380
Aroclor 1254
  hepatocytomegaly with exposure to, 198
  testicular or sperm damage with, 192
Aroclor, effects of on marine molluscs, 303
Aromatic hydrocarbons, see also Polycyclic aromatic hydrocarbons

laboratory studies of effects of, 371
liver lesions and, 464–465
pathological disorders produced by, 362–363
substituted, 371–372
symbiont influence on, 142
Arsenic
histopathology of female reproductive tissues with, 192
pathological disorders produced by, 361
toxicologic characterization of in mollusc organs, 380
Arthropoda, 319–320
Arthropods
molting cycle in, 247
resistance of to disease, 336
studies on diseases of, 320
Ascardidoids
environmental effects on, 127–130
hosts of, 127–129
Ascomycetes, on marine crustaceans, 325
Ascorbic-acid deficiency syndrome, 345
Aspergillosis, 228
Asphyxiation, with algal ichthyotoxins, 5–6
Asteroidea, 423
Atlantic cod adenovirus, 43
Atlantic cod ulcus syndrome, 42
Atlantic croaker, lymphocystis on, 41
Atlantic menhaden
early ulcerative mycosis lesion in, 90
spinning disease in, 8, 42
ulcerative lesions with fungal infection in, 7
Atlantic salmon, 43–44
Aureobasidium, 106
Azo compounds
pathological disorders produced by, 363
toxic effects of, 372

B

B lymphocytes, of fish, 495–496
Bacillary necrosis, 260
Bacteria
agglutinins of in bivalve serum, 491
iron effects on, 137
opsonization of, 499
Bacterial cold-water disease, 74
Bacterial diseases, 53–77
conditions of propagation of, 53
in crustaceans, 321–325
in intensive aquaculture systems, 12
of marine mammals, 223–227
in molluscs, 262–264
number of affecting fish, 53
in sponges, 396–399
in wild fishes, 7–8
Bacterial pathogens, 54–55
Baculovirus infections
definitive experiments on, 337
penaei, 321, 253
Bald sea urchin disease, 427, 428
Baltic Sea, pollution of and mussel pathology, 378–379
Basophilic foci cells, 199
Beluga whales, neoplasms in, 234
Benedenia monticelli, 120
Bent spines, pollution-induced, 467
Benzo[a]pyrene
carcinogenic effect of, 382
liver lesions with exposure to, 465–466
in mussel tissues, 378
and neoplasm formation, 381
pathological disorders produced by, 371
Bile ductular/ductal hyperplasia, 197
Bilobular epidermal neoplasms, 163
Bioconcentration factor (BCF), 305
Bioindicators, for parasites, 147–149
Biological organization levels, 528
Biomedical research, 13–15
Birnaviruses
division of serotypes of, 29
infecting molluscs, 256
isolated from epizootic ulcerative syndrome, 104
in wild fishes, 8
Bivalves
bacterial diseases in, 262–263
cell surface receptors of, 491–492
chemical effects on resistance to infections in, 492–494
crude oil exposures in, 373–374
fungal disease in, 264
hemocytic and serum lysosomal enzymes of, 488–489
hemocytic neoplasia in, 291, 295
humoral factors of, 490–491
immunomodulation in, 494–495
paramyxean pathogens in, 272–276
petroleum-contaminated sediment effect on, 281
phagocytic activity in, 486–488
pollution-induced structural modifications in, 310
reactive oxygen intermediates in, 489–490
shell disease of, 264, 468
structural anomalies in, 290
temperature and salinity effects on hemocyte activity in, 492

total and differential hemocyte counts in, 484–486
vibriosis of larvae of, 260
xenobiotic-induced mesotheliomas and myxomas in, 382
Black band disease, 401–404
Black death disease, 345
Black gill disease, 326
cadmium and, 370
noninfectious causes of, 354
in penaeid shrimp, 353–354
pollution-associated, 471
Black Mat Disease, 326
Blastomycosis, 228
Bleaching, of corals, 407–410
Blood flukes
eggs of in red hind ventricle, 133
heavy metal effects on, 138
mortality related to, 130–132
Blue crab, 246
Blue-green algae, shrimp disease with, 351–352
Bonamia ostreae, 276–277
Bone development, 185
Botulinum toxin, 66–67
Brain coral, bleaching of, 408
Brain damage, with environmental pollution, 470
Braunina cordiformis, in cetaceans, 233
Breeding strategies, 193
British Columbia study, 378
Brittlestars, 423–425
Broken back syndrome, 185
Bronchostitial pneumonia, 223
Bulbous epidermal neoplasms, 169

C

C-type retrovirus, 44
Cadmium
and chemoluminescence response, 501
concentration of in molluscs, 306
effects of on phagocytosis, 487
histopathology of female reproductive tissues with, 192
immunosuppressive properties of, 4
laboratory studies of effects of, 360–370
leukocyte count with exposure to, 498
pathological disorders produced by, 361
protective disinfecting role of on parasite, 138
testicular or sperm damage with, 192
vitamin C depletion with, 185
Cage studies, 150
Calcium carbonate exoskeleton, , 411
Calcium oxalate crystal accumulation, 13
Calcium phosphate, 466
Calici virus, 219–220
Caligid copepod, 136
Calyptospora funduli infection
environmental effects on, 121–124
low temperature waters and, 3
Camobacterium piscicola, 68
Candida albicans, 228
Candidiasis in marine mammals, 228
Captive fishes, 8–13
Carbamate insecticide, 303
Carbazoles, in sediments, 374
Carcinogenesis, 380–381
and chemical pollution, 382–383
complexities in phenomena associated with, 521
genetic alteration leading to, 297–299
in molluscs, 297
Carcinogens
broad occurrence of in aquatic ecosystems, 383
complexities in exposure to, 521
and liver neoplasms in fish, 160
in neoplasm formation, 381
sensitivity to in medaka, 465–466
serial progression studies in exposure to, 199
uptake and accumulation of by aquatic animals, 300
Carcinoma, 202–204
Cardiac neoplasms, 189
Cardiovascular system, toxicologic pathology of in fish, 188–189
Caribbean sea urchin mass mortalities, 429–432
Carnivora, 217
Cataracts, pollution-induced, 469–470
Catarrhal enteritis, 28
Catastrophic events, 235–236
Catecholamine release, 496–497
Cause-effect relationships, 524–526
Cell-mediated immunity, 296–297
Cell metabolism, 466–468
Cell surface receptors, 491–492
Cellular tissue responses, 516–517
Central Long Island Sound, dredging and disposal in, 376
Central nervous system, 186–187
Cercaria, 413
Ceroid deposition, 13
Cestodes, as indicators of environmental conditions, 132–134
Cetaceans, 217
candidiasis in, 228
endoparasites in, 232–233

neoplasms in, 234
pox viruses in, 219
Sarcocystis species in, 229–230
Chaetoceros graeilis, senescent blooms of, 350
Chambering, 468
Channel catfish, *Edwardsiella tarda* infection in, 56
Chattonella antiqua, detrimental algal blooms of, 5
Chemical carcinogens, 297–299
Chemical injury, 251–252
Chemical stressors, 495
Chemically induced tumors, 252–253
Chemicals
causing histopathology in aquatic invertebrates, 359–383
cellular phenomena associated with, 252
effects of on resistance to infection in bivalves, 492–494
field studies of effects of, 374–379, 381–383
laboratory studies of effects of, 360–374, 379–381
neoplasms associated with, 169–170
pathological disorders produced by, 361–366
physiological effects of, 238
Chemoautotrophic bacteria, mutualistic symbioses of, 394
Chemoluminescence
in fish, 499–501
in oysters, 489–490
Chemotherapeutics, external, 9
Chinook salmon
hematopoietic tumors of, 165
metacestode infection in smolts, 9–10
viral hemorrhagic septicemia in, 31
Chitinoclastic microorganisms, 328
Chitinous layers, 329
Chlamydia-like organisms, in crustaceans, 324
Chlamydiae, causing disease in molluscs, 262–263
Chlordane, effects of on marine molluscs, 303
Chloride cell hyperplasia, 190
Chlorinated hydrocarbons
and early oocyte atresia and low fertilization rates, 456–457
effects of on marine molluscs, 302–303
larval mortality with, 455
pathological disorders produced by, 363
in sediments, 374
skeletal deformities with, 467
tissue levels of, 376, 455–456
Chlorinated organic compounds, in Puget Sound sediment, 378
Chlorine, concentration of in molluscs, 306–307
Chlorine bleach effluent, effect of on parasitic diseases, 138–139
Chloroacetones, in chromatophoroma etiology, 166
Cholangiocarcinomas, 203, 463
Cholangioma, 203
Chromatoblastomas, epizootic, 166–167
Chromatophoromas, 182–184
Chromium, 361–370
Chromosomal anomalies, 453
Chronic soft-shell syndrome, in penaeid shrimp, 347–349
Chrysochromulina polylepis, fish kills caused by, 6
Chum salmon reovirus (CSR), 37
Ciliate infestions, kraft mill effluent effects on, 139
Cirripedia, polycyclic aromatic hydrocarbons in, 380
Cladosporium, 106
Clams
acid phosphatase release in after copper exposure, 489
bifurcation of foot of, 290
crude oil exposures in, 373–374
hemocyte lysis in, 484
neoplasms in, 294
pollution-induced vacuolar lesions in, 379
RNA viruses of, 256
studies of anomalies in, 375
Clear cell foci, 200, 201
Clostridium botulinum, 66–67
Clostridium perfringens, water pollution with and sea urchin mortalities, 431
Cnidaria
Black Band Disease in, 401–404
bleaching of, 407–410
important recent epizootics in, 434–435
mutualistic symbiotic relationship of with zooxanthellae, 407–408
species of, 400–401
tumors in, 410–412
white band disease in, 404–406
Coagulative hepatic necrosis, 196
Coccidian infections
environmental effects on, 121–124

and low temperature waters, 3
Coccidiodomycosis, 228
Cockles, pollution-induced tumors in, 379
Coho salmon
rickettsial infection of, 63–66
viral hemorrhagic septicemia in, 31
Cold water, fish kills associated with, 2–3
Combjellies, diseases of, 412–415
Comparative approach, 520
Concanavalin A, binding of to bivalve hemocyte membranes, 491–492
Concanavalin A-receptor complexes, 492
Connective tissue neoplasms, clusters of, 157
Contaminated sites, pathological disorders at, 367–369
Contracecum, in pinnipeds, 231
Copepods, 134–136
Copper
and chemoluminescence response, 500–501
concentration of in oysters, 306
effects of on bivalve hemocyte count, 485
effects of on phagocytosis, 487
immunosuppressive properties of, 4
laboratory studies of effects of, 370
pathological disorders produced by, 361–362
and reduced acid phosphatase release, 489
research on toxic effects of, 380
sublethal effects of on molluscs, 307–309
Copper sulfate pollution, acute fish mortality with, 4
Coral
black band disease in, 401–404
bleaching of, 407–410
loss of brownish pigments from tissues of, 407–410
mortalities of with environmental changes, 409–410
mutualistic symbiotic relationship of with zooxanthellae, 407–408
prostaglandins synthesized from, 394
sloughing of tissues of, 404, 405–406
tumors in, 410–412
white band disease in, 404–406
Coral reefs, 393, 395
Crabs
chlamydia-like and rickettsia-like infections in, 324
fungal infections of, 326
Hematodinium perezi infection in, 327–328
shell disease in claw of, 461
shell lesions in, 330
viral disease in, 321
Cramped muscle syndrome, 346–347
Crayfish
Aphanomyces astaci infection, 329–330
DMNA in, 373
fungal infections in, 325–326
internal defense mechanism of, 336
Crayfish plague, course of, 334–335
Crude oil
detrimental effects of on neurosensory system, 187
effects of on bivalve hemocyte count, 484–485
and hemic neoplasms, 378
laboratory studies of effects of, 373–374
Crustacea, 319–320
Crustacean symbionts, environmental effects on, 134–135
Crustaceans
agglutininlectin capacities of, 335
bacterial and rickettsial diseases in, 321–325
course of infections in by specific pathogens, 330–335
course of non-specific infection in, 328–330
defense mechanisms of, 334–335
diseases of, 246, 320
fungal infections in, 325–327
infectious diseases of, 319–338
noninfectious diseases of, 343–354
phenoloxidase system in, 335–336
physiology of, 462
protozoans, helminths, and parasites in, 327–328
shell disease in, with environmental pollution, 461–463
toxic and carcinogenic properties of insecticides in, 372
viral families recorded in, 322
viruses in, 321
Cryptocaryon irritans
heavy metal effects on, 137
infection of skin and gills with, 12
life cycles and environmental effects on, 118–119
Cryptococcoses, 228
Crystalloid formation, and pollution gradients, 381
Ctenophora, 412–415, 435
Cutaneous neoplasms, 165–166, 182
Cyanobacterial infections, 258, 401–404
Cyanobacterial symbionts, of sponges, 396–397, 399–400

Cyclorichis campula, in cetaceans, 233
Cymothoids, 134
Cytochrome P450 oxidases, 300
Cytochrome P450IA1, in pillar cells, 191
Cytofluorographic analysis, 496
Cytophaga-like external bacteria, 35

D

2,4-D, lesions induced by, 188
DDE, larval mortality with, 455–456
DDT
- carcinogenic effects of, 383
- chronic exposure to and pathogen growth, 281
- effects of on marine molluscs, 303
- larval mortality with, 455–456
- toxic effects of, 372
- toxicologic effects of in endocrine system, 189–190

Debilitation, disease secondary to, 181
Deep Water Dumpsite 106, pollution of, 380
Defense mechanism impairment, pollution-induced, 4–5
Demodex zalophi, 230
Demosponges, 393–394, 399–400
DENA, see N-nitrosodiethylamine
Dental epithelial neoplasms, 163
Dermal pigment cell neoplasms, chemical etiology of, 183
Dermatologic disease, in dolphins, 224
Dermatophilosis, 226
Dermatophilus congolensis, 226
Dermatophytosis, 228
Dermo, description and diagnosis of, 245
Detergents, effects of on marine molluscs, 304
Deuteromycetes, in marine crustaceans, 325, 327
Developmental stage, effect of on parasitic infections, 114
Developmental toxicants, 455
Diadema antillarum, 429–432
Dieldrin, effects of on marine molluscs, 303
Diet
- artificial in intensive systems, 13
- effect of on parasitic infections, 114
- in epicuticular repair with shell disease, 329–330
- pathological conditions attributed to, 13

Digestive diverticular epithelial atrophy, 515–516
Digestive gland tubules, atrophied, 310
Digestive system, toxicologic pathology of in fish, 194–204
Dimecron, toxicity of, 147
Dinoflagellates
- algal blooms caused by, 5
- blooms of causing disease in shrimp, 350–351
- highly ichthyotoxic, 99–100
- in marine crustaceans, 327–328
- toxins of, 350–351

Dioxin, immunosuppressive properties of, 4
Dipetalonema spirocauda, in pinnipeds, 232
Diphyllobothrium latum, 7, 144
Diplostomum species, 143
Dirofilaria immitis, in pinnipeds, 232
Disciplinary interaction view, 524
Disease
- of captive fishes, 8–13
- caused by physical and chemical changes, 2–5
- definition of, 112, 255
- of fishes in biomedical research, 13–15
- importance of in marine fish, 1–15
- interactive factors determining, 343, 344
- and pollutants, 451–474
- of wild fishes, 2–5

Dissolved oxygen, fish kills with low levels of, 3
DMBA, 183, 194
DMNA, toxic effects of, 372–373
DNA/RNA hybridization methods, 523
Dolphin pox virus, 219
Dolphins
- erysipelas in, 224
- lobomycosis in, 227–228
- mass die-off of due to chronic pollution, 237
- parasites of, 232–233
- *Sarcocystis* in, 230

Dyes, laboratory studies of effects of, 372
Dylox, fish sensitivity to, 147
Dysgerminoma, 193

E

Early life history stages, pollutant effects on, 453–457
Echinoderms
- associations and tumors of, 423–425
- bald sea urchin disease in, 427
- Caribbean mass mortalities of, 429–432
- defense mechanisms of, 423

green sea urchin amoebiasis in, 427–429
important recent epizootics in, 435–436
parasitic castration in, 425–427
species of, 422–423
Echinorhynchus gadi, 141–142
Ecosystems, 471–472, 525–526
Ectoparasites
as bioindicators, 148
of echinoderms, 423–424
examinations for, 180–181
lesions of resembling neoplasms, 184
in marine mammals, 230
protozoan, 12
Ectoplasmic nets, 277–278, 279
Edwardsiella infections, 55–56, 225–226
Electron microscopy, 43–45, 180
Endocrine system, 189–190
Endocuticle, 329
Endoparasites, 148, 230–233
Endotheliomas, 516
Endpoint measurements, 528
Endrin, 190, 303
Enteric red mouth disease, 35, 56–58
Enteritis, necrotizing, 226
Environment
interaction of with neoplasms in wild fish from marine ecosystem, 157–170
natural and man-made changes of, 343
studies of using fish as sentinel organisms, 177–178
and viral diseases in marine fish, 26
wild fishes as indicators of contamination of, 3–4
Environmental factors
in coral neoplasms, 410–412
in parasitic diseases, 111–112
anthropogenic, 135–147
natural, 118–135
Environmental research, 45–46
Environmental stress
and bleaching of corals, 408–410
in invertebrate disease, 433
macrophage impairment assay as biomarker of, 503
Environmental stressors
critical role of in disease, 471
immunomodulation by, 483–503
in marine molluscs, 291–310
and parasitic disease, 229
Eosinophilic cell foci, 199–201
Epibiontic growth, in crustaceans, 325
Epicuticle, 329–330
Epidermal neoplasms, 161–163, 169
Epidermal papillomas, globular type of, 162
Epidermoid carcinomas, prevalence of, 182–183
Epithelial neoplasms, 163, 184
Epithelioma papillosum cyprini (EPC) cell line, 161
Epithelium, irritation of by sedimentation, 146
Epizootic neoplasms
cutaneous fibrocytic, 165–166
epidermal, 161–163
hematopoietic, 163–165
liver, 158–160
peripheral nerve sheath cell, 168, 169–170
pigment cell, 166–167, 169
Epizootic ulcerative syndrome, 100–106
Erysipelas, 224
Erysipelothrix rhusiopathiae, 224
Erythematous dermatitis lesions, 102
Erythrocytic inclusion body syndrome (EIBS), 34–35
Erythrocytic necrosis virus (ENV), 37–39
Erythromycin, for bacterial kidney disease, 70–71
Estuarine zone, 96, 157
Eumycota, 264–265
Eutrophication, 11–12, 144
Excretory system, toxicologic pathology of in fish, 193–194
Exocuticle, 329
Exophiala, 106
Exophthalmia, with streptococcal infections, 76
Exoskeleton, 461–463
Extensive culture methods, 9–10
Exxon Valdez oil spill, 235
Eye, 187–188

F

Facultative parasites, 9–10
Faviid corals, black band disease in, 401–404
Fecundity rates, 456–457
Feeding, effect of parasites on, 114
Fibrocytic neoplasms, 169
Fibromas, 145, 165–166
Field studies
of chemical effects, 374–379, 381–383
of parasitic infections, 148
Filamentous algae, 400
Filamentous blue-green algae blooms, 351–352

Filamentous microepibionts, 325
Filtering/detoxifying systems, 463–466
Fin erosion, 458–459
Fin lesions, 4
Fin ray deformities, 467–468
Fin rot, 181–182
Finfish mariculture, 9
Finfishes, status of sewage sludge relative to, 145
Fish
  bacterial diseases of, 53–77
  bacterial pathogens of, 54–55
  chemoluminescence in, 499–501
  concern about disease in farming of, 8–9
  disease-pollution interaction in, 451–474
  diseases of as bioindicators of anthropogenic contamination, 1
  effects of pollution on populations of, 27–28
  fungal diseases of, 85–107
  infection of humans and animals by pathogens of, 1, 7
  parasitic diseases of and relationship with toxicants and environmental factors, 111–150
  peripheral blood of, 495–496
  phagocytosis in, 498–499
  primary stress responses in, 496–497
  as sentinel organisms, 177–178
  skeletal deformities in with environmental pollution, 467–468
  toxicologic histopathology of, 177–204
  viral diseases of, 25–46
  white blood cell populations in, 497–498
Fish kills, 2–3, 5
Fish parasites, 6–7, see also Parasites
Fish pox, viral etiology of, 169
Flagellate infestation, with sedimendation, 146
Flatfish, liver tumors in, 451
Flavobacterium species, 74–75
FLDV-1 strain, 40
FLDV-2 strain, 40
Flexibacter species, 72–74
Flounder, 43, 463–465
Flukes, in cetaceans, 233; see also Blook flukes
Focal fatty vacuolation, 200
Focal necrosis disease, 258–260
Folithion, toxicity of, 147
Follicular atresia, 191–192
Forebrain malformations, 470
Fungal disease, 85–107
  in crustaceans, 325–327
  in molluscs, 264–265
  of sponges, 396
  in wild fishes, 7
Fungi, 85
Fungicides, carcinogenic effects of, 382
Furunculosis, 60–61
Fusariomycosis, 228
Fusarium, 106, 326
Fused vertebrae, pollution-induced, 467

G

Gaffkemia, 323–324, 330–334
Gaffkya homari, 323
Gametogenesis, 193
Ganglioneuromas, 168
Gas-bubble disease, 349–350
Gastroenteropathy, sea otter hemorrhagic, 226
Gastropods, 262
Genetic abnormalities, 519
  in marine molluscs, 290–291
  pollution-associated, 471
Genetic disarray, 453
Germinomas, agrochemicals and, 382
Gill necrosis virus, molluscan, 257
Gill tuberculosis, 72
Gills
  black or brown pigment in, 353–354
  cadmium and melanization of, 370
  deformation of with iron and steel factory pollution, 379
  deformities of with vertebral defects, 185–186
  disease of
    bacterial, 75
    free-living protozoans causing, 119
    Paramoeba pemaquidensis associated with, 9
    pollution-induced, 382
  effect of sewage and waste on, 145
  electrophoretic changes in mucus of, 6
  exposure of to kraft mill waste, 139
  hyperplasia of with encapsulated blood flukes, 130–131
  as indicator of environmental stress, 191
  inflammation and hyperplasia of with pollution, 375–376
  irritation of by sedimentation, 145–146
  lesions of attributed to anthropogenic contamination, 4
  reduced carbonic anhydrase activity of, 5–6

toxicologic histopathology of, 190–191
Gliding bacteria, 262
Globular epidermal neoplasms, 169
Globular epitheliomas, 162
Glomerular filtration, 193
Glycera dibranchiata, 416
Gorgonia flabellum tissue, 403
Gram-negative bacteria, 7, 323
Gram-positive bacteria, in wild marine fishes, 7–8
Granulocytes, 310, 485
Granuloma
in *Calyptospora funduli*-infected liver, 123
in cetaceans, 233
Pearsonellum corventum egg fragments in, 131
Granulomatous disease, 13, 15
Green sea urchin amoebiasis, 427–429
Gregarines
association of annelids, 417–418
co-occurring in *Echinocardium cordatum*, 424
in marine crustaceans, 327
in shrimp ponds, 418–419
Gregarinid sporozoa, in oligochaetes, 420
Gulf Coast, 377
Gulf killifish, 3
Gusathion A, 348–349
Gyrodinium aureolum, 6

H

Habitat degradation, 451–474
Halocerus, in cetaceans, 232–233
Haplosporidian, 268–272
challenge of to researchers, 512
hyperparisitism in, 280
parasitic, hosts of, 273
spore formation of, 270
Haplosporidium nelsoni, 268–272, 512
Haplosporosomes, 272–274
Harbor seals
herpes virus in, 221–222
seal pox virus in, 218–219
Toxoplasma gondii in, 229
Health, 524–529
Heart, toxicologic injury to, 188
Heartworm disease, in pinnipeds, 232
Heated effluents, and parasite life cycles, 142–143
Heavy metals, see also Metals; specific elements
effects of on bivalve hemocyte count, 485
effects of on fish immune and other systems, 136–137
effects of on phagocytosis, 487
histopathology of female reproductive tissues with, 192
and parasitic diseases, 137–138
specific target tissues of, 513
Helminths
causing mortality, 130–132
hydrocarbon effects on in digestive tract, 141–142
as indicators of environmental conditions, 132–134
infestations of in crustaceans, 327–328
internal, 127–130
life cycle dynamics of, 132
toxic and carcinogenic properties of insecticides in, 372
Hemagglutinins, 490–491
Hemangioendotheliomas, 189
Hemangioendotheliosarcomas, 189
Hemangiomas, 189
Hemangiopericytic sarcomas, 189
Hemangiopericytoma, 189, 516
Hematodinium perezi, in marine crustaceans, 327–328
Hematopoietic neoplasms
epizootic, 163–165
progression of, 164
viral etiology of, 164–165
virus-induced, 157
Hematopoietic virus, 13
Hemic neoplasms, 294
Hemocyte count in bivalves, 484–486
Hemocyte membranes, 491–492
Hemocytes
in pollutant detoxification, 485–486
temperature and salinity effects on activity of, 492
Hemocytic enteritis (HE), 351–352
Hemocytic enzymes, 488–489
Hemocytic neoplasia, 291, 295
Hemolymph, 322
Hemolymph factors, 494
Hemolysin, 490–491
Hepatic adenoma, 202
Hepatic megalocytosis, 4
Hepatic neoplasms, 13–14, 199
Hepatitis virus, of sea lions, 220–221
Hepato-cholangiocellular carcinoma, 203–204
Hepatocarcinomas, 463

Hepatocellular carcinoma, 202
Hepatocellular hypertrophy, 197–198
Hepatocellular necrosis, 196
Hepatocellular nodule, 202
Hepatocellular vacuolation, 198–199
Hepatocytes, 199–202
Hepatocytomegaly, 197–199
Hepatoma, 202
Hepatopancreas, 331–334, 352–353
Heptachlor, 188, 303
Heptachlor epoxide, larval mortality with, 455
Herbicides
  carcinogenic effects of, 382–383
  causing hyperplastic lesions of brain, 186
  and disordered cardiac structure, 381
  effects of on marine molluscs, 303–304
Hermaphroditism, 193
Herpesvirus
  causing plaque epidermal neoplasms, 161–162
  of marine mammals, 221–222
  Pacific cod, 44
  turbot, 45
Herpesvirus salmonis disease, 35
Herring
  Ichthyophonus hoferi infection of, 7
  viral erythrocytic necrosis in, 39
Heterophyid trematodes, infection of man by, 7
Hexachlorobenzene
  effects of on bivalve hemocyte count, 484
  effects of on resistance to infection, 493
  and intracellular lysosome levels, 489
  reduced hemocyte uptake and killing of bacteria with, 486–487
Hexachlorobutadiene, nephrotoxic effects of, 193–194
β-Hexachlorocyclohexane (HCH), 189
High water temperatures, 3
Himasthla quisetensis, 419
Hippospongia communis, 396–398
Hirame rhabdovirus (HRV), 43
Hirudinea, associates of, 420–422
Histiocytic dermatitis, in dolphins, 227
Histopathologic analysis, 178–181, 252
Histopathological changes, 4
Histopathology, 177–178
Histoplasmosis, 228
Holothurians, 425
Homerange, 114–115
Hookworms, in pinnipeds, 231
Hormonal balance, 457, 496
Host-environment-pathogen interactions, 343–344
Host immune systems, 120–121, 132
Host migration, and parasitic infections, 114–115
Hosts, 113–115, 148
Human parasites, 144–145
Humoral factors, 490–491
Hydrocarbons, see also Aromatic hydrocarbons; Chlorinated hydrocarbons; Polycyclic aromatic hydrocarbons
  aromatic, laboratory studies of effects of, 371–372
  kidney abnormalities with exposure to, 466
  petroleum, 139–142, 309, 469–470
Hydrogen peroxide, with xenobiotic exposure in bivalves, 489
Hydrothol 191, lesions induced by, 188
Hydroxyl radical, with xenobiotic exposure in bivalves, 489
Hyperiid amphipods, association of with ctenophores, 413–414
Hyperoche species, association of with ctenophores, 413–414
Hyperparisitism, in metazoan parasites of molluscs, 280
Hyperpigmentation, in *Ichthyophonus* infection, 85
Hypothalamic-pituitary-interrenal (HPI) axis, 496
Hypoxia, 99–100
Hysterothylacium reliquens, 129–130

I

Ichthyophonus, 106
  diagnosis of, 86–88
  ecological and economic impact of, 88
  future research needs for, 89
  general characteristics of, 85–88
  histopathology of resting spore of, 87
  hoferi, 7, 85
Ichthyophthirius multifiliis, 118–119, 137
Ichthyosporidium giganteum, 124–126
Icthyobodo necator infestation, with sedimendation, 146
Immune deficiencies, ulcerative lesions with, 181–182
Immune response
  chemical effects on, 492–494
  modulation of by environmental stressors, 483–503

and parasites, 113, 132
Immune system
in protection from bacterial diseases, 223–224
specificity of, 483–484
Immunocompromise, 487–488
Immunoglobulins, 483
Immunological recognition factors, 491
Immunosuppression
caused by pollution, 4–5
with chronic environmental pollution, 237
with cold water temperatures, 3
with crowding, 8
disease secondary to, 181
and reduced chemoluminescence, 500–501
xenobiotic-induced, 493–494
Immunosuppressive chemicals, 4–5
Incertae sedis, 276–277
Indian River, 377
Industrial toxicants, 146–147
Infection, 112, 492–494
Infectious agents, causing neoplasms, 14–15
Infectious disease
challenges of to researchers, 518
gap in knowledge of, 517
of marine crustaceans, 319–338
of molluscs, 255–281
pitfalls of studies of in invertebrates, 253
pollution's role in, 337
residual lesions from, 181
in wild fishes, 6–8
Infectious hematopoietic necrosis (IHNV), 32–33
Infectious pancreatic necrosis virus (IPNV), 28–30, 162
Infestation, definition of, 112
Inflammation
in invertebrates, 248
parasites and processes of, 132
Inflammatory disorders, 519
Influenza virus, 221
Inorganic toxicants, 306–307
Insecticides, 147, 372
Integument, 458–463
Intensive culture systems, 9, 12–13
International Council for the Exploration of the Sea, 452
International Mussel Watch Program, 308–309
Intestinal degeneration, diet-associated, 13
Invertebrates
chemically induced histopathology in, 359–383
commercial uses and functions of, 393–394
death and postmortem change in, 248–249
diseases of, 245–247, 393–437
organismic information on, 247–249
pitfalls in studies of, 250–253
and population information, 249–250
environmental tolerances of, 249–250
inflammation and wound repair in, 248
normal histology of, 247–248
normal life cycle of, 247
normal microbial flora and parasite load of, 249
normal mortality of, 249
phyla of, 393
role of as disease vectors, 433
symbiotic associations among, 394
viral and infectious diseases of, 253
Iridovirus, 8, 39
Iron, 137
Islet cell hyperplasia, 190
Isopods, 134

J

Jet fuel, 380

K

Kepone toxicity, scoliosis with, 468
Kidney stones, in bivalves, 290
Kidneys
abnormalities of with environmental pollution, 466
bacterial disease of, 35, 68–71
function of, 193
neoplasms of, 194
pollution-induced diseases of, 382
toxicologic histopathology of, 193–194
Killifish, 122
Kraft pulp mill effluent
in chromatophoroma etiology, 166–167, 183–184
effect of on parasitic diseases, 138–139
Kudoa infections, 6–7
Kyphosis, 184–185

L

Laboratory studies
of chemical effects, 360–374, 379–381
of parasitic infections, 148
Labyrinthomorpha, 277–279
Labyrinthuloides haliotidis, 277–279

Lactic acid bacteria, 67–68
Lactobacilli, 67–68
Lactobacillus piscicola, 68
Lagenidium callinectes, 326
Lamellar epithelial cells, 146, 190
Larvae, 114, 454–457
Larval mycosis, of crustaceans, 326
Larval vibriosis, 260
Lectins, 491–492, 503
Leeches, 415, 420–422
Lens tissue damage, pollution-induced, 469–470
Lepeophtheirus salmonis, 10
Leptospira pomona, 224–225
Leptospirosis, 224–225
Leucocidin, 61
Leucothrix mucor, in crustaceans, 322, 325
Leukocytes, 497–498, 502
Lice, in sea lion pups, 230
Light microscopic examination, 180
Lindane, 303
Lip neoplasms, 182–183
Liver
  adenoma of, 202
  coagulative necrosis of, 196
  diet-associated anomalies of, 13
  focal hepatocellular changes in, 199–202
  function in, 195
  hyperplasia of regeneration of, 196–197
  necrosis and cell degeneration of, 122–123
  neoplasms of, 158–160, 195–196
  regenerative foci in, 196–197
  tumors of, 463–466, 471
Liver fluke, of sea lions, 231
Lobomycosis, 227–228
Lobsters
  fungal diseases of, 326
  gaffkemia in, 323–324, 330–334
  phycomycete infections in, 326–327
  shell lesions in, 330
Lordosis, 184–185
Lung mites, in pinnipeds, 230
Lungworms, 230–233
Lymphocystis, 39–41, 134–135
Lymphocyte responses, 483–484
Lymphocytes, in fish stress response, 497
Lymphocytopenia, stress-related in fish, 497
Lysosomal membrane, 308

M

Macrophages, 181, 502–503
Malathion, 147
Malformations, 454–455
MAMA, papillary adenoma with, 194
Manatees, *Toxoplasma gondii* in, 229
Marine diseases
  as complex response processes, 515
  endpoints in study of, 522–524
  of fish, importance of, 1–15
  linking cause and effect in study of, 520–529
  of mammals, pathobiology of, 217–239
  progress in understanding problems of, 511–514
  recent successes in research on, 511–512
  significant unsolved problems of, 515–519
  state of studies of, 511–529
  subcellular levels of, 522
Marine mammals
  bacterial diseases of, 223–227
  environmental pollution-associated diseases of, 235–238
  immune system of, 223–224
  mycotic diseases of, 227–228
  neoplasms in, 233–235
  parasitic diseases of, 228–233
  stressors that affect health of, 217–218
  types of, 217
  viral diseases of, 218–223
Marine mortalities, recently recognized causes of, 514
Marteilia refringens, 272–276
Mass mortalities
  of Caribbean sea urchins, 429–432
  ecological impact of, 394–395
  fish, 302
Medulloepitheliomas, in medullo, 14
Megalocytosis, 4, 198
Melanization, 335
Melanized gills, 353–354
Melanoma, in platyfish, 14
Melanoma associated antigens, 168
Menhaden, Atlantic, 91–94, 98
Meninx, hyperplastic lesions of, 186–187
Mercury
  concentration of in oysters, 306
  laboratory studies of effects of, 370
  pathological disorders produced by, 362
  testicular or sperm damage with, 192
Meront, 267
Mesotheliomas, 381–382
Metacercariae, 150, 414
Metacestodes, in aquaculture systems, 9–10

Metals, see also Heavy metals; specific elements
diseases associated with, 519
laboratory studies of effects of, 360–370
sublethal effects of on molluscs, 307–309
Metazoan parasites, 255
challenges of to researchers, 518
infecting molluscs, 279–280
as monitors of dynamic environmental conditions, 127
in polychaetes, 419
Methoxychlor, 187
Methylazoxymethanol acetate
kidney neoplasms with, 194
microophthalmia with, 188
neoplasms induced by, 14
3-Methylcholanthrene, 197–198
Methylcholanthrene (MCA), 371
METRO outfalls, 145
Microbial infections, 150, 414–415
Microepibionts, 325
Microfilaria, in pinnipeds, 232
Microophthalmia, 188, 470
Microspora, in marine crustaceans, 327
Migration, and parasitic infections, 114–115
Minchinia
in clams, 270
nelsoni, 268–269
teredinis, immature spore of, 271
MNNG
and chromatophoromas, 183
kidney neoplasms with, 194
papillary adenoma with, 194
toxic effects of, 373
MNU, toxic effects of, 373
Moldy feeds, disease with, 352
Molluscs
disordered cardiac structure in, 381
environmental insults to, 291–310
eumycota infecting, 264–265
genetic abnormalities in, 290–291
hemagglutinins and serum agglutinins of, 491
hemocytes in pollutant detoxification in, 485–486
infectious diseases of, 255–281
inflammatory response in, 245
kidney abnormalities in with environmental pollution, 466
metazoan parasites of, 279–280
neoplasms of, 381, 383
noninfectious diseases of, 289–311
nutritional deficiencies of, 289
PAHs in, 380
paramyxea infecting, 272–276
pollution and infectious disease of, 280–281
pollution-related tumors in, 375
prokaryotes of, 258–264
Protista infecting, 265–279
reduced hemocytes in hemic proliferative disease of, 487
shell abnormalities in, 468
toxic and carcinogenic properties of insecticides in, 372
viruses of, 256–258
xenobiotic-induced mesotheliomas and myxomas in, 382
Molting cycle, 247
Monoclonal antibodies, 503
Monocytes, of fish, 495–496
Monogeneans
with agricultural toxicants, 146
epizootics of, 12
and petroleum hydrocarbons, 140
prevalence and intensity of on gills of Atlantic cod, 141
Morbillivirus, 222–223
MSX (multinucleated sphere unknown), 246, 512
Mucous cell hyperplasia, 190–191
Mucus, skin secretion of, 182
Mullet, 101–104
Muscle necrosis, in penaeid shrimp, 350
Muscular tetany, 185
Musculoskeletal system, 184–186
Mussels, 378–379, 484–485
Mutagens, in chromatophoroma etiology, 166–167
Mycelial fungus, 281
Mycobacteriosis, 71, 226–227
Mycobacterium species, 71
marinum, in wild fishes, 7–8
in seals, 226–227
Mycoplasms, 262–263
Mycoses, inflammatory, 106–107
Mycotic diseases, 227–228
Mycotic granulomatosis, 105–106
Mytilicola, 280
Myxobolus lintoni, 125, 128
Myxobolus cerebralis infection, 13
Myxomas, and pollution gradients, 382
Myxosporeans, 6–7, 421–422
Myxosporidans, 126–128

N

N-nitrosodiethylamine
and growth of pathogens, 280
immunosuppressive effect of, 493–494
spongiosis hepatis, 466
toxic effects of, 372–373
N-nitrosodipropylamine (NDPA), 373
Nacrezation, 290
Narragansett Bay, 375–376
Nasitrema, in cetaceans, 232
Natural cytotoxic cells (NCC), 502
Necropsy-based approach, 178
Necrotizing dermatitis, 102–103
Nematodes
in cetaceans, 232–233
in echinoderms, 425
environmental effects on, 127–130
heavy metal effects on, 138
in oligochaetes, 420–421
Nematopsis infection, in shrimp, 417–418
Neodiplostomum species, 143
Neoplasms, see also Tumors
and agrochemicals, 382–383
cardiac, 381
cardiovascular, 189
causes of, 234–235
chemically induced, 13–15, 157, 252–253, 359
with chronic pollution, 4
in clams, 294
complex pathogenesis of, 521
connective tissue, 157
and contaminated sediments, 381
of corals, 410–412
definition of, 291–292
of echinoderms, 423–425
epizootic liver, 158–160
first recorded in marine mollusc, 292
hematopoietic, 157, 292
hemocytic, 291, 295, 296–297
hepatic, 195–196, 463–466
histopathology of, 513
immunological compromise, 487–488
infectious agents causing, 14–15
of invertebrate phyla, 433–437
in kidneys, 194
of marine mammals, 233–235
in marine molluscs, 291–301
with *Oncorhynchus masou* virus, 36–37
in oysters, 293
petroleum contamination and, 281
pollution-associated, 471
of reproductive organs, 193
of scleractians, 411
types of, 516
virus-induced, 180, 299
in wild fish, 157–170
13Nephrocalcinosis, 381
Neritic zone, 157
Nervous system, 186–188
Netpens, 10–12
Neural damage, 470
Neurilemmoma, 168
Neurofibromas, 168–170
Neurosensory system, 187, 469–470
Neutrocytophilia, 497
Neutrophilic granulocytes, 497
Neutrophils, 495–496
New Bedford Harbor, 376–377
New York Bight, 377, 451
Nifurpirinol, 167
Nitrite, 354
Nitrosamines, 381
Nitroso compounds
laboratory studies of effects of, 372–373
pathological disorders produced by, 364–365
research on toxic effects of, 380
Nitrotriacetic acid (NTA), 373
Nocardia species, 72, 260
Nocardiosis, 72
Noninfectious agents, 515
Noninfectious disease
challenges of to researchers, 519
of crustaceans, 343–354
pitfalls of studies of, 251
Nonlysozyme bactericidal factors, 490
Nonself material, detection of, 483
Nuclear power plant, 143
Nutrition, 114, 177
Nutritional deficiencies, 289
Nutritional disorders, 13, 345–349
Nuvan, toxicity of, 147

O

Oil, see also Crude oil; Petroleum
chemical components of, 236
research on toxic effects of, 380
Oil spills, 235–236
effects of in temperate and arctic regions, 374
histopathologic effects of, 379

Olfactory epithelium, 469
Olfactory organ, 187
Oligochaetes, 415, 420–422
Oncogenes, in molluscs, 301
Oncogenesis, 300–301
Oncorhynchus masou herpesvirus, 169
Oncorhynchus masou virus (OMV), 35–37
Oocyte atresia, 191–192
Oomycete mycoses, 106–107
Oomycete skin infections, 106
Oomycete water molds, 95
Opaleye fish, SMSLV in, 220
Opportunistic pathogens, 343–344
Optic malformations, 470
Oral papillomas, 183
Orchitophyra stellarum, 425–426
Organic chemicals
  diseases associated with, 519
  effects of on marine molluscs, 302–304
  in fish liver
    neoplasms, 160
  sublethal effects of in molluscs, 309
Organic herbicides, 303–304
Organochlorines
  bone structure changes with, 185
  effects of on marine molluscs, 303
  physiological effects of on marine mammals, 238
Organophosphates
  causing chronic soft-shell syndrome, 348–349
  causing spinal fractures, 185
  effects of on isopods, 147
  skeletal deformities with, 467–468
Organotoxicants, 281
Orthodichlorobenzene, 302
Orthohalarachne species, in pinnipeds, 230
Oscillatoriaceae, 351–352
Ostracoblabe implexa, 264
Otostrongylus circultitus, in pinnipeds, 231
Ovacystis disease, 377
Ovary, 5
Ovoid basophilic bodies, 404, 406
Oxolinic acid, for furunculosis, 61
Oxygen-caused gas-bubble disease, 349
Oxygen depletion, 144–145
Oxytetracycline, for furunculosis, 61
Oyster Mortality Workshops, 246
Oyster velar virus disease (OVVD), 258
Oysters
  Bonamia ostreae in, 276–277
  concanavalin A receptors on hemocytes of, 492
  crude oil exposures in, 373–374
  digestive diverticular epithelial atrophy in, 515–516
  focal necrosis disease of, 258–260
  fungal disease in shells of, 264
  Haplosporidia of, 268–269
  Haplosporidium nelsoni in, 512
  infectious disease in with organotoxicant exposure, 281
  inorganic toxicants in, 306–307
  Marteilia refringens in, 272–276
  neoplasms in, 293
  nonlysozyme bactericidal factors in, 490
  PAHs in, 380–381
  Perkinsus marinus in, 266–267, 511–512
  pollution-induced vacuolar lesions in, 379
  pollution-related disease in, 377
  rate and pattern of postmortem changes in, 248
  shell abnormalities in, 468
  studies of mortalities of, 245–246
  Vibrio diseases in, 260–261
  viruses of, 257
Ozone sterilization, 12

P

Pacific cod herpesvirus, 44
Pacific salmon, 71
PAHs, see Polycyclic aromatic hydrocarbons
Pamlico River, 99–100
Pancreatic neoplasms, 13–14, 195
Papillary adenomas, 194–195
Papilliform tumors, 371
Papillomas
  associated with pollution, 163
  PAH-induced, 183
  pleuronectid, 44
  with toxic etiology, 182
  winter flounder, 45
Papillomatosis, Atlantic salmon, 43–44
Paradiplozoon homoion, 142
Parafilaroides decorus, in pinnipeds, 230–231
Paramoeba pemaquidensis, 9, 119
Paramoeba perniciosa, in marine crustaceans, 327
Paramyxea, 272–276
Paranophrys infection, 327
Parasites
  of ctenophores, 414–415
  of echinoderms, 423–425
  facultative, 9–10

in heated European lakes and reservoirs, 143–144
of polychaetes, 416–419
seasonality of, 112
temperature effects on life cycles of, 142–143
type of, 113
Parasitic castration, in echinoderms, 425–427
Parasitic copepods, 114
Parasitic disease
anthropogenic environmental effects on, 135–147
bioindicators for, 147–149
and disturbance of host-symbiont-environment equilibrium, 149–150
future research on, 147–150
host effects in, 113–118
as indicators of environmental stresses, 111–112
of marine mammals, 228–233
natural environmental effects on, 118–135
types of, 112, 113
in wild fishes, 6–7
Parasitic infections
in crustaceans, 327–328
and environmental stresses and factors, 148–149
models for, 149
seasonal patterns of, 148
sources of, 228–229
Parasitic surveys, 147–149
Parasitosis, 412–414
Paratenesis, 412–414
Parathion
acute fish mortality with exposure to, 4
chronic exposure to and pathogen growth, 281
toxic effects of, 372
Pasteurella species, 7, 62–63
Pathobiological endpoints, 529
Pathogenesis, 521–522
Pathogens
accurate identification of, 253
interaction of with host and environment, 343–344
life cycle stages of, 517
methods of identification of, 523
recent recognized causes of marine mortalities, 514
Pathologic pigmentation, 379–380
PCBs, see Polychlorinated biphenyls
Pearl formation, 290
Pearsonellum corventum, 131–132
Pelecypod infections, 262
Penaeid shrimp, 343–354
Pentachlorophenol
effect of on resistance to infection, 493
effects of on bivalve hemocyte count, 484
and intracellular lysosome levels, 489
reduced hemocyte uptake and killing of bacteria with, 486–487
toxic effects of, 372
Pericardial mesotheliomas, in clams, 375
Peripheral blood, of fish, 495–496
Peripheral nerve sheath cell neoplasms, epizootic, 167–170
Perkinsiosis, 245–246
Perkinsus marinus, 265–269, 511–512
Pesticides
carcinogenic effects of, 382–383
causing chronic soft-shell syndrome, 348–349
causing hyperplastic lesions of brain, 186
contaminating Puget Sound, 378
histopathology, 188–189
pathological disorders, 363–364
toxicologic effects of on endocrine system, 189–190
Petroleum
in arctic marine environments, 378
and hemic disorders, 375
laboratory studies of effects of, 373–374
pathogenesis induced by, 382
pathological disorders produced by, 365
Petroleum hydrocarbons
effect of on parasitic diseases, 139–142
eye lens tissue damage with, 469–470
fish mortalities with, 140–141
neural damage with, 470
olfactory epithelium damage with, 469–470
sublethal effects of in molluscs, 309
Phagocytes, 483, 494–502
Phagocytosis, 486–488, 495–499
Phenol
effects of on bivalve hemocyte count, 484
laboratory studies of effects of, 372
pathological disorders produced by, 364
reducing phagocytic blood cell number, 486–487
Phenoloxidase, 334–336
Phenoloxidase system, 335–336
Phenoxyacetate herbicides, 188
Pholeter gastrophilus, 233
Phorbol esters, 300

Phorbol myristate acetate (PMA), 499
Phormidium corallyticum, 401–404
Phosphorus pollution, 4
Phycomycetes, 325–327
Physical extremes, 349–350
Phytohemagglutinin (PHA), 498
Phytoplankton, 5–6
Picornavirus, 162
Pigment cell neoplasms, 166–170
Pike, 165
Pinctada maxima, 259, 261
Pinnipedia, 217
Pinnipeds
  candidiasis in, 228
  cutaneous disease in, 226
  distemper virus of, 222–223
  ectoparasites in, 230
  endoparasites in, 230–232
  leptospirosis in, 224–225
  neoplasms in, 234
  pox viruses in, 218–219
  Sarcocystis species in, 229–230
Pinocytosis, 486
Plankton blooms, 3
Plaque epidermal neoplasms, 161–162, 169
Plasma cell tumors, 15
Pleuronectid papilloma, 44
Pneumonia, 222
Pollutant-disease relationship, 472–473
Pollutants, see also Chemicals; Pollution; Toxicants; specific pollutants
  categories of, 135
  causing histopathology in aquatic invertebrates, 359–383
  in chromatophoroma etiology, 166–167
  effects of on cell and tissue metabolic pathways, 466–468
  effects of on early life history stages, 453–457
  effects of on filtering/detoxifying systems, 463–466
  effects of on integument, 458–463
  effects of on neurosensory systems, 469–470
  effects of on parasites, 135–147
  hemocytes in detoxification of, 485–486
  interaction of with disease, 451–474
  and liver cancer in fish, 158–160
  neoplasms due to, 234
  sources of, 237
  specific target tissues of, 513
  sublethal effects of in molluscs, 307–309
  vulnerability to, 452–453
Pollution
  carcinogenic effects of, 380–383
  from catastrophic events, 235–236
  from chronic discharge of waste, 236–237
  crustacean exposure to, 336
  in diseases of marine mammals, 235–238
  effects of on fish, 27–28
  epidermal neoplasms associated with, 163
  in etiology of epizootic neoplasms, 169–170
  and infectious disease of molluscs, 280–281
  in invertebrate disease, 433
  organ and species specificity of, 379–382
  physiological effects of, 238
  as precipitator of fish diseases, 27–28
  role of in development of infections, 337
  skin lesions associated with, 181–182
Polychaetes, 393–394, 415–419
Polychlorinated biphenyls, 359
  contaminating Massachusetts waters, 376–377
  effects of on marine molluscs, 303
  levels of in gonads and embryo mortality, 456–457
  physiological effects of on marine mammals, 238
  in Puget Sound sediment, 378
  research on toxic effects of, 380
  in sediments, 374
Polycyclic aromatic hydrocarbons
  carcinogenic effect of, 380–383
  cataract prevalence with exposure to, 470
  contaminating Massachusetts waters, 376–377
  effect of on phagocytic function in oysters, 492
  effect of on phagocytosis, 498–599
  indicator of in environment, 191
  pathological disorders produced by, 371
  in sediments, 374
  tumor-inducing potential for, 359
Polycyclic aromatic ketones, 374
Polycyclic aromatic quinones, 374
Polynuclear aromatic hydrocarbons, 183
Porifera, 395–400
Porpoises, 232–233
Postlarval stage, 114
Power plant thermal effluent, 3
Pox viruses, 218–219
Prokaryotes, 258–264
Prophenoloxidase-activating system, 335, 462
Protista, 255, 265–279

Protogonyaulax tamarensis, 6
Protozoa
  causing gill-disease, 119
  in crustaceans, 327–328
  life cycles of and environmental effects on, 124–126
  parasitic, in oligochaetes, 420
Protozoan disease, 229–230, 518
Pseudocoelomate bilateria, 248
Pseudomonas infection, 75
  enalia, in focal necrosis disease, 260
  in marine mammals, 225
  pseudomallei, 225
  ulceration with, 459
Pseudoneoplastic conditions, 180
Pseudotuberculosis, 62–63
Puget Sound, 378–380, 458–459
Pytchodiscus brevis, 6

Q

Quality state, 526–529
Quantitative structure-activity relationships (QSAR), 304–305
Quincy Bay, 376–377

R

Rainbow trout
  enteric redmouth in, 56–57
  with infectious pancreatic necrosis virus, 30
  social stress in, 497–498
  viral hemorrhagic septicemia in, 30, 32
Rappahannock River, 99
Reactive oxygen intermediates, 489–490, 499–500
Red eye, 225
Red sore disease, 143
Red spherule cells, 427
Red spot disease, 100–106
Red tide, 350–351, 409–410
Registry of Tumors in Lower Animals (RTLA), 157, 359
Regulatory genes, 297–299
Renibacterium salmoninarum, 68–71
  granulomatous lesions induced by, 15
  natural hosts of, 70
  physiological characteristic of, 69
  susceptibility to with erythrocytic inclusion body syndrome, 35
  in wild fishes, 7
Reoviridae, 37
Reproductive system
  disturbance in with anthropogenic contamination, 5
  failure of and environmental contamination, 455–456
  toxicologic pathology of in fish, 191–193
Research disciplines, 522–524
Respiratory system, 190–191
Retina, 13–14, 470
Retroviruses
  in dermal sarcoma of freshwater walleye, 166
  infecting molluscs, 256
  in plaque epidermal neoplasms, 162
Rhabdovirus, 8, 104
Rickettsia
  causing disease in molluscs, 262–263
  challenges of to researchers, 518
  in crustaceans, 321–325
Rickettsia-like organisms, 324
Rickettsiales, 63–67
Rickettsiosis, 376
RNA viruses, 256
Romet, 56
Rotenone, 348–349

S

Sabellid worm, 416
Sagenogenetosomes, 277
Salinity
  effect of on symbionts, 115
  and hemocyte activity, 492
  tolerance of and ulcerative mycosis prevalence, 93
Salmon, 7, 10–11
Salmonella species, 225–226
Salmonids, 8, 28–37
San Miguel Sea Lion Virus (SMSLV), 219–220
Sand dollars, 424
Saponin, 348–349
Saprolegnia, 92–93, 106
Saprolegniaceous water molds, 92
Sarcinomyces species, 106
Sarcocystis species, 229–230
Sarcomas, 166, 294
Schizont, 267
Schwannomas, 168, 170
Scleractinian corals, 401–403, 410–411
Scoliosis, 184–185, 468
Sea cucumbers, 425–427
Sea lice, on sockeye salmon head, 10

Sea lions, 218–225, 230–234
Sea otter, 226
Sea stars, 425–427
Sea urchins
accumulation of sediment on spines of, 430
diseases of, 427–429
function of, 393
mass mortalities of in Caribbean, 429–432
parasitic castration in, 425–427
Seafood, 383
Seal pox virus, 218–219
Seals
effects of environmental chemicals on, 238
influenza virus in, 221
Mycobacterium species in, 226–227
Sarcocystis species in, 229–230
Secondary sex characteristics, 183
Sediment
compounds bound to and tumor formation, 381
contaminated, 177, 366, 380
Sediment-associated chemicals, 464
Sedimentation, 145–146
Selenium deficiency, in penaeid shrimp, 347
Seminoma, 193
Sense organs, 186–187
Septic hepatopancreatic necrosis (SHN), 352–353
Septicemic disease, 322–323
Serum lysosomal enzymes, 488–489
Serum lysosomes, 488
Sevin, 372, 383
Sewage
contamination of Quincy Bay with, 376–377
effect of on parasitic diseases, 144–145
Sex, effect of on parasitic infections, 114
Sheepshead minnow, 128
Shell disease
and chromium, 370
with environmental pollution, 461–463
in penaeid shrimp, 353–354
pollution-associated, 471
prevalence of, 462
Shell disease syndrome, 328–330, 337
Shell erosion, 461–463
Shell malformation, 290–291, 468
Shellfish
disease-pollution interaction in, 451–474
diseases of, recent successes in research on, 511–512
pollution-induced neoplasms in, 383
Shrimp, see also Penaeid shrimp
black death disease in, 345
fungal infections of, 326
gregarines in ponds of, 418–419
Nematopsis infection in, 417–418
pollution-related disease in, 377
Shut-down-reaction (SDR), 409
Silver, 360–361, 380
Sirenia, 217
Sirolpidium zoophthorum, 264
Skeletal deformities, 467–468
Skin, 4, 181–184
Snails, 262
Snieszko sphere model, 343
Social stress, effects of on blood cell composition in rainbow trout, 497–498
Sodium arsenite, 188
Soft-shell shrimp, 347–349
Sperm, toxicants damaging or impairing, 192–193
Spinal deformities, 33
Spleen, 188–189
Sponge gemmules, 370
Sponge-wasting disease, 396
Sponges, 247, 393–400
Spongia officinalis, 397–398
Spongin fibers, 397–400
Spongin microfibrils, 400
Spongiophaga, 396
Spongiosis hepatis, 466
Sporotrichosis, 228
Spot, 124–125
Squamous cell carcinomas, 169, 182
Staphylococcus species, 75
Starfish, 423–427
Stenurus species, 232
Steringophorus furciger, 141–142
Stony corals, 393, 401
Streptococcus infections, 7, 75–76
Stress, 27–28, 149–150, 495–502
Stress-induced mucus, 144–145
Stress-related necrosis (SRN), 406
Stressors, 217–218; see also Environmental stressors
Strongylocentrotus droebachiensis, 427–429
Sulfamerazine, 61
Sulfonimide, 56
Supernumerary siphons, 290
Superoxide anion, 489–490
Superoxide dismutase (SOD), 500
Swimbladder, 194
Symbionts, 112
as bioindicators, 148

and effect of aromatic hydrocarbons, 142
natural environmental effects on, 116–135
relationship of with host, 113–115
temperature and salinity effects on, 115

T

T lymphocytes, of fish, 495–496
Tar balls, 236
Temperature
associated with red spot disease and epizootic ulcerative syndrome outbreaks, 104
and chemoluminescence response, 499–500
effect of on symbionts, 115
and hemocyte activity, 492
increase in and bleaching of corals, 408–409
influence of on toxic effects, diseases, and parasites, 142–144
Teratoid siphon anomalies, in clams, 375
Terramycin, 56
Testis, 192–193
Tetrachlorocyclopentene-1,3-dione, 166
Tetradecanoylphorbol acetate (TPA), 300
Thecamoeba hoffmani, 119
Thermal effluent, 3
Thermal pollution, 142–144
Thermal stress, 408–409, 433
Thermoregulation, 236
Thioglycolate Diagnostic Technique, 245–246
Thyroid adenocarcinomas, 190
Thyroid follicles, 189–190
Tissue damage, 251
Tissue metabolic pathways, 466–468
Toxaphene, 303, 372
Toxic disease, of crustacea and penaeid shrimp, 350–354
Toxicants
cellular tissue responses to, 516–517
disease conditions associated with, 452–453
effects of on parasites, 135–147
Toxicologic histopathology, 177–204
Toxicologic pathology
of cardiovascular system and spleen, 188–189
of digestive system, 194–204
of endocrine system, 189–190
of excretory system, 193–194
of musculoskeletal system, 184–186
need for research on, 516–517
of nervous system, 186–188
of reproductive system, 191–193
of respiratory system, 190–191
of skin, 181–184
Toxicological diseases, 301–309
Toxigenic algae, 350–351
Toxaphene, 281
Toxoplasma gondii, 229
Trematodes, 233, 279–280
Triactinomyxon, 422
Tributyltin
and chemoluminescence response, 501
immunosuppressive properties of, 4
shell abnormalities with, 468
suppression of chemoluminescence by, 490
Trichloroethylene, 193
Trichodinid ciliate infestations, 119, 140
Trichomaris invadens, 326
Trifluralin, 189
Trypanoplasma bullocki, 120
Trypanosomes, 120–121, 141
TTS assay, 492
Tubifex, as host for myxosporean agent, 421–422
Tumor regulatory gene complex (Tu-complex), 14
Tumors, see also Neoplasms
of Atlantic salmon swim bladder sarcoma virus, 44
chemically induced, 252–253
clusters of, 157
in cnidarians, 410–412
of echinoderms, 423–425
examination of, 180
liver, 463–466
malignant, 253
with *Oncorhynchus masou* virus, 36–37
pollution-associated, 471
types of, 516
Turbellarians, PAHs in, 380
Turbidity, 145–146
Turbot herpesvirus, 45

U

Ulcerations, pollution-associated, 459–460, 471

Ulcerative dermal necrosis (UDN), 106
Ulcerative lesions, 181–182
Ulcerative mycosis
  accurate assessment of impacts of, 98
  challenges of to researchers, 518
  data linking water quality to, 99–100
  documented occurrences of, 94
  ecological and economic impact of, 97–98
  epidemiology of, 93–97
  in estuarine fishes, 96
  experimental model of, 98–99
  occurrences of associated with oomycete water molds, 95
  pathogenesis of, 89–93
  similarity of to red spot disease and epizootic ulcerative syndrome, 101–102
Ulcus syndrome, 8
Ultraviolet irradiation, and coral bleaching, 409
Ultraviolet sterilization, 12
Uncinaria lucasi, in pinnipeds, 231
Urosporidium spisuli, 280

V

Vacuolar foci, 200
Vagococcus salmoninarum, 68
Velvet disease, 117
Vertebral deformities, 184–186
Vesicular exanthema of swine (VESV), 220
Vibrio, 58–59
  alginolyticus, 260
  anguillarum, 58–59
    in focal necrosis disease, 260
    in sea urchin, 427
    ulceration with, 459, 460
    in wild marine fishes, 7
  in crustaceans, 322–323
  ordalii, 58–59
  in oysters, 260–261
  salmonicida, 58
  salmoninarum, 59
Vibriosis
  agents causing, 58–59
  control of, 59
  in crustaceans, 322–323
  in intensive aquaculture systems, 12
  in wild marine fishes, 7
Viral disease
  classification of, 27
  and environments, 26
  fate of surviving fish populations, 46
  in intensive systems, 13
  known only by electron microscopy, 43–45
  literature on, 25–27
  of marine fish, 25–46
  of marine mammals, 218–223
  of molluscs, 256–258
  of non-salmonid fishes, 37–43
  pollution and, 27–28
  research needs, 45–46
  of salmonids, 28–37
  in wild fishes, 8
Viral erythrocytic necrosis (VEN), 37–39
Viral hemorrhagic septicemia (VHSV), 30–32
Viruses
  causing epidermal neoplasms in fish, 161–163
  causing epithelial tumors, 184
  challenges of to researchers, 518
  of crustaceans, 321
  difficulty of accurate identification of, 253
  in epizootic neoplasm etiologies, 169, 170
  groupings of recorded in crustaceans, 322
  in hematopoietic neoplasm etiology, 164–165
  with malignant neoplasms, 299
  neoplasms caused by, 157, 180
  shortcomings of studies of, 337
  susceptibility to and water quality, 45
Vitamin C, in bone development, 185
VR-299 strain, 28, 29

W

Waste, health effects of chronic discharge of, 236–237
Water molds
  oomycete, 95
  saprolegniaceous, 92–93
  in ulcerative mycosis lesions, 97–98
Water quality, and ulcerative mycosis, 99–100
Whales
  candidiasis in, 228
  effect of oil spills on, 236
  herpesvirus in, 222
  neoplasms in, 234
  Sarcocystis species in, 229–230
Whirling disease, 421–422
White band disease, 404–406
White blood cells, in fish, 497–498
White muscle lesions, in penaeid shrimp, 347
White plague, see White band disease
Winter flounder papilloma, 45
Wound repair, in invertebrates, 248

X

X-cell pseudotumors, 180

Xenobiotics
and chemoluminescence response, 499–501
effect of on resistance to infection, 493–494
neoplastic disease epizootics with, 4
oocyte or follicular atresia with exposure to, 191–192
sublethal concentrations of, 483
threat of to human nutrition, 177
Xenomas, with protozoan infections, 124

Y

Yaquina Bay mussel study, 378
Yersinia, 35, 56–58

Z

Zalophotrema hepaticum, 231
Zinc, 306
Zooxanthellae, 407–408, 411
Zygomycosis, 228
Zymosan, opsonization of, 499

T - #0151 - 101024 - C0 - 229/152/31 [33] - CB - 9780849386626 - Gloss Lamination